Set Valued Mappings with Applications in Nonlinear Analysis

SERIES IN MATHEMATICAL ANALYSIS AND APPLICATIONS

Series in Mathematical Analysis and Applications (SIMAA) is edited by Ravi P. Agarwal, National University of Singapore and Donal O'Regan, National University of Ireland, Galway, Ireland.

The series is aimed at reporting on new developments in mathematical analysis and applications of a high standard and of current interest. Each volume in the series is devoted to a topic in analysis that has been applied, or is potentially applicable, to the solutions of scientific, engineering and social problems.

Volume 1
Method of Variation of Parameters for Dynamic Systems
V. Lakshmikantham and S.G. Deo

Volume 2
Integral and Integrodifferential Equations: Theory, Methods and Applications
edited by *Ravi P. Agarwal and Donal O'Regan*

Volume 3
Theorems of the Leray-Schauder Type and Applications
Donal O'Regan and Radu Precup

Volume 4
Set Valued Mappings with Applications in Nonlinear Analysis
edited by *Ravi P. Agarwal and Donal O'Regan*

Set Valued Mappings with Applications in Nonlinear Analysis

Edited by

Ravi P. Agarwal

National University of Singapore

and

Donal O'Regan

National University of Ireland, Galway, Ireland

CRC Press
Taylor & Francis Group
Boca Raton London New York

CRC Press is an imprint of the
Taylor & Francis Group, an **informa** business
A CHAPMAN & HALL BOOK

CRC Press
Taylor & Francis Group
6000 Broken Sound Parkway NW, Suite 300
Boca Raton, FL 33487-2742

First issued in paperback 2019

ISBN-13: 978-0-415-28424-0 (hbk)
ISBN-13: 978-0-367-39578-0 (pbk)

Visit the Taylor & Francis Web site at
http://www.taylorandfrancis.com

and the CRC Press Web site at
http://www.crcpress.com

Contents

Preface

This book is a collection of research articles related to the mathematical analysis of multifunctions. By a set valued map $F : X \to 2^Y$ we simply mean a map that assigns to each $x \in X$ a subset $F(x) \subseteq Y$. The theory of set valued maps is a beautiful mixture of analysis, topology and geometry. Over the last thirty years or so there has been a huge interest in this area of research. This is partly due to the rich and plentiful supply of applications in such diverse fields as for example Biology, Control theory and Optimization, Economics, Game theory and Physics. This book titled 'set valued mappings with applications in nonlinear analysis' contains 29 research articles from leading mathematicians in this area from around the world. Topological methods in the study of nonlinear phenomena is the central theme. As a result the chapters were selected accordingly and no attempt was made to cover every area in this vast field. The topics covered in this book can be grouped in the following major areas: integral inclusions, ordinary and partial differential inclusions, fixed point theorems, boundary value problems, variational inequalities, game theory, optimal control, abstract economics, and nonlinear spectra.

In particular the theory of set valued maps is used in the chapters of Agarwal, Meehan and O'Regan, Andres, Candito, Kamenski and Nistri, Kryszewski, Matzakos and Papageorgiou, and Palmucci and Papalini to present results for differential and integral inclusions in various settings. The Baire category method is used by De Blasi and Pianigiani to discuss existence problems for partial differential inclusions. Structure of solution sets is addressed by Agarwal and O'Regan, and Obukhovskii and Zecca. The chapter of Matzakos, Papageorgiou and Yannakakis contains results on optimal control for nonlinear parabolic partial differential equations. Many new fixed point theorems for set valued maps are contained in the contributions of Agarwal and O'Regan, Daffer and Kaneko, Frigon, Morales, Ricceri, and Takahashi. Nonlinear spectral theory is discussed by Appell, Conti and Santucci, random fixed point theory by Shahzad, and fuzzy mappings by Cho, Shim, Huang and Kang. In a long survey chapter Milojević presents new results in the theory of A-proper maps. Variational inequalities are discussed in the long survey article of Chowdhury and Tarafdar, and in the chapters of Isac, Tarafdar and Yuan, and Park. Maximal element principles are presented in the contributions of Ding, and Isac and Yuan. Applications of fixed point theory in abstract economies and game theory appear in the chapter of Tan and Wu. Some interesting fixed point algorithms are contained in the chapters of Reich and Zaslavski, and Verma.

We wish to express our appreciation to all the contributors. Without their cooperation this book would not have been possible.

Ravi P Agarwal
Donal O'Regan

SIMAA 4(2002) 1–9

1. Positive L^p and Continuous Solutions for Fredholm Integral Inclusions

Ravi P. Agarwal[1], Maria Meehan[2] and Donal O'Regan[3]

[1]*Department of Mathematics, National University of Singapore, 10 Kent Ridge Crescent, Singapore 119260*
[2]*School of Mathematical Sciences, Dublin City University, Glasnevin, Dublin 9, Ireland*
[3]*Department of Mathematics, National University of Ireland, Galway, Ireland*

Abstract: In this chapter a multivalued version of Krasnoselski's fixed point theorem in a cone is used to discuss the existence of $C[0,T]$ and $L^p[0,T]$ solutions to the nonlinear integral inclusion $y(t) \in \int_0^T k(t,s)\, f(s,y(s))\, ds$. Throughout we will assume k: $[0,T] \times [0,T] \to \mathbf{R}$ and f: $[0,T] \times \mathbf{R} \to 2^{\mathbf{R}}$.

1. INTRODUCTION

In this chapter we present new results which guarantee that the Fredholm integral inclusion

$$y(t) \in \int_0^T k(t,s) f(s,y(s)) ds \tag{1.1}$$

has a positive solution $y \in L^p[0,T]$, $1 \le p < \infty$, or has a nonnegative solution $y \in C[0,T]$. Throughout this chapter $T > 0$ is fixed, k: $[0,T] \times [0,T] \to \mathbf{R}$ and f: $[0,T] \times \mathbf{R} \to 2^{\mathbf{R}}$; here $2^{\mathbf{R}}$ denotes the family of nonempty subsets of $\mathbf{R}$. It is only recently [6] that a general theory has been developed which guarantees that the operator equation, $y(t) = \int_0^T k(t,s) g(s,y(s))\, ds$ for a.e. $t \in [0,T]$, has a positive solution $y \in L^p[0,T]$ (note by a positive solution we mean $y(t) > 0$ for a.e. $t \in [0,T]$); here g: $[0,T] \times \mathbf{R} \to \mathbf{R}$ is single valued. In Section 2 using the 1991 paper of Cellina *et al.* [3] we are able to establish criteria which guarantees that (1.1) has a positive solution $y \in L^p[0,T]$. Section 3 discusses $C[0,T]$ solutions to (1.1); the results here improve those in [1].

The main idea in this chapter relies on the multivalued analogue [1] of Krasnoselski's fixed point theorem in a cone. Let $E = (E, \|\cdot\|)$ be a Banach space and $C \subseteq E$. For $\rho > 0$ let

$$\Omega_\rho = \{x \in E : \|x\| < \rho\} \quad \text{and} \quad \partial\Omega_\rho = \{x \in E : \|x\| = \rho\}.$$

Theorem 1.1: *Let $E = (E, \|\cdot\|)$ be a Banach space, $C \subseteq E$ a cone and let $\|\cdot\|$ be increasing with respect to C. Also r, R are constants with $0 < r < R$. Suppose $A: \overline{\Omega_R} \cap C \to K(C)$ (here $K(C)$ denotes the family of nonempty, convex, compact subsets of C) is an upper semicontinuous, compact map and assume* one *of the following conditions*

(A) $\|y\| \le \|x\|$ *for all* $y \in A(x)$ *and* $x \in \partial\Omega_R \cap C$ *and* $\|y\| > \|x\|$ *for all* $y \in A(x)$ *and* $x \in \partial\Omega_r \cap C$

or

(B) $\|y\| > \|x\|$ *for all* $y \in A(x)$ *and* $x \in \partial\Omega_R \cap C$ *and* $\|y\| \le \|x\|$ *for all* $y \in A(x)$ *and* $x \in \partial\Omega_r \cap C$

hold. Then A has a fixed point in $C \cap (\overline{\Omega_R} \backslash \Omega_r)$.

2. $L^p[0,T]$ SOLUTIONS

In this section we discuss the nonlinear Fredholm integral inclusion

$$y(t) \in \int_0^T k(t,s)f(s,y(s))ds \quad \text{a.e. } t \in [0,T], \tag{2.1}$$

where $k: [0,T] \times [0,T] \to \mathbf{R}$ and $f: [0,T] \times \mathbf{R} \to K(\mathbf{R})$. We would like to know what conditions one requires on k and f in order that the inclusion (2.1) has a positive solution $y \in L^p[0,T]$, where $1 \le p < \infty$. Here by a positive solution y we mean $y(t) > 0$ for a.e. $t \in [0,T]$. Throughout this section $\|\cdot\|_q$ denotes the usual norm on L^q for $1 \le q \le \infty$.

Theorem 2.1: *Let* $k: [0,T] \times [0,T] \to \mathbf{R}$ *and* $f: [0,T] \times \mathbf{R} \to K(\mathbf{R})$ *and suppose the following conditions hold*:

the map $u \mapsto f(t,u)$ *is upper semicontinuous for a.e.* $t \in [0,T]$; (2.2)

the graph of f *belongs to the* σ*-field* $\mathcal{L} \otimes \mathcal{B}(\mathbf{R} \times \mathbf{R})$ *(here* $\mathcal{L}$ *denotes the Lebesgue* σ*-field on* $[0,T]$ *and* $\mathcal{B}(\mathbf{R} \times \mathbf{R}) = \mathcal{B}(\mathbf{R}) \otimes \mathcal{B}(\mathbf{R})$ *is the Borel* σ*-field in* $\mathbf{R} \times \mathbf{R}$); (2.3)

$\exists p_2, 1 \le p_2 < \infty, a_1 \in L^{p_2}[0,T]$ *and* $a_2 > 0$ *a constant, with* $|f(t,y)| = \sup\{|z|: z \in f(t,y)\} \le a_1(t) + a_2|y|^{\frac{p}{p_2}}$ *for a.e.* $t \in [0,T]$ *and all* $y \in \mathbf{R}$; (2.4)

$(t,s) \mapsto k(t,s)$ *is measurable*; (2.5)

$\exists 0 < M \le 1, k_1 \in L^p[0,T], k_2 \in L^{p_1}[0,T]$, *here* $\frac{1}{p_1} + \frac{1}{p_2} = 1$, *such that* $0 < k_1(t), k_2(t)$ *a.e.* $t \in [0,T]$ *and* $Mk_1(t)k_2(s) \le k(t,s) \le k_1(t)k_2(s)$ *a.e.* $t \in [0,T]$, *a.e.* $s \in [0,T]$; (2.6)

for a.e. $t \in [0,T]$ *and all* $y \in (0,\infty)$, $u > 0$ *for all* $u \in f(t,y)$; (2.7)

$$\begin{aligned}&\exists q \in L^{p_2}[0,T] \text{ and } \psi\colon [0,\infty) \to [0,\infty), \psi(u) > 0 \text{ for } u > 0,\\ &\text{continuous and nondecreasing with for a.e. } t \in [0,T] \text{ and } y > 0,\\ &u \geq q(t)\psi(y) \text{ for all } u \in f(t,y);\end{aligned} \tag{2.8}$$

$$\exists \alpha > 0 \quad \text{with} \quad 1 < \frac{\alpha}{2^{\frac{p_2-1}{p_2}} \|k_1\|_p \|k_2\|_{p1} \left(\|a_1\|_{p_2}^{p_2} + [a_2]^{p_2}\, \alpha^p\right)^{\frac{1}{p_2}}} \tag{2.9}$$

and

$$\exists \beta > 0,\ \beta \neq \alpha \quad \text{with} \quad 1 > \frac{\beta}{M\|k_1\|_p \int_0^T k_2(s)q(s)\psi(a(s)\beta)ds}, \tag{2.10}$$

where

$$a(t) = M\frac{k_1(t)}{\|k_1\|_p}. \tag{2.11}$$

Then (2.1) has at least one positive solution $y \in L^p[0,T]$ and either

(A) $0 < \alpha < \|y\|_p < \beta$ *and* $y(t) \geq a(t)\alpha$ *a.e.* $t \in [0,T]$ *if* $\alpha < \beta$

or

(B) $0 < \beta < \|y\|_p < \alpha$ *and* $y(t) \geq a(t)\beta$ *a.e.* $t \in [0,T]$ *if* $\beta < \alpha$

holds.

Proof: Let $E = (L^p[0,T], \|\cdot\|_p)$ and

$$C = \{y \in L^p[0,T] : y(t) \geq a(t)\|y\|_p \text{ a.e. } t \in [0,T]\}.$$

It is easy to see that $C \subseteq E$ is a cone. Next let $A = K \circ N_f\colon C \to 2^E$, where the linear integral (single valued) operator K is given by

$$Ky(t) = \int_0^T k(t,s)y(s)\,ds,$$

and the multivalued Nemytskij operator N_f is given by

$$N_f u = \{y \in L^{p_2}[0,T] : y(t) \in f(t,u(t)) \text{ a.e. } t \in [0,T]\}.$$

Remark 2.1: Note A is well defined since if $x \in C$ then (2.2)–(2.4) and [3] guarantee that $N_f x \neq \emptyset$.

We first show $A\colon C \to 2^C$. To see this let $x \in C$ and $y \in Ax$. Then there exists a $v \in N_f x$ with

$$y(t) = \int_0^T k(t,s)v(s)ds \quad \text{for a.e. } t \in [0,T].$$

Now

$$|y(t)|^p \leq [k_1(t)]^p \left(\int_0^T k_2(s)v(s)ds \right)^p \quad \text{for a.e. } t \in [0,T]$$

so

$$\|y\|_p \leq \|k_1\|_p \int_0^T k_2(s)v(s)ds. \tag{2.12}$$

Combining this with (2.6) gives

$$y(t) \geq M \int_0^T k_1(t)k_2(s)v(s)ds \geq M \frac{k_1(t)}{\|k_1\|_p} \|y\|_p = a(t)\|y\|_p \quad \text{for a.e. } t \in [0,T].$$

Thus $y \in C$ so $A\colon C \to 2^C$. Also notice [3,6] guarantees that

$$A\colon C \to 2^C \text{ is upper semicontinuous.} \tag{2.13}$$

In addition note [8,9,10:pp. 47–49] implies $K\colon L^{p_2}[0,T] \to L^p[0,T]$ is completely continuous, and $N_f\colon L^p[0,T] \to 2^{L^{p_2}[0,T]}$ maps bounded sets into bounded sets. Consequently

$$A\colon C \to K(C) \text{ is completely continuous.} \tag{2.14}$$

Let

$$\Omega_\alpha = \{y \in L^p[0,T] : \|y\|_p < \alpha\} \quad \text{and} \quad \Omega_\beta = \{y \in L^p[0,T] : \|y\|_p < \beta\}.$$

Assume that $\beta < \alpha$ (a similar argument holds if $\alpha < \beta$). It is immediate from (2.13) and (2.14) that

$$A\colon C \cap \overline{\Omega_\alpha} \to K(C) \text{ is upper semicontinuous and compact.}$$

If we show

$$\|y\|_p < \|x\|_p \quad \text{for all } y \in Ax \text{ and } x \in C \cap \partial\Omega_\alpha \tag{2.15}$$

and

$$\|y\|_p > \|x\|_p \quad \text{for all } y \in Ax \text{ and } x \in C \cap \partial\Omega_\beta \tag{2.16}$$

are true, then Theorem 1.1 guarantees that the operator A has a fixed point in $C \cap (\overline{\Omega_\alpha} \backslash \Omega_\beta)$. This in turn implies that (2.1) has at least one solution $y \in L^p[0,T]$ with $\beta \leq \|y\|_p \leq \alpha$ and $y(t) \geq a(t)\beta$ for a.e. $t \in [0,T]$.

Suppose $x \in C \cap \partial\Omega_\alpha$, so $\|x\|_p = \alpha$, and $y \in Ax$. Then there exists a $v \in N_f x$ with

$$y(t) = \int_0^T k(t,s)v(s)ds \quad \text{for a.e. } t \in [0,T].$$

Now (2.4) and (2.6) guarantee that

$$|y(t)| \leq k_1(t) \int_0^T k_2(s)\left[|a_1(s)| + a_2|x(s)|^{\frac{p}{p_2}}\right] ds \quad \text{for a.e. } t \in [0,T].$$

This together with (2.9) yields

$$\begin{aligned}
\|y\|_p &\leq \|k_1\|_p \|k_2\|_{p_1} \left(\int_0^T \left[|a_1(s)| + a_2 \, |x(s)|^{\frac{p}{p_2}}\right]^{p_2} ds \right)^{\frac{1}{p_2}} \\
&\leq \|k_1\|_p \|k_2\|_{p_1} \left(2^{p_2 - 1} \int_0^T [|a_1(s)|^{p_2} + [a_2]^{p_2} |x(s)|^p] ds \right)^{\frac{1}{p_2}} \\
&= 2^{\frac{p_2-1}{p_2}} \|k_1\|_p \, \|k_2\|_{p_1} \left(\|a_1\|_{p_2}^{p_2} + [a_2]^{p_2} \, \|x\|_p^p \right)^{\frac{1}{p_2}} \\
&= 2^{\frac{p_2-1}{p_2}} \, \|k_1\|_p \, \|k_2\|_{p_1} \left(\|a_1\|_{p_2}^{p_2} + [a_2]^{p_2} \, \alpha^p \right)^{\frac{1}{p_2}} \\
&< \alpha = \|x\|_p
\end{aligned}$$

and so (2.15) is satisfied.

Now suppose $x \in C \cap \partial\Omega_\beta$, so $\|x\|_p = \beta$ and $x(t) \geq a(t)\beta$ for a.e. $t \in [0,T]$, and $y \in Ax$. Then there exists a $v \in N_f x$ with

$$y(t) = \int_0^T k(t,s)v(s)ds \quad \text{for a.e. } t \in [0,T].$$

Notice (2.8) guarantees that $v(s) \geq q(s)\psi(x(s))$ for a.e. $s \in [0,T]$ and this together with (2.6) yields

$$y(t) \geq M k_1(t) \int_0^T k_2(s) q(s) \psi(x(s)) ds \quad \text{for a.e. } t \in [0,T].$$

Combining with (2.10) gives

$$\begin{aligned}
\|y\|_p &\geq M \, \|k_1\|_p \int_0^T k_2(s) q(s) \psi(x(s)) ds \\
&\geq M \|k_1\|_p \int_0^T k_2(s) q(s) \psi(a(s)\beta) ds \\
&> \beta = \|x\|_p
\end{aligned}$$

and thus (2.16) is satisfied. Now apply Theorem 1.1. □

3. $C[0,T]$ SOLUTIONS

In this section we discuss the Fredholm integral inclusion

$$y(t) \in \int_0^T k(t,s) f(s, y(s)) ds \quad \text{for } t \in [0,T], \tag{3.1}$$

where k: $[0,T] \times [0,T] \to \mathbf{R}$ and f: $[0,T] \times \mathbf{R} \to K(\mathbf{R})$. We will use Theorem 1.1 to establish the existence of a nonnegative solution $y \in C[0,T]$ to (3.1). We will let $|\cdot|_0$ denote the usual norm on $C[0,T]$ i.e., $|u|_0 = \sup_{[0,T]}|u(t)|$ for $u \in C[0,T]$.

Theorem 3.1: *Let $1 \le p < \infty$ and q, $1 < q \le \infty$, the conjugate to p, k: $[0,T] \times [0,T] \to \mathbf{R}$, f: $[0,T] \times \mathbf{R} \to K(\mathbf{R})$ and assume the following conditions are satisfied*:

$$\text{for each } t \in [0,T], \text{ the map } s \mapsto k(t,s) \text{ is measurable}; \tag{3.2}$$

$$\sup_{t \in [0,T]} \left(\int_0^T |k(t,s)|^q \, ds \right)^{\frac{1}{q}} < \infty; \tag{3.3}$$

$$\int_0^T |k(t',s) - k(t,s)|^q \, ds \to 0 \quad \text{as } t \to t', \text{ for each } t' \in [0,T]; \tag{3.4}$$

$$\text{for each } t \in [0,T], \quad k(t,s) \ge 0 \quad \text{for a.e. } s \in [0,T]; \tag{3.5}$$

$$\begin{aligned}&\text{for each measurable } u\text{: } [0,T] \to \mathbf{R}, \text{ the map } t \mapsto f(t,u(t)) \\ &\text{has measurable single valued selections};\end{aligned} \tag{3.6}$$

$$\text{for a.e. } t \in [0,T], \text{ the map } u \mapsto f(t,u) \text{ is upper semicontinuous}; \tag{3.7}$$

$$\begin{aligned}&\text{for each } r > 0, \exists\, h_r \in L^p[0,T] \text{ with } |f(t,y)| \le h_r(t) \\ &\text{for a.e. } t \in [0,T] \text{ and every } y \in \mathbf{R} \text{ with } |y| \le r;\end{aligned} \tag{3.8}$$

$$\text{for a.e. } t \in [0,T] \text{ and all } y \in (0,\infty), u > 0 \text{ for all } u \in f(t,y); \tag{3.9}$$

$$\begin{aligned}&\exists\, g \in L^q[0,T] \text{ with } g\text{: } [0,T] \to (0,\infty) \text{ and} \\ &\text{with } k(t,s) \le g(s) \text{ for } t \in [0,T];\end{aligned} \tag{3.10}$$

$$\begin{aligned}&\exists\, \delta, \epsilon, 0 \le \delta < \epsilon \le T \text{ and } M, 0 < M < 1, \\ &\text{with } k(t,s) \ge M\, g(s) \text{ for } t \in [\delta,\epsilon];\end{aligned} \tag{3.11}$$

$$\begin{aligned}&\exists\, h \in L^p[0,T] \text{ with } h\text{: } [0,T] \to (0,\infty), \text{ and } w \ge 0 \text{ continuous} \\ &\text{and nondecreasing on } (0,\infty) \text{ with } |f(t,y)| \le h(t)\, w(y) \\ &\text{for a.e. } t \in [0,T] \text{ and all } y \in (0,\infty);\end{aligned} \tag{3.12}$$

$$\begin{aligned}&\exists\, \tau \in L^p[\delta,\epsilon] \text{ with } \tau > 0 \text{ a.e. on } [\delta,\epsilon] \text{ and with for a.e.} \\ &t \in [\delta,\epsilon] \text{ and } y \in (0,\infty), u \ge \tau(t)\, w(y) \text{ for all } u \in f(t,y);\end{aligned} \tag{3.13}$$

$$\exists\, \alpha > 0 \quad \text{with} \quad 1 < \frac{\alpha}{w(\alpha) \sup_{t \in [0,T]} \int_0^T k(t,s) h(s) ds} \tag{3.14}$$

and

$$\exists \beta > 0,\ \beta \neq \alpha \quad \textit{with} \quad 1 > \frac{\beta}{w(M\beta) \int_\delta^\epsilon \tau(s)\, k(\sigma, s)\, ds}; \tag{3.15}$$

here $\sigma \in [0, T]$ *is such that*

$$\int_\delta^\epsilon \tau(s) k(\sigma, s) ds = \sup_{t \in [0,T]} \int_\delta^\epsilon \tau(s) k(t, s) ds. \tag{3.16}$$

Then (3.1) has at least one nonnegative solution $y \in C[0, T]$ *and either*

(*A*) $0 < \alpha < |y|_0 < \beta$ *and* $y(t) \geq M\alpha$ *for* $t \in [\delta, \epsilon]$ *if* $\alpha < \beta$

or

(*B*) $0 < \beta < |y|_0 < \alpha$ *and* $y(t) \geq M\beta$ *for* $t \in [\delta, \epsilon]$ *if* $\beta < \alpha$

holds.

Proof: Let $E = (C[0, T], |\cdot|_0)$ and

$$C = \left\{ y \in C[0, T] : y(t) \geq 0 \text{ for } t \in [0, T] \text{ and } \min_{t \in [\delta, \epsilon]} y(t) \geq M|y|_0 \right\}.$$

Also let $A = K \circ N_f \colon C \to 2^E$, where $K \colon L^p[0, T] \to C[0, T]$ and $N_f \colon C[0, T] \to 2^{L^p[0,T]}$ are given by

$$Ky(t) = \int_0^T k(t, s) y(s) ds$$

and

$$N_f u = \{ y \in L^p[0, T] : y(t) \in f(t, u(t)) \quad \text{a.e. } t \in [0, T] \}.$$

Remark 3.1: Note A is well defined since if $x \in C$ then [4,5] guarantee that $N_f x \neq \emptyset$.

We first show $A \colon C \to 2^C$. To see this let $x \in C$ and $y \in Ax$. Then there exists a $v \in N_f x$ with

$$y(t) = \int_0^T k(t, s) v(s) ds \quad \text{for } t \in [0, T].$$

This together with (3.10) yields

$$|y(t)| \leq \int_0^T g(s) v(s) ds \quad \text{for } t \in [0, T]$$

and so

$$|y|_0 \leq \int_0^T g(s) v(s) ds. \tag{3.17}$$

On the other hand (3.11) and (3.17) yields

$$\min_{t \in [\delta, \epsilon]} y(t) = \min_{t \in [\delta, \epsilon]} \int_0^T k(t, s) v(s) ds \geq M \int_0^T g(s) v(s) ds \geq M|y|_0,$$

so $y \in C$. Thus $A\colon C \to 2^C$. A standard result from the literature [5,7,8,10] guarantees that

$$A\colon C \to K(C) \text{ is upper semicontinuous and completely continuous.}$$

Let

$$\Omega_\alpha = \{u \in C[0,T] : |u|_0 < \alpha\} \quad \text{and} \quad \Omega_\beta = \{u \in C[0,T] : |u|_0 < \beta\}.$$

Without loss of generality assume $\beta < \alpha$. If we show

$$|y|_0 < |x|_0 \quad \text{for all } y \in Ax \text{ and } x \in C \cap \partial\Omega_\alpha \tag{3.18}$$

and

$$|y|_0 > |x|_0 \quad \text{for all } y \in Ax \text{ and } x \in C \cap \partial\Omega_\beta \tag{3.19}$$

are true, then Theorem 1.1 guarantees the result.

Suppose $x \in C \cap \partial\Omega_\alpha$, so $|x|_0 = \alpha$, and $y \in Ax$. Then there exists $v \in N_f x$ with

$$y(t) = \int_0^T k(t,s)v(s)\,ds \quad \text{for } t \in [0,T].$$

Now (3.12) implies that for $t \in [0,T]$ we have

$$\begin{aligned} |y(t)| &\le \int_0^T k(t,s)h(s)w(x(s))\,ds \le w(|x|_0)\int_0^T k(t,s)h(s)\,ds \\ &\le w(\alpha) \sup_{t\in[0,T]} \int_0^T k(t,s)h(s)\,ds. \end{aligned}$$

This together with (3.14) yields

$$|y|_0 \le w(\alpha) \sup_{t\in[0,T]} \int_0^T k(t,s)h(s)ds < \alpha = |x|_0,$$

so (3.18) holds.

Next suppose $x \in C \cap \partial\Omega_\beta$, so $|x|_0 = \beta$ and $M\beta \le x(t) \le \beta$ for $t \in [\delta,\epsilon]$, and $y \in Ax$. Then there exists $v \in N_f x$ with

$$y(t) = \int_0^T k(t,s)v(s)ds \quad \text{for } t \in [0,T].$$

Notice (3.13) and (3.15) imply

$$\begin{aligned} y(\sigma) &= \int_0^T k(\sigma,s)v(s)ds \ge \int_\delta^\epsilon k(\sigma,s)v(s)ds \\ &\ge \int_\delta^\epsilon k(\sigma,s)\tau(s)w(x(s))ds \ge w(M\beta)\int_\delta^\epsilon k(\sigma,s)\tau(s)ds \\ &> \beta = |x|_0. \end{aligned}$$

Thus $|y|_0 > |x|_0$, so (3.19) holds. Now apply Theorem 1.1. □

REFERENCES

[1] R.P. Agarwal and D.O'Regan (2000). A note on the existence of multiple fixed points for multivalued maps with applications, *Jour. Differential Equations*, **160**, 389–403.

[2] C.D. Aliprantis and K.C. Border (1985). *Infinite dimensional analysis*. Berlin: Springer-Verlag.

[3] A. Cellina, A. Fryszkowski and T. Rzezuchowski (1991). Upper semicontinuity of Nemytskij operators, *Annali di Matematica Pura ed Applicata*, **160**, 321–330.

[4] K. Deimling (1992). *Multivalued differential equations*. Berlin: Walter De Gruyter.

[5] M. Frigon (1995). Thoremes d'existence de solutions d'inclusions diffrentilles. In A. Granas and M. Frigon (Eds), *Topological methods in differential equations and inclusions*, NATO ASI Series C, Dordrecht: Kluwer Academic Publishers, **472**, 51–87.

[6] M. Meehan and D. O'Regan. *Positive L^p solutions of Fredholm integral equations*, Archiv der Mathematik (to appear).

[7] D. O'Regan (1996). Integral inclusions of upper semicontinuous or lower semicontinuous type, *Proc. Amer. Math. Soc.*, **124**, 2391–2399.

[8] D. O'Regan (1997). A topological approach to integral inclusions, *Proc. Royal Irish Acad.*, **97A**, 101–111.

[9] D. O'Regan (1997). A note on solutions in $L^1[0,1]$ to Hammerstein integral equations, *Jour. Integral Equations and Applications*, **9**, 165–178.

[10] D. O'Regan and M. Meehan (1998). *Existence theory for nonlinear integral and integro-differential equations*. Dordrecht: Kluwer Academic Publishers.

SIMAA 4(2002) 11–15

2. A Note on the Structure of the Solution Set for the Cauchy Differential Inclusion in Banach Spaces

R.P. Agarwal[1] and Donal O'Regan[2]

[1]*Department of Mathematics, National University of Singapore, 10 Kent Ridge Crescent, Singapore 119260*
[2]*Department of Mathematics, National University of Ireland, Galway, Ireland*

Abstract: In this chapter we investigate the topological structure of the solution set of the Cauchy differential inclusion in Banach spaces. Our multivalued map will be assumed to satisfy a "local" integrably boundedness assumption.

1. INTRODUCTION

Let E be a Banach space, $T > 0$ and F: $[0,T] \times E \to C(E)$ a multivalued map; here $C(E)$ denotes the family of nonempty, closed, convex subsets of E. In this chapter we discuss the topological structure of the solution set of the Cauchy differential inclusion

$$\begin{aligned} &y'(t) \in F(t, y(t)) \quad \text{a.e. } t \in I \equiv [0,T] \\ &y(0) = x_0, \end{aligned} \tag{1.1}$$

using a recent result of Cichon and Kubiaczyk [1]. Throughout this chapter E is a real Banach space with norm $\|\cdot\|$. We denote by $C([0,T],E)$ the space of continuous functions y: $[0,T] \to E$. Let u: $[0,T] \to E$ be a measurable function. By $\int_0^T u(t)dt$ we mean the Bochner integral of u, assuming it exists. We define the Sobolev class $W^{1,1}([0,T],E)$ as the set of continuous functions u such that there exists $v \in L^1([0,T],E)$ with $u(t) - u(0) = \int_0^t v(s)ds$ for all $t \in [0,T]$. Notice if $u \in W^{1,1}([0,T],E)$ then u is differentiable almost everywhere on $[0,T]$, $u' \in L^1([0,T],E)$ and $u(t) - u(0) = \int_0^t u'(s)ds$ for all $t \in [0,T]$. By a solution to (1.1) we mean a function $y \in W^{1,1}([0,T],E)$ satisfying the differential equation in (1.1). Let $S(x_0)$ denote the solution set of (1.1). Recently [1] Cichon and Kubiaczyk proved the following result concerning the topological structure of $S(x_0)$.

Theorem 1.1: *Let* $E = (E, \|\cdot\|)$ *be a real Banach space,* $T > 0$, F: $[0,T] \times E \to C(E)$ *with the following conditions satisfied*:

$$F(\cdot, x) \text{ has a strongly measurable selection for each } x \in E; \tag{1.2}$$

$$F(t,\cdot) \text{ is upper semicontinuous for each } t \in [0,T]; \tag{1.3}$$

$$\exists \eta \in L^1(I) \text{ with } \|F(t,x)\| = \sup\{|y| : y \in F(t,x)\} \le \eta(t) \text{ on } I \times E \tag{1.4}$$

and

for any bounded set $\Omega \subseteq E$, $\lim_{h \to 0^+} \alpha(F(I_{t,h} \times \Omega)) \le w(t, \alpha(\Omega))$ *a.e. on* I; *here* $I_{t,h} = [t-h, t] \cap I$, $\alpha(\cdot)$ *the Kuratowski measure of noncompactness and* w: $I \times [0,\infty) \to [0,\infty)$ *is a Kamke function (i.e.,* w *is a Carathéodory map with* $\max_{s \in [0,r]} w(t,s) \in L^1(I)$ *for all* $r > 0$, *and* $\rho \equiv 0$ *is the only absolutely continuous function satisfying* $\rho(0) = 0$ *and* $\rho'(t) = w(t, \rho(t))$ *a.e. on* I. (1.5)

Then $S(x_0)$ *is nonempty, compact and connected in* $C([0,T],E)$. *In fact* $S(x_0)$ *is a* R_δ*-set.*

Remark 2.1: A set is a R_δ-set if it is the intersection of a decreasing sequence of nonempty, compact, absolute retracts.

Remark 2.2: Deimling and Rao [3] and Tolstonogov [8] showed that $S(x_0)$ is nonempty, compact and connected. Recently Cichon and Kubiaczyk [1] established that $S(x_0)$ is a R_δ-set.

The main goal of this chapter is to remove the "global" integrably boundedness assumption (1.4) on F. By using Theorem 1.1 and a trick involving the Urysohn function we are able to accomplish this if we assume a "local" integrably boundedness assumption on F. This is exactly what is needed from an application viewpoint.

2. SOLUTION SET

First we establish a general existence principle for (1.1). We assume (1.2), (1.3) and (1.5) hold. In addition suppose the following two conditions are satisfied:

$$\begin{aligned} &\text{for each } r > 0 \text{ there exists } h_r \in L^1[0,T] \text{ such that } \|F(t,x)\| \le h_r(t) \\ &\text{for a.e. } t \in [0,T] \text{ and all } x \in E \text{ with } \|x\| \le r \end{aligned} \tag{2.1}$$

and

$$\begin{aligned} &\exists M > \|x_0\| \text{ with } \|y\|_0 = \sup_{t \in [0,T]} \|y(t)\| < M \\ &\text{for any possible solution to (1.1).} \end{aligned} \tag{2.2}$$

Let $\epsilon > 0$ be given and let τ_ϵ: $E \to [0,1]$ be the Urysohn function for

$$(\overline{B}(0,M), \quad E \backslash B(0, M+\epsilon))$$

such that $\tau_\epsilon(x) = 1$ if $\|x\| \le M$ and $\tau_\epsilon(x) = 0$ if $\|x\| \ge M + \epsilon$. Let $\tilde{F}(t,x) = \tau_\epsilon(x) F(t,x)$ and consider the differential inclusion

$$\begin{aligned} y'(t) &\in \tilde{F}(t, y(t)) \quad \text{a.e. } t \in [0,T] \\ y(0) &= x_0. \end{aligned} \tag{2.3}$$

We will let $S_\epsilon(x_0)$ denote the solution set of (2.3).

Theorem 2.1: *Let $E = (E, \|\cdot\|)$ be a real Banach space, F: $[0,T] \times E \to C(E)$, and assume (1.2), (1.3), (1.5), (2.1) and (2.2) hold. Let $\epsilon > 0$ and suppose*

$$\|u\|_0 < M \quad \textit{for any possible solution } u \in W^{1,1}([0,T], E) \textit{ to (2.3)} \tag{2.4}$$

is true. Then $S(x_0)$ is a R_δ-set.

Proof: Notice $S(x_0) = S_\epsilon(x_0)$. Also (2.1) and the definition of τ_ϵ guarantees that $\tilde{F}$ satisfies (1.4) (with F replaced by $\tilde{F}$). Notice also if Ω is a bounded subset of E then for a.e. $t \in I$ we have

$$\tilde{F}(I_{t,h} \times \Omega) \subseteq \text{co}(F(I_{t,h} \times \Omega) \cup \{0\}).$$

Now from a standard property of α we have for a.e. $t \in I$ that

$$\alpha(\tilde{F}(I_{t,h} \times \Omega)) \leq \alpha(F(I_{t,h} \times \Omega))$$

As a result for a.e. $t \in I$ we have

$$\lim_{h \to 0^+} \alpha(\tilde{F}(I_{t,h} \times \Omega)) \leq \lim_{h \to 0^+} \alpha(F(I_{t,h} \times \Omega)) \leq w(t, \alpha(\Omega)).$$

Thus (1.5) is true with F replaced by $\tilde{F}$. Now Theorem 1.1 (applied to $\tilde{F}$) guarantees that $S_\epsilon(x_0)$ is a R_δ-set. □

We now use the existence principle, Theorem 2.1, to establish two applicable results for (1.1). First however recall the following three Lemma's from the literature [5,6,7].

Lemma 2.2: *Let $E = (E, \|\cdot\|)$ be a real Banach space. If $x \in W^{1,1}([0,T], E)$ then $\|x\| \in W^{1,1}([0,T], \mathbf{R})$.*

Lemma 2.3: *Let $E = (E, \|\cdot\|)$ be a real Banach space. Then the following properties hold:*

(*i*) $|\langle x, y\rangle_-| \leq \|x\|\,\|y\|$; *here* $x, y \in E$ *and* $\langle x, y\rangle_- = \|x\| \lim_{t \to 0^-} \frac{\|x+ty\| - \|x\|}{t}$;
(*ii*) $\langle \alpha x, \beta y\rangle_- = \alpha\beta\langle x, y\rangle_-$ *for all* $\alpha\beta \geq 0$ *and* $x, y \in E$;
(*iii*) *if* x: $[0,T] \to E$ *is differentiable at t then* $\|x(t)\|\, D^- \|x(t)\| = \langle x(t), x'(t)\rangle_-$; *here* D^- *is the left Dini derivative.*

Lemma 2.4: *Let $R \geq 0$, $r \in L^1([0,T], [0,\infty))$ and ψ: $[0,\infty) \to (0,\infty)$ be a Borel function such that*

$$\int_0^T r(s)ds < \int_R^\infty \frac{dx}{\psi(x)}.$$

Let M_0 be such that $\int_0^T r(s)ds = \int_R^{M_0} \frac{dx}{\psi(x)}$. Then for any $[t_0, t_1] \subseteq [0,T]$ and $z \in W^{1,1}([t_0, t_1], [0,\infty))$ with $z'(t) \leq r(t)\psi(z(t))$ for a.e. $t \in [t_0, t_1]$ and $z(t_0) \leq R$, we have $z(t) \leq M_0$ for all $t \in [t_0, t_1]$.

Proof: If $z(s) \leq R$ for all $s \in [t_0, t_1]$ then the lemma holds. On the other hand if $z(t) > R$ for some $t \in [t_0, t_1]$ then since $z(t_0) \leq R$ there exists $\mu \in [t_0, t_1]$ with $z(\mu) = R$ and $z(s) \geq R$ for $s \in [\mu, t]$. Now since $z'(t) \leq r(t)\,\psi(z(t))$ for a.e. $t \in [t_0, t_1]$ we may divide by $\psi(z(t))$ and integrate from μ to t to obtain

$$\int_R^{z(t)} \frac{dx}{\psi(x)} \leq \int_\mu^t r(s)ds \leq \int_0^T r(s)ds = \int_R^{M_0} \frac{dx}{\psi(x)}.$$

Thus $z(t) \leq M_0$. □

Theorem 2.5: *Let $E = (E, \|\cdot\|)$ be a real Banach space, F: $[0,T] \times E \to C(E)$, and assume (1.2), (1.3), (1.5) and (2.1) hold. In addition suppose the following conditions are satisfied:*

$\exists\, q \in L^1([0,T],[0,\infty))$ and ϕ: $[0,\infty) \to (0,\infty)$ a Borel measurable function such that for a.e. $t \in [0,T]$ and all $v \in E$ we have $\langle v, z\rangle_- \leq q(t)\phi(\|v\|)$ for all $z \in F(t,v)$ (2.5)

and

$$\int_0^T q(s)ds < \int_{\|x_0\|}^\infty \frac{x}{\phi(x)}\, dx. \tag{2.6}$$

Then $S(x_0)$ is a R_δ-set.

Proof: Let $\epsilon > 0$ be given,

$$I(z) = \int_{\|x_0\|}^z \frac{x}{\phi(x)} dx, \quad M_0 = I^{-1}\left(\int_0^T q(s)ds\right) \quad \text{and} \quad M = M_0 + 1.$$

We will show any solution u of (1.1) satisfies $\|u\|_0 < M$ and any possible solution y of (2.3) satisfies $\|y\|_0 < M$. If this is true, then Theorem 2.1 guarantees the result.

Suppose u is a possible solution of (1.1). Then

$$\|u(t)\|' = D^-\,\|u(t)\| \leq q(t)\,\frac{\phi(\|u(t)\|)}{\|u(t)\|} \quad \text{a.e. on } \{t : \|u(t)\| > 0\};$$

here we used Lemma 2.2 and Lemma 2.3. Now Lemma 2.4, applied with $R = \|x_0\|$, $\psi(x) = \frac{\phi(x)}{x}$ and $z(t) = \|u(t)\|$, implies $\|u(t)\| \leq M_0$ for all $t \in [0,T]$. Consequently $\|u(t)\| < M$ for all $t \in [0,T]$.

Next let y be a possible solution of (2.3). Now for a.e. $t \in [0,T]$ and all $v \in E$ we have, since τ_ϵ: $E \to [0,1]$, that

$$\langle v, z\rangle_- \leq q(t)\,\phi(\|v\|)$$

for all $z \in \tilde{F}(t,v) = \tau_\epsilon(v)F(t,v)$. Thus

$$\|y(t)\|' \leq q(t)\frac{\phi(\|y(t)\|)}{\|y(t)\|} \quad \text{a.e. on } \{t : \|y(t)\| > 0\}.$$

Now Lemma 2.4 implies $\|y(t)\| \leq M_0 < M$ for all $t \in [0, T]$. □

Corollary 2.6: *Let $E = (E, \|\cdot\|)$ be a real Banach space, F: $[0, T] \times E \to C(E)$, and assume (1.2), (1.3), (1.5) and (2.1) hold. In addition suppose the following conditions are satisfied:*

$$\begin{array}{l} \exists\, q \in L^1([0,T],[0,\infty)) \text{ and } \mu\colon [0,\infty) \to (0,\infty) \text{ a Borel} \\ \text{measurable function such that for a.e. } t \in [0,T] \\ \text{and all } v \in E \text{ we have } \|F(t,v)\| \leq q(t)\,\mu(\|v\|) \end{array} \tag{2.7}$$

and

$$\int_0^T q(s)ds < \int_{\|x_0\|}^{\infty} \frac{dx}{\mu(x)}. \tag{2.8}$$

Then $S(x_0)$ is a R_δ-set.

REFERENCES

[1] M. Cichon and I. Kubiaczyk (1999). Some remarks on the structure of the solution set for differential inclusions in Banach spaces, *Jour. Math. Anal. Appl.*, **233**, 597–606.

[2] K. Deimling (1992). *Multivalued differential equations*. Berlin: Walter de Gruyter.

[3] K. Deimling and M. Rao (1988). On solution set of multivalued differential equations, *Appl. Anal.*, **30**, 129–135.

[4] R. Kannan and D. O'Regan (2000). A note on the solution set of integral inclusions, *Jour. Integral Eqns. Applications*, **12**, 85–94.

[5] V. Lakshmikantham and S. Leela (1981). *Nonlinear differential equations in abstract spaces*. New York: Pergamon Press.

[6] M. Frigon and D. O'Regan (1994). Existence results for initial value problems in Banach spaces, *Differential Equations Dynam. Systems*, **2**, 41–48.

[7] M. Frigon and D. O'Regan (1997). Nonlinear first order initial and periodic problems in Banach spaces, *Applied Math. Letters*, **10**(4), 41–46.

[8] A.A. Tolstonogov (1983). On the structure of the solution set for differential inclusions in Banach spaces, *Math. USSR Sbornik*, **46**, 1–15.

SIMAA 4(2002) 17–26

3. Fixed Point Theory for Acyclic Maps between Topological Vector Spaces having Sufficiently many Linear Functionals, and Generalized Contractive Maps with Closed Values between Complete Metric Spaces

Ravi P. Agarwal[1] and Donal O'Regan[2]

[1]*Department Of Mathematics, National University of Singapore, 10 Kent Ridge Crescent, Singapore 119260*
[2]*Department Of Mathematics, National University of Ireland, Galway, Ireland*

Abstract: Two new fixed point theorems for acyclic maps defined on Hausdorff topological spaces with sufficiently many linear functionals and two fixed point theorems for multivalued contractions in the sense of Bose and Mukherjee are presented.

1. INTRODUCTION

This chapter has two main sections. In Section 2 we present two fixed point results for acyclic maps defined on Hausdorff topological vector spaces having sufficiently many linear functionals. In particular we will present an acyclic version of Mönch's fixed point theorem. Recall a map is acyclic if it is a upper semicontinuous multifunction with nonempty, compact, acyclic values. The results of this section contain as a special case the results of Ky Fan (see [5,6] and the references therein). In fact if the maps considered were Kututani maps (i.e., upper semicontinuous multifunction with nonempty, compact, convex values) then the results of this section could be improved considerably. The Kututani maps will be discussed in greater details in a future paper of the authors. In Section 3, we present two fixed point results for multivalued generalized contractive maps (in the sense of Bose and Mukherjee) with closed values defined on complete metric spaces.

2. FIXED POINT THEORY FOR ACYCLIC MAPS

In this section we present two fixed point results for acyclic multivalued maps on Hausdorff topological vector spaces having sufficiently many linear functionals. Recall a topological vector space E is said to have sufficiently many linear functionals if for every $x \in E$ with $x \neq 0$ there exists $l \in E^*$ (the dual space of E) with $l(x) \neq 0$ (Notice from the Hahn-Banach theorem that every Hausdorff locally convex linear topological space has sufficiently many linear functionals). The proof of our results rely on the following result in the literature due to Park [5,6].

Theorem 2.1: *Let Q be a compact, convex subset of a Hausdorff topological vector space having sufficiently many linear functionals and let $F\colon Q \to AC(Q)$ be an upper semicontinuous map (here $AC(Q)$ denotes the family of nonempty, compact, acyclic subsets of Q). Then F has a fixed point in Q.*

Remark 2.1: Recall a nonempty topological space is acyclic if all of its reduced Čech homology groups over rationals vanish.

Theorem 2.2: *Let Ω be a closed, convex subset of a Hausdorff topological vector space having sufficiently many linear functionals and $x_0 \in \Omega$. Suppose there is a closed map (i.e., has closed graph) $F\colon \Omega \to AC(\Omega)$ with the properties:*

$$A \subseteq \Omega,\ A = \mathrm{co}(\{x_0\} \cup F(A)) \text{ implies } \bar{A} \text{ is compact} \tag{2.1}$$

and

$$\text{for any } A \subseteq \Omega \text{ with } \bar{A} \text{ compact, then } F(\bar{A}) \subseteq \overline{F(A)}. \tag{2.2}$$

Then F has a fixed point in Ω.

Remark 2.2: *If* F is lower semicontinuous then (2.2) holds. Indeed if F is lower semicontinuous then for any $A \subseteq \Omega$ we have $F(\bar{A}) \subseteq \overline{F(A)}$. To see this let $x \in \bar{A}$. We wish to show $F(x) \subseteq \overline{F(A)}$. Let $z \in F(x)$, and let U be an open neighborhood of z. Then since F is lower semicontinuous we have that $F^{-1}(U)$ is an open set containing x. Now since $x \in \bar{A}$ we have $F^{-1}(U) \cap A \neq \emptyset$. Consequently $U \cap F(A) \neq \emptyset$, so $z \in \overline{F(A)}$. We can do this for all $z \in F(x)$ so $F(x) \subseteq \overline{F(A)}$.

Proof: Let

$$D_0 = \{x_0\}, \quad D_n = \mathrm{co}(\{x_0\} \cup F(D_{n-1})) \quad \text{for } n = 1, 2, \ldots \text{ and } D = \bigcup_{n=0}^{\infty} D_n.$$

Now for $n = 0, 1, \ldots$ notice D_n is convex. Also by induction we see that

$$D_0 \subseteq D_1 \subseteq \ldots \subseteq D_{n-1} \subseteq D_n \subseteq \ldots \subseteq \Omega.$$

Consequently D is convex. It is also immediate since (D_n) is increasing that

$$D = \bigcup_{n=1}^{\infty} \mathrm{co}(\{x_0\} \cup F(D_{n-1})) = \mathrm{co}(\{x_0\} \cup F(D)). \tag{2.3}$$

This together with (2.1) implies that $\bar{D}$ is compact. Also from (2.3) we have $F(D) \subseteq D \subseteq \bar{D}$ and this together with (2.2) gives $F(\bar{D}) \subseteq \overline{F(D)} \subseteq \bar{D}$. Consequently $F\colon \bar{D} \to AC(\bar{D})$ is a closed map. Now [2:pp. 465] implies $F\colon \bar{D} \to AC(\bar{D})$ is upper semicontinuous and Theorem 2.1 implies that there exists $x \in \bar{D} \subseteq \Omega$ with $x \in F(x)$. □

Next we present a Mönch fixed point theorem for acyclic maps defined on Hausdorff topological vector spaces having sufficiently many linear functionals.

Theorem 2.3: *Let Ω be a closed, convex subset of a Hausdorff topological vector space E having sufficiently many linear functionals and $x_0 \in \Omega$. Suppose there is an upper semicontinuous map $F\colon \Omega \to AC(\Omega)$ with (2.2) holding. In addition assume the following properties hold:*

$$\begin{array}{l} A \subseteq \Omega, A = \text{co}(\{x_0\} \cup F(A)) \textit{ with } \bar{A} = \bar{C} \\ \textit{and } C \subseteq A \textit{ countable, implies } \bar{A} \textit{ is compact;} \end{array} \tag{2.4}$$

$$\begin{array}{l} \textit{for any relatively compact subset } A \textit{ of } E \textit{ there} \\ \textit{exists a countable set } B \subseteq A \textit{ with } \bar{B} = \bar{A} \end{array} \tag{2.5}$$

and

$$\textit{if } A \textit{ is a compact subset of } E \textit{ then } \overline{\text{co}}(A) \textit{ is compact.} \tag{2.6}$$

Then F has a fixed point in Ω.

Remark 2.3: If E is metrizable then (2.5) holds since compact metric spaces are separable.

Remark 2.4: If E is a quasicomplete locally convex linear topological space then (2.6) holds.

Proof: Let D_n, $n = 0, 1, \ldots$, and D be as in Theorem 2.2. Now

$$D_0 \subseteq D_1 \subseteq \ldots \subseteq D_{n-1} \subseteq D_n \ldots \subseteq \Omega$$

and D is convex. In addition

$$D = \text{co}(\{x_0\} \cup F(D)). \tag{2.7}$$

We now show D_n is relatively compact for $n = 0, 1, \ldots$ Suppose D_k is relatively compact for some $k \in \{1, 2, \ldots\}$. Then [2] guarantees that $F(\bar{D}_k)$ is compact since F is upper semicontinuous. This together with (2.6) guarantees that D_{k+1} is relatively compact.

Now (2.5) implies that for each $n \in \{0, 1, \ldots\}$ there exists C_n with C_n countable, $C_n \subseteq D_n$, and $\bar{C}_n = \bar{D}_n$. Let $C = \cup_{n=0}^{\infty} C_n$. Now since

$$\bigcup_{n=0}^{\infty} D_n \subseteq \bigcup_{n=0}^{\infty} \bar{D}_n \subseteq \overline{\bigcup_{n=0}^{\infty} D_n},$$

we have

$$\overline{\bigcup_{n=0}^{\infty} \bar{D}_n} = \overline{\bigcup_{n=0}^{\infty} D_n} = \bar{D} \quad \textit{and} \quad \overline{\bigcup_{n=0}^{\infty} \bar{D}_n} = \overline{\bigcup_{n=0}^{\infty} \bar{C}_n} = \overline{\bigcup_{n=0}^{\infty} C_n} = \bar{C}.$$

Thus $\bar{C} = \bar{D}$. This together with (2.4) and (2.7) implies that $\bar{D}$ is compact. Also from (2.7) we have $F(D) \subseteq D \subseteq \bar{D}$ and this together with (2.2) gives $F(\bar{D}) \subseteq \overline{F(D)} \subseteq \bar{D}$. Consequently $F\colon \bar{D} \to AC(\bar{D})$ is upper semicontinuous. Now apply Theorem 2.1. □

Remark 2.5: Suppose in Theorem 2.3 we assumed

$$\text{for any } A \subseteq \Omega \quad \text{we have} \quad F(\bar{A}) \subseteq \overline{F(A)} \tag{2.8}$$

instead of (2.2). Then we could replace (2.4) with

$$C \subseteq \Omega \text{ countable}, \quad \bar{C} = \overline{\text{co}}(\{x_0\} \cup F(C)) \quad \text{implies } \bar{C} \text{ is compact.} \tag{2.9}$$

To see this let $A \subseteq \Omega$, $A = \text{co}\,(\{x_0\} \cup F(A))$ with $\bar{A} = \bar{C}$ and $C \subseteq A$ countable. We must show $\bar{A}(= \bar{C})$ is compact. Now (2.8) implies

$$F(A) \subseteq F(\bar{A}) = F(\bar{C}) \subseteq \overline{F(C)} \subseteq \overline{\text{co}}(\{x_0\} \cup F(C))$$

and so

$$\overline{\text{co}}(\{x_0\} \cup F(A)) \subseteq \overline{\text{co}}(\{x_0\} \cup F(C)).$$

Of course trivially

$$\overline{\text{co}}(\{x_0\} \cup F(C)) \subseteq \overline{\text{co}}(\{x_0\} \cup F(A)),$$

so

$$\overline{\text{co}}(\{x_0\} \cup F(C)) = \overline{\text{co}}(\{x_0\} \cup F(A)).$$

Consequently,

$$\bar{C} = \bar{A} = \overline{\text{co}}(\{x_0\} \cup F(A)) = \overline{\text{co}}(\{x_0\} \cup F(C)),$$

so (2.9) guarantees that $\bar{C}(= \bar{A})$ is compact.

Remark 2.6: It is possible to remove (2.2) in Theorem 2.3 if we assume

$$\begin{array}{l}\text{for any acyclic subset } A \text{ of } \Omega \text{ we have } A \cap \bar{D} \\ \text{is acyclic (where } D \text{ is as in Theorem 2.2).}\end{array} \tag{2.2*}$$

To see this proceed as in Theorem 2.2 to obtain that $\bar{D}$ is compact and $F(D) \subseteq \bar{D}$. Now let $F^*\colon \bar{D} \to AC(\bar{D})$ (see (2.2)*) be given by

$$F^*(x) = F(x) \cap \bar{D}.$$

(Note to check that $F^*(x) \neq \emptyset$ for $x \in \bar{D}$ it is enough to show $\bar{D} \subseteq F^{-1}(\bar{D})$. We have $D \subseteq F^{-1}(\bar{D})$ and since $F^{-1}(\bar{D})$ is closed, since F is upper semicontinuous, we have $\bar{D} \subseteq F^{-1}(\bar{D})$). Now F^* is upper semicontinuous since for any closed set A of $\bar{D}$ its easy to check that $(F^*)^{-1}(A) = F^{-1}(A) \cap \bar{D}$ so $(F^*)^{-1}(A)$ is closed. Now apply Theorem 2.1 (with $F = F^*$ and $Q = \bar{D}$).

Remark 2.7: It is easy to see that the upper semicontinuity assumption in Theorem 2.3 (and Remark 2.6) could be replaced by *graph* (F) is closed and F maps compact sets into relatively compact sets.

3. FIXED POINT THEORY FOR GENERALIZED CONTRACTIVE MAPS

This section presents two new fixed point results (Theorems 3.1 and 3.3) for multivalued maps with closed values defined on a complete metric space X. Our first result (Theorem 3.1) is a local version of a result of Bose and Mukherjee [3] (we will note also in this section that the result of Bose and Mukherjee follows immediately from Theorem 3.1). Our other new result (Theorem 3.3) extends some ideas of Frigon and Granas [4] to the maps considered in this section.

Let (X, d) be a metric space. By $B(x, r)$ we denote the open ball in X centered at x of radius r and by $B(C, r)$ we denote $\cup_{x\in C} B(x, r)$ where C is a subset of X. For C and K two nonempty closed subsets of X we define the generalised Hausdorff distance D to be

$$D(C, K) = \inf\{\epsilon : C \subseteq B(K, \epsilon),\ K \subseteq B(C, \epsilon)\} \in [0, \infty].$$

Theorem 3.1: *Let (X, d) be a complete metric space, $x_0 \in X$ and F: $\overline{B(x_0, r)} \to C(X)$; here $r > 0$ and $C(X)$ denotes the family of nonempty closed subsets of X. Suppose for $x, y \in \overline{B(x_0, r)}$ we have*

$$D(F(x), F(y)) \leq a_1 \operatorname{dist}(x, F(x)) + a_2 \operatorname{dist}(y, F(y)) + a_3 \operatorname{dist}(y, F(x)) \\ + a_4 \operatorname{dist}(x, F(y)) + a_5\, d(x, y),$$

where $a_1, \ldots, a_5$ are nonnegative real numbers with $a_1 + a_2 + a_3 + a_4 + a_5 < 1$, $a_1 + a_4 + a_5 > 0$, $a_2 + a_3 + a_5 > 0$, and with either $a_1 = a_2$ or $a_3 = a_4$. In addition assume

$$\operatorname{dist}(x_0, F(x_0)) < \left(\frac{1 - A_1 A_2}{1 + A_1}\right) r,$$

where

$$A_1 = \frac{a_1 + a_4 + a_5}{1 - a_2 - a_4} \quad \textit{and} \quad A_2 = \frac{a_2 + a_3 + a_5}{1 - a_1 - a_3}.$$

Then F has a fixed point (i.e., there exists $x \in \overline{B(x_0, r)}$ with $x \in F(x)$).

Remark 3.1: Note if $a_3 = a_4$ then $0 < A_1 < 1$ and $0 < A_2 < 1$, whereas if $a_1 = a_2$ we have $0 < A_1 A_2 < 1$. We may if we wish (because of symmetry) take $a_3 = a_4$ and $a_1 = a_2$ (in this case $A_1 = A_2$).

Proof: Choose $x_1 \in F(x_0)$ such that

$$d(x_1, x_0) < \left(\frac{1 - A_1 A_2}{1 + A_1}\right) r,$$

so $x_1 \in \overline{B(x_0, r)}$.

Next choose $\epsilon > 0$ such that

$$A_1\, d(x_1, x_0) + \frac{\epsilon}{1 - a_2 - a_4} < A_1 \left(\frac{1 - A_1 A_2}{1 + A_1} \right) r. \tag{3.1}$$

Then choose $x_2 \in F(x_1)$ with

$$\begin{aligned} d(x_1, x_2) &\le D(F(x_0), F(x_1)) + \epsilon \\ &\le a_1 \operatorname{dist}(x_0, F(x_0)) + a_2 \operatorname{dist}(x_1, F(x_1)) + a_3 \operatorname{dist}(x_1, F(x_0)) \\ &\quad + a_4 \operatorname{dist}(x_0, F(x_1)) + a_5\, d(x_0, x_1) + \epsilon \\ &\le a_1\, d(x_0, x_1) + a_2\, d(x_1, x_2) + a_4\, d(x_0, x_2) + a_5\, d(x_0, x_1) + \epsilon \end{aligned}$$

and so

$$d(x_1, x_2) \le A_1\, d(x_0, x_1) + \frac{\epsilon}{1 - a_2 - a_4}.$$

Now with ϵ chosen as in (3.1) we have

$$d(x_1, x_2) < A_1 \left(\frac{1 - A_1 A_2}{1 + A_1} \right) r.$$

Notice

$$x_2 \in \overline{B(x_0, r)}$$

since (we give an argument here which can be used in the general step)

$$\begin{aligned} d(x_0, x_2) &\le \left(\frac{1 - A_1 A_2}{1 + A_1} \right) r + \left(\frac{1 - A_1 A_2}{1 + A_1} \right) A_1\, r \\ &\le \frac{1 - A_1 A_2}{1 + A_1}\, r \{ 1 + A_1 A_2 + (A_1 A_2)^2 + \cdots + A_1 [1 + A_1 A_2 + (A_1 A_2)^2 + \cdots] \} \\ &= \left(\frac{1 - A_1 A_2}{1 + A_1} \right) r \left[\frac{1 + A_1}{1 - A_1 A_2} \right] = r. \end{aligned}$$

Next choose $\delta > 0$ such that

$$A_2\, d(x_1, x_2) + \frac{\delta}{1 - a_1 - a_3} < A_2 A_1 \left(\frac{1 - A_1 A_2}{1 + A_1} \right) r.$$

Then choose $x_3 \in F(x_2)$ with

$$d(x_2, x_3) \le D(F(x_2), F(x_1)) + \delta.$$

A similar reasoning as above yields

$$d(x_2, x_3) \leq A_2\, d(x_1, x_2) + \frac{\delta}{1 - a_1 - a_3}$$

and so

$$d(x_3, x_2) < A_2 A_1 \left(\frac{1 - A_1 A_2}{1 + A_1}\right) r.$$

Note

$$x_3 \in \overline{B(x_0, r)}$$

since (see the reasoning above)

$$\begin{aligned} d(x_0, x_3) &\leq \left(\frac{1 - A_1 A_2}{1 + A_1}\right) r\, [1 + A_1 + A_1 A_2] \\ &\leq \left(\frac{1 - A_1 A_2}{1 + A_1}\right) r \left[\frac{1 + A_1}{1 - A_1 A_2}\right] = r. \end{aligned}$$

Proceed inductively to obtain $x_n \in F(x_{n-1})$, $n = 4, 5, \cdots$ with $x_n \in \overline{B(x_0, r)}$ and

$$d(x_{2j+1}, x_{2j+2}) \leq (A_1 A_2)^j\, A_1 \left(\frac{1 - A_1 A_2}{1 + A_1}\right) r, \quad j = 1, 2, \ldots$$

and

$$d(x_{2j}, x_{2j+1}) \leq (A_1 A_2)^j \left(\frac{1 - A_1 A_2}{1 + A_1}\right) r, \quad j = 2, 3, \ldots.$$

Now it is immediate since $0 < A_1 A_2 < 1$ that (x_n) is Cauchy. Also since X is complete there exists $x \in \overline{B(x_0, r)}$ with $\lim_{n\to\infty} x_n = x$. It remains to show $x \in F(x)$. Notice

$$\begin{aligned} \text{dist}(x, F(x)) &\leq d(x, x_{2n+1}) + \text{dist}(x_{2n+1}, F(x)) \\ &\leq d(x, x_{2n+1}) + D(F(x_{2n}), F(x)) \end{aligned}$$

and so

$$\begin{aligned} D(F(x_{2n}), F(x)) &\leq a_1\, \text{dist}(x_{2n}, F(x_{2n})) + a_2\, \text{dist}(x, F(x)) + a_3\, \text{dist}(x, F(x_{2n})) \\ &\quad + a_4\, \text{dist}(x_{2n}, F(x)) + a_5\, d(x_{2n}, x) \\ &\leq a_1\, d(x_{2n}, x_{2n+1}) + a_2\, [d(x, x_{2n+1}) + D(F(x_{2n}), F(x))] \\ &\quad + a_3\, d(x, x_{2n+1}) + a_4 [d(x_{2n}, x_{2n+1}) + D(F(x_{2n}), F(x))] \\ &\quad + a_5\, d(x_{2n}, x). \end{aligned}$$

Consequently,

$$\begin{aligned} D(F(x_{2n}), F(x)) &\leq \left(\frac{a_1 + a_4}{1 - a_2 - a_4}\right) d(x_{2n}, x_{2n+1}) + \left(\frac{a_2 + a_3}{1 - a_2 - a_4}\right) d(x, x_{2n+1}) \\ &\quad + \left(\frac{a_5}{1 - a_2 - a_4}\right) d(x_{2n}, x). \end{aligned}$$

As a result we have

$$\begin{aligned}\operatorname{dist}(x, F(x)) &\leq \left(\frac{a_1 + a_4}{1 - a_2 - a_4}\right) d(x_{2n}, x_{2n+1}) + \left(\frac{1 + a_3 - a_4}{1 - a_2 - a_4}\right) d(x, x_{2n+1}) \\ &\quad + \left(\frac{a_5}{1 - a_2 - a_4}\right) d(x_{2n}, x) \\ &\to 0 \quad \text{as } n \to \infty.\end{aligned}$$

Thus $x \in \overline{F(x)} = F(x)$ and we are finished. □

We next note that we obtain Bose and Mukherjee's result [3] as a Corolllary of Theorem 3.1.

Theorem 3.2: *Let (X, d) be a complete metric space, $F\colon X \to C(X)$. Suppose for $x, y \in X$ we have*

$$\begin{aligned}D(F(x), F(y)) \leq a_1 \operatorname{dist}(x, F(x)) + a_2 \operatorname{dist}(y, F(y)) + a_3 \operatorname{dist}(y, F(x)) \\ + a_4 \operatorname{dist}(x, F(y)) + a_5\, d(x, y)\end{aligned}$$

where $a_1, \ldots, a_5$ are nonnegative real numbers with $a_1 + a_2 + a_3 + a_4 + a_5 < 1$, $a_1 + a_4 + a_5 > 0$, $a_2 + a_3 + a_5 > 0$, and with either $a_1 = a_2$ or $a_3 = a_4$. Then F has a fixed point.

Proof: Fix $x_0 \in X$. Choose $r > 0$ so that

$$\operatorname{dist}(x_0, F(x_0)) < \left(\frac{1 - A_1 A_2}{1 + A_1}\right) r.$$

Now Theorem 3.1 guarantees that there exists $x \in \overline{B(x_0, r)}$ with $x \in F(x)$. □

Theorem 3.3: *Let (X, d) be a complete metric space with U an open subset of X. Suppose $H\colon \bar{U} \times [0,1] \to C(X)$ is a closed map (i.e., has closed graph) with the following satisfied:*

(a) *$x \notin H(x, t)$ for $x \in \partial U$ and $t \in [0,1]$;*
(b) *for all $t \in [0,1]$ and $x, y \in \bar{U}$, $D(H(x,t), H(y,t)) \leq a_1 \operatorname{dist}(x, H(x,t)) + a_2 \operatorname{dist}(y, H(y,t)) + a_3 \operatorname{dist}(y, H(x,t)) + a_4 \operatorname{dist}(x, H(y,t)) + a_5\, d(x,y)$ (here $a_1, \ldots, a_5$ are nonnegative real numbers with $a_1 + a_2 + a_3 + a_4 + a_5 < 1$, $a_1 + a_4 + a_5 > 0$, $a_2 + a_3 + a_5 > 0$, and with either $a_1 = a_2$ or $a_3 = a_4$); and*
(c) *there exists a continuous increasing function $\phi\colon [0,1] \to \mathbf{R}$ such that $D(H(x,t), H(x,s)) \leq |\phi(t) - \phi(s)|$ for all $t, s \in [0,1]$ and $x \in \bar{U}$.*

Then $H(\cdot, 0)$ has a fixed point iff $H(\cdot, 1)$ has a fixed point.

Proof: Suppose $H(\cdot, 0)$ has a fixed point. Consider

$$Q = \{(t, x) \in [0,1] \times U : x \in H(x, t)\}.$$

Now Q is nonempty since $H(\,\cdot\,, 0)$ has a fixed point. On Q define the partial order

$$(t, x) \leq (s, y) \quad \text{iff} \quad t \leq s \quad \text{and} \quad d(x, y) \leq 2\left(\frac{1 + A_1}{1 - A_1 A_2}\right) [\phi(s) - \phi(t)],$$

where

$$A_1 = \frac{a_1 + a_4 + a_5}{1 - a_2 - a_4} \quad \text{and} \quad A_2 = \frac{a_2 + a_3 + a_5}{1 - a_1 - a_3}.$$

Let P be a totally ordered subset of Q and let

$$t^* = \sup\{t : (t, x) \in P\}.$$

Take a sequence $\{(t_n, x_n)\}$ in P such that $(t_n, x_n) \leq (t_{n+1}, x_{n+1})$ and $t_n \to t^*$. We have

$$d(x_m, x_n) \leq 2\left(\frac{1 + A_1}{1 - A_1 A_2}\right) [\phi(t_m) - \phi(t_n)] \quad \text{for all } m > n,$$

and so (x_m) is a Cauchy sequence, which converges to some $x^* \in \bar{U}$. Now since H is a closed map we have $(t^*, x^*) \in Q$ (note $x^* \in H(x^*, t^*)$ by closedness and (a) implies $x^* \in U$). It is also immediate from the definition of t^* and the fact that P is totally ordered that

$$(t, x) \leq (t^*, x^*) \quad \text{for every } (t, x) \in P.$$

Thus (t^*, x^*) is an upper bound of P. By Zorn's Lemma Q admits a maximal element $(t_0, x_0) \in Q$.

We *claim* $t_0 = 1$ (if our claim is true then we are finished). Suppose our claim is false. Then, choose $r > 0$ and $t \in (t_0, 1]$ with

$$\overline{B(x_0, r)} \subseteq U \quad \text{and} \quad r = 2\left(\frac{1 + A_1}{1 - A_1 A_2}\right) [\phi(t) - \phi(t_0)].$$

Notice

$$\begin{aligned} \operatorname{dist}(x_0, H(x_0, t)) &\leq \operatorname{dist}(x_0, H(x_0, t_0)) + D(H(x_0, t_0)), H(x_0, t))) \\ &\leq \phi(t) - \phi(t_0) = \frac{1}{2}\left(\frac{1 - A_1 A_2}{1 + A_1}\right) r < \left(\frac{1 - A_1 A_2}{1 + A_1}\right) r. \end{aligned}$$

Now Theorem 3.1 guarantees that $H(\cdot, t)$ has a fixed point $x \in \overline{B(x_0, r)}$. Thus $(x, t) \in Q$ and notice since

$$d(x_0, x) \leq r = 2\left(\frac{1 + A_1}{1 - A_1 A_2}\right) [\phi(t) - \phi(t_0)] \quad \text{and} \quad t_0 < t,$$

we have $(t_0, x_0) < (t, x)$. This contradicts the maximality of (t_0, x_0). □

REFERENCES

[1] R.P. Agarwal and D. O'Regan (2001). Essentiality and Mönch type maps, *Proc. Amer. Math. Soc.*, **129**, 1015–1020.

[2] C.D. Aliprantis and K.C. Border (1994). *Infinite dimensional analysis*. Berlin: Springer-Verlag.

[3] R.K. Bose and R.N. Mukherjee (1977). Common fixed points of some multivalued maps, *Tamkang Jour. Math.*, **8**, 245–249.

[4] M. Frigon and A. Granas (1994). Résultats du type de Leray-Schauder pour des contractions multivoques, *Topolo. Methods Nonlinear Anal.*, **4**, 197–208.

[5] S. Park (1988). Fixed point theorems on compact convex sets in topological vector spaces, *Contemp. Math.*, **72**, 183–191.

[6] S. Park (1996). *Fixed points of acyclic maps on topological vector space*, World Congress of Nonlinear Analysts '92, Walter de Gruyter (V. Lakshmikantham (Ed.)), Berlin, 2171–2177.

SIMAA 4(2002) 27–38

4. Using the Integral Manifolds to Solvability of Boundary Value Problems*

Jan Andres

Department of Math. Analysis, Fac. of Science, Palacký University, Tomkova 40, 779 00 Olomouc-Hejčín, Czech Republic. E-mail: andres@risc.upol.cz

Abstract: Acyclic solution sets of quasi-linearized differential systems with constraints are employed for solving nonlinear boundary value problems. A special attention is paid to periodic and anti-periodic solutions. The sufficient criteria are obtained in terms of (not necessarily smooth) bounding functions. The intersection of sublevel sets of these Liapunov-like functions forms a desired bound set with a transversality behaviour on its boundary.

Keywords and Phrases: Solution sets, boundary value problems, topological structure, multi-valued method, bound sets, bounding functions

AMS Subject Classification: 34A60, 34B15, 47H04

1. INTRODUCTION

The aim of our chapter consists in showing how methods of set-valued analysis can be adapted to solve nontraditionally boundary value problems for ordinary differential equations or, more generally, inclusions. More precisely, the solution sets (i.e., integral manifolds, whence the title) of linearized differential systems, satisfying given boundary conditions, give rise to multivalued operators with suitable properties. Thus, the original problem turns out to be equivalent with a fixed-point problem for these operators (cf. [2,3,14,15,25] and the references therein).

This approach requires, besides another, verifying the topological structure of solution sets to linearized systems (in Chapter 3) and the transversality behaviour of solutions on the boundary of a certain set, called the bound set, (in Chapter 5). The latter is guaranteed by constructing a special sort of Liapunov-like functions, called bounding functions, which is always a difficult task. So, the multivalued method developed in [3] (and recalled in Chapter 4) is appropriately elaborated in Chapter 6, where the main results are formulated.

We believe that this method deserves some future interest, because new results might be obtained in this way (cf. [7]), having no analogy by means of a standard (single-valued) manner.

* Supported by the Council of Czech Government (J 14/98: 153100011)

2. SOME PRELIMINARIES

In the entire text, all topological spaces are metric. Let us recall that a set is *acyclic* (w.r.t. any continuous theory of cohomology) if it is homologically same as a one point space. A nonempty space Y is called an *absolute retract* (AR) if, for any metrizable X and any closed $A \subset X$, every continuous map $F: A \to Y$ is extendable over X. By an *R_δ-set* we mean the intersection of a decreasing sequence of compact AR-spaces. Let us note that any R_δ-set is well-known to be acyclic.

Furthermore, let us recall that a multivalued map with closed values $\varphi: X \rightsquigarrow Y$ (i.e., $\varphi: X \to 2^Y \setminus \{\emptyset\}$) is *measurable* if, for any open $U \subset Y$, the set $\{x \in X : \varphi(x) \cap U \neq \emptyset\}$ is measurable. It is *upper semi-continuous* (*u.s.c.*) if $\{x \in X : \varphi(x) \subset U\}$ is open in X, for every open subset $U \subset Y$. Obviously, all u.s.c. maps are measurable. A multivalued map is called *acyclic* (or, in particular, R_δ) if it is an u.s.c. map with nonempty, compact, acyclic (or, in particular, R_δ) values.

At last, all boundary value problems (BVPs) under our consideration will take the form (in general)

$$\begin{aligned} &x' \in F(t,x) \quad \text{for a.a. } t \in I, \\ &x \in S, \end{aligned} \tag{0}$$

where I is a given real compact interval and $F: I \times \mathbb{R}^n \rightsquigarrow \mathbb{R}^n$ is an (upper) *Carathéodory function*, i.e.,

(i) the set of values of F is nonempty, compact and convex for all $(t,x) \in I \times \mathbb{R}^n$;
(ii) $F(t,\,\cdot\,)$ is u.s.c. for a.a. $t \in I$;
(iii) $F(\,\cdot\,,x)$ is measurable for all $x \in \mathbb{R}^n$.

By a *solution* $x(t)$ of (0) we always mean the Carathéodory one, namely an absolutely continuous function $x(t) \in AC(I,\mathbb{R}^n)$ satisfying (0), for a.a. $t \in I$.

For more details and information, we recommend the monograph [21].

3. TOPOLOGICAL STRUCTURE OF SOLUTION SETS

Although the problem of the investigation of structure of solution sets to differential systems comes back to H. Kneser in 1923, they are only rare results related to boundary value problems (see [2,4,5,10–13,17–19,22,26–28]). On the other hand, Cauchy (initial value) problems are treated with this respect rather frequently (see e.g., [3–5,9,18,21] and the references therein). Nevertheless, because of our interest (i.e., BVPs), we recall here only those in [10] and [13] which seem to be the most appropriate for our goal.

In [10], a rather large family of multivalued BVPs has been examined as follows.

Proposition 1: *Consider the problem*

$$\begin{aligned} &x' + A(t)x \in F(t,x) \quad \textit{for a.a. } t \in I, \\ &Lx = \Theta \quad (\Theta \in \mathbb{R}^n), \end{aligned} \tag{1}$$

on a compact interval I, where $A: I \to \mathbb{R}^n \times \mathbb{R}^n$ is a single-valued essentially bounded Lebesgue measurable $(n \times n)$-matrix and $F: I \times \mathbb{R}^n \rightsquigarrow \mathbb{R}^n$ is a Carathéodory function *which is lipschitzian in x, for a.a. $t \in I$, with a sufficiently small Lipschitz constant k (i.e., $h(F(t,x), F(t,y)) \leq k|x-y|$ for all $x, y \in \mathbb{R}^n$ and a.a. $t \in I$, where $h(.,.)$ stands for the Hausdorff metric.*

Let, furthermore, $L\colon C(I, \mathbb{R}^n) \to \mathbb{R}^n$ be a linear operator such that the associated homogeneous problem,

$$\begin{aligned} &x' + A(t)x = 0 \quad \text{for a.a. } t \in I, \\ &Lx = 0, \end{aligned}$$

has only the trivial solution on I.

Then the set of solutions of (1) is a (nonempty) compact AR-space (i.e., more than R_δ-set).

Remark 1: In fact, Proposition 1 represents only a particular case of a more general result in [10], where the functional dependence has been also taken into account. Moreover, the covering dimension of solution sets was studied there. On the other hand, because of the applied contraction principle, the assertion reduces just to the uniqueness property for ODEs.

Therefore, we add still the following statements in [13] (cf. Theorems 4 and 5 in [13]).

Proposition 2: *Consider the Floquet problem*

$$\begin{aligned} &x' = f(t,x) \quad \text{for a.a. } t \in [a,b], \\ &x(a) + \mu x(b) = \xi \quad (\mu > 0, \xi \in \mathbb{R}^n), \end{aligned} \tag{2}$$

where $f\colon I \times \mathbb{R}^n \to \mathbb{R}^n$ is a bounded Carathéodory function. Assume, furthermore, that f satisfies

$$|f(t,x) - f(t,y)| \leq p(t)|x-y| \quad \text{for a.a. } t \in [a,b] \text{ and } x, y \in \mathbb{R}^n, \tag{3}$$

where $p\colon [a,b] \to [0,\infty)$ is a Lebesgue integrable function such that

$$\int_a^b p(t)dt \leq \sqrt{\pi^2 + \ln^2 \mu}. \tag{4}$$

Then the set of solutions of (2) is an R_δ-set.

Proposition 3: *Consider the Cauchy–Nicoletti problem*

$$\begin{aligned} &x_i' = f_i(t, x_1, \ldots, x_n) \quad \text{for a.a. } t \in [a,b], (i = 1, \ldots, n), \\ &x_i(t_i) = \xi_i \quad (\xi_i \in \mathbb{R}, t_i \in [a,b], i = 1, \ldots, n;\ \xi = (\xi_1, \ldots, \xi_n)), \end{aligned} \tag{5}$$

where $f = (f_1, \ldots, f_n)\colon [a,b] \times \mathbb{R}^n \to \mathbb{R}^n$ is a bounded Carathéodory function. Assume, furthermore, that f satisfies (3), where $p\colon [a,b] \to [0,\infty)$ is a Lebesgue integrable function satisfying this time

$$\int_a^b p(t)dt \leq \frac{\pi}{2}. \tag{6}$$

Then the set of solutions of (5) is an R_δ-set.

Remark 2: As pointed out in [13], if the sharp inequalities hold in (4) or (6), then problem (2) or (5) has a unique solution, respectively. On the other hand, for non-sharp inequalities (4) or (6), problem (2) or (5) can possess more solutions, respectively.

To conclude this section, in order to apply the above propositions appropriately in the sequel, let us note that, for a Carathéodory function $G(t, x, y)$: $I \times \mathbb{R}^n \times \mathbb{R}^n \rightsquigarrow \mathbb{R}^n$, the composed multifunction $G(t, x, q(t))$, with $q \in C(I, \mathbb{R}^m)$, becomes Carathéodory, provided G is product-measurable (see e.g., [8], p. 36). In the single-valued case, it is well-known that this property is satisfied automatically.

Definition 1: For a Carathéodory function $G(t, x, q(t))$, where $q \in C(I, \mathbb{R}^m)$ and $G(t, x, y)$: $I \times \mathbb{R}^n \times \mathbb{R}^n \rightsquigarrow \mathbb{R}^n$ is product-measurable and Carathéodory, the sufficient conditions in Propositions 1–3 will be denoted, for the sake of simplicity, as (P1), (P2), (P3), respectively.

4. GENERAL MULTIVALUED METHOD

Now, we shall see how the information about the structure of solution sets to quasi-linearized systems (when applying the Schauder linearization device) can be employed (see condition (i) below) for solving given boundary value problems. The appropriately modified Theorem 2.33 in [3] reads as follows.

Theorem 1: *Consider the boundary value problem*

$$\begin{aligned} &x' \in F(t, x) \quad \text{for a.a. } t \in I, \\ &x \in S, \end{aligned} \tag{7}$$

where I is a given real compact interval, F: $I \times \mathbb{R}^n \rightsquigarrow \mathbb{R}^n$ is a Carathéodory function and S is a subset of $AC(I, \mathbb{R}^n)$.

Let G: $I \times \mathbb{R}^n \times \mathbb{R}^n \times [0,1] \rightsquigarrow \mathbb{R}^n$ be a Carathéodory function such that

$$G(t, c, c, 1) \subset F(t, c), \quad \text{for all } (t, c) \in I \times \mathbb{R}^n.$$

Assume that

(*i*) *There exists a bounded retract Q of $C(I, \mathbb{R}^n)$ such that $Q \backslash \partial Q$ is nonempty open and a closed bounded subset S_1 of S such that the associated problem*

$$\begin{aligned} &x' \in G(t, x, q(t), \lambda) \quad \text{for a.a. } t \in I, \\ &x \in S_1, \end{aligned} \tag{8}$$

is solvable with R_δ-sets of solutions, for each $(q, \lambda) \in Q \times [0, 1]$;

(*ii*) *There exists a locally integrable function α: $I \to \mathbb{R}$ such that*

$$|G(t, x(t), q(t), \lambda)| \leq \alpha(t) \quad \text{a.e. in } I,$$

for any $(x, q, \lambda) \in \Gamma_T$ (i.e., from the graph Γ of T), where T denotes the set-valued map which assigns to any $(q, \lambda) \in Q \times [0, 1]$ the set of solutions of (8);

(*iii*) $T(Q \times \{0\}) \subset Q$;
(*iv*) *The map T has no fixed points on the boundary ∂Q of Q, for every $(2, \lambda) \in Q \times [0, 1]$.*

Then problem (7) admits a solution.

Remark 3: Theorem 1 extends many of its analogies (cf. e.g., [2,14,15] and the references therein). On the other hand, often "only" acyclicity of solution sets is to our disposal (see e.g., [4,12]).

Hence, we add still another statement based on the application of the Eilenberg–Montgomery fixed-point theorem (cf. Corollary 2.35 in [3]).

Theorem 2: *Consider problem (7). Let $G: I \times \mathbb{R}^n \times \mathbb{R}^n \rightsquigarrow \mathbb{R}^n$ be a Carathéodory function such that*

$$G(t, c, c) \subset F(t, c), \quad \textit{for all } (t, c) \in I \times R^n.$$

Assume that

(*i*) *There exists a convex closed subset Q of $C(I, \mathbb{R}^n)$ such that the associated problem*

$$\begin{aligned} &x' \in G(t, x, q(t)) \quad \textit{for a.a. } t \in I, \\ &x \in S \cap Q, \end{aligned} \tag{9}$$

has an acyclic set of solutions, for each $q \in Q$;
(*ii*) *There exists a locally integrable function $\alpha: I \to \mathbb{R}$ such that*

$$|G(t, x(t), q(t))| \le \alpha(t) \quad \textit{a.e. in } I,$$

for any pair $(q, x) \in \Gamma_T$;
(*iii*) *$T(Q)$ is bounded in $C(I, \mathbb{R}^n)$ and $\overline{T(Q)} \subset S$.*

Then problem (7) admits a solution.

Of course, both theorems can be easily specified w.r.t. concrete BVPs (1) or (2) or (3), when applying properties (P1) or (P2) or (P3) (see Definition 1) in condition (i), respectively. However, condition (iv) in Theorem 1 (or, in particular, condition (iii) in Theorem 2) are still rather implicit. Therefore, in the next section, this condition will be expressed more explicitly in terms of locally lipschitzian bounding functions.

5. BOUND SETS FOR BVPs

In this part, we follow the ideas in [20,24,30] and especially in [29], in order to elaborate them, by means of the techniques in [1,9,23], in a multivalued way. Unlike in all mentioned chapters, but [1], solutions are understood in the Carathéodory sense, which brings some technical difficulties. The details will be published elsewhere (see [6]).

Let us recall some appropriately modified definitions.

Definition 2: By a bound set to problem (7) (or (8)) we mean a bounded subset $\mathcal{B} = \bigcup_{A \in J} \mathcal{B}(A) \subset \mathbb{R}^n$ such that $\mathcal{B}(A)$ is nonempty and open, for each $A \in J$, for which there is no solution $x(t)$ of (7) (or (8)) such that if $x(t) \in \overline{\mathcal{B}(A)}$, for every $t \in J$, then $x(t_0) \in \partial\mathcal{B}(A_0)$, for some $t_0 \in J$.

Obviously, if $\mathcal{B}$ is a bound set to (7) (or (8)) and

$$\Omega := \{x \in S : x(t) \in \mathcal{B}(A), \text{for all } t \in J\},$$

$$\partial\Omega := \{x \in S : x(t) \in \overline{\mathcal{B}(A)}, \text{for all } t \in J, \text{and } x(t_0) \in \partial\mathcal{B}(A_0), \text{for some } t_0 \in J\},$$

then no solution $x(t)$ of (7) (or (8)) can belong to the boundary $\partial\Omega$ of Ω.

As usual, a function $V(t,x) \in C(J \times \mathbb{R}^n, \mathbb{R})$ is said to be *locally lipschitzian at a point* $x_0 \in \mathbb{R}^n$ if there exist a positive constant L and a neighbourhood U of x_0 such that

$$|V(t,x) - V(t,y)| \le L|x-y| \quad \text{for all } x,y \in U, \text{uniformly w.r.t. } t \in J.$$

For such a function V, we can define in a standard manner the *upper right* and the *lower left Dini derivatives* of V at $x_0 \in \mathbb{R}^n$, calculated in $x_1 \in \mathbb{R}^n$, by

$$D^+V(t,x_0)(x_1) := \limsup_{h\to 0^+} \frac{1}{h}[V(t+h, x_0+hx_1) - V(t,x_0)],$$

and

$$D_-V(t,x_0)(x_1) := \liminf_{h\to 0^-} \frac{1}{h}[V(t+h, x_0+hx_1) - V(t,x_0)],$$

respectively.

In order to find desired bound sets to (7) (or (8), for verifying condition (iv) in Theorem (1), bounding functions $V(t,x)$ can be constructed as follows.

Lemma 1: *Assume that a nonempty, open, bounded subset $\mathcal{B} \subset \mathbb{R}^n$ exists jointly with a one-parameter family of bounding functions $V_u(t,x) \in C(J \times \mathbb{R}^n, \mathbb{R})$, which are locally lipschitzian in $u \in \bigcup_{A\in J} \partial\mathcal{B}(A)$, such that the following conditions are satisfied:*

$$\begin{aligned} &\forall u \in \cup\partial\mathcal{B}(s)\, \exists r_u > 0 \ \textit{such that} \\ &V_u(t,y) \le 0\, \forall y \in \overline{\mathcal{B}} \cap B_u^{r_u}, \ \textit{uniformly w.r.t. } t \in J, \\ &\textit{and } V_u(t,u) \equiv 0, \end{aligned} \tag{B1}$$

where $B_u^r = \{x \in \mathbb{R}^n : |x-u| < r\}$,

$$\begin{aligned} &\forall u \in \bigcup_{A\in J} \partial\mathcal{B}(A), \quad \textit{for a.a. } t \in J, \forall w \in \{F(t,u)\}: \\ &(-\varepsilon, \varepsilon) \not\subset [D^+V_u(t,u)(w), D_-V_u(t,u)(w)], \end{aligned} \tag{B2}$$

where ε is a suitable positive constant.

Then every solution $x(t)$ of a Carathéodory inclusion $x' \in F(t,x)$ with $x(t) \in \overline{\mathcal{B}(A)}$, for every $t \in J$, satisfies $x(t) \in \mathcal{B}(A)$, for every $t \in \text{int}J = \overline{J}\backslash\partial J$.

Proof: Can be done similarly as in [29], where only classical solutions of ODEs have been considered, by means of the arguments used in a slightly different context in [9], (cf. also [23]).

Remark 4: For C^1-functions $V_u(t,x)$, we have

$$D^+V_u(t,u)(w) = D_-V_u(t,u)(w) = \langle \mathbf{grad}V_u(t,u),(1,w)\rangle,$$
for every $w \in \{F(t,u)\}$.

Thus, condition (B2) takes the simple form

$$\forall u \in \partial\mathcal{B}(A), \text{ for a.a. } t \in J, \langle \mathbf{grad}V_u(t,u),(1,F(t,u))\rangle \geq \varepsilon > 0 \text{ or}$$
$$\langle \mathbf{grad}V_u(t,u),(1,F(t,u))\rangle \leq -\varepsilon < 0,$$

where $\langle \cdot,\cdot \rangle$ stands for an inner product.

In order $\mathcal{B}$ to be a bound set to (7) (or (8), we must still show that, for each solution $x(t)$ of (7) (or (8)) such that $x(t) \in \overline{\mathcal{B}(A)}$, for every $t \in J = [a,b]$, $x(a) \notin \partial\mathcal{B}(a)$ and $x(b) \notin \partial\mathcal{B}(b)$.

For this, the most convenient problem for our consideration seems to be

$$\begin{aligned} &x' \in F(t,x) \quad \text{for a.a. } t \in [a,b], \\ &x(b) = Mx(a), \end{aligned} \tag{10}$$

where F is a Carathéodory function and M is a regular (nonsingular) $n \times n$-matrix. Observe that, in particular, for $M = -E$, (10) becomes an anti-periodic problem (cf. (2)) and, for $M = E$, a periodic problem.

Let us also recall that a subset $\mathcal{S} \subset \mathbb{R}^n$ is said to be *invariant w.r.t. a subgroup $\mathcal{H}$* of the group $GL_n(\mathbb{R})$ of the real nonsingular $(n \times n)$-matrices if

$$\forall u \in \mathcal{S}, \forall H \in \mathcal{H} : Hu \in \mathcal{S}.$$

Lemma 2: *Let the assumptions of Lemma 1 be satisfied. Assume, furthermore, that $M\partial\mathcal{B}(a) = \partial\mathcal{B}(b)$ takes place and sufficiently small positive constants δ, ε exist such that*

$$\begin{aligned} &\forall u \in \partial\mathcal{B}(A_a), \textit{for a.a. } t_a \in [a, a+\delta), \textit{for a.a. } t_b \in (b-\delta, b], \\ &\forall w_a \in \{F(t_a,u)\}, \forall w_b \in \{F(t_b, Mu)\} : \\ &(-\varepsilon,\varepsilon) \not\subset [D^+V_u(a,u)(w_a), D_-V_{Mu}(b,Mu)(w_b)]. \end{aligned} \tag{B3}$$

Then $\mathcal{B}$ is a bound set to (10).

Proof: Can be done similarly as in [29], by means of the arguments used in a slightly different context in [9].

Remark 5: For C^1-functions $V_u(t,x)$, condition (B3) takes the form

$$\forall u \in \partial\mathcal{B}(A_a), \text{ for a.a. } t_a \in [a,a+\delta), \text{ for a.a. } t_b \in (b-\delta,b] : (-\varepsilon,\varepsilon) \not\subset$$
$$[\langle \mathbf{grad}V_u(a,u),(1,F(t_a,u))\rangle, \langle \mathbf{grad}V_{Mu}(b,Mu),(1,F(t_b,Mu))\rangle].$$

Unfortunately, problems like (5) are much less convenient to handle for the same goal. Therefore, in the remaining part, we restrict ourselves only to (10) and its particular cases $M = \pm E$.

6. MAIN RESULTS

Hence, let us consider the problem

$$\begin{aligned} &x' + A(t)x \in F(t,x) \quad \text{for a.a. } t \in [a,b], \\ &x(b) = Mx(a), \end{aligned} \tag{11}$$

where A: $[a,b] \to \mathbb{R}^n \times \mathbb{R}^n$ is a single-valued essentially bounded Lebesgue measurable $(n \times n)$-matrix, F: $[a,b] \times \mathbb{R}^n \rightsquigarrow \mathbb{R}^n$ is a Carathéodory function and M is a regular $(n \times n)$-matrix.

Summarizing the information from the foregoing sections, we are ready to give the first main result.

Theorem 3: *Consider problem (11) and let G: $[a,b] \times \mathbb{R}^n \times \mathbb{R}^n \times [0,1] \rightsquigarrow \mathbb{R}^n$ be a product-measurable Carathéodory function such that*

$$G(t,c,c,1) \subset F(t,c) \quad \textit{for all } (t,c) \in [a,b] \times \mathbb{R}^n.$$

Assume that

(i) *The associated homogeneous problem*

$$\begin{aligned} &x' + A(t)x = 0 \quad \textit{for a.a. } t \in [a,b], \\ &x(b) = Mx(a), \end{aligned}$$

has only the trivial solution;

(ii) *There exists a bounded retract Q of $C([a,b], \mathbb{R}^n)$ with the nonempty open $Q \backslash \partial Q$ such that $G(t,x,q(t),\lambda)$ is lipschitzian in x with a sufficiently small Lipschitz constant, for a.a. $t \in [a,b]$ and each $(q,\lambda) \in Q \times [0,1]$*;

(iii) *There exists a Lebesgue integrable function α: $[a,b] \to \mathbb{R}$ such that*

$$|G(t,x(t),q(t),\lambda)| \le \alpha(t) \quad \textit{a.e. in } [a,b],$$

for any $(x,q,\lambda) \in \Gamma_T$, where T denotes the set-valued map which assigns to any $(q,\lambda) \in Q \times [0,1]$ the set of solutions of

$$\begin{aligned} &x' + A(t)x \in G(t,x,q(t),\lambda) \quad \textit{for a.a. } t \in [a,b], \\ &x(b) = Mx(a); \end{aligned}$$

(iv) *$T(Q \times \{0\}) \subset Q$ and ∂Q is fixed-point free*;

(v) *For each $u \in \bigcup_{A \in J} \partial \mathcal{B}(A)$, where $\mathcal{B} = \operatorname{int}\{x(t) \in \mathbb{R}^n : x \in Q\} \neq \phi$ is the same, for every $A \in [a,b]$, there exists a bounding function $V_u(t,x) \in C([a,b] \times \mathbb{R}^n, \mathbb{R}^n)$, which is locally lipschitzian in $u \in \partial \mathcal{B}$, such that conditions (B1)–(B3) hold for any $(q,\lambda) \in Q \times (0,1]$, where $F := G(t,x,q(t),\lambda) - A(t)x$ and $I := [a,b]$*;

(*vi*) $M\partial\mathcal{B}(a) = \partial\mathcal{B}(b)$.

Then problem (11) admits a solution.

Proof: Follows immediately from Theorem 1, Proposition 1 and Lemma 2.

If, in particular $M = E$ or $M = -E$, then Theorem 3 significantly simplifies as follows.

Corollary 1: *Consider problem (11), where $M = E$ and $F(t,x) \equiv F(t+(b-a),x)$. Let $G\colon [a,b] \times \mathbb{R}^n \times \mathbb{R}^n \rightsquigarrow \mathbb{R}^n$ be a product-measurable Carathéodory function such that*

$$G(t,c,c) \subset F(t,c) \quad \textit{for all } (t,c) \in [a,b] \times \mathbb{R}^n.$$

Assume that

(*i*) *A is a piece-wise continuous single-valued bounded $(b-a)$-periodic $(n \times n)$-matrix whose Floquet multipliers lie off the unit circle;*

(*ii*) *There exists a bounded retract Q of $C([a,b],\mathbb{R}^n)$, with $0 \in Q\backslash\partial Q$, where $Q\backslash\partial Q$ is nonempty open, such that $G(t,x,q(t))$ is lipschitzian in x with a sufficiently small Lipschitz constant, for a.a. $t \in [a,b]$ and each $q \in Q$;*

(*iii*) *There exists a Lebesgue integrable function $\alpha\colon [a,b] \to \mathbb{R}$ such that*

$$|G(t,x(t),q(t))| \leq \alpha(t) \quad \textit{a.e. in } [a,b],$$

for any $(x,q,\lambda) \in \Gamma_T$, where T denotes the set-valued map which assigns, to any $(q,\lambda) \in Q \times [0,1]$, the set of solutions of

$$\begin{aligned} &x' + A(t)x \in \lambda G(t,x,q(t)) \quad \textit{for a.a. } t \in [a,b], \\ &x(a) = x(b); \end{aligned}$$

(*iv*) *For each $u \in \bigcup_{A\in J} \partial\mathcal{B}(A)$, where $\mathcal{B} = \operatorname{int}\{x(t) \in \mathbb{R}^n : x \in Q\} \neq \phi$ is the same, for every $A \in [a,b]$, there exists a bounding function $V_u(t,x) \in C([a,b] \times \mathbb{R}^n, \mathbb{R}^n)$, which is locally lipschitzian in $u \in \partial\mathcal{B}$, such that conditions* (B1)–(B3) *hold for any $(q,\lambda) \in Q \times (0,1]$, where $F := \lambda G(t,x,q(t)) - A(t)x$ and $I := [a,b]$;*

Then the inclusion $x' + A(t)x \in F(t,x)$ admits a $(b-a)$-periodic solution.

Proof: Follows from Theorem 3, when realizing that condition (i) is implied by the given properties of the related Floquet multipliers (see e.g., [2]). Subsequently, the relation $T(Q \times \{0\}) \subset Q$ reduces to the requirement that $0 \in Q$. Condition (vi) holds trivially.

Corollary 2: *Consider problem* (11)*, where $M = -E$, $A(t) \equiv 0$ and $F(t,x) \equiv -F(t+(b-a),-x)$. Let $G\colon [a,b] \times \mathbb{R}^n \times \mathbb{R}^n \times [0,1] \rightsquigarrow \mathbb{R}^n$ be a product-measurable Carathéodory function such that*

$$G(t,c,c,1) \subset F(t,c) \quad \textit{for all } (t,c) \in [a,b] \times \mathbb{R}^n.$$

Assume that

(*i*) *There exists a bounded retract Q of $C([a,b],\mathbb{R}^n)$ with the nonempty open $Q\backslash\partial Q$, symmetrical w.r.t. the origin $0 \in Q$, such that $G(t,x,q(t),\lambda)$ is lipschitzian in x with a sufficiently small Lipschitz constant, for* a.a. $t \in [a,b]$ *and each $(q,\lambda) \in Q \times [0,1]$;*

(ii) *There exists a Lebesgue integrable function* α: $[a,b] \to \mathbb{R}$ *such that*

$$|G(t, x(t), q(t), \lambda)| \le \alpha(t) \quad \text{a.e. in } [a,b],$$

for any $(x, q, \lambda) \in \Gamma_T$, *where* T *denotes the set-valued map which assigns, to any* $(q, \lambda) \in Q \times [0,1]$, *the set of solutions of*

$$x' \in G(t, x, q(t), \lambda), \quad \text{for a.a. } t \in [a,b],$$
$$x(a) = -x(b);$$

(iii) $T(Q \times \{0\}) \subset Q$ *and* ∂Q *is fixed-point free;*

(iv) *For each* $u \in \partial\mathcal{B}$, *where* $\mathcal{B} = \text{int}\{x(t) \in \mathbb{R}^n : x \in Q\} \neq \phi$ *is the same, for every* $A \in [a,b]$, *there exists a bounding function* $V_u(t,x) \in C([a,b] \times \mathbb{R}^n, \mathbb{R}^n)$, *which is locally lipschitzian in* $u \in \partial\mathcal{B}$, *such that conditions (B1)–(B3) hold for any* $(q, \lambda) \in Q \times (0,1]$, *where* $F := G(t, x, q(t), \lambda)$ *and* $I := [a,b]$;

Then the inclusion $x' \in F(t,x)$ *admits a* $2(b-a)$*-periodic solution* $x(t) \equiv -x(t + (b-a))$.

Proof: Follows from Theorem 3, when realizing that condition (i) and (vi) hold trivially, provided additionally $A(t) \equiv 0$ and a symmetry of Q w.r.t. the origin $0 \in Q$.

In the single-valued case, Corollary 2 can be improved in view of Proposition 2, where $\mu = 1$ and $\xi = 0$, as follows.

Theorem 4: *Consider problem (2), where* $\mu = 1, \xi = 0$ *and* $f(t,x) \equiv -f(t + (b-a), -x)$. *Let* g: $[a,b] \times \mathbb{R}^n \times \mathbb{R}^n \times [0,1] \to \mathbb{R}^n$ *be a Carathéodory function such that*

$$g(t, c, c, 1) = f(t, c) \quad \text{for all } (t, c) \in [a,b] \times \mathbb{R}^n.$$

Assume that

(i) *There exists a bounded retract* Q of $C([a,b], \mathbb{R}^n)$ *with the nonempty open* $Q \backslash \partial Q$, *symmetrical w.r.t. the origin* $0 \in Q$, *such that*

$$|g(t, x, q(t), \lambda) - g(t, y, q(t), \lambda)| \le p(t)$$

holds for a.a. $t \in [a,b]$; $x, y, \in \mathbb{R}^n$ *and each* $(q, \lambda) \in Q \times [0,1]$, *where* p: $[a,b] \to [0, \infty)$ *is a Lebesgue integrable function with* (cf. (4))

$$\int_a^b p(t)dt \le \pi;$$

(ii) *There exists a positive constant* α *such that*

$$|g(t, x(t), q(t), \lambda)| \le \alpha \quad \text{a.e. in } [a,b],$$

for any $(x, q, \lambda) \in \mathbb{R}^n \times Q \times [0,1]$;

(iii) $T(Q \times \{0\}) \subset Q$ *and* ∂Q *is fixed-point free, where* T *denotes the set-valued map which assigns, to any* $(q, \lambda) \in Q \times [0,1]$, *the set of solutions of*

$$x' = g(t, x, q(t), \lambda) \quad \text{for a.a. } t \in [a,b],$$
$$x(a) = -x(b);$$

(*iv*) *For each* $u \in \partial\mathcal{B}$, *where* $\mathcal{B} = \text{int}\{x(t) \in \mathbb{R}^n : x \in Q\} \neq \phi$ *is the same, for every* $A \in [a, b]$, *there exists a bounding function* $V_u(t, x) \in C([a, b] \times \mathbb{R}^n, \mathbb{R}^n)$, *which is locally lipschitzian in* $u \in \partial\mathcal{B}$, *such that conditions (B1)–(B3) hold for any* $(q, \lambda) \in Q \times (0, 1]$, *where* $f := g(t, x, q(t), \lambda)$, $M = -E$ *and* $I := [a, b]$;

Then the equation $x' = f(t, x)$ *admits a* $2(b-a)$*-periodic solution* $x(t) \equiv -x(t + (b - a))$.

Proof: Is quite analogous to the one of Corollary 2, but when for condition (i) we apply Proposition 2, instead of Proposition 1, i.e., when appropriately modified property (P1) is replaced by (P2).

7. CONCLUDING REMARKS

The usage of integral manifolds in the above spirit is typical for differential inclusions, which are mostly fully (Schauder-like) linearized and then the invariantness of sets Q, under the associated operators T, is ensured, as required in Theorem 2. Those operators are usually at least acyclic (see e.g. [2,16]). Thus, for instance, the Floquet problems can be immediately solved, provided at most linear growth restrictions, with sufficiently small coefficients, of product-measurable Carathéodory right-hand sides F.

On the other hand, to find nonstandard existence criteria for differential equations (e.g., by means of Theorem 4) seems to be still far from obvious (cf. [7]). Nevertheless, we believe (and encourage the reader to this goal) that the application of the above technique can really bring some very delicate new results.

REFERENCES

[1] J. Andres (1997). On the multivalued Poincaré operators, *Topol. Meth. Nonlin. Anal.*, **10**(1), 171–182.

[2] J. Andres (1999). Almost-periodic and bounded solutions of Carathéodory differential inclusions, *Diff. Integral Eqns.*, **12**(6), 887–912.

[3] J. Andres, G. Gabor and L. Górniewicz (1999). Boundary value problems on infinite intervals, *Trans. Amer. Math. Soc.*, **351**(12), 4861–4903.

[4] J. Andres, G. Gabor and L. Górniewicz (2000). Topological structure of solution sets to multivalued asymptotic problems, *Zeitschrift für Anal. und ihre Anw.*, **19**(1), 35–60.

[5] J. Andres, G. Gabor and L. Górniewicz (2001). *Acyclity of solution sets to functional inclusions*. Nonlin. Anal., to appear.

[6] J. Andres, L. Malaguti and V. Taddei. (2001). Flogues boundary value problems for differential inclusions: a bound sets approach, *Zeitschrift für Anal. und ihre Anw.*, to appear.

[7] G. Anichini, G. Conti and P. Zecca (1991). Using solution sets for solving boundary value problems for ordinary differential equations, *Nonlin. Anal.*, **17**(5), 465–472.

[8] J. Appel, E. DePascale, N.H. Thái and P.P. Zabreiko (1995). Multi-valued superpositions, *Dissertationes Math.*, **345**, 1–97.

[9] J.-P. Aubin and A. Cellina (1984). *Differential Inclusions*. Berlin: Springer.

[10] A. Augustynowicz, Z. Dzedzej and B.D. Gelman (1998). The solution set to BVP for some functional differential inclusions, *Set-Valued Anal.*, **6**(3), 257–263.

[11] G. Bartuzel and A. Fryszkowski (1995). A topological property of the solution set to the Sturm–Liouville differential inclusions, *Demonstr. Math.*, **28**(4), 903–914.

[12] J. Bebernes and M. Martelli (1980). On the structure of the solution set to periodic boundary value problems, *Nonlin. Anal.*, **4**(4), 821–830.

[13] D. Bielawski and T. Pruszko (1991). On the structure of the set of solutions of a functional equation with applications to boundary value problems, *Anal. Polon. Math.*, **53**(3), 201–209.

[14] M. Cecchi, M. Furi and M. Marini (1985). On continuity and compactness of some nonlinear operators associated with differential equations in noncompact intervals, *Nonlin. Anal.*, **9**(2), 171–180.

[15] M. Cecchi, M. Furi and M. Marini (1985). About the solvability of ordinary differential equations with asymptotic boundary conditions, *Boll. U.M.I. (Anal. Funz. Appl.)*, **6**, **4-C**(1), 329–345.

[16] M. Cecchi, M. Marini and P. Zecca (1985). Existence of bounded solutions for multivalued differential systems, *Nonlin. Anal.*, **9**(8), 775–786.

[17] F.S. DeBlasi and G. Pianigiani (1993). Solution sets of boundary value problems for nonconvex differential inclusions, *Topol. Meth. Nonlin. Anal.*, **1**, 303–314.

[18] R. Dragoni, J.W. Macki, P. Nistri and P. Zecca (1996). *Solution Sets of Differential Equations in Abstract Spaces*. Pitman RNMS 342, Harlow: Longman.

[19] M. Furi and M.-P. Pera (1979). On the existence of an unbounded connected set of solutions for nonlinear equations in Banach spaces, *Atti Accad. Naz. Lincei – Rend. Sci. fis. mat. nat.*, **67**(1–2), 31–38.

[20] R.E. Gaines and J. Mawhin (1997). *Coincidence Degree and Nonlinear Differential Equations*. LNM 586, Berlin: Springer.

[21] L. Górniewicz (1999). Topological Fixed Point Theory of Multivalued Mappings. Dordrecht: Kluwer.

[22] T. Kaminogo (1978). Kneser's property and boundary value problems for some retarded functional differential equations, *Tôkoku Math. J.*, **30**, 471–486.

[23] M. Lewicka (1998). Locally lipschitzian guiding function method for ODEs, *Nonlin. Anal.*, **33**, 747–758.

[24] J. Mawhin (1998). Bound sets and Floquet boundary value problems for nonlinear differential equations, *Iagell. Univ. Acta Math.*, **36**, 41–53.

[25] A. Margheri and P. Zecca (1993). Solution sets and boundary value problems in Banach spaces, *Topol. Meth. Nonlin. Anal.*, **2**, 179–88.

[26] J.J. Nieto (1986). Aronszajn's theorem for some nonlinear Dirichlet problems with unbounded nonlinearities, *Proceed. Edinburgh Math. Soc.*, **35**, 346–351.

[27] M.P. Pera (1983). A topological method for solving nonlinear equations in Banach spaces and some related global results on the structure of the solution sets, *Rend. Sem. Mat. Univ. Politecn. Torino*, **41**(3), 9–30.

[28] R. Reissig (1979). Continual of periodic solutions of the Liénard equation. In J. Albrecht, L. Collatz and K. Kirchgässner (Eds), *Constructive Methods for Nonlinear Boundary Value Problems and Nonlinear Oscillations*, Basel: Birkhäuser, 126–133.

[29] V. Taddei (2000). *Bound sets for Floquet boundary value problems: the nonsmooth case*, *Discr. Cons. Dynam. Syst.*, **6**, 459–473.

[30] F. Zanolin (1987). Bound sets, periodic solutions and flow-invariance for ordinary differential equations in $\mathbb{R}^n$: some remarks, *Rend. Ist. Mat. Univ. Trieste*, **19**, 76–92.

SIMAA 4(2002) 39–47

5. On the Semicontinuity of Nonlinear Spectra

J. Appell[1], G. Conti[2] and P. Santucci[3]

[1]*Universität Würzburg, Mathematisches Institut, Am Hubland, D-97074 Würzburg, Germany. E-mail: appell@mathematik.uni-wuerzburg.de*
[2]*Università di Firenze, Dipartimento di Matematica, Via dell'Agnolo 14, I-50122 Firenze, Italy. E-mail: gconti@cesit1.unifi.it*
[3]*Università di Roma "La Sapienza", Dipartimento di Metodi e Modelli Matematici, Via A. Scarpa 16, I-00161 Roma, Italy. E-mail: santucci@dmmm.uniroma1.it*

Abstract: We discuss a fairly general approach to proving the closedness and upper semicontinuity of the multivalued map which associates to each continuous nonlinear operator from a certain operator class a natural spectrum. In this way, we show that the Kachurovskij spectrum for Lipschitz continuous operators, the Furi–Martelli–Vignoli spectrum for quasibounded operators, and the Feng spectrum for so-called *k*-epi operators are all upper semicontinuous with respect to a suitable normed or locally convex topology. On the other hand, we give a simple counterexample which shows that the Dörfner spectrum for linearly bounded operators has not a closed graph and is not upper semicontinuous either.

Keywords: Nonlinear operator, spectrum, closed multivalued map, upper semicontinuous multivalued map

Classification: 47H04, 47H12

Let X be a Banach space over $\mathbb{K} = \mathbb{R}$ or $\mathbb{K} = \mathbb{C}$. Denote by $\mathfrak{L}(X)$ the algebra of all bounded linear operators in X, and by

$$\sigma(L) = \{\lambda : \lambda \in \mathbb{K},\ (\lambda - L)^{-1} \notin \mathfrak{L}(X)\} \tag{1}$$

the spectrum of $L \in \mathfrak{L}(X)$. It is well known (see e.g., [13] or [11, Th. 3.1]) that the map σ: $\mathfrak{L}(X) \to 2^{\mathbb{K}}$ which associates to each operator L its spectrum $\sigma(L)$ is upper semicontinuous but not lower semicontinuous. Roughly speaking, the spectrum $\sigma(L)$ may "collaps", but not "blow up" if the operator L changes continuously.

To see that σ need not be lower semicontinuous, consider, for example, for $\varepsilon \in \mathbb{R}$ the operator $L_\varepsilon \in \mathfrak{L}(l_1(\mathbb{Z}))$ defined on the canonical basis $\{e_k : k \in \mathbb{Z}\}$ in $l_1(\mathbb{Z})$ by $L_\varepsilon e_k = e_{k-1}$ for $k \neq 0$ and $L_\varepsilon e_0 = \varepsilon e_{-1}$. Then we have $\sigma(L_0) = [-1, 1]$, on the one hand, but $\sigma(L_\varepsilon) = \{-1, 1\}$ for $\varepsilon \neq 0$, on the other. Consequently, the map $\varepsilon \mapsto \sigma(L_\varepsilon)$ is not lower semicontinuous at $\varepsilon = 0$.

In view of the importance of spectral theory for linear operators in various fields of mathematics, physics, or engineering, it is not surprising at all that various attempts have been made to define and study spectra also for *nonlinear operators*. Here we mention the

Kachurovskij spectrum [10] for Lipschitz continuous operators, the Neuberger spectrum [12] for Fréchet differentiable operators, the Rhodius spectrum [14] for continuous operators, the Dörfner spectrum [5] for operators with sublinear growth, the Furi–Martelli–Vignoli spectrum [8] for quasibounded operators, and the Feng spectrum [6] for so-called k-epi operators (see [9,15]). In particular, the Furi–Martelli–Vignoli spectrum and the Feng spectrum are upper semicontinuous with respect to a suitable topology. For the other spectra no semicontinuity properties have been studied so far. A comparison of the various advantages and drawbacks of all these spectra may be found in the recent survey [2].

In this note we propose a unified (and quite elementary) approach to proving the upper semicontinuity of the spectrum for several classes of continuous *nonlinear* operators. This approach consists in two parts. First we show that, for f belonging to some class $\mathfrak{M}(X)$ of continuous operators in X, the graph of the multivalued map $\sigma: f \mapsto \sigma(f)$, is *closed* in the product $\mathfrak{M}(X) \times \mathbb{K}$. Afterwards it suffices to use the *local boundedness* of this map, by means of simple upper estimates for the spectral radius, to deduce its upper semicontinuity.

In the first section we recall some semicontinuity definitions for multivalued maps, as well as their connections with closed graphs and local boundedness. This is applied in the second section to give a simple proof of the upper semicontinuity (which is already known) of the Furi–Martelli–Vignoli spectrum and Feng spectrum. Using the same method we show in the third section that also a certain modification of the Furi–Martelli–Vignoli spectrum which was introduced in [3] is upper semicontinuous. In the fourth section we prove the upper semicontinuity of the Kachurovskij spectrum; this gives the upper semicontinuity of the familiar spectrum of a bounded linear operator as a straightforward consequence. Finally, we give a simple counterexample which shows that the Dörfner spectrum has not a closed graph, and is not upper semicontinuous either.

1. SEMICONTINUITY OF MULTIVALUED MAPS

Let $\mathfrak{M}$ be a linear space over $\mathbb{K} = \mathbb{R}$ or $\mathbb{K} = \mathbb{C}$ with seminorm p, and let $\sigma\colon \mathfrak{M} \to 2^{\mathbb{K}}$ be a multivalued functional on $\mathfrak{M}$. For $f \in \mathfrak{M}$ and $\delta > 0$ we denote by $U_\delta(f)$ the p-neighbourhood $\{g : g \in \mathfrak{M},\ p(g-f) < \delta\}$ of f. Recall that σ is called *upper semicontinuous* [resp. *lower semicontinuous*] if, for any $f \in \mathfrak{M}$ and open $V \subseteq \mathbb{K}$ with $\sigma(f) \subseteq V$ [resp. $\sigma(f) \cap V \neq \emptyset$], one can find a $\delta > 0$ such that $\sigma(U_\delta(f)) \subseteq V$ [resp. $\sigma(U_\delta(f)) \cap V \neq \emptyset$]. Moreover, σ is called *closed* if the graph of σ is closed in $\mathfrak{M} \times \mathbb{K}$, i.e., $\lambda_n \in \sigma(f_n)$, $\lambda_n \to \lambda$ and $p(f_n - f) \to 0$ imply that $\lambda \in \sigma(f)$. Obviously, σ is closed if and only if, for every $f \in \mathfrak{M}$ and $\lambda \in \mathbb{K}\backslash\sigma(f)$, there exist $\delta > 0$ and $V_\lambda \subset \mathbb{K}$ open such that $\lambda \in V_\lambda$ and $\sigma(U_\delta(f)) \cap V_\lambda = \emptyset$.

It is easy to see that every upper semicontinuous map with closed values is closed, but not vice versa. For example, the map $\sigma\colon \mathbb{R} \to 2^{\mathbb{R}}$ defined by $\sigma(t) = [-1/t, 1/t]$ for $t \neq 0$ and $\sigma(0) = 0$ is closed, but not upper semicontinuous at 0. The following proposition provides a simple condition which implies the upper semicontinuity of a closed map.

Lemma 1: *Let $(\mathfrak{M}, p)$ be a seminormed linear space and $\sigma\colon \mathfrak{M} \to 2^{\mathbb{K}}$ a closed map with bounded values. If*

$$\sup_{\lambda \in \sigma(f)} |\lambda| \leq p(f) \quad (f \in \mathfrak{M}), \tag{2}$$

then σ is upper semicontinuous.

Proof: Let $f \in \mathfrak{M}$, and let $V \subseteq \mathbb{K}$ be open with $\sigma(f) \subseteq V$. Choose $\eta > 0$ with $\sigma(U_\eta(f))\backslash V \neq \emptyset$. For $g \in U_\eta(f)$ we have

$$\sup_{\lambda \in \sigma(g)} |\lambda| \leq p(g) \leq p(f) + \eta.$$

Consequently, $\sigma(U_\eta(f))$ is bounded, and so $C := \overline{\sigma(U_\eta(f))}\backslash V$ is compact.

Fix $\lambda \in C$. Since $\lambda \notin \sigma(f)$ and the map σ is closed, we find $\delta(\lambda) > 0$ and $V_\lambda \subseteq \mathbb{K}$ open with $\lambda \in V_\lambda$ and $\sigma(U_{\delta(\lambda)}(f)) \cap V_\lambda = \emptyset$. Obviously, $\{V_\lambda : \lambda \in C\}$ is an open covering of C. From the compactness of C we get $C \subseteq V_{\lambda_1} \cup \ldots \cup V_{\lambda_m}$ for suitable $\lambda_1, \ldots, \lambda_m \in C$. Putting $\delta := \min\{\eta, \delta(\lambda_1), \ldots, \delta(\lambda_m)\}$ we see that $\sigma(U_\delta(f)) \subseteq V$ as claimed. □

In what follows we apply Lemma 1 to various subsets of the linear space $\mathfrak{C}(X)$ of all continuous operators $f: X \to X$. To this end, we have to recall some metric characteristics for such operators. Given $f \in \mathfrak{C}(X)$, let

$$[f]_{\text{Lip}} = \sup_{x \neq y} \frac{||f(x) - f(y)||}{||x - y||}, \quad [f]_{\text{lip}} = \inf_{x \neq y} \frac{||f(x) - f(y)||}{||x - y||}, \tag{3}$$

$$[f]_B = \sup_{x \neq 0} \frac{||f(x)||}{||x||}, \quad [f]_b = \inf_{x \neq 0} \frac{||f(x)||}{||x||} \tag{4}$$

and

$$[f]_Q = \limsup_{||x|| \to \infty} \frac{||f(x)||}{||x||}, \quad [f]_q = \liminf_{||x|| \to \infty} \frac{||f(x)||}{||x||}. \tag{5}$$

Moreover, we will use the topological characteristics

$$[f]_A = \inf\{k : k \geq 0, \alpha(f(M)) \leq k\alpha(M)\} \tag{6}$$

and

$$[f]_a = \sup\{k : k \geq 0, \alpha(f(M)) \geq k\alpha(M)\}, \tag{7}$$

where $\alpha(M)$ denotes the Kuratowski measure of noncompactness (see e.g. [1,4]) of a bounded set $M \subset X$.

Obviously, $[f]_b \leq [f]_q \leq [f]_Q \leq [f]_B$ for all $f \in \mathfrak{C}(X)$, and $[f]_{\text{lip}} \leq [f]_b \leq [f]_B \leq [f]_{\text{Lip}}$ if $f(0) = 0$. Moreover, $[f]_{\text{lip}} \leq [f]_a \leq [f]_A \leq [f]_{\text{Lip}}$ if $\dim X = \infty$. The following lemma contains an estimate for the characteristics (3)–(7) which will be used several times in the sequel.

Lemma 2: *For $f, g \in \mathfrak{C}(X)$ and $(K, k) \in \{(A, a), (B, b), (Q, q), (Lip, lip)\}$ the estimates*

$$[f]_k - [g]_K \leq [f + g]_k \leq [f]_k + [g]_K \tag{8}$$

are true.

Proof: We prove (8) only for $(K, k) = (A, a)$. In case $\dim X < \infty$ the estimate (8) is trivial. In case $\dim X = \infty$ we have, for all noncompact bounded sets $M \subset X$,

$$
\begin{aligned}
[f+g]_a &= \inf_{\alpha(M)>0} \frac{\alpha((f+g)(M))}{\alpha(M)} \\
&\le \inf_{\alpha(M)>0} \frac{\alpha(f(M)) + \alpha(g(M))}{\alpha(M)} \\
&\le \inf_{\alpha(M)>0} \frac{\alpha(f(M))}{\alpha(M)} + \sup_{\alpha(M)>0} \frac{\alpha(g(M))}{\alpha(M)} = [f]_a + [g]_A.
\end{aligned}
$$

This proves the second estimate in (8); the first estimate follows replacing g by $-g$. The proof for the other values of (K, k) is analogous. □

2. THE FURI–MARTELLI–VIGNOLI AND FENG SPECTRA

Following [7] we call an operator $f \in \mathfrak{C}(X)$ *stably solvable* if, for any compact operator $g \in \mathfrak{C}(X)$ with $[g]_Q = 0$, the equation $f(x) = g(x)$ has a solution $x \in X$. Putting, in particular, $g(x) \equiv y$ one sees that every stably solvable operator f is onto; the converse is true for linear f [8].

The *Furi–Martelli–Vignoli spectrum* $\sigma_{\text{FMV}}(f)$ contains, by definition, all $\lambda \in \mathbb{K}$ such that either $\min\{[\lambda - f]_a, [\lambda - f]_q\} = 0$ or $\lambda - f$ is *not* stably solvable [8]. In spite of its technical definition, this spectrum is quite natural. For instance, in the linear case it coincides precisely with the familiar spectrum (see [8, Prop. 6.1.2]).
In order to study the semicontinuity properties of this spectrum put

$$
p_{\text{FMV}}(f) = \max\{[f]_A, [f]_Q\}, \quad \mathfrak{M}_{\text{FMV}}(X) = \{f : f \in \mathfrak{C}(X), p_{\text{FMV}}(f) < \infty\}. \tag{9}
$$

It is evident that p_{FMV} is a seminorm on the linear space $\mathfrak{M}_{\text{FMV}}(X)$.

Theorem 1: *The multivalued map* $\sigma_{\text{FMV}}\colon \mathfrak{M}_{\text{FMV}}(X) \ni f \mapsto \sigma_{\text{FMV}}(f) \in 2^{\mathbb{K}}$ *is closed.*

Proof: Let $(f_n)_n$ and $(\lambda_n)_n$ be sequences with $\lambda_n \in \sigma_{\text{FMV}}(f_n)$, $p_{\text{FMV}}(f_n - f) \to 0$, and $\lambda_n \to \lambda$. Then

$$
p_{\text{FMV}}((\lambda_n - f_n) - (\lambda - f)) \le |\lambda_n - \lambda| + p_{\text{FMV}}(f_n - f) \to 0 \quad (n \to \infty),
$$

and so the sequence $(\lambda_n - f_n)_n$ tends to $\lambda - f$ in $\mathfrak{M}_{\text{FMV}}(X)$. We have to show that $\lambda \in \sigma_{\text{FMV}}(f)$.

Suppose that $\lambda \notin \sigma_{\text{FMV}}(f)$, hence $[\lambda - f]_a > 0$, $[\lambda - f]_q > 0$, and $\lambda - f$ is stably solvable. Choose n so large that

$$
\max\{p_{\text{FMV}}(f_n - f), |\lambda_n - \lambda|\} < \frac{1}{2}\min\{[\lambda - f]_a, [\lambda - f]_q\}.
$$

Applying the estimate (8) twice to (A, a) and (Q, q) we get

$$
[\lambda_n - f_n]_a \ge [\lambda - f_n]_a - |\lambda_n - \lambda| \ge [\lambda - f]_a - [f - f_n]_A - |\lambda_n - \lambda| > 0
$$

and

$$
[\lambda_n - f_n]_q \ge [\lambda - f_n]_q - |\lambda_n - \lambda| \ge [\lambda - f]_q - [f - f_n]_Q - |\lambda_n - \lambda| > 0.
$$

Moreover, $\lambda_n - f_n$ is stably solvable, by the perturbation theorem of Furi–Martelli–Vignoli for stably solvable operators (see [8, Prop. 6.1.3]). We conclude that $\lambda_n \notin \sigma_{\mathrm{FMV}}(f_n)$, a contradiction. □

In [8, Prop. 8.1.2] the authors prove that

$$\sup_{\lambda \in \sigma_{\mathrm{FMV}}(f)} |\lambda| \leq p_{\mathrm{FMV}}(f)$$

for all $f \in \mathfrak{M}_{\mathrm{FMV}}(X)$. Consequently, Lemma 1 implies the following

Corollary 1 [8]: *The multivalued map $\sigma_{\mathrm{FMV}}\colon \mathfrak{M}_{\mathrm{FMV}}(X) \ni f \mapsto \sigma_{\mathrm{FMV}}(f) \in 2^{\mathbb{K}}$ is upper semicontinuous.*

Now we pass to the spectrum which has been introduced by Feng in [6]. From now on we write $B_r(X) = \{x : x \in X, \|x\| \leq r\}$ for the closed ball, and $S_r(X) = \partial B_r(X) = \{x : x \in X, \|x\| = r\}$ for the sphere of radius $r > 0$ in X. Following [15] we call a continuous operator $f\colon B_r(X) \to X$ *k-epi* ($k \geq 0$) if $f(x) \neq 0$ on $S_r(X)$ and, for any continuous operator $g\colon B_r(X) \to X$ with $[g]_A \leq k$ and $g(x) \equiv 0$ on $S_r(X)$, the equation $f(x) = g(x)$ has a solution $x \in B_r(X)$.

The *Feng spectrum* $\sigma_{\mathrm{F}}(f)$ consists, by definition, of all $\lambda \in \mathbb{K}$ such that either $\min\{[\lambda - f]_a, [\lambda - f]_b\} = 0$ or $\lambda - f$ is *not k-epi* on $B_r(X)$ for any $k > 0$ and $r > 0$ [6]. Also this spectrum coincides in the linear case precisely with the usual spectrum (see [6, Th. 3.3]).

The definition of the Feng spectrum suggests to consider now the functional

$$\|f\|_{\mathrm{F}} = \max\{[f]_A, [f]_B\}, \quad \mathfrak{M}_{\mathrm{F}}(X) = \{f : f \in \mathfrak{C}(X), \|f\|_{\mathrm{F}} < \infty\}. \tag{10}$$

Since $[f]_B = 0$ implies $f(x) \equiv 0$, the class $\mathfrak{M}_{\mathrm{F}}(X)$ is, in contrast to the class $\mathfrak{M}_{\mathrm{FMV}}(X)$, even a normed space.

Theorem 2: *The multivalued map $\sigma_{\mathrm{F}}\colon \mathfrak{M}_F(X) \ni f \mapsto \sigma_{\mathrm{F}}(f) \in 2^{\mathbb{K}}$ is closed.*

Proof: Let $(f_n)_n$ and $(\lambda_n)_n$ be sequences with $\lambda_n \in \sigma_{\mathrm{F}}(f_n)$, $\|f_n - f\|_{\mathrm{F}} \to 0$, and $\lambda_n \to \lambda$. Then

$$\|(\lambda_n - f_n) - (\lambda - f)\|_{\mathrm{F}} \leq |\lambda_n - \lambda| + \|f_n - f\|_{\mathrm{F}} \to 0 \quad (n \to \infty),$$

and so the sequence $(\lambda_n - f_n)_n$ tends to $\lambda - f$ in $\mathfrak{M}_{\mathrm{F}}(X)$. We have to show that $\lambda \in \sigma_{\mathrm{F}}(f)$.

Suppose again that $\lambda \notin \sigma_{\mathrm{F}}(f)$, hence $[\lambda - f]_a > 0$, $[\lambda - f]_b > 0$, and $\lambda - f$ is *k-epi* for some $k > 0$ on every ball $B_r(X)$. Choose n so large that

$$\max\{\|f_n - f\|_{\mathrm{F}}, |\lambda_n - \lambda|\} < \frac{1}{2}\min\{[\lambda - f]_a, [\lambda - f]_b, k\}.$$

Applying the estimate (8) twice to (A, a) and (B, b) we get $[\lambda_n - f_n]_a > 0$ and $[\lambda_n - f_n]_b > 0$ as before. Consider the set

$$\Sigma_n := \{x : x \in B_r(X), \lambda x - f(x) + h_n(x, t) = 0 \text{ for some } t \in [0, 1]\},$$

where h_n is the homotopy

$$h_n(x, t) = t(\lambda_n x - \lambda x - f_n(x) + f(x)).$$

We claim that $\Sigma_n = \{0\}$. Indeed, for $x \in \Sigma_n$, $x \neq 0$ we have

$$\lambda x - f(x) = t(\lambda - \lambda_n)x + t(f_n(x) - f(x)),$$

hence

$$[\lambda - f]_b\|x\| \leq \|\lambda x - f(x)\| \leq |\lambda - \lambda_n|\,\|x\| + \|f_n(x) - f(x)\| < \frac{1}{2}[\lambda - f]_b\|x\| + \frac{1}{2}[\lambda - f]_b\|x\| = [\lambda - f]_b\|x\|$$

which is impossible. We conclude that $x = 0$, hence $\Sigma_n \cap S_r(X) = \emptyset$ for n sufficiently large. Moreover, $[h_n]_A \leq |\lambda_n - \lambda| + [f_n - f]_b \to 0$ for $n \to \infty$. From the homotopy invariance of k-epi operators (see e.g., [15]) it follows that $\lambda - f + h_n(\cdot, 1) = \lambda_n - f_n$ is k_n-epi on $B_r(X)$ for $k_n = k - |\lambda_n - \lambda| \geq \frac{1}{2}k > 0$. We conclude that $\lambda_n \notin \sigma_F(f_n)$, a contradiction. □

In [6, Th. 3.6] the author proves that

$$\sup_{\lambda \in \sigma_F(f)} |\lambda| \leq \|f\|_F$$

for all $f \in \mathfrak{M}_F(X)$. Consequently, Lemma 1 implies the following

Corollary 2 [6]: *The multivalued map* $\sigma_F\colon \mathfrak{M}_F(X) \ni f \mapsto \sigma_F(f) \in 2^{\mathbb{K}}$ *is upper semi-continuous.*

3. A MODIFICATION OF THE FURI–MARTELLI–VIGNOLI SPECTRUM

The Furi–Martelli–Vignoli spectrum has been slightly modified in the recent paper [3] in order to remove the artificial condition $\min\{[\lambda - f]_a, [\lambda - f]_q\} = 0$ in the definition of the spectrum.

For $k \geq 0$, we call an operator $f \in \mathfrak{C}(X)$ *k-stably solvable* if, for any operator $g \in \mathfrak{C}(X)$ with $[g]_A \leq k$ and $[g]_Q \leq k$, the equation $f(x) = g(x)$ has a solution $x \in X$. Clearly, 0-stable solvability is just stable solvability in the sense of [7], and every k-stably solvable operator is also k'-stably solvable for $k' \leq k$. It seems therefore natural to call the number

$$\mu(f) = \sup\{k : k \geq 0,\ f \text{ is } k\text{-stably solvable}\} \tag{11}$$

the *measure of stable solvability* of f. For example, the Darbo fixed point Theorem [4] implies that the measure of stable solvability of the identity is equal to 1.

The *modified Furi–Martelli–Vignoli spectrum* $\hat{\sigma}_{FMV}(f)$ contains, by definition, all $\lambda \in \mathbb{K}$ such that $\lambda - f$ is *not* k-stably solvable for some $k > 0$. Again, in the linear case this spectrum coincides with the familiar spectrum (see [3, Prop. 11]).

The following lemma establishes a simple Rouché type perturbation result for k-stably solvable operators.

Lemma 3: *Let* $f, g \in \mathfrak{C}(X)$, *where* $\mu(f) > 0$ *and* $p_{FMV}(g) < \mu(f)$ *with* $p_{FMV}(g)$ *given by (9). Then* $f + g$ *satisfies*

$$\mu(f + g) \geq \mu(f) - p_{FMV}(g). \tag{12}$$

Proof: Fix $k < \mu(f) - p_{\text{FMV}}(g)$, and let $h \in \mathfrak{C}(X)$ be given with $p_{\text{FMV}}(h) \leq k$. We have to show that the equation $f(x) + g(x) = h(x)$ has a solution in X. From the definition (9) it follows that $p_{\text{FMV}}(h - g) \leq p_{\text{FMV}}(h) + p_{\text{FMV}}(g) \leq k + p_{\text{FMV}}(g) < \mu(f)$. Consequently, there exists $x \in X$ such that $h(x) - g(x) = f(x)$, and we are done. □

Theorem 3: *The multivalued map* $\hat{\sigma}_{\text{FMV}}$: $\mathfrak{M}_{\text{FMV}}(X) \ni f \mapsto \hat{\sigma}_{\text{FMV}}(f) \in 2^{\mathbb{K}}$ *is closed.*

Proof: We prove that the complement is open. So, let $f \in \mathfrak{M}_{\text{FMV}}(X)$ be given and $\lambda \notin \hat{\sigma}_{\text{FMV}}(f)$, i.e., $\mu(\lambda - f) > 0$. Applying Lemma 3 to $\lambda - f$ instead of f and $g(x) = (\lambda' - \lambda)x$ we see that $\mu(\lambda' - f) > 0$ provided we choose $|\lambda' - \lambda| < \mu(\lambda - f)$. □

From the definition it follows immediately that $\hat{\sigma}_{\text{FMV}}(f) \subseteq \sigma_{\text{FMV}}(f)$; an example for strict inclusion may be found in [3]. This shows that every estimate for the spectral radius of $\sigma_{\text{FMV}}(f)$ holds also for the spectral radius of $\hat{\sigma}_{\text{FMV}}(f)$, and so the following is true.

Corollary 3 [3]: *The multivalued map* $\hat{\sigma}_{\text{FMV}}$: $\mathfrak{M}_{\text{FMV}}(X) \ni f \mapsto \hat{\sigma}_{\text{FMV}}(f) \in 2^{\mathbb{K}}$ *is upper semicontinuous.*

4. THE KACHUROVSKIJ AND DÖRFNER SPECTRA

One of the most "natural" spectra is that defined by Kachurovskij [10]. Consider the class

$$\mathfrak{M}_{\text{K}}(X) = \{f : f \in \mathfrak{C}(X),\ [f]_{\text{Lip}} < \infty\} \tag{13}$$

with the seminorm $p_{\text{K}}(f) = [f]_{\text{Lip}}$. The *Kachurovskij spectrum* of f [10] is defined by

$$\sigma_{\text{K}}(f) = \{\lambda : \lambda \in \mathbb{K}, (\lambda - f)^{-1} \notin \mathfrak{M}_{\text{K}}(X)\}. \tag{14}$$

Obviously, this definition is modelled on the classical definition (1) for the "linear" spectrum. Indeed, in the linear case one has $[L]_{\text{Lip}} = ||L||$, hence $\sigma_{\text{K}}(L) = \sigma(L)$.

Theorem 4: *The multivalued map* σ_{K}: $\mathfrak{M}_{\text{K}}(X) \ni f \mapsto \sigma_{\text{K}}(f) \in 2^{\mathbb{K}}$ *is closed.*

Proof: The proof is similar to that of Theorem 1 and Theorem 2. In fact, applying the estimate (8) twice to $(K, k) = (\text{Lip}, \text{lip})$ we get

$$[\lambda_n - f_n]_{\text{lip}} \geq [\lambda - f_n]_{\text{lip}} - |\lambda_n - \lambda| \geq [\lambda - f]_{\text{lip}} - [f - f_n]_{\text{Lip}} - |\lambda_n - \lambda| > 0,$$

contradicting the hypothesis $\lambda_n \in \sigma_{\text{K}}(f_n)$. □

Corollary 4: *The multivalued map* σ_{K}: $\mathfrak{M}_{\text{K}}(X) \ni f \mapsto \sigma_{\text{K}}(f) \in 2^{\mathbb{K}}$ *is upper semi-continuous.*

Proof: Let $f \in \mathfrak{M}_{\text{K}}(X)$ and $|\lambda| > p_{\text{K}}(f) = [f]_{\text{Lip}}$. For fixed $y \in X$, the operator $f_y(x) := \frac{1}{\lambda}(f(x) + y)$ is continuous and satisfies $[f_y]_{\text{Lip}} < 1$. The Banach-Caccioppoli fixed point principle implies the existence of a unique $x \in X$ with $f_y(x) = x$, hence $\lambda x - f(x) = y$. Moreover, $[(\lambda - f)^{-1}]_{\text{Lip}} = [\lambda - f]_{\text{lip}}^{-1} < \infty$. So we have proved that $\lambda \notin \sigma_{\text{K}}(f)$, and thus

$$\sup_{\lambda \in \sigma_{\text{K}}(f)} |\lambda| \leq p_{\text{K}}(f) \quad (f \in \mathfrak{M}_{\text{K}}(X)).$$

The assertion follows again from Lemma 1. □

As a particular case we get from Corollary 4 the semicontinuity property of the linear spectrum mentioned at the beginning.

Corollary 5 [13]: *The multivalued map σ: $\mathfrak{L}(X) \ni L \mapsto \sigma(L) \in 2^{\mathbb{K}}$ which associates to each bounded linear operator its spectrum is upper semicontinuous.*

To conclude let us consider the class introduced by Dörfner

$$\mathfrak{M}_{\mathrm{D}}(X) = \{f : f \in \mathfrak{C}(X),\, [f]_B < \infty\}, \tag{15}$$

equipped with the norm $[\cdot]_B$. Given $f \in \mathfrak{M}_{\mathrm{D}}(X)$, the *Dörfner spectrum* [5] of f is defined by

$$\sigma_{\mathrm{D}}(f) = \{\lambda : \lambda \in \mathbb{K},\, (\lambda - f)^{-1} \notin \mathfrak{M}_{\mathrm{D}}(X)\}. \tag{16}$$

Although this definition is rather similar to that of Kachurovskij, one gets a completely different spectrum. In particular, the spectrum (16) does not have a closed graph in the product $\mathfrak{M}_{\mathrm{D}}(X) \times \mathbb{K}$, as is shown by the following

Example: Take $X = \mathbb{R}$, put

$$c_n = \sum_{k=1}^{n} \frac{1}{k^2(k+1)}, \quad c = \sum_{k=1}^{\infty} \frac{1}{k^2(k+1)},$$

and define a piecewise linear function $f \in \mathfrak{C}(\mathbb{R})$ by $f(0) = 0$ and

$$f(x) = \begin{cases} \frac{1}{n}x + c_n - \frac{1}{n+1} & \text{for } \frac{n}{n+1} \le x \le \frac{n+1}{n+2}, \\ \frac{1}{n}x - c_n + \frac{1}{n+1} & \text{for } -\frac{n+1}{n+2} \le x \le -\frac{n}{n+1}, \\ cx & \text{for } |x| \ge 1. \end{cases}$$

It is easy to see that $f \in \mathfrak{M}_{\mathrm{D}}(X)$ with $[f]_B = 1$. Since f is a homeomorphism with $[f]_b = c > 0$ we have $0 \notin \sigma_{\mathrm{D}}(f)$. On the other hand, every $\lambda \in \mathbb{R}$ with $c \le |\lambda| \le 1$ is certainly an eigenvalue of f. Moreover, every scalar $\lambda_n^{\pm} = \pm\frac{1}{n}$ belongs to $\sigma_{\mathrm{D}}(f)$, since the function $f_n(x) = f(x) - \frac{1}{n}x$ is constant on the interval $[\frac{n}{n+1}, \frac{n+1}{n+2}]$.

Obviously, we have $[f - f_n]_B \to 0$ as $n \to \infty$. So, if we take $\lambda_n \equiv 0$, we have $\lambda_n \in \sigma_{\mathrm{D}}(f_n)$, $\lambda_n \to 0$, but $0 \notin \sigma_{\mathrm{D}}(f)$. This shows that the multivalued map σ_{D} is *neither closed nor upper semicontinuous* on $\mathfrak{M}_{\mathrm{D}}(\mathbb{R})$. So, in contrast to all the preceding spectra, the Dörfner spectrum $\sigma_{\mathrm{D}}(f)$ may "blow up" when f changes continuously.

Acknowledgement: This chapter was written in the framework of a DFG project (DFG-Gz. Ap 4015-1). Financial support by the DFG is gratefully acknowledged.

REFERENCES

[1] R.R. Akhmerov, M.I. Kamenskij, A.S. Potapov, A.E. Rodkina and B.N. Sadovskij (1992). *Measures of Noncompactness and Condensing Operators* [in Russian]. Nauka, Novosibirsk, 1986; Engl. transl.: Birkhäuser, Basel.

[2] J. Appell, E. De Pascale and A. Vignoli (2000). A comparison of different spectra for nonlinear operators, *Nonlin. Anal. TMA*, **40**, 73–90.

[3] J. Appell, E. Giorgieri and M. Väth. *On a class of maps related to the Furi–Martelli–Vignoli spectrum. Annali Mat. Pura Appl.*

[4] G. Darbo (1955). *Punti uniti in trasformazioni a codominio non compatto*, Rend. Sem. Mat. Univ. Padova, **24**, 84–92.

[5] M. Dörfner (1997). *Beiträge zur Spektraltheorie nichtlinearer Operatoren*, Ph. D. thesis.

[6] W. Feng (1997). A new spectral theory for nonlinear operators and its applications, *Abstr. Appl. Anal.*, **2**, 163–183.

[7] M. Furi, M. Martelli and A. Vignoli (1976). Stably solvable operators in Banach spaces, *Atti Accad. Naz. Lincei Rend. Cl. Sci. Fis. Mat. Nat.*, **60**, 21–26.

[8] M. Furi, M. Martelli and A. Vignoli (1978). Contributions to the spectral theory for nonlinear operators in Banach spaces, *Annali Mat. Pura Appl.*, **118**, 229–294.

[9] M. Furi, M. Martelli and A. Vignoli (1980). On the solvability of nonlinear operator equations in normed spaces, *Annali Mat. Pura Appl.*, **128**, 321–343.

[10] R.I. Kachurovskij (1969). Regular points, spectrum and eigenfunctions of nonlinear operators, *Soviet Math. Dokl.*, **10**, 1101–1105.

[11] T. Kato (1966). *Perturbation Theory for Linear Operators*, Berlin: Springer.

[12] J.W. Neuberger (1969). Existence of a spectrum for nonlinear transformations, *Pacific J. Math.*, **31**, 157–159.

[13] J.D. Newburgh (1951). The variation of spectra, *Duke Math. J.*, **18**, 165–176.

[14] A. Rhodius (1977). Der numerische Wertebereich und die Lösbarkeit linearer und nichtlinearer Operatorgleichungen, *Math. Nachr.*, **79**, 343–360.

[15] E.U. Tarafdar and H.B. Thompson (1987). On the solvability of nonlinear noncompact operator equations, *J. Austral. Math. Soc.*, **43**, 103–114.

SIMAA 4(2002) 49–61

6. Existence Results for Two-point Boundary Value Problems via Set-valued Analysis

Pasquale Candito*

Dipartimento di Matematica, Universita di Messina, contrada Papardo, salita Sperone, N° 31, 98166 Sant' Agata (ME), Italy. E-mail: candito@dipmat.unime.it

Abstract: The aim of this chapter is to prove some existence results for two-point boundary value problems with possibly discontinuous non-linearities. In addition, it points out a recent approach to the study of the same problems for second-order implicit equations. We emphasize that the whole chapter is based on Theorem 1 below (namely, Theorem 2.1 [13]). Moreover, set-valued analysis furnishes the abstract framework inside which these problems are investigated.

Keywords and Phrases: Boundary value problem, discontinuous right-hand side, convexification, multi-valued boundary value problem, implicit equation

1991 Mathematics Subject Classification: Primary – 34A60, 34B15

1. INTRODUCTION

Let $[a, b]$ be a compact real interval with the Lebesgue measure structure; n a positive integer; $\mathbb{R}^n$ the Euclidean n-space, whose zero element is denoted by $\theta_{\mathbb{R}^n}$; $p \in [1, +\infty]$; $W^{2,p}([a, b], \mathbb{R}^n)$ the space of all $u \in C^1([a, b], \mathbb{R}^n)$ such that u' is absolutetly continuous in $[a, b]$ and $u'' \in L^p([a, b], \mathbb{R}^n)$. Consider the problem

$$\begin{cases} u'' = f(t, u, u') \\ u(a) = u(b) = \theta_{\mathbb{R}^n}, \end{cases} \qquad (P_f)$$

where f: $[a, b] \times \mathbb{R}^n \times \mathbb{R}^n \to \mathbb{R}^n$ is a possibly discontinuous function satisfying a suitable integrable boundedness type condition. A function u: $[a, b] \to \mathbb{R}^n$ is said to be a *solution* of (P_f) if $u \in W^{2,p}([a, b], \mathbb{R}^n)$, $u(a) = u(b) = \theta_{\mathbb{R}^n}$, and $u''(t) = f(t, u(t), u'(t))$ a.e. in $[a, b]$. The approach we develop to investigate (P_f) is strictly based on set-valued analysis. We first define an appropriate regularization G_f: $[a, b] \times \mathbb{R}^n \times \mathbb{R}^n \to 2^{\mathbb{R}^n}$ of f

* Current Address: DIMET Facoltà di Ingegneria, Università di Reggio Calabria, via Graziella (Feo di Vito) 89100 Reggio Calabria Italy, E-mail: candito@ns.ing.unirc.it

(see [9,11,12]) and, through Theorem 1 below (Theorem 2.1 [13]), we get a solution $u \in W^{2,p}([a,b],\mathbb{R}^n)$ to the problem

$$\begin{cases} u''(t) \in G_f(t,u(t),u'(t)) \text{ a.e. in } [a,b] \\ u(a) = u(b) = \theta_{\mathbb{R}^n}. \end{cases} \qquad (P_{G_f})$$

Next, under suitable assumptions, we come back to problem (P_f) by proving that u turns out also a solution of (P_f). To do this, we follow two ways:

(I) Using a technical relationship between f and its convexification G_f, firstly introduced in [12], we obtain Theorem 3, which plays a key role in solving problem (P_f) via set-valued analysis. To be precise, it claims that if a solution of (P_{G_f}) enjoys a right property, then it is also a solution of (P_f). For instance, when f doesn't depend on t, we simply ask that at least one of the projections of the set of the discontinuities of f has measure zero.

(II) Requiring that f is "locally" directionally continuous with respect to a suitable cone of $\mathbb{R} \times \mathbb{R}^n \times \mathbb{R}^n$ (see [6,8,9]).

Our approach is slightly different from the usual ones [9,12], in the sense that the regularization $G_f(t,x,z)$ considered here involves only the variables x and z and we don't need to guarantee *a priori* that f be bounded.
Here is the plain of the present chapter.

After some notations and preliminary results, we establish a technical lemma (Proposition 5), which reveals very useful to apply Theorem 1 in solving problem (P_{G_f}). Roughly speaking, it states that, under suitable hypotheses on f, the multifunction G_f satisfies the assumptions of Theorem 1. In particular, when $f(t,\cdot,\cdot)$ has a closed graph, we immediately get an existence result for problem (P_f) (Theorem 2). The main results are presented in Section 3. Finally, in Section 4, we consider the implicit problems

$$\begin{cases} u''(t) \in Y \\ f(t,u,u',u'') = 0 \\ u(a) = u(b) = \theta_{\mathbb{R}^n}, \end{cases} \qquad (P_f^i)$$

where Y denotes a non-empty, connected, and locally connected subset of $\mathbb{R}^n$ while $f\colon [a,b] \times \mathbb{R}^n \times \mathbb{R}^n \times Y \to \mathbb{R}$ is a continuous function, and

$$\begin{cases} u''(t) \in Z \\ g(u'') = f(t,u,u') \\ u(a) = u(b) = \theta_{\mathbb{R}^n}, \end{cases} \qquad (P_{g,f}^i)$$

where Z denotes an arcwise connected subset of $\mathbb{R}^n$, $g\colon Z \to \mathbb{R}$ is a continuous function, and $f\colon [a,b] \times \mathbb{R}^n \times \mathbb{R}^n \to \mathbb{R}$ is an essentially bounded function. In order to show how these problems can be reduced to an appropriate multivalued problem, to which Proposition 5 and Theorem 1 apply, we give both an "ordinary" version of Theorem 3.1 in [15] (Theorem 6) and a direct proof of a simpler form of Theorem 4 in [2] (Theorem 7). This result is essentially derived form the corresponding one of [17]. The crucial fact in proving this latter is that Theorem 1.1 of [19] yields a suitable multifunction $Q\colon [a,b] \times \mathbb{R}^n \times \mathbb{R}^n \to 2^Y$, involving f, which turns out lower semicontinuous. Next, using standard arguments, we construct another multifunction $F\colon [a,b] \times \mathbb{R}^n \times \mathbb{R}^n \to 2^{\mathbb{R}^n}$

with the same regularity as Q but having values in a suitable ball. By Theorem 2.1 in [9] (see [2,17]) and Theorem 5, we achieve the conclusion. To establish Theorem 6, we exploit the approach previously developed in [20] to reduce problem $(P^i_{g,f})$ to (P_{G_h}), where h: $[a,b] \times \mathbb{R}^n \times \mathbb{R}^n \to \mathbb{R}^n$ is an appropriate function related with g^{-1} and f. Next, through the results obtained in the preceding section, we solve problem (P_{G_h}). Let us finally emphasize that Theorem 1 represents the key to give an application of Theorem 11 in [21] for solving $(P^i_{g,f})$ (see Theorem 3 of [2,5]), and that the papers [9,18,19–21], basically represent the starting point for our investigations involving problems (P_f), (P^i_f) and $(P^i_{g,f})$ (see also [3]). As classical works on the subject treated here, we also mention [1,4,10,11].

2. BASIC DEFINITIONS AND PRELIMINARY RESULTS

Let X, Y be two non-empty sets. A *multifunction* $F: X \to 2^Y$ is a function from X into the family of all subsets of Y. The *graph* of F is the set $gr(F) = \{(x,y) \in X \times Y : y \in F(x)\}$. For $W \subseteq Y$, define $F^-(W) = \{x \in X : F(x) \cap W \neq \emptyset\}$. If V is a non-empty subset of X, we denote by $F_{|V}$ the restriction of F to V. Let (X, Ξ) be a measurable space and let Y be a topological space, F is called *measurable*, when for each open subset W of Y one has $F^-(W) \in \Xi$. From now on, "measurable" always means Lebesgue measurable and $m(E)$ stands for the measure of E. If Y is a topological space and $W \subseteq Y$, the simbols $\text{int}(W), \bar{W}$, $\text{co}(W)$ will denote, respectively, the interior, the clousure, and the convex hull of the set W. Let (X, d) be a metric space. For every $x \in X$ and $\rho > 0$, we write $B(x,\rho) := \{y \in X : d(x,y) \leq \rho\}$, $B^\circ(x,\rho) := \{y \in X : d(x,y) < \rho\}$.

Let $\|\cdot\|, \|\cdot\|_1, \|\cdot\|_2$ be three fixed norms on $\mathbb{R}^n$; d the metric induced by $\|\cdot\|$; $\|\cdot\|_{\mathbb{R}^n \times \mathbb{R}^n}$ the norm on $\mathbb{R}^n \times \mathbb{R}^n$ defined by putting, for every $(x,z) \in \mathbb{R}^n \times \mathbb{R}^n$,

$$\|(x,z)\|_{\mathbb{R}^n \times \mathbb{R}^n} = \begin{cases} \max\{\frac{4}{b-a}\|x\|_1, \ \|z\|_2\} & \text{if } b-a \leq 4 \\ \max\{\|x\|_1, \ \frac{b-a}{4}\|z\|_2\} & \text{if } b-a > 4. \end{cases}$$

If c_1, c_2 are two positive constants such that

$$\|x\|_1 \leq c_1\|x\|, \quad \|x\|_2 \leq c_2\|x\|, \quad \forall x \in \mathbb{R}^n,$$

we set

$$\gamma = \max\{c_1, c_2\}\gamma',$$

where

$$\gamma' = \begin{cases} 1 & \text{if } p = 1, \\ \left[\frac{(b-a)(p-1)}{2p-1}\right]^{1-\frac{1}{p}} & \text{if } 1 < p < +\infty, \\ \frac{(b-a)}{2} & \text{if } p = +\infty, \end{cases}$$

or

$$\gamma' = \begin{cases} \frac{(b-a)}{4} & \text{if } p = 1, \\ \frac{(b-a)}{4}\left[\frac{(b-a)(p-1)}{2p-1}\right]^{1-\frac{1}{p}} & \text{if } 1 < p < +\infty, \\ \frac{(b-a)^2}{8} & \text{if } p = +\infty, \end{cases}$$

according to whether $b-a \leq 4$ or $b-a > 4$. One clearly has

$$\|(x,z)\|_{\mathbb{R}^n \times \mathbb{R}^n} \leq \max\left\{\frac{4c_1}{b-a}\|x\|, c_1\|x\|, \frac{(b-a)c_2}{4}\|z\|, c_2\|z\|\right\}$$

for every $(x,z) \in \mathbb{R}^n \times \mathbb{R}^n$.

Let us now introduce the classes of functions to which the right-hand side of problem (P_f) belongs. Denote by $\mathbb{H}(p,s,r)$ the family of functions f: $[a,b] \times \mathbb{R}^n \times \mathbb{R}^n \to \mathbb{R}^n$ for which there exist $p, s \in [1, +\infty]$, with $p \leq s$, and $r \in]0, +\infty[$ such that the real-valued function

$$S_r(t) = \sup_{\|(x,z)\|_{\mathbb{R}^n \times \mathbb{R}^n} \leq \gamma r} \|f(t,x,z)\|, \quad t \in [a,b],$$

belongs to $L^s([a,b], \mathbb{R})$ and $\|S_r\|_{L^p([a,b],\mathbb{R})} \leq r$.

Furthermore, write $\mathbb{A}$ for the class of functions f: $[a,b] \times \mathbb{R}^n \times \mathbb{R}^n \to \mathbb{R}^n$ such that the set

$$D_f := \{(t,x,z) \in [a,b] \times \mathbb{R}^n \times \mathbb{R}^n \ : \ f \text{ is discontinuous at } (t,x,z)\}$$

has measure zero and indicate with $\mathbb{B}$ the family of functions f: $[a,b] \times \mathbb{R}^n \times \mathbb{R}^n \to \mathbb{R}^n$ for which there exists a set $\Omega_f \subseteq [a,b]$, with $m(\Omega_f) = 0$, such that the set

$$D_{\Omega_f} = \cup_{t \in [a,b] \setminus \Omega_f} \{(x,z) \in \mathbb{R}^n \times \mathbb{R}^n : f(t,\cdot,\cdot) \text{ is discontinuous at } (x,z)\}$$

has measure zero and $f(\cdot, x, z)$ is measurable for every $(x,z) \in \mathbb{R}^n \times \mathbb{R}^n \setminus D_{\Omega_f}$. Let us note that $\mathbb{A} \cap \mathbb{B} \neq \emptyset$ and moreover $\mathbb{A} \not\subseteq \mathbb{B}, \mathbb{B} \not\subseteq \mathbb{A}$.

Finally, whenever $f \in \mathbb{H}(p,s,r)$, we denote by $S(f)$ the class of the solutions u to problem (P_f) belonging to $W^{2,p}([a,b], \mathbb{R}^n)$ and such that $\|u''(t)\| \leq S_r(t)$ a.e. in $[a,b]$. Evidently, if $u \in S(f)$, then $\|u''\|_{L^p([a,b],\mathbb{R}^n)} \leq r$.

Here and in the sequel, the multifunction G_f: $[a,b] \times \mathbb{R}^n \times \mathbb{R}^n \to 2^{\mathbb{R}^n}$ defined by putting, for every $(t,x,z) \in [a,b] \times \mathbb{R}^n \times \mathbb{R}^n$,

$$G_f(t,x,z) = \cap_{\varrho > 0} \overline{co}(\{f(t,v,w) : \|(v,w) - (x,z)\|_{\mathbb{R}^n \times \mathbb{R}^n} < \varrho\})$$

is called the convexification (or regularization) of f. In order to come back from problem (P_{G_f}) to problem (P_f), we need the following technical results.

Proposition 1: *Let u: $[a,b] \to \mathbb{R}$ be absolutely continuous and let $E \subseteq \mathbb{R}$, with $m(E) = 0$. Then, $u'(t) = 0$ for almost every $t \in u^{-1}(E)$.*

Proof: Since u is absolutely continuous, Jordan's Theorem yields $u = u_1 - u_2$, with u_1, u_2 absolutely continuous and increasing. Therefore, it is not restrictive to suppose u increasing. Now, if B is any measurable subset of $u^{-1}(E)$, by Lemma 2.3 of [4], we have $\int_B u'(t)dt = m(u(B)) = 0$. This completes the proof. □

Arguing in the same way it is possible to prove

Proposition 2: *Let $u \in W^{2,1}([a,b], \mathbb{R})$ and let $E \subseteq \mathbb{R}$, with $m(E) = 0$. Then $u''(t) = 0$ for almost every $t \in u^{-1}(E)$.*

Recall that, if $M > 0$ is given and Γ^M denotes the cone

$$\Gamma^M := \{(t, x, z) \in \mathbb{R} \times \mathbb{R}^n \times \mathbb{R}^n : \|(x, z)\|_{\mathbb{R}^n \times \mathbb{R}^n} \leq Mt\},$$

a function $f\colon E \to \mathbb{R}^n, E \subseteq \mathbb{R} \times \mathbb{R}^n \times \mathbb{R}^n$, is said to be Γ^M-continuous at the point $(t, x, z) \in E$, whenever to every $\varepsilon > 0$ there corresponds $\delta > 0$ such that if $(\tau, v, w) \in E \cap ((t, x, z) + B(0, \delta) \cap \Gamma^M)$ then $\|f(\tau, v, w) - f(t, x, z)\| < \varepsilon$. The function f is called Γ^M-continuous (or simply directionally continuous) when it is Γ^M-continuous at each point of E.

Assuming that the right-hand side of problem (P_f) is directionally continuous, we obtain the following

Proposition 3: *Let $f\colon E \to \mathbb{R}^n$ be a Γ^M-continuous function. Then $f \in \mathbb{A}$.*

Proof: It is not restrictive to equip $\mathbb{R}^n$ with the Euclidean norm. So, if

$$f(t, x, z) = (f_1(t, x, z), f_2(t, x, z), \ldots, f_n(t, x, z))$$

with $f_i\colon E \to \mathbb{R}$, one clearly has

$$|f_i(t, x, z) - f_i(t, v, w)| \leq \|f_i(t, x, z) - f_i(t, v, w)\|$$

for every $(t, x, z) \in E$ and every $i = 1, 2, \ldots n$. Using the fact that f is Γ^M-continuous in E, it is easy to verify that each f_i enjoys the same property. Arguing as in Lemma 1.4 of [6] we then see that D_{f_i}, the set of discontinuities of f_i, has measure zero. Since $D_f = \cup_{i=1}^{n} D_{f_i}$, the conclusion follows. □

To simplify notations, in the next proposition we assume the indeterminate expressions, when $p = 1$ or $p = \infty$, to be read as $\lim_{p \to 1^+}$ or $\lim_{p \to \infty}$, respectively.

Proposition 4: (Lemma 3 [2]). *If $u \in W^{2,p}([a, b], \mathbb{R}^n)$, $p \in [1, +\infty]$, and $u(a) = u(b) = \theta_{\mathbb{R}^n}$, then*

(*i*) $\|u(t)\| \leq \frac{(b-a)}{4}\left[\frac{(b-a)(p-1)}{2p-1}\right]^{1-\frac{1}{p}}\|u''\|_{L^p([a,b],\mathbb{R}^n)} \forall t \in [a, b]$;

(*ii*) $\|u'(t)\| \leq \left[\frac{(b-a)(p-1)}{2p-1}\right]^{1-\frac{1}{p}}\|u''\|_{L^p([a,b],\mathbb{R}^n)} \forall t \in [a, b]$;

(*iii*) $\|u(t^*) - u(t)\| \leq \left[\frac{(b-a)(p-1)}{2p-1}\right]^{1-\frac{1}{p}}\|u''\|_{L^p([a,b],\mathbb{R}^n)}(t^* - t) \,\forall\, t, t^* \in [a, b]$ *with* $t \leq t^*$.

For the convenience of the reader we now state the basic result of S.A. Marano that represents our main tool.

Theorem 1: (Theorem 2.1 [13]). *Let $F\colon [a, b] \times \mathbb{R}^n \times \mathbb{R}^n \to 2^{\mathbb{R}^n}$ be a multifunction with non-empty, closed, and convex values. Assume that:*

(*i*) *For almost every $t \in [a, b]$ the multifunction $F(t, \cdot, \cdot)$ has a closed graph.*
(*ii*) *The set $\{(x, z) \in \mathbb{R}^n \times \mathbb{R}^n : F(\cdot, x, z)$ is measurable$\}$ is dense in $\mathbb{R}^n \times \mathbb{R}^n$.*
(*iii*) *There exist $p, s \in [1, +\infty]$, with $p \leq s$, and $r \in]0, +\infty[$ such that the real function*

$$M_r(t) = \sup_{\|(x,z)\|_{\mathbb{R}^n \times \mathbb{R}^n} \leq \gamma r} \{d(0, F(t, x, z))\}, \quad t \in [a, b]$$

belongs to $L^s([a,b],\mathbb{R})$ and its norm in $L^p([a,b],\mathbb{R})$ is less than or equal to r. Then, the problem

$$\begin{cases} u''(t) \in F(t,u(t),u'(t)) & a.e.\ in\ [a,b] \\ u(a) = u(b) = \theta_{\mathbb{R}^n}, \end{cases}$$

admits at least one solution $u \in W^{2,p}([a,b],\mathbb{R}^n)$ such that $\|u''(t)\| \le M_r(t)$ a.e. in $[a,b]$.
To solve (P_{G_f}) through Theorem 1, we will use the next proposition.

Proposition 5: *Let f: $[a,b] \times \mathbb{R}^n \times \mathbb{R}^n \to \mathbb{R}^n$. Assume that either $f \in H(p,s,r) \cap \mathbb{A}$ or $f \in H(p,s,r) \cap \mathbb{B}$.*
Then, the convexification G_f enjoys the following properties:

(δ_1) *$G_f(t,x,z)$ in non-empty, convex, and closed.*
(δ_2) *$G_f(t,\cdot,\cdot)$ has a closed graph for every $t \in [a,b]$.*
(δ_3) *The set $\{(x,z) \in \mathbb{R}^n \times \mathbb{R}^n : G_f(\cdot,x,z)$ is measurable$\}$ is dense in $\mathbb{R}^n \times \mathbb{R}^n$.*

Moreover, the real-valued function

$$M_r(t) = \sup_{\|(x,z)\|_{\mathbb{R}^n\times\mathbb{R}^n} \le \gamma r} \{d(0, G_f(t,x,z))\}, \quad t \in [a,b]$$

belongs to $L^s([a,b],\mathbb{R})$ and its norm in $L^p([a,b],\mathbb{R})$ is less than or equal to r.

Proof: Obviously, $G_f(t,x,z)$ is non-empty, because $f(t,x,z) \in G_f(t,x,z)$, convex, and closed. Furthermore, by definition of G_f, it is a simple matter to verify (δ_2). Now, we distinguish two cases. If $f \in H(p,s,r) \cap \mathbb{B}$ then, arguing in a standard way, it is easy to prove (δ_3) (see for instance [7,16:p. 171]). Otherwise, namely when $f \in H(p,s,r) \cap \mathbb{A}$, there exists a set $\Omega \subset \mathbb{R}^n \times \mathbb{R}^n$, with $m(\Omega) = 0$, such that

$$D_f^{(x,z)} := \{t \in [a,b] : (t,x,z) \in D_f\}$$

has measure zero for every $(x,z) \in (\mathbb{R}^n \times \mathbb{R}^n)\backslash\Omega$. Evidently, $(\mathbb{R}^n \times \mathbb{R}^n)\backslash\Omega$ is dense in $\mathbb{R}^n \times \mathbb{R}^n$ and by continuity, $t \in [a,b]\backslash D_f^{(x,z)}$ implies $G_f(t,x,z) = \{f(t,x,z)\}$. Since for any $(x,z) \in (\mathbb{R}^n \times \mathbb{R}^n)\backslash\Omega$ and any open set $W \subseteq \mathbb{R}^n$ one has that

$$\begin{aligned} &G_f^-(t,x,z)(W) = \\ &= \left\{t \in [a,b]\backslash D_f^{(x,z)} : f(t,x,z) \in W\right\} \cup \left\{t \in D_f^{(x,z)} : G_f(t,x,z) \cap W \neq \emptyset\right\} \end{aligned}$$

condition (δ_3) holds. Next, bearing in mind (δ_2), (δ_3), and Lemma 6 of [2], we infer that the function M_r is measurable (see [16:p. 171]). Combining this with the obvious inequality

$$M_r(t) \le S_r(t), \ t \in [a,b],$$

completes the proof. □

Remark 1: It is worthwhile to note that putting together Proposition 5 and Theorem 1 gives a solution of problem (P_{G_f}) belonging to $S(f)$.

3. EXISTENCE RESULTS FOR PROBLEM (P_f)

The next result is an immediate consequence of Proposition 5 and Theorem 1.

Theorem 2: *Let* f: $[a,b] \times \mathbb{R}^n \times \mathbb{R}^n \to \mathbb{R}^n$. *Assume that:*

(*i*) *Either* $f \in \mathbb{H}(p,s,r) \cap \mathbb{A}$ *or* $f \in \mathbb{H}(p,s,r) \cap \mathbb{B}$.
(*ii*) *For almost every* $t \in [a,b]$ *the function* $f(t,\cdot,\cdot)$ *has a closed graph.*

Then, problem (P_f) *admits at least one solution* u *lying in* $S(f)$.

Remark 2: We explicitly observe that any Carathéodory function belonging to $\mathbb{H}(p,s,r)$ satisfies the hypotheses of Theorem 2.

We now wish to prove the following

Theorem 3: *Let* f: $[a,b] \times \mathbb{R}^n \times \mathbb{R}^n \to \mathbb{R}^n$ *belong to* $H(p,s,r) \cap \mathbb{B}$. *Assume that:*

(δ_4) *There exists a set* $\Omega_1 \subseteq [a,b]$, *with* $m(\Omega_1) = 0$, *such that for every* $t \in [a,b]\backslash\Omega_1$ *and* $(x,z) \in D_{\Omega_f}$, $0 \in G_f(t,x,z) \Rightarrow f(t,x,z) = 0$.
(δ_5) *There is at least one solution* u *of problem* (P_{G_f}) *such that* $u''(t) = 0$ *almost everywhere in* $\Omega := \{t \in [a,b] : (u(t), u'(t)) \in D_{\Omega_f}\}$.

Then, u *is a solution to* (P_f) *lying to* $S(f)$.

Proof: Let us observe that by (δ_5) there exist two subsets Ω_2 and Ω_3 of $[a,b]$, with $m(\Omega_2) = m(\Omega_3) = 0$, such that $u''(t) \in G_f(t,u(t),u'(t))$ for every $t \in [a,b]\backslash\Omega_2$, and $u''(t) = 0$ for every $t \in \Omega\backslash\Omega_3$. Set $\Omega^* = (\cup_{i=1}^{3} \Omega_i)$. Obviously, Ω^* has measure zero in $[a,b]$. If $t \in [a,b]\backslash\Omega^*$, then

$$G_f(t,u(t),u'(t)) = \{f(t,u(t),u'(t))\} \text{ whenever } t \notin \Omega,$$

whereas, by (δ_4),

$$u''(t) = 0 = f(t,u(t),u'(t)), \text{ every time that } t \in \Omega.$$

Therefore, $u''(t) = f(t,u(t),u'(t))$ a.e in $[a,b]$. □

Indicate with p_i: $[a,b] \times \mathbb{R}^k \to \mathbb{R}$ the projection onto the i-th axis of $\mathbb{R}^k$ if $i = 1,2,\ldots,k$ and reserve p_0 for that onto $[a,b]$. The following corollary states that if $n = 1$ and at least one of the projection of D_{Ω_f} has measure zero, then assumption (δ_5) in Theorem 3 holds.

Corollary 1: *Let* f: $[a,b] \times \mathbb{R} \times \mathbb{R} \to \mathbb{R}$ *be a function beloging to* $H(p,s,r)$ *and satisfying assertion* (δ_4). *Assume that:*

(δ_6) *There exists* $i \in \{1,2\}$ *such that* $m(p_i(D_{\Omega_f})) = 0$.

Then, problem (P_f) *admits at least one solution* $u \in S(f)$.

Proof: We claim that there exist $u \in W^{2,p}([a,b],\mathbb{R})$ and $\Omega_2 \subset [a,b]$, with $m(\Omega_2) = 0$, such that $u(a) = u(b) = 0$ and $u''(t) \in G_f(t,u(t),u'(t))$ for every $t \in [a,b]\backslash\Omega_2$. To see this, we note that, because of (δ_6), one has $m(D_{\Omega_f}) = 0$ and consequently

$$f \in H(p,s,r) \cap \mathbb{B}.$$

So, owing to Proposition 5 and Theorem 1, the asserted u and Ω_2 actually exist. Of course, it clearly results

$$\Omega \subseteq u^{-1}(p_1(D_{\Omega_f})) \cap u'^{-1}(p_2(D_{\Omega_f})).$$

Then, owing to Propositions 1 and 2, there exists a set $\Omega_3 \subseteq [a,b]$, with $m(\Omega_3) = 0$, such that $u''(t) = 0$ for every $t \in \Omega \backslash \Omega_3$. Therefore, all the assumptions of Theorem 3 are satisfied, and the conclusion follows. □

Theorem 4: *Let f: $[a,b] \times \mathbb{R} \times \mathbb{R} \to \mathbb{R}$ be a function beloging to $H(p,s,r)$. Assume that:*

(δ_7) $m(p_i(D_f)) = 0$ *for some* $i \in \{0,1,2\}$.
(δ_8) $\sup_{[a,b]\times B(\vartheta_{\mathbb{R}^n\times\mathbb{R}^n},\gamma r)} f(t,x,z) < 0$.

Then, problem (P_f) admits at least one positive solution $u \in S(f)$.

Proof: Thanks to our assumptions, Proposition 5 and Theorem 1 hold. Then there exist $u \in W^{2,p}([a,b],\mathbb{R})$ and $\Omega_2 \subseteq [a,b]$, with $m(\Omega_2) = 0$, such that $u''(t) \in G_f(t,u(t),u'(t))$ for every $t \in [a,b]\backslash\Omega_2$. Moreover,

$$\widetilde{\Omega} = \{t \in [a,b] : (t,u(t),u'(t)) \in D_f\} \subseteq$$

$$\subseteq p_0(D_f) \cap u^{-1}(p_1(D_f)) \cap u'^{-1}(p_2(D_f)).$$

We claim that $m(\widetilde{\Omega}) = 0$. If $i = 0$, the assertion is trivial. Otherwise, by Propositions 1 and 2 one has $u''(t) = 0$ a.e. in $u^{-1}(p_1(D_f))$ or $u'^{-1}(p_2(D_f))$, respectively. Since Proposition 4 implies $(u(t),u'(t)) \in B(\vartheta_{\mathbb{R}^n\times\mathbb{R}^n},\gamma r)$, by ($\delta_8$), in both cases it results $0 \notin G_f(t,u(t),u'(t))$ for evey $t \in [a,b]$. Thus $m(u^{-1}(p_1(D_f))) = 0$ or $m(u'^{-1}(p_2(D_f))) = 0$ according to whether $i = 1$ or $i = 2$. Now, elementary arguments yield the conclusion. □

In order that each solution of (P_{G_f}) also solves (P_f), a quite different approach consists in requiring that the right-hand side of problem (P_f) be "locally" Γ^M-continuous, as the next theorem shows.

Theorem 5: *Let f: $[a,b] \times \mathbb{R}^n \times \mathbb{R}^n \to \mathbb{R}^n$ belong to $\mathbb{H}(p,s,r)$. Assume that:*

(δ_9) *There exists a sequence $\{E_k\}_{k\in N}$ of pairwise disjoint, non-empty, closed subsets of $[a,b]$ such that $\cup_{k=0}^{+\infty} E_k = [a,b]$, $m(E_0) = 0$, and $S_k = \max_{t\in E_k} S_r(t) < +\infty$ for which $f_{|E_k\times\mathbb{R}^n\times\mathbb{R}^n}$ is Γ^{M_k}-continuous, where*

$$M_k > \max\left\{c_1\left[\left(\frac{(b-a)(p-1)}{2p-1}\right)^{1-\frac{1}{p}}\right]r, \left[\frac{4c_1}{(b-a)^{1\backslash p}}\left(\frac{p-1}{2p-1}\right)^{1-\frac{1}{p}}\right]r,\right.$$
$$\left. c_2(1+S_k), c_2\left(\frac{b-a}{4}\right)(1+S_k)\right\}.$$

Then, problem (P_f) admits at least one solution $u \in S(f)$.

Proof: For simplicity of notation we write h_k instead of $f_{|E_k \times \mathbb{R}^n \times \mathbb{R}^n}$, $k \geq 1$, and indicate with D_{h_k} the set of discontinuity points of h_k. Since h_k is Γ^{M_k}-continuous, Proposition 3 implies $m(D_{h_k}) = 0$. Taking into account that $D_f = \cup_{k=0}^{+\infty} D_{h_k}$, we get

$$f \in H(p, s, r) \cap \mathbb{A}.$$

Thanks to Remark 1 and arguing in a way by now evident, we achieve that problem (P_{G_f}) admits at least one solution $u \in S(f)$. We claim that u also solves (P_f). This is achieved once we prove that

$$u''(t) = h_k(t, u(t), u'(t))$$

a.e. in E_k. We will verify the preceding formula for every $t \in T_k$, where T_k denotes the set of points of $[a, b]$ having the following properties:

(1) $u''(t) \in G_f(t, u(t), u'(t))$.
(2) There exists a strictly decreasing sequence $(t_j)_{j \in \mathbb{N}} \subseteq E_k$ such that

$$t_j \to t, u''(t_j) \to u''(t), u''(t_j) \in G_f(t_j, u(t_j), u'(t_j)) \, \forall t \in T_k.$$

This is clearly enough because, by Lemma 4 of [2], one has $T_k \neq \emptyset$ and $m(T_k) = m(E_k)$. Fix $\varepsilon > 0$ and $t \in T_k$. Since h_k is Γ^{M_k}-continuous at $(t, u(t), u'(t))$, there exists $\delta > 0$ such that for every $(\tau, v, w) \in \{(t, u(t), u'(t)) + B(0, \delta) \cap \Gamma^{M_k}\}$ one has

$$\|h_k(\tau, v, w) - h_k(t, u(t), u'(t))\| < \frac{\varepsilon}{2}.$$

Taking j sufficiently large, it results

$$\left\| \frac{u'(t_j) - u'(t)}{t_j - t} - u''(t) \right\| < 1$$

and so, by Proposition 4,

$$\|u'(t_j) - u'(t)\| \leq$$

$$\leq \left(\left\| \frac{u'(t_j) - u'(t)}{t_j - t} - u''(t) \right\| + \|u''(t)\| \right)(t_j - t) \leq (1 + S_k)(t_j - t).$$

Therefore,

$$\|(u(t_j), u'(t_j)) - (u(t), u'(t))\|_{\mathbb{R}^n \times \mathbb{R}^n} < M_k(t_j - t).$$

This implies that there exists $\rho > 0$ such that

$$\overset{\circ}{B}[(t_j, u(t_j), u'(t_j)), \rho] \subset \{(t, u(t), u'(t)) + B(0, \delta) \cap \Gamma^{M_k}\},$$

which leads to

$$u''(t_j) \in G_f(t_j, u(t_j), u'(t_j)) \subset B\Big(h_k(t, u(t), u'(t)), \frac{\varepsilon}{2}\Big).$$

Hence, as an easy computation shows,

$$\|u''(t) - h_k(t, u(t), u'(t))\|_{\mathbb{R}^n} \leq$$

$$\leq \|u''(t) - u''(t_j)\|_{\mathbb{R}^n} + \|u''(t_j) - h_k(t, u(t), u'(t))\|_{\mathbb{R}^n} < \varepsilon.$$

Since ε was arbitrary, the conclusion follows. □

4. EXISTENCE RESULTS FOR IMPLICIT EQUATIONS

In this section we state two existence results concernig the Dirichlet problem for second-order implicit equations. For more general case we refer the reader to [2], and [17] as previous results on the subject we mention [2], [15], and [17].

Theorem 6: *Let Z be an arcwise connected subset of $\mathbb{R}^n$, let g: $Z \to \mathbb{R}$ be continuous and let f: $[a, b] \times \mathbb{R}^n \times \mathbb{R}^n \to \mathbb{R}$. Suppose that:*

(k_1) $\inf p_j(\overline{\text{co}}(Z)) \cdot \sup p_j(\overline{\text{co}}(Z)) > 0$ *for each* $j \in \{1, 2, \ldots, n\}$.
(k_2) *f is essentially bounded.*
(k_3) $m(p_j(D_f)) = 0$ *for some* $j \in \{0, 1, 2, \ldots, 2n\}$.
(k_4) $m(p_j(f^{-1}(r)\backslash \text{int}(f^{-1}(r)))) = 0$ *for some* $j \in \{0, 1, 2, \ldots, 2n\}$ *and for every* $r \in g(Z)$.
(k_5) $\overline{f(([a, b] \times \mathbb{R}^n \times \mathbb{R}^n)\backslash D_f)} \subseteq g(Z)$.

Then problem $(P^i_{g,f})$ admits at least one solution $u \in S(f)$. Further, u'' is essentially bounded and almost everywhere equal to a function whose set of discontinuity points has measure zero.

Proof: Write $S = [a, b] \times \mathbb{R}^n \times \mathbb{R}^n$, and observe that, by ($k_2$), there exists $c > 0$ such that

$$S\backslash D_f \subseteq \{(t, x, z) \in S : f(t, x, z) \leq c\}.$$

Now, set

$$a = \min \overline{f(S\backslash D_f)} \quad \text{and} \quad b = \max \overline{f(S\backslash D_f)}.$$

Assumption (k_5) ensures that $g(y_1) = a$ and $g(y_2) = b$ for some $y_1, y_2 \in Z$. Pick a continuous function λ: $[0, 1] \to Z$ compling with $\lambda(0) = y_1, \lambda(1) = y_2$ and define, for every $t \in [0, 1], \tilde{g}(t) = g(\lambda(t))$. We now distinguish two cases. First, suppose that $\tilde{g}$ is nonconstant and choose $t_1, t_2 \in [0, 1]$ fulfilling

$$\tilde{g}(t_1) = \min_{t \in [0,1]} \tilde{g}(t), \quad \tilde{g}(t) = \max_{t \in [0,1]} \tilde{g}(t).$$

Obviously, $t_1 \neq t_2$ and it is not restrictive to assume $t_1 < t_2$. Let ψ: $\tilde{g}([0, 1]) \to [0, 1]$ be defined by $\psi(r) = \min\{\tilde{g}^{-1}(r) \cap [t_1, t_2]\}$ for every $r \in \tilde{g}([0, 1])$. We claim that ψ is strictly increasing. Indeed, pick $r_1, r_2 \in \tilde{g}([0, 1])$ with $r_1 < r_2$. Then $\psi(r_1) \neq \psi(r_2)$ and $\psi(r_1) > t_1$. From $\tilde{g}(\psi(r_2)) = r_2 > r_1$ and $\tilde{g}(t_1) \leq r_1$, taking into account the continuity of

$\widetilde{g}$, we immediately infer $\psi(r_1) < \psi(r_2)$. Therefore, the set D_K of all discontinuity points of the function $K\colon \mathbb{R} \to Z$ given by

$$K(r) = \begin{cases} \lambda(\psi(\widetilde{g}(t_1))) & if\ r \in]-\infty, \widetilde{g}(t_1)[, \\ \lambda(\psi(r)) & if\ r \in \widetilde{g}([0,1]), \\ \lambda(\psi(\widetilde{g}(t_2))) & if\ r \in]\widetilde{g}(t_2), +\infty[, \end{cases}$$

is at most countable. Now, suppose $\widetilde{g}$ is constant and write $K(r) = y_1$ for every $r \in \mathbb{R}$. In both cases, we define $h\colon S \to Z$ by putting, for every $(t, x, z) \in S$,

$$h(t, x, z) = K(f(t, x, z)).$$

It is easy to prove that if D_h denotes the set of discontinuities of h then one has

$$D_h = D_f \cup \left[\cup_{r \in D_K} (f^{-1}(r) \setminus \text{int}(f^{-1}(r)))\right]$$

or $D_h = D_f$, according to whether $\widetilde{g}$ is nonconstant or not. Therefore, by (k_3) and (k_4), one has $h \in \mathbb{A}$. Since h is bounded, we also have

$$h \in H(+\infty, +\infty, r) \cap \mathbb{A}$$

for some $r \in]0, +\infty[$. Hence, thanks to Remark 1, there exist $u \in W^{2,p}([a, b], \mathbb{R}^n)$ and $\Omega_2 \subseteq [a, b]$, with $m(\Omega_2) = 0$ such that $u''(t) \in G_h(t, u(t), u'(t))$ for every $t \in [a, b] \setminus \Omega_2$. Now, we observe that the set

$$\widetilde{\Omega} = \{t \in [a, b] : (t, u(t), u'(t)) \in D_h\}$$

is contained in

$$p_0(D_h) \cap \left(\bigcap_{j=1}^{n} u_j^{-1}(p_j(D_h)) \right) \cap \left(\bigcap_{j=n+1}^{2n} \left(u_j'^{-1}(p_j(D_h)) \right) \right).$$

Bearing in mind (k_1), for every $y = (y_1, y_2, \ldots, y_n) \in Z$ one has $y_i > 0$ for $i = 1, 2, \ldots, n$. Using Propositions 1 and 2 as in the proof Theorem 5 we achieve $m(\widetilde{\Omega}) = 0$. Hence, $u''(t) = h(t, u(t), u'(t))$ a.e. in $[a, b]$. Consequently, the set of discontinuity points of u'' has measure zero and

$$g((u''(t)) = \widetilde{g}(\psi(f(t, u(t), u'(t)))) = f(t, u(t), u'(t)) \text{ a.e. in } [a, b].$$

This completes the proof. □

Theorem 7: (Theorem 4 [2]). *Let Y be a non-empty, connected and locally connected subset of $\mathbb{R}^n$ and let $f\colon [a, b] \times \mathbb{R}^n \times \mathbb{R}^n \times Y \to \mathbb{R}$ be continuous. Assume that there exist $p, s \in [1, +\infty]$, with $p \leq s, r \in]0, +\infty[$ and a non-negative function $d \in L^s([a, b], \mathbb{R})$, with $r \geq \|d\|_{L^p([a,b],\mathbb{R})}$, such that:*

(k_6) *For almost every $t \in [a, b]$ and every $(x, z) \in B(\theta_{\mathbb{R}^n \times \mathbb{R}^n}, \gamma r)$, the set*

$$\{y \in Y : f(t, x, z, y) = 0\}$$

has empty interior in Y and

$$0 \in \mathrm{int}\{f(t,x,z,Y \cap B^{\circ}(\theta_{\mathbb{R}^n}, d(t)))\}.$$

(k_7) *For almost every $t \in [a,b]$, the set $Y \cap B^{\circ}(\theta_{\mathbb{R}^n}, d(t))$ is connected.*

Then, problem (P_f^i) admits at least one solution $u \in W^{2,p}([a,b],\mathbb{R}^n)$ such that $\|u''(t)\| \leq d(t)$ a.e in $[a,b]$.

Proof: Define Q: $[a,b] \times \mathbb{R}^n \times \mathbb{R}^n \to 2^{\mathbb{R}^n}$ by putting, for every $(t,x,z) \in [a,b] \times \mathbb{R}^n \times \mathbb{R}^n, Q(t,x,z) = \{y \in Y : f(t,x,z,y) = 0$ and y is not a local extremum point for$f(t,x,z,\cdot)\}$. Thanks to our assumptions, Theorem 1.1 of [19] holds and, by standard arguments, it is easy to prove that the multifunction Q has non-empty closed values and is lower semicontinuous (see [2,17]). Using Vitali's covering theorem and Lusin's theorem, yields a sequence $\{E_k\}_{k\in N}$ of pairwise disjoint, non-empty, closed subsets of $[a,b]$ such that $\cup_{k=0}^{+\infty} E_k = [a,b], m(E_0) = 0$, and $d_{|E_k}$ is continuous. Now, we set

$$F(t,x,z) = \begin{cases} \overline{Q(t,x,z) \cap B^{\circ}(\theta_{\mathbb{R}^n}, d(t))} & \text{if } (t,x,z) \in \{E_k \times B(\theta_{\mathbb{R}^n \times \mathbb{R}^n}, \gamma r)\}, \\ B(\theta_{\mathbb{R}^n}, d(t)) & \text{if } (t,x,z) \notin \{[a,b] \times B(\theta_{\mathbb{R}^n \times \mathbb{R}^n}, \gamma r)\}. \end{cases}$$

Owing to Lemma 5 in [2], the multifunction F: $[a,b] \times \mathbb{R}^n \times \mathbb{R}^n \to 2^{\mathbb{R}}$ is lower semicontinuous as a simple computation shows. So, by Theorem 2.1 of [9], for every $k \geq 1$ there exists a Γ^{M_k}-continuous selection h_k of $F_{|E_k \times \mathbb{R}^n \times \mathbb{R}^n}$, where

$$M_k > \max\left\{ c_1 \left[\left(\frac{(b-a)(p-1)}{2p-1} \right)^{1-\frac{1}{p}} \right] r, \left[\frac{4c_1}{(b-a)^{1\backslash p}} \left(\frac{p-1}{2p-1} \right)^{1-\frac{1}{p}} \right] r, \right.$$
$$\left. c_2(1+d_k), c_2\left(\frac{b-a}{4}\right)(1+d_k) \right\}$$

and $d_k = \max_{t \in E_k} d_k(t)$. Of course, we adopt here the same convention stated at the beginning of Proposition 4 holds. Define g: $[a,b] \times \mathbb{R}^n \times \mathbb{R}^n \to \mathbb{R}^n$ by putting, for every $(t,x,z) \in [a,b] \times \mathbb{R}^n \times \mathbb{R}^n$,

$$g(t,x,z) = g_k(t,x,z) \text{ if } t \in E_k, k \in \mathbb{N}^+.$$

Evidently, the function g satisfies all the hypotheses of Theorem 5 and, then there exists $u \in S(f)$ such that $u''(t) = g(t,u(t),u'(t))$, and $u(a) = u(b) = \theta_{\mathbb{R}^n}$. Taking into account Proposition 4 one has $(u(t),u'(t)) \in B(\theta_{\mathbb{R}^n \times \mathbb{R}^n}, \gamma r)$ for every $t \in [a,b]$, thus $u''(t) \in Q(t,u(t),u'(t))$ a.e in $[a,b]$. This completes the proof. □

ACKNOWLEDGMENT

The author wishes to thank Professor S.A. Marano for introducing him to the topics treated in this chapter and for many stimulating conversations.

REFERENCES

[1] Ravi P. Agarwal (1986). Boundary Value Problems for Higher Order Differential Equations. Singapore: *World Sci. Publ.*

[2] D. Averna and G. Bonanno (1999). Existence of solutions for a multivalued boundary value problem with non-convex and unbounded right-hand side, *Ann. Polon. Math.*, **71**(3), 253–271.

[3] D. Averna and S.A. Marano (1999). An existence theorem for inclusions of the type $\Psi(u)(t) \in F(t, \Phi(t))$ and application to a multivalued boundary value problem, *Appl. Anal.*, **72**, 449–458.

[4] P. Binding (1979). The Differential Equation $x' = f \circ x$, *J. Differential Equations*, **31**, 183–199.

[5] G. Bonanno (1996). Differential inclusions with nonconvex right-hand side and applications to implicit integral and differential equations, *Rend. Accad. Naz. Sci. XL Mem. Mat. Appl.*, **114**(20), 193–203.

[6] G. Bonanno and S.A. Marano. Elliptic problems in R^n with discontinuous non-linearities, *Proc. Edinburgh Math. Soc.*, (to appear).

[7] G. Bonanno and S.A. Marano (1996). Positive solutions of elliptic equations with discontinuous nonlinearities, *Topol. Methods Nonlinear Anal.*, **8**, 263–273.

[8] A. Bressan (1998). Directionally continuous selections and differential inclusions, *Funkcial. Ekvac.*, **31**, 459–470.

[9] A. Bressan (1989). Upper and lower semicontinuous differential inclusion: A unified approach. In H. Sussmann (Ed.), *Controllability and Optimal Control*, New York: Dekker, 21–31.

[10] K. Deimling (1992). Multivalued Differential Equations, de Gruyter Ser. *Nonlinear Anal. Appl.*, 1 de Gruyter, Berlin.

[11] A.F. Filippov (1988). Differential Equations with Discontinuous Righthand Sides. Dordrecht: *Kluwer Acad. Publ.*

[12] S. Hu (1991). Differential equations with discontinuous right-hand sides, *J. Math. Anal. Appl.*, **154**, 377–390.

[13] S.A. Marano (1992). Existence theorems for a multivalued boundary value problem, *Bull. Austral. Math. Soc.*, **45**, 249–260.

[14] S.A. Marano (1994). Implicit elliptic differential equations, *Set-Valued Anal.*, **2**, 545–558.

[15] S.A. Marano (1996). Implicit elliptic boundary-value problems with discontinuous nonlinearities, *Set-Valued Anal.*, **4**, 287–300.

[16] S.A. Marano (1995). Elliptic boundary-value problems with discontinuous nonlinearities, *Set-Valued Anal.*, **3**, 167–180.

[17] S.A. Marano (1994). On a boundary value problem for the differential equation $f(t, x, x', x'') = 0$, *J. Math. Anal. Appl.*, **182**, 309–319.

[18] O. Naselli Ricceri and B. Ricceri (1990). An existence theorem for inclusions of the type $\Psi(u)(t) \in F(t, \Phi(t))$ and application to a multivalued boundary value problem, *Appl. Anal.*, **38**, 259–270.

[19] B. Ricceri (1982). Applications de théorèmes de semi-continuité inférieure, *C.R. Acad. Sci. Paris Sér. I*, **295**, 75–78.

[20] B. Ricceri (1985). Lipschitzian solution of the implicit Cauchy problem $g(x') = f(t, x)$, $x(0) = 0$, with f discontinuous in x, *Rend. Circ. Mat. Palermo*, **34**(2), 127–135.

[21] B. Ricceri (1987). On multifunctions of one real variable, *J. Math. Anal. Appl.*, **295**, 225–236.

SIMAA 4(2002) 63–77

7. Generalized Strongly Nonlinear Implicit Quasi-variational Inequalities for Fuzzy Mappings*

Y.J. Cho[1], S.H. Shim[1], N.J. Huang[2] and S.M. Kang[1]

[1]*Department of Mathematics, Gyeongsang National University, Chinju 660-701, Korea*
[2]*Department of Mathematics, Sichuan University, Chengdu, Sichuan 610064, People's Republic of China*

Abstract: In this chapter, we study a class of generalized strongly nonlinear implicit quasi-variational inequality problems for fuzzy mappings and prove that the generalized strongly nonlinear implicit quasi-variational inequality problem for fuzzy mappings is equivalent to solving the fixed point problem. Using the quivalence, a new iterative algorithm for finding the approximate solutions of the generalized strongly nonlinear implicit quasi-variational inequality problem for fuzzy mappings is suggested. We also discuss the existence of solutions for this kind of generalized strongly nonlinear implicit quasi-variational inequality problem for fuzzy mappings without compactness and the convergence of iterative sequences generated by the algorithms.

Keywords and Phrases: Nonlinear implicit quasi-variational inequality, set-valued mapping, fuzzy mapping, nonlinear implicit quasi-complementarity problem, algorithm

1991 AMS Mathematics Subject Classification: 49J40, 47H04, 46S40

1. INTRODUCTION

Variational inequality and complementarity problem theory play an important and fundamental role in the study of a wide class of problems arising in mechanics, physics, nonlinear programming, optimization and control, economics and transportation equilibrium, contact problems in elasticity, fluid flow through porous media, and many other branches of mathematical and engineering sciences (see, for example, [1–5,10–13, 16,17,24,27–29,37,38,43] and the references therein).

In recent years, they have been extended and generalized in several directions to study a large number of unrelated problems in a unified and general framework. Among these generalizations of the variational inequality and complementarity problems, the (implicit) quasi-variational inequality and complementarity problems are important and useful generalizations in which the constraint set depends on the solution.

---

* This work was supported by Korea Research Foundation Grant (KRF-99-005-D0003). The third author was supported by '98 APEC Post-Doctor Fellowship (KOSEF) while he visited Gyeongsang National University.

In 1991, Chang and Huang [5,6] introduced and studied some new classes of (implicit) quasi-complementarity problems for set-valued mappings with compact values in Hilbert spaces, which include many complementarity and quasi-complementarity problems studied by Noor [31,32], Isac [23] and Pang [39,40] as special cases.

Recently, Huang [18] introduced a new class of generalized nonlinear implicit quasi-variational inequalities for set-valued mappings and proved the existence of solutions for this class of generalized nonlinear implicit quasi-variational inequalities for set-valued mappings without compactness and the convergence of iterative sequences generated by the algorithms, which improves and extends the earlier and recent results of Chang [4], Chang and Huang [5,6], Ding [14], Huang [19], Noor [33], Siddiqi and Ansari [41,42], and Zeng [44].

On the other hand, in 1989, Chang and Zhu [9] first introduced and studied a class of variational inequalities for fuzzy mappings. Recently, several kinds of variational inequalities and complementarity problems for fuzzy mappings were considered and studied by Chang [4], Chang and Huang [7,8], Huang [19–21], Noor [37] and Lee *et al.* [25,26]. These works may lead to new and significant results in these areas [38].

In this chapter, motivated and inspired by recent research work in this field, we study a class of generalized strongly nonlinear implicit quasi-variational inequality problems for fuzzy mappings and prove that the generalized strongly nonlinear implicit quasi-variational inequality problem is equivalent to solving the fixed point problem. By using the quivalence, a new iterative algorithm for finding the approximate solutions of the generalized strongly nonlinear implicit quasi-variational inequality problem for fuzzy mappings is suggested. We also discuss the existence of solutions for this kind of generalized strongly nonlinear implicit quasi-variational inequality problem for fuzzy mappings without compactness and the convergence of iterative sequences generated by the algorithm. Our results improve and extend some known results in this field.

2. PRELIMINARIES AND FORMULATIONS

Let H be a real Hilbert space endowed with the norm $\|\cdot\|$ and inner product $(\cdot,\cdot)$. Let D be a nonempty subset of H and $\mathcal{F}(E)$ denote a collection of all fuzzy sets over E for any $E \subset H$. A mapping $\tilde{F}$ from D into $\mathcal{F}(H)$ is called a *fuzzy mapping* on D. If $\tilde{F}$ is a fuzzy mapping on D, then $\tilde{F}(x)$ (denote it by $\tilde{F}_x$, in the sequel) is a fuzzy set on H and $\tilde{F}_x(y)$ is the membership function of y in $\tilde{F}_x$.

A fuzzy mapping $\tilde{F}$: $D \to \mathcal{F}(H)$ is said to be *closed* if, for any $x \in D$, the function $\tilde{F}_x(y)$ is upper semi-continuous with respect to y (i.e., for any given point $y_0 \in H$ and any net $\{y_\alpha\} \subset H$, when $y_\alpha \to y_0$, we have $\tilde{F}_x(y_0) \geq \limsup_\alpha \tilde{F}_x(y_\alpha)$).

Let $B \in \mathcal{F}(H), q \in [0,1]$. Then the set

$$(B)_q = \{x \in H : B(x) \geq q\}$$

is called a q-cut set of B.

Let $\tilde{F}, \tilde{G}$: $D \to \mathcal{F}(H)$ and $\tilde{A}$: $D \to \mathcal{F}(D)$ be closed fuzzy mappings satisfying the following condition (I):

(I) There exist mappings a, b, c: $D \to [0,1]$ such that, for all $x \in D$, the sets $(\tilde{F}_x)_{a(x)}, (\tilde{G}_x)_{b(x)}$ and $(\tilde{A}_x)_{c(x)}$ are nonempty and bounded.

Obviously, if $\tilde{F}, \tilde{G}$: $D \to \mathcal{F}(H)$ and $\tilde{A}$: $D \to \mathcal{F}(D)$ are closed fuzzy mappings satisfying the condition (I), then for all $x \in D$, the sets $(\tilde{F}_x)_{a(x)} \in CB(H), (\tilde{G}_x)_{b(x)}$

$\in CB(H)$ and $(\tilde{G}_x)_{b(x)} \in CB(D)$, where $CB(E)$ denotes the family of all nonempty bounded closed subsets of E for any $E \subset H$. In fact, for each $x \in D$, let $\{y_j\}_{j \in I} \subset (\tilde{F}_x)_{a(x)}$ be a net in H and $y_j \to y_0 \in H$. Then $(\tilde{F}_x)(y_j) \geq a(x)$ for $j \in I$. Since $\tilde{F}$ is closed, we have

$$\tilde{F}_x(y_0) \geq \limsup_{j \in I} \tilde{F}_x(y_j) \geq a(x).$$

This implies that $y_0 \in (\tilde{F}_x)_{a(x)}$ and so $(\tilde{F}_x)_{a(x)} \in CB(H)$. Similarly, $(\tilde{G}_x)_{b(x)} \in CB(H)$ and $(\tilde{A}_x)_{c(x)} \in CB(D)$.

Let $\tilde{F}, \tilde{G}$: $D \to \mathcal{F}(H)$ and $\tilde{A}$: $D \to \mathcal{F}(D)$ be closed fuzzy mappings satisfying the condition (I). Then we can define set-valued mappings F, G and A as follows:

$$\begin{aligned} F&: D \to CB(H), \quad x \longmapsto (\tilde{F}_x)_{a(x)}, \\ G&: D \to CB(H), \quad x \longmapsto (\tilde{G}_x)_{b(x)}, \\ A&: D \to CB(D), \quad x \longmapsto (\tilde{A}_x)_{c(x)}. \end{aligned}$$

In the sequel, F, G and A are called the set-valued mappings induced by the fuzzy mappings $\tilde{F}, \tilde{G}$: $D \to \mathcal{F}(H)$ and $\tilde{A}$: $D \to \mathcal{F}(D)$, respectively.

Let a, b, c: $H \to [0, 1]$ be mappings and $\tilde{F}, \tilde{G}$: $D \to \mathcal{F}(H)$ and $\tilde{A}$: $D \to \mathcal{F}(D)$ be fuzzy mappings. Let S, T: $H \to H, N$: $H \times H \to H$ and g: $D \to H$ be single-valued mappings. Let K: $D \to 2^H$ be a set-valued mapping such that, for each $x \in D, K(x)$ is a nonempty closed convex subset of H. We consider the following problem:

Find $u, w \in D$ and $x, y \in H$ such that

$$\begin{cases} g(u) \in K(w), \\ \tilde{F}_u(x) \geq a(u), \tilde{G}_u(y) \geq b(u), \tilde{A}_u(w) \geq c(u), \\ (N(Sx, Ty), g(v) - g(u)) \geq 0 \end{cases} \tag{2.1}$$

for all $g(v) \in K(w)$. The problem (2.1) is called the *generalized strongly nonlinear implicit quasi-variational inequality* for fuzzy mappings.

If $K(x)$ is a closed convex cone in H and

$$K^*(x) = \{u \in H : (u, v) \geq 0 \text{ for all } v \in K(x)\}$$

is a polar cone of $K(x)$ in H for each $x \in D$, then the problem (2.1) is equivalent to finding $u, w \in D$ and $x, y \in H$ such that

$$\begin{cases} g(u) \in K(w), \\ \tilde{F}_u(x) \geq a(u), \tilde{G}_u(y) \geq b(u), \tilde{A}_u(w) \geq c(u), \\ N(Sx, Ty) \in K^*(w), \\ (g(u), N(Sx, Ty)) = 0, \end{cases} \tag{2.2}$$

which is called the *generalized strongly nonlinear implicit quasi-complementarity problem* for fuzzy mappings.

If F, G and A are classical set-valued mappings, the problems (2.1) and (2.2) are reduced to finding $u \in D, x \in Fu, y \in Gu$ and $w \in Au$ such that

$$\begin{cases} g(u) \in K(w), \\ (N(Sx, Ty), g(v) - g(u)) \geq 0 \end{cases} \tag{2.3}$$

for all $g(v) \in K(w)$ and

$$\begin{cases} g(u) \in K(w), \\ N(Sx, Ty) \in K^*(w), \\ (g(u), N(Sx, Ty)) = 0, \end{cases} \tag{2.4}$$

respectively. The problems (2.3) and (2.4) are called the *generalized set-valued strongly nonlinear implicit quasi-variational inequality* and the *generalized set-valued strongly nonlinear implicit quasi-complementarity problem*, respectively.

If $N(x, y) = x - y, Sx = Tx = x$ for all $x, y \in H$ and $Ax = x$ for all $x \in D$, then the problems (2.3) and (2.4) are reduced to finding $u \in D, x \in Fu$ and $y \in Gu$ such that

$$\begin{cases} g(u) \in K(u), \\ (x - y, g(v) - g(u)) \geq 0 \end{cases} \tag{2.5}$$

for all $g(v) \in K(w)$ and

$$\begin{cases} g(u) \in K(u), \\ x - y \in K^*(u), \\ (g(u), x - y) = 0, \end{cases} \tag{2.6}$$

respectively. The problems (2.3) and (2.4) are called the *generalized strongly nonlinear quasi-variational inequality* and the *generalized strongly nonlinear quasi-complementarity problem*, respectively (see Ding [15]).

Obviously, a number of known classes of variational inequalities, complementarity problems, (implicit) quasi-variational inequalities and (implicit) quasi-complementarity problems studied previously by many authors including Chang [4], Chang and Huang [5,6], Ding [14,15], Huang and Hu [22], Noor [31–36], Siddiqi and Ansari [41,42], and Zeng [44] can be obtained as special cases of the problems (2.1) and (2.2).

3. EQUIVALENCE AND ITERATIVE ALGORITHMS

We first introduce the following results for our main theorems.

Lemma 3.1: ([29]) *If K is a closed convex subset of a Hilbert space H and z is a given point in H, then $u \in K$ satisfies the inequality*

$$(u - z, v - u) \geq 0$$

for all $v \in K$ if and only if

$$u = P_K z, \tag{3.1}$$

where P_K is the projection of H onto K.

Lemma 3.2: ([29]) *The mapping P_K defined by (3.1) is nonexpansive, i.e.,*

$$\|P_K u - P_K v\| \leq \|u - v\|$$

for all $u, v \in H$.

Lemma 3.3: ([36]) If *$K(u) = m(u) + K$ and K is a closed convex subset of H, then for any $u, v \in H$,*

$$P_{K(u)} v = m(u) + P_K(v - m(u)). \tag{3.2}$$

Theorem 3.1: *Let D be a nonempty subset of a Hilbert space H and $K\colon D \to 2^H$ be set-valued mappings such that, for each $x \in D$, $K(x)$ is a nonempty closed convex cone in H and $K(D) \subset g(D)$. Then the generalized strongly nonlinear implicit quasi-variational inequality problem (2.1) and the generalized strongly nonlinear implicit quasi-complementarity problem (2.2) have the same set of solutions.*

Proof: If (u, x, y, w) is a solution of the problem (2.1), then $u, w \in D$ and $x, y \in H$ such that

$$\begin{cases} g(u) \in K(w), \\ \tilde{F}_u(x) \geq a(u), \tilde{G}_u(y) \geq b(u), \tilde{A}_u(w) \geq c(u), \\ (N(Sx, Ty), g(v) - g(u)) \geq 0 \end{cases} \tag{3.3}$$

for all $g(v) \in K(w)$. Since $2g(u) \in K(w)$, it follows from (3.3) that $(N(Sx, Ty), g(u)) \geq 0$. Since $0 \in K(w) \subset g(D)$, there exists $z \in D$ such that $0 = g(z) \in K(w)$. By (3.3), we have $(N(Sx, Ty), g(u)) \leq 0$ and so we have

$$(N(Sx, Ty), g(u)) = 0.$$

Since $K(D) \subset g(D)$, for any $z \in K(w)$, there exists $v \in D$ such that $z = g(v) \in K(w)$ and hence we have $z + g(u) \in K(w)$ and

$$\begin{aligned} (N(Sx, Ty), z) &= (N(Sx, Ty), z + g(u) - g(u)) \\ &= (N(Sx, Ty), g(v) + g(u)) - (N(Sx, Ty), g(u)) \\ &= (N(Sx, Ty), g(v) - g(u)) \geq 0, \end{aligned}$$

which implies that $N(Sx, Ty) \in K^*(w)$. This prove that (u, x, y, w) is a solution of the problem (2.2).

Conversely, if (u, x, y, w) is a solution of the problem (2.2), then $u, w \in D$ and $x, y \in H$ are such that

$$\begin{cases} g(u) \in K(w), \\ \tilde{F}_u(x) \geq a(u), \tilde{G}_u(y) \geq b(u), \tilde{A}_u(w) \geq c(u), \\ N(Sx, Ty) \in K^*(w), \\ (g(u), N(Sx, Ty)) = 0 \end{cases}$$

and so it follows that

$$\begin{aligned}(N(Sx,Ty),g(v)-g(u)) &= (N(Sx,Ty),g(v))-(N(Sx,Ty),g(u))\\ &= (N(Sx,Ty),g(v))\\ &\geq 0\end{aligned}$$

for each $g(v) \in K(w)$. Therefore (u,x,y,w) is a solution of the problem (2.1). This completes the proof.

Remark 3.1: Theorem 3.1 is a generalization of Theorem 3.1 of Ding [15].

Theorem 3.2: *Let D be a nonempty subset of a Hilbert space H and $K\colon D \to 2^H$ be set-valued mappings such that, for each $x \in D, K(x)$ is a nonempty closed convex subset in H and $K(D) \subset g(D)$. Then (u,x,y,w) is a solution of the generalized strongly nonlinear implicit quasi-variational inequality problem (2.1) if and only if $u \in D, \tilde{F}_u(x) \geq a(u)$, $\tilde{G}_u(y) \geq b(u), \tilde{A}_u(w) \geq c(u)$ and*

$$g(u) = P_{K(w)}(g(u) - \rho N(Sx,Ty)), \tag{3.4}$$

where $\rho > 0$ is a constant.

Proof: Let (u,x,y,w) be a solution of the problem (2.1). Then $u,w \in D$ and $x,y \in H$ are such that

$$\begin{cases} g(u) \in K(w), \\ \tilde{F}_u(x) \geq a(u), \tilde{G}_u(y) \geq b(u), \tilde{A}_u(w) \geq c(u), \\ (N(Sx,Ty),g(v)-g(u)) \geq 0 \end{cases}$$

for all $g(v) \in K(w)$. Since $K(D) \subset g(D)$, for any $\rho > 0$, we have

$$(g(u) - [g(u) - \rho(N(Sx,Ty)], z - g(u)) \geq 0$$

for all $z \in K(w)$. Therefore, it follows from Lemma 2.1 that

$$g(u) = P_{K(w)}(g(u) - \rho N(Sx,Ty)).$$

Conversely, if $u \in D, \tilde{F}_u(x) \geq a(u), \tilde{G}_u(y) \geq b(u)$ and $\tilde{A}_u(w) \geq c(u)$ are such that (3.4) holds, then it follows from Lemma 2.1 that $g(u) \in K(w)$ and

$$(g(u) - [g(u) - \rho N(Sx,Ty)], z - g(u)) \geq 0$$

for all $z \in K(w)$. Thus we have

$$(N(Sx,Ty), g(v) - g(u)) \geq 0$$

for all $g(v) \in K(w)$ and so (u,x,y,w) is a solution of the problem (2.1). This completes the proof.

Based on Theorem 3.2, we are now in a position to propose the following generalized and unified new algorithm for solving the problem (2.1).

Let D be a nonempty subset of a Hilbert space $H, S, T: H \to H, N: H \times H \to H$ and $g: D \to H$ be single-valued mappings. Let $K: D \to 2^H$ be a set-valued mapping such that, for each $x \in D, K(x)$ is a nonempty closed convex subset in H and $K(D) \subset g(D)$. Suppose that $\tilde{F}, \tilde{G}: D \to \mathcal{F}(H)$ and $\tilde{A}: D \to \mathcal{F}(D)$ are closed fuzzy mappings satisfying the condition (I). Let F, G and A be the set-valued mappings induced by the fuzzy mappings $\tilde{F}, \tilde{G}$ and $\tilde{A}$, respectively. For given $u_0 \in D, x_0 \in Fu_0, y_0 \in Gu_0$ and $w_0 \in Au_0$, by the assumption $K(D) \subset g(D)$, there is $u_1 \in D$ such that

$$g(u_1) = P_{K(w_0)}(g(u_0) - \rho N(Sx_0, Ty_0)),$$

where $\rho > 0$ is a constant. Since $x_0 \in Fu_0 \in CB(H), y_0 \in Gu_0 \in CB(H)$ and $w_0 \in Au_0 \in CB(D)$, by the result of Nadler [30], there exist $x_1 \in Fu_1$, $y_1 \in Gu_1$ and $w_1 \in Au_1$ such that

$$\begin{cases} \|x_0 - x_1\| \le (1+1)H(Fu_0, Fu_1), \\ \|y_0 - y_1\| \le (1+1)H(Gu_0, Gu_1), \\ \|w_0 - w_1\| \le (1+1)H(Au_0, Au_1), \end{cases}$$

where $H(\cdot,\cdot)$ is the Hausdorff metric. By the assumption $K(D) \subset g(D)$, there is $u_2 \in D$ such that

$$g(u_2) = P_{K(w_1)}(g(u_1) - \rho N(Sx_1, Ty_1)).$$

By induction, we have the following algorithm:

Algorithm 3.1: *Let D be a nonempty subset of a Hilbert space $H, S, T: H \to H, N: H \times H \to H$ and $g: D \to H$ be single-valued mappings. Let $K: D \to 2^H$ be a set-valued mapping such that, for each $x \in D, K(x)$ is a nonempty closed convex subset in H and $K(D) \subset g(D)$. Suppose that $\tilde{F}, \tilde{G}: D \to \mathcal{F}(H)$ and $\tilde{A}: D \to \mathcal{F}(D)$ are closed fuzzy mappings satisfying the condition (I). Let F, G and A be the set-valued mappings induced by the fuzzy mappings $\tilde{F}, \tilde{G}$ and $\tilde{A}$, respectively. Then, for given $u_0 \in D$, we can obtain sequences $\{x_n\}, \{y_n\}, \{w_n\}$ and $\{u_n\}$ such that*

$$\begin{cases} x_n \in Fu_n, \; \|x_n - x_{n+1}\| \le \left(1 + \dfrac{1}{1+n}\right) H(Fu_n, Fu_{n+1}), \\ y_n \in Gu_n, \; \|y_n - y_{n+1}\| \le \left(1 + \dfrac{1}{1+n}\right) H(Gu_n, Gu_{n+1}), \\ w_n \in Au_n, \; \|w_n - w_{n+1}\| \le \left(1 + \dfrac{1}{1+n}\right) H(Au_n, Au_{n+1}), \\ g(u_{n+1}) = P_{K(w_n)}(g(u_n) - \rho N(Sx_n, Ty_n)) \end{cases} \tag{3.5}$$

for all $n = 0, 1, 2, \ldots,$ where $\rho > 0$ is a constant.

From Algorithm 3.1, we have the following algorithm:

Algorithm 3.2: *Let D be a nonempty subset of a Hilbert space $H, S, T\colon H \to H, N\colon H \times H \to H$ and $g\colon D \to H$ be single-valued mappings. Let $K\colon D \to 2^H$ be a set-valued mapping such that, for each $x \in D, K(x)$ is a nonempty closed convex subset in H and $K(D) \subset g(D)$. Suppose that $A\colon D \to CB(D)$ and $F, G\colon D \to CB(H)$ are set-valued mappings. Then, for given $u_0 \in D$, we can obtain sequences $\{x_n\}, \{y_n\}, \{w_n\}$ and $\{u_n\}$ such that*

$$\begin{cases} x_n \in Fu_n, \ \|x_n - x_{n+1}\| \le \left(1 + \dfrac{1}{1+n}\right) H(Fu_n, Fu_{n+1}), \\ y_n \in Gu_n, \ \|y_n - y_{n+1}\| \le \left(1 + \dfrac{1}{1+n}\right) H(Gu_n, Gu_{n+1}), \\ w_n \in Au_n, \ \|w_n - w_{n+1}\| \le \left(1 + \dfrac{1}{1+n}\right) H(Au_n, Au_{n+1}), \\ g(u_{n+1}) = P_{K(w_n)}(g(u_n) - \rho N(Sx_n, Ty_n)) \end{cases} \tag{3.6}$$

for all $n = 0, 1, 2, \ldots,$ where $\rho > 0$ is a constant.

From Algorithm 3.2, we have the following algorithm:

Algorithm 3.3: *Let D be a nonempty subset of a Hilbert space $H, g\colon D \to H$ be single-valued mapping and $K\colon D \to 2^H$ be a set-valued mapping such that, for each $x \in D, K(x)$ is a nonempty closed convex subset in H and $K(D) \subset g(D)$. Suppose that $F, G\colon D \to CB(H)$ are set-valued mappings. Then, for given $u_0 \in D$, we can obtain sequences $\{x_n\}, \{y_n\}$ and $\{u_n\}$ such that*

$$\begin{cases} x_n \in Fu_n, \ \|x_n - x_{n+1}\| \le \left(1 + \dfrac{1}{1+n}\right) H(Fu_n, Fu_{n+1}), \\ y_n \in Gu_n, \ \|y_n - y_{n+1}\| \le \left(1 + \dfrac{1}{1+n}\right) H(Gu_n, Gu_{n+1}), \\ g(u_{n+1}) = P_{K(u_n)}(g(u_n) - \rho(x_n - y_n)) \end{cases} \tag{3.7}$$

for all $n = 0, 1, 2, \ldots,$ where $\rho > 0$ is a constant.

It is easy to see that, if we suppose that the mapping $g\colon D \to H$ is expansive, then the inverse mapping g^{-1} of g exists and each u_n is computable.

4. EXISTENCE AND CONVERGENCE THEOREMS

In this section, we consider some conditions under which the solution of the generalized strongly nonlinear implicit quasi-variational inequality problem (2.1) exists and the sequences of approximate solutions obtained from Algorithm 3.1 converge strongly to the exact solution of the generalized strongly nonlinear implicit quasi-variational inequality problem (2.1).

Definition 4.1: Let D be a nonempty subset of a Hilbert space $H, N: H\times H \to H$, $S: H \to H$ and $g: D \to H$ three single-valued mappings. Let $F: D \to CB(H)$ be a set-valued mapping. Then

(1) F is said to be *strongly monotone* on D with respect to the first argument of N and (S, g) if there exists a constant $\alpha > 0$ such that, for all $u, v \in D, x \in Fu$ and $y \in Fv$,

$$(g(u) - g(v), N(Sx, \cdot) - N(Sy, \cdot)) \geq \alpha\|g(u) - g(v)\|^2.$$

(2) F is said to be *strongly monotone* on D with respect to S and g if there exists a constant $\alpha > 0$ such that, for all $u, v \in D, x \in Fu$ and $y \in Fv$,

$$(g(u) - g(v), Sx - Sy) \geq \alpha\|g(u) - g(v)\|^2.$$

If $Su = u$ for all $u \in H$, then F is said to be *strongly monotone* on D with respect to g, and if $Su = u$ for all $u \in H$ and $g(u) = u$ for all $u \in D$, then F is said to be *strongly monotone* on D.

(3) S is said to be *Lipschitz continuous* on H with respect to the first argument of N if there exists a constant $\beta > 0$ such that, for all $u, v \in H$,

$$\|N(Su, \cdot) - N(Sv, \cdot)\| \leq \beta\|u - v\|.$$

(4) F is said to be *H-Lipschitz continuous* on D with respect to g if there exists a constant $\beta > 0$ such that, for all $u, v \in D$,

$$H(Fu, Fv) \leq \beta\|g(u) - g(v)\|.$$

If $g(u) = u$ for all $u \in D$, then F is said to be *H-Lipschitz continuous* on D.

In a similar way, we can define Lipschitz continuity of S with respect to the second argument of N.

Definition 4.2: Let D be a nonempty subset of a Hilbert space H and $K: D \to 2^H$ be a set-valued mapping such that, for each $x \in D, K(x)$ is a nonempty closed convex subset of H. The projection $P_{K(x)}$ is said to be *Lipschitz continuous* if there exists a constant $\eta > 0$ such that for all $z \in H$ and $x, y \in D$,

$$\|P_{K(x)}z - P_{K(y)}z\| \leq \eta\|x - y\|.$$

Remark 4.1: If $K(x)$ is defined as in Lemma 3.3, and $m: D \to H$ is Lipschitz continuous on D, then $P_{K(x)}$ is Lipschitz continuous. In fact, for all $x, y \in D$ and $z \in H$, it follows from Lemmas 3.3 and 3.2 that

$$\begin{aligned}\|P_{K(x)}z - P_{K(y)}z\| &= \|m(x) + P_K(z - m(x)) - m(y) - P_K(z - m(y))\| \\ &\leq \|m(x) - m(y)\| + \|P_K(z - m(x)) - P_K(z - m(y))\| \\ &\leq 2\|m(x) - m(y)\|.\end{aligned}$$

Theorem 4.1: *Let D be a nonempty subset of a Hilbert space H and $g: D \to H$ be such that $g(D)$ is a closed in H. Suppose that $N: H \times H \to H$ is a single-valued mapping and $K: D \to 2^H$ is a set-valued mapping such that, for each $x \in D, K(x)$ is a nonempty* *closed convex subset in H and $K(D) \subset g(D)$. Let $\tilde{F}, \tilde{G}$: $D \to \mathcal{F}(H)$ and $\tilde{A}$: $D \to \mathcal{F}(D)$ be closed fuzzy mappings satisfying the condition (I). Let F, G and A be the set-valued mappings induced by the fuzzy mappings $\tilde{F}, \tilde{G}$ and $\tilde{A}$, respectively. Suppsoe that F, G and A are H-Lipschitz continuous on D with respect to g with H-Lipschitz constants ϵ, η and γ, respectively, and F strongly monotone on D with respect to the first argument of N and (S, g) with constant α. Let S, T: $H \to H$ be Lipschitz continuous on D with respect the first and second arguments of N with Lipschitz constants β, ξ, respectively. If $P_{K(x)}$ is Lipschitz continuous with constant μ for each $x \in D$ and the following conditions hold:*

$$\begin{cases} \left|\rho - \dfrac{\alpha - (1-\mu\gamma)\xi\eta}{\beta^2\epsilon^2 - \xi^2\eta^2}\right| < \dfrac{\sqrt{[\alpha - (1-\mu\gamma)\xi\eta]^2 - (\beta^2\epsilon^2 - \xi^2\eta^2)\mu\gamma(2-\mu\gamma)}}{\beta^2\epsilon^2 - \xi^2\eta^2}, \\ \alpha > (1-\mu\gamma)\xi\eta + \sqrt{(\beta^2\epsilon^2 - \xi^2\eta^2)\mu\gamma(2-\mu\gamma)}, \\ \rho\xi\eta < 1 - \mu\gamma, \quad \xi\eta < \beta\epsilon, \end{cases} \tag{4.1}$$

then there exist $u \in D$, $x \in Fu$, $y \in Gu$ and $w \in Au$ such that (u, x, y, w) is a solution of the generalized strongly nonlinear implicit quasi-variational inequality problem (2.1) and

$$g(u_n) \to g(u), \quad x_n \to x, \quad y_n \to y, \quad w_n \to w \quad \text{as} \quad n \to \infty,$$

where $\{u_n\}, \{x_n\}, \{y_n\}$ and $\{w_n\}$ are sequences defined in Algorithm 3.1.

Proof: From Algorithm 3.1, Lemma 3.2, Definition 4.2, the Lipschitz continuity of $P_{K(u)}$ and T, it follows that

$$\begin{aligned} &\|g(u_{n+1}) - g(u_n)\| \\ &= \|P_{K(w_n)}(g(u_n) - \rho N(Sx_n, Ty_n)) \\ &\quad - P_{K(w_{n-1})}(g(u_{n-1}) - \rho N(Sx_{n-1}, Ty_{n-1}))\| \\ &\le \|P_{K(w_n)}(g(u_n) - \rho N(Sx_n, Ty_n)) \\ &\quad - P_{K(w_n)}(g(u_{n-1}) - \rho N(Sx_{n-1}, Ty_{n-1}))\| \\ &\quad + \|P_{K(w_n)}(g(u_{n-1}) - \rho N(Sx_{n-1}, Ty_{n-1})) \\ &\quad - P_{K(w_{n-1})}(g(u_{n-1}) - \rho N(Sx_{n-1}, Ty_{n-1}))\| \\ &\le \|g(u_n) - g(u_{n-1}) - \rho(N(Sx_n, Ty_n) - N(Sx_{n-1}, Ty_n))\| \\ &\quad + \rho\|N(Sx_n, Ty_n) - N(Sx_n, Ty_{n-1})\| + \mu\|w_n - w_{n-1}\| \\ &\le \|g(u_n) - g(u_{n-1}) - \rho(N(Sx_n, Ty_n) - N(Sx_{n-1}, Ty_n))\| \\ &\quad + \rho\xi\|y_n - y_{n-1}\| + \mu\|w_n - w_{n-1}\|. \end{aligned} \tag{4.2}$$

By Algorithm 3.1 and the H-Lipschitz continuity of F with respect to g, we obtain

$$\begin{aligned} \|x_n - x_{n-1}\| &\le \left(1 + \frac{1}{n}\right) H(Fu_n, Fu_{n-1}) \\ &\le \left(1 + \frac{1}{n}\right) \epsilon \|g(u_n) - g(u_{n-1})\|. \end{aligned} \tag{4.3}$$

Since F is strongly monotone on D with respect to the first argument of N and (S, g) and S is Lipschitz continuous with respect to the first argument of N, it follows from (4.3) that

$$\begin{aligned}\|g(u_n)-g(u_{n-1}) - \rho(N(Sx_n, Ty_n) - N(Sx_{n-1}, Ty_n))\|^2 &\\ \leq \|g(u_n) - g(u_{n-1})\|^2 - 2\rho\alpha\|g(u_n) - g(u_{n-1})\|^2 &\\ + \rho^2\beta^2\|x_n - x_{n-1}\|^2 & \qquad (4.4)\\ \leq \left(1 - 2\rho\alpha + \rho^2\beta^2\left(1+\frac{1}{n}\right)^2\epsilon^2\right)\|g(u_n) - g(u_{n-1})\|^2. &\end{aligned}$$

Further, since G and A are H-Lipschitz continuous on D with respect to g, we have

$$\begin{aligned}\|y_n - y_{n-1}\| &\leq \left(1+\frac{1}{n}\right)H(Gu_n, Gu_{n-1})\\ &\leq \left(1+\frac{1}{n}\right)\eta\|g(u_n) - g(u_{n-1})\|\end{aligned} \qquad (4.5)$$

and

$$\begin{aligned}\|w_n - w_{n-1}\| &\leq \left(1+\frac{1}{n}\right)H(Au_n, Au_{n-1})\\ &\leq \left(1+\frac{1}{n}\right)\gamma\|g(u_n) - g(u_{n-1})\|.\end{aligned} \qquad (4.6)$$

From (4.2) and (4.4) $\sim$ (4.6), it follows that

$$\|g(u_n) - g(u_{n+1})\| \leq \theta_n\|g(u_n) - g(u_{n-1})\|, \qquad (4.7)$$

where

$$\theta_n = \sqrt{1 - 2\rho\alpha + \rho^2\beta^2\left(1+\frac{1}{n}\right)^2\epsilon^2} + \rho\xi\left(1+\frac{1}{n}\right)\eta + \mu\left(1+\frac{1}{n}\right)\gamma.$$

Let

$$\theta = \sqrt{1 - 2\rho\alpha + \rho^2\beta^2\epsilon^2} + \rho\xi\eta + \mu\gamma.$$

Then $\theta_n \to \theta$ as $n \to \infty$. It follows from (4.1) that $\theta < 1$. Hence, $\theta_n < 1$ for n sufficiently large. From (4.7), we know that $\{g(u_n)\}$ is a Cauchy sequence in $g(D)$. Since $g(D)$ is closed in H, there exists $u \in D$ such that

$$g(u_n) \to g(u) \quad \text{as} \quad n \to \infty.$$

By (4.3), (4.5) and (4.6), we know that $\{x_n\}$, $\{y_n\}$ and $\{w_n\}$ are all Cauchy sequence in H. Let

$$x_n \to x, \quad y_n \to y, \quad w_n \to w \quad \text{as} \quad n \to \infty.$$

Further we have

$$\begin{aligned} d(x, Fu) &= \inf\{\|x - z\|: \in Fu\} \\ &\leq \|x - x_n\| + d(x_n, Fu) \\ &\leq \|x - x_n\| + H(Fu_n, Fu) \\ &\leq \|x - x_n\| + \epsilon\|g(u_n) - g(u)\| \to 0 \end{aligned}$$

and hence $x \in Fu$. Similarly, $y \in Gu$ and $w \in Au$. Therefore, $\tilde{F}_u(x) \geq a(u), \tilde{G}_u(y) \geq b(u)$ and $\tilde{A}_u(w) \geq c(u)$.

Now we prove that

$$g(u) = P_{K(w)}(g(u) - \rho N(Sx, Ty)).$$

In fact, from Lemma 3.2 and the Lipschitz continuity of $P_{K(u)}$, we have

$$\begin{aligned} &\|g(u_{n+1}) - P_{K(w)}(g(u) - \rho N(Sx, Ty))\| \\ &\quad = \|P_{K(w_n)}(g(u_n) - \rho N(Sx_n, Ty_n)) - P_{K(w)}(g(u) - \rho N(Sx, Ty))\| \\ &\quad \leq \|P_{K(w_n)}(g(u_n) - \rho N(Sx_n, Ty_n)) - P_{K(w_n)}(g(u) - \rho N(Sx, Ty))\| \\ &\qquad + \|P_{K(w_n)}(g(u) - \rho N(Sx, Ty)) - P_{K(w)}(g(u) - \rho N(Sx, Ty))\| \\ &\quad \leq \|g(u_n) - g(u) - \rho(N(Sx_n, Ty_n) - N(Sx, Ty))\| + \mu\|w_n - w\| \\ &\quad \leq \|g(u_n) - g(u)\| + \rho\|N(Sx_n, Ty_n) - N(Sx, Ty_n)\| \\ &\qquad + \rho\|N(Sx, Ty_n) - N(Sx, Ty)\| + \mu\|w_n - w\| \\ &\quad \leq \|g(u_n) - g(u)\| + \rho\beta\|x_n - x\| + \rho\xi\|y_n - y\| + \mu\|w_n - w\|, \end{aligned}$$

which implies that

$$g(u_{n+1}) \to P_{K(w)}(g(u) - \rho N(Sx, Ty)).$$

Therefore

$$g(u) = P_{K(w)}(g(u) - \rho N(Sx, Ty)).$$

This completes the proof.

From Theorems 3.1 and 4.1, we have the following:

Theorem 4.2: *Let $D, g, \tilde{F}, \tilde{G}, \tilde{A}, F, G, A, S$ and T be the same as in Theorem 4.1. Suppose that $K: D \to 2^H$ be a set-valued mapping such that $K(D) \subset g(D), K(x)$ is nonempty closed convex cone in H and $P_{K(x)}$ is Lipschitz continuous with constant μ for each $x \in D$. If the condition (4.1) of Theorem 4.1 holds, then there exist $u \in D$, $x \in Fu$, $y \in Gu$ and $w \in Au$ such that (u, x, y, w) is a solution of the generalized strongly nonlinear implicit quasi-complementarity problem (2.2) and*

$$g(u_n) \to g(u), \quad x_n \to x, \quad y_n \to y, \quad w_n \to w \quad \text{as} \quad n \to \infty,$$

where $\{u_n\}, \{x_n\}, \{y_n\}$ *and* $\{w_n\}$ *are sequences defined in Algorithm 3.1.*

From Theorems 4.1 and 4.2, we have the following results:

Theorem 4.3: *Let* D *be a nonempty subset of a Hilbert space* H *and* $g\colon D \to H$ *be a sigle-valued mapping such that* $g(D)$ *is closed in* H. *Let* $F, G\colon D \to CB(H)$ *and* $A\colon D \to CB(D)$ *be* H*-Lipschitz continuous on* D *with respect to* g *with* H*-Lipschitz constants* ϵ, η *and* γ, *respectively. Let* $N\colon H \times H \to H$ *and* F *be strongly monotone on* D *with respect to the first argument of* N *and* (S, g) *with constant* α. *Let* $S, T\colon H \to H$ *be Lipschitz continuous on* D *with respect to the first and second arguments of* N *with Lipschitz constants* β, ξ, *respectively. Suppose that* $K\colon D \to 2^H$ *is a set-valued mapping such that* $K(D) \subset g(D), K(x)$ *is nonempty closed convex subset of* H *and* $P_{K(x)}$ *is Lipschitz continuous with constant* μ *for each* $x \in D$. *If the conditions (4.1) in Theorem 4.1 hold, then there exist* $u \in D, x \in Fu, y \in Gu$ *and* $w \in Au$ *such that* (u, x, y, w) *is a solution of the generalized set-valued strongly nonlinear implicit quasi-variational inequality problem (2.3) and*

$$g(u_n) \to g(u), \quad x_n \to x, \quad y_n \to y, \quad w_n \to w \quad \text{as} \quad n \to \infty,$$

where $\{u_n\}, \{x_n\}, \{y_n\}$ *and* $\{w_n\}$ *are sequences defined in Algorithm 3.2.*

Theorem 4.4: *Let* D, g, F, G, A, S *and* T *be the same as in Theorem 4.3. Suppose that* $K\colon D \to 2^H$ *is a set-valued mapping such that* $K(D) \subset g(D), K(x)$ *is nonempty closed convex cone in* H *and* $P_{K(x)}$ *is Lipschitz continuous with constant* μ *for each* $x \in D$. *If the condition (4.1) of Theorem 4.1 holds, then there exist* $u \in D, x \in Fu, y \in Gu$ *and* $w \in Au$ *such that* (u, x, y, w) *is a solution of the generalized set-valued strongly nonlinear implicit quasi-complementarity problem (2.4) and*

$$g(u_n) \to g(u), \quad x_n \to x, \quad y_n \to y, \quad w_n \to w \quad \text{as} \quad n \to \infty,$$

where $\{u_n\}, \{x_n\}, \{y_n\}$ *and* $\{w_n\}$ *are sequences defined in Algorithm 3.2.*

Remark 4.2: Our results improve and extend some known results. For example, Theorems 4.3 and 4.4 improve and extend Theorems 4.1 and 4.2 of Ding [15], respectively.

REFERENCES

[1] C. Baiocchi and A. Capelo (1984). *Variational and Quasivariational Inequalities, Application to Free Boundary Problems*. New York: Wiley.

[2] A. Bensoussan (1982). *Stochastic Control by Functional Analysis Method*, Amsterdam: North-Holland.

[3] A. Bensoussan and J.L. Lions (1984). *Impulse Control and Quasivariational Inequalities*. Bordas, Paris: Gauthiers-Villers.

[4] Shih-sen Chang (1991). *Variational Inequality and Complementarity Problem Theory with Applications*. Shanghai: Shanghai Scientific and Tech. Literature Publishing House.

[5] Shih-sen Chang and Nan-jing Huang (1991). Generalized strongly nonlinear quasi-complementarity problems in Hilbert spaces, *J. Math. Anal. Appl.* **158**, 194–202.

[6] Shih-sen Chang and Nan-jing Huang (1991). Generalized multivalued implicit complementarity problems in Hilbert spaces, *Math. Japonica*, **36**, 1093–1100.

[7] Shih-sen Chang and Nan-jing Huang (1993). Generalized complementarity problems for fuzzy mappings, *Fuzzy sets and Systems*, **55**, 227–234.

[8] Shih-sen Chang and Nan-jing Huang (1996). Generalized quasi-complementarity problems for a pair of fuzzy mappings, *J. Fuzzy Math.*, **4**, 343–354.

[9] Shih-sen Chang and Yuanguo Zhu (1989). On variational inequalities for fuzzy mappings, *Fuzzy sets and Systems*, **32**, 359–367.

[10] R.W. Cottle (1976). Complementarity and variational problems, *Sympos. Math.*, **19**, 177–208.

[11] R.W. Cottle and G.B. Dantzig (1968). Complementarity pivot theory of mathematical programming, *Linear Algebra Appl.*, **1**, 163–185.

[12] R.W. Cottle, J.P. Pang and R.E. Stone (1992). *The Linear Complementarity Problem*. London: Academic Press.

[13] J. Crank (1984). *Free and Moving Boundary Problems*. Oxford: Clarendon.

[14] X.P. Ding (1993). Generalized strongly nonlinear quasivariational inequalities, *J. Math. Anal. Appl.*, **173**, 577–587.

[15] X.P. Ding (1998). Set-valued implicit Wiener-Hopf equations and generalized strongly nonlinear quasivariational inequalities, *J. Appl. Anal.*, **4**, 59–71.

[16] F. Giannessi and A. Maugeri (1995). *Variational Inequalities and Network Equilibrium Problems*. New York: Plenum.

[17] P.T. Harker and J.S. Pang (1990). Finite-dimensional variational inequality and nonlinear complementarity problems: A survey of theory, algorithms and applications, *Math. Programming*, **48**, 161–220.

[18] Nan-jing Huang (1997). On the generalized implicit quasivariational inequalities, *J. Math. Anal. Appl.*, **216**, 197–210.

[19] Nan-jing Huang (1997). Completely generalized strongly nonlinear quasi-complementarity problems for fuzzy mappings, Indian *J. Pure Appl. Math.*, **28**, 23–32.

[20] Nan-jing Huang (1997). A new method for a class of nonlinear variational inequalities with fuzzy mappings, *Appl. Math. Lett.*, **10**(6), 129–133.

[21] Nan-jing Huang (1998). A general class of nonlinear variational inclusions for fuzzy mappings, *Indian J. Pure Appl. Math.*, **29**, 957–964.

[22] Nan-jing Huang and Xin-qi Hu (1994). Generalized multi-valued nonlinear implicit quasi-complementarity problems in Hilbert spaces, *J. Sichuan Univ.*, **31**, 306–310.

[23] G. Isac (1990). A special variational inequality and the implicit complementarity problems, *J. Fac. Sci. Univ. Tokyo*, **37**, 109–127.

[24] G. Isac (1992). *Complementarity problems, Lecture Notes in Math. 1528*. Berlin: Springer-Verlag.

[25] B.S. Lee, G.M. Lee, S.J. Cho and D.S. Kim (1993). A variational inequality for fuzzy mappings, Proc. of Fifth Internat. Fuzzy Systems Association World Congress, Seoul, 326–329.

[26] G.M. Lee, D.S. Kim, B.S. Lee and S.J. Cho (1993). Generalized vector variational inequality and fuzzy extension, *Appl. Math. Lett.*, **6**, 47–51.

[27] C.E. Lemke (1965). Bimatrix equilibrium points and mathematical programming, *Management Sci.*, **11**, 681–689.

[28] Y. Lin and C. Cryer (1985). An alternating direction implicit algorithm for the solution of linear complementarity problems arising from boundary problems, *Appl. Math. Optim.*, **13**, 1–17.

[29] U. Mosco (1976). Implicit variational problems and quasi-variational inequalities in Lecture Notes in *Math.* **543**, Berlin: Springer-Verlag, 83–156.

[30] S.B. Nadler, Jr. (1969). Multi-valued contraction mappings, *Pacific J. Math.*, **30**, 475–488.

[31] M.A. Noor (1986). Iterative method for quasi-complementarity problems, *Methods Oper. Res.*, **56**, 75–83.

[32] M.A. Noor (1988). The quasi-complementarity problems, *J. Math. Anal. Appl.*, **130**, 344–353.

[33] M.A. Noor (1987). On the nonlinear complementarity problems *J. Math. Anal. Appl.*, **123**, 455–460.

[34] M.A. Noor (1988). Iterative methods for a class of complementarity, *J. Math. Anal. Appl.*, **133**, 366–382.

[35] M.A. Noor (1988). Fixed point approach for complementarity problems, *J. Math. Anal. Appl.*, **133**, 437–448.

[36] M.A. Noor (1985). An iterative scheme for a class of quasi-variational inequalities, *J. Math. Anal. Appl.*, **110**, 463–468.

[37] M.A. Noor (1997). Some recent advances in variational inequalities, I, II, *New Zealand J. Math.*, **26**, 53–80, 229–255.

[38] M.A. Noor, K.I. Noor and T.M. Rassias (1993). Some aspects of variational inequalities, *J. Comput. Appl. Math.*, **47**, 285–312.

[39] J.P. Pang (1982). On the convergence of a basic iterative method for the implicit complementarity problems, *J. Optim. Theory Appl.*, **37**, 149–162.

[40] J.P. Pang (1981). The implicit complementarity problem. In O.L. Mangasarian, R. Meyer, and S.M. Robinson (Eds), *Nonlinear Programming, 4*, New York/London: Academic Press, 487–518.

[41] A.H. Siddiqi and Q.H. Ansari (1990). Strongly nonlinear quasivariational inequalities, *J. Math. Anal. Appl.*, **149**, 444–450.

[42] A.H. Siddiqi and Q.H. Ansari (1992). General strongly nonlinear variational inequalities, *J. Math. Anal. Appl.*, **166**, 386–392.

[43] George Xian-Zhi Yuan (1999). *KKM Theory and Applications in Nonlinear Analysis*, Marcel Dekker, Inc.

[44] L.C. Zeng (1995). Completely generalized strongly nonlinear quasicomplementarity problems in Hilbert spaces, *J. Math. Anal. Appl.*, **193**, 706–714.

SIMAA 4(2002) 79–127

8. Vector Variational Inequalities, Multi-objective Optimizations, Pareto Optimality and Applications

Mohammad S.R. Chowdhury and Enayet Tarafdar

Department of Mathematics, The University of Queensland, Brisbane, Queensland 4072, Australia

Abstract: Over the past two decades, there has been a rapid development of the field of vector variational inequality and multi-objective optimization problems in its theory of existence, uniqueness and sensitivity of solutions, in the theory of algorithms, and in the application of these techniques to mathematical economics. This chapter provides a partial state-of-the-art review of these developments.

Keywords and Phrases: Generalized vector variational-like inequality, cone, $L - \eta$-condition, $C_x - \eta$-pseudomonotone, weakly $C_x - \eta$-pseudomonotone, multi-objective optimizations, Pareto optimality, topological vector space

2000 Mathematics Subject Classification: 47H04, 47H10, 47N10, 49A29, 49J35, 49J40, 49J45

"... the members of a collectivity enjoy maximum ophelimity in a certain position when it is impossible to find a way of moving from that position very slightly in such a manner that the ophelimity enjoyed by each of the individuals of that collectivity increases or decreases. That is to say, any small displacement in departing from that position necessarily has the effect of increasing the ophelimity which certain individuals enjoy, and decreasing that which others enjoy, of being agreeable to some and disagreeable to others." – V. Pareto, *Manuel d'economic politique*, Girard and Briere, 1909. English Translation by A.S. Schwier, Augustus M. Kelley Publishers,* New York, 1971.

1. INTRODUCTION

Since Professor Giannessi first introduced Vector Variational Inequality in a finite dimensional Euclidean space $\mathbb{R}^n$ [35, 1980] there have been numerous generalizations of vector variational inequalities (in short, VVI) in abstract spaces. Chen and Cheng [10] studied the VVI in infinite dimensional spaces and applied it to Vector Optimization Problem (in short, VOP). Since then, Chen *et al.* [12–15,17], Siddiqi *et al.* [69], Lai and Yao [51] and Yu and Yao [83] have intensively studied the VVI on different assumptions in abstract spaces. More recently, Lee *et al.* [52,53], Lin *et al.* [57], Danilidis and

* The authors are grateful to Augustus M. Kelley Publishers for granting their permission to reproduce this quotation.

Hadjisavvas [24] and Konnov and Yao [50] considered the generalized VVI problem involving set-valued mapping with monotonicity and obtained existence results under different assumptions.

Our plan is to begin with the very fundamental results and to proceed gradually toward the frontier. Due to space limitations, much of our discussion will be brief. The details of proof, specialized results, examples, etc. will be omitted which can be found in the references.

The organization of the remainder of the chapter will be as follows. In Section 2, we present some theorems on alternatives and formally state vector variational inequality problems and also present some fundamental definitions and facts concerning these problems. Section 3 develops the theory on the existence and uniqueness of solutions for the vector variational inequality problems, and vector optimization problems are presented in Section 4. Multi-objective optimization problems will be discussed in Section 5, and Section 6 will present the Pareto Equilibria for multi-objective games. Pareto solution of a cone variational inequality and Pareto optimality of a mapping with applications to mathematical economics will be surveyed in Section 7.

2. ALTERNATIVES AND VECTOR VARIATIONAL INEQUALITIES

2.1. Some Theorems on Alternatives

We shall begin with some theorems of alternatives for non-linear systems in topological vector spaces.

Theorem 2.1.1: *Suppose U and V are topological vector spaces over $\mathbb{R}$ and X is an arbitrary non-empty set. Let C_1 and C_2 be closed and convex cones in U and V, respectively, such that $\operatorname{int} C_1 \neq \emptyset$. Consider two functions $f: X \to C_1$ and $g: X \to C_2$. Define the set $H =: \{(u, v) \in U \times V : u \in \operatorname{int} C_1$ and $v \in C_2\}$.*

Suppose $G: U \times V \to \mathbb{R}$ is a real-valued function such that $G(u, v) > 0$ iff $(u, v) \in H$. Then the system

$$f(x) \in \operatorname{int} C_1, \ g(x) \in C_2 \quad \textit{for some } x \in X \tag{2.1}$$

is not possible iff *the inequalities*

$$G(f(x), g(x)) \leq 0, \quad \textit{for all } x \in X \tag{2.2}$$

hold.

Proof: If (2.1) is possible, then there exists $\bar{x} \in X$ such that $f(\bar{x}) = \bar{u} \in \operatorname{int} C_1$ and $g(\bar{x}) = \bar{v} \in C_2$, i.e., $(\bar{u}, \bar{v}) \in H$. Thus $G(\bar{u}, \bar{v}) > 0$, i.e., $G(f(\bar{x}), g(\bar{x})) > 0$. Hence (2.2) is false.

If (2.1) is impossible, i.e., $(u, v) = (f(x), g(x)) \notin H$, for all $x \in X$, then $G(u, v) = G(f(x), g(x)) \leq 0$, for all $x \in X$. Hence (2.2) is true. This completes the proof. □

As a corollary we have the following result which is Theorem 1 in [35]:

Corollary 2.1.1: [35]: *Consider a set $X \subset \mathbb{R}^n$, and two vector-valued functions $f: X \to \mathbb{R}^l$, $g: X \to \mathbb{R}^m$. Define the set*

$$H =: \{(u, v) \in \mathbb{R}^l \times \mathbb{R}^m = \mathbb{R}^{l+m} : u > 0, v \geq 0\}.$$

Suppose G: $\mathbb{R}^l \times \mathbb{R}^m \to \mathbb{R}$ is a real-valued function such that $G(u, v) > 0$ iff $(u, v) \in H$. Then the system

$$f(x) > 0, \ g(x) \geq 0 \quad \textit{for some } x \in X \tag{2.3}$$

is not possible iff *the inequalities*

$$G(f(x), \ g(x)) \leq 0, \quad \textit{for all } x \in X \tag{2.4}$$

hold.

Theorem 2.1.2: [35]: *Consider a set $X \subset \mathbb{R}^n$. Suppose that λ: $X \to \mathbb{R}^l_+$ and μ: $X \to \mathbb{R}^m_+$ are two functions. Then the system (2.1) is not possible* iff$(\lambda(x), \mu(x)) \neq 0$ *for all $x \in X$ and such that*

$$\langle \lambda(x), f(x) \rangle + \langle \mu(x), g(x) \rangle \leq 0 \quad \textit{for all } x \in X, \tag{2.5}$$

where $\langle \lambda(x), f(x) \rangle + \langle \mu(x), g(x) \rangle = 0$ when $\lambda(x) = 0$.

Theorem 2.1.3: [35]: *Consider the same assumptions as in Corollary 2.1.1 and Theorem 2.1.2. Suppose, in addition, that f and g are concave and $X \subset \mathbb{R}^n$ is convex. Then system (2.1) is impossible* iff *there exists $\Lambda = (\lambda, \mu) \in \mathbb{R}^l_+ \times \mathbb{R}^m_+$ with $(\lambda, \mu) \neq 0$ such that*

$$\langle \lambda, f(x) \rangle + \langle \mu, g(x) \rangle \leq 0, \quad \textit{for all } x \in X, \tag{2.6}$$

where the inequality must be strictly satisfied if $\lambda = 0$.

Proof: For each $x \in X$, consider $E(x) = \{(u, v) \in \mathbb{R}^l \times \mathbb{R}^m = \mathbb{R}^{l+m} : u \leq f(x); \, v \leq g(x)\}$ and let $E = \bigcup_{x \in X} E(x)$.

Since f and g are concave and X is convex, it can be shown that E is convex. Clearly, E is the intersection of supporting half-spaces of $K = \{(u, v) \in \mathbb{R}^l \times \mathbb{R}^m = \mathbb{R}^{l+m} : u = f(x); \, v = g(x); \quad \text{for all } x \in X\}$ of the kind $\langle \lambda, u \rangle + \langle \mu, v \rangle \leq k$ with $(\lambda, \mu) \geq 0$ and $(\lambda, \mu) \neq 0$. Thus by using a well known theorem of separation for convex sets we can establish (2.6). Hence the proof is complete. □

Corollary 2.1.2: [35]: *Consider the same assumptions as in Corollary 2.1.1 and Theorem 2.1.2. Suppose that f and g are concave and X is convex. Then system (2.1) is impossible iff there exists $\lambda \in \mathbb{R}^l_+$ and $\mu \in \mathbb{R}^m_+$ with $(\lambda, \mu) \neq 0$ such that*

$$\langle \lambda, f(x) \rangle + \langle \mu, g(x) \rangle \leq 0, \quad \textit{for all } x \in X, \tag{2.7}$$

where the inequality must be strictly satisfied if $\lambda = 0$.

2.2. Basic Definitions and Results on Vector Variational Inequalities

Now we shall introduce extremum problems and Vector Variational Inequalities (VVI).

Definition 2.2.1: Suppose X is an arbitrary non-empty set and ϕ: $X \to \mathbb{R}$ is a real-valued function. Let C_1 and C_2 (as defined before) be closed and convex cones in topological vector spaces U and V, respectively such that int $C_1 \neq \emptyset$. Consider a function g: $X \to C_2$ and define the set

$$R = \{x \in X : g(x) \in C_2\}.$$

Then an extremum problem is to find an $\hat{x} \in R$ such that

$$\phi(\hat{x}) = \inf_{x \in R} \phi(x). \tag{2.8}$$

Theorems 2.1.1 and 2.1.2 may be used to state a necessary and sufficient optimality condition for problem (2.8). Let us set $U = \mathbb{R}$ and define f: $X \to C_1 = \mathbb{R}$ by $f(x) = \phi(\hat{x}) - \phi(x)$. Then Theorem 2.1.1 (respectively, Theorem 2.1.2) may be read this way: $\hat{x} \in R$ is an optimal solution of (2.8) iff (2.2) (respectively, (2.5)) holds. Similarly, when X is convex and ϕ and g are concave, we can obtain from Theorem 2.1.3 that, $\hat{x} \in R$ is an optimal solution of (2.8) iff (2.6) holds.

Giannessi first introduced VVI (Vector Variational Inequalities) in a general finite dimensional setting (for details see [35]). Since then there have been various generalizations of VVI. We shall present VVI first in the following form.

Definition 2.2.2: Suppose F: $X \to \mathbb{R}^n$. Then a VVI problem is to find an $\hat{x} \in R = \{x \in X : g(x) \in C_2\}$ such that

$$\langle F(\hat{x}), x - \hat{x} \rangle \geq 0, \quad \text{for all } x \in R. \tag{2.9}$$

We shall now define VVI in a more general setting as follows:

Definition 2.2.3: Suppose that U and V are topological vector spaces over $\mathbb{R}$ and V is endowed with an associated partial order induced by a convex cone S (i.e., if $A, B \in V$ then, $A \geq B$ means that $A - B \in S$). Now consider the set $L(U, V)$ of all linear operators from U into V. We shall denote the value of $A \in L(U, V)$ at $u \in U$ by $\langle A, u \rangle$. Let X be a set in U. Suppose that F: $X \to L(U, V)$ is a mapping from X into $L(U, V)$. Then the VVI is to find a point $\hat{x} \in X$ such that

$$\langle F(\hat{x}), x - \hat{x} \rangle \geq 0, \quad \text{for all } x \in X. \tag{2.10}$$

We shall now discuss applications to vector extremum problems.

Definition 2.2.4: Consider mappings ϕ: $X \to W$ and θ: $X \to Z$ where W and Z are topological vector spaces. Define $\mathcal{R} = \{x \in X : \theta(x) \in Z\}$ and

$$\min_{x \in \mathcal{R}} \phi(x) = \phi(\hat{x}) \tag{2.11}$$

for some $\hat{x} \in \mathcal{R}$ for which there exists no $x \in \mathcal{R}$ such that $\phi(x) \leq \phi(\hat{x})$ and $\phi(x) \neq \phi(\hat{x})$.

Giannessi [35] defined vector extremum problems as follows:

Definition 2.2.5: Let ϕ: $X \to \mathbb{R}^h$ and θ: $X \to \mathbb{R}^k$ be two functions. Define $\mathcal{R} = \{x \in X : \theta(x) \geq 0\}$. Then the vector extremum problem is defined as

$$\min_{x \in \mathcal{R}} \phi(x) = \phi(\hat{x}), \tag{2.12}$$

for some $\hat{x} \in \mathcal{R}$ for which there exists no $x \in \mathcal{R}$ such that $\phi(x) \leq \phi(\hat{x})$ and $\phi(x) \neq \phi(\hat{x})$.

We can use Corollary 2.1.1 and Theorem 2.1.2 to state a necessary and sufficient (vector) optimality condition for problem (2.12). Let us set

$$f_i(x) = \phi_i(\hat{x}) - \phi_i(x), \quad g(x) = [\phi(\hat{x}) - \phi(x), \theta(x)],$$

and consider the system

$$\begin{aligned} f_i(x) > 0 \quad & \text{for at least one } i = 1, 2, \ldots, h \\ g(x) \geq 0 \quad & \text{for some } x \in X. \end{aligned} \tag{2.13}$$

We can show that $\hat{x} \in \mathcal{R}$ is (vector) optimal for (2.12) iff (2.13) is impossible. For every fixed i let us identify (2.13) with (2.1) at $l = 1$. Let $u_i = f_i(x) = \phi_i(\hat{x}) - \phi_i(x)$ and $v_i = g(x) = [\phi(\hat{x}) - \phi(x), \theta(x)]$. Then, applying Corollary 2.1.1, we see that (2.13) is impossible iff $G(u_i, v_i) \leq 0$; i.e., $(u_i, v_i) \notin H = \{(u, v) \in \mathbb{R}^{1+m} : u > 0;\ v \geq 0\}$. Thus, $u_i \leq 0$ or $v_i \leq 0$. If $u_i \leq 0$ then, $u_i = \phi_i(\hat{x}) - \phi_i(x) \leq 0$. Therefore, $\phi_i(\hat{x}) \leq \phi_i(x)$ for all $x \in X$. Hence, $\hat{x}$ is (vector) optimal for (2.12). Thus, $\hat{x} \in \mathcal{R}$ is (vector) optimal for (2.12) iff (2.2) holds for every $i = 1, 2, \ldots, h$.

Again, applying Theorem 2.1.2, we see that system (2.13) is impossible iff there exist functions λ: $X \to \mathbb{R}_+$ and μ: $X \to \mathbb{R}^m_+$ with $(\lambda(x), \mu(x)) \neq 0$ for all $x \in X$ such that

$$\langle \lambda(x), f_i(x) = \phi_i(\hat{x}) - \phi_i(x) \rangle + \langle \mu(x), g(x) \rangle \leq 0 \quad \text{for all } x \in X.$$

Thus $\lambda(x)(\phi_i(\hat{x}) - \phi_i(x)) + \langle \mu(x), g(x) \rangle \leq 0$ for all $x \in X$. Hence $\lambda(x)(\phi_i(\hat{x}) - \phi_i(x)) \leq 0$. Therefore, $(\phi_i(\hat{x}) - \phi_i(x)) \leq 0$ as $\lambda(x) \in \mathbb{R}_+$. Consequently, $\phi_i(\hat{x}) \leq \phi_i(x)$ for all $x \in X$. Hence, $\hat{x}$ is (vector) optimal for (2.12). Thus, $\hat{x} \in \mathcal{R}$ is (vector) optimal for (2.12) iff (2.5) holds for every $i = 1, 2, \ldots, h$.

Then, Corollary 2.1.1 (respectively, Theorem 2.1.2) may be read this way:

Proposition 2.2.1: *$\hat{x} \in \mathcal{R}$ is (vector) optimal for (2.12)* iff *(2.2) (respectively, (2.5)) holds for every $i = 1, 2, \ldots, h$.*

We shall now give an application to Proposition 2.2.1.

Consider the particular case in which ϕ is linear, θ is affine and $X = \mathbb{R}^n$. Let us set

$$\phi(x) = Dx, \quad \theta(x) = Ax - b, \tag{2.14}$$

where D, A, and b are matrices of orders $h \times n$, $k \times n$ and $k \times 1$, respectively. Let D_i be the i-th row of D and let $y = D\hat{x}$. We shall now apply Corollary 2.1.2. We have that $\hat{x} \in \mathcal{R}$ is an optimal solution of (2.12) iff there exist non-negative vector ω^i, μ^i, such that, for every $i = 1, 2, \ldots, h$, either

$$y_i - \langle D_i, x \rangle + \langle \omega^i, y - Dx \rangle + \langle \mu^i, Ax - b \rangle \leq 0, \quad \text{for all } x \in \mathbb{R}^n \tag{2.15}$$

or

$$\langle \omega^i, y - Dx \rangle + \langle \mu^i, Ax - b \rangle < 0, \quad \text{for all } x \in \mathbb{R}^n. \tag{2.16}$$

Without loss of generality we assume that $\mathcal{R} \neq \emptyset$. Then, $g(x) \geq 0$ is possible. Thus, (2.16) is impossible. Hence, (2.14) is equivalent to (2.15). We see that (2.15) is possible iff the system

$$\mu A - \omega D = D_i, \quad y_i + \langle \omega, y \rangle \leq \langle \mu, b \rangle, \quad \omega \geq 0, \quad \mu \geq 0 \tag{2.17}$$

is possible for every $i = 1, 2, \ldots, h$. Let $\mathcal{Y}$ be the set all y, such that system (2.17) is possible for every $i = 1, 2, \ldots, h$. Thus, we are able to consider the following vector extremum problem

$$\max(y), \quad y \in \mathcal{Y}. \tag{2.18}$$

According to what was stated in [40:p. 34], (2.18) is called *dual* to (2.12). In [40] y was replaced by Ub, U being an $h \times n$ matrix of (dual) unknowns. Duality for (2.12) is contained also in [39:p. 33]. □

We shall briefly discuss the concept of vector variational inequalities in the field of vector extremum problems introduced in [35]. Let us consider the functions $F_i: X \to \mathbb{R}^n, i = 1, 2, \ldots, h$. Now consider the problem: to find an $\hat{x} \in \mathcal{R}$ for which there exists no $x \in \mathcal{R}$ such that

$$\langle F_i(\hat{x}), x - \hat{x}\rangle \leq 0, \; i = 1, \ldots, h \quad \text{and} \quad \langle F_i(\hat{x}), x - \hat{x}\rangle \neq 0 \quad \text{for at least one } i. \quad (2.19)$$

This problem of vector variational inequality becomes (2.9) at $h = 1$ and for $R = \{x \in X : g(x) \geq 0\}$. We note that $\hat{x} \in \mathcal{R}$ is a solution of (2.19), iff the system

$$f_i(x) = \langle F_i(\hat{x}), \hat{x} - x > 0 \quad \text{for at least one} \quad i = 1, \ldots, h$$

$$f(x) \geq 0, \quad \theta(x) \geq 0, \quad \text{for some } x \in X; \quad \text{where } (f = (f_1, f_2, \ldots, f_h)) \quad (2.20)$$

is impossible.

Among other applications, Giannessi discussed the following concepts in his chapter [35]: (i) Concave auxiliary functions for quadratic programs; (ii) Decomposition of convex quadratic programs, (iii) Decomposition of non-convex quadratic programs, (iv) Decomposition of a class of complementarity problems. For details see [35:pp. 167–184].

3. EXISTENCE AND UNIQUENESS THEORY

Over the past two decades, a large body of literature has developed on the existence and uniqueness of solutions to vector variational inequalities problems. Due to the diversity of results, it is not possible to list them all. In this section, we have chosen to provide the most general results and present the most fundamental results as their consequences; many other results can be found in the references listed at the end of the chapter. In the following two sections (3.1 and 3.2) we summarize the work of Ding and Tarafdar [28,29].

3.1. Generalized Vector Variational-like Inequalities without Monotonicity

In this section, we shall present a class of generalized vector variational-like inequalities without monotonicity which generalizes and unifies generalized vector variational inequalities, vector variational inequalities as well as various extensions of the classical variational inequalities in the literature. Some existence theorems for the generalized vector variation-like inequality problem without monotonicity are obtained in noncompact setting of topological vector spaces.

The Vector Variational Inequality (in short, VVI) in a finite dimensional Euclidean space has been introduced [35] and some applications have been given. Chen and Cheng [10] studied the VVI in infinite dimensional space and applied it to Vector Optimization Problem (in short, VOP). Since then, Chen *et al.* [12–15,17], Siddiqi *et al.* [69], Lai and Yao [51] and Yu and Yao [83] have intensively studied the VVI on different assumptions in abstract spaces. More recently, Lee *et al.* [52,53], Lin *et al.* [57] and Konnov and Yao

[50] considered the generalized VVI problem involving set-valued mapping with monotonicity and obtained existence results under different assumptions.

Let X be a topological vector space. A nonempty subset P of X is called a convex cone iff $P+P\subset P$ and $\lambda P\subseteq P, \forall\lambda\geq 0$. A cone P is said to be pointed iff $P\cap(-P)=\{0\}$ and solid iff it has nonempty interior, i.e., $\operatorname{int} P\neq\emptyset$. Throughout this chapter, we assume that X and Y are real Hausdorff topological vector spaces, $L(X, Y)$ is the space of all continuous linear operators from X into Y and σ is the family of all bounded subsets of X whose union is total in X, i.e., the linear hull of $\cup\{S\colon S\in\sigma\}$ is dense in X. Let $\mathcal{B}$ be a neighbourhood base of 0 in Y. When S runs through σ, V through $\mathcal{B}$, the family

$$M(S,V)=\{l\in L(X,Y):\bigcup_{x\in S}\langle l,x\rangle\subset V\}$$

is a neighbourhood base of 0 in $L(X, Y)$ for a unique translation-invariant topology, called the topology of uniform convergence on the sets $S\in\sigma$, or, briefly, the σ-topology where $\langle l,x\rangle$ denotes the evaluation of the linear operator $l\in L(X,Y)$ at $x\in X$ (see, [68:pp. 79–80]). By the Corollary of Schaefer [68:p. 80], $L(X, Y)$ becomes a locally convex topological vector space under the σ-topology, where Y is assumed a locally convex topological vector space.

In order to prove the main results, we need the following very useful result.

Lemma 3.1.1: *Let X and Y be real Hausdorff topological vector spaces and $L(X, Y)$ be the topological vector space under the σ-topology. Then the bilinear mapping $\langle\cdot,\cdot\rangle$: $L(X,Y)\times X\to Y$ is continuous on $L(X,Y)\times X$ where $\langle l,x\rangle$ denotes the evaluation of the linear operator $l\in L(X,Y)$ at $x\in X$.*

Proof: Let $(l_\alpha,x_\alpha)_{\alpha\in\Gamma}$ be a net in $L(X,Y)\times X$ and $(l_\alpha,x_\alpha)\to(l_0,x_0)$, then we have $l_\alpha\to l_0$ under the σ-topology of $L(X, Y)$ and $x_\alpha\to x_0$ in X. Let $V\in\mathcal{B}$ be an arbitrary given neighbourhood of 0 in Y, we can choose a neighbourhood V_1 of 0 in Y such that $V_1+V_1+V_1\subset V$. Since $x_\alpha\to x_0$ in $X, \{x_\alpha-x_0\}_{\alpha\in\Gamma}\cup\{x_0\}$ must be bounded and hence $\{x_\alpha-x_0\}_{\alpha\in\Gamma}\cup\{x_0\}\subset S_0$ for some $S_0\in\sigma$. Let

$$M(S_0,V_1)=\{l\in L(X,Y):\bigcup_{x\in S_0}\langle l,x\rangle\subset V_1\}.$$

Then $M(S_0,V_1)$ is a neighbourhood of 0 under the σ-topology of $L(X, Y)$. Since $l_\alpha-l_0\to 0$ under the σ-topology of $L(X, Y)$, there exists $\alpha_1\in\Gamma$ such that $l_\alpha-l_0\in M(S_0,V_1)$ for all $\alpha\geq\alpha_1$. It follows that

$$\langle l_\alpha-l_0,x_\alpha-x_0\rangle\in V_1 \text{ and } \langle l_\alpha-l_0,x_0\rangle\in V_1, \forall\alpha\geq\alpha_1.$$

Since $x_\alpha\to x_0$ in X and $l_0\in L(X,Y)$ is a continuous linear operator, there exists $\alpha_2\in\Gamma$ such that

$$\langle l_0,x_\alpha-x_0\rangle\in V_1, \forall\alpha\geq\alpha_2.$$

Hence there exists $\alpha_3\in\Gamma$ such that

$$\begin{aligned}\langle l_\alpha,x_\alpha\rangle-\langle l_0,x_0\rangle &= \langle l_\alpha-l_0,x_\alpha-x_0\rangle+\langle l_\alpha-l_0,x_0\rangle+\langle l_0,x_\alpha-x_0\rangle\\ &\in V_1+V_1+V_1\subset V, \ \forall\alpha\geq\alpha_3.\end{aligned}$$

This proves that the bilinear mapping $\langle\cdot,\cdot\rangle$: $L(X,Y)\times X\to Y$ is continuous. □

Let K be a nonempty and convex subset of $X, T: K \Rightarrow L(X, Y)$ be a set-valued mapping, $\eta: K \times K \to X$ be a single-valued mapping and $C: K \Rightarrow Y$ be a set-valued mapping such that $\forall x \in K$, $C(x)$ is a closed, pointed, convex and solid cone in Y with apex at 0.

In this section, we study the Generalized Vector Variational-like Inequality (in short GVVLI): find $y \in K$, such that there exists an $\hat{v} \in T(y)$ satisfying

$$\langle \hat{v}, \eta(x, y) \rangle \nleq_{\operatorname{int} C(y)} 0, \quad \forall x \in K, \tag{3.1}$$

where $A \nleq_D B$ means $B - A \notin D$.

We note that $\hat{v}$ depends on x generally and any vector $\hat{x}$ with $0 \in T(\hat{x})$ is a solution of the GVVLI.

Special Cases

(1) If T is a single-valued mapping and $\eta(x, y) = x - g(y)\ \forall x, y \in K$ where $g: K \to X$ is a single-valued mapping, then the problem (3.1) reduces to finding $y \in K$, such that

$$\langle T(\hat{x}), x - g(\hat{x}) \rangle \nleq_{\operatorname{int} C(y)} 0, \quad \forall x \in K. \tag{3.2}$$

The problem (3.2) was considered by Siddiqi *et al.* [69].

(2) If X and Y are both Banach spaces and $\eta(x, y) = x - y\ \forall x, y \in K$, then (3.1) reduces to finding $y \in K$, such that there exists a $\hat{v} \in T(y)$ satisfying

$$\langle \hat{v}, x - \hat{x} \rangle \nleq_{\operatorname{int} C(y)} 0, \quad \forall x \in K. \tag{3.3}$$

(3.3) was introduced and studied by Lin *et al.* [57] and Konnov and Yao [50]. If furthermore C is a constant mapping, then (3.3) was considered by Lee *et al.* [52,53].

(3) If T is a single-valued mapping then (3.3) reduces to finding $y \in K$ such that

$$\langle T(y), x - y \rangle \nleq_{\operatorname{int} C(y)} 0, \quad \forall x \in K. \tag{3.4}$$

(3.4) was studied by Lai and Yao [51], Yu and Yao [83] and Chen *et al.* [10–15].

(4) If $Y = \mathbb{R}$, $C(x) = [0, \infty[\ \forall x \in K$ and T is a single-valued mapping then (3.1) reduces to finding $y \in K$, such that

$$\langle T(y), \eta(x, y) \rangle \geq 0, \quad \forall x \in K. \tag{3.5}$$

(3.5) is called Variational-like Inequality; it was introduced and studied by Parida *et al.* [63].

The main purpose of this section is to derive some existence results for the GVVLI involving set-valued mapping without monotonicity under noncompact setting of topological vector spaces.

3.1.1. Existence of Solutions for the GVVLI

In this section, we prove some existence results of solutions for the GVVLI.

If X and Y are both topological spaces. A set-valued mapping $T: X \to 2^Y$ is said to be upper semicontinuous (in short, u.s.c) on X if, for each open set $U \subseteq Y$, the set $T^{-1}(U) = \{x \in X : T(x) \subset U\}$ is open in X.

Definition 3.1.1: Let X, Y be two real topological vector spaces, K be a nonempty and convex subset of X, T: $K \Rightarrow L(X, Y)$, η: $K \times K \to X$ and C: $K \Rightarrow Y$ be such that, $\forall x \in K$, $C(x)$ is a closed, pointed and convex cone with apex at 0. T is said to satisfy generalized $L - \eta$-condition iff for any finite set $\{x_1, \ldots, x_n\} \subseteq K$, $\bar{x} = \sum_{i=1}^{n} \lambda_i x_i$ with $\lambda_i \geq 0$ and $\sum_{i=1}^{n} \lambda_i = 1$, there exists $\bar{v} \in T(\bar{x})$ such that

$$\left\langle \bar{v}, \sum_{i=1}^{n} \lambda_i \eta(x_i, \bar{x}) \right\rangle \not\leq_{\operatorname{int} C(\bar{x})} 0.$$

Remark 3.1.1: If $\eta(x, y)$ is affine in first argument and, $\forall x \in K$, there exists $v \in T(x)$, such that

$$\langle v, \eta(x, x) \rangle \not\leq_{\operatorname{int} C(x)} 0,$$

then T satisfies the generalized $L - \eta$-condition.

If $\eta(x, y) = x - y$ $\forall x, y \in K$, then we have that for any $v \in T(\bar{x})$

$$\left\langle v, \sum_{i=1}^{n} \lambda_i (x_i - \bar{x}) \right\rangle = \langle v, \bar{x} - \bar{x} \rangle = 0 \not\leq_{\operatorname{int} C(x)} 0,$$

and hence the generalized $L - \eta$-condition is satisfied trivially.

The following result is a special case of Theorem 3 of Ding and Tan [27]:

Lemma 3.1.2: *Let K be a nonempty and convex subset of a topological vector space X. Let G: $K \Rightarrow K$ be such that*

(a) $\forall x \in K, G(x)$ *is compactly closed, i.e., for each compact subset D of K, $G(x) \cap D$ is closed in D,*
(b) *G is a KKM mapping, i.e., for any finite set $\{x_1, \ldots, x_n\} \subseteq K$, $\operatorname{conv}\{x_1, \ldots, x_n\} \subseteq \bigcup_{i=1}^{n} G(x_i)$, where* conv *denotes convex hull,*
(c) *there exist a nonempty compact convex subset L of K and a nonempty and compact subset D of K, such that $\forall y \in K \backslash D$, there is $x \in \operatorname{co}(L \cup \{y\})$ satisfying $y \notin G(x)$.*

Then $D \cap (\bigcap_{x \in K} G(x)) \neq \emptyset$.

The following result is Lemma 2.1 of Chen [17].

Lemma 3.1.3: *Let (X, P) be an ordered topological vector space equipped with a closed, pointed, convex and solid cone P. Then we have that $\forall x, y \in X$,*

(i) $y - z \in \operatorname{int} P$ *and* $y \notin \operatorname{int} P \Rightarrow z \notin \operatorname{int} P$.
(ii) $y - z \in P$ *and* $y \notin \operatorname{int} P \Rightarrow z \notin \operatorname{int} P$.
(iii) $y - z \in -\operatorname{int} P$ *and* $y \notin -\operatorname{int} P \Rightarrow z \notin -\operatorname{int} P$.
(iv) $y - z \in -P$ *and* $y \notin \operatorname{int} P \Rightarrow z \notin -\operatorname{int} P$.

Theorem 3.1.1: *Let X and Y be two real topological vector spaces, K be a nonempty and convex subset of X and $L(X, Y)$ be equipped with the σ-topology. Let T: $K \Rightarrow L(X, Y)$ be u.s.c. in K with compact values and η: $K \times K \to X$ be continuous with respect to the second argument, such that T satisfies the generalized $L - \eta$-condition. Let C: $K \Rightarrow Y$ be such that $\forall x \in K$, $C(x)$ is a closed pointed convex and solid cone and the mapping W: $K \Rightarrow Y$ defined by $W(x) = Y \backslash (-\operatorname{int} C(x))$ is u.s.c. on K. Suppose that the following condition is satisfied.*

(A) There exist a nonempty compact convex subset L of K and a compact subset D of K such that for each $y \in K\backslash D$, there is $x \in \text{co}(L \cup \{y\})$ satisfying

$$\langle v, \eta(x, y)\rangle \not\leq_{\text{int}\, C(y)} 0, \quad \forall v \in T(y).$$

Then, the GVVLI has a solution $\hat{x} \in D$.

Corollary 3.1.1: *Let X, Y be two real Hausdorff topological vector spaces and K be a nonempty, compact and convex subset of X. Suppose that $L(X, Y)$ is equipped with the σ-topology. Let T: $K \Rightarrow L(X, Y)$ be u.s.c. with compact values and η: $K \times K \to X$ be affine with respect to the first argument and continuous with respect to the second argument, such that $\forall x \in K$, there is a $v \in T(x)$ satisfying $\langle v, x - g(x)\rangle \not\leq_{\text{int}\, C(x)} 0$. If the set-valued mapping $W(x) = Y\backslash(-\text{int}\, C(x))$ is u.s.c. on K, then there exists $\hat{x} \in K$ such that $\forall x \in K$, there is an $\hat{u} \in T(\hat{x})$, such that*

$$\langle \hat{v}, \eta(x, \hat{x})\rangle \not\leq_{\text{int}\, C(\hat{x})} 0.$$

Remark 3.1.2: If T is a single-valued mapping and $\eta(x, y) = x - g(y)$ where g: $K \to X$ is continuous, the Corollary 3.1.1 reduces to Theorem 2.1 of Siddiqi *et al.* [69]. Hence Theorem 2.1.1 and Corollary generalizes Theorem 2.1 in [69] to generalized Vector Variational-like Inequalities.

Corollary 3.1.2: *Let X, Y be two real Banach spaces, K be a nonempty and convex subset of X and $L(X, Y)$ be equipped with the norm topology. Let T: $K \Rightarrow L(X, Y)$ be u.s.c. with compact values and η: $K \times K \to X$ be continuous in second argument such that T satisfies the generalized $L - \eta$-condition. Suppose W: $K \Rightarrow Y$, defined by $W(x) = Y\backslash(-\text{int}\, C(x))$, is u.s.c. on K. Suppose that the condition* (A) *in Theorem 3.1.1 holds. Then, there exists $\hat{x} \in D$, such that $\forall x \in K$, there is a $\hat{v} \in T(\hat{x})$ satisfying*

$$\langle \hat{v}, \eta(x, \hat{x})\rangle \not\leq_{\text{int}\, C(\hat{x})} 0.$$

Remark 3.1.3: The result in Theorem 3.1.1 and Corollary 3.1.2 are new and interesting. As they do not depend on any monotonicity of mappings.

Theorem 3.1.2: *Let X, Y be two real Hausdorff topological vector spaces, K be a nonempty convex subset of X and $L(X, Y)$ be equipped with the σ-topology. Let T: $K \Rightarrow L(X, Y)$ be u.s.c. with compact values and η: $K \times K \to X$ be continuous in second argument. Let C: $K \Rightarrow Y$ be such that $\forall x \in K$, $C(x)$ is a closed, pointed convex and solid cone and W: $K \to 2^Y$, defined by $W(x) = Y\backslash(-\text{int}\, C(x))$, is u.s.c. on K. Suppose that there exists a mapping h: $K \times K \to Y$, such that*

(i) $\forall x, y \in K$, *there exists* $v \in T(y)$, *such that*

$$h(y, x) - \langle v, \eta(x, y)\rangle \in -\text{int}\, C(y).$$

(ii) *for any finite set* $\{x_1, \ldots, x_n\} \subset K$ *and any* $\bar{x} = \sum_{i=1}^{n} \lambda_i x_i$ *with* $\lambda_i \geq 0$ *and* $\sum_{i=1}^{n} \lambda_i = 1$, *there is an* $i_0 \in \{1, \ldots, n\}$ *such that* $h(\bar{x}, x_{i_0}) \notin -\text{int}\, C(\bar{x})$.

(iii) *There exist a nonempty, compact and convex subset L of K and a nonempty compact subset D of K, such that $\forall y \in K\backslash D$, there is an $x \in \text{co}(L \cup \{y\})$ satisfying*

$$\langle v, \eta(x, y)\rangle \not\leq_{\text{int}\, C(y)} 0, \quad \forall v \in T(y).$$

Then, the GVVLI has a solution $\hat{x} \in D$.

Corollary 3.1.3: *Let X, Y be two real Hausdorff topological vector spaces, K be a nonempty and convex subset of X and $L(X, Y)$ be equipped with the σ-topology. Let $T: K \to L(X, Y)$ be continuous and $\eta: K \times K \to X$ be continuous with respect to second argument. Let $C: K \Rightarrow Y$ be such that $\forall x \in K$, $C(x)$ is a closed, pointed, convex and solid cone and $W: K \Rightarrow Y$ defined by $W(x) = Y \backslash (-\text{int}\, C(x))$ is u.s.c. on K. Suppose that there exists a mapping $h: K \times K \to Y$ such that*

(i) $h(y, x) - \langle T(y), \eta(x, y)\rangle \in -\text{int}\, C(y)$
(ii) *the set* $\{x \in K: h(y, x) \in -\text{int}\, C(y)\}$ *is convex* $\forall y \in K$,
(iii) $h(x, x) \notin -\text{int}\, C(x)\ \forall x \in K$
(iv) *there exist a nonempty compact convex subset L of K and a nonempty compact subset D of K, such that $\forall y \in K \backslash D$, there is a $x \in \text{co}\{L \cup \{y\}\}$ satisfying*

$$\langle T(y), \eta(x, y)\rangle \not\leq_{\text{int}\, C(y)} 0.$$

Then, there exists $\hat{x} \in D$, such that

$$\langle T(\hat{x}), \eta(x, \hat{x})\rangle \not\leq_{\text{int}\, C(\hat{x})} 0, \quad \forall x \in K.$$

Remark 3.1.4: Theorem 3.1.2 and Corollary 3.1.3 improve and generalize Theorem 2.2 of Siddiqi *et al.* [69].

3.2. Generalized Vector Variational-like Inequalities with $C_x - \eta$-Pseudomonotone Set-valued Mappings

In this section, we shall present a class of generalized vector variational-like inequalities involving $C_x - \eta$-pseudomonotone and weakly $C_x - \eta$-pseudomonotone set-valued mappings. The generalized vector variational-like inequality problem unifies and generalizes the generalized vector variational inequalities, vector variational inequalities and various extensions of the classic variational inequalities involving single-valued and set-valued mappings of various monotone types in the literature. Several existence theorems are established under noncompact setting in topological vector spaces. These new results unify and generalize many recent known results in the literature.

The vector variational inequality problem in a finite dimensional Educlidean space has been introduced in [35]. Since then, Chen *et al.* [15–17], Lai and Yao [51] and Yu and Yao [83] have intensively studied the vector variational inequalities (in short VVI) on different monotonicity assumptions in abstract spaces. Recently, Lee *et al.* [52,53], Lin *et al.* [57], Danilidis and Hadjsavvas [24] and Konnev and Yao [50] introduced and studied the generalized vector variational inequalities (in short GVVI) involving monotone type set-valued mappings in Banach spaces.

In this section, we shall present a new class of generalized vector variation-like inequalies (in short, GVVLI) inequalities involving $C_x - \eta$-pseudomonotone (weakly $C_x - \eta$-pseudomonotone) set-valued mappings in topological vector spaces. Several existence results for solutions of the GVVI are presented under non-compact setting in topological vector spaces. These results are new and unify and extend many known results mentioned above.

Let X, Y be real Hausdorff topological vector spaces, $L(X, Y)$ be the space of all continuous linear operators from X into Y and σ be the family of all bounded subsets of X. Let $\mathcal{B}$ be a neighbourhood base of 0 in Y. When S runs through σ, V through $\mathcal{B}$, the family

$$M(S, V) = \{l \in L(X, Y) : \bigcup_{x \in S} \langle l, x\rangle \subset V\}$$

is a neighbourhood base of 0 in $L(X, Y)$ for a unique translation-invariant topology, called the topology of uniform convergerence on the sets $S \in \sigma$, or, briefly, the σ-topology where $\langle l, x\rangle$ denotes the evaluation of the linear operator $l \in L(X, Y)$ at $x \in X$ (see, [68:p.79–80). By the Corollary in [68:p. 80), if Y is locally convex space and the set $\bigcup\{S\colon S \in \sigma\}$ is total in X, then $L(X, Y)$ becomes a locally convex topological vector space under the σ-topology. A nonempty subset P of Y is called a convex cone if $P + P \subset P$ and $\lambda P \subset P$ for all $\lambda > 0$. A cone P is said to be a pointed cone if $P \cap (-P) = \{0\}$ where 0 is the zero vector in Y. A cone P is said to be proper if it is properly contained in Y.

Let X, Y be real Hausdorff topological vector spaces and K be a nonempty convex subsets of X. Let $T\colon K \Rightarrow L(X, Y)$ and $C\colon K \Rightarrow Y$ be two set-valued mappings such that for each $x \in K, C(x)$ is a proper closed convex cone in Y with $\operatorname{int} C(x) \neq \emptyset$. Let $\eta\colon K \times K \to X$ be a single-valued mapping. We consider the following (GVVLI) problem: find $x \in K$ such that for each $y \in K$, there is a $l \in T(x)$ satisfying

$$\langle l, \eta(y, x)\rangle \not\leq_{\operatorname{int} C(x)} 0, \tag{3.7}$$

where $A \not\leq_D B$ means $B - A \notin D$.

Special Cases

(I) If $\eta(x, y) = x - y$ for all $x, y \in K$ and X, Y are both Banach spaces, (3.7) reduces to finding $x \in K$ such that for each $y \in K$, there is a $l \in T(x)$ satisfying

$$\langle l, y - x\rangle \not\leq_{\operatorname{int} C(x)} 0. \tag{3.8}$$

(3.8) was introduced and studied by Lin *et al.* [57], Daniilidis and Hadjisavvas [24] and Konnov and Yao [50]. If C is also a constant mapping, then (3.8) was considered by Lee *et al.* [52,53].

(II) If T is a single-valued mapping and $\eta(x, y) = x - g(y)$ for all $x, y \in K$ where $g\colon K \to X$ is a single-valued mapping, then (3.7) reduces to finding $x \in K$ such that

$$\langle T(x), y - g(x)\rangle \not\leq_{\operatorname{int} C(x)} 0, \quad \forall y \in K \tag{3.9}$$

(3.9) was introduced and studied by Siddiqi *et al.* [69].

(III) If T is a single-valued mapping, $\eta(x, y) = x - y$ for all $x, y \in K$ and X, Y are both Banach spaces, (3.7) reduces to finding $x \in K$ such that

$$\langle T(x), y - x\rangle \not\leq_{\operatorname{int} C(x)} 0, \quad \forall y \in K. \tag{3.10}$$

Then (3.10) and its special cases were extensively studied by Chen *et al.* [15–17], Lai and Yao [50] and Yu and Yao [83].

(IV) If $Y = \mathbb{R}, C(x) = [0, \infty)$ for all $x \in K$ and T is a single-valued mapping, then (3.7) reduces to finding $x \in K$ such that

$$\langle T(x), \eta(y, x)\rangle \geq 0, \quad \forall y \in K. \tag{3.11}$$

(3.11) is called variational-like inequality problem which was introduced and studied by Parida *et al.* [63] and others.

Vector variational inequalities and vector variational-like inequalities have shown to be a particularly useful tool in the geometric features of optimization, (see [10–15,45]). The main purpose of this chapter is to derive several existence results of solutions for the (GVVLI) involving $C_x - \eta$-pseudomonotone and weakly $C_x - \eta$-pseudomonotone set-valued mappings under non-compact setting in topological vector spaces. These results unify and extend many known results mentioned above.

In order to prove our main results, we need the following Definitions and Lemmas.

Definition 3.2.1: Let X, Y be real topological vector spaces and K be a nonempty subset of X. Let T: $K \rightrightarrows L(X, Y)$ and C: $K \rightrightarrows Y$ such that for each $x \in K, C(x)$ is a proper closed convex cone with $\operatorname{int} C(x) \neq \emptyset$. Let η: $K \times K \to X$ and P is a convex cone in Y.

(i) T is $P - \eta$-monotone on K if for each pair of points $x, y \in K$ and for all $l \in T(x), l' \in T(y)$, we have

$$\langle l' - l, \eta(y, x)\rangle \in P.$$

(ii) T is $P - \eta$-pseudomonotone on K if for each pair of points $x, y \in K$ and for all $l \in T(x), l' \in T(y)$, we have

$$\langle l, \eta(y, x)\rangle \in P \Rightarrow \langle l', \eta(y, x)\rangle \in P.$$

(iii) T is $C_x - \eta$-pseudomonotone on K if for each pair of points $x, y \in K$ and for all $l \in T(x), l' \in T(y)$, we have

$$\langle l, \eta(y, x)\rangle \not\leq_{\operatorname{int} C(x)} 0 \Rightarrow \langle l', \eta(x, y)\rangle \not\leq_{\operatorname{int} C(x)} 0.$$

(iv) T is weakly P-monotone on K if for each pair of points $x, y \in K$ and for each $l \in T(x)$, there exists $l' \in T(y)$ such that

$$\langle l' - l, \eta(y, x)\rangle \in P.$$

(v) T is weakly P-pseudomonotone on K if for each pair of points $x, y \in K$ and for each $l \in T(x)$, we have

$$\langle l, \eta(y, x)\rangle \in P \Rightarrow \langle l', \eta(y, x)\rangle \in P \quad \text{for some } l' \in T(y).$$

(vi) T is weakly $C_x - \eta$-pseudomonotone on K if for each pair of points $x, y \in K$ and for each $l \in T(x)$, we have

$$\langle l, \eta(y, x)\rangle \not\leq_{\operatorname{int} C(x)} 0 \Rightarrow \langle l', \eta(y, x)\rangle \not\leq_{\operatorname{int} C(x)} 0 \quad \text{for some } l' \in T(y).$$

Remark 3.2.1: The $P - \eta$-pseudomonotone mappings can be regarded as some extensions of P-pseudomonotone and pseudomonotone mappings introduced by Konnov and Yao [50] and Karamardian [44] respectively. The $C_x - \eta$-pseudomonotone and weakly

$C_x - \eta$-pseudomonotone mapping can be regarded as some generalizations of C_x-pseudomonotone and weakly C_x-pseudomonotone mappings introduced by Konnov and Yao [50] (also see [83]) respectively.

From Definition 3.2.1 it is easy to show that the following relationship between the mappings in Definition 3.2.1 hold.

Proposition 3.2.1: *Let X, Y, K, T, η, C and P as in Definition 3.2.1.*

(*i*) *If T is $P - \eta$-monotone, then T is weakly $P - \eta$-monotone and $P - \eta$-pseudomonotone.*
(*ii*) *If T is $P - \eta$-pseudomonotone (resp., $C_x - \eta$-pseudomonotone), then it is weakly $P - \eta$-pseudomonotone (resp., weakly $C_x - \eta$-pseudomonotone).*
(*iii*) *If T is weakly $P - \eta$-monotone, then it is weakly $P - \eta$-pseudomonotone.*
(*iv*) *If T is $C_- - \eta$-monotone) (resp. weakly $C_- - \eta$-monotone), then it is C_x-η-pseudomonotone (resp., weakly $C_x - \eta$-pseudomonotone) where $C_- = \bigcap_{x \in K} C(x)$.*

If X, Y are both topological spaces. A set-valued mapping $T: X \Rightarrow 2^Y$ is called upper semicontinuous (in short, u.s.c.) on X if, for each open set $U \subset Y$, the set $T^{-1}(U) = \{x \in X: T(x) \subset U\}$ is open in X.

Definition 3.2.2: Let X, Y be real topological vector spaces and K be a nonempty subset of X. Let $T: K \Rightarrow L(X, Y)$ be a set-valued mapping and $\eta: K \times K \to X$ be a single-valued mapping. T is said to be η-hemicontinuous on K if for any $x, y \in K$ and $\alpha \in [0, 1]$, the mapping

$$\alpha \mapsto \langle T(x + \alpha(y - x)), \eta(y, x)\rangle = \bigcup_{l \in T(x+\alpha(y-x))} \langle l, \eta(y, x)\rangle$$

is u.s.c. at 0^+.

Remark 3.2.2: If X, Y are Banach spaces and $\eta(x, y) = x - y$ for all $x, y \in K$, then the η-hemicontinuity of T reduces to the u-hemicontinuity of T introduced by Konnov and Yao [50] (also see, [57,83]).

Lemma 3.2.1: *Let X, Y be real Hausdorff topological vector spaces and K be a nonempty convex subset of X. Let T: $K \Rightarrow L(X, Y)$ be a set-valued mapping and η: $K \times K \to X$ be a single-valued mapping. Let C: $K \Rightarrow Y$ be a set-valued mapping such that for each $x \in K$, $C(x)$ is a closed convex cone with* int $C(x) \neq \emptyset$. *We consider the following problems:*

(I) $x \in K$ *such that* $\forall y \in K, \exists l \in T(x) : \langle l, \eta(y, x)\rangle \nleq \operatorname{int} C(x) 0;$

(II) $x \in K$ *such that* $\forall y \in K, \exists l' \in T(y) : \langle l', \eta(y, x)\rangle \nleq \operatorname{int} C(x) 0;$

(III) $x \in K$ *such that* $\langle l', \eta(y, x)\rangle \nleq \operatorname{int} C(x) 0$, *for all* $y \in K$ *and* $l' \in T(y)$.

Then

(*i*) *Problem (III) implies Problem (II).*
(*ii*) *Problem (II) implies Problem (I) if T is η-hemicontinuous on K, $\eta(x, y)$ is affine in first argument and $\eta(x, x) = 0$ for each $x \in K$.*
(*iii*) *Problem (I) implies Problem (III) if T is $C_x - \eta$-pseudomonotone, and it implies Problem (II) if T is weakly $C_x - \eta$-pseudomonotone.*

Remark 3.2.3: If $\eta(x, y) = x - y$ for all $x, y \in K$, Lemma 2.1 reduces to Lemma 2.1 of Konnov and Yao [50] which extends Lemma 2.1 of Yu and Yao [83].

The following result is a special case of Theorem 3 of Ding and Tan [27].

Lemma 3.2.2: *Let K be a nonempty convex subset of a topological vector space X and G: $K \rightrightarrows K$ be such that*

(*a*) *for each $x \in K, G(x)$ is compactly closed, i.e., for each compact subset D of $K, G(x) \cap D$ is closed in D,*
(*b*) *G is a KKM mapping, i.e., for any finite set $\{x_1, \ldots, x_n\} \subset K, \mathrm{co}\{x_1, \ldots, x_n\} \subset \bigcup_{i=1}^{n} G(x_i)$,*
(*c*) *there exist a nonempty compact convex subset L of K and a nonempty compact subset D of K such that for each $y \in K \backslash D$, there is $x \in \mathrm{co}(L \cup \{y\})$ satisfying $y \notin G(x)$.*

Then $D \cap \left(\bigcap_{x \in K} G(x) \right) \neq \emptyset$.

3.2.1. Existence Theorems for Solutions of the (GVVLI)

In this section, we present several existence theorems for solutions of the (GVVLI) involving $C_x - \eta$-pseudomonotone and weakly $C_x - \eta$-pseudomonotone set-valued mappings. These results are presented under noncompact setting in locally convex topological vector spaces.

Theorem 3.2.1: *Let X, Y be real locally convex Hausdorff topological vector spaces and K be a nonempty convex subset of X. Let C: $K \rightrightarrows Y$ be such that for each $x \in K, C(x)$ is a proper closed convex cone with $\operatorname{int} C(x) \neq \emptyset$, and W: $K \rightrightarrows Y$ be defined by $W(x) = Y \backslash (-\operatorname{int} C(x))$ such that the graph $\mathrm{Gr}(W)$ of W is weakly closed in $X \times Y$. Let T: $K \rightrightarrows L(X, Y)$ and η: $K \times K \to X$ be such that*

(*i*) *T is $C_x - \eta$-pseudomonotone and η-hemicontinuous on K,*
(*ii*) *$\eta(x, y)$ is affine in first argument and weakly continuous in second argument such that $\eta(x, x) = 0$ for all $x \in K$,*
(*iii*) *there exist a nonempty weakly compact convex subset L of K and a nonempty weakly compact subset D of K such that for each $x \in K \backslash D$, there is $y \in \mathrm{co}(L \cup \{x\})$ satisfying*

$$\langle l', \eta(y, x) \rangle \in -\operatorname{int} C(x), \quad \text{for some } l' \in T(y)$$

then the (GVVLI) has a solution $\hat{x} \in D$.

Remark 3.2.4: If K is also weakly compact, then the condition (iii) is satisfied automatically. If $\eta(x, y) = x - y$ for all $x, y \in K$, then the condition (ii) is satisfied trivially. Hence Theorem 3.2.1 generalizes Theorem 3.1 of Konnov and Yao [50] from the (GVVI) to the (GVVLI) and from compact setting of Banach spaces to noncompact setting of locally convex topological vector spaces. Theorem 3.1, in turn, generalizes Theorem 3.1 of Lin *et al.* [57].

Theorem 3.2.2: *Let X, Y, C, W, T and η be as in Theorem 3.2.1 and K be a closed convex (not necessary bounded) subset of X. Suppose that conditions (i) and (ii) of Theorem 3.2.1 hold and the condition (iii) is replaced by the following*

(*iii*)′ *there exist a nonempty weakly compact convex subset L of K and a nonempty weakly compact subset D of K such that for each $x \in K \backslash D$, there is $y \in \mathrm{co}(L \cup \{x\})$ satisfying $x \notin \overline{F(y)}^{w}$ where $F(y) = \{x \in K : \exists l \in T(x), \langle l, (y, x) \rangle \not\leq_{\operatorname{int} C(x)} 0\}$ and $\overline{F(y)}^{w}$ is the weak closure of $F(y)$.*

Then the (GVVLI) has a solution $\hat{x} \in D$.

Remark 3.2.5: Note that the condition (iii) is weaker than the generalized v-coercive condition on T in Theorem 3.2.2 of Konnov and Yao [50]. Hence Theorem 3.2 extends Theorem 3.2 of Konnov and Yao [50] to the (GVVLI) and locally convex topological vector spaces.

In the following, we present two existence theorems of solutions for the (GVVLI) with weakly $C_x - \eta$-pseudomonotone set-valued mappings.

Theorem 3.2.3: *Let X, Y be real locally convex Hausdorff topological vector spaces, K be a nonempty convex subset of X and $L(X, Y)$ be equipped with the σ-topology. Let $C\colon K \Rightarrow Y$ be such that for each $x \in K, C(x)$ is a proper closed convex cone with $\operatorname{int} C(x) \neq \emptyset$, and $W\colon K \Rightarrow Y$ be defined by $W(x) = Y \backslash (-\operatorname{int} C(x))$ such that the graph $\operatorname{Gr}(W)$ of W is weakly closed in $X \times Y$. Let $T\colon K \Rightarrow L(X, Y)$ and $\eta\colon K \times K \to X$ be such that*

(*i*) *T is weakly $C_x - \eta$-pseudomonotone and η-hemicontinuous on K with nonempty compact values,*
(*ii*) *$\eta(x, y)$ is affine in first argument and weakly continuous in second argument such that $\eta(x, x) = 0$ for all $x \in K$,*
(*iii*) *there exist a nonempty weakly compact convex subset L of K and a nonempty weakly compact subset D of K such that for each $x \in K \backslash D$, there is $y \in \operatorname{co}(L \cup \{x\})$ satisfying*

$$\langle l', \eta(y, x) \rangle \underset{\operatorname{int} C(x)}{\leq} 0, \quad \forall l' \in T(y).$$

Then the (GVVLI) has a solution $\hat{x} \in D$.

Remark 3.2.6: If K is also weakly compact, then the condition (iii) is satisfied automatically. Hence Theorem 3.2.3 generalizes Theorem 3.3 (a) of Konnov and Yao [50] to the (GVVLI) and noncompact setting in locally convex topological vector spaces.

Theorem 3.2.4: *Let $X, Y, L(X, Y), C, W, T$ and η be as in Theorem 3.3 and K be a nonempty closed convex subset of X. Suppose that the conditions (i) and (ii) of Theorem 3.3 are satisfied and the condition (iii) is replaced by*

(*iii*)$'$ *there exist a nonempty weakly compact subset D of X and $y_0 \in X$ such that for each $x \in K \backslash D$ and $l \in T(x)$,*

$$\langle l, \eta(y_0, x) \rangle \underset{\operatorname{int} C(x)}{\leq} 0.$$

Then the (GVVLI) has a solution $\hat{x} \in D$.

Remark 3.2.7: If X is a reflexive Banach space, it is easy to show that the assumption (iii)$'$, is equivalent to the generalized d-coerciveness of T introduced by Konnov and Yao ([50]:p. 52). Hence Theorem 3.2.4 generalizes Theorem 3.3 (b) of Konnov and Yao [50] to (GVVLI) and locally convex topological vector spaces.

4. THEORY OF VECTOR OPTIMIZATION

4.1. Cones, Efficient Points and Existence Theorems

Let E be a real topological vector space and C a convex cone in E. Then $\leq$ is a partial order in E generated by $\leq$, i.e., $x \leq y \;\Leftrightarrow\; y - x \in C$. Throughout this chapter we shall use the following notations for a set C in E: $\mathrm{cl}\,C, \mathrm{int}\,C, \mathrm{ri}\,C, C^c$, and $\mathrm{conv}(C)$ denote the closure, interior, relative interior, complement and convex hull of C in E, respectively. We shall use the notation $\ell(C)$ to denote the set $C \cap -C$.

We shall now present the following definitions (see [59:p. 39]):

Definition 4.1.1: Let A be a non-empty subset of E. Then,

(1) $x \in A$ is an ideal efficient (or ideal minimum) point of A with respect to C if $x \leq y$ for every $y \in A$; The set of all ideal minimal points of A will be denoted by $\mathrm{I}\min(A|C)$;
(2) $x \in A$ is an efficient (or Pareto-minimal, or non-dominated) point of A with respect to C if $y \leq x$ for some $y \in A$, then $x \leq y$; The set of all efficient points of A will be denoted by $\min(A|C)$;
(3) $x \in A$ is a (global) properly efficient point of A with respect to C if there exists a convex cone K which is not the whole space and contains $C \backslash \ell(C)$ in its interior so that $x \in \min(A|K)$; The set of all proper efficient points of A will be denoted by $\mathrm{Pr}\min(A|C)$;
(4) supposing that $\mathrm{int}\,C$ is non-empty, $x \in A$ is a weakly efficient point of A with respect to C if $x \in \min(A|\{0\} \cup \mathrm{int}\,C)$; The set of all weakly efficient points of A will be denoted by $\mathrm{W}\min(A|C)$.

Proposition 4.1.1: [59:p. 41]: *We have the inclusions:*

$$\mathrm{Pr}\min(A) \subset \min(A) \subset \mathrm{W}\min(A).$$

Moreover, if $\mathrm{I}\min(A)$ *is non-empty, then*

$$\mathrm{I}\min(A) = \min(A)$$

and it is a point whenever C is pointed.

Proposition 4.1.2: [59:pp. 41,42]: *An equivalent definition of efficiency can be stated as follows:*

(1) $x \in \mathrm{I}\min(A)$ *if and only if* $x \in A$ *and* $A \subset x + C$;
(2) $x \in \min(A)$ *if and only if* $A \cap (x - C) \subset x + \ell(C)$, *or equivalently, there is no* $y \in A$ *such that* $x > y$. *In particular, when C is pointed,* $x \in \min(A)$ *if and only if* $A \cap (x - C) = \{x\}$;
(3) *when C is not the whole space,* $x \in \mathrm{W}\min(A)$ *if and only if* $A \cap (x - \mathrm{int}\,C) = \emptyset$, *or equivalently, there is no* $y \in A$ *such that* $y << x$.

Proof: This is immediate from the definition. □

Proposition 4.1.3: [59:p. 42]: *Suppose that there exists a convex pointed cone K containing C. Then*

(1) $\mathrm{I}\min(A|K) = \mathrm{I}\min(A|C)$ *in case* $\mathrm{I}\min(A|C)$ *exists,*
(2) $\mathrm{Pr}\min(A|K) \subset \mathrm{Pr}\min(A|C)$,
(3) $\min(A|K) \subset \min(A|C)$,
(4) $\mathrm{W}\min(A|K) \subset \mathrm{W}\min(A|C)$.

A counterexample for Proposition 4.1.3, in the case where the pointedness of K is violated, is obtained when K is the whole space. In that case every point of a set, in particular, the points which are not efficient with respect to C, is efficient with respect to K. However, the following proposition provides a useful exception.

Proposition 4.1.4: [59:pp. 42,43]: *Assume that there is a closed homogeneous half space H which contains $C\backslash\ell(C)$ in its interior. Then*

(1) $\text{I}\min(A|H) \subset \text{I}\min(A|C)$ *in case right hand side set is nonempty,*
(2) $\min(A|H) \subset \min(A|C)$,
(3) $\text{W}\min(A|H) \subset \text{W}\min(A|C)$.

Proposition 4.1.5: [59:p. 43]: *Let B and A be two sets in E with $B \subset A$. Then*

(1) $\text{I}\min(A) \cap B \subset \text{I}\min(B)$;
(2) $\text{Pr}\min(A) \cap B \subset \text{Pr}\min(B)$;
(3) $\min(A) \cap B \subset \min(B)$;
(4) $\text{W}\min(A) \cap B \subset \text{W}\min(B)$.

Definition 4.1.2: [59:p. 2]: Suppose that C is a convex cone in a real topological vector space E. Then C is said to be

(i) pointed if $\ell(C) = \{0\}$,
(ii) acute if its closure is pointed,
(iii) strictly supported if $C\backslash\ell(C)$ is contained in an open homogeneous half space, and
(iv) correct if $(\text{cl}\, C) + C\backslash\ell(C) \subset C$.

Definition 4.1.3: [59:pp. 8,9]: Let $\mathcal{B}$ denote the filter of neighbourhoods of ∞ and X be a non-empty subset of E. The the recession cone of X is the cone

$$\text{Rec}(X) = \cap\ \text{cl}\,\text{cone}(\text{X} \cap \text{V}) : \text{V} \in \mathcal{B}\}.$$

If the intersection $X \cap V$ is empty then we shall set $\text{cl}\,\text{cone}\ (X \cap V) = \{0\}$.

Let C be a non-empty convex cone in a separated topological space E. We shall now define the sets which are closed or compact not in the usual sense, but with respect to the cone C.

Definition 4.1.4: [59:pp. 13–14]: Let X be a subset of E. We say that it is

(i) C-bounded if for each neighbourhood U of zero in E, there is some positive t such that $X \subset tU + C$,
(ii) C-closed if $X+\ \text{cl}C$ is closed,
(iii) C-compact if any cover of X of the form $\{U_\alpha + C : \alpha \in I, U_\alpha \text{ are open}\}$ admits a finite subcover,
(iv) C-semi-compact if any cover of X of the form $\{(x_\alpha - \text{cl}C)^c : \alpha \in I, x_\alpha \in X\}$ admits a finite subcover.

The last definition was first given by Corley [20]. Clearly, when $C = \{0\}$, every C-notion becomes the corresponding ordinary one.

Definition 4.1.5: [59:p. 43]: Let $x \in E$. The set $A \cap (x - C)$ is called a section of A at x and denoted by A_x

Proposition 4.1.6: [59:p. 44]: *For any $x \in E$ with A_x being non-empty, we have*

(i) $\text{I}\min(A_x) \subset \text{I}\min(A)$ *in case the right hand side set is non-empty*;
(ii) $\min(A_x) \subset \min(A)$;
(iii) $\text{W}\min(A_x) \subset \text{W}\min(A)$.

We note that there is nothing stated in the proposition above about the proper efficiency. Ordinarily, the inclusion $\Pr\min(A_x) \subset \Pr\min(A)$ is not true. For instance, an efficient point of A which is not proper is a proper efficient point of the section at itself. Nevertheless, if the point where the section is taken is well chosen, then a positive result can be expected.

Proposition 4.1.7: [59:p. 44]: *Let E be a finite dimensional Euclidean space and C an acute convex cone with non-empty interior. Suppose that* $\text{Rec}(A) \cap -\text{cl}\, C = \{0\}$, *then* $x \in \Pr\min(A)$ *if and only if* $x \in \Pr\min(A_e)$ *for some* $e \in E$ *with* $x << e$, *that is with* $e \in x - \text{int}\, C$.

Definition 4.1.6: [59:p. 46]: A net $\{x_\alpha : \alpha \in I\}$ in a real topological vector space E is said to be decreasing with respect to the convex cone C if $x_\beta < x_\alpha$ for each $\alpha, \beta \in I$, and $\alpha < \beta$.

Definition 4.1.7: [59:p. 46]: Let A be a subset of a real topological vector space E and C a convex cone in E. Then A is said to be C-complete (respectively, strongly C-complete) if it has no covers of the form $\{(x_\alpha - \text{cl}\, C)^c : \alpha \in I\}$ (respectively, $\{(x_\alpha - C)^c : \alpha \in I\}$) with $\{x_\alpha\}$ being a decreasing net in A.

Clearly, when C is closed, C-completeness and strong C-completeness coincide. Now we present some results on the existence of efficient points.

Theorem 4.1.1: [59:p. 46]: *Let E be a real topological vector space and C a convex correct cone in E. Suppose that A is a non-empty set in E. Then* $\min(A|C)$ *is non-empty if and only if A has a non-empty C-complete section.*

Theorem 4.1.2: [59:p. 47]: *Let E be a real topological vector space and C a convex cone in E. Suppose that A is a non-empty set in E. Then* $\min(A|C)$ *is non-empty if and only if A has a non-empty strongly C-complete section.*

We shall present below some criteria for a set to be C-complete and by means of these criteria several results on the existence of efficient points have been established in the literature.

If E is a real topological vector space and C is a convex cone in E, then recall that the cone C is said to be Daniell if any decreasing net having a lower bound converges to its infimum. Also the space E with the given cone C is boundedly order complete if any bounded decreasing net has an infimum [64].

Lemma 4.1.1: [59:p. 47]: *Let E be a real topological vector space and C a convex cone in E. Then a set $A \subset E$ is C-complete in the following cases*:

(*i*) *A is C-semi-compact, in particular, A is C-compact or compact*;
(*ii*) *A is weakly compact and E is a locally convex space*;
(*iii*) *A is closed bounded and C is Daniell, E is boundedly order complete*;
(*iv*) *A is closed minorized (i.e., there is $x \in E$ such that $A \subset x + C$) and C is Daniell.*

We conclude this section by presenting a condition for the existence of efficiency in terms of recession cones.

Theorem 4.1.3: [59:p. 52]: *Suppose that E is a real topological vector space and C is not the whole space. If A is a non-empty set in E and* $\text{W}\min(A|C)$ *is non-empty, then* $\text{Rec}(A) \cap -\text{int}\, C = \emptyset$, *and if* $\Pr\min(A|C)$ *is non-empty, then* $\text{Rec}(A) \cap -\text{cl}\, C \subset \ell(\text{cl}\, C)$.

It is worthwhile noticing that the conditions stated in Theorem 4.1.3 are necessary, but not sufficient for the existence of efficient points. However, the following result is sometimes helpful when the sets are polyhedral. We recall that a set is polyhedral if it is the sum of a polyhedron and a polyhedral cone.

Theorem 4.1.4: [59:p. 53]: *Suppose that E is a real topological vector space and C is a convex cone in E. Let A be a polyhedral set in E. Then $\min(A|C)$ is non-empty if and only if* $\mathrm{Rec}(A) \cap -C \subset \ell(C)$.

Suppose that E_1 and E_2 are two real topological vector spaces and C is a convex cone in E_2. Let T be a set-valued map from E_1 to E_2 which means that $T(x)$ is a set in E_2 for each $x \in E_1$. We shall use the following notations for set-valued maps:

$$\operatorname{dom} T = \{x \in E_1 : T(x) \neq \emptyset\}$$

$$\operatorname{graph} T = G(T) = \{(x, y) \in E_1 \times E_2 : y \in T(x), x \in \operatorname{dom} T\}$$

$$\operatorname{epi} T = \{(x, y) \in E_1 \times E_2 : y \in T(x) + C, x \in \operatorname{dom} T\}$$

Definition 4.1.8: [59:p. 33]: Let X be subset of $\operatorname{dom} T$. We say that

(i) T is upper C-continuous at $x_0 \in X$ if for each neighbourhood V of $T(x_0)$ in E_2, there is a neighbourhood U of x_0 in E_1 such that

$$T(x) \subset V + C, \quad \text{for all } x \in U \cap \operatorname{dom} T;$$

(ii) T is lower C-continuous at $x_0 \in X$ if for any $y \in T(x_0)$, any neighbourhood V of y in E_2, there is a neighbourhood U of x_0 in E_1 such that

$$T(x) \cap (V + C) \neq \emptyset, \quad \text{for each } x \in U \cap \operatorname{dom} T;$$

(iii) T is C-continuous at x_0 if it is upper and lower C-continuous at that point; and T is upper (respectively, lower) C-continuous on X if it is upper (respectively, lower) C-continuous at every point of X;

(iv) T is C-closed if $\operatorname{epi} T$ is closed;

(v) whenever "N" denotes some property of sets in E_2, we say that T is "N"-valued on X if $T(x)$ has the property "N", for every $x \in X$.

We now state the following result on upper C-continuous maps.

Theorem 4.1.5: [59:p. 34]: *Suppose that X is a compact set in E_1 and T is an "N"-valued, upper C-continuous map from E_1 to E_2 with $X \subset \operatorname{dom} T$, where "$N$" may be C-closed, C-bounded, C-compact or C-semi-compact. Then $T(x)$ has the property "N" in E_2.*

4.2. Vector Optimization Problems

Let X be a non-empty subset of a topological vector space F and T: $X \to 2^E$, a set-valued mapping from X into E, where E is a real topological vector space which is ordered by a convex cone C.

The general vector optimization problem corresponding to X and T will be denoted by (VP) and we shall wrtite it as follows:

$$\min T(x)$$
$$\text{s.t. } x \in X.$$

This amounts to finding a point $x \in X$, called an optimal (or minimal, or efficient) solution of (VP), such that

$$T(x) \cap \min(T(X)|C) \neq \emptyset,$$

where $T(X)$ is the union of $T(x)$ over X. The elements of $\min(T(X)|C)$ are also called optimal values of (VP). The set of all solutions of (VP) will be denoted by $S(X;\,T)$. We shall get the notions of $IS(X;\,T)$, $\Pr S(X;\,T)$ and $W\,S(X;\,T)$ replacing $\min(T(X)|C)$ by $\mathrm{I}\min$, $\Pr\min$, and $\mathrm{W}\min$ respectively. The set X is sometimes called the set of alternatives and $T(X)$ is the set of outcomes.

Problems with set-valued data arose originally in the theory of vector optimization. For a vector problem, its dual constructed by several means, is a problem whose objective function is set-valued, whatever might be the objective of the primal problem. On the other hand, several objects appeared in the theory of optimization, non-smooth analysis etc., have the set-valuedness as an inherent property. For instance, the sets of subdifferentials, tangent cones in non-smooth analysis, or the sets of feasible solutions, optimal solutions in parametric programming are all set-valued maps. Therefore optimization theory for set-valued maps are of increasing interest.

As in mathematical programming, additional constraints are often imposed on the set X. Namely, let E_1 be a topological vector space and K a convex cone in it. Let G and H be two set-valued maps from X to E_1. Two kinds of constraints generalize the inequality constraint in scalar programming:

$$\{x \in X, G(x) \cap -K \neq \emptyset\}, \tag{4.2.1}$$
$$\{x \in X, H(x) \subset -K\}. \tag{4.2.2}$$

These constraints may be obtained from each other by an appropriate re-definition of the maps G and H. We can show this by supposing first that (4.2.1) is given. Then define H by the rule:

$$H(x) = G(x) \cap -K \quad \text{if} \quad G(x) \cap -K \neq \emptyset, \tag{4.2.3}$$
$$H(x) = G(x) \quad \text{otherwise.}$$

It is easy to show that $\operatorname{dom} H$ and $\operatorname{dom} G$ coincide and the sets defined by (4.2.1) and (4.2.2) are the same. Conversely, suppose that (4.2.2) is given. The G can be defined as follows:

$$G(x) = H(x) \quad \text{if} \quad H(x) \subset -K, \tag{4.2.4}$$
$$G(x) = H(x) \backslash -K \quad \text{otherwise.}$$

The same conclusion is true for the sets of (4.2.1) and (4.2.2) with G being defined by (4.2.4). The only disadvantage of this construction is that neither convexity nor continuity properties are preserved, for instance if $H(x)$ is convex, then it is not necessary for $G(x)$ of (4.2.4) to be convex.

The following proposition gives a helpful relationship between efficient, properly efficient, and weakly efficient solutions of (VP).

Proposition 4.2.1: [59:p. 58]: *We have the following inclusions for a (VP):*

$$\operatorname{Pr} S(X;\ T) \subset S(X;\ T) \subset \operatorname{W}S(X;\ T).$$

Furthermore, if IS(X; T) is non-empty, then

$$\operatorname{I}S(X;\ T) = S(X;\ T).$$

Proof: These inclusions are immediate from the definition of the sets above and from Proposition 4.1.1. □

We shall present below some existence results for optimal solutions of (VP). We shall first state a lemma concerning C-complete sets.

Lemma 4.2.1: [59:p. 58]: *Suppose that X is a non-empty compact subset of F, C is a convex cone in E and T: $X \to 2^E$ is upper C-continuous (see Definition 4.1.8) on X with $T(x) + C$ being closed C-complete for every $x \in X$. Then $T(X)$ is C-complete.*

Theorem 4.2.1: [59:p. 59]: *Assume all the hypotheses of Lemma 4.2.1. If, in addition, C is correct, then the set $S(X;\ T)$ is non-empty.*

Proof: By Lemma 4.2.1, $T(X)$ is a C-complete set. The result follows by applying Theorem 4.1.1 to the set $T(X)$. □

Theorem 4.2.2: [59:p. 59]: *Suppose that X is a non-empty compact subset of F, C is a correct cone in E and T: $X \to 2^E$ is upper C-continuous (see Definition 4.1.8) and C-semi-compact-valued on X. Then the set $S(X;\ T)$ is non-empty.*

Proof: Invoke the theorem to Theorem 4.1.5 and Theorem 4.1.1. □

We return now to the case where the problem is with explicit constraint of the form in (4.2.1):

$$\begin{aligned} &\min T(x) \\ &\text{s.t. } x \in X,\ G(x) \cap -K \neq \emptyset. \end{aligned} \tag{GP}$$

Definition 4.2.1: [59:p. 59]: A point $x_0 \in X$ is said to be a feasible solution of (GP) if $G(x_0) \cap -K \neq \emptyset$.

The set of feasible solutions of (GP) is denoted by X_0.

Proposition 4.2.2: [59:pp. 59–60]: *Let x_0 be a feasible solution of (GP). Then $a_0 \in T(x_0)$ is an optimal value of (GP) if and only if*

$$(a_0 - C \setminus \ell(C), -K) \cap (T(x), G(x)) = \emptyset, \tag{4.2.5}$$

for all $x \in X$.

In the rest of this section, we shall take up the special case where T is a point-valued map. We shall use f instead of T to indicate this case.

Definition 4.2.2: [59:p. 61]: We say that (VP) is

(i) a linear problem if X is a polyhedral set, C is a polyhedral cone and f is a linear map;
(ii) a convex problem (respectively, strictly convex, quasi-convex, strictly quasi-convex problem) if X is convex and f is C-convex (respectively, strictly C-convex, C-quasi-convex, strictly C-quasi-convex) (see Definition 6.1 in [59:p. 29]).

Proposition 4.2.3: [59:p. 61]: *Let (VP) be a linear problem. Then $S(X; f)$ is non-empty if and only if* $\text{Rec}(f(X)) \cap -C = \{0\}$.

Proof: Invoke this proposition to Theorem 4.1.4. □

Proposition 4.2.4: [59:p. 61]: *For (VP) being strictly convex or strictly quasi-convex, the sets $WS(X; f)$ and $S(X; f)$ coincide.*

4.3. A Vector Variational Inequality and Optimization Over an Efficient Set

Multiobjective optimization problems arise in many applications. Usually, only efficient solutions need be considered as possible optima. One natural approach is to optimize a suitable utility function over the set of efficient solutions. This set is not generally convex, even when the given multiobjective problem is convex.

Philip [65] has considered some special multiobjective problems, and proposed an algorithm. More recently, Benson [4,5] has discussed the optimization of a linear utility function over the set of efficient solutions, and has proposed an algorithm.

4.3.1. Definitions and Preliminary Results

Let X be a real linear topological space, and let (Y, S) be a real topological linear space with a partial order $\geq$ induced by a pointed closed convex cone S, with non-empty interior int S; thus $y_1 \geq y_2 \Leftrightarrow y_1 - y_2 \in S$. Let $S_0 = S\backslash 0\}$. Let $L(X, Y)$ be the space of continuous linear operators from X into Y. Let $C \subset X$ be a non-empty convex set. Let f: $C \to Y$ be a mapping, and let G: $C \to 2^Y$ be a point-to-set mapping. If $\phi \in X'$, the dual space of X, then ϕx denotes the evaluation of ϕ at $x \in X$.

Definition 4.3.1: The mapping f: $C \to Y$ is S-convex if for all $x_1, x_2 \in C$ and for each $\alpha \in (0, 1), \alpha f(x_1) + (1 - \alpha)f(x_2) \in f(\alpha x_1 + (1 - \alpha)x_2) + S$.

The point-to-set mapping G: $C \to 2^Y$ is S-convex if for all $x_1, x_2 \in C$ and for each $\alpha \in (0, 1), \alpha G(x_1) + (1 - \alpha)G(x_2) \subset G(\alpha x_1 + (1 - \alpha)x_2) + S$.

Definition 4.3.2: Let $C \subset X$ be a non-empty set, $x_0 \in C$, and G: $C \to L(X, Y)$ a point-to-set mapping. A generalized vector variational inequality is the problem of finding a vector $x_0 \in C$ and a linear operator $A \in G(x_0)$ such that for all $x \in C$, $A(x - x_0) \notin -\text{int}\, S$.

Consider now a multi-objective optimization problem:

$$W\min f(x) \quad \text{subject to } x \in C, \tag{4.3.1}$$

where $C \subset X$ is a non-empty set, f: $X \to Y$ is a mapping, and $W\min$ denotes weak minimum.

Definition 4.3.3: A point x_0 is a weak minimum, or weak efficient point, for problem (4.3.1) if for all $x \in C$, $f(x) - f(x_0) \notin -\operatorname{int} S$. The set of all weak minimum points for (4.3.1) is denoted by C_E.

Definition 4.3.4: A linear operator $A \in L(X, Y)$ is a weak subgradient of $f: C \to Y$ (where $C \subset X$) at $x_0 \in C$ if for all $x \in C$, $f(x) - f(x_0) - A(x - x_0) \notin -\operatorname{int} S$.

The weak subdifferential of f at x_0 is the set $\partial_W f(x_0)$ of all weak subgradients of f at x_0. A linear operator $A \in L(X, Y)$ is a strong subgradient of $f: C \to Y$ at $x_0 \in C$ if for all $x \in C$, $f(x) - f(x_0) - A(x - x_0) \in S$. The set of all strong subgradients of f at x_0 is denoted by $\partial_S f(x_0)$.

Since S is a pointed cone with non-empty interior, $\partial_S f(x_0) \subset \partial_W f(x_0)$ for all $x_0 \in C$.

Definition 4.3.5 [42]: A topological vector space Y, partially ordered by a convex cone S, is order-complete if every subset A which has an upper bound b in terms of the ordering (that is, for all $y \in A, b - y \in S$) then has a supremum $\hat{b}$ (that is, there exists $\hat{b} \in Y$ such that $\hat{b}$ is an upper bound to A, and each upper bound b to A satisfies $-\hat{b} \in S$).

Remark 4.3.1: From Definition 4.3.5, a similar statement holds, replacing upper bound by lower bound, and supremum by infimum. It is well known that $\mathbb{R}^n$, with an order cone S having exactly n generators, is thus order-complete; but that $C(I)$ (the space of continuous functions on an interval I, with the uniform norm) is not order-complete.

4.3.2. Existence of Subgradients

In this section, the existence of weak and strong subgradients is presented. In order to present the existence of strong subgradients, a generalized Hahn-Banach extension theorem is introduced. This result (Theorem 4.3.2) is a generalized version of Giles' "Hahn-Banach dominated extension theorem" for functionals [36].

Lemma 4.3.1: *Let $C \subset X$ be a convex set, with $\operatorname{int} C \neq \emptyset$; let $f: C \to Y$ be an S-convex mapping, continuous at some point $x_0 \in \operatorname{int} C$; let $\operatorname{int} S \neq \emptyset$. Then the set*

$$\operatorname{epi} f := \{(x, y) \in X \times Y : x \in C, y - f(x) \in S\}$$

is convex, and $\operatorname{int} \operatorname{epi} f \neq \emptyset$.

For proof we refer to [Lemma 1 [13]].

Theorem 4.3.1: *Let $C \subset X$ be convex, with $\operatorname{int} C \neq \emptyset$; let the cone S be pointed, with $\operatorname{int} S \neq \emptyset$; let $F: C \to Y$ be an S-convex mapping, continuous at $x_0 \in \operatorname{int} C$. Then there exists a weak subgradient B of F at x_0, satisfying the further condition that $Bz \notin -\operatorname{int} S \Leftrightarrow Bz \in S$. (Continuity of F need not be assumed in finite dimensions.)*

Theorem 4.3.2: *Let X be a real linear topological space, and let (Y, S) be a real order complete linear topological space, with order cone S. Let $C \subset X$ be convex, with $\operatorname{int} C \neq \emptyset$. Let the mapping $F: C \to Y$ be S-convex, and let X_0 be a proper subspace of X, with $X_0 \cap \operatorname{cor} C \neq \emptyset$. Let $h: X_0 \to Y$ be a continuous affine mapping such that $h(x) \leq F(x)$ (in terms of S), for each $x \in X_0 \cap C$. Then there exists a continuous affine $k: X \to Y$ such that $k(x) = h(x)$ for all $x \in X_0 \cap C$, and $k(x) \leq F(x)$ (in terms of S), for all $x \in C$.*

For proof we refer to [Theorem 2 [13]].

The following result can be proved by following similar arguments as in the proof of Theorem 4.3.2.

Theorem 4.3.3: *Let X be a real linear topological space; let (Y, S) be an order-complete real partially ordered linear topological space; let $C \subset X$ be convex, with non-empty interior, and let $x_0 \in \operatorname{int} C$; let F: $X \to Y$ be a S-convex mapping. Then there exists a continuous affine mapping h: $X \to Y$ such that $h(x_0) = F(x_0)$, and $h(x) \leq F(x)$ for all $x \in C$.*

Theorem 4.3.4: *Let X, Y, S, C, x_0 and F be as in Theorem 4.3.2. Then there exists a strong subgradient of F at $x_0 \in \operatorname{int} C$. If S is a pointed cone, then there also exists a weak subgradient of F at x_0.*

4.3.3. Equivalence of the Weak Minimization Problem (4.3.1) and a Vector Variational Inequality

In this section, we shall present that the weak minimization problem (4.3.1) is equivalent, under some restrictions, to a generalized vector variational inequality.

Consider the following *generalized variational inequality*: Given non-empty $C \subset X$ and S-convex f: $C \to Y$, find $x_0 \in C$ and $B \in \partial_W f(x_0)$ such that

$$B(x - x_0) \notin -\operatorname{int} S, \quad \text{for all } x \in C. \tag{4.3.2}$$

Theorem 4.3.5: *Let $C \subset X$ be convex with $\operatorname{int} C \neq \emptyset$; let the convex cone S be pointed, and let $\operatorname{int} S \neq \emptyset$; let f: $C \to Y$ be S-convex, and continuous at a point $x_0 \in \operatorname{int} C$. If x_0 is a weak minimum of the multi-objective optimization problem (4.3.1), then x_0 solves the generalized variational inequality (4.3.2). Conversely, if (x_0, A) solves (4.3.2), where also $A \in \partial_S f(x_0)$, then x_0 is a weak minimum of (4.3.1).*

Remark 4.3.2: If (x_0, B) solves (4.3.2), but $B \in \partial_W f(x_0) \backslash \partial_S f(x_0)$, then x_0 is not necessarily a weak minimum for (4.3.1). The above proof does not extend, since $W + W$ is not contained in W.

If $B \in \partial_W f(x_0)$ and $B(x - x_0) \in S$ for all $x \in C$ then, by a similar proof, x_0 is a weak minimum of (4.3.1).

If x_0 is a weak minimum of (4.3.1), then usually $0 \in \partial_S f(x_0)$ does not hold, i.e., when the constraint $x_0 \in C$ is inactive, thus when $x_0 \in \operatorname{int} C$. However, suppose also that C is convex, and f is (Fréchet or linear Gáteaux) differentiable at x_0 (as f is S-convex). A weak minimum of (4.3.1) at x_0 implies that $f(x) - f(x_0) \in W := Y \backslash (-\operatorname{int} S)$. Let $x_0 + v \in C$; since C is convex, $x_0 + \alpha v \in C$ for all $\alpha \in (0, 1)$. Hence

$$[f(x_0 + \alpha v) - f(x)]/\alpha \in W.$$

Since W is closed, and f is differentiable at x_0, it follows that $f'(x_0)(x - x_0) \in W$; x_0 and $f'(x_0)$ satisfy the generalized variational inequality (4.3.2), and $f'(x_0) \in \partial_S f(x_0)$.

Theorem 4.3.6: *Let $C \subset X$ be convex; let f: $C \to Y$ be S-convex and linearly Gáteaux differentiable at x_0. Then $x_0 \in C$ is a weak minimum of (4.3.1) if and only if x_0 and $f'(x_0)$ solves (4.3.2).*

4.3.4. Optimization Over an Efficient Set

Given $C \subset C, f\colon C \to Y$, and $\phi\colon X \to \mathbb{R}$, consider the following problems:

(I) *Multi-objective optimization*: W min $f(x)$ subject to $x \in C$. Denote the set weak minimum (= efficient) points for this problem by E.

(II) *Generalized vector variational inequality*: Find $x_0 \in C$ and $A \in \partial_W f(x_0)$ such that $A(x - x_0) \notin -\text{int}\, S$ for all $x \in C$. Denote the set of optima for this problem by V.

(III) *Optimization over an efficient set*: Minimize $\phi(x)$ subject to $x \in E$.

(IV) *Linearized problem*: W min $f'(x_0)(x - x_0)$ subject to $x \in C$.

(V) W min$[\phi(x), f'(x_0)(x - x_0)]$ subject to $x \in C$.

(VI) Minimize $\phi(x)$ subject to $x \in V$.

(VII) Minimize $\phi(x)$ subject to $f'(x_0)(x - x_0) \in W \equiv Y\backslash(-\text{int}\, S)$ for all $x \in C$.

For (IV), (V) and (VII), f is assumed (linearly Gâteaux) differentiable. The weak minimization is with respect to the cone S, or $\mathbb{R}_+ \times S$ for (V).

Theorem 4.3.7: *Let $C \subset X$ be closed convex, let* int S *be non-empty, and let $f\colon X \to Y$ be S-convex and (Fréchet or linearly Gâteaux) differentiable, with derivative $f'(x_0)$ at $x_0 \in X$. Then:*

(*a*) *Problem (I) is equivalent to problem (IV); if also f is (linearly Gâteaux) differentiable at x_0, then (I) is equivalent to (II).*

(*b*) *If f is (linearly Gâteaux) differentiable at x_0, then (III) is equivalent to (VI).*

(*c*) *If x_0 is an optimum for (III), then x_0 is an optimum for (V). (So, any necessary conditions for an optimum of (V) hold also for (III).)*

(*d*) *Problem (III) is equivalent to problem (VI).*

4.3.5. Kuhn–Tucker Necessary Conditions for Optimization Over an Efficient Set

Theorem 4.3.8: *Kuhn–Tucker necessary conditions for the point x_0 to be an optimum for problem (III) are:*

$$\alpha^T \phi'(x_0) + \zeta^T f'(x_0) \in N_C(x_0),\ \ \alpha \in \mathbb{R}_+,\ \ \zeta \in S^*,\ \ (\alpha, \zeta) \neq (0,0),$$

where $N_C(x_0)$ denotes the normal cone to C at x_0.

Consider now the constraint $-g(x) \in T$, where T is a closed convex cone, and g is a differentiable vector function. Replacing $x \in C$ by $-g(x) \in T$, and assuming a constraint qualification holds for this constraint at x_0, necessary Kuhn–Tucker conditions for $x = z$ to be a weak minimum of $f(x)$, subject to $-g(x) \in T$, are:

$$(Q)\quad \tau^T f'(z) + \rho^T g'(z) = 0,\ \tau \in S^*,\ \rho \in T^*,\ \tau^T e = 1,\ -g(z) \in T,\ \rho^T g(z) = 0,$$

where e is any constant vector in int S, so that $\tau^T e = 1$ ensures that $\tau \neq 0$. (If $S = \mathbb{R}^p_+$ then $e = (1, 1, \ldots, 1)^T$ may be chosen.) Denote now by K the set of weak minima for $f(x)$, subject to $-g(x) \in T$.

Consider now the problem:

$$(H)\quad \text{Minimize } \phi(z) \quad \text{subject to } z \in K.$$

Assume now also that f is S-convex, and g is T-convex; then the necessary conditions (Q) for $x \in K$ are also sufficient. (note that the hypotheses could be reduced; it suffices if f is S-pseudo-convex and g is T-quasi-convex.) Under these assumptions, problem (H) is equivalent to minimizing $\phi(z)$ subject to constraints (Q). A Langrangian function for this latter problem is:

$$\phi(z) + \theta[\tau^T f'(z) + \rho^T g'(z)] + \lambda^T g(z) - \tau^T\sigma - \rho^T\omega + \beta[\tau^T e - 1] + \delta\rho^T g(z).$$

Consequently, Kuhn–Tucker necessary conditions (assuming a constraint qualification) for z_0 to minimize $\phi(z)$ over the efficient set K are that Lagrange multipliers $\lambda \in T^*, \theta \in \mathbb{R}, \sigma \in S, \omega \in T, \beta \in \mathbb{R}, \delta \in \mathbb{R}$ exist, satisfying the conditions:

$$\begin{aligned}
&\phi'(z_0) + \theta[\tau^T f''(z_0) + \rho^T g''(z_0)] + \lambda^T g'(z_0) + \delta\rho^T g(z_0) = 0;\\
&\theta f'(z_0) - \sigma + \beta e = 0;\\
&\theta g'(z_0) - \omega + \delta g(z_0) = 0;\\
&\lambda^T g(z) = 0;\\
&\lambda^T \sigma = 0;
\end{aligned}$$

Consider, in particular, the *multi-linear* (linear multi-objective) problem:

$$\text{W min } Mx \quad \text{subject to } Ax - b \leq 0. \tag{L}$$

Here M is an $r \times n$ matrix, and A is an $m \times n$ matrix; $x \in \mathbb{R}^n$. Denote by E_L the set of weak minima for problem (L), minimizing with respect to the cone $\mathbb{R}^r_+$. Then

$$\begin{aligned}
x \in E_L \Leftrightarrow &[Ax - b \leq 0,\ \tau^T M + \rho^T M + \rho^T A = 0,\ \tau \in \mathbb{R}^r_+,\ \rho \in \mathbb{R}^m_+,\\
&\tau^T e = 1,\ \rho^T(Ax - b) = 0].
\end{aligned}$$

The requirement $\tau^T e = 1$ ensures that $\tau \neq 0$. If $c^T x$ is another (real) objective function, then the problem of minimizing $c^T x$ over $x \in E_L$ is equivalent to the problem:

Minimize $c^T x$ subject to

$$Ax - b \leq 0,\ \tau^T M + \rho^T A = 0,\ \tau \in \mathbb{R}^r_+,\ \rho \in \mathbb{R}^m_+,\ \tau^T e = 1,\ \rho^T(Ax - b) = 0].$$

This last problem fails to be a linear program, because of the complementary slackness constraint $\rho^T(Ax - b) = 0$. The minimization is with respect to the variables x, τ, ρ. Kuhn–Tucker necessary conditions for a minimum are that Lagrange multipliers $\lambda \in \mathbb{R}^m_+, \theta \in \mathbb{R}, \sigma \in \mathbb{R}^r_+, \beta \in \mathbb{R}, \delta \in \mathbb{R}$ exist, satisfying the constraints:

$$\begin{aligned}
&C^T + \lambda^T A = 0,\ \theta M - \sigma + \beta e = 0,\ \theta A - \omega + \gamma(Az_0 - b) = 0, \lambda^T(Az_0 - b) = 0,\\
&\tau^T\sigma = 0, \rho^T\omega = 0.
\end{aligned}$$

5. ESSENTIAL WEAK EFFICIENT SOLUTION IN MULTIOBJECTIVE OPTIMIZATION PROBLEMS

5.1. Introduction

Let X be a Hausdorff space, 2^X a space of all non-empty compact subsets of X, and $C^m(X)$ a space of all bounded continuous functions from X to R^m. Suppose that 2^X is endowed with the Vietoris topology (see Definition 5.2.2 below) and $C^m(X)$ is topologized by the usual uniform convergence norm. Let $C \subset \mathbb{R}^m$ be a convex cone with $\text{int}\, C \neq \emptyset$. Consider $P = 2^X \times C^m(X)$. Then every pair $p = (A, f) \in P$ determines a multiobjective optimization problem with respect to C:

$$\min_{x \in A} f(x).$$

If $f(x) \in \min(f(A)|\{0\} \cup \text{int}\, C)$, then $x \in A$ is a weakly efficient solution with respect to C as mentioned in [59]. In other words, $x \in A$ is a weakly efficient solution if there is no $x' \in A$ such that $f(x') - f(x) \in \text{int}\, C$. Denote by $F(p)$, the set of all weakly efficient solutions of $p \in P$. $x \in F(p)$ is said to be an essential weak efficient solution if for each open neighbourhood $N(x)$ of x, there exists an open neighbourhood $O(p)$ such that for any $p' \in O(p)$, there exists $x' \in F(p')$ and $x' \in N(x)$. A problem $p \in P$ is said to be almost essential (respectively, essential) if there is an $x \in F(p)$ which is essential (respectively, if all $x \in F(p)$ are essential).

In this section, we shall present the results which will show that the multi-valued mapping F is almost lower semi-continuous (respectively, lower semi-continuous) at $p \Leftrightarrow$ the problem p is almost essential (respectively, essential). When X is a Cech-complete space from the class $\mathcal{L}$, there exists a dense G_δ subset Q of P such that F is almost lower semi-continuous on Q; in other words, most of the multi-valued optimization problems (in the sense of Baire category) are almost essential, i.e., they have at least one essential weak efficient solution. When X is a complete metric space, there exists a dense G_δ subset Q' of P such that F is lower semi-continuous on Q'; in other words, most of the multi-valued optimization problems (in the sense of Baire category) are essential, i.e., their weakly efficient solutions are all essential.

Consequently, most of the optimization problems will have unique solutions [3,48].

5.2. Preliminaries

Let X be a Hausdorff space and $\mathcal{P}_0(X)$ be the space of all non-empty subsets of X.

Lemma 5.2.1: [43] *Let $A_n \in 2^X (n = 1, 2, 3 \ldots)$, and $A \in 2^X$ be such that $A_n \to A$. Then we have the following:*

(1) *If $G \supset A$ is open, then there exists N such that for all $n \geq N, G \supset A_n$.*
(2) *If $x_n \in A_n (n = 1, 2, 3 \ldots)$, and $x_n \to x$, then $x \in A$.*
(3) *If $x' \in A$, then for any open neighbourhood G' of x', there exists N' such that for all $n \geq N', G' \cap A_n \neq \emptyset$.*
(4) $\bigcup_{n=1}^{\infty} A_n \cup A \in 2^X$.

Definition 5.2.2: [49:pp. 7,8] Let $(X, \mathcal{T})$ be a topological space and let G denote an arbitrary non-empty open subset of X (i.e., member of $\mathcal{T}$). Consider $\mathcal{P}_0(X) = \mathcal{P}(X) \setminus \{\emptyset\}$. Then the family

$$\mathcal{U} = \{[\cdot, G] = \{U \in \mathcal{P}_0(X) : U \subset G\} : G \in \mathcal{T}\},$$

consisting of all the closed initial segments (see Definition 1.1.3 in [49:p. 5]) determined by such G, is a base for a topology. Also the family

$$\mathcal{L} = \{I_G = \{U \in \mathcal{P}_0(X) : U \cap G \neq \emptyset\} : G \in \mathcal{T}\}$$

form a subbase for a topology on $\mathcal{P}_0(X)$.

From the above two topologies, we shall now define the following three topologies on $\mathcal{P}_0(X)$:

(a) The upper topology on $\mathcal{P}_0(X)$ is that generated by the base $\mathcal{U}$ where $\mathcal{U} = \{[\cdot, G] : G \in \mathcal{T}\}$. This topology is denoted by $\mathcal{T}_{\mathcal{U}}$.
(b) The lower topology on $\mathcal{P}_0(X)$ is that generated by the subbase $\mathcal{L}$. This topology is denoted by $\mathcal{T}_{\mathcal{L}}$.
(c) The Vietoris topology $\mathcal{T}_{\mathcal{V}}$ on $\mathcal{P}_0(X)$ is that generated by $\mathcal{U}$ and $\mathcal{L}$ together.

Definition 5.2.3: [43] X is said to be a Cech-complete space if it can be embedded as a G_δ subset of some compact Hausdorff space.

Lemma 5.2.4: [43] *If X is a Cech-complete space, then $P = 2^X \times C^m(X)$ is Cech-complete.*

Definition 5.2.5: Let Y be a Hausdorff space and F: $Y \to \mathcal{P}_0(X)$. Then

(1) F is said to be upper semi-continuous (u.s.c) at $y \in Y$, if for each open set G with $F(y) \subset G$, there exists an open neighbourhood $O(y)$ of y such that for each $y' \in O(y), F(y') \subset G$.
(2) F is said to be lower semi-continuous (l.s.c) at $y \in Y$, if for each open set G with $F(y) \cap G \neq \emptyset$, there exists an open neighbourhood $O(y)$ of y such that for each $y' \in O(y), F(y') \cap G \neq \emptyset$.
(3) F is said to be almost lower semi-continuous (a.l.s.c) at $y \in Y$, if there exists $x \in F(y)$ such that for each open neighbourhood $N(x)$ of x there exists an open neighbourhood $O(y)$ of y with $N(x) \cap F(y') \neq \emptyset$, for each $y' \in O(y)$.
(4) F is said to be an usco map, if F is upper semi-continuous on Y and $F(y)$ is compact for each $y \in Y$.

Lemma 5.2.6: (Theorem 2 of [33]) *Let X be a metric space, Y a Baire space, and F: $Y \to \mathcal{P}_0(X)$ an usco map. Then there is a dense G_δ subset Q' of Y such that F is l.s.c. on Q'.*

Definition 5.2.7: [3,48] Let X be a Hausdorff space. Then X is said to belong to the class $\mathcal{L}$ if for every Cech-complete space Y and for every usco map F: $Y \to \mathcal{P}_0(X)$, F is l.s.c. on some dense G_δ subset of Y.

Each completely metrizable space is in $\mathcal{L}$ (Lemma 5.2.6) and each Banach space with its weak topology (which is non-metrizable) is in $\mathcal{L}$ (Theorem 2 of [19]).

5.3. Some Results on Essential Weak Efficient Solution in Multi-objective Optimization Problems

Lemma 5.3.1: [43] *For each $p \in P$, $F(p) \in 2^X$ where F is as defined in Definition 5.2.5.*

Lemma 5.3.2: [43] *Let F be defined as in Definition 5.2.5. Then F is u.s.c. on P.*

Theorem 5.3.3: [43] *Let F be defined as in Definition 5.2.5. Then*

(1) *F is a.l.s.c at $p \in P \Leftrightarrow p$ is almost essential.*
(2) *F is l.s.c at $p \in P \Leftrightarrow p$ is essential.*

Theorem 5.3.4: [43] *Let F be defined as in Definition 5.2.5. Suppose $F(p) = \{x\}$. Then p is essential.*

Theorem 5.3.5: [43] *Suppose F is defined as in Definition 5.2.5. Let X be a Cech-complete space from the class $\mathcal{L}$. Then there is a G_δ subset Q of P such that F is a.l.s.c. on Q; in other words, most of the multi-objective optimization problems (in the sense of Baire category) are almost essential; i.e., they have at least one essential weak efficient solution.*

Remark 5.3.6: [43] If X is a compact Hausdorff space and belongs to the class $\mathcal{L}$, then Theorem 5.3.5 holds.

Theorem 5.3.7: [43] *Suppose F is defined as in Definition 5.2.5. Let X be a complete metric space. Then there is a dense G_δ subset Q' of P such that F is l.s.c. on Q'; in other words, most of the multi-objective optimization problems (in the sense of Baire category) are essential; i.e., their weakly efficient solutions are all essential.*

5.4. Applications of "Essential Weak Efficient Solution in Multi-objective Optimization Problems"

Let $m = 1$, and $C = (-\infty, 0)$ be a convex cone in $\mathbb{R}$. Then $\operatorname{int} C = (-\infty, 0)$. Clearly, we obtain the following lemma.

Lemma 5.4.1: [43] *x is a weakly efficient solution with respect to $C \Leftrightarrow x$ is a solution of* $\min_{x \in A} f(x)$.

Theorem 5.4.2: [43] *If* $\min_{x\in A}$ *$f(x)$ has a unique solution, then p is essential. If X is Cech-complete and p is almost essential, then* $\min_{x\in A}$ *$f(x)$ has a unique solution.*

Theorem 5.4.3: [3] *Let X be a Cech-complete space which belongs to the class $\mathcal{L}$. Then there is a dense G_δ subset Q of P each of whose elements has a unique solution.*

Remark 5.4.4: [43] If X is a compact Hausdorff space which belongs to the class $\mathcal{L}$, then Theorem 5.4.3 holds ([48]).

5.5. Approximate Dual and Approximate Vector Variational Inequality for Multiobjective Optimization

An approximate dual has been presented for a multi-objective optimization problem. The approximate dual has a finite feasible set, and is constructed without using a perturbation. An approximate weak duality theorem and an approximate strong duality theorem are presented, and also an approximate variational inequality condition for efficient multi-objective solutions are surveyed.

5.5.1. Introduction

Consider the finite dimensional linear space $\mathbb{R}^p$, equipped with a partial ordering $\geq_S$, defined by a closed pointed convex cone S, with interior $\operatorname{int} S \neq \emptyset$.

Definition 5.5.1: (See [58]) For a given $e \in S$, a point $c \in A \subset \mathbb{R}^p$ is said to be an e-[weak] minimum of A if there exists no $x \in A$ satisfying $0 \neq c - x - e \in S[\in \text{int}\, S]$. A point $c \in A$ is said to be an e-[weak] maximum if there exists no $x \in A$ such that $0 \neq x - c - d \in S[\in \text{int}\, S]$. When $e = 0$, an e-[weak] minimum (respectively, e-[weak] maximum) is said to be a [weak] minimum (respectively, [weak] maximum). A set W is said to e-upper dominate a set V if

$$(\forall v \in V)(\exists w \in W) w + e - v \in S.$$

A set W is said to e-strongly upper dominate a set V if

$$(\forall v \in V)(\exists w \in W) w + e - v \in \text{ int}\, S.$$

The order-interval $[-e, e] := \{x \in \mathbb{R}^p : e \geq_S x \geq_S -e\}$. Denote its interior by $(-e, e)$. Denote by $\mathbb{R}^{p \times m}$ the space of all $p \times m$ real matrices Λ, and by $\|\Lambda\|$ the norm of Λ in this space. Let $D \subset \mathbb{R}^m$ be a closed convex cone. Let $M := \{\Lambda \in \mathbb{R}^{p \times m} : \|\Lambda\| = 1, \Lambda(D) \subset S\}$. Then M is compact.

A function h: $C \to \mathbb{R}^p$ is S-convex if $C \subset \mathbb{R}^n$ is convex, and

$$(\forall x, y \in C, \forall \alpha \in (0,1)) \quad h(\alpha x + (1-\alpha)y) \leq_S \alpha h(x) + (1-\alpha) h(y).$$

Consider a nonlinear multi-objective optimization problem:

$$\text{W}\min\, f(x) \quad \text{subject to } x \in X := \{x \in E, -g(x) \in D\}, \tag{P}$$

where f: $\mathbb{R}^n \to \mathbb{R}^p$, g: $\mathbb{R}^n \to \mathbb{R}^m$ are vector functions, $E \subset \mathbb{R}^n$, and weak minimum (W min; and W max later) are as in Definition 5.5.1.

Definition 5.5.2: A vector valued Lagrangian for (P) is $L(x, \Lambda) := f(x) + \Lambda g(x)$. Following Sawaragi *et al.* [67], it will be assumed that

(i) E is non-empty compact,
(ii) f is continuous, and
(iii) g is continuous.

Under these assumptions, it is readily shown that X and $f(X)$ are compact.

Definition 5.5.3: The dual map Φ: $M \to \mathbb{R}^p$ is defined by

$$\Phi(\Lambda) := \text{W}\min\, L(E, \Lambda),$$

for each $\Lambda \in M$. A dual problem [67] to (P) is

$$\text{W}\max \bigcup_{\Lambda \in M} \Phi(\Lambda). \tag{D_{TS}}$$

Lemma 5.5.1: *Under the above assumptions, for each $\Lambda \in M$, the sets $L(E, \Lambda)$ and $\Phi(\Lambda)$ are compact.*

Lemma 5.5.2: *The map Φ: $M \to \mathbb{R}^p$ is upper semi-continuous, and $\Phi(M)$ is compact.*

Remark 5.5.1: The conclusions of Lemma 5.5.1 and Lemma 5.5.2 still hold if $\Phi(\Lambda)$ is changed to $e - \text{W}\min\, L(E, \Lambda)$.

5.5.2. Approximate Dual

We now present an approximate dual for the multi-objective problem (P). No perturbation map is required.

Lemma 5.5.3: *Let $A \subset \mathbb{R}^p$ be compact, and $e \in \operatorname{int} S$. Then there exists a finite subset $K \subset A$ such that K e-strongly upper dominates A.*

Remark 5.5.2: Similar results hold with upper replaced by lower.

Since f is continuous and X is compact, $f(X)$ is compact. By Lemma 5.5.3, there exists a finite subset $U \subset f(X)$, such that U e-strongly lower dominates $f(X)$, and thus $(\forall v \in f(X))$ $(\exists u \in U) u - e - v \in -\operatorname{int} S$. So a *primal approximate problem* may be defined as

$$\text{W}\min U. \qquad (P^{\#})$$

For each $\Lambda \in M$, define $\Phi^{\#}(\Lambda)$ to be the set of e-weak minima of $L(E, \Lambda)$. Then $\Phi^{\#}(\Lambda) \neq \emptyset$. Set $Q := \bigcup_{\Lambda \in M} \Phi^{\#}(\Lambda)$. By Lemma 5.5.2, the set Q is compact. By Lemma 5.5.3, there exists a finite subset $W \subset Q$ such that $(\forall q \in Q)(\exists w \in W)\ w - e - q \in -S$. So an *approximate dual problem* may be defined as

$$\text{W}\max W. \qquad (D^{\#})$$

Two obvious corollaries follow.

Corollary 5.5.1: *If u_0 is a weak minimum of $(P^{\#})$, then u_0 is an e-weak minimum of (P).*

Corollary 5.5.2: *If w_0 is a weak maximum of $(D^{\#})$ then w_0 is an e-weak maximum of (D_{TS}).*

Theorem 5.5.1: *(Approximate Weak Duality). For each $u \in U$ and each $w \in W$, $w - u - e \notin S \backslash \{0\}$.*

Proof: If $w \in W$, then $w \in \Phi^{\#}(\Lambda)$ for some $\Lambda \in M$. Then

$$(\forall x \in X) w - e - [f(x) + \Lambda g(x)] \in H := \mathbb{R}^p \backslash \operatorname{int} S.$$

Theorem 5.5.2: *(Approximate Strong Duality). Let $x^* \in X$ and $\Lambda^* \in M$ satisfying $w^* := f(x^*) \in \Phi^{\#}(\Lambda^*) \cap U \cap W$. Then x^* is an e-weak minimum of $f(X)$, and w^* is an e-weak maximum of $\Phi(M)$.*

5.5.3. Approximate Vector Variational Inequality

Let $C \subset \mathbb{R}^n$ be a non-empty convex set, $S \subset \mathbb{R}^p$ a closed convex cone with $\operatorname{int} S \neq \emptyset$, and $f\colon C \to \mathbb{R}^p$ a vector valued function. Denote by $\mathbf{L}(\mathbb{R}^n, \mathbb{R}^p)$ the space of all linear mappings from $\mathbb{R}^n$ into $\mathbb{R}^n$. Let $G\colon C \to \mathbf{L}(\mathbb{R}^n, \mathbb{R}^p)$ be a mapping. A *generalized vector variational inequality* is the problem of finding $x_0 \in C$ and $A \in G(x_0)$ such that $(\forall x \in C) A(x - x_0) \notin -\operatorname{int} S$.

Theorem 5.5.3: *Let $C \subset \mathbb{R}^n$ be compact convex; let $S \subset \mathbb{R}^p$ be a closed convex pointed cone with $\operatorname{int} S \neq \emptyset$; let $f\colon C \to \mathbb{R}^p$ be S-convex, (Fréchet) continuously differentiable at*

$x_0 \in C$, and differentiable on C; let x_0 be a weak minimum of $f(x)$ subject to $x \in C$. Then, for each $e \in \operatorname{int} S$, there exists a finite subset $W \subset C$ such that $(\forall w \in W)$ $f'(x_0)(w - x_0) + e \notin -\operatorname{int} S$.

Consider now the generalized inequality system, for $x \in C$:

$$(\forall u \in U(x)) f'(x)(u - x) - e \notin -\operatorname{int} S. \qquad (GI)$$

Theorem 5.5.4: *Let $f\colon C \to \mathbb{R}^p$ be S-convex and continuously differentiable on C, where $C \subset \mathbb{R}^n$ is a compact convex set; let $e \in \operatorname{int} S$. If x_0 is a solution (GI), then x_0 is a weak minimum of $f(x)$ subject to $x \in C$.*

6. NON-COMPACT PARETO EQUILIBRIA FOR MULTI-OBJECTIVE GAMES

In this section we shall present the results on the existence of weighted Nash-equilibria and Pareto equilibria for non-compact multi-objective games with multi-criteria as applications of the Ky Fan's minimax principle in 1972. Several existence theorems for non-compact weighted Nash-equilibria and Pareto equilibria will also be presented. These will include the corresponding results in the literature as special cases.

6.1. Introduction

Recently, the study of the existence of Pareto equilibria in game theory with vector payoffs has been attracted by many authors, for example, see Bergstresser and Yu [7], Borm *et al.* [8], Ghose and Prasad [34], Szidarovszky *et al.* [70], Tanaka [71], Wang [79], Yu [86], Yu and Yuan [84], Zeleny [89], and references theirin. The motivation for the study of multi-criteria models can be found in Bergstresser and Yu [7], Szidarovszky *et al.* [70], and Zeleny [89]. The existence of Pareto equilibria is one of the fundamental problems in game theory. In this section, we shall present some existence results of Pareto equilibria as applications of the Ky Fan's minimax inequality. The results presented in this section could be regarded as a unified improvement of the corresponding existence results given in the literature.

Now, we shall recall and present some notations and definitions. In this section, we shall consider a finite-players game with multi-criteria in its strategic form (also called a normal form) $G := (X_i, F^i)_{i \in N}$, where $N := \{1, 2, \ldots, n\}$. For each player $i \in N$, X_i is the set of strategies in $\mathbb{R}^{k_i}$ and each F^i is a mapping from $X := \prod_{i \in N} X_i$ into $\mathbb{R}^{k_i}$, which is called the payoff function (or say, multi-criteria) of the i-th player (here k_i is a positive integer). If a strategy $x := (x^1, x^2, \ldots, x^n) \in X$ is played, each player i is trying to maximize his/her payoff function $F^i := (f_1^i(x), f_2^i(x), \ldots, f_{k_i}^i(x))$, which consists of non-commensurable outcomes. Each player i has a preference $\succeq_i$ over the outcome space $\mathbb{R}^{k_i}$. For each player $i \in N$, its preference $\succeq_i$ is given as

$$z^1 \succeq_i z^2 \quad \text{if and only if} \quad z_j^i \geq z_j^2$$

for each $j = 1, 2, \ldots, k_i$, where $z^1 := (z_1^1, z_2^1, \ldots, z_{k_i}^1)$ and $z^2 = (z_1^2, z_2^2, \ldots, z_{k_i}^2)$ are any element in $\mathbb{R}^{k_i}$. The players' preference relations induce the preference on X, defined for each player i, and chooses $x = (x^1, x^2, \ldots, x^n)$ and $y = (y^1, y^2, \ldots, y^n) \in X$ by

$$x \succeq_i y \quad \text{whenever } F^i(x) \succeq_i F^i(y).$$

Also we assume that the model of a game in this section is a non-cooperative game, i.e., there is no communication between players, so players act as free agents; each player is trying to maximize his/her own payoff according to his/her preferences.

For the games with vector payoff functions (or say, multi-criteria), it is well known that in general there does not exist a strategy $\hat{x} \in X$ to maximize (or equivalently, minimize) all f_j^i's for each player; for example, see the reference of Yu [86]. Hence we need to recall some solution concepts for multi-criteria games.

Throughout this section, for each given $m \in N$, we shall denote by $\mathbb{R}_+^m$ the non-negative orthant of $\mathbb{R}^m$, i.e.,

$$\mathbb{R}_+^m := \{u := (u^1, \ldots, u^m) \in \mathbb{R}^m \quad \text{such that } u^j \geq 0 \quad \text{for } j := 1, \ldots, m\},$$

so that the non-negative orthant $\mathbb{R}_+^m$ of $\mathbb{R}^m$ has a non-empty interior with the topology induced in terms of convergence of vectors with respect to the Euclidean metric. That is,

$$\text{int}\,\mathbb{R}_+^m := \{u := (u^1, \ldots, u^m) \in \mathbb{R}^m : u^j > 0 \quad \text{for all } j := 1, \ldots, m\}.$$

For each $i \in N$, denote $X^{\hat{i}} := \prod_{j \in N \setminus \{i\}} X_j$. If $x = (x^1, \ldots, x^n) \in X$, we shall write $x^{\hat{i}} = (x^1, \ldots, x^{i-1}, x^{i+1}, \ldots, x^n) \in X^{\hat{i}}$. If $x^i \in X_i$ and $x^{\hat{i}} \in X^{\hat{i}}$, we shall use $(x^{\hat{i}}, x^i)$ to denote $y = (y^1, \ldots, y^n) \in X$ such that $y^i = x^i$ and $y^{\hat{i}} = x^{\hat{i}}$. Let $\hat{x} = (\hat{x}^1, \ldots, \hat{x}^n) \in X$. Now we have the following definition.

Definition 6.1.1: [87] A strategy $\hat{x}^i \in X_i$ of player i is said to be a Pareto efficient strategy (respectively, a weak Pareto efficient strategy) with respect to $\hat{x}$ if there is no strategy $x^i \in X_i$ such that

$$F^i(\hat{x}) - F^i(\hat{x}^{\hat{i}}, x^i) \in \mathbb{R}_+^{k_i} \backslash 0\} \text{ (respectively, } F^i(\hat{x}) - F^i(\hat{x}^{\hat{i}}, x^i) \in \text{int}\,\mathbb{R}_+^{k_i}).$$

Definition 6.1.2: [87] A strategy $\hat{x} \in X$ is said to be a Pareto equilibrium (respectively, a weak Pareto equilibrium) of a game $G := (X_i, F^i)_{i \in N}$ if, for each player i, $\hat{x}^i \in X_i$ is a Pareto efficient strategy with respect to $\hat{x}$ (respectively, a weak Pareto efficient strategy with respect to $\hat{x}$).

From the above definition, it is clear that each Pareto equilibrium is a weak Pareto equilibrium, but the converse is not always true. We also need the following definition which was first given by Wang in [79].

Definition 6.1.3: [87] A strategy $\hat{x} \in X$ is said to be a Weighted Nash-equilibrium with respect to the Weighted vector $W := (W^1, \ldots, W^n)$ of a game $G := (X_i, F^i)_{i \in N}$ if, for each player $i \in N$, we have that

(1) $W^i \in \mathbb{R}_+^{k_i} \setminus \{0\}$; and
(2) $W^i \cdot F^i(\hat{x}) \leq W^i \cdot (F^i(\hat{x}^{\hat{i}}, x^i))$ for each $x^i \in X_i$, where $\cdot$ denotes the inner product.

Remark 6.1.1: In particular, when $W^i \in T_+^{k_i}$ for all $i \in N$, the strategy $\hat{x} \in X$ is said to be a normalized weighted Nash-equilibrium with respect to W, where $T_+^{k_i}$ is a simplex of $\mathbb{R}^{k_i}$, i.e.,

$$T_+^{k_i} := \left\{ u := (u^1, \ldots, u^{k_i}) \in \mathbb{R}_+^{k_i} \quad \text{such that } \sum_{j=1}^{k_i} u^j = 1 \right\}.$$

Let $W^i \in \mathbb{R}_+^{k_i}\backslash\{0\}$ be fixed for each $i \in N$. From the definition above, it is not difficult to verify that a strategy $\hat{x} \in X$ is a weighted Nash-equilibrium with respect to the weighted vector $W = (W^1, \ldots, W^n)$ of a game $(X_i, F^i)_{i\in N}$ if and only if for each $i \in N, \hat{X}^i$ is an optimal solution of the optimization problem:

$$\min_{x^i \in X_i} W^k \cdot F^i(\hat{x}^{\hat{i}}, x^i). \qquad (P(i, W^i))$$

6.2. The Existence of Weighted Nash-Equilibria

In order to study the existence of Pareto equilibria as applications of Ky Fan's minimax inequality in [32], we first need the following non-compact version of the Ky Fan's minimax inequality which is due to Ding and Tan in [26:p. 235].

Theorem 6.2.1: (Ky Fan's Minimax Principle) *Let X be a non-empty convex subset of a topological vector space and f an extended real-valued function defined on $X \times X$. Suppose that*

(1) *for each fixed $x \in X, f(x, y)$ is a lower semi-continuous function of y on each non-empty compact subset C of X;*
(2) *for each $A \in \mathcal{F}(X)$ and each $y \in \text{co}(A)$, we have $\min_{x\in A} f(x, y) \leq 0$ where $\mathcal{F}(X)$ denotes the family of all non-empty finite subsets of X (in particular, condition (2) is satisfied if we assume that for each fixed $y \in X, f(x, y)$ is a quasi-concave function of x on X and $f(x, x) \leq 0$ for each $x \in X$); and*
(3) *there exist a non-empty compact convex subset X_0 of X and a non-empty compact subset K of X such that for each $y \in X\backslash K$, there exists an $x \in \text{co}(X_0 \cup \{y\})$ with $f(x, y) > 0$.*

Then there exists $\hat{y} \in K$ such that $f(x, \hat{y}) \leq 0$ for all $x \in X$. Equivalently, we can say that there exists $\hat{y} \in K$ such that

$$\sup_{x\in X} f(x, \hat{y}) \leq \sup_{x\in X} f(x, x).$$

Remark 6.2.1: Indeed, the famous Ky Fan's minimax inequality above has been generalized in many ways. For more details, interested readers can find other generalizations of Ky Fan's minimax inequalities and their applications in Chowdhury and Tan [18], Tarafdar [75], Yuan [88], and many other references therein.

As an application of Theorem 6.2.1, we have the following:

Theorem 6.2.2: [87:pp. 159–160]: *Let $G := (X_i, F^i)_{i\in N}$ be a given multi-objective game, where $F^i = (f_1^i, \ldots, f_{k_i}^i)$ for $i \in N$. For each $i \in N$, suppose X_i is a non-empty convex subset of a Hausdorff topological vector space E_i. Suppose X_i^0 is a non-empty compact and convex subset of X_i, and K_i is a non-empty compact (not necessarily convex) subset of X_i, for each $i \in I$. If there is a weighted vector $W = (W^1, \ldots, W^n)$ with $W^i \in \mathbb{R}_+^{k_i}\backslash\{0\}$ such that the following assumptions are satisfied.*

(1) *For each fixed $y \in X$, the mapping $x \mapsto \sum_{i\in N} W^i \cdot F^i(x^{\hat{i}}, y^i)$ is upper semi-continuous on X;*
(2) *The mapping $(x, y) \mapsto \sum_{i\in N} W^i \cdot F^i(x^{\hat{i}}, y^i)$ is jointly lower semi-continuous on X;*
(3) *For each fixed $x \in X$, the mapping $y \mapsto \sum_{i\in N} W^i \cdot F^i(x^{\hat{i}}, y^i)$ is quasi-convex; and*

(4) *For each $x \in X \setminus K$, there exists $y \in \mathrm{co}(X_0 \cup \{x\})$ such that*

$$\sum_{i=1}^{n} W^i \cdot F^i(x^{\hat{i}}, x^i) > \sum_{i=1}^{n} W^i \cdot F^i(x^{\hat{i}}, y^i),$$

where $X_0 := \prod_{i=1}^{n} X_i^0$ and $K := \prod_{i=1}^{n} K_i$. Then G has at least one weighted Nash-equilibria.

Remark 6.2.2: Theorem 6.2.2 does not only improve corresponding existence results of weighted Nash-equilibria for multi-objective games with multi-criteria given by Wang [Theorem 3.1 [79]], but it also extends Theorem 1.11 of Szidarovszky *et al.* in [70] and Theorem 1 of Borm *et al.* in [8] to Hausdorff topological vector spaces under weaker continuous hypotheses.

6.3. The Existence of Pareto Equilibria

In this section, as applications of weighted Nash-equilibria, we shall derive some existence of Pareto equilibrium for multi-objective games. In order to do so, we first need the following lemma which tells us that the existence problem for Pareto equilibria can be reduced to the existence of the weighted Nash-equilibria under certain circumstances.

Lemma 6.3.1: [87:p. 161]: *Each normalized weighted Nash-equilibrium $\hat{x} \in X$ with a weight $W = (W^1, \ldots, W^n) \in \mathrm{int}\, T_+^{k_1} \times \cdots \times \mathrm{int}\, T_+^{k_n})$ is a weak Pareto equilibrium (respectively, a Pareto equilibrium) of the game $(X_i, F^i)_{i \in N}$.*

Proof: See Lemma 2.1 of Wang in [79:pp. 376–377]. □

Remark 6.3.1: We should note that the conclusion of above Lemma 6.3.1 still holds if $\hat{x}$ is a weighted Nash-equilibrium with a weight $W = (W^1, \ldots, W^n)$ satisfying that $W^i \in \mathbb{R}_+^{k_i}$ (respectively, $W^i \in \mathrm{int}\, \mathbb{R}_+^{k_i}$)for each $i \in N$. Secondly, we note that a Pareto equilibria is not necessarily a weighted Nash-equilibrium.

Now by combining Lemma 6.3.1 and Theorem 6.2.2, we have the following existence of Pareto equilibria for multi-objective games in Hausdorff topological vector spaces.

Theorem 6.3.1: [87:p. 161]: *Let $G = (X_i, F^i)_{i \in N}$ be a multi-objective game, where $F^i = (f_1^i, \ldots, f_{k_i}^i)$ for $i \in N$. For each $i \in N$, suppose X_i is a non-empty compact and convex subset of a Hausdorff topological vector space E_i. Suppose that for each $i \in N$, X_i^0 and K_i are a non-empty compact and convex subset of X_i and a non-empty compact (not necessarily convex) subset of X_i, respectively. If there is a weighted vector $W = (W^1, \ldots, W^n)$ with $W^i \in \mathbb{R}_+^{k_i} \setminus \{0\}$ such that the following are satisfied:*

(1) *for each fixed $y \in X$, the mapping $x \mapsto \sum_{i \in N} W^i \cdot F^i(x^{\hat{i}}, y^i)$ is upper semi-continuous on X;*
(2) *the mapping $(x, y) \mapsto \sum_{i \in N} W^i \cdot F^i(x^{\hat{i}}, y^i)$ is jointly lower semi-continuous on $X \times X$;*
(3) *for each fixed $x \in X$, the mapping $y \mapsto \sum_{i \in N} W^i \cdot F^i(x^{\hat{i}}, y^i)$ is quasi-convex; and*
(4) *for each $x \in X \setminus K$, there exists $y \in \mathrm{co}(X_0 \cup \{x\})$ such that*

$$\sum_{i=1}^{n} W^k \cdot F^i(x^{\hat{i}}, x^i) > \sum_{i=1}^{n} W^i \cdot F^i(x^{\hat{i}}, y^i),$$

where $X_0 := \prod_{i=1}^{n} X_i^0$ and $K := \prod_{i=1}^{n} K_i$,

The G has at least one weak Pareto equilibrium in K. In addition, if $W^i \in \operatorname{int} T_+^{k_i}$ for each $i = 1, \ldots, n$, then G has at least one Pareto equilibrium.

As an immediate consequence of Theorem 6.3.1, we have the following result:

Theorem 6.3.2: [87:p. 162]: *Let $G = (X_i, F^i)_{i\in N}$ be a multi-objective game, where $F^i = (f_1^i, \ldots, f_{k_i}^i)$ for $i \in N$. For each $i \in N$, suppose X_i is a non-empty convex subset of a Hausdorff topological vector space E_i, and X_i^0 and K_i are a non-empty compact and convex subset of X_i and a non-empty compact (not necessarily convex) subset of X_i, respectively. Moreover we assume that the following conditions are satisfied for $j = 1, \ldots, k_i$:*

(1) *the f_j^i is jointly lower semicontinuous on X;*
(2) *for each fixed $y^i \in X_i$, the function $f_j^i(x^{\hat{i}}, y^i)$ is upper semicontinuous on $X^{\hat{i}}$;*
(3) *for each fixed $x^{\hat{i}} \in X^{\hat{i}}$, the function $f_j^i(x^{\hat{i}}, \cdot)$ is convex on X_i; and*
(4) *for each $x \in X\backslash K$, there exists $y \in \mathrm{co}(X_0 \cup \{x\})$ such that for all $i = 1, \ldots, n$, we have $f_j^i(x^{\hat{i}}, x^i) > f_j^i(x^{\hat{i}}, y^i)$ where $X_0 := \prod_{i=1}^n X_i^0$ and $K := \prod_{i=1}^n K_i$.*

Then the multi-objective game G has at least one Pareto equilibrium in K.

Before we close this section, we state the following result which is a compact version of Theorem 6.3.2.

Corollary 6.3.1: *Let $G = (X_i, F^i)_{i\in N}$ be a multi-objective game. For each $i \in N$, suppose X_i is a non-empty compact and convex subset of a Hausdorff topological vector space E_i and the following conditions are satisfied for $j = 1, \ldots, k_i$:*

(1) *the f_j^i is jointly semicontinuous on X;*
(2) *for each fixed $y^i \in X_i$, the function $f_j^i(x^{\hat{i}}, y^i)$ is upper semicontinuous on $X^{\hat{i}}$; and*
(3) *for each fixed $x^{\hat{i}} \in X^{\hat{i}}$, the function $f_j^i(x^{\hat{i}}, \cdot)$ is convex on X_i.*

Then the multi-objective game G has at least one Pareto equilibrium.

Remark 6.3.2: In all these subsections of Section 6, we studied the existence of weighted Nash-equilibria and Pareto equilibria as applications of some nonlinear analysis principle, i.e., mainly the Ky Fan minimax principle which was established by Fan himself in 1972. As the Ky Fan minimax inequality plays a very important role in the study of applied nonlinear analysis (e.g., see Aubin [2], Aubin and Ekeland [1], Lin and Simons [56], Yu and Yuan [84], Yuan [88], and others), hopefully, the Ky Fan minimax inequality method would also play another important role in the study of optimization, in particular for the investigation of the existence for Pareto equilibria in the vector-valued optimization problems and multi-objective games; in this way, the pioneer work has been carried out by Yu and Yuan in [84].

7. PARETO SOLUTION OF A CONE VARIATIONAL INEQUALITY AND PARETO OPTIMALITY OF A MAPPING WITH APPLICATIONS TO MATHEMATICAL ECONOMICS

The concepts of cone monotone, cone upper semicontinuous and cone hemicontinuous mapping and Pareto solution of cone variational inequality are introduced. Also the notion of a Pareto optimality of a mapping is introduced. Theorems on the existence of Pareto solution of cone variational inequality and on the existence of Pareto optimality of a mapping are proved. Moreover, applications to Mathematical Economics, in particular, existence of a Pareto optimum of a Pure exchange economy as well as a private ownership economy are discussed.

7.1. Pareto Solution of a Cone Variational Inequality and Pareto Optimality of a Mapping

The notion of a Pareto optimum is of fundamental importance in all aspects of economics and game theory. It evaluates allocation and optimizes pay off. Hence it is of great interest that this notion is put into the frame work of a mathematical theory. In this chapter we have made an attempt to do that.

Let $F: X \times X \to V$ be a mapping where X is a nonempty convex subset of a Hausdorff topological vector space E and V be an ordered Banach space ordered by a cone P of V. Then a point $x_0 \in X$ is called a Pareto solution of the cone variational inequaltiy for F if $F(x_0, y) \in P \cup [P \cup (-P)]^c$ for all $y \in X$, where $[P \cup (-P)]^c$ denotes the complement of $P \cup (-P)$ in V. Given a mapping $f: X \to V$, a point $x_0 \in X$ is called a Pareto optimum of f if it is a Pareto solution of the cone variational inequaltiy for the mapping $F: X \times X \to V$ defined by $F(x, y) = f(x) - f(y), x, y \in X$. In this chapter we proved theorems on the existence of both. In [76] we have shown that a Pareto optimum of a private ownership economy in general and a pure exchange economy in particular is always a Pareto optimum of a mapping and vice versa. In here we have also introduced the concepts of cone monotone, cone upper semicontinuous and cone hemicontinuous mappings.

7.1.1. Pareto Solution of a Cone Inequality

A nonempty subset P of a real Banach space V is called a *cone* if $\bar{P} = P, P + P \subset P, \mathbb{R}_+ P \subset P$ and $P \cap (-P) = \{0\}$, where $\bar{P}$ is the closure of P and $\mathbb{R}_+ = [0, \infty)$.

Each cone P induces in V an ordering $\preceq$ defined by $x \preceq y$ if and only if $y - x \in P$. This relation $\preceq$ is evidently reflexive, antisymmetric and transitive. The pair (V, P) is called an ordered Banach space with the ordering $\preceq$ induced by P, the positive cone of V. $P^* = \{f \in V^* : f(x) \geq 0 \text{ for all } x \in P\}$ is called the dual cone where V^* is the continuous dual of V, that is P^* is the set of order preserving continuous linear functional on V.

A cone P of V is said to be *normal* if and only if there exists a positive number e such that if $x, y \in P$ with $\|x\| \geq 1$ and $\|y\| \geq 1$, then $\|x + y\| \geq e$ (for other equivalent definitions see [41] and [47]). In what follows V will always denote an ordered Banach space and we write $x \prec y$ if $y - x \in P \backslash \{0\}$.

Let X be a nonempty convex subset of a Haudorff real topological vector space. A mapping $F: X \to V$ is said to be *order or cone convex* if $F(\lambda x + \mu y) \preceq \lambda F(x) + \mu F(y)$ for all $x, y \in X$ and $\lambda \geq 0, \mu \geq 0$ with $\lambda + \mu = 1$, i.e. for such x, y, λ, μ we have

$$\lambda F(x) + \mu F(y) - F(\lambda x + \mu y) \in P.$$

We note that F is cone convex if and only if $f \cdot F$ is convex for all $f \in P^*$. [If F is cone convex, then $f \cdot F$ is convex for $f \in P^*$ as f is order preserving. Next, let fF be convex for each $f \in V^*$. If possible, let for some $x, y \in X$ and $\lambda \geq 0, \mu \geq 0$ with

$\lambda + \mu = 1, [\lambda F(x) + \mu F(y) - F(\lambda x + \mu y)] \notin P$, then by a consequence of the separation theorem (see [47:p. 225]) there exists a $f \in P^*$ such that $f[\lambda F(x) + \mu F(y) - F(\lambda x + \mu y)] < 0$ which implies that fF is not convex].

A mapping F: $X \to V$ is said to be *order or cone lower (upper) semicontinuous* if $f \cdot F$ is lower (upper) semicontinuous for each $f \in P^*$.

We recall from [61] and [74] that a function G: $X \times X \to \mathbb{R}$ is called *monotone* if $G(x, y) + G(y, x) \geq 0$ for all $x, y \in X$ and is called *hemicontinuous* if the function $k(t) = G(x + t(y - x), y)$ of the real variable $t \in [0, 1]$ is lower semicontinuous on X as $t \downarrow 0^+$ for arbitrary given vectors x and y in X.

A mapping F: $X \times X \to V$ is said to be *order or cone monotone* if $F(x, y) + F(y, x) \succeq 0$ for all $x, y \in X$ and is said to be *strictly cone monotone* if in addition $F(x, y) + F(y, x) = 0$ implies $x = y$. A mapping F: $X \times X \to V$ is said to be *order or cone hemicontinuous* if fF is hemicontinuous for each $f \in P^*$.

Lemma 7.1.1: *If P is a normal cone of a real Banach space V and $x \in V$ with $x \succ 0(\prec 0)$, then there exists $f \in P^*$ such that $f(x) > 0$ $(f(x) < 0)$.*

In what follows the following fixed point theorem [73] which is an extension of an earlier theorem in [72] will be the main tool:

Theorem 7.1.1: *Let X be a nonempty convex subset of a real Hausdorff topological vector space. Let T: $X \to 2^X$ be a set valued mapping such that*

(*i*) *for each $x \in X, T(x)$ is a nonempty convex subset of X;*
(*ii*) *for each $y \in X, T^{-1}(y) = \{x \in X : y \in T(x)\}$ contains a relatively open subset O_y of X (O_y may be empty for some y);*
(*iii*) $\bigcup_{x \in X} O_x = X$; *and*
(*iv*) *X contains a nonempty subset X_0 contained in a compact convex subset X_1 of X such that the set $D = \bigcap_{x \in X_0} O_x^c$ is either empty or compact (O_x^c denotes the complement of O_x in X).*

Then there exists a point $x_0 \in X$ such that $x_0 \in T(x_0)$.

With X a nonempty convex subset of a real Hausdorff topological vector space $E, (V, P)$ an ordered Banach space and P a normal cone, we consider the following order or cone variational problem (to see connection with variational problem we refer to the work of Tarafdar in [74]).

We consider the following problem:

(7.1) F: $X \times X \to V$ is a mapping with $F(x, x) = 0$ for each $x \in X$ and satisfying

(i) for each $x \in X, F(x, \cdot)$ is (a) cone concave and (b) cone upper semicontinuous on E;
(ii) F is cone monotone and cone upper hemicontinuous; and
(iii) for $x, y \in X, F(x, y) \in [P \cup (-P)]^c$ if and only if

$$F(y, x) \in [P \cup (-P)]^c.$$

We are interested to find the existence of a point $x_0 \in X$ such that

$$F(x_0, y) \in (-P) \cup [P \cup (-P)]^c \quad \text{for all } y \in X. \tag{7.2}$$

x_0 will be called a Pareto solution of the variational cone inequality (7.2).

We first prove the following lemma which is a generalization of a well-known lemma due to Minty (see [9]) (see also [61,72,74]).

Lemma 7.1.2: *If $X \times X \to V$ is a mapping as defined above and satisfying (i)(a), (ii) and (iii), then x_0 is a solution of the variational cone inequality*

$$F(y, x_0) \in P \cup [P \cup (-P)]^c \quad \text{for all } y \in X \tag{7.3}$$

if and only if x_0 is a solution of the variational cone inequality (7.2).

Theorem 7.1.2: *If X is a closed convex subset of E and F: $X \times X \to V$ is a mapping as defined as in (7.1) and satisfy (7.1) (i), (ii) and (iii), then for each $x \in X$, the set*

$$\begin{aligned} G(x) &= \overline{\{y \in X : F(y,x) \preceq 0\}} \cup \overline{\{y \in X : F(y,x) \in [P \cup (-P)]^c\}} \\ &= \overline{\{y \in X : F(y,x) \succ 0\}^c} \end{aligned}$$

is a subset of the set

$$\begin{aligned} B(x) &= \overline{\{y \in X : F(x,y) \succeq 0\}} \cup \overline{\{y \in X : F(x,y) \in [P \cup (-P)]^c\}} \\ &= \overline{\{y \in X : F(y,x) \prec 0\}^c} \end{aligned}$$

Further assume that

(a) $\bigcup_{x \in X} \{y \in X : F(x,y) \prec 0\} = X$ *implies* $\bigcup_{x \in X} [B(x)]^c = X$;

(b) there is a nonempty subset X_0 contained in a compact convex subset X_1 of X such that one of the followings holds:

(A) the set $L = \bigcap_{x \in X_0} G(x)$ *is either empty or compact and*

(B) the set $K = \bigcap_{x \in X_0} B(x)$ *is either empty or compact.*

Then there is a solution of the variational cone inequality (7.2).

Corollary 7.1.1: *Let X be a closed convex subset of E and F: $X \times X \to V$ be a mapping as defined in (7.1) and satisfying (7.1) (i), (ii) and (iii). Further assume that the condition (a) of Theorem 7.1.2 holds and that there exists at least one point $\bar{x} \in X$ such that either $G(\bar{x})$ or $B(\bar{x})$ is empty or compact. Then there is a Pareto solution of the cone variational inequaltiy (7.2).*

Corollary 7.1.2: *If X is a compact convex subset of E and F: $X \times X \to V$ is a mapping as defined in (7.1) and satisfying (7.1) (i), (ii) and (iii) and condition (a) of Theorem 7.1.2, then there is a Pareto solution of cone inequality (7.2).*

Remark 7.1.1: If the mapping F: $X \times X \to V$ is such that $F(x,y) = -F(y,x)$ for all $x, y \in X$, then F satisfies (7.1)(i), (7.1)(iii) and F is cone monotone. Moreover it is trivial that Lemma 7.1.2 holds without cone semicontinuity and cone hemicontinuity. Also it is easy to see that for each $x \in X, B(x) = G(x)$ where $B(x)$ and $G(x)$ are as defined in Theorem 7.1.2. Thus exactly as above we can prove the following results.

Theorem 7.1.3: *If X is a closed convex subset of E and F: $X \times X \to V$ is a mapping such that $F(x,y) = -F(y,x)$ for all $x, y \in X$ and for each $x \in X, F(x, \cdot)$ is cone concave. Further assume that*

(a) the condition (a) of Theorem 7.1.2 holds;
(b) the condition (b) of Theorem 7.1.2 holds.

Then there is a Pareto solution of the variational cone inequality (7.2).

Corollary 7.1.3: *Let L be a closed convex subset of E and F: $X \to V$ a mapping such that for all $x, y \in X, F(x,y) = -F(y,x)$, and for each $x \in X, F(x, \cdot)$ is cone concave.*

Further assume that the condition (a) of Theorem 7.1.2 holds and there exists a point $\bar{x} \in X$ such that $G(\bar{x})$ is empty or compact.

Then there exists a Pareto solution of cone variational inequality (7.2).

Corollary 7.1.4: *If X is compact convex subset of E and F: $X \times X \to V$ is a mapping such that for all $x, y \in X, F(x,y) = -F(y,x)$, and for each $x \in X, F(x, \cdot)$ is cone concave.*

Further assume that the condition (a) of Theorem 7.1.2 holds. Then there is a Pareto solution of the cone variational inequality (7.2).

7.1.2. Generalized Pareto Optimum

In this section we present the concept of Pareto optimality of a mapping. If a point $x_0 \in X$ satisfies the cone inequality

$$F(x_0, y) \in P \cup [P \cup (-P)]^c \quad \text{for all } y \in X, \tag{7.8}$$

then x_0 is said to be a Pareto maximal of F and if $x_0 \in X$ satisfies the cone inequality

$$F(x_0, y) \in (-P) \cup [P \cup (-P)]^c \quad \text{for all } y \in X, \tag{7.9}$$

then x_0 is said to be a Pareto minimal of F. In either case x_0 is said to be a generalized Pareto optimum, or simply a Pareto optimum of F. Throughout the rest of the chapter by a generalized Pareto optimum or Pareto optimum we always mean a Pareto maximal. In order to apply our concept to more specific problems such as the problems of economics and game theory we need to formalize further our concepts.

With a mapping f: $X \to V$, we can define a preordering $\preceq_f$ in X as follows: for $x, y \in X, x \preceq_f y$ if and only if $f(x) \preceq f(y)$ where $\preceq$ is the order in V induced by the cone P. Thus $\preceq_f$ is reflexive and transitive as $\preceq$ is so. Further we define $x \prec_f y$ if and only if $f(x) \prec f(y)$. Obviously the mapping f: $(X, \preceq_f) \to (V, \preceq)$ is an increasing mapping.

A point $\bar{x} \in X$ is called a maximal element with respect to $\preceq_f$ if there is no point $y \in X$ such that $x \prec_f y$. It readily follows that $\bar{x} \in X$ is maximal element with respect to $\preceq_f$ if and only if $[f(\bar{x}) - f(y)] \in P \cup [P \cup (-P)]^c$ for all $y \in X$, that is if and only if $\bar{x}$ is a Pareto solution of the cone inequality $F(\bar{x}, y) \in P \cup [P \cup (-P)]^c$ for all $y \in X$ where F: $X \times X \to V$ is defined by $F(x, y) = f(x) - f(y), x, y \in X$.

A maximal element $\bar{x} \in X$ with respect to $\preceq_f$ is said to be generalized Pareto optimum or simply a Pareto optimum for f.

We note that if x_1 and x_2 are two Pareto optima of f, then x_1 and x_2 are either incomparable or indifferent.

Theorem 7.1.4: *Let X be a closed convex subset of E and $f\colon X \to V$ be a cone upper semicontinuous and cone concave mapping. Further assume that the conditions (a) and (b) of Theorem 7.1.2 for the mapping $F\colon X \times X \to V$ defined by $F(x,y) = f(y) - f(x), x, y \in X$, holds. Then there is a point $x_0 \in X$ such that*

$$[f(x_0) - f(y)] \in P \cup [P \cup (-P)]^c \quad \textit{for all} \quad y \in X$$

i.e. there is Pareto optimum $x_0 \in X$ of f.

Corollary 7.1.5: *Let X be a closed convex subset of E and $f\colon X \to V$ be a cone upper semicontinuous and cone concave mapping. Further assume that the condition (a) of Theorem 7.1.2 for the mapping $F\colon X \times X \to V$ defined by $F(x,y) = f(y) - f(x), x, y \in X$ holds and there is at least one point $\bar{x} \in X$ such that $G(\bar{x})$ or $B(\bar{x})$ is either empty or compact, where for each $x \in X, G(x)$ and $B(x)$ are as defined as in Theorem 7.1.2 but with $F(x,y) = f(y) - f(x)$. Then there is a Pareto optimum $x_0 \in X$ of f.*

Corollary 7.1.6: *Let X be a compact convex subset of E and $f\colon X \to V$ be a cone upper semicontinuous and cone concave mapping satisfying the condition (a) of Theorem 7.1.2 for F defined by $F(x,y) = f(y) - f(x), x, y \in X$. Then there is a Pareto optimum x_0 of f.*

7.2. Applications of Pareto Optimality of a Mapping to Mathematical Economics

The concept of a Pareto optimum of mapping, which has the root in the quotation given in the very beginning of this chapter, has been introduced and its existence has been proved in a chapter [77] appeared in the Proceedings of the First World Congress of Nonlinear Analysts, Walter de Gruyter, Berlin, 1992. In the present section we have presented some results on the existence of a Pareto optimum of a Pure exchange economy as well as a private ownership economy.

In the chapter [77] appeared in this proceedings we have introduced the concept of a Pareto solution of cone variational inequality with a mapping $F\colon X \times X \to V$ and a Pareto optimum of a mapping $f\colon X \to V$ have been introduced, where X is a nonempty convex subset of a Hausdorff topological vector space and V is an ordered Banach space ordered by a cone P. Theorems on the existence of Pareto solution of cone variational inequality and Pareto optimum of a mapping have also been proved. In the present chapter we have applied the results of [77] to prove the existence of Pareto optimum of a pure exchange economy and a private ownership economy. Our main result (Theorem 7.2.2) in this chapter is different from that of Debreu [9] in that we have assumed condition (P) (see Section 3) instead of insatiability conditon.

7.2.1. Economy and related concept

In this section we define an economy and all other related terms that we need for this chapter.

An economy is described by:

$$m \text{ consumers indexed by } i = 1, 2, \ldots, m; \; n \text{ produces}$$
$$\text{indexed by } j = 1, 2, \ldots, n; \quad \text{for each } i = 1, 2, \ldots, m$$

a consumption set $(X_i, \preceq_i)$ where X_i is a nonempty subset of $\mathbb{R}^l$ and $\preceq_i$ is a preference preordering, i.e., a reflexive and transitive relation on X_i; for each $j = 1, 2, \ldots, n$, a nonempty subset Y_j, of $\mathbb{R}^l$, called the production set for the producer j, and an *a priori* given vector $\omega \in \mathbb{R}^l$, called the total resources of the economy $\mathcal{E}$.

A pair $(x, y) = ((x_i), (y_i))$ where $x_i \in X_i$ and $y_j \in Y_j, i = 1, 2, \ldots, m$ and $j = 1,2, \ldots, n$ is called a state of the economy $\mathcal{E}$. Thus a state of the economy $\mathcal{E}$ is an $(m + n)$ tuple of $\mathbb{R}^l$ which can be represented by a point of $\mathbb{R}^{(m+n)l}$. Given a state $(x, y) = ((x_i), (y_j))$ of $\mathcal{E}$, the point $x - y = \sum_{i=1}^m x_i - \sum_{j=1}^n y_j$ is called the net demand and the point $z = x - y - \omega$ is called the excess demand. Thus every point of the set $Z = X - Y - \{\omega\}$ represents an excess demand corresponding to a state, where $X = \sum_{i=1}^m X_i$ and $Y = \sum_{j=1}^n Y_j$.

A state $(x, y) = ((x_i), (y_j))$ of $\mathcal{E}$ is called a market equilibrium if $x - y = \sum_{i=1}^m x_i - \sum_{j=1}^n y_j = \omega$, that is if excess demand is 0. A state $(x, y) = ((x_i), (y_j))$ of $\mathcal{E}$ is said to be attainable if $x_i \in X_i$ for $i = 1, 2, \ldots, m, y_j \in Y_j$ for $j = 1, 2, \ldots, n$ and $x - y = \sum_{i=1}^m x_i - \sum_{j=1}^n y_j = \omega$. Note that the two different notations $x = (x_i)$ and $y = (y_i)$ and $x = \sum_{i=1}^m x_i$ and $y = \sum_{j=1}^n y_j$ have been used.

The set of all attainable states of $\mathcal{E}$ is denoted by A. If X_i is the commodity space, a subset of $\mathbb{R}^l$ and $\preceq_i$ is the preference preordering of the ith consumer, an increasing function $u_i: X_i \to \mathbb{R}$ (i.e., $x_i, x_i' \in X_i$ with $x_i \preceq_i x_i' \Rightarrow u_i(x_i) \leq u_i(x_i')$) is called the utility function of the ith consumer. Without loss of generality we may assume u_i to be nonnegative.

Let us consider the following condition on $(X_i, \preceq_i)$:

(a) for each $x_i' \in X_i$, the set $\{x_i \in X_i : x_i \preceq_i x_i'\}$ is closed.

It is well known that $\preceq_i$ satisfying (a) represents an increasing upper semicontinuous utility function $u_i: X_i \to \mathbb{R}$. Conversely given a upper semicontinuous function $u_i: X_i \to \mathbb{R}$, we can define a preordering $\preceq_{u_i}$ on X_i as follows: with x_i and x_i' in $X_i, x_i \preceq_{u_i} x_i'$ if and only if $u_i(x_i) \leq u_i(x_i')$. Clearly $\preceq_{u_i}$ satisfies (a) stated above and u_i is increasing with respect to $\preceq_{u_i}$.

Throughout the rest of the chapter we will assume that the economy $\mathcal{E}$ is given by $\mathcal{E} = \big((X_i, \preceq_i), (Y_j)\big)$ where $\preceq_i = \preceq_{u_i}$ and u_i is upper semicontinuous and concave for each $i = 1,2, \ldots, m$. That is, each preference can be represented by an upper semicontinuous concave function.

In the sequel we will need the following result ([9]:p. 77) which we write as lemma:

Lemma 7.2.1: *(A priori bound of the economy): Let $\mathcal{E}$ be an economy such that $X = \sum_{i=1}^m X_i$ has a lower bound for $\leq$ (i.e., for each i, there is a point $a_i \in \mathbb{R}^l$ such that $a_i \leq x_i$ for all $x_i \in X_i$ (coordinate wise) so that $a = \sum_{i=1}^m a_i$ is a lower bound of X for $\leq$), Y is closed, convex and $Y \cap \Omega = \{0\}$, where Ω is the nonnegative orthant of $\mathbb{R}^l$. If $Y \cap (-Y) \subset \{0\}$, then A is bounded.*

Private Ownership Economies (The Arrow Debreu Model)

A private ownership economy is defined by $\mathcal{E} = \big((X_i, \preceq_i), (Y_j), \omega\big)$ where for each i, there is a point ω_i (resources of the ith consumer) of $\mathbb{R}^l$ such that $\omega = \sum_{i=1}^m \omega_i$.

Pareto Optimum

Let us consider an economy $\mathcal{E} = \big((X_i, \preceq_i), (Y_j), \omega\big)$. We define a preordering $\preceq_a$ on A, the set of all attainable states of $\mathcal{E}$ as follows:

Given two attainable states or allocations $((x_i), (y_j))$ and $((x_i'), (y_j')), ((x_i), (y_j)) \preceq_a ((x_i'), (y_j'))$ if and only if $x_i \preceq_i x_i'$ for each i, i.e., each consumer i desires his consumption

x_i' at least as much as his consumption x_i. Given two comparable attainable states $((x_i), (y_j))$ and $((x_i'), (y_j'))$, the second one is said to be better or more efficient than the first one, if $x_i \preceq_i x_i'$ for each i and $x_i \prec_i x_i'$ for at least one i. In this case we write $((x_i), (y_j)) \prec_a ((x_i'), (y_j'))$. It should be noted that two attainable states $((x_i), (y_j))$ and $((x_i'), (y_j'))$ may be incomparable with respect to $\preceq_a$.

A maximal element of A with respect to $\preceq_a$ is called a Pareto optimum of the economy $\mathcal{E}$ (quotation at the beginning).

(A point $\bar{x}_i$ is a satiation point if there is no consumption $x_i' \in X_i$ such that $x_i' \succ_i \bar{x}_i$, i.e., $\bar{x}_i$ is a greatest element of X_i with respect to $\preceq_i$ which is a complete preordering.)

A real valued function f defined in a convex set X is called quasiconvex if for each real number t, the set $\{x \in X : f(x) > t\}$ is either empty or convex. The condition (ii) above or equivalently condition (a″) in ([9:p. 59]) can easily be seen to represent a quasiconcave utility function u_i. Thus Debreu's two fundamental results on the existence of "equilibrium points" and "Pareto optimum" (respectively Theorem 5.7 and Theorems 6.3 and 6.4 [9]) concern economies with continuous and quasiconcave utilities or preferences. In this chapter we will prove by using the abstract results of our work in [77] of Pareto optimum of economies with upper semicontinuous and concave utilities with insatiability condition replaced by a new condition. Thus while we weaken the continuity condition, we need to strengthen the convexity condition. Nevertheless our results are different from those of Debreu.

7.2.2. Applications to Economy

Pure Exchange Economy

We first consider the pure exchange economy, the simplest case of an economy where no production is considered, i.e., the total production set $Y = \{0\}$. Thus it is a very special case of the economy considered in Section 1. The model of a pure exchange economy $\mathcal{E}$ can be described by: m consumers indexed by $i = 1, 2, \ldots, m$; the commodity space $\Omega = \{(x^1, x^2, \ldots, x^l) \in \mathbb{R}^l : x^i \geq 0 \text{ for } i = 1, 2, \ldots, l\}$ which is a positive cone of $\mathbb{R}^l$; the space of attainable states of the economy $= \Omega^m = \{x = (x_1, x_2, \ldots, x_m) \in \mathbb{R}^{lm} : x_i \in \Omega,\ i = 1, 2, \ldots, m \text{ and } \sum_{i=1}^m x_i = w\}$, where $w \in \mathbb{R}^l$ is the resources of the economy $\mathcal{E}$ and for each $i = 1, 2, \ldots, m, \tilde{u}_i : \Omega \to \mathbb{R}_+$ is a utility function (not necessarily continuous) for the ith consumer.

The prefence $\preceq_i$ for the ith consumer is defined by: for $x, x' \in \Omega, x \preceq_i x'$ if and only if $\tilde{u}_i(x) \leq \tilde{u}_i(x')$. We define a preordering $\preceq_a$ on Ω^m as follows: given two attainable states $(x_i) = (x_1, x_2, \ldots, x_m) \in \Omega^m$ and $(x_i') = (x_1', x_2', \ldots, x_m') \in \Omega^m, (x_i) \preceq_a (x_i')$ if and only if $x_i \preceq_i x_i'$ for each $i = 1, 2, \ldots, m$. Given two attainable states (x_i) and (x_i') of $\mathcal{E}, (x_i)$ may not be comparable to (x_i').

Pareto optimum of the pure exchange economy is a maximal element of Ω^m with respect to $\preceq_a$.

Let $X = \Omega^m$ which is a compact convex subset of $\mathbb{R}^{lm}$ and $V = (\mathbb{R}^m, \Omega')$ be an ordered Banach space where $\Omega' = \{(x_1, x_2, \ldots, x_m) \in \mathbb{R}^m \colon x_i \geq 0 \text{ for } i = 1, 2, \ldots, m\}$ is the positive cone of $\mathbb{R}^m$.

Next we define the mapping $f \colon X \to V$ by

$$f(x) = (\tilde{u}_1(x_1), \tilde{u}_2(x_2), \ldots, \tilde{u}_m(x_m))$$

where $x = (x_1, x_2, \ldots, x_m) \in \Omega^m = X$ and $x_i \in \Omega$ for each $i = 1, 2, \ldots, m$. Then it is evident that given two states x and y in X, $f(x) \preceq f(y)$ if and only if $\tilde{u}_i(x_i) \leq \tilde{u}_i(y_i)$ for each $i = 1, 2, \ldots, m$ where $x = (x_i)$ and $y = (y_i)$ and $\preceq$ is the order in $\mathbb{R}^m$ induced by the cone Ω'. Thus $x \preceq_f y$ if and only if $f(x) \preceq f(y)$ if and only if $\tilde{u}_i(x_i) \leq \tilde{u}_i(y_i)$ for each i, if and only if $x_i \preceq_i y_i$ for each i, if and only if $(x_i) \preceq_a (y_i)$, i.e., $x \preceq_a y$. Thus the preordering $\preceq_a$ in X is equivalent to the preordering $\preceq_f$. Thus from the definition of Pareto optimum of a mapping we have the following proposition:

Proposition 7.2.1: *A point $x_0 \in X = \Omega^m$ is a Pareto optimum of the pure exchange economy $\mathcal{E}$ if and only if x_0 is a Pareto optimum of the mapping f: $X \to V$ defined above.*

Existence Theorem

We consider the following assumption in our economy

$$\mathcal{E} = \left((X_i, \underset{i}{\preceq}), (Y_j), (w_i) \right) :$$

(P) If $(x, y) = ((x_i), (y_j))$ and $(x', y') = ((x'_i), (y'_j))$ are two attainable states of the economy $\mathcal{E}$ such that $u_i(x_i) \geq u_i(x'_i)$ for all i and $u_i(x_i) > u_i(x'_i)$ for at lease one i, then there exists an attainable state $(\bar{x}, \bar{y}) = ((\bar{x}_i), (\bar{y}_i))$ such that $u_i(\bar{x}_i) > u_i(x'_i)$ for all i.

Note that in the above assumption (P) for the pure exchange $\mathcal{E}$, the second coordinate does not appear.

Theorem 7.2.1: *If for each consumer $i = 1, 2, \ldots, m$, the utility function $\bar{u}_i$ is a upper semicontinuous and concave function, then the prure exchange economy $\mathcal{E}$ satisfying the above condition (P) has a Pareto optimum.*

Private Ownership Economy (Arrow–Debreu Model)

Proposition 7.2.2: *Let $\mathcal{E} = ((X_i, \preceq_i), (Y_j), (w_i))$ be a private ownership economy and $X = A$ be the set of all attainable states of $\mathcal{E}$. If f: $X \to \mathbb{R}^m$ be the mapping defined by*

$$f((x_i), (y_j)) = (u_1(x_1), u_2(x_2), \ldots, u_m(x_m)),$$
$$((x_i), (y_j)) \in A = X.$$

Then $((\bar{x}_i), (\bar{y}_j)) \in A$ is a Pareto optimum of f if and only if $((\bar{x}_i), (\bar{y}_j))$ is a Pareto optimum of $\mathcal{E}$.

Theorem 7.2.2: *Let $\mathcal{E} = \big((X_i, \preceq_i), (Y_j), (w_i)\big)$ be a private ownership economy such that*

(*a*) for each $i = 1, 2, \ldots, m$,
  (*i*) *X_i is closed, convex and has lower bound for $\leq$ (see Lemma 7.2.1 for definition)*;
  (*ii*) *u_i is upper semicontinuous*;
  (*iii*) *u_i is concave*;
  (*iv*) *there exists $x_i^0 \in X_i$ such that $x_i^0 \ll w_i$* (i.e., *each coordinate of x_i^0 is strictly less than the corresponding coordinate of w_i*);
  (v) *the economy $\bar{\mathcal{E}} = \big((X_i \preceq_i), (\overline{\mathrm{co}\, Y_j}), (w_j)\big)$ satisfies the condition* (P), *where $\overline{\mathrm{co}\, Y_j}$ denotes the closed convex hull of Y_j*;

(*b*) *for each $j = 1, 2, \ldots, n, 0 \in Y_j$*;
(*c*) *Y is closed and convex*;
(*d*) *$Y \cap (-Y) \subset \{0\}$*;
(*e*) *$Y \supset (-\Omega)$.*

Then there exists a Pareto optimum of the economy $\mathcal{E}$.

For further ideas, results and applications we refer to [14], [16], [23], [37–38], [46], [54–55], [60], and [80–82], and references therein.

REFERENCES

[1] J.P. Aubin and I. Ekeland (1984). *Applied Nonlinear Analysis*. New York: John Wiley and Sons.

[2] J.P. Aubin (1982). *Mathematical Methods of Game Theory and Economic Theory*. Amsterdam: North-Holland.

[3] G. Beer (1988). On a generic optimization theorem of Kenderov, *Nonlinear Anal. T. M. A.*, **6**, 647–655.

[4] H.P. Benson (1984). Optimization over the efficient set, *J. Math. Anal. Appl.*, **98**, 562–580.

[5] H.P. Benson (1986). An algorithm for optimizing over the weakly efficient set, *European J. Operational Research*, **25**(2), 192–199.

[6] C. Berge (1963). *Topological Spaces*. New York: Macmillan.

[7] K. Bergstresser and P.L. Yu (1977). Domination structures and multicriteria problems in N-person games, *Theory and Decision*, **8**, 5–48.

[8] P.E. Borm, S.H. Tijs and J. Van Der Aarssen (1990). Pareto equilibrium in multi-objective games, *Methods Oper. Res.*, **60**, 303–312.

[9] F.E. Browder (1965). Nonlinear monotone operators and convex sets in Banach spaces, *Bull. Amer. Math. Soc.*, **71**, 780–785.

[10] G.Y. Chen and G.M. Cheng (1987). Vector variational inequalities and vector optimization. *Lecture Notes in Econom. and Math. System*, Berlin: Springer-Verlag, **1**(285), 408–416.

[11] G.-Y. Chen and G.-M. Cheng, Vector variational inequality and vector optimization problems, Proceedings of 7th MCDM Conference, Kyoto, Japan. Berlin: Springer-Verlag, 408–416.

[12] G.Y. Chen and B.D. Craven (1989). Approximate dual and approximate vector variational inequality for multiobjective optimization, *J. Austral. Math. Soc.*, ser. A **47**, 418–423.

[13] G.-Y. Chen and B.D. Craven (1990). A vector variational inequality and optimization over an efficient set, *Z. Oper. Res.*, **34**, 1–12.

[14] G.-Y. Chen and S.-H. Hou (1999). Existence of solutions for vector variational inequalities. In F. Giannessi (Ed.), *Vector Variational Inequalities and Vector Equilibria, Mathematical Theories*, Kluwer Acad. Pub., (December), 73–86.

[15] G.Y. Chen and X.Q. Yang (1990). The vector complementarity problem and its equivalences with the weak minimal element in ordered spaces, *J. Math. Anal. Appl.*, **153**, 136–158.

[16] G.Y. Chen and X.Q. Yang (1999). On the existence of solutions to vector complementarity problems, In F. Giannessi (Ed.), *Vector Variational Inequalities and Vector Equilibria, Mathematical Theories*, Kluwer Acad. Pub. (December), 87–95.

[17] G.Y. Chen (1992). Existence of solutions for a vector variational inequality: an extension of Hartman-Stampacchia Theorem, *J. Optim. Theory Appls.*, **74**, 445–456.

[18] M.S.R. Chowdhury and K.-K. Tan (1996). Generalization of Ky Fan's minimax inequality with applications to generalized variational inequalities for pseudo-monotone operators and fixed point theorems, *J. Math. Anal. Appl.*, **204**, 910–929.

[19] J.P.R. Christensen (1982). Theorems of Namioka and Johnson type upper semi-continuous and compact valued set-valued mappings, *Proc. Amer. Math. Soc.*, **86**, 649–655.

[20] H.W. Corley (1980). An existence result for maximization with respect to cones, *J. Optim. Theory Appl.*, **31**, 277–281.

[21] B.D. Craven (1977). Lagrangean conditions and quasiduality, *Bull. Austral. Math. Soc.*, **16**, 325–339.

[22] B.D. Craven (1978). *Mathematical Programming and Control Theory*. London: Chapman & Hall.

[23] P. Daniele and A. Maugeri (1999). Vector variational inequalities and modelling of a continuum traffic equilibrium problem. In F. Giannessi (Ed.), *Vector Variational Inequalities and Vector Equilibria, Mathematical Theories*, Kluwer Acad. Pub. (December), 97–111.

[24] A. Danilidis and N. Hadjisavvas (1996). Existence theorems for vector variational inequalities, *Bull. Austral. Math. Soc.*, **54**, 473–481.

[25] G. Debreu (1959). *Theory of Value*. New York: Wiley.

[26] X.P. Ding and K.-K. Tan (1992). A minimax inequality with applications to existence of equilibrium points and fixed point theorems, *Colloq. Math.*, **63**, 233–247.

[27] X.P. Ding and K.K. Tan (1993). Generalizations of KKM theorem and applications to best approximations and fixed point theorems, *SEA Bull. Math.*, **17**(2), 139–150.

[28] X.P. Ding and E. Tarafdar (1999). Generalized vector variational-like inequalities without monotonicity, In F. Giannessi (Ed.), *Vector Variational Inequalities and Vector Equilibria, Mathematical Theories*. Kluwer Acad. Pub. (December), 113–124.

[29] X.P. Ding and E. Tarafdar (1999). Generalized vector variational-like inequalities with $C_x - \eta$-pseudomonotone set-Valued mappings, In F. Giannessi (Ed.), *Vector Variational Inequalities and Vector Equilibria, Mathematical Theories*. Kluwer Acad. Pub. (December), 125–140.

[30] R. Engleking (1977). *General Topology*. Warsaw: Polish Scientific.

[31] K. Fan (1961). A generalization of Tychonoff's fixed point theorem, *Math. Ann.*, **142**, 305–310.

[32] K. Fan (1972). Minimax inequality and application. In O. Shisha (Ed.), *Inequalities, III*, New York: Academic Press.

[33] M.K. Fort, Jr. (1951). Points of continuity of semi-continuous functions, *Publ. Math. Debrecen*, **2**, 100–102.

[34] D. Ghose and U.R. Prasad (1989). Solution concepts in two-person multicriteria games, *J. Optim. Theory Appl.*, **63**, 167–189.

[35] F. Giannessi (1980). Theorems of the Alternative, Quadratic Programs, and Complementarity Problems. In R.W. Cottle, F. Giannessi and J.L. Lions (Eds), *Variational Inequalities and Complementarity Problems*, New York: John Wiley and Sons, 151–186.

[36] J.R. Giles (1982). *Convex Analysis with Application in Differentiation of Convex Functions*. Boston: Pitman Advanced Publishing Program.

[37] C.J. Goh and X.Q. Yang (1999). Scalarization methods for vector variational inequality. In F. Giannessi, *Vector Variational Inequalities and Vector Equilibria, Mathematical Theories*. Kluwer Acad. Pub. (December), 217–232.

[38] X.H. Gong, W.T. Fu and W. Liu (1999). Super efficiency for a vector equilibrium in locally convex topological vector spaces, In F. Giannessi (Ed.), *Vector Variational Inequalities and Vector Equilibria, Mathematical Theories*. Kluwer Acad. Pub. (December), 233–252.

[39] E.L. Hannan (1978). Using duality theory for identification of primal efficient points and for sensitivity analysis in multiple objective linear programming, *J. Oper. Res. Soc.*, **29**(7), 643–649.

[40] H. Isermann (1978). On some relations between a dual pair of multiple objective linear programs, *Z. Oper. Res.*, **22**, 33–41.

[41] G. Jameson (1970). *Ordered Linear Spaces*, Lecture Notes. New York: Springer-Verlag.

[42] G. Jameson (1970). *Ordered linear spaces*, Lecture Notes in Mathematics 141. Berlin: Springer-Verlag.

[43] Y. Jian (1992). Essential weak efficient solution in multi-objective optimization problems, *J. Math. Anal. Appl.*, **166**(1), 230–235.

[44] S. Karamadian (1976). Complementarity over cones with monotone and pseudomonotone maps, *J. Optim. Theory Appl.*, **18**, 445–454.

[45] K.K. Kazmi (1996). Existence of solutions for vector optimization, *Appl. Math. Lett.*, **9**(6), 19–22.

[46] K.R. Kazmi (1999). Existence of solutions for vector saddle-point problems, In F. Giannessi (Ed.), *Vector Variational Inequalities and Vector Equilibria, Mathematical Theories*. Kluwer Acad. Pub. (December), 267–275.

[47] J.L. Kelley and I. Namioka (1963). *Linear Topological Spaces*. Princeton, N.J.: Van Nostrand.

[48] P.S. Kenderov (1984). Most of the optimization problems have unique solution. In B. Brosowski and F. Deutsch (Eds), *Proceedings, Oberwolhfach Conference on Parametric Optimization*, Birkhauser International Series of Numerical Mathematics, Basel: Birkhauser, **72**, 203–216.

[49] E. Klein and A. Thompson (1984). *Theory of Correspondences*. New York: Wiley.

[50] I.V. Konnov and J.C. Yao (1997). On the generalized vector variational inequality problem, *J. Math. Anal. Appl.*, **206**, 42–58.

[51] T.C. Lai and J.C. Yao (1996). Existence results for VVIP, *Appl. Math. Lett.*, **9**(3), 17–19.

[52] G.M. Lee, D.S. Kim and B.S. Lee (1996). Generalized vector variational inequality, *Appl. Math. Lett.* **9**(1), 39–42.

[53] G.M. Lee, D.S. Kim and S.J. Cho (1993). Generalized vector variational inequality and fuzzy extension, *Appl. Math. Lett.*, **6**(6), 47–51.

[54] G.M. Lee, D.S. Kim, B.S. Lee and N.D. Yen (1999). Vector variational inequality as a tool for studying vector optimization problems. In F. Giannessi (Ed.), *Vector Variational Inequalities and Vector Equilibria, Mathematical Theories*. Kluwer Acad. Pub. (December), 277–305.

[55] S.J. Li, X.Q. Yang and G.-Y. Chen (1999). Vector Ekeland variational principle. In F. Giannessi (Ed.), *Vector Variational Inequalities and Vector Equilibria, Mathematical Theories*. Kluwer Acad. Pub. (December), 321–333.

[56] B. Lin and S. Simons (1987). *Nonlinear and Convex Analysis – Proceeding in Honor of Ky Fan*. New York: Dekker.

[57] K.L. Lin, D.P. Yang and J.C. Yao (1997). Generalized vector variational inequalities, *J. Optim. Theory Appl.*, **92**(1), 117–125.

[58] P. Loridan (1984). ϵ-solutions in vector minimization problems, *J. Optim. Theory Appl.*, **43**, 265–276.

[59] D.T. Luc (1989). Theory of vector optimization. In M. Beckmann and W. Krelle (Eds), *Lecture Notes in Econom. and Math. Systems*, Vol. **319**, Berlin: Springer-Verlag, 173 pages.

[60] G. Mastroeni (1999). On Minty vector variational inequality, In F. Giannessi (Ed.), *Vector Variational Inequalities and Vector Equilibria, Mathematical Theories*. Kluwer Acad. Pub. (December), 351–361.

[61] U. Mosco (1976). Implicit variational prablems and quasi variational inequalities. In J.P. Gossez, E.J. Lami Dozo, J. Mawhin and L. Waebroeck (Eds), *Nonlinear Operators and the Calculus of Variations*, Lecture Notes in Math., New York: Springer-Verlag, **543**, 83–156.

[62] H. Nikaido (1963). *Convex Structures and Economic Theory*. New York: Academic Press.

[63] I. Parida, M. Sahoo and A. Kumar (1989). A variational-like inequality problem, *Bull. Austral. Math. Soc.*, **39**, 225–231.

[64] A.L. Peressini (1967). *Ordered Topological Vector Spaces*. New York: Harper and Row Publ.

[65] J. Philip (1972). Algorithms for one vector maximization problem, *Math. Programming*, **2**, 207–229.

[66] W. Rudin (1973). *Functional Analysis*. New York: McGraw-Hill Book Company.

[67] Y. Sawaragi, H. Nakayama and T. Tanino (1985). Theory of Multiobjective Optimization, *Math. Sci. Engrg.*, **176**, New York: Academic Press, 137–138.

[68] H.H. Schaefer (1971). *Topological Vector Spaces*. New York: Springer-Verlag.

[69] A.H. Siddiqi, Q.H. Ansari and A. Khaliq (1995). On vector variational inequality, *J. Optim. Theory Appl.*, **84**, 171–180.

[70] F. Szidarovszky, M.E. Gershon, and L. Duckstein (1986). *Techniques for Multiobjective Decision Making in Systems Management*. Amsterdam: Elsevier.

[71] T. Tanaka (1990). A characterization of generalized saddle points of vector-valued functions via scalarization, *Nihonkai Math. J.*, **1**, 209–227.

[72] E. Tarafdar (1977). On nonlinear variational inequalities, *Proc. Amer. Math. Soc.*, **67**, 95–98.

[73] E. Tarafdar (1987). A fixed point theorem equivalent to Fan-Knaster-Kuratowski-Mazurkiewicz's theorem, *J. Math. Anal. Appl.*, **128**, 475–479.

[74] E. Trafdar (1990). Nonlinear variational inequality with application to the boundary value problem for quasilinear operator in generalized diversence form, *Funkcialaj Ekvacioj* **33**, 441–453.

[75] E. Tarafdar (1990). A fixed point theorem in H-space and related results, *Bull. Austral. Math. Soc.*, **42**, 133–140.

[76] E. Tarafdar (1992). Applications of Pareto optimality of a mapping to economics, *Proceedings of World Congress of Nonlinear Analysis*.

[77] E. Tarafdar (1992). Pareto solution of cone variational inequality and Pareto Optimality of a mapping, to appear in the Proceedings of First World Congress of Nonlinear Analysts.

[78] J. Vályi (1987). Approximate saddle-point theorems in vector optimization, *J. Optim. Theory Appl.*, **55**, 436–448.

[79] S.Y. Wang (1993). Existence of a Pareto equilibrium, *J. Optim. Theory Appl.*, **79**, 373–384.

[80] X.Q. Yang and C.-J. Goh (1999). Vector variational inequalities, vector equilibrium flow and vector optimization. In F. Giannessi (Ed.), *Vector Variational Inequalities and Vector Equilibria, Mathematical Theories*. Kluwer Acad. Pub. (December), 447–465.

[81] N.D. Yen and G.M. Lee (1999). On monotone and strongly monotone vector variational inequalities. In F. Giannessi (Ed.), *Vector Variational Inequalities and Vector Equilibria, Mathematical Theories*. Kluwer Acad. Pub. (December), 467–478.

[82] H. Yin and C. Xu (1999). Vector variational inequality and implicit vector complementarity problems, In F. Giannessi (Ed.), *Vector Variational Inequalities and Vector Equilibria, Mathematical Theories*. Kluwer Acad. Pub. (December), 491–505.

[83] S.J. Yu and J.C. Yao (1996). On vector variational inequalities, *J. Math. Anal. Appl.*, **89**, 749–769.

[84] J. Yu and X.Z. Yuan (1995). The study of Pareto equilibria for multiobjective games by fixed point and Ky Fan minimax inequality methods. Research Report No. 1/95, Department of Institute of Mathematics, Guizhou Institute of Technology, China.

[85] P.L. Yu (1974). Cone convexity, cone extreme points, and nondominated solutions in decision problems with multiobjectives, *J. Optim. Theory Appl.*, **14**, 318–323.

[86] P.L. Yu (1979). Second-order game problems: Decision dynamics in gaming phenomena, *J. Optim. Theory Appl.*, **27**, 147–166.

[87] X.Z. Yuan and E. Tarafdar (1996). Non-compact Pareto equilibria for multiobjective games, *J. Math. Anal. Appl.*, **204**, 156–163.

[88] X.Z. Yuan (1995). KKM principle, Ky Fan minimax inequalities and fixed point theorems, *Nonlinear World*, **2**, 131–169.

[89] M. Zeleny (1976). Game with multiple payoffs, *Internat. J. Game Theory*, **4**, 179–191.

SIMAA 4(2002) 129–136

9. Variational Principle and Fixed Points

Peter Z. Daffer[1], Hideaki Kaneko[2] and Wu Li[2]

[1]*Information Systems Planning and Analysis, 1718 Peachtree St., N.W. Suite 560, South Tower, Atlanta, Georgia 30309*
[2]*Department of Mathematics and Statistics, Old Dominion University, Norfolk, Virginia 23529-0077*

Abstract: In this chapter, we extend the study that was initiated by Hamel [7] on the relationship among theorems of Ekeland, Caristi and Takahashi. We shall give a new proof of Takahashi's theorem and subsequently a new proof to show that the aforementioned theorems are equivalent which was done in [7]. A series of fixed point theorems for multivalued maps are presented that are equivalent to a well known theorem of Caristi.

1991 Mathematics Subject Classification: 47H10

1. INTRODUCTION

We begin by listing theorems of Takahashi, Caristi and Ekeland. We give an alternative proof to Takahashi's theorem. The main theorem of [12] is the following ("proper" means not identically equal to ∞):

Theorem 1.1 [12]: *Let (X, d) be a complete metric space and let φ: $X \to (-\infty, \infty]$ be a proper lower semicontinuous function, bounded from below. Suppose that, for each $u \in X$ with $\varphi(u) > \inf_{x\in X} \varphi(x)$, there is a $v \in X$ such that $v \neq u$ and $\varphi(v) + d(u, v) \leq \varphi(u)$. Then there exists an $x_0 \in X$ such that $\varphi(x_0) = \inf_{x\in X} \varphi(x)$.*

Proof: We let $m = \inf\{\varphi(v)|v \in X\}$, and suppose that $\varphi(u) > m$ for every $u \in X$. We define inductively a sequence $\{u_n\}$ as follows; take $u_0 \in X$ with $\varphi(u_0) < \infty$ and let

$$S_0 = \{v \in X \,|\, d(v, u_0) \leq \varphi(u_0) - \varphi(v)\}.$$

$S_0 \neq \emptyset$ since $u_0 \in S_0$, and S_0 is closed since φ is lower semi-continuous. Also, by hypothesis, S_0 is not a singleton set, since $\varphi(u_0) > m$ there is $u_1 \in S_0$, $u_1 \neq u_0$. Clearly $\varphi(u_1) < \varphi(u_0)$ and we are free to choose u_1 such that $\varphi(u_1) = \inf\{\varphi(v) \,|\, v \in S_0\}$. This minimum is attained because S_0 is a bounded set and φ is lower semi-continuous, bounded from below.

If $u_0, u_1, \ldots, u_n$ have been chosen, choose $u_{n+1} \in S_n = \{v \in X \,|\, d(u_n, v) \leq \varphi(u_n) - \varphi(v)\}$ with $u_{n+1} \neq u_n$, $\varphi(u_{n+1}) < \varphi(u_n)$ and $\varphi(u_{n+1}) = \inf\{\varphi(v) \,|\, v \in S_n\}$. We claim that $\{u_n\}$ is Cauchy. To see this, note that $d(u_{n+k}, u_n) \leq \sum_{i=1}^{k} d(u_{n+i}, u_{n+i-1}) \leq \sum_{i=1}^{k} (\varphi(u_{n+i-1}) - \varphi(u_{n+i})) = \varphi(u_n) - \varphi(u_{n+k})$, and since $\varphi(u_n)$ is a decreasing sequence converging to c for some c, for any $\epsilon > 0$, we have $d(u_{n+k}, u_n) < \epsilon$ for all k if n is sufficiently large.

Let $\{u_n\}$ converge to $u \in X$. We make a point here that u_n are all different. We claim that $u \in S_n$ for every n. We have

$$\begin{aligned} d(u_{n-k}, u_n) &\leq \sum_{i=0}^{k-1} d(u_{n-k+i}, u_{n-k+i+1}) \\ &\leq \sum_{i=0}^{k-1} [\varphi(u_{n-k+1}) - \varphi(u_{n-k+i+1})] \\ &= \varphi(u_{n-k}) - \varphi(u_n), \end{aligned}$$

which shows that $u_n \in S_{n-k}$, for $k = 1, 2, \ldots, n$, so that $u_n \in \cap_{i=0}^{n-1} S_i$. We thus have, for $k > n$, $u_k \in \cap_{i=0}^{n} S_i$, and since $\cap_{i=0}^{n} S_i$ is closed, $u \in \cap_{i=0}^{n} S_i$. Thus, in particular, $u \in S_n$, and since $u \neq u_n$, we get $\varphi(u) < \varphi(u_n)$. This contradicts the choice of u_n in S_n. This completes the proof. □

We note that the formulation that Hamel uses to describe Takahashi's condition is the following:

Condition (T): There exists an $\alpha > 0$ such that for each $u \in X$ such that $\varphi(u) > \inf_{x \in X} \varphi(x)$, there is a $v \in X$ such that $v \neq u$ and $\varphi(v) + \alpha d(u, v) \leq \varphi(u)$.

Condition (T) is different from the condition that was originally given by Takahashi. The condition used in Theorem 1.1 is the original definition of Takahashi's condition. It should be pointed out that condition (T) encompasses a wider class of lower semicontinuous functions than the condition in Theorem 1.1. For example, consider $f\colon R \to R$ defined by $f(x) = 1$ for $x \in (0, 4)$ and $f(x) = 0$ elsewhere. The function f does not satisfy the condition in Theorem 1.1: for $x = 2$, there is no y such that $|y - 2| \leq 1 - f(y)$ holds. However, condition (T) is satisfied by f by the metric $\alpha d(x, y) = \frac{1}{2}|x - y|$.

Takahashi [12] observed that Theorem 1.1 includes as corollaries three well known theorems. They are the fixed point theorem of Caristi [2], the ϵ-variational principle of Ekeland [6] and the fixed point theorem of Nadler [9]. Caristi's theorem is the following:

Theorem 1.2 [2]: *Let X be a complete metric space and let $\varphi\colon X \to R$ be a lower semicontinuous function bounded from below. Let $T\colon X \to X$ be a mapping satisfying*

$$d(x, Tx) \leq \varphi(x) - \varphi(Tx) \tag{1.2}$$

for every $x \in X$. Then there exists an $x_0 \in X$ with $Tx_0 = x_0$.

Much attention was drawn to Caristi's theorem since its publication because it requires no continuity on the mapping T. It also contains as a special case the fixed point theorem for multivalued contractions of Nadler [9] that is a generalization of the classical Banach fixed point principle. Nadler proved that if T is a mapping of a complete metric space X to $CB(X)$ (= the family of all closed bounded subsets of X) that satisfies

$$H(Tx, Ty) \leq kd(x, y) \quad \text{for all } x, y \in X,\ 0 \leq k < 1,$$

then T has a fixed point in X. Recently, some work has been made toward generalizing this theorem by relaxing the condition on the contractive constant k. In [11:p. 40], Reich proved that a mapping T: $X \to K(X)$ (=the family of all compact subsets of X) has a fixed point in X if it satisfies $H(Tx, Ty) \leq k(d(x, y))d(x, y)$ for all $x, y \in X$, where k: $(0, \infty) \to [0, 1)$ satisfies $\limsup_{r \to t^+} k(r) < 1$ for every $t \in (0, \infty)$. This result generalizes the fixed point theorem for single-valued mappings that was proved by Boyd and Wong [1]. Some attempts [3,5,8] were made to replace $K(X)$ by $CB(X)$ thereby generalizing Reich's theorem. Theorem 1.3 of Caristi, although it is of extremely general character, does not seem to have an immediate relationship with these theorems.

Now we return to the main scope of this chapter and state the variational principle of Ekeland.

Theorem 1.3 [6]: *Let (X, d) be a complete metric space and φ: $X \to (-\infty, \infty]$ a proper lower semicontinuous function bounded from below. Let $\epsilon > 0$ be given and a point $u \in X$ such that*

$$\varphi(u) \leq \inf_{x \in X} \varphi(x) + \epsilon.$$

Then there exists some point $v \in X$ such that

$$\varphi(v) \leq \varphi(u)$$

$$d(u, v) \leq 1$$

$$\varphi(w) > \varphi(v) - \epsilon d(v, w) \quad \textit{for all } w \neq v.$$

A number of useful applications of Theorem 1.3 are described in [6]. It is well documented that Theorems 1.2 and 1.3 are equivalent, e.g., see [7,12]. In [7], Hamel observed that Theorem 1.1 can be derived from Theorem 1.2, making Theorems 1.1, 1.2 and 1.3 equivalent. The main purpose of Section 2 is to provide the reader with additional equivalent formulations of these theorems. This provides us with an alternative perspective to the approach of Hamel in [7]. In addition, a series of fixed point theorems are given for lower semicontinuous multifunctions. In the final section, Section 3, the argument used to prove Theorem 1.1 is used to prove the existence of the weak sharp minima for a class of lower semicontinuous functions. This confirms and expands the result obtained in Theorem 2(ii) of [7].

2. VARIATIONAL PRINCIPLE AND FIXED POINT THEOREMS

First, in this section, we make an observation that the sequence $\{u_n\}$ generated in the proof of theorem 1.1 [12] actually converges to a minimizer of φ. Theorem 2.1 below is Theorem 1.1 slightly modified to reflect the observed point above. The proof is essentially the one given in [12].

Theorem 2.1: *Let (X, d) be a complete metric space and let φ: $X \to (-\infty, \infty]$ be a proper lower semicontinuous function, bounded from below. Suppose that, for each $u \in X$ with $\varphi(u) > \inf_{x \in X} \varphi(x)$, there is a $v \in X$ such that $v \neq u$ and $\varphi(v) + d(u, v) \leq \varphi(u)$. Then a sequence $\{u_n\}$ can be constructed that converges to a minimizer x_0 of φ, i.e., $x_0 \in X$ is such that $\varphi(x_0) = \inf_{x \in X} \varphi(x)$.*

Proof: Let $u_0 \in X$. If $\varphi(u_0) = \inf_{x\in X} \varphi(x) \equiv c$, then we are done. If $\varphi(u_i) > c$ for $i = 0, 1, \ldots, n-1$ ($n \geq 1$), then find $u_n \in S_n$ where

$$S_n = \{w \in X : \varphi(w) + d(u_{n-1}, w) \leq \varphi(u_{n-1})\}$$

such that

$$\varphi(u_n) \leq \inf_{w\in S_n} \varphi(w) + \frac{1}{2}\left\{\varphi(u_{n-1}) - \inf_{w\in S_n} \varphi(w)\right\}.$$

Arguing as in [12], we see that $\{u_n\}$ is a Cauchy sequence and that, with $u_n \to x_0 \in X$,

$$d(u_n, x_0) \leq \varphi(u_n) - \varphi(x_0).$$

We claim that x_0 is a minimizer for φ. If not, by hypothesis, there exists $z \in X$, $z \neq x_0$ such that

$$\begin{aligned}\varphi(z) &\leq \varphi(x_0) - d(x_0, z)\\ &\leq \varphi(x_0) - d(x_0, z) + \varphi(u_n) - \varphi(x_0) - d(u_n, x_0)\\ &\leq \varphi(u_n) - d(u_n, z).\end{aligned}$$

Hence $z \in S_n$. The definition of S_n implies that

$$2\varphi(u_n) - \varphi(u_{n-1}) \leq \inf_{w\in S_n} \varphi(w) \leq \varphi(z).$$

Using the above inequalities, we obtain the following contradiction,

$$\varphi(z) < \varphi(x_0) \leq \varphi(z).$$

Hence x_0 is a minimizer of φ. □

Takahashi demonstrated the generality of Theorem 1.1 by including as corollaries three well known theorems. They are the fixed point theorem of Caristi [2], the ϵ-variational principle of Ekeland [6] and the fixed point theorem of Nadler [9]. Our next task is to show that Theorems 1.1 and 1.3 of Caristi are indeed equivalent, providing an alternative proof to Theorem 1 of [7].

Proposition 2.2: *Theorem 1.1 of Takahashi and Theorem 1.2 of Caristi are equivalent. Hence they are equivalent to Theorem 1.3 of Ekeland.*

Proof: As was stated before this Proposition, Takahashi [12] showed that his Theorem 1.1 contains Theorem 1.2 as a corollary. For a converse, suppose that there is no $x_0 \in X$ such that $\varphi(x_0) = \inf_{x\in X} \varphi(x)$. Now define $S\colon X \to 2^X\backslash\emptyset$ by $Sx = \{y\colon \varphi(y) + d(x, y) \leq \varphi(x)\}$ and $T\colon X \to X$ by $Tx \in Sx\backslash\{x\}$. This is possible by the hypothesis to Takahashi's theorem. But T is fixed point free, and this is impossible by Caristi's theorem. Hence there is $x_0 \in X$ with $\varphi(x_0) = \inf_{x\in X} \varphi(x)$.

It is well known that Theorems 1.2 and 1.3 are equivalent [12]. □

We conclude this section by describing four other formulations in terms of the fixed points of lower semicontinuous multifunctions that can be shown to be equivalent to Theorem 1.1 of Takahashi. A similar development was made by Park [10].

Theorem 2.3: *Let X be a complete metric space and let φ: $X \to (-\infty, \infty]$ be a proper lower semicontinuous function bounded from below. Let T: $X \to 2^X \setminus \{\emptyset\}$. Suppose that, for all $x \notin T(x)$, there exists $y \neq x$ that satisfies*

$$\varphi(y) + d(y, x) \leq \varphi(x).$$

Then T has a fixed point in X.

Theorem 2.4: *Let X be a complete metric space and let φ: $X \to (-\infty, \infty]$ be a proper lower semicontinuous function bounded from below. Let T: $X \to 2^X \setminus \{\emptyset\}$ be such that for all $x \in X$ with $T(x) \neq \emptyset$ and for all $y \in T(x)$,*

$$\varphi(y) + d(y, x) \leq \varphi(x).$$

Then there exists $u \in X$ such that $T(u) = \{u\}$.

Theorem 2.5: *Let X be a complete metric space and let φ: $X \to (-\infty, \infty]$ be a proper lower semicontinuous function bounded from below. Let T: $X \to 2^X \setminus \{\emptyset\}$ be such that for all $x \in X$, there exists $y \in T(x)$ such that*

$$d(y, x) \leq \varphi(x) - \varphi(y).$$

Then T has a fixed point in X.

Theorem 2.6: *Let X be a complete metric space and let φ: $X \to (-\infty, \infty]$ be a proper lower semicontinuous function bounded from below. Let T: $X \to 2^X \setminus \{\emptyset\}$ be closed. Suppose that*

$$d(x, T(x)) \leq \varphi(x) - \sup_{y \in T(x)} \varphi(y) \quad \text{for } x \in X.$$

Then T has a fixed point in X.

The main result of the section is the following:

Proposition 2.7: *Theorems 2.3, 2.4, 2.5, 2.6 and 1.1 are equivalent.*

Proof: *Theorem 1.1 $\Rightarrow$ Theorem 2.3*
Suppose that T has no fixed point. Then for all $x \in X$, $x \notin T(x)$ so that by assumption there exists $y \neq x$ such that $\varphi(y) + d(y, x) \leq \varphi(x)$. Then by Theorem 1.1, there exists $x_0 \in X$ such that $\varphi(x_0) = \inf_{x \in X} \varphi(x)$. Let $y_0 \in T(x_0)$ be such that $y_0 \neq x_0$ and $\varphi(y_0) + d(y_0, x_0) \leq \varphi(x_0)$. Then

$$0 < d(x_0, y_0) \leq \varphi(x_0) - \varphi(y_0) \leq \varphi(y_0) - \varphi(y_0) = 0.$$

This contradiction proves the implication.

Theorem 2.3 $\Rightarrow$ Theorem 2.4
Suppose that there is no element $u \in X$ for which $T(u) = \{u\}$. For each $x \in X$, define f by $f(x) \in T(x) \setminus \{x\}$ so that f is fixed point free. Then $\varphi(f(x)) + d(f(x), x) \leq \varphi(x)$.

By Theorem 3.3, T must have a fixed point and it must be a fixed point of f also by Theorem 1.3 of Caristi. This contradiction proves the impication.

Theorem 2.4 ⇒ Theorem 1.1

Define T: $X \to 2^X$ by $T(x) = \{y: \varphi(y) + d(x,y) \leq \varphi(x)\}$. Suppose that there is no $x_0 \in X$ such that $\varphi(x_0) = \inf_{x\in X} \varphi(x)$. By the assumption in Theorem 1.1, $T(x) \neq \emptyset$. Then for each $x \in X$ and $y \in T(x)$, $\varphi(y) + d(y,x) \leq \varphi(x)$. Hence by Theorem 3.4, there exists $u \in X$ such that $T(u) = \{u\}$. This shows that there is no $v \neq u$ for which $\varphi(v) + d(u,v) \leq \varphi(u)$.

Theorem 2.4 ⇒ Theorem 2.5

This is obvious, since the condition in Theorem 2.4 implies the condition in Theorem 2.5.

Theorem 2.5 ⇒ Theorem 2.6

Let $\psi(x) \equiv \frac{1}{2}\varphi(x)$. Let $x \in X$. If $d(x, T(x)) = 0$, then $x \in T(x)$ since $T(x)$ is closed in X. Thus if T has no fixed point, then $d(x, T(x)) > 0$ for any $x \in X$. Let $y \in T(x)$ be such that $d(x,y) < \frac{1}{2}d(x, T(x))$. Then

$$d(x,y) \leq \frac{1}{2}d(x,T(x)) \leq \frac{1}{2}\left(\varphi(x) - \sup_{y\in T(x)} \varphi(y)\right) \leq \psi(x) - \psi(y).$$

Since ψ is a proper lower semicontinuous function bounded from below, by Theorem 2.5, T has a fixed point.

Theorem 2.6 ⇒ Theorem 2.4

Note that $T(x) \equiv \{y\}$ is closed. Hence Theorem 2.4 is a special case of Theorem 2.6.

This completes the proof of Proposition 2.7. □

3. APPLICATION TO WEAK SHARP MINIMA

In this section, we make use of the argument employed to demonstrate Theorem 1.1 to prove the existence of weak sharp minima for a class of lower semicontinuous functions. This expands Theorem 2(ii) of Hamel by giving an alternative approach. As before, let X be a complete metric space and φ: $X \to (-\infty, \infty]$ be lower semicontinuous. We define

$$m \equiv \inf\{\varphi(u) \mid u \in X\}$$

and

$$M \equiv \{v \in X \mid \varphi(v) = m\}. \tag{3.1}$$

Then we say that φ has weak sharp minima if, for any $u \in X$, we have

$$d(u, M) \leq \varphi(u) - m.$$

Hamel begins Section 2 of his paper [7] by stating that "we want to characterize functions which satisfy the condition of Takahashi" and that

Theorem 3.1 (Theorem 2 [7]): *Let X be a complete metric space and φ: $X \to (-\infty, \infty]$ be a lower semicontinuous function that is bounded from below.*

(*i*) *If there exists an* $\alpha > 0$ *and a minimizer* $u \in X$ *of the function* φ *such that*

$$\varphi(v) - \varphi(u) \geq \alpha d(v,u), \quad \textit{for all } v \in X, \tag{1}$$

then f *satisfies the condition of Takahashi (as in Condition (T) defined earlier) with same* $\alpha > 0$.

(*ii*) *Suppose that* φ *satisfies Condition (T) with* $\alpha > 0$. *Then*

$$\varphi(u) - \varphi(v) \geq \alpha\, \mathrm{dist}(x, M), \quad \textit{for all } u \in X \quad \text{and} \quad v \in M. \tag{2}$$

We note that Condition (T) does indeed imply (2) above, but (2) does not in turn imply (1). The fact that (2) can hold without (1) holding may be seen by letting $f\colon R \to R$ be defined by $f(x) = 0$ for all $x \in R$. Hence Theorem 2 of [7] does not provide a complete characterization of functions which satisfy Condition (T). Our final goal of this chapter is to provide a new proof of the existence of weak sharp minima for lower semicontinuous functions that satisfy the condition of Takahashi (the condition described in Theorem 1.1).

Theorem 3.2 (Compare to Theorem 3.1(*ii*): *Let* X *be a complete metric space and* $\varphi\colon X \to (\infty, \infty]$ *be a lower semicontinuous function that is bounded from below. Suppose that, for any* $u \in X$ *with* $\inf_{x \in x} \varphi(x) < \varphi(u)$, *there exists* $v \in X$ *such that* $v \neq u$ *and*

$$d(u,v) \leq \varphi(u) - \varphi(v).$$

Then M *defined in (3.1) is nonempty and* φ *has weak sharp minima.*

Proof: For $u \in X$, define

$$A(u) \equiv \{v \in X \mid d(u,v) \leq \varphi(u) - \varphi(v)\}.$$

Since φ is lower semicontinuous, $A(u)$ is a closed set. By Theorem 1.1, $M \neq \emptyset$, where M is defined in (3.1). Note that $\varphi(v) \leq \varphi(u)$ for every $v \in A(u)$. Now, by way of contradiction, let us assume that there is $u_0 \in X$ with

$$\varphi(u_0) - m < d(u_0, M). \tag{3.2}$$

Clearly, $u_0 \notin M$ and this is true for every $v \in A(u_0)$. For if there were $v \in A(u_0)$ with $\varphi(v) = m$, then we get $d(u_0, M) \leq d(u_0, v) \leq \varphi(u_0) - m$, which contradicts (3.2). We also note that (3.2) holds for every $v \in A(u_0)$. To see this, take $v \in A(u_0)$, $w \in M$, so that $d(u_0, w) \leq d(u_0, v) + d(v, w) \leq \varphi(u_0) - \varphi(v) + d(v, w)$ and this yields $d(u_0, M) \leq \varphi(u_0) - \varphi(v) + d(v, M)$. But $d(u_0, M) > \varphi(u_0) - m$, from (3.2), and together this gives $\varphi(u_0) - m < \varphi(u_0) - \varphi(v) + d(v, M)$, which is $\varphi(v) - m < d(v, M)$, giving (3.2) with v in place of u_0.

Since $\varphi(u_0) > m$, by hypothesis there is $u_1 \in A(u_0)$ with $u_1 \neq u_0$. Since $\varphi(u_1) - m < d(u_1, M)$, again it is clear that $u_1 \notin M$, that $\varphi(u_1) < \varphi(u_0)$ and that we can again show as above that $\varphi(v) - m < d(v, M)$ for every $v \in A(u_1)$ and $A(u_1) \cap M = \emptyset$. In addition, we select u_1 such that $\varphi(u_1) = \inf\{\varphi(v) \mid v \in A(u_1)\}$. This is possible since X is complete, φ is lower semicontinuous, and $A(u_1)$ is closed and nonempty. Continuing in this way, we generate a sequence $\{u_n\}$ with the above properties. Namely, if $u_0, u_1, \ldots, u_n$ have been chosen so that at $u_i \in A(u_{i-1})$, $\varphi(u_i) < \varphi(u_{i-1})$, $\varphi(u_i) = \inf\{\varphi(v) \mid v \in A(u_i)\}$, $A(u_{i-1}) \cap M = \emptyset$, $i = 1, 2, \ldots, n$ and $\varphi(v) - m < d(v, M)$ for every $v \in \bigcup_{i=1}^{n} A(u_{i-1})$, then, since $u_n \notin M$, we can choose $u_{n+1} \in A(u_n)$, $u_{n+1} \neq u_n$, with $\varphi(u_{n+1}) = \inf\{\varphi(v) \mid v \in A(u_n)\}$, and as above we will have $\varphi(u_{n+1}) > m$, $\varphi(u_{n+1}) < \varphi(u_n)$ and $\varphi(v) - m < d(v, M)$ for each $v \in A(u_{n+1})$. To see the latter, we write again, just as above, $\varphi(u_n) - m < d(u_n, M) \leq \varphi(u_n) - \varphi(u_{n+1}) + d(u_{n+1}, M)$, giving $\varphi(u_{n+1}) - m < d(u_{n+1}, M)$. Hence, $A(u_{n+1}) \cap M = \emptyset$.

We now have our sequence $\{u_n\}$ consisting of all different elements and $\varphi(u_{n+1}) < \varphi(u_n)$. Since $d(u_{n+k}, u_n) \leq \sum_{i=1}^{k} d(u_{n+i}, u_{n+i-1}) \leq \sum_{i=1}^{k} (\varphi(u_{n+i-1}) - \varphi(u_{n+i})) = \varphi(u_n) - \varphi(u_{n+k})$, and noting that $\varphi(u_n)$ monotonically decreases to some c, $\{u_n\}$ must be Cauchy. Let u_n converge to $u \in X$. We now show that $u \in \bigcap_{i=0}^{\infty} A(u_i)$. We first show that, for every n, $u_n \in \bigcap_{i=0}^{n-1} A(u_i)$. This follows from the following:

$$\begin{aligned} d(u_{n-k}, u_n) &\leq \sum_{j=0}^{k-1} d(u_{n-k+j}, u_{n-k+j+1}) \\ &\leq \sum_{j=0}^{k-1} [\varphi(u_{n-k+j}) - \varphi(u_{n-k+j+1})] \\ &= \varphi(u_{n-k}) - \varphi(u_n), \end{aligned}$$

which shows (recall that all the u_i are outside M) that $u_n \in A(u_{n-k})$, $k = 1, \ldots, n$, hence $u_n \in \bigcap_{i=0}^{n-1} A(u_i)$. It follows immediately from this that $u_k \in \bigcap_{i=0}^{n-1} A(u_i)$ for all $k \geq n$. Since $\bigcap_{i=0}^{n-1} A(u_i)$ is a closed set, $u \in \bigcap_{i=0}^{\infty} A(u_i)$. Thus $u \in A(u_n)$, and $u \neq u_n$; hence $\varphi(u) < \varphi(u_n)$, and this is a contradiction, since $\varphi(v) \geq \varphi(u_n)$ for every $v \in A(u_n)$. □

REFERENCES

[1] D.W. Boyd and J.S.W. Wong (1968). On Nonlinear Contractions, *Proc. A.M.S.*, **89**, 458–464.

[2] J. Caristi (1976). Fixed Point Theorems for Mappings Satisfying Inwardness Conditions, *Trans. Amer. Math. Soc.*, **215**, 241–251.

[3] P.Z. Daffer and H. Kaneko (1995). Fixed Points of Generalized Contractive Multi-Valued Mappings, *Jl. Math. Anal. Appl.*, **192**, 655–666.

[4] P.Z. Daffer and H. Kaneko (1998). A variational principle of Ekeland. In P.K. Jain (Ed.), *Functional AnalysisSelected Topics*, New Delhi, India: Narosa Publishing House, 152–157.

[5] P.Z. Daffer, H. Kaneko and W. Li (1996). On a Conjecture of S. Reich, *Proc. Amer. Math. Soc.*, **124**(10), 3159–3162.

[6] I. Ekeland (1979). Nonconvex Minimization Problems, *Bull. Amer. Math. Soc.* (New Series), **1**, 443–474.

[7] A. Hamel (1994). Remarks to an equivalent formulation of Ekeland's variational principle, *Optimization*, **31**, 233–238.

[8] N. Mizoguchi and W. Takahashi (1989). Fixed Points Theorems for Multi-Valued Mappings on Complete Metric Spaces, *Jl. Math. Anal. Appl.*, **141**, 177–188.

[9] S.B. Nadler Jr. (1969). Multivalued Contraction Mappings, *Pacific Jl. Math.*, **30**, 475–488.

[10] S. Park (1983). Equivalent Formulations of Ekeland's Variational Principle and Their Applications. Tech. Report No. 1, The Math. Sci. Res. Inst. of Korea.

[11] S. Reich (1972). Fixed Points of Contractive Functions, *Boll. Un. Mat. Ital.*, **5**(4), 26–42.

[12] W. Takahashi (1989). Existence Theorems Generalizing Fixed Point Theorems for Multi-valued Mappings. In M.A. Thra and J.B. Baillon (Eds), *Fixed Point Theory and Applications*, Longman Scientific and Technical, 397–406.

SIMAA 4(2002) 137–148

10. On the Baire Category Method in Existence Problems for Ordinary and Partial Differential Inclusions

F.S. De Blasi[1] and G. Pianigiani[2]

[1]*Dipartimento di Matematica, Università di Roma II (Tor Vergata), Via della Ricerca Scientifica, 00133 Roma, Italy*
[2]*Dipartimento di Matematica per le Decisioni, Università di Firenze, Via Lombroso, 6/17, 50134 Firenze, Italy*

1. INTRODUCTION

Let $\mathbb{E}$ be a Banach space and let f: $I \times \mathbb{E} \to \mathbb{E}$ be a map belonging to a set, say $\mathcal{M}$. For $f \in \mathcal{M}$ consider the differential equation $x'(t) = f(t, x(t))$. The question whether this equation has properties like existence, uniqueness, continuous dependence, periodic solutions etc. is in general difficult to be answered, especially in infinite dimensions. In this context a number of mathematicians including Orlicz, Peixoto, Markus, Smale started to investigate whether the above properties were, perhaps, generically satisfied in $\mathcal{M}$. Here, if $\mathcal{M}$ is a complete metric space of maps, Baire category enters naturally if we agree to say that a property is *generic* in $\mathcal{M}$ provided that it is satisfied by all f belonging to a residual subset $\mathcal{M}_0$ of $\mathcal{M}$, i.e., with $\mathcal{M}\backslash\mathcal{M}_0$ a set of Baire first category. It is worthwhile to point out that knowing that a property is generic in $\mathcal{M}$ does not furnish, in general, any information whether a specific $f \in \mathcal{M}$ belongs or not to the generic set.

In this note we are concerned with existence results for Cauchy problems for ordinary differential inclusions and Dirichlet problems for partial differential inclusions of the form

$$x'(t) \in F(t, x(t)) \quad x(0) = x_0 \tag{1}$$

$$\begin{aligned} \nabla u(x)) &\in F(x, u(x)) && x \in \Omega \text{ a.e.} \\ u(x) &= \varphi(x) && x \in \partial\Omega. \end{aligned} \tag{2}$$

The method we discuss is based on the Baire category; yet, unlike the situation considered before, it enables us to establish existence results for specific differential inclusions.

This area of research was started by De Blasi and Pianigiani [14,15] in the framework of existence problems for non convex valued differential inclusions of the form (1). In this development a stimulus was offered by a paper of Cellina [7] in which it was shown that the solution set to the Cauchy problem $x'(t) \in \{-1, 1\}$, $x(0) = 0$, is a residual subset of the solution set of $x'(t) \in [-1,1]$, $x(0) = 0$, if the latter is endowed with the metric of uniform convergence.

In the present chapter we give a partial review of existence results, for differential inclusions, whose proofs are based on the Baire category method. Significant developments and applications are due, among others, to Bressan [2], Bressan and Colombo [3], Bressan and Flores [4], Dacorogna and Marcellini [12,13], Papageorgiou [24,25], Suslov [27,28], Tolstonogov [29,30].

In Sections 2,3,4 we are concerned with existence of solutions to the Cauchy problem (1) when the values of F are assumed to have nonempty interior and then when the values of F are compact. In Section 5 we study the existence of solutions to the Dirichlet problem (2) in the scalar and vectorial case.

2. CAUCHY PROBLEM

Throughout $\mathbb{E}$ stands for a reflexive real Banach space and $K(\mathbb{E})$ for the space of all nonempty bounded closed convex subsets of $\mathbb{E}$ endowed with the Hausdorff metric.

Let X be a metric space with distance d. By $B(a,r)$ or $B[a,r]$ we mean an open or closed ball with center a and radius r. For a non empty set $A \subset X$, we put $B[A,r] = \{x \in X\colon d(x,A) \leq r\}$, where $d(x,A) = \inf\{d(x,a) : a \in A\}$; furthermore we denote by diam A the diameter of A.

For $A \subset \mathbb{E}$ we denote by ∂A, $\operatorname{int}A$, $\operatorname{ext}A$, $\overline{co}A$ respectively, the boundary, the interior, the set of the extreme points, and the closed convex envelope of A. We set $I = [0,1]$.

By a *solution* of the Cauchy problem (1) we mean a Lipschitzean function $x\colon I \to \mathbb{E}$ satisfying (1) a.e. Observe that $x'(t)$ exists a.e. for $\mathbb{E}$ is reflexive. Set

$$\mathcal{M}_F = \{x \in C(I,\mathbb{E}) : x \text{ is solution of } (1)\}.$$

In the sequel $\mathcal{M}_F$ will always be endowed with the metric of $C(I,\mathbb{E})$.

In order to give an idea of the method we start with a simple situation. Consider the following differential inclusion

$$x'(t) \in \partial F(t,x(t)) \quad x(0) = x_0. \tag{3}$$

Theorem 1: [14] *Let $\mathbb{E}$ be a reflexive Banach space. Let $F\colon I \times B(x_0,r) \to K(\mathbb{E})$ be Hausdorff continuous and bounded by a constant $M < r$. Moreover, assume that $\operatorname{int}F(t,x) \neq \emptyset$ for every $(t,x) \in I \times B(x_0,r)$. Then the Cauchy problems* (1) *and* (3) *have solutions and the solution set $\mathcal{M}_{\partial F}$ of* (3) *is a residual subset of $\mathcal{M}_F$.*

Proof: (Sketch) The space $\mathcal{M}_F$ is nonempty, as F has locally Lipschitzean selections f, and for any such f the Cauchy problem $x'(t) = f(t,x(t))$, $x(0) = x_0$ has a unique solution defined on I. Further, under our assumptions, the metric space $\mathcal{M}_F$ is complete. For $n \in N$ set

$$\mathcal{M}_n = \left\{x \in \mathcal{M}_F\colon \int_I d(x'(t), \partial F(t,x(t))dt < 1/n\right\}.$$

The main step is to prove that each $\mathcal{M}_n$ is an open and dense subset of $\mathcal{M}_F$. Then, by the Baire category theorem, it follows that the set $\cap\mathcal{M}_n$ is residual in $\mathcal{M}_F$, hence nonempty. As $\mathcal{M}_{\partial F} = \cap\mathcal{M}_n$ the statement follows.

Let F: $I \times B(x_0, r) \to K(\mathbb{E})$ be a Hausdorff continuous multifunction bounded by a constant $M < r$. Consider the Cauchy problem

$$x'(t) \in \text{ext}F(t, x(t)) \quad x(0) = x_0. \tag{4}$$

Note that the set $\text{ext}F(t, x)$ is not necessarily closed and the map $(t, x) \to \text{ext}F(t, x)$ not necessarily Hausdorff continuous. We associate to (4) the convex Cauchy problem

$$x'(t) \in F(t, x(t)) \quad x(0) = x_0. \tag{5}$$

By a well known counterexample due to Plis the set $\mathcal{M}_{\text{ext}F}$ of all solutions of (4) may fail to be dense in $\mathcal{M}_F$. Nevertheless in Theorems 2 and 3 it is proved that, under suitable assumptions on F, the set $\mathcal{M}_{\text{ext}F}$ is nonempty and approximates some significant subsets of $\mathcal{M}_F$.

To study the Cauchy problem (4) we need to introduce the Choquet function $d_F(t, x, z)$ (see [10,5,26]) which "measures" the distance of a point $z \in F(t, x)$ from the set of the extreme points of $F(t, x)$. Some useful properties of d_F are listed below:

(i) d_F is non negative and bounded by M^2 on the graph of F;
(ii) $d_F(t, x, z) = 0$ if and only if $z \in \text{ext}F(t, x)$;
(iii) if $\{x_n\} \subset \mathcal{M}_F$ is a sequence converging to $x \in \mathcal{M}_F$, then

$$\limsup \int_I d_F(t, x_n(t), x_n'(t))dt \leq \int_I d_F(t, x(t), x'(t))dt.$$

3. *F* WITH NONEMPTY INTERIOR

Suppose that F: $I \times B(x_0, r) \to K(\mathbb{E})$ is a Hausdorff continuous multifunction, bounded by a constant $M < r$, taking values with nonempty interior. Set

$$S_F = \{f\colon I \times B(x_0, r) \to \mathbb{E}\colon f \text{ is a locally Lipschitzean selection of } F\}.$$

By using locally constant selections of F (which exist as the values of F have nonempty interior) and a locally Lipschitzean partition of unity, it can be shown that S_F is nonempty. For $f \in S_F$ we denote by x_f: $I \to \mathbb{E}$ the solution, which exists and is unique, of the Cauchy problem

$$x'(t) = f(t, x(t)) \quad x(0) = x_0. \tag{6}$$

Under the above assumptions on F the space $\mathcal{M}_F$ is nonempty and complete under the metric of $C(I, \mathbb{E})$.

The following lemma holds.

Lemma 1: *Let F be as above, $f \in S_F$, and let $\eta, \theta > 0$. Then there exists $g \in S_F$ such that*

$$x_g \in \mathcal{M}_\theta \cap B(x_f, \eta).$$

The proof is rather technical and can be found in [17]. The meaning of the Lemma is that for each selection $f \in S_F$ there exists another selection $g \in S_F$ such that x_g is close to x_f, and $x_g'(t)$ takes values near the extreme points of $F(t, x_g(t))$, in the sense that $\int_I d_F(t, x_g(t), x_g'(t))dt < \theta$.

Theorem 2: [17] *Let* $\mathbb{E}$ *be a reflexive separable Banach space, and let* $F\colon I \times B(x_0, r) \to K(\mathbb{E})$ *be a Hausdorff continuous multifunction, bounded by a constant* $M < r$, *taking values with nonempty interior. Then for each* $f \in S_F$ *and* $\eta > 0$ *the set* $\mathcal{M}_{\text{ext}F} \cap B(x_f, \eta)$ *is nonempty. In particular* $\mathcal{M}_{\text{ext}F}$ *is nonempty.*

Proof: (sketch) For $\theta > 0$ set

$$\mathcal{M}_\theta = \{x \in \mathcal{M}_F : \int_I d_F(t, x(t), x'(t))dt < \theta\}.$$

By property (iii) of the Choquet function the set $\mathcal{M}_\theta$ is open in $\mathcal{M}_F$. Put $\theta_n = 1/(n+1)$. By Lemma 1 there exists $g_1 \in S_F$ such that

$$x_{g_1} \in \mathcal{M}_{\theta_0} \cap B(x_f, \eta).$$

Hence there exists $0 < \eta_1 < \theta_1$ such that

$$B[x_{g_1}, \eta_1] \subset \mathcal{M}_{\theta_0} \cap B(x_f, \eta).$$

Arguing by induction, we construct a decreasing sequence of balls $B[x_{g_n}, \eta_n] \subset \mathcal{M}_F$, $0 < \eta_n < \theta_n$, satisfying

$$B[x_{g_{n+1}}, \eta_{n+1}] \subset \mathcal{M}_{\theta_n} \cap B(x_{g_n}, \eta_n). \tag{7}$$

As each $B[x_{g_n}, \eta_n]$ is closed and $\eta_n \to 0$, by Cantor intersection theorem it follows that $\bigcap B[x_{g_n}, \eta_n]$ is nonempty. If x is in this intersection then, by (7), $x \in M_{\theta_n}$ for all n, hence

$$\int_I d_F(t, x(t), x'(t))dt = 0.$$

Property (ii) of the Choquet function implies that $x'(t) \in \text{ext}F(t, x(t))$ a.e. As $B[x_{g_n}, \eta_n] \subset B(x_f, \eta)$ for all n, it follows that $x \in B(x_f, \eta)$, which completes the proof.

4. *F* COMPACT

Let $F\colon I \times B(x_0, r) \to K(\mathbb{E})$ be a Hausdorff continuous multifunction, bounded by a constant $M < r$, whose values are contained in a compact subset of $\mathbb{E}$. Let

$$S_F = \{f\colon I \times B(x_0, r) \to \mathbb{E}\colon f \text{ is a continuous selection of } F\}.$$

Clearly S_F is non empty by Michael's selection theorem (see [1]). Under the above assumptions on F, the space $\mathcal{M}_F$ is nonempty and compact. For $f \in S_F$ we denote by P_f the set (which is nonempty and compact) of all solutions of (6).

Lemma 2: *Let F be as above, $f \in S_F$, and let $\eta, \theta > 0$. Then there exists $g \in S_F$ such that*

$$P_g \subset \mathcal{M}_\theta \cap B(P_f, \eta).$$

The proof of this lemma is quite long and can be found in [16]. The main steps are the following. First we define a suitable partition $\mathcal{R}$ and using it we approximate f locally by functions taking values near the extreme points of F. Hence we use a refinement $\mathcal{R}'$ of $\mathcal{R}$ to construct a selection g of F with the properties stated in the lemma.

Theorem 3: [16] *Let $\mathbb{E}$ be a reflexive separable Banach space. Let $F: I \times B(x_0, r) \to K(\mathbb{E})$ be a Hausdorff continuous and compact multifunction, bounded by a constant $M < r$. Then for each $f \in S_F$ and $\eta > 0$ the set $\mathcal{M}_{\text{ext}F} \cap B(P_f, \eta)$ is nonempty. In particular $\mathcal{M}_{\text{ext}F}$ is nonempty.*

Proof: (sketch) Let $\mathcal{M}_\theta$ be as in the proof of Theorem 2, and observe that $\mathcal{M}_\theta$ is open by property (iii) of the Choquet function d_F. Set $\theta_n = 1/(n+1)$. By Lemma 2 there exists $g_1 \in S_F$ such that

$$P_{g_1} \subset \mathcal{M}_{\theta_0} \cap B(P_f, \eta)$$

thus, for some $0 < \eta_1 < \theta_1$,

$$B[P_{g_1}, \eta_1] \subset \mathcal{M}_{\theta_0} \cap B(P_f, \eta).$$

By induction one can construct a decreasing sequence of nonempty closed sets $B[P_{g_n}, \eta_n] \subset \mathcal{M}_F$, where $g_n \in S_F$, $0 < \eta_n < \theta_n$, such that

$$B[P_{g_{n+1}}, \eta_{n+1}] \subset \mathcal{M}_{\theta_n} \cap B(P_{g_n}, \eta_n).$$

As $\mathcal{M}_F$ is compact, it follows that $\bigcap B[P_{g_n}, \eta_n]$ is nonempty. The conclusion is as in the proof of Theorem 2.

The Baire approach works equally well under Caratheodory type assumptions on F. Let

$$S'_F = \{f: I \times B(x_0, r) \to \mathbb{E}: f \text{ is a Caratheodory selection of } F\}.$$

Theorem 4: [16] *Let $\mathbb{E}$ be a reflexive separable Banach space. Let $F: I \times B(x_0, r) \to K(\mathbb{E})$ be a compact Caratheodory multifunction, bounded by a constant $M < r$. Then for each $f \in S'_F$ and $\eta > 0$ the set $\mathcal{M}_{\text{ext}F} \cap B(P_f, \eta)$ is nonempty. In particular $\mathcal{M}_{\text{ext}F}$ is nonempty.*

Proof: (Sketch) The analog of Lemma 2 remains valid with S'_F in place of S_F. To see that, let $f \in S'_F$ and let $\eta, \theta > 0$. By Scorza-Dragoni's theorem there is a closed set $J \subset I$, with measure close enough to 1, such that the restrictions of f and F to $J \times B(x_0, r)$ are both continuous. By Dugundji's theorem these maps have continuous extensions, say $f^*: I \times B(x_0, r) \to E$ and $F^*: I \times B(x_0, r) \to K(\mathbb{E})$, respectively. By Lemma 2 there is $g^* \in S_{F^*}$ such that

$$P_{g^*} \subset \mathcal{M}_\theta \cap B(P_{f^*}, \eta)$$

and then from g^* we construct a Caratheodory selection $g \in S'_F$ such that

$$P_g \subset \mathcal{M}_\theta \cap B(P_f, \eta).$$

From this point the proof runs as that of Theorem 3.

5. DIRICHLET PROBLEM

Recently Dacorogna and Marcellini [12] (see also [4]) have used the Baire method to establish existence of solutions to the Dirichlet problem for Hamilton-Jacobi equations of the form

$$\begin{aligned} &H(\nabla u(x)) = 0 && x \in \Omega \text{ a.e.} \\ &u(x) = \varphi(x) && x \in \partial\Omega. \end{aligned}$$

where $\Omega \subset \mathbb{R}^n$ is open. They consider both, the scalar and the vectorial problem. After rewriting the above problem in the equivalent form

$$\begin{aligned} &\nabla u(x) \in C && x \in \Omega \text{ a.e.} \\ &u(x) = \varphi(x) && x \in \partial\Omega, \end{aligned}$$

where $C = \{z \in \mathbb{R}^n \mid H(z) = 0\}$, they prove an existence theorem by using a Baire category approach very much in the spirit of Theorems 1 and 2 above. With reference to the scalar case the basic assumptions are:

(j) φ is C^1, or piecewise C^1;
(jj) $\nabla\varphi(x) \in \operatorname{int}\overline{co}\, C$.

Now we discuss some generalizations of the above mentioned results. In the scalar case, which we consider first, we show that assumption (j) can be relaxed. More details can be found in [18]. We retain notation and terminology of [22] and [11].

Let $\Omega \subset \mathbb{R}^n$ be open and bounded and let $\varphi \in C(\overline{\Omega}) \cap W^{1,\infty}(\Omega)$. Consider the following Dirichlet problem

$$\begin{aligned} &\nabla u(x)) \in F(x, u(x)) && x \in \Omega \text{ a.e.} \\ &u(x) = \varphi(x) && x \in \partial\Omega. \end{aligned} \qquad (D_F)$$

where F is a multifunction, defined on $\Omega \times \mathbb{R}$, taking values in the space of the nonempty bounded subsets of $\mathbb{R}^n$.

As before, problem (D_F) corresponds to the Dirichlet problem for a Hamilton-Jacobi equation of the form

$$\begin{aligned} &H(x, u(x), \nabla u(x)) = 0 && x \in \Omega \text{ a.e.} \\ &u(x) = \varphi(x) && x \in \partial\Omega. \end{aligned}$$

By a *solution* of (D_F) we mean a function $u \in C(\overline{\Omega}) \cap W^{1,\infty}(\Omega)$ satisfying $\nabla u(x)) \in F(x, u(x))$, $x \in \Omega$ a.e. and $u(x) = \varphi(x)$, $x \in \partial\Omega$.

Now we state a technical approximation lemma which will play an important role in the proof of the next theorem. This lemma is a sharper version of approximation results of the type considered by Ekeland and Temam [20], Dacorogna [11], and Dacorogna and Marcellini [12].

Lemma 3: *Let Ω be a nonempty open and bounded subset of $\mathbb{R}^n$. Let $F\colon \Omega \times \mathbb{R} \to K(\mathbb{R}^n)$ be a Hausdorff continuous multifunction bounded by a constant M. Moreover, assume that* $\operatorname{int}F(x,u) \neq \emptyset$ *for every $(x,u) \in \Omega \times \mathbb{R}$. Let u be a solution of the Dirichlet problem*

$$\begin{aligned} &\nabla u(x) \in \operatorname{int}F(x,u(x)) && x \in \Omega \text{ a.e.} \\ &u(x) = \varphi(x) && x \in \partial\Omega, \end{aligned} \qquad (D_{\operatorname{int}F})$$

where $\varphi \in C(\overline{\Omega}) \cap W^{1,\infty}(\Omega)$. Let $\varepsilon, \alpha > 0$. Then there exists a compact set $\Omega_0 \subset \Omega$, $m(\Omega \backslash \Omega_0) < \alpha$, where m denotes the Lebesgue measure, and for each $x_0 \in \Omega_0$ there exist a family $\{A_\sigma : \sigma \in \Sigma_{x_0}\}$ of compact sets $A_\sigma \subset \Omega$ and a family $\{u_\sigma : \sigma \in \Sigma_{x_0}\}$ of functions $u_\sigma : \overline{\Omega} \to \mathbb{R}$, where $\Sigma_{x_0} = (0, \sigma_{x_0})$ for some $\sigma_{x_0} > 0$, satisfying for each $\sigma \in \Sigma_{x_0}$ the following properties:

(*i*) $B(x_0, \eta\sigma) \subset A_\sigma \subset B(x_0, 2\sigma)$, *where $0 < \eta < 1$ is independent of σ;*
(*ii*) $u_\sigma \in C(\overline{\Omega}) \cap W^{1,\infty}(\Omega)$ *is solution of the Dirichlet problem* $(D_{\operatorname{int}F})$;
(*iii*) $\|u_\sigma - u\|_{C(\overline{\Omega})} < \varepsilon$;
(*iv*) $d_F(x, u_\sigma(x), \nabla u_\sigma(x)) < \alpha, \quad x \in A_\sigma$ *a.e.*;
(*v*) $u_\sigma(x) = u(x)$, $x \in \overline{\Omega} \backslash A_\sigma$.

The proof is technical and can be found in [18].
In order to prove the next existence result we introduce an appropriate space $\mathcal{M}$. Set

$$\mathcal{M} = cl\{u : \overline{\Omega} \to \mathbb{R} \mid \nabla u(x) \in \operatorname{int}F(x,u(x)), x \in \Omega \text{ a.e.} \quad \text{and} \quad u(x) = \varphi(x), x \in \partial\Omega\},$$

where the closure is understood in the metric of uniform convergence.

Theorem 5: *Let Ω be a nonempty open bounded subset of $\mathbb{R}^n$, and let $F\colon \Omega \times \mathbb{R} \to K(\mathbb{R}^n)$ be a Hausdorff continuous multifunction bounded by a constant M. Moreover, assume that* $\operatorname{int}F(x,u) \neq \emptyset$ *for every $(x,u) \in \Omega \times \mathbb{R}$. Let $\varphi \in C(\overline{\Omega}) \cap W^{1,\infty}(\Omega)$ be such that* $\nabla\varphi(x) \in \operatorname{int}F(x, \varphi(x))$, *$x \in \Omega$ a.e. Then the Dirichlet problem*

$$\begin{aligned} &\nabla u(x) \in \operatorname{ext}F(x,u(x)) && x \in \Omega \text{ a.e.} \\ &u(x) = \varphi(x) && x \in \partial\Omega, \end{aligned} \qquad (D_{\operatorname{ext}F})$$

has a solution $u \in C(\overline{\Omega}) \cap W^{1,\infty}(\Omega)$. Furthermore the set of all solutions of $(D_{\operatorname{ext}F})$ is dense in $\mathcal{M}$.

Proof: (Sketch) Under our assumptions the space $\mathcal{M}$, endowed with the metric of uniform convergence, is a nonempty complete metric space. For $\alpha > 0$ set

$$\mathcal{M}_\alpha = \{u \in \mathcal{M} : \int_\Omega d_F(x, u(x), \nabla u(x))dx < \alpha\}.$$

By using the properties of the Choquet function d_F one can prove that $\mathcal{M}_\alpha$ is open in $\mathcal{M}$. To see that $\mathcal{M}_\alpha$ is dense in $\mathcal{M}$ it suffices to show that, for every $\varepsilon > 0$ and every solution u of the Dirichlet problem $(D_{\text{int}F})$, there exists $v \in \mathcal{M}_\alpha$ such that

$$\|v - u\|_{C(\overline{\Omega})} < \varepsilon.$$

Let $u: \overline{\Omega} \to \mathbb{R}$ be a solution of the Dirichlet problem $(D_{\text{int}F})$ and let $\varepsilon > 0$. Set $\alpha' = \alpha/(2M^2 + m(\Omega))$. By virtue of Lemma 3 (with α' in the place of α) there exists a compact $\Omega_0 \subset \Omega$, with $m(\Omega \backslash \Omega_0) < \alpha'$, such that for each $x_0 \in \Omega_0$ there exist a family $\{A_\sigma^{x_0}: \sigma \in \Sigma_{x_0}\}$ of compact sets $A_\sigma^{x_0} \subset \Omega$ and a family $\{u_\sigma^{x_0}: \sigma \in \Sigma_{x_0}\}$ of functions $u_\sigma^{x_0}: \overline{\Omega} \to \mathbb{R}$, where $\Sigma_{x_0} = (0, \sigma_{x_0})$ for some $\sigma_{x_0} > 0$, satisfying properties (i)–(v) of Lemma 3. Since $\operatorname{diam}(A_\sigma^{x_0}) \to 0$ as $\sigma \to 0$, the family $\mathcal{V} = \{A_\sigma^{x_0}: x_0 \in \Omega_0, \sigma \in \Sigma_{x_0}\}$ is a Vitali covering of Ω_0. Since $m(\Omega_0) < +\infty$, by Vitali's theorem there exists a finite family of pairwise disjoint sets $\{A_{\sigma_i}^{x_i}: x_i \in \Omega_0, \ \sigma_i \in \Sigma_{x_i}, \ i = 1, \ldots, k\}$ such that

$$m\left(\Omega_0 \backslash \bigcup_{i=1}^{k} A_{\sigma_i}^{x_i}\right) < \alpha'.$$

Now, define $v: \overline{\Omega} \to \mathbb{R}$ by

$$v(x) = \begin{cases} u_{\sigma_i}^{x_i}(x) & \text{if } x \in A_{\sigma_i}^{x_i} \text{ for some } 1 \leq i \leq k \\ u(x) & \text{if } x \in \overline{\Omega} \backslash \bigcup_{i=1}^{k} A_{\sigma_i}^{x_i}. \end{cases}$$

The compact sets $A_{\sigma_i}^{x_i}$ are pairwise disjoint thus, by Lemma 3 (ii),(iii),(v), we have that $v \in C(\overline{\Omega}) \cap W^{1,\infty}(\Omega)$ is solution of the Dirichlet problem $(D_{\text{int}F})$, hence $v \in \mathcal{M}$ and

$$\|v - u\|_{C(\overline{\Omega})} < \varepsilon.$$

Furthermore $v \in \mathcal{M}_\alpha$. In fact, in view of Lemma 3 (iv), we have

$$\int_\Omega d_F(x, v(x), \nabla v(x))dx \leq \int_{\bigcup_{i=1}^{k} A_{\sigma_i}^{x_i}} d_F(x, v(x), \nabla v(x))dx +$$

$$\int_{\Omega_0 \backslash \bigcup_{i=1}^{k} A_{\sigma_i}^{x_i}} d_F(x, v(x), \nabla v(x))dx + \int_{\Omega \backslash \Omega_0} d_F(x, v(x), \nabla v(x))dx \leq$$

$$\sum_{i=1}^{k} \int_{A_{\sigma_i}^{x_i}} d_F(x, v(x), \nabla v(x))dx + M^2 m(\Omega \backslash \bigcup_{i=1}^{k} A_{\sigma_i}^{x_i}) + M^2 m(\Omega \backslash \Omega_0) \leq$$

$$\alpha' m(\Omega) + M^2 \alpha' + M^2 \alpha' < \alpha.$$

As the solution u of $(D_{\text{int}F})$ and $\varepsilon > 0$ were arbitrary, we have actually proved that $\mathcal{M}_\alpha$ is dense in $\mathcal{M}$.

Now set $\alpha_k = 1/k$ and let

$$\mathcal{M}_0 = \bigcap_{k=1}^{+\infty} \mathcal{M}_{\alpha_k}.$$

By the Baire theorem it follows that $\mathcal{M}_0$ is nonempty and dense in $\mathcal{M}$. Each $v \in \mathcal{M}_0$ is a solution of the Dirichlet problem $(D_{\text{ext}F})$. In fact $v \in \mathcal{M}_{\alpha_k}$, $k \in \mathbb{N}$, implies that $d_F(x, v(x), \nabla v(x)) = 0$, $x \in \Omega$ a.e. and hence $\nabla v(x) \in \text{ext}F(x, v(x))$, $x \in \Omega$ a.e. Furthermore, $v(x) = \varphi(x)$, $x \in \partial\Omega$, and clearly $v \in C(\overline{\Omega}) \cap W^{1,\infty}(\Omega)$. Consequently v is solution of the Dirichlet problem $(D_{\text{ext}F})$. This completes the proof.

In the previous Theorem 5 the assumption that φ satisfies $\nabla\varphi(x) \in \text{int}F(x, \varphi(x))$, $x \in \Omega$ a.e. is crucial as it is shown by the following example.

Example 1: Let $\Omega = (0,1) \times (0,1)$. Let F: $\Omega \times \mathbb{R} \to \mathbb{R}^2$ be defined by $F(x,u) = [(-1, |u|), (1, |u|)]$ and let $\varphi = 0$. Then the Dirichlet problem

$$\begin{aligned} &\nabla u(x) \in \text{ext}F(x, u(x)) \quad && x \in \Omega \text{ a.e.} \\ &u(x) = \varphi(x) && x \in \partial\Omega \end{aligned} \qquad (\text{D}_{\text{extF}})$$

has no solution.

Proof: Note that $\text{ext}F(x,u)$ is the set consisting of the endpoints of the closed interval $[(-1, |u|), (1, |u|)])$. As $0 \in F(x, 0)$, it is evident that $\nabla\varphi(x)$ belongs to $F(x, \varphi(x))$, actually to its relative interior, but not to the interior of $F(x, \varphi(x))$ which is empty.

Suppose that the problem $(D_{\text{ext}F})$ has a solution $u \in C(\overline{\Omega}) \cap W^{1,\infty}(\Omega)$. Then $u_x = \pm 1$ which implies $|u| > 0$ a.e. and so $u_y = |u| > 0$ a.e. Since, clearly, the equation

$$\begin{aligned} &u_y = |u| \quad && x \in \Omega \text{ a.e.} \\ &u = 0 && x \in \partial\Omega \end{aligned}$$

has no solution, we have a contradiction.

We consider now the vectorial problem

$$\begin{aligned} &\nabla u(x)) \in F(x, u(x)) \quad && x \in \Omega \text{ a.e.} \\ &u(x) = \varphi(x) && x \in \partial\Omega \end{aligned} \qquad (D_F)$$

where F: $\Omega \times \mathbb{R}^n \to K(\mathbb{R}^{n\times n})$. As in the scalar case the convexity plays a central role. However, in the vectorial case, the usual notion of convexity is not appropriate to establish a result in the line of theorem 5, as it will be shown in the following example.

Let M_2 be the set of the 2×2 real matrices. Recall that the singular values $\sigma_1(A), \sigma_2(A)$ of $A \in M_2$ are the square roots of the eigenvalues of the matrix AA^* (A^* the transpose of A). Let

$$F = \{A \in M_2 \colon \sigma_1(A) \leq 1, \sigma_1(A) + \sigma_2(A) \leq 1 + 1/2\}.$$

It is known [21] that

(i) F is convex
(ii) $\mathrm{int}F = \{A \in M_2\colon \sigma_1(A) < 1, \sigma_1(A) + \sigma_2(A) < 1 + 1/2\}$
(iii) $\mathrm{ext}F = \{A \in M_2\colon \sigma_1(A) = 1, \sigma_2(A) = 1/2\}$.

Example 2: Let $\Omega = (0,1) \times (0,1)$ and let $\varphi\colon \Omega \to \mathbb{R}^2$ be defined by $\varphi(x) = \alpha x$, where $3/4 > \alpha > \sqrt{2}/2$. Then φ is a solution of the Dirichlet problem

$$\begin{aligned} &\nabla u(x) \in \mathrm{int}\, F \quad x \in \Omega \text{ a.e.} \\ &u(x) = \varphi(x) \qquad x \in \partial\Omega \end{aligned} \tag{D$_{\mathrm{intF}}$}$$

yet the Dirichlet problem ($D_{\mathrm{ext}F}$) has no solution.

Proof: Suppose $u \in C(\overline{\Omega}) \cap W^{1,\infty}(\Omega)$ is a solution of ($D_{\mathrm{ext}F}$). Then $\sigma_1(\nabla u(x)) = 1$, $\sigma_2(\nabla u(x)) = 1/2$ a.e. and so $|\det \nabla u(x)| = \sigma_1(\nabla u(x))\sigma_2(\nabla u(x)) = 1/2$ a.e. On the other hand it is well known [11] that the determinant is a quasiconvex function, thus $\int_\Omega \det(A + \nabla\psi(x))dx \geq \det A$ for all $\psi \in W_0^{1,\infty}(\Omega)$ and for all $A \in M_2$. As $u - \varphi \in W_0^{1,\infty}(\Omega)$ it follows that $\int_\Omega \det \nabla(u(x) - \varphi(x))dx \geq \det 0 = 0$. On the other hand

$$0 \leq \int_\Omega \det \nabla(u(x) - \varphi(x))dx = \int_\Omega \det \nabla u(x)dx - \alpha \int_\Omega \mathrm{div}(u(x) - \varphi(x))dx - \alpha^2.$$

By the divergence theorem we have

$$\int_\Omega \mathrm{div}(u(x) - \varphi(x))dx = 0$$

hence $\int_\Omega \det \nabla u(x)dx \geq \alpha^2$. Thus

$$1/2 = \int_\Omega |\det \nabla u(x)|dx \geq \int_\Omega \det \nabla u(x)dx \geq \alpha^2.$$

Since $\alpha^2 > 1/2$ we have a contradiction.

In the vectorial problems a natural substitute of assumption (jj) is the stronger requirement that $\nabla\varphi(x)$ be contained in the interior of $\overline{co}^q C$, the closed quasiconvex envelope of C (see [12]).

Furthermore one would expect that the smoothness assumption (j) could be relaxed as in the scalar case. The next theorem 6 is a partial result in this direction.

A "simple" and important vectorial problem, coming from elasticity and optimal design theory, is the prescribed singular values problem. This problem has been studied by several authors (see [9,6,12,13,23] and the references therein). The following theorem holds:

Theorem 6: *Let $a > 0$ and let $\Omega \subset \mathbb{R}^n$ be open. Let $\varphi \in C(\overline{\Omega}) \cap W^{1,\infty}(\Omega)$ be such that $|\varphi(x) - \varphi(y)| < a|x - y|$, for all $x, y \in \Omega$, $x \neq y$. Then there exists a solution $u \in C(\overline{\Omega}) \cap W^{1,\infty}(\Omega)$ of the Dirichlet problem*

$$\begin{aligned} &\sigma_i(\nabla u(x)) = a, \quad i = 1, \ldots, n \quad x \in \Omega \text{ a.e.} \\ &u(x) = \varphi(x) \quad x \in \partial\Omega. \end{aligned}$$

As in the scalar case, the proof of the above theorem relies on an approximation result of the type of Lemma 4, see [19].

We remark that if we set $F = \{A \in M_n : \sigma_i(A) = a, i = 1, \ldots, n\}$ then the convex envelope and the quasiconvex envelope are equal.

The general case, where not all singular values coincide, has not been solved in the full generality. For significant progresses see [13].

REFERENCES

[1] J.P. Aubin and A. Cellina (1984). Differential Inclusions, Berlin: Springer-Verlag.

[2] A. Bressan (1990). Differential inclusions with non-closed, non-convex right-hand side, *Differential Integral Equations*, **3**, 633–638.

[3] A. Bressan and G. Colombo (1988). Generalized Baire category and differential inclusions in Banach spaces, *J. Differential Equations*, **76**, 135–158.

[4] A. Bressan and F. Flores (1994). On total differential inclusions, *Rend. Sem. Mat. Univ. Padova*, **92**, 9–16.

[5] C. Castaing and M. Valadier (1977). Convex Analysis and Measurable Multifunctions, Lecture, *Notes in Math.*, **580**, Berlin: Springer-Verlag.

[6] P. Celada and S. Perrotta (1998). Functions with prescribed singular values of the gradient, *NoDEA Nonlinear Differential Equations and Applications*, **5**, 383–396.

[7] A. Cellina (1980). On the differential inclusion $x' \in [-1, 1]$, *Atti Accad. Naz. Lincei. Rend. Sc. Fis. Mat. Nat.*, **69**, 1–6.

[8] A. Cellina (1993). On minima of a functional of the gradient: sufficient conditions, *Nonlinear Analysis Theory Meth. Appl.*, **20**, 343–347.

[9] P. Cellina and S. Perrotta (1995). On a problem of potential wells, *J. Convex Analysis*, **2**, 103–115.

[10] G. Choquet (1969). Lectures on Analysis, Mathematics Lecture Notes Series, Massachusetts: Benjamin Reading.

[11] B. Dacorogna (1989). Direct Methods in the Calculus of Variations, Berlin: Springer-Verlag.

[12] B. Dacorogna and P. Marcellini (1997). General existence theorems for Hamilton-Jacobi equations in the scalar and vectorial cases, *Acta Mathematica*, **178**, 1–37.

[13] B. Dacorogna and P. Marcellini (1998). Cauchy-Dirichlet problem for first order nonlinear systems, *J. Funct. Anal.*, **152**, 404–446.

[14] F.S. De Blasi and G. Pianigiani (1982). A Baire category approach to the existence of solutions of multivalued differential equations in Banach spaces, *Funkcial. Ekvac.*, **25**, 153–162.

[15] F.S. De Blasi and G. Pianigiani (1987). Differential inclusions in Banach spaces, *J. Differential Equations*, **66**, 208–229.

[16] F.S. De Blasi and G. Pianigiani (1991). Non convex valued differential inclusions in Banach spaces, *J. Math. Anal. Appl.*, **157**, 469–494.

[17] F.S. De Blasi and G. Pianigiani (1992). On the density of extremal solutions of differential inclusions, *Ann. Polon. Math.*, **56**, 133–142.

[18] F.S. De Blasi and G. Pianigiani (1999). On the Dirichlet problem for first order partial differential equations. A Baire category approach, *NoDEA Nonlinear Differential Equations and Applications*, **6**, 13–34.

[19] F.S. De Blasi and G. Pianigiani. A remark on the Baire approach to the prescribed singular values problem, (a preprint).

[20] I. Ekeland and R. Temam (1974). Analyse Convexe et Problemes Variationnels, Paris: Dunod Gauthier-Villars.

[21] R.A. Horn and C.R. Johnson (1991). Topics in Matrix Analysis, Cambridge, UK: Cambridge University Press.

[22] P.L. Lions (1982). Generalized Solutions of Hamilton-Jacobi Equations, Research Notes in Math. Vol. **69**, London: Pitman.

[23] S. Muller and V. Sverak (1998). Unexpected solutions of first and second order partial differential equations, In Proceedings of the International Congress of Mathematicians, Documents mathematica, Vol II, Berlin, 691–702.

[24] N.S. Papageorgiou (1993). On the "bang-bang" principle for nonlinear evolution inclusions, *Aequationes Math.*, **45**, 267–280.

[25] N.S. Papageorgiou (1994). On the solution set of nonconvex subdifferential evolution inclusions, *Czech. Math. J.*, **44**, 481–500.

[26] G. Pianigiani (1990). Differential inclusions. The Baire category method, In A. Cellina (Ed.), *Methods of Nonconvex Analysis*, Lecture Notes in Math., Berlin: Springer-Verlag, **1446**, 104–136.

[27] S.I. Suslov (1991). Nonlinear "bang-bang" principle in $\mathbb{R}^n$, *Math. Notes*, **49**, 518–523.

[28] S.I. Suslov (1992). Nonlinear "bang-bang" principle in Banach spaces, *Siberian Math. J.*, **33**, 675–685.

[29] A.A. Tolstonogov (1991). Extremal selections of multivalued mappings and the "bang-bang" principle for evolution inclusions, *Sov. Math., Dokl.*, **43**, 589–593.

[30] A.A. Tolstonogov (1995). Extreme continuous selectors of multivalued maps and their applications, *J. Differential Equations*, **122**, 161–180.

SIMAA 4(2002) 149–174

11. Maximal Element Principles on Generalized Convex Spaces and Their Applications*

Xie Ping Ding

Department of Mathematics, Sichuan Normal University, Chengdu, Sichuan 610066, P.R. China

Abstract: Some existence theorems of maximal elements involving admissible set-valued mappings and the set-valued mappings with compactly local intersection property are first proved in generalized convex spaces. Next, as applications, some new coincidence theorems, fixed point theorems, minimax inequalities, section theorems and existence theorems of solutions for quasi-equilibrium problems are given in generalized convex spaces. Finally, noncompact infinite optimization problems and equilibrium existence problems for constrained games are also discussed. These theorems improve and generalize many important known results in recent literature.

Keywords and Phrases: Maximal elements, Coincidence theorem, minimax inequality, section theorem, quasi-equilibrium, infinite optimization, constrained game, generalized convex space

1991 Mathematics Subject Classification: 90A14, 54H25, 47H19, 49J35

1. INTRODUCTION

Recently, Park and Kim [1,2] introduced the concepts of admissible set-valued mappings and generalized convex (or G-convex) spaces which include many classes of known set-valued mappings and the topological spaces with various convex structure of appearing in nonlinear analysis, respectively as special cases. Theses concepts have become the adequate and important tools to studies various problems in nonlinear analysis.

In this chapter, we first introduce a new class of set-valued mappings with compactly local intersection property and establish some existence theorems of maximal elements involving admissible set-valued mappings and the set-valued mappings with compactly local intersection property on G-convex spaces. Next, as applications of our results, some new minimax inequalities, section theorems and existence theorems of solutions for quasi-equilibrium problems are proved in G-convex spaces. Finally, noncompact infinite optimization problems and equilibrium existence problems for constrained games are also discussed. These theorems improve and generalize many important known results in recent literature.

* This project supported by the National Natural Science Foundation of China (19871059) and the Natural Science Foundation of Education Department of Sichuan province ([2000] 25)

2. PRELIMINARIES

Let X and Y be two nonempty sets. We denote by 2^Y and $\mathcal{F}(X)$ the family of all subsets of Y and the family of all nonempty finite subsets of X respectively. For any $A \in \mathcal{F}(X)$, we denote by $|A|$ the cardinality of A. Let Δ_n be the standard n-dimensional simplex with vertices $e_0, e_1, \ldots, e_n$. If J is a nonempty subset of $\{0, 1, \ldots, n\}$, we denote by Δ_J the convex hull of the vertices $\{e_j: j \in J\}$. A topological space X is said to be contractible if the identity mapping I_X of X is homotopic to a constant function. A topological space X is said to be an acyclic space if all of its reduced Cech homology groups over the rationals vanish. In particular, any contractible space is acyclic, and hence any convex or star-shaped set in a topological vector space is acyclic.

For a topological space X, we shall denote by $C(X)$ and $ka(X)$ the family of all nonempty compact subsets of X and the family of all compact acyclic subsets of X respectively. The following notions were introduced by Ding [3]. A subset A of X is said to be compactly open (resp., compactly closed) in X if for any nonempty compact subset K of X, $A \cap K$ is open (resp., closed) in K. For any given subset A of X, we define the compact closure and the compact interior of A, denoted by ccl (A) and cint (A) as

$$\text{ccl}\,(A) = \bigcap\{B \subset X : A \subset B \text{ and } B \text{ is compactly closed in } X\}, \text{ and}$$

$$\text{cint}\,(A) = \bigcup\{B \subset X : B \subset A \text{ and } B \text{ is compactly open in } X\}$$

respectively. It is easy to see that cint (A) (resp., ccl (A)) is compactly open (resp., compactly closed) in X and for each nonempty compact subset K of X, we have $\text{ccl}(A) \cap K = \text{cl}_K(A \cap K)$ and $\text{cint}(A) \cap K = \text{int}_K(A \cap K)$ where $\text{cl}_K(A \cap K)$ and $\text{int}_K(A \cap K)$ denote the closure and the interior of $A \cap K$ in K respectively. It is clear that a subset A of X is compactly open (resp., compactly closed) in X if and only if cint $(A) = A$ (resp., ccl $(A) = A$). Let X and Y be two topological spaces. A mapping $G: X \to 2^Y$ is said to be transfer compactly open-valued (resp., transfer compactly closed-valued) on X if for $x \in X$ and for each nonempty compact subset K of Y, $y \in G(x) \cap K$ (resp., $y \notin G(x) \cap K$) implies that there exists a point $x' \in X$ such that $y \in \text{int}_K(G(x') \cap K)$ (resp., $y \notin \text{cl}_K(G(x') \cap K)$). Clearly, each open-valued (resp., closed-valued) mapping $G: X \to 2^Y$ is transfer open-valued (resp., transfer closed-valued) (see the definitions 6 and 7 of Tian [4]) and is also compactly open-valued (resp., compactly closed-valued). Each transfer open-valued (resp., transfer closed-valued) mapping $G: X \to 2^Y$ is transfer compactly open-valued (resp., transfer compactly closed-valued) and the inverse is not true in general.

Let X and Y be two topological spaces. A mapping $G: X \to 2^Y$ is said to be compact if $G(X)$ is included in a compact subset of Y. G is said to have the local intersection property on X if for each $x \in X$ with $G(x) \neq \emptyset$, there exists an open neighborhood $N(x)$ of x in X such that $\bigcap_{z \in N(x)} G(z) \neq \emptyset$ (see [5]). The example in [5:p. 63] shows that a set-valued mapping with the local intersection property may not have the property of open inverse values. Now, we introduce the following new notion for set-valued mappings. $G: X \to 2^Y$ is said to have the compactly local intersection property on X if for each nonempty compact subset K of X and for each $x \in K$ with $G(x) \neq \emptyset$, there exist a open neighborhood $N(x)$ of x in X such that $\bigcap_{z \in N(x) \cap K} G(z) \neq \emptyset$. Clearly, if G has the compactly local intersection property, then for any compact subset K of X, the restriction $G|_K$: $K \to 2^Y$ of G on K has the local intersection property. It is also clear that each set-valued mapping with the local intersection property have the compactly local intersection property and the inverse is not true in general.

Let X and Y be two topological spaces. For a given class $\mathbf{L}$ of set-valued mappings, $\mathbf{L}(X, Y)$ denotes the set of set-valued mappings T: $X \to 2^Y$ belonging to $\mathbf{L}$, and $\mathbf{L}_c$ the set of finite composites of set-valued mappings in $\mathbf{L}$.

Let $\mathcal{U}$ denote the class of set-valued mappings satisfying the following properties:

(1) $\mathcal{U}$ contains the class $\mathbf{C}$ of (single-valued) continuous mappings;
(2) each $F \in \mathcal{U}_c(X, Y)$ is u.s.c. with nonempty compact values;
(3) for any standard n-simplex Δ_n, each $F \in \mathcal{U}_c(\Delta_n, \Delta_n)$ has a fixed point.

Examples of the class $\mathcal{U}$ are $\mathbf{C}$, the Kakutani mappings $\mathbf{K}$ (with convex values), the Aronszajn mappings $\mathbf{M}$ (with $\mathbf{R}_\delta$ values), the acyclic mappings $\mathbf{V}$ (with acyclic values), the approachable mappings $\mathbf{A}$ and the admissible mappings in the sense of Gorniewicz and others. For details, see Park [1,2].

A class $\mathcal{U}_c^\kappa$ of set-valued mappings is defined as follows: $F \in \mathcal{U}_c^\kappa(X, Y)$ if and only if for any compact subset K of X there exists a $F^* \in \mathcal{U}_c(K, Y)$ such that $F^*(x) \subset F(x)$ for all $x \in K$. The class $\mathcal{U}_c^\kappa$ of set-valued mappings was introduced by Park [1,2] and is called the class of admissible set-valued mappings. The class $\mathcal{U}_c^\kappa$ includes many important classes of set-valued mappings in nonlinear analysis as special cases, see [1,2] and the references therein.

The following notion of a generalized convex (or G-convex) space was introduced by Park and Kim [1,2]. $(X, D; \Gamma)$ is said to be a G-convex space if X is a topological space, D is a nonempty subset of X and Γ: $\mathcal{F}(D) \to 2^X$ is such that

(1) for each $A \in \mathcal{F}(D)$ with $|A| = n+1$, there exists a continuous mapping ϕ_A: $\Delta_n \to \Gamma(A)$ such that $B \in \mathcal{F}(A)$ with $|B| = J+1$, implies $\phi_A(\Delta_J) \subset \Gamma(B)$, where Δ_J denotes the face of Δ_n corresponding $B \in \mathcal{F}(A)$.

When $D = X$, we write $(X; \Gamma)$ in place of $(X, X; \Gamma)$. Let $(X, D; \Gamma)$ be a G-convex space and $K \subset X$. K is said to be G-convex if for each $N \in \mathcal{F}(D)$, $N \subset K$ implies $\Gamma(N) \subset K$. A function f: $K \to \mathbf{R} \cap \{\pm\infty\}$ is called G-quasiconcave (resp., G-quasiconvex) if for each $\lambda \in \mathbf{R}$, the set $\{x \in K : f(x) > \lambda\}$ (resp., $\{x \in K : f(x) < \lambda\}$) is G-convex. The concept of G-convex space is a generalization of many topological spaces with various convex structure, and it includes convex spaces due to Komiya [6] and Lassonde [7], pseudo-convex space due to Horvath [8], simplicial convexity due to Bielawski [9], H-spaces due to Horvath [10,11] etc., as special cases. For details, see Park and Kim [1,2].

In order to prove our main theorems, we need the following results.

Lemma 2.1: *Let X and Y be topological spaces, K be a nonempty compact subset of X and G: $X \to 2^Y$ be a set-valued mapping such that $G(x) \neq \emptyset$ for each $x \in K$. Then the following conditions are equivalent:*

(I) *G has the compactly local intersection property,*
(II) *for each $y \in Y$, there exists a open subset O_y of X (which may be empty) such that $O_y \cap K \subset G^{-1}(y)$ and $K = \bigcup_{y\in Y} (O_y \cap K)$,*
(III) *there exists a set-valued mapping F: $X \to 2^Y$ such that for each $y \in Y$, $F^{-1}(y)$ is open or empty in X; $F^{-1}(y) \cap K \subset G^{-1}(y)$, $\forall\, y \in Y$, and $K = \bigcup_{y\in Y} (F^{-1}(y) \cap K)$,*

(IV) *for each* $x \in K$*, there exists* $y \in Y$ *such that* $x \in \operatorname{cint} G^{-1}(y) \cap K$ *and*

$$K = \bigcup_{y \in Y} (\operatorname{cint} G^{-1}(y) \cap K) = \bigcup_{y \in Y} (G^{-1}(x) \cap K),$$

(V) $G^{-1}: Y \to 2^X$ *is transfer compactly open-valued on* X.

Proof: (I) ⇒ (II). Since for each $x \in K$, $G(x) \neq \emptyset$, by (I), there exists a open neighborhood $N(x)$ of x in X such that

$$M(x) = \bigcap_{z \in N(x) \cap K} G(z) \neq \emptyset.$$

It follows that there exists $y \in M(x) \subset Y$ such that $N(x) \cap K \subset G^{-1}(y)$, and hence we have

$$K = \bigcup_{x \in K} (N(x) \cap K) \subset \bigcup_{y \in Y} (G^{-1}(y) \cap K) \subset K.$$

For each $y \in Y$, if $y \in M(x)$ for some $x \in K$, let $O_y = N(x)$ and if $y \notin M(x)$ for all $x \in K$, let $O_y = \emptyset$. Then the family $\{O_y\}_{y \in Y}$ of open sets satisfies the condition (II).
(II) ⇒ (III). Suppose the condition (II) holds. Define a mapping $F: X \to 2^Y$ by

$$F(x) = \{y \in Y : x \in O_y\}, \quad \forall x \in X.$$

Then for each $y \in Y$, we have

$$F^{-1}(y) = \{x \in X : y \in F(x)\} = \{x \in X : x \in O_y\} = O_y.$$

By (II), we have that $F^{-1}(y)$ is open X, $F^{-1}(y) \cap K \subset G^{-1}(y)$ and

$$K = \bigcup_{y \in Y} (O_y \cap K) = \bigcup_{y \in Y} (F^{-1}(y) \cap K).$$

(III) ⇒ (IV). Suppose the condition (III) holds. Then for each $y \in Y$, $F^{-1}(y)$ is open in X and hence

$$\begin{aligned} F^{-1}(y) \cap K &= \operatorname{int}_K (F^{-1}(y) \cap K) \subset \operatorname{int}_K (G^{-1}(y) \cap K) \\ &= \operatorname{cint} G^{-1}(y) \cap K \subset G^{-1}(y). \end{aligned}$$

Therefore, we have

$$\begin{aligned} K &= \bigcup_{y \in Y} (F^{-1}(y) \cap K) \subset \bigcup_{y \in Y} (\operatorname{cint} G^{-1}(y) \cap K) \\ &\subset \bigcup_{y \in Y} (G^{-1}(y) \cap K) \subset K. \end{aligned}$$

and for each $x \in K$, there exists $y \in Y$ such that $x \in \operatorname{cint} G^{-1}(y) \cap K$.

(IV) $\Rightarrow$ (V). Suppose the condition (IV) holds. Then

$$K = \bigcup_{y \in Y} (\operatorname{cint} G^{-1}(y) \cap K) = \bigcup_{y \in Y} \operatorname{int}_K (G^{-1}(y) \cap K) = \bigcup_{y \in Y} (G^{-1}(y) \cap K).$$

It follows that for each $y \in Y$, $x \in G^{-1}(y) \cap K$ implies that there exists a point $y' \in Y$ such that $x \in \operatorname{int}_K(G^{-1}(y') \cap K)$. This shows that G^{-1} is transfer compactly open-valued on Y.

(V) $\Rightarrow$ (I). Suppose the condition (V) holds. We first show that

$$\bigcup_{y \in Y} \operatorname{int}_K (G^{-1}(y) \cap K) = \bigcup_{y \in Y} (G^{-1}(y) \cap K).$$

It is enough to show $\cup_{y \in Y}(G^{-1}(y) \cap K) \subset \cup_{y \in Y} \operatorname{int}_K(G^{-1}(y) \cap K)$. If it is false, then there exists $x \in \cup_{y \in Y}(G^{-1}(y) \cap K)$ such that $x \notin \cup_{y \in Y} \operatorname{int}_K(G^{-1}(y) \cap K)$. Hence there exists $y \in Y$ such that $x \in G^{-1}(y) \cap K$ and $x \notin \operatorname{int}_K(G^{-1}(z) \cap K)$ for all $z \in Y$. Since G^{-1} is transfer compactly open-valued on Y, there must is a $y' \in Y$ such that $x \in \operatorname{int}_K(G^{-1}(y') \cap K)$ which is a contradiction. Noting $G(x) \neq \emptyset$ for each $x \in K$, we have

$$K = \bigcup_{y \in Y} (G^{-1}(y) \cap K) = \bigcup_{y \in Y} \operatorname{int}_K (G^{-1}(y) \cap K).$$

It follows that for each $x \in K$, there exists $y \in Y$ such that $x \in \operatorname{int}_K(G^{-1}(y) \cap K)$ and so there is a relatively open neighborhood $N_1(x)$ of x in K such that $N_1(x) \subset \operatorname{int}_K(G^{-1}(y) \cap K)$. Therefore there exists an open neighborhood $N(x)$ of x in X such that

$$N(x) \cap K = N_1(x) \subset \operatorname{int}_K (G^{-1}(y) \cap K) \subset G^{-1}(y).$$

It follows that $y \in \bigcap_{z \in N(x) \cap K} G(x)$ and $\bigcap_{z \in N(x) \cap K} G(z) \neq \emptyset$. This shows that G has the compactly local intersection property on X.

Remark 2.1: Lemma 2.1 improves and generalizes Lemma 1 of Ding [12,13] and Lemma 2.1 of Ding [14].

3. MAXIMAL ELEMENTS AND COINCIDENCE THEOREMS

Theorem 3.1: *Let (X, D, Γ) be a G-convex space, Y be a topological space and $F \in \mathcal{U}_c^{\kappa}(X, Y)$. Let $G: Y \to 2^D$ be such that*

(*i*) *for each $N \in \mathcal{F}(D)$ and $x \in N$, $F(\Gamma(N)) \cap G^{-1}(x)$ is relatively open in $F(\Gamma(N))$,*
(*ii*) *for each $N \in \mathcal{F}(N)$, $F(\Gamma(N)) \subset \bigcup_{x \in N} (Y \backslash G^{-1}(x))$.*

Then we have

(1) *for any $N \in \mathcal{F}(D)$, $F(\Gamma(N)) \cap (\bigcap_{x \in N} (Y \backslash G^{-1}(x))) \neq \emptyset$,*
(2) *for any $N \in \mathcal{F}(D)$, there exists a $y \in F(\Gamma(N))$ such that $G(y) \cap N = \emptyset$.*

Proof: Clearly, the conclusions (1) and (2) are equivalent. It is enough to show that the conclusion (2) holds. Suppose the conclusion (2) is false. Then there exists a set $A = \{x_0, \ldots, x_n\} \in \mathcal{F}(D)$ with $|A| = n+1$ such that $G(y) \cap A \neq \emptyset$ for each $y \in F(\Gamma(A))$. It follows that

$$F(\Gamma(A)) \subset \bigcup_{x \in A} G^{-1}(x). \tag{3.1}$$

Since $(X, D; \Gamma)$ is a G-convex space, there exists a continuous mapping ϕ_A: $\Delta_n \to \Gamma(A)$ such that for any $B \in \mathcal{F}(A)$ with $|B| = J+1$,

$$\phi_A(\Delta_J) \subset \Gamma(B). \tag{3.2}$$

Since $\phi_A(\Delta_n)$ is compact in X and $F \in \mathcal{U}_c^{\kappa}(X, Y)$, there exists a $\tilde{F} \in \mathcal{U}_c(\phi_A(\Delta_n), Y)$ such that $\tilde{F}(x) \subset F(x)$ for each $x \in \phi_A(\Delta_n)$. Since $\tilde{F}$ is u.s.c. with compact values and $\phi_A(\Delta_n)$ is compact, by Proposition 3.1.11 of Aubin and Ekeland [15], $\tilde{F}(\phi_A(\Delta_n))$ is compact in Y. By (3.1) and (3.2), we have

$$\tilde{F}(\phi_A(\Delta_n)) = \bigcup_{i=0}^{n} (G^{-1}(x_i) \cap \tilde{F}(\phi_A(\Delta_n))).$$

Let $\{\psi_i\}_{i=0}^n$ is the continuous partition of unity subordinated to the open cover

$$\{G^{-1}(x_i) \cap \tilde{F}(\phi_A(\Delta_n))\}_{i=0}^n,$$

i.e., for each $i \in N = \{0, 1, \ldots, n\}$, ψ_i: $\tilde{F}(\phi_A(\Delta_n)) \to [0,1]$ is continuous,

$$\{y \in \tilde{F}(\phi_A(\Delta_n)) : \psi_i(y) \neq 0\} \subset G^{-1}(x_i) \cap \tilde{F}(\phi_A(\Delta_n)) \subset G^{-1}(x_i), \tag{3.3}$$

and $\sum_{i=1}^n \psi_i(y) = 1$ for each $y \in \tilde{F}(\phi_A(\Delta_n))$. Define a mapping ψ: $\tilde{F}(\phi_A(\Delta_n)) \to \Delta_n$ by

$$\psi(y) = \sum_{i=0}^{n} \psi_i(y) e_i, \quad \forall\, y \in \tilde{F}(\phi_A(\Delta_n)).$$

Then we have that $\psi \tilde{F} \phi_A \in \mathcal{U}_c(\Delta_n, \Delta_n)$ has a fixed point $z_0 \in \Delta_n$, that is $z_0 \in \psi \tilde{F} \phi_A(z_0)$. Hence there exists a $y_0 \in \tilde{F}\phi_A(z_0)$ such that

$$z_0 = \psi(y_0) = \sum_{j \in J(y_0)} \psi_j(y_0) e_j \in \Delta_{J(y_0)},$$

where $J(y_0) = \{j \in N : \psi_j(y_0) \neq 0\}$ and $N = \{0, 1, \ldots, n\}$. It follows from (3.2) and the condition (ii) that

$$\begin{aligned} y_0 \in \tilde{F}\phi_A(z_0) &\subset \tilde{F}(\phi_A(\Delta_{J(y_0)})) \subset \tilde{F}(\Gamma(\{x_j\}_{j \in J(y_0)})) \\ &\subset F(\Gamma(\{x_j\}_{j \in J(y_0)})) \subset \bigcup_{j \in J(y_0)} (Y \backslash G^{-1}(x_j)). \end{aligned}$$

Therefore there exists a $j_0 \in J(y_0)$ such that $y_0 \notin G^{-1}(x_{j_0})$. On the other hand, by the definition of $J(y_0)$, we have $\psi_{j_0}(y_0) \neq 0$ and it follows from (3.3) that $y_0 \in G^{-1}(x_{j_0})$ which is a contradiction. This completes the proof.

Theorem 3.2: *Let $(X, D; \Gamma)$ be a G-convex space and Y be a topological space. Let $F \in \mathcal{U}_c^\kappa(X, Y)$ and $G: Y \to 2^D$ be such that,*

(i) *for each $x \in D$, $G^{-1}(x)$ is compactly open in Y,*
(ii) *for each $N \in \mathcal{F}(D)$, $F(\Gamma(N)) \subset \bigcup_{x\in N}(Y\backslash G^{-1}(x))$,*
(iii) *there exists a nonmempty compact subset K of Y such that either*
  (a) *for some $M \in \mathcal{F}(D)$, $Y\backslash K \subset \bigcup_{x\in M} G^{-1}(x)$, or*
  (b) *for each $N \in \mathcal{F}(D)$, there exists a compact G-convex subset L_N of X containing N such that $F(L_N)\backslash K \subset \bigcup_{x\in L_N\cap D} G^{-1}(x)$.*

Then we have

(1) $\overline{F(X)} \cap K \cap (\bigcap_{x\in D}(Y\backslash G^{-1}(x))) \neq \emptyset$;
(2) *there exists a point $\hat{y} \in \overline{F(X)} \cap K$ such that $G(\hat{y}) = \emptyset$.*

Proof: Clearly, the conclusions (1) and (2) are equivalent. It is enough to show that the conclusion (1) holds. Suppose the conclusion (1) is not true, then we have

$$\overline{F(X)} \cap K \subset \bigcup_{x\in D}(G^{-1}(x) \cap K). \tag{3.4}$$

Since $\overline{F(X)} \cap K$ is compact in Y. By the condition (i) and (3.4), there exists a finite set $N \in \mathcal{F}(D)$ such that

$$\overline{F(X)} \cap K \subset \bigcup_{x\in N}(G^{-1}(x) \cap K). \tag{3.5}$$

Case (iii)(a): By the assumption (iii) (a) and (3.5), we have that there exists a finite set $N_1 = N \cup M \in \mathcal{F}(D)$ such that $\overline{F(X)} \subset \bigcup_{x\in N_1} G^{-1}(x)$. Theorem 3.1 implies the conclusion (2) holds.

Case(iii)(b): Let L_N be the compact G-convex subset of X in the condition (iii)(b) and $\tilde{F} \in \mathcal{U}_c(L_N, Y)$ be such that $\tilde{F}(x) \subset F(x)$ for all $x \in L_N$. Then we have

$$A \in \mathcal{F}(L_N \cap D) \Rightarrow \tilde{F}(\Gamma(A)) \subset F(\Gamma(A)) \subset \bigcup_{x\in A}(Y\backslash G^{-1}(x)).$$

Since $\tilde{F}(L_N)$ is compact and for each $x \in D$, $Y\backslash G^{-1}(x)$ is compactly closed, $\{\tilde{F}(L_N) \cap (Y\backslash G^{-1}(x)) : x \in L_N \cap D\}$ is a family of compact subsets in $\tilde{F}(L_N)$. For any $A \in \mathcal{F}(L_N \cap D)$, by Theorem 3.1 there exists $y \in \tilde{F}(\Gamma(A))$ such that $G(y) \cap A = \emptyset$. It follows that $y \notin G^{-1}(x)$ for each $x \in A$. Hence we have

$$y \in \tilde{F}(\Gamma(A)) \cap \left(\bigcap_{x\in A}(Y\backslash G^{-1}(x)\right) = \bigcap_{x\in A}[\tilde{F}(\Gamma(A)) \cap (Y\backslash G^{-1}(x))].$$

Since L_N is G-convex, we have $\Gamma(A) \subset L_N$ and

$$y \in \bigcap_{x\in A}[\tilde{F}(L_N) \cap (Y\backslash G^{-1}(x))].$$

Therefore the family $\{\tilde{F}(L_N) \cap (Y\backslash G^{-1}(x)) : x \in L_N \cap D\}$ has the finite intersection property. It follows that

$$\tilde{F}(L_N) \cap \left(\bigcap_{x \in L_N \cap D} (Y\backslash G^{-1}(x)) \right) \neq \emptyset.$$

Take any $z \in \tilde{F}(L_N) \cap (\bigcap_{x \in L_N \cap D}(Y\backslash G^{-1}(x))) = \tilde{F}(L_N)\backslash \bigcup_{x \in L_N \cap D} G^{-1}(x)$, i.e., $z \in \tilde{F}(L_N)$ and $z \notin \bigcup_{x \in L_N \cap D} G^{-1}(x)$. By (iii) (b), we have $z \in K$. It follows from (3.5) that

$$z \in \overline{F(X)} \cap K \subset \bigcup_{x \in N} G^{-1}(x).$$

Hence there exists $x^* \in N \subset L_N \cap D$ such that $z \in G^{-1}(x^*)$ which contradicts the fact $z \notin \bigcup_{x \in L_N \cap D} G^{-1}(x)$. This completes the proof.

Theorem 3.3: *Let $(X, D; \Gamma)$ be a G-convex space and Y be a topological space. Let $F \in \mathcal{U}_c^\kappa(X, Y)$ and $G: Y \to 2^D$ be such that*

(*i*) *G satisfies one of the conditions (I)–(V) in Lemma 2.1,*
(*ii*) *for each $N \in \mathcal{F}(D)$, $F(\Gamma(N)) \subset \bigcup_{x \in N}(Y\backslash \text{cint}(G^{-1}(x)))$,*
(*iii*) *there exists a nonempty compact subset K of Y such that either*
 (*a*) *for some $M \in \mathcal{F}(D)$, $Y\backslash K \subset \bigcup_{x \in M} \text{cint}(G^{-1}(x))$; or*
 (*b*) *for each $N \in \mathcal{F}(D)$, there exists a compact G-convex subset L_N of X containing N such that*

$$F(L_N)\backslash K \subset \bigcup_{x \in L_N \cap D} \text{cint}(G^{-1}(x)).$$

Then we have

(1) $\overline{F(X)} \cap K \cap (Y\backslash \bigcup_{x \in D} G^{-1}(x)) \neq \emptyset$,
(2) *there exists a point $\hat{y} \in \overline{F(X)} \cap K$, such that $G(\hat{y}) = \emptyset$.*

Proof: By Lemma 2.1, the conditions (I)–(V) are equivalent. Without loss of generality, we can assume that G^{-1} is transfer compactly open values. Define a mapping $G_1: Y \to 2^D$ by $G_1(y) = \{x \in D : y \in \text{cint}(G^{-1}(x))\}$, then we have $G_1^{-1}(x) = \text{cint}(G^{-1}(x))$ and G_1^{-1} has compact open values. It is easy to see that F and G_1 satisfy all conditions of Theorem 3.2. It follows from Theorem 3.2 that

$$\overline{F(X)} \cap K \cap \bigcap_{x \in D}(Y\backslash \text{cint}(G^{-1}(x))) \neq \emptyset.$$

Now we claim that

$$K \cap \bigcap_{x \in D}(Y\backslash \text{cint}(G^{-1}(x))) = K \cap \bigcap_{x \in D}(Y\backslash G^{-1}(x)). \tag{3.6}$$

Since $\text{cint}(G^{-1}(x)) \subset G^{-1}(x)$ for each $x \in D$, it is clear that

$$K \cap \bigcap_{x \in D}(Y\backslash G^{-1}(x)) \subset K \cap \bigcap_{x \in D}(Y\backslash \text{cint}(G^{-1}(x))).$$

If the inverse inclusion relation is not true, then there exists a point

$$\begin{aligned} y &\in K \cap \bigcap_{x\in D}(Y\backslash \mathrm{cint}(G^{-1}(x))) = K\backslash \bigcup_{x\in D}(K \cap \mathrm{cint}(G^{-1}(x))) \\ &= K\backslash \bigcup_{x\in D} \mathrm{int}_K(K\cap G^{-1}(x)) \end{aligned} \tag{3.7}$$

such that

$$y \notin K \cap \bigcap_{x\in D}(Y\backslash G^{-1}(x)) = K\backslash \bigcup_{x\in D}(K\cap G^{-1}(x)). \tag{3.8}$$

By (3.8), there exists a point $x \in D$ such that $y \in K\cap G^{-1}(x)$. Since G^{-1} is transfer compactly open values, there exists a point $x' \in D$ such that $y \in \mathrm{int}_K(K\cap G^{-1}(x'))$ which contradicts with (3.7). Hence (3.6) holds and so we have

$$\overline{F(X)} \cap K \cap (\bigcap_{x\in D}(Y\backslash G^{-1}(x)) = \overline{F(X)} \cap K \cap (\bigcap_{x\in D}(Y\backslash \mathrm{cint}(G^{-1}(x))) \neq \emptyset.$$

Therefore there exists a point $\hat{y} \in \overline{F(X)} \cap K$ such that $G(\hat{y}) = \emptyset$. This completes the proof.

Theorem 3.3 is equivalent to the following KKM type theorem.

Theorem 3.4: *Let $(X, D; \Gamma)$ be a G-convex space and Y be a topological space. Let $F \in \mathcal{U}_c^{\kappa}(X, Y)$ and S: $D \to 2^Y$ be such that*

(i) S is transfer compactly closed-valued on D,
(ii) for any $N \in \mathcal{F}(D)$, $\Gamma(N) \subset \bigcap_{x\in N} \mathrm{ccl}(S(x))$,
(iii) there exists a nonempty compact subset K of Y such that either
(a) for some $M \in \mathcal{F}(D)$, $\bigcap_{x\in M} \mathrm{ccl} S(x) \subset K$,
(b) for each $N \in \mathcal{F}(D)$, there exists a compact G-convex subset L_N of X containing N such that $F(L_N) \cap \bigcap_{x\in L_N\cap D} \mathrm{ccl} S(x) \subset K$.

Then $\overline{F(X)} \cap K \cap \bigcap_{x\in D} S(x) \neq \emptyset$.

Proof: Define a mapping G: $Y \to 2^D$ by

$$G(y) = \{x \in D : y \in Y\backslash S(x)\}.$$

Then, for each $x \in D$, $G^{-1}(x) = Y\backslash S(x)$. By (i), G^{-1} is transfer compactly open-valued on D and hence G satisfies the condition (i) of Theorem 3.3 by Lemma 2.1. It is easy to check that the conditions (ii), (iii)(a) and (iii)(b) imply the conditions (ii), (iii)(a) and (iii)(b) of Theorem 3.3, respectively. From the conclusion (1) of Theorem 3,3 we obtain

$$\overline{F(X)} \cap K \cap (Y\backslash \bigcup_{x\in D} G^{-1}(x)) = \overline{F(X)} \cap K \cap \bigcap_{x\in D} S(x) \neq \emptyset.$$

This completes the proof.

Remark 3.1: Theorem 3.4 improves Theorems 3.2 and 3.3 of Ding [3]. If for each $x \in D$, $S(x)$ is compactly closed in Y, then S is transfer compactly closed-valued on D

and for each $x \in D$, $\text{ccl}S(x) = S(x)$. Hence Theorem 3.4 also generalizes Theorems 3 and 4 of Park and Kim [2] and, in turn, generalizes many known KKM type theorems, for example, see the Remarks 3.3 and 3.4 in [3] and the Remark in [2:p. 557].

By applying Theorem 3.3 we can obtain the following coincidence theorem.

Theorem 3.5: *Let $(X, D; \Gamma)$ be a G-convex space and K be nonempty compact subset of a topological space Y. Let $F \in \mathcal{U}_c^\kappa(X,Y)$, $G: Y \to 2^D$ and $T: Y \to 2^X$ be such that*

(i) *G satisfies one of the conditions (I)–(V) in Lemma 2.1,*
(ii) *for $y \in F(X)$, $N \in \mathcal{F}((\text{cint}\, G^{-1})^{-1}(y))$ implies $\Gamma(N) \subset T(y)$,*
(iii) *$\overline{F(X)} \cap K \subset \bigcup_{x \in D} G^{-1}(x)$,*
(iv) *one of the following conditions hold:*
 (a) *for some $M \in \mathcal{F}(D)$, $Y \backslash K \subset \bigcup_{x \in M} \text{cint}(G^{-1}(x))$;*
 (b) *for each $N \in \mathcal{F}(D)$, there exists a compact G-convex subset L_N of X containing N such that*

$$F(L_N) \backslash K \subset \bigcup_{x \in L_N \cap D} \text{cint}(G^{-1}(x)).$$

Then there exists $(x_0, y_0) \in X \times Y$ such that $x_0 \in T(y_0)$ and $y_0 \in F(x_0)$.

Proof: By the condition (iii), we have $\overline{F(X)} \cap K \cap (Y \backslash \bigcup_{x \in D} G^{-1}(x)) = \emptyset$ and hence the conclusion (1) of Theorem 3.3 does not hold. By Theorem 3.3, there exists $N \in \mathcal{F}(D)$ such that $F(\Gamma(N)) \not\subset \bigcup_{x \in N} (Y \backslash \text{cint}(G^{-1}(x)))$. Hence there exist $y_0 \in F(\Gamma(N))$ and $x_0 \in \Gamma(N)$ such that $y_0 \in \text{cint}(G^{-1}(x))$ for all $x \in N$ and $y_0 \in F(x_0)$. It follows that $N \in \mathcal{F}((\text{cint}G^{-1})^{-1}(y_0))$. By the condition (ii), we obtain $x_0 \in \Gamma(N) \subset T(y_0)$. This completes the proof.

Remark 3.2: Theorem 3.5 generalizes Theorems 4.2 and 4.3 of Ding [3], Theorem 3.1 of Ding and Tarafdar [16], Theorem 1 of Park and Kim [2] and many recent known results, see the Remarks 4.2 and 4.3 in [3] and the Remarks in [1:p. 178].

Theorem 3.6: *Let $(X, D; \Gamma)$ be a G-convex space and Y be a topological space. Let $F \in \mathcal{U}_c^\kappa(X,Y)$ be a compact mapping, $G: Y \to 2^D$ and $T: Y \to 2^X$ be such that*

(i) *G satisfies one of the conditions (I)–(V) in Lemma 2.1,*
(ii) *for each $y \in F(X)$, $N \in \mathcal{F}(G(y))$ implies $\Gamma(N) \subset T(y)$,*
(iii) *$\overline{F(X)} \subset \bigcup_{x \in D} G^{-1}(x)$.*

Then there exists $(x_0, y_0) \in X \times \overline{F(X)}$ such that $x_0 \in T(y_0)$ and $y_0 \in F(x_0)$.

Proof: Since $F \in \mathcal{U}_c^\kappa(X,Y)$ is a compact mapping, $\overline{F(X)}$ is compact in Y. Let $K = \overline{F(X)}$ and $G|_K: K \to 2^D$ be the restriction of G on K. Then, by the condition (i), we can assume that $G|_K$ has the local intersection property. Noting that for each $y \in F(X)$, $(\text{cint}G^{-1})^{-1}(y) \subset G(y)$ and $G|_K(y) = G(y)$, From the condition (ii) it follows that for each $y \in F(X)$, $N \in \mathcal{F}((\text{cint}(G|_K)^{-1})^{-1}(y)) \subset G|_K(y) = G(y)$ implies $\Gamma(N) \subset T(y)$ and hence $G|_K$ satisfies the condition (ii) of Theorem 3.4. By the condition (iii), we have

$$\overline{F(X)} \cap K \subset \bigcup_{x \in D} (G^{-1}(x) \cap K) = \bigcup_{x \in D} (G|_K)^{-1}(x),$$

and so $G|_K$ satisfies the condition (iii) of Theorem 3.4. Putting $Y = K = \overline{F(X)}$, the condition (iv)(a) of Theorem 3.4 is satisfied trivially. By Theorem 3.4 with $Y = K = \overline{F(X)}$ and $G = G|_K$, there exists $(x_0, y_0) \in X \times \overline{F(X)}$ such that $x_0 \in T(y_0)$ and $y_0 \in F(x_0)$.

Theorem 3.7: *Let $(X, D; \Gamma)$ be a compact G-convex space and Y be a topological space. Let $F \in \mathcal{U}_c^\kappa(X,Y)$, $G\colon Y \to 2^D$ and $T\colon Y \to 2^X$ be such that the conditions (i)–(iii) of Theorem 3.5 hold. Then there exists $(x_0, y_0) \in X \times Y$ such that $x_0 \in T(y_0)$ and $y_0 \in F(x_0)$.*

Proof: Since X is compact, we may assume that $F \in \mathcal{U}_c(X,Y)$ and hence F is upper semicontinuous with compact values. By Proposition 3.1.11 of Aubin and Ekeland [15], $F(X)$ is compact in Y. The conclusion of Theorem 3.7 follows from Theorem 3.6.

Theorem 3.8: *Let $(X, D; \Gamma)$ be a G-convex space and K be a nonempty compact subset of X. Let $G\colon X \to 2^D$ and $T\colon X \to 2^X$ be such that*

(i) *G satisfies one of the conditions (I)–(V) in Lemma 2.1,*
(ii) *for each $x \in X$, $N \in \mathcal{F}((\mathrm{cint}\, G^{-1})^{-1}(x))$ implies $\Gamma(N) \subset T(x)$,*
(iii) *for each $x \in K$, $G(x) \neq \emptyset$,*
(iv) *for each $N \in \mathcal{F}(D)$, there exists nonempty compact G-convex subset L_N of X containing N such that*

$$L_N \backslash K \subset \bigcup_{x \in L_N \cap D} \mathrm{cint}(G^{-1}(x)).$$

Then there exists a point $\hat{x} \in D$ such that $\hat{x} \in T(\hat{x})$.

Proof: Let $Y = X$ and $F(x) = \{x\}$ be the identity mapping. Then $F \in \mathcal{U}_c(X,X) \subset \mathcal{U}_c^\kappa(X,X)$ and the conditions (iii) and (iv) imply that the conditions (iii) and (iv)(b) of Theorem 3.4 hold. It is easy to check that all conditions of Theorem 3.4 are satisfied. The conclusion follows from Theorem 3.4.

Corollary 3.1: *Let $(X, D; \Gamma)$ be a G-convex space and K be a nonempty compact G-convex subset of X. Let $G\colon X \to 2^D$ be such that the conditions (i), (iii) and (iv) of Theorem 3.8 hold and the condition (ii) is replaced by the following condition*

(ii)′ for each $x \in X$, $G(x)$ is G-convex.

Then there exists $\hat{x} \in D$ such that $\hat{x} \in G(\hat{x})$.

Proof: Let $T(x) = G(x)$ for each $x \in X$ in Theorem 3.5. Noting that for each $x \in X$, $(\mathrm{cint} G^{-1})^{-1}(x) \subset G(x)$, it is easy to see that the condition (ii)′ implies that the condition (ii) of Theorem 3.8 holds. By Theorem 3.8, the conclusion holds.

4. MINIMAX INEQUALITIES AND SECTION THEOREMS

Let $(X, D; \Gamma)$ be a G-convex space, Y be a topological spaces and $\lambda \in \mathbf{R}$. A function $f\colon X \times Y \to \mathbf{R} \cup \{\pm\infty\}$ is said to be λ-transfer compactly lower (or, upper) semicontinuous in y if for each compact subset K of Y and for each $y \in K$, there exists $x \in X$

such that $f(x, y) > \lambda$ (or, $f(x, y) < \lambda$) implies that there exist a open neighborhood $N(y)$ of y in Y and a point $x' \in Y$ such that $f(x', z) > \lambda$ (or, $f(x, z) < \lambda$) for all $z \in N(y) \cap K$ (see, [3]). It is easy to see that if we define a mapping $G: Y \to 2^X$ by $G(y) = \{x \in X : f(x, y) > \lambda\}$ (or, $G(y) = \{x \in X : f(x, y) < \lambda\}$) for some $\lambda \in \mathbf{R}$, then G has the compactly local intersection property if and only if $f(x, y)$ is λ-transfer compactly lower (or, upper) semicontinuous in y.

By applying our coincidence theorems obtained in the above section, we first show the following alternative theorems and minimax inequalities.

Theorem 4.1: *Let $(X, D; \Gamma)$ be a G-convex space, K be a nonempty compact subset of a topological space Y and $A, B \subset Z$ be three sets. Let $F \in \mathcal{U}_c^\kappa(X, Y)$, $g: D \times Y \to Z$ and $f: X \times Y \to Z$ be such that*

(i) *the mapping $G: Y \to 2^D$ defined by $G(y) = \{x \in D : g(x, y) \in A\}$ satisfies one of the conditions (I)–(V) in Lemma 2.1 with X and Y being replaced by Y and D,*
(ii) *for each $y \in F(X)$, $N \in \mathcal{F}(G(y))$ implies $\Gamma(N) \subset \{x \in X : f(x, y) \in B\}$,*
(iii) *one of the following conditons holds:*
 (a) *there exists $M \in \mathcal{F}(D)$ such that for each $y \in Y \backslash K$, there is a $x \in M$ satisfying $y \in \text{cint}(\{y \in Y : g(x : y) \in A\})$,*
 (b) *for each $N \in \mathcal{F}(D)$, there exists a compact G-convex subset L_N of X containing N such that for each $y \in F(L_N) \backslash K$, there is a $x \in L_N \cap D$ satisfying $y \in \text{cint}(\{y \in Y : g(x, y) \in A\}$.*

Then either

(1) *there exists a $\hat{y} \in \overline{F(X)} \cap K$ such that $g(x, \hat{y}) \notin A$ for all $x \in D$; or*
(2) *there exists $(\hat{x}, \hat{y}) \in Gr(F)$ such that $f(\hat{x}, \hat{y}) \in B$, where $Gr(F)$ is the graph of F.*

Proof: Define a mapping $T: Y \to 2^X$ by $T(y) = \{x \in X : f(x, y) \in B\}$. Suppose that the conclusion (1) does not hold. Then, for each $y \in \overline{F(X)} \cap K$, there exists $x \in D$ such that $g(x, y) \in A$ and hence for each $y \in \overline{F(X)} \cap K$, $G(x) \neq \emptyset$. It follows that $\overline{F(X)} \cap K \subset \bigcup_{x \in D} G^{-1}(x)$ and the condition (iii) of Theorem 3.5 is satisfied. Since for each $y \in F(X)$, we have $(\text{cint}\, G^{-1})^{-1}(y) \subset G(y)$, the condition (ii) implies that the condition (ii) of Theorem 3.5 holds. It is easy to see that all conditions of Theorem 3.5 are satisfied. By Theorem 3.5, there exists $(\hat{x}, \hat{y}) \in X \times Y$ such that $\hat{y} \in F(\hat{x})$ and $\hat{x} \in T(\hat{y})$. Hence we have $(\hat{x}, \hat{y}) \in Gr(F)$ and $f(\hat{x}, \hat{y}) \in B$.

Remark 4.1: Theorem 4.1 improves Theorem 5 of Park and Kim [2] and generalizes many known results, see the Particular Forms in [2:p. 559].

Theorem 4.2: *Let $(X, D; \Gamma)$ be a G-convex space, Y be a topological space and $A, B \subset Z$ be three sets. Let $F \in \mathcal{U}_c^\kappa(X, Y)$ be a compact mapping, and $g: D \times Y \to Z$ and $f: X \times Y \to Z$ be such that the conditions (i) and (ii) of Theorem 4.1 hold. Then either*

(1) *there exists a point $\hat{y} \in \overline{F(X)}$ such that $g(x, y) \notin A$ for all $x \in D$; or*
(2) *there exists $(\hat{x}, \hat{y}) \in Gr(F)$ such that $f(\hat{x}, \hat{y}) \in B$.*

Proof: Define $G: Y \to 2^D$ and $T: Y \to 2^X$ by

$$G(y) = \{x \in D : g(x, y) \in A\}, \quad \text{and} \quad T(y) = \{x \in X : f(x, y) \in B\},$$

respectively. Suppose that the conclusion (i) does not hold. Then, for each $y \in \overline{F(X)}$, $G(x) \neq \emptyset$. It follows that $\overline{F(X)} \subset \bigcup_{x \in D} G^{-1}(x)$ and hence the condition (iii) of Theorem 3.6 holds. It is easy to see that the conditions (i) and (ii) imply the conditions (i) and (ii) of Theorem 3.6 hold. By Theorem 3.6, there exists $(\hat{x}, \hat{y}) \in X \times \overline{F(X)}$ such that $\hat{x} \in T(\hat{y})$ and $\hat{y} \in F(\hat{x})$. Hence we have $(\hat{x}, \hat{y}) \in Gr(F)$ and $f(\hat{x}, \hat{y}) \in B$.

Remark 4.2: Theorem 4.2 improves Theorem 6 of Park and Kim [2] and generalizes many known results, see the Particular Forms in [2:p. 560].

Theorem 4.3: *Let $(X, D; \Gamma)$ be a G-convex space, K be a nonempty compact subset of a topological space Y and $\alpha, \beta \in \mathbf{R}$. Let $F \in \mathcal{U}_c^\kappa(X, Y)$, g: $D \times Y \to \mathbf{R} \cup \{\pm\infty\}$ and f: $X \times Y \to \mathbf{R} \cup \{\pm\infty\}$ be such that*

(i) $g(x, y)$ is *α-transfer compactly lower semicontinuous in y,*
(ii) *for each $y \in F(X)$, $N \in \mathcal{F}(\{x \in D : g(x, y) > \alpha\})$ implies $\Gamma(N) \subset \{x \in X : f(x, y) > \beta\}$,*
(iii) *one of the following conditions holds:*
 (a) *there exists $M \in \mathcal{F}(D)$ such that for each $y \in Y \backslash K$, there is a $x \in M$ satisfying $y \in \text{cint}(\{y \in Y : g(x, y) > \alpha\}$,*
 (b) *for each $N \in \mathcal{F}(D)$, there exists a compact G-convex subset L_N of X containing N such that for each $y \in F(L_N) \backslash K$, there is a $x \in L_N \cap D$ satisfying $y \in \text{cint}(\{y \in Y : g(x, y) > \alpha\}$.*

Then either

(1) *there exists a $\hat{y} \in \overline{F(X)} \cap K$ such that $g(x, \hat{y}) \leq \alpha$ for all $x \in D$; or*
(2) *there exists $(\hat{x}, \hat{y}) \in Gr(F)$ such that $f(\hat{x}, \hat{y}) > \beta$.*

Proof: Define mappings G: $Y \to 2^D$ and T: $Y \to 2^X$ by $G(y) = \{x \in D : g(x, y) > \alpha\}$ and $T(y) = \{x \in X : f(x, y) > \beta\}$ respectively. Then, by (i), G has the compactly local intersection property. Putting $Z = \mathbf{R} \cup \{\pm\infty\}$, $A = (\alpha, \infty]$ and $(\beta, \infty]$ in Theorem 4.1, the conclusion of Theorem 4.3 follows from Theorem 4.1.

Remark 4.3: Theorem 4.3 generalizes Corollary 4.1 of Ding [3], Theorem 4 of Tian [4], Theorem 6 of Park and Kim [2] and many known results, see the Special Forms in [2:p. 560].

Theorem 4.4: *Let $(X, D; \Gamma)$ be a G-convex space, Y be a topological space and $\alpha, \beta \in \mathbf{R}$. Let $F \in \mathcal{U}_c^\kappa(X, Y)$ be a compact mapping, g: $D \times Y \to \mathbf{R} \bigcup \{\pm\infty\}$ and f: $X \times Y \to \mathbf{R} \cup \{\pm\infty\}$ be such that the conditions (i) and (ii) of Theorem 4.3 hold.*

Then either

(1) *there exists $\hat{y} \in \overline{F(X)}$ such that $g(x, \hat{y}) \leq \alpha$ for all $x \in D$; or*
(2) *there exists $(\hat{x}, \hat{y}) \in Gr(F)$ such that $f(\hat{x}, \hat{y}) > \beta$.*

Proof: Putting $Z = \mathbf{R} \cup \{\pm\infty\}$, $A = (\alpha, \infty]$ and $B = (\beta, \infty]$ in Theorem 4.2, the conclusion of Theorem 4.4 follows from Theorem 4.2.

By Theorems 4.3 and 4.4, we easily obtain the following generalization of Ky Fan type minimax inequality.

Theorem 4.5: *Under assumptions of Theorem 4.3, if $\alpha = \beta = \sup_{(x,y)\in Gr(F)} f(x,y)$, then*

(1) *there exists $\hat{y} \in \overline{F(X)} \cap K$ such that*

$$g(x,\hat{y}) \leq \sup_{(x,y)\in Gr(F)} f(x,y), \quad \forall x \in D;$$

(2) *we have the following minimax inequality*

$$\inf_{y\in K} \sup_{x\in D} g(x,y) \leq \sup_{(x,y)\in Gr(F)} f(x,y).$$

Theorem 4.6: *Under assumptions of Theorem 4.4, if $\alpha = \beta = \sup_{(x,y)\in Gr(F)} f(x,y)$, then*

(1) *there exist $\hat{y} \in \overline{F(X)}$ such that*

$$g(x,\hat{y}) \leq \sup_{(x,y)\in Gr(F)} f(x,y), \quad \forall x \in D;$$

(2) *the conclusion (2) of Theorem 4.5 holds.*

Remark 4.4: It is clear that the conclusion (2) of Theorem 4.5 can be written as

$$\inf_{y\in K} \sup_{x\in D} g(x,y) \leq \inf_{F\in\mathcal{U}(X,Y)} \sup_{(x,y)\in Gr(F)} f(x,y).$$

Theorems 4.5 and 4.6 improve and generalize Theorem 4.5 of Ding [3], Theorem 7 of Park and Kim [2], Theorem 0 of Park and Chen [17], Theorem 3 of Yuan [18], Theorem 1 of Ha [19] and many other known resluts, see [2:p. 561].

Theorem 4.7: *Let $(X, D; \Gamma)$ be a G-convex space and Y be a topological space. Suppose that g: $D \times Y \to \mathbf{R} \cup \{\pm\infty\}$ and f: $X \times Y \to \mathbf{R} \cup \{\pm\infty\}$ be such that*

(i) *$g(x,y) \leq f(x,y)$ for all $(x,y) \in D \times Y$,*
(ii) *for any $\lambda \in \mathbf{R}$, $g(x,y)$ is λ-transfer compactly lower semicontinuous in y,*
(iii) *for each $y \in Y$, $g(\cdot, y)$ is G-quasiconcave,*
(iv) *$f(x,y)$ is lower semicontinuous on $X \times Y$ such that for each $K \in C(Y)$, $y \in Y$ and for each $\lambda \in \mathbf{R}$, the set $\{y \in K : f(x,y) \leq \lambda\}$ is acyclic.*

Then we have

$$\inf_{y\in Y} \sup_{x\in D} g(x,y) \leq \inf_{K\in C(X)} \sup_{x\in D} \inf_{y\in K} f(x,y).$$

If in addition, Y is compact and $f(x,y) = g(x,y)$ for all $(x,y) \in X \times Y$, then we have

$$\min_{y\in Y} \sup_{x\in D} g(x,y) = \sup_{x\in D} \min_{y\in Y} g(x,y).$$

Proof: If the conclusion is false, then there exist $\alpha, \beta \in \mathbf{R}$ such that

$$\inf_{y\in Y} \sup_{x\in D} g(x,y) > \alpha > \beta > \inf_{K\in C(Y)} \sup_{x\in D} \inf_{y\in K} f(x,y).$$

Hence there exists a $K \in C(Y)$ such that $\beta > \sup_{x\in D} \inf_{y\in K} f(x,y)$. Define two mappings F: $X \to 2^K$ and G:$K \to 2^D$ by

$$F(x) = \{y \in K : f(x,y) \leq \beta\} \quad \text{and} \quad G(y) = \{x \in D : g(x,y) > \alpha\}$$

respectively. Then $F(x) \neq \emptyset$ and $F(x) \subset K$ for each $x \in X$ and $G(y) \neq \emptyset$ for each $y \in K$. From the condition (iv) it follows that $F \in \mathbf{V}(X, K) \subset \mathcal{U}_c^\kappa(X, K)$. By (ii), G has the compactly local intersection property. By (iii) and the definition of G, $G(y)$ is G-convex for each $y \in Y$. Hence, for each $y \in F(X) \subset K$ and $N \in \mathcal{F}((\text{cint}G^{-1})^{-1}(y) \subset \mathcal{F}(G(y))$, we have $\Gamma(N) \subset G(y)$ and the condition (ii) of Theorem 3.5 with $G = T$ is satisfied. Since $F(X) \subset K$ and $G(y) \neq \emptyset$ for each $x \in K$, it is easy to see that the condition (iii) of Theorem 3.5 is satisfied. If $Y = K$, the condition (iv) of Theorem 3.5 is satisfied trivially. Hence, by Theorem 3.5 with $Y = K$ and $G = T$, there exists $(\hat{x}, \hat{y}) \in D \times K$ such that $\hat{y} \in F(\hat{x})$ and $\hat{x} \in G(\hat{y})$. It implies $f(\hat{x}, \hat{y}) \leq \beta$ and $g(\hat{x}, \hat{y}) > \alpha$. It follows from $\alpha > \beta$ that $f(\hat{x}, \hat{y}) < g(\hat{x}, \hat{y})$ which contradicts the condition (i). Hence the first conclusion holds.

Now assume that Y is compact and $f(x, y) = g(x, y)$ for all $(x, y) \in X \times Y$. By the above conclusion, we have

$$\begin{aligned} \min_{y\in Y} \sup_{x\in D} f(x,y) &\leq \inf_{K\in C(Y)} \sup_{x\in D} \min_{y\in K} f(x,y) \\ &= \sup_{x\in D} \min_{y\in Y} f(x,y). \end{aligned}$$

It is clear that $\sup_{x\in D} \inf_{y\in Y} f(x,y) \leq \inf_{y\in Y} \sup_{x\in D} f(x,y)$. Hence we must have

$$\min_{y\in Y} \sup_{x\in D} f(x,y) = \sup_{x\in D} \min_{y\in Y} f(x,y).$$

Remark 4.5: Theorem 4.7 improves and develops Theorem 3 of Wu [20], Theorem 5 of Yuan [18] and Theorem 4 of Ha [21] to noncompact topological spaces without linear structure.

In the following, we establish two new Ky Fan type section theorems.

Theorem 4.8: *Let $(X, D; \Gamma)$ be a G-convex space, K be a nonempty compact subset of a topological space Y and $F \in \mathcal{U}_c^\kappa(X, Y)$. Let $A \subset B \subset X \times Y$ and $C \subset D \times Y$ be such that*

(i) *the mapping $G\colon Y \to 2^D$ defined by $G(y) = \{x \in D : (x, y) \notin C\}$ satisfies one of the conditions (I)–(V) in Lemma 2.1,*
(ii) *for each $y \in F(X)$, $N \in \mathcal{F}(G(y))$ implies $\Gamma(N) \subset \{x \in X : (x, y) \notin B\}$,*
(iii) *A contains the graph $Gr(F)$ of F,*
(iv) *one of the following conditions holds:*
    (a) *there exists $M \in \mathcal{F}(D)$ such that for each $y \in Y \backslash K$, there is a $x \in M$ satisfying $y \in \text{cint}(\{z \in Y : (x, z) \notin C\})$,*
    (b) *for each $N \in \mathcal{F}(D)$, there exists a nonempty compact G-convex subset L_N of X containing N such that for each $y \in F(L_N) \backslash K$, there is a $x \in L_N \cap D$ satisfying $y \in \text{cint}(\{z \in Y : (x, z) \notin C\})$.* Then there exists a point $\hat{y} \in \overline{F(X)} \cap K$ such that $D \times \{\hat{y}\} \subset C$.

Proof: It is easy to see that the condition (iv) implies the condition (iii) of Theorem 3.3 holds. We claim that the condition (ii) of Theorem 3.3 is also satisfied. If it is false, then there exists a $N \in \mathcal{F}(D)$ such that $F(\Gamma(N)) \not\subset \bigcup_{x\in N} (Y \backslash \text{cint}(G^{-1}(x)))$, and hence there exists $y^* \in F(\Gamma(N)) \subset F(X)$ such that $y^* \notin \bigcup_{x\in N} (Y \backslash \text{cint}(G^{-1}(x)))$. It follows that $y^* \in G^{-1}(x)$ for all $x \in N$ and $N \subset G(y^*)$. By (ii), we have $\Gamma(N) \subset \{x \in X : (x, y^*) \notin B\}$, i.e., $(x, y^*) \notin B$ for all $x \in \Gamma(N)$. On the other hand, by $y^* \in F(\Gamma(N))$, there exists

$x^* \in \Gamma(N)$ such that $(x^*, y^*) \in Gr(F) \subset A \subset B$ by (iii) which contradicts the fact $(x, y^*) \notin B$ for all $x \in \Gamma(N)$. Therefore the condition (ii) of Theorem 3.3 is satisfied. By Theorem 3.3, there exists a point $\hat{y} \in \overline{F(X)} \cap K$ such that $G(\hat{y}) = \emptyset$. It follows that $D \times \{\hat{y}\} \subset C$.

Remark 4.6: Theorem 4.8 improves Theorem 4.1 of Ding [3], Theorem 9 of Park and Kim [2] and many other known results, see the Remark 4.1 in [3] and the Particular Forms in [2:p. 564].

Theorem 4.9: *Let $(X, D; \Gamma)$ be a G-convex space and Y be a topological space. Let $A, B \subset X \times Y$ be such that*

(i) *the mapping $G: Y \to 2^D$ defined by $G(y) = \{x \in D: (x, y) \notin A\}$ satisfies one of the conditions* (I)–(V) *in Lemma 2.1,*
(ii) *for each $y \in \bigcup_{x \in X}\{y \in K: (x, y) \in C\}$, $N \in \mathcal{F}(G(y))$ implies $\Gamma(N) \subset \{x \in X: (x, y) \notin B\}$,*
(iii) *there exist a subset C of B and a compact subset K of Y such that C is closed in $D \times Y$ and for each $x \in D$, the set $\{y \in K: (x, y) \in C\}$ is acyclic.*

Then there exists point $\hat{y} \in Y$ such that $D \times \{\hat{y}\} \subset A$.

Proof: Define a mapping $F: X \to 2^K$ by $F(x) = \{y \in K: (x, y) \in C\}$. Since K is compact in Y and C is closed in $D \times Y$, therefore F has closed graph and by Corollary 3.1.9 of Aubin and Ekeland [15], F is upper semicontinuous. By (iii), we have $F \in \mathbf{V}(X, K) \subset \mathcal{U}_c^\kappa(X, Y)$. We claim that the condition (ii) of theorem 3.3 is satisfied. If it is not true, then there exists $N \in \mathcal{F}(D)$ such that $F(\Gamma(N)) \not\subset \bigcup_{x \in N}(Y \backslash \text{cint}(G^{-1}(x))$. Hence there exists $y^* \in F(\Gamma(N))$ such that $y^* \in G^{-1}(x)$ for all $x \in N$ and so $N \in \mathcal{F}(G(y^*))$. By (ii) we have $(x, y^*) \notin B$ for all $x \in \Gamma(N)$. On the other hand, since $y^* \in F(\Gamma(N)) = \bigcup_{x \in \Gamma(N)}\{y \in K: (x, y) \in C\}$, there exist $x^* \in \Gamma(N)$ such that $(x^*, y^*) \in C \subset B$ which is a contradiction and hence the condition (ii) of Theorem 3.3 holds. From Theorem 3.3 with $Y = K$ it follows that there exists a point $\hat{y} \in \overline{F(X)} \cap K \subset K$ such that $G(\hat{y}) = \emptyset$. Hence we have $D \times \{\hat{y}\} \subset A$.

Remark 4.7: Theorem 4.9 improves and generalizes Theorem 4 of Wu [20], Theorem 2 of Yuan [18] and Theorem 3 of Ha [21] in several aspects, and in turn generalizes Fan's section theorem in [22,23].

5. EXISTENCE OF QUASI-EQUILIBRIUM PROBLEMS

Let X and Y be nonempty sets and 2^X be the family of all subsets of X. Let $T: X \to Y$ be a single-valued mapping, $A: X \to 2^X$ be a set-valued mapping and $f: X \times Y \to \mathbf{R} \cup \{\pm\infty\}$ be a function. The quasi-equilibrium problem $QEP(T, A, f)$ is to find $\hat{x} \in X$ such that

$$\begin{aligned} &\hat{x} \in A(\hat{x}), \\ &f(\hat{x}, T\hat{x}) \le f(y, T\hat{x}), \quad \forall y \in A(\hat{x}). \end{aligned} \tag{5.1}$$

The $QEP(T, A, F)$ was introduced and studied by Noor and Oettli [24]. Cubiotti [25] and Ding [26] proved some existence theorems of equilibrium points for the $QEP(T, A, f)$(5.1) in finite demensional space $\mathbf{R}^n$ and topological vector spaces.

Recently, Lin and Park [27] considered the following quasi-equilibrium problem $QEP(A, f)$ which is to find $\hat{x} \in X$ such that

$$\begin{aligned} &\hat{x} \in A(\hat{x}), \\ &f(y, \hat{x}) \geq 0, \quad \forall y \in A(\hat{x}), \end{aligned} \tag{5.2}$$

where $A: X \to 2^X$ and $f: X \times X \to \mathbf{R} \cup \{\pm\infty\}$. They proved some equilibrium existence theorems for the $QEP(A, f)$(5.2) in compact G-convex spaces without the linear structure.

The $QEP(T, A, f)$(5.1) and $QEP(A, f)$(5.2) include many Optimization problems, Nash type equilibrium problems, quasi-variational inequality problems, quasi-complementarity problems and others as special cases, see [24–27] and the references therein.

In this section, by applying Corollary 3.1, some new existence theorems of equilibrium points for the $QEP(T, A, f)$ and $QEP(A, f)$ are proved in noncompact G-convex spaces. These results include many key known results in the fields as special cases.

Lemma 5.1: *Let X and Y be topological spaces, E be a nonempty closed subset of X and $\Phi, \Psi: X \to 2^Y$ be two set-valued mappings with nonempty values such that $\Phi(x) \subset \Psi(x)$ for each $x \in X$. Suppose that $\Phi^{-1}, \Psi^{-1}: Y \to 2^X$ are both transfer compactly open-valued on Y. Then the mapping $G: X \to 2^Y$ defined by*

$$G(x) = \begin{cases} \Phi(x), & \text{if } x \in E, \\ \Psi(x), & \text{if } x \in X \backslash E \end{cases}$$

is such that $G^{-1}: Y \to 2^X$ is also transfer compactly open-valued on Y.

Proof: For any given $y \in Y$, we have $\Phi^{-1}(y) \subset \Psi^{-1}(y)$ since $\Phi(x) \subset \Psi(x)$ for each $x \in X$. It follows that

$$\begin{aligned} G^{-1}(y) &= \{x \in X : y \in G(x)\} \\ &= \{x \in E : y \in \Phi(x)\} \cup \{x \in X \backslash E : y \in \Psi(x)\} \\ &= (E \cap \Phi^{-1}(y)) \cup [(X \backslash E) \cap \Psi^{-1}(y)] \\ &= [(E \cap \Phi^{-1}(y)) \cup (X \backslash E)] \cap [(E \cap \Phi^{-1}(y)) \cup \Psi^{-1}(y)] \\ &= [X \cap (\Phi^{-1}(y) \cup (X \backslash E))] \cap [(E \cup \Psi^{-1}(y)) \cap (\Phi^{-1}(y) \cup \Psi^{-1}(y))] \\ &= [\Phi^{-1}(y) \cup (X \backslash E)] \cap \Psi^{-1}(y) \\ &= \Phi^{-1}(y) \cup [(X \backslash E) \cap \Psi^{-1}(y)]. \end{aligned}$$

Hence for each nonempty compact subset K of X, we have

$$G^{-1}(y) \cap K = (\Phi^{-1}(y) \cap K) \cup [(X \backslash E) \cap \Psi^{-1}(y) \cap K].$$

If $x \in G^{-1}(y) \cap K$, then we have that either $x \in \Phi^{-1}(y) \cap K$ or $x \in (X \backslash E) \cap \Psi^{-1}(y) \cap K$. If $x \in \Phi^{-1}(y) \cap K$, noting that Φ^{-1} is transfer compactly open-valued on Y, there exists $y' \in Y$ such that

$$x \in \text{int}_K(\Phi^{-1}(y') \cap K) \subset \text{int}_K(G^{-1}(y') \cap K).$$

If $x \in (X\backslash E) \cap \Psi^{-1}(y) \cap K$, noting that $X\backslash E$ is an open set and Ψ^{-1} is transfer compactly open-valued on Y, by using similar argument, we can show that there exists $y' \in Y$ such that $x \in \text{int}_K(G^{-1}(y') \cap K)$. This prove that G^{-1} is transfer compactly open-valued on Y.

Remark 5.1: Lemma 5.1 improves and generalizes Lemma 1.4 of Ding [26] and Proposition 4.1 of Cubiotti [25].

We first prove the following equilibrium existence theorem of *QEP*(T, A, f).

Theorem 5.1: *Let (X, Γ) be a G-convex space, K be a nonempty compact subset of X and Y be a nonempty set. Let T: $X \to Y$, A: $X \to 2^X$ and f: $X \times Y \to \mathbf{R} \cup \{\pm\infty\}$ be such that*

(i) *A has nonempty G-convex values and satisfies one of the conditions (I)–(V) in Lemma 2.1,*
(ii) *the set $E = \{x \in X : x \in A(x)\}$ is closed in X.*
(iii) *the mappings $P, B: X \to 2^X$ defined by*

$$P(x) = \{y \in X : f(x, Tx) - f(y, Tx) > 0\},$$
$$B(x) = \{y \in A(x) : f(x, Tx) - f(y, Tx) > 0\}$$

both have the compactly local intersection property,
(iv) *for each $x \in X$, $y \mapsto f(y, Tx)$ is G-quasiconvex,*
(v) *for each $N \in \mathcal{F}(X)$, there exists a nonempty compact G-convex subset L_N of X containing N such that for each $x \in L_N\backslash K$, if $x \notin E$, then there exists $y \in L_N$ such that $x \in \text{cint}A^{-1}(y)$; if $x \in E$, then there is $y \in L_N$ such that $x \in \text{cint}(\{x \in A^{-1}(y) : f(x, Tx) - f(y, Tx) > 0\})$.*

Then there exists $\hat{x} \in X$ such that

$$\hat{x} \in A(\hat{x}),$$
$$f(\hat{x}, T\hat{x}) \leq f(y, \hat{x}), \quad \forall y \in A(\hat{x}),$$

i.e., $\hat{x}$ is a equilibrium point of the *QEP*(T, A, f).

Proof: Define a mapping G: $X \to 2^X$ by

$$G(x) = \begin{cases} B(x), & \text{if } x \in E, \\ A(x), & \text{if } x \in X\backslash E. \end{cases}$$

From the condition (iii) and Lemma 2.1 it follows that the mappings B^{-1}, A^{-1}: $X \to 2^X$ are both transfer compactly open-valued on X. Note that $B(x) \subset A(x)$ for each $x \in X$, by Lemma 5.1, G^{-1}: $X \to 2^X$ is also transfer compactly open-valued on X. By the assumptions (i) and (iv), for each $x \in X$, $G(x)$ is G-convex. Now assume that for each $x \in E$, $B(x) = A(x) \cap P(x) \neq \emptyset$, then for each $x \in X$, $G(x) \neq \emptyset$ by (i). It is easy to see that the condition (v) implies that the condition (iv) of Theorem 3.8 holds. Hence all conditions of Corollary 3.1 with $D = X$ are satisfied. By Corollary 3.1, there exists $\hat{x} \in X$ such that $\hat{x} \in G(\hat{x})$. By the definition of E and G, we must have $\{x \in X : x \in G(x)\} \subset E$. It follows that $\hat{x} \in A(\hat{x}) \cap P(\hat{x}) \cap E$. In particular, we obtain $f(\hat{x}, T\hat{x}) - f(\hat{x}, T\hat{x}) > 0$ which is impossible. Therefore there exists $\hat{x} \in E$ such that $A(\hat{x}) \cap P(\hat{x}) = \emptyset$, that is $\hat{x} \in A(\hat{x})$ and $f(\hat{x}, T\hat{x}) \leq f(y, T\hat{x})$ for all $y \in A(\hat{x})$. This completes the proof.

Remark 5.2: Theorem 5.1 improves and generalized Theorem 2.1 of Ding [26] and Theorem 4.2 of Cubiotti [25] from topological vector spaces to noncompact G-convex spaces without linear structure under much weaker assumptions.

Theorem 5.2: *Let (X,Γ) be a G-convex space, K be a nonempty compact subset of X and Y be a topological space. Let T: $X \to Y$, A: $X \to 2^X$ and f: $X \times Y \to \mathbf{R} \cup \{\pm\infty\}$ be such that*

(i) *A has nonempty G-convex values such that A^{-1}: $X \to 2^X$ is compactly open-valued on X,*
(ii) *the set $E = \{x \in X : x \in A(x)\}$ is closed in X,*
(iii) *T and f are continuous such that for each $x \in X$, $y \mapsto f(y,Tx)$ is G-quasiconvex,*
(iv) *for each $N \in \mathcal{F}(X)$ there exists a nonempty compact G-convex subset L_N of X containing N such that for each $x \in L_N \backslash K$, if $x \notin E$, then $A(x) \cap L_N \neq \emptyset$ if $x \in E$, then there is $y \in A(x) \cap L_N$ satisfying $f(y,Tx) < f(x,Tx)$.*

Then the QEP(T, A, f) has an equilibrium point $\hat{x} \in X$.

Proof: Define P: $X \to 2^X$ by

$$P(x) = \{y \in X : f(x,Tx) - f(y,Tx) > 0\}, \quad \forall\, x \in X.$$

Since T and f are both continuous, we have that for each $y \in X$, $P^{-1}(y) = \{x \in X : f(x,Tx) - f(y,Tx) > 0\}$ is open in X and hence P^{-1}: $X \to 2^X$ have compactly open values. By the assumption, A^{-1} has compactly open values and hence the mapping $B^{-1} = (A \cap P)^{-1} = A^{-1} \cap P^{-1}$ also have compactly open values. By Lemma 2.1, the condition (iii) of Theorem 5.1 is satisfied. By the condition (iv), for each $x \in L_N \backslash K$, if $x \notin E$, we have $A(x) \cap L_N \neq \emptyset$ and hence there exists $y \in L_N$ such that $x \in A^{-1}(y) = \text{cint}A^{-1}(y)$ since $A^{-1}(y)$ is compactly open; if $x \in E$, we have $y \in L_N$ and $x \in A^{-1}(y) \cap P^{-1}(y) = \text{cint}(A^{-1}(y) \cap P^{-1}(y)) = \text{cint}(\{x \in A^{-1}(y) : f(x,Tx) - f(y,Tx) > 0\})$, since the set $A^{-1}(y) \cap P^{-1}(y)$ is compactly open. Hence the condition (v) of Theorem 5.1 is satisfied. It is easy to see that all conditions of Theorem 5.1 are satisfied. By Theorem 5.1, the conclusion of Theorem 5.2 holds.

Corollary 5.1: *Let (X,Γ) be a G-convex space and K be a nonempty compact subset of X. Let A: $X \to 2^X$ and f: $X \times X \to \mathbf{R} \cup \{\pm\infty\}$ be such that*

(i) *A has nonempty G-convex values such that A^{-1}: $X \to 2^X$ is compactly open-valued on X and the set $E = \{x \in X : x \in A(x)\}$ is closed in X,*
(ii) *f is continuous such that for each $x \in X$, $y \mapsto f(x,y)$ is G-quasiconvex,*
(iii) *for each $N \in \mathcal{F}(X)$, there exists a nonempty compact G-convex subset L_N of X containing N such that for each $x \in L_N \backslash K$, if $x \notin E$, then $A(x) \cap L_N \neq \emptyset$; if $x \in E$, then there is a $y \in A(x) \cap L_N$ satisfying $f(y,x) < f(x,x)$.*

Then there exists a $\hat{x} \in X$ such that $\hat{x} \in A(\hat{x})$ and $f(\hat{x},\hat{x}) \leq f(y,\hat{x})\ \forall y \in A(\hat{x})$.

Proof: Putting $X = Y$ and T being the identity mapping, it is easy to see that all conditions of Theorem 5.2 are satisfied. The conclusion follows from Theorem 5.2.

If f is replaced $-f$ in Corollary 5.1, then we have the following result.

Corollary 5.2: *Let (X,Γ) be a G-convex space and K be a nonempty compact subset of X. Let A: $X \to 2^X$ and f: $X \times X \to \mathbf{R} \cup \{\pm\infty\}$ be such that*

(*i*) *the condition (i) of Corollary 5.1 holds,*
(*ii*) *f is continuous such that for each $x \in X$, $y \mapsto f(y,x)$ is G-quasiconcave,*
(*iii*) *for each $N \in \mathcal{F}(X)$, there exists a nonempty compact G-convex subset L_N of X containing N such that for each $x \in L_N \backslash K$, if $x \notin E$, then $A(x) \cap L_N \neq \emptyset$; if $x \in E$, then there is a $y \in A(x) \cap L_N$ satisfying $f(x,x) < f(y,x)$.*

Then there exists a $\hat{x} \in X$ such that $\hat{x} \in A(\hat{x})$ and

$$f(\hat{x},\hat{x}) = max_{y \in A(\hat{x})} f(y,\hat{x}).$$

Theorem 5.3: *Let (X,Γ) be a G-convex space, K be a nonempty compact subset of X. Let A: $X \to 2^X$ and f: $X \times X \to \mathbf{R} \cup \{\pm\infty\}$ be such that*

(*i*) *A has nonempty G-convex values such that A^{-1}: $X \to 2^X$ has compactly open values and the mapping $\mathrm{cl}\, A$: $X \to 2^X$, defined by $(\mathrm{cl}\, A)(x) = \mathrm{cl}\, A(x)$, is upper semicontinuous,*
(*ii*) *f is a continuous function such that for each $x \in X$, $y \mapsto f(y,x)$ is G-quasiconvex,*
(*iii*) *for each $N \in \mathcal{F}(X)$, there exists nonempty compact G-convex subset L_N of X containing N such that for each $x \in L_N \backslash K$, if $x \notin \overline{E} = \{x \in X : x \in \mathrm{cl}\, A(x)\}$, then $A(x) \cap L_N \neq \emptyset$; if $x \in \overline{E}$, then there is $y \in A(x) \cap L_N$ satisfying $f(y,x) < f(x,x)$.*

Then there exists $\hat{x} \in X$ such that $\hat{x} \in \mathrm{cl}\, A(\hat{x})$ and $f(\hat{x},\hat{x}) \leq f(y,\hat{x})$, $\forall y \in A(\hat{x})$. If further assume that $f(x,x) \geq 0$ for all $x \in X$, then we have that $\hat{x} \in \mathrm{cl} A(\hat{x})$ and $f(y,\hat{x}) \geq 0$, $\forall y \in A(\hat{x})$, i.e., $\hat{x}$ is an equilibrium point of the QEP(A, f) (5.2).

Proof: Since $\mathrm{cl}\, A$: $X \to 2^X$ is upper semicontinuous with closed values, the set $\overline{E} = \{x \in X$: $x \in \mathrm{cl}\, A(x)\}$ must be closed in X. By letting $Y = X$, T being the identity mapping and $\overline{E}$ in place of E, it is easy to see that all conditions of Theorem 5.2 are satisfied. By Theorem 5.2, there exists $\hat{x} \in X$ such that $\hat{x} \in \mathrm{cl}\, A(\hat{x})$ and $f(\hat{x},\hat{x}) \leq f(y,\hat{x})$, $\forall y \in A(\hat{x})$. If $f(x,x) \geq 0$ for all $x \in X$, then we must have $\hat{x} \in \mathrm{cl}\, A(\hat{x})$ and $f(y,\hat{x}) \geq 0$, $\forall y \in A(\hat{x})$, i.e., $\hat{x}$ is an equilibrium point of the *QEP*(*A*, *f*)(2).

Remark 5.3: If (X,Γ) is a compact G-convex space, by letting $X = K = L_N$ for each $N \in \mathcal{F}(X)$, then the condition (iii) of Theorem 5.3 is satisfied trivially. Hence Theorem 5.3 generalizes Theorem 4 of Lin and Park [27] to noncompact setting.

From Theorem 5.3, we obtain the following existence result for generalized quasi-equilibrium problems.

Theorem 5.4: *Let (X,Γ) be a G-convex space, K be a nonempty compact subset of X and Y be a topological space. Let T: $X \to 2^Y$ have a continuous selection g: $X \to Y$. Let A: $X \to 2^X$ and ϕ: $X \times Y \times X \to \mathbf{R} \cup \{\pm\infty\}$ be such that*

(*i*) *A satisfies the conditions (i) of Theorem 5.3,*
(*ii*) *ϕ is a continuous function such that for each $(x,y) \in X \times Y$, $z \mapsto \phi(x,y,z)$ is G-quasiconvex,*
(*iii*) *for each $N \in \mathcal{F}(X)$, there exists a nonempty compact G-convex subset L_N of X containing N such that for each $x \in L_N \backslash K$, if $x \notin \overline{E} = \{X \in X : x \in \mathrm{cl}\, A(x)\}$ then $A(x) \cap L_N \neq \emptyset$; if $x \in \overline{E}$, then there is a $y \in A(x) \cap L_N$ satisfying $\phi(x, g(x), y) < \phi(x, g(x), x)$.*

Then there exist $\hat{x} \in X$ *and* $\hat{y} = g(\hat{x}) \in T(\hat{x})$ *such that*

$$\hat{x} \in \operatorname{cl} A(\hat{x}),$$

$$\phi(\hat{x}, \hat{y}, \hat{x}) \leq \phi(\hat{x}, \hat{y}, z), \quad \forall\, z \in A(\hat{x}).$$

If further assume that $\phi(x, g(x), x) \geq 0$ *for all* $x \in X$*, then we obtain*

$$\hat{x} \in \operatorname{cl} A(\hat{x}),$$

$$\phi(\hat{x}, \hat{y}, z) \geq 0, \quad \forall\, z \in A(\hat{x}).$$

Proof: Define f: $X \times X \to \mathbf{R} \cup \{\pm\infty\}$ by

$$f(z, x) = \phi(x, g(x), z), \quad \forall\, (z, x) \in X \times X.$$

Then the conclusion of Theorem 5.4 holds from Theorem 5.3.

Remark 5.4: Theorem 5.4 generalizes Corollary 5 of Lin and Park [27] to noncompact setting and Theorem 3.1 of Chang *et al.* [28] in several aspects.

6. OPTIMIZATION PROBLEM AND CONSTRAINED GAMES

Let I be any (finite or infinite) index set and for each $i \in I$, X_i be a topological space. We use the notation

$$X = \prod_{i \in I} X_i \quad \text{and} \quad X^i = \prod_{j \in I,\, j \neq i} X_j$$

For each $x \in X$, x_i denotes its ith coordinate and x^i the projection of x on X^i. Write $x = (x_i, x^i)$.

For each $i \in I$, let F_i: $X^i \to 2^{X_i}$ be a set-valued mapping and f_i: $X \to \mathbf{R} \cup \{\pm\infty\}$ be a function. The infinite optimization problem is to find $\hat{x} = (\hat{x}_i, \hat{x}^i) \in X$ such that for each $i \in I$,

$$\hat{x}_i \in F_i(\hat{x}^i),$$

$$f_i(\hat{x}) = \max_{y_i \in F_i(\hat{x}^i)} f_i(y_i, \hat{x}^i). \tag{6.1}$$

The problem have been studied by Kaczynski and Zeidan [29], Park [30], Ding [31,32] and others in topological vector spaces. Some existence theorems of solutions for the finite or infinite optimization problem (6.1) were established in [29,32] under various different assumptions in topological vector spaces.

If I is the set of players. Each player $i \in I$ has a strategy set X_i, a constraint correspondence F_i: $X^i \to 2^{X_i}$ and a loss function f_i: $X \to \mathbf{R} \cup \{\pm\infty\}$. A constrained game

$\Gamma = (X_i, F_i, f_i)_{i\in I}$ is defined as a family of ordered triples (X_i, F_i, f_i). A point $\hat{x} \in X$ is called an equilibrium point of Γ if for each $i \in I$,

$$\hat{x}_i \in F_i(\hat{x}^i),$$
$$f_i(\hat{x}) \le f_i(y_i, \hat{x}^i), \quad \forall\, y_i \in F_i(\hat{x}^i). \tag{6.2}$$

If $F_i(x^i) = X_i$ for each $i \in I$ and for each $x^i \in X^i$, then the constrained game $\Gamma = (X_i, F_i, f_i)_{i\in I}$ reduces to the conventional game $\Gamma = (X_i, f_i)_{i\in I}$ and its an equilibrium point is said to be a Nash equilibrium point.

The constrained game problem (6.2) have been studied by Aubin and Ekeland [15], Ding [31,32], Yuan *et al.* [33] and many others. Some equilibrium existence theorems were established in [15,31–33] under various different assumptions in topological vector spaces.

In this section, by using Corollary 5.1 and Corollary 5.2, several existence theorems of solutions for infinite optimization problem (6.1) and the constrained game problem (6.2) are proved under noncompact setting of generalized convex spaces without linear structure. These results are new and interesting which improve and generalize a number of known results in literature to generalized convex spaces.

The following result is Lemma 4.3 of Tan [34].

Lemma 6.1: *Let $(X_i, \Gamma_i)_{i\in I}$ be any family of G-convex spaces and $X = \prod_{i\in I} X_i$ be equipped with the product topology. For each $i \in I$, let π_i: $X \to X_i$ be the projection. Define Γ: $X \to 2^X\backslash\{\emptyset\}$ by*

$$\Gamma(A) = \prod_{i\in I} \Gamma_i(\pi_i(A)), \quad \forall\, A \in \mathcal{F}(X).$$

Then (X, Γ) is a G-convex space.

Theorem 6.1: *Let I be (finite or infinite) index set, $(X_i, \Gamma_i)_{i\in I}$ be a family of G-convex spaces, $X = \prod_{i\in I} X_i$, each $K_i \subset X_i$ be a nonempty compact set and $K = \prod_{i\in I} K_i$. For each $i \in I$, let F_i: $X^i \to 2^{X_i}$ be a set-valued mapping and f_i: $X \to \mathbf{R} \cup \{\pm\infty\}$ be a function such that the mapping A: $X \to 2^X$ and the function f: $X \times X \to \mathbf{R} \cup \{\pm\infty\}$ defined by*

$$A(x) = \prod_{i\in I} F_i(x^i) \quad \textit{and} \quad f(y, x) = \sum_{i\in I} f_i(y_i, x^i)$$

respectively satisfy the following conditions:

(i) *for each $i \in I$, F_i has nonempty G-convex valuex such that the inverse mapping A^{-1}: $X \to 2^X$ of A is compactly open-valued on X and the set $D = \{x \in X : x \in A(x)\}$ is closed in X,*
(ii) *f is continuous such that for each $x \in X$, $y \mapsto f(y, x)$ is* G-quasiconcave,
(iii) *for each $N \in \mathcal{F}(X)$, there exists nonempty compact G-convex subset L_N of X containing N such that for each $x \in L_N\backslash K$, if $x \notin D$, then $A(x) \cap L_N \neq \emptyset$; if $x \in D$, then there is> $y \in A(x) \cap L_N$ satisfying $f(x, x) < f(y, x)$.*

Then there exists $\hat{x} \in X$ such that for each $i \in I$,

$$\hat{x}_i \in F_i(\hat{x}^i),$$
$$f_i(\hat{x}) = \max\nolimits_{y_i\in F_i(\hat{x}^i)} f_i(y_i, \hat{x}^i),$$

i.e., $\hat{x} \in X$ is a solution of the noncompact infinite optimization problem (6.1).

Proof: Let $X = \prod_{i\in I} X_i$ and $\Gamma = \prod_{i\in I} \Gamma_i$. By Lemma 6.1, (X, Γ) is a G-convex space and $K = \prod_{i\in I} K_i$ be a nonempty compact subset of X. From the conditions (i)–(iii) it is easy to see that the A and f satisfy all conditions of Corollary 5.2. Therefore there exists $\hat{x} \in X$ such that $\hat{x} \in A(\hat{x})$ and

$$f(\hat{x}, \hat{x}) \geq f(y, \hat{x}), \ \forall\, y \in A(\hat{x}).$$

Hence we have that for each $i \in I$, $\hat{x}_i \in F_i(\hat{x}^i)$ and

$$\sum_{i\in I} f_i(\hat{x}_i, \hat{x}^i) \geq \sum_{i\in I} f_i(y_i, \hat{x}^i), \quad \forall\, y \in A(\hat{x}). \tag{6.3}$$

Choose $\hat{y} \in X$ such that $\hat{y}_i = y_i \in F_i(\hat{x}^i)$ and $\hat{y}_j = \hat{x}_j$ for all $j \in I$ with $j \neq i$, then $\hat{y} \in A(\hat{x})$. It follows from (6.3) that for each $i \in I$,

$$\begin{aligned} f_i(\hat{x}_i, \hat{x}^i) &= \sum_{i\in I} f_i(\hat{x}_i, \hat{x}^i) - \sum_{j\in I, j\neq i} f_j(\hat{x}_j, \hat{x}^j) \\ &\geq \sum_{i\in I} f_i(\hat{y}_i, \hat{x}^i) - \sum_{j\in I, j\neq i} f_j(\hat{x}_j, \hat{x}^i) = f_i(y_i, \hat{x}^i), \quad \forall\, y_i \in F_i(\hat{x}^i). \end{aligned}$$

Hence we obtain that for each $i \in I$,

$$\hat{x}_i \in F_i(\hat{x}^i),$$
$$f_i(\hat{x}) = \max_{y_i \in F_i(\hat{x}^i)} f_i(y_i, \hat{x}^i),$$

i.e., $\hat{x}$ is a solution of the noncompact infinite optimization problem (6.1).

Remark 6.1: If for $i \in I$, (X_i, Γ_i) is a compact G-comvex space, then the condition (iii) of Theorem 6.1 is satisfied trivially. Theorem 6.1 is a improving variant of Theorem 4 and Theorem 5 of Park [30], Theorem 3.1 of Ding [31,32] and Theorem of Kaczynski and Zeidan [29] in G-convex spaces without linear structure. The example of a set-valued mapping A satisfying the condition (i) of Theorem 6.1 was given by Cubiotti [25:p. 20].

Theorem 6.2: *Let I be a (finite or infinite) set of players, $\Gamma = (X_i, F_i, f_i)_{i\in I}$ be a constrained game where each (X_i, Γ_i) be a G-convex spaces and $X = \prod_{i\in I} X_i$. For each $\in I$, let $K_i \subset X_i$ be a nonempty compact set and $K = \prod_{i\in I} K_i$. Suppose that the mapping A: $X \to 2^X$ and the function f: $X \times X \to \mathbf{R} \cup \{\pm\infty\}$ defined by*

$$A(x) = \prod_{i\in I} F_i(x^i) \quad \text{and} \quad f(y, x) = \sum_{i\in I} f(y_i, x^i)$$

respectively satisfy the following conditions:

(i) *F_i and A satisfies the conditions (i) of Theorem 6.1,*
(ii) *f is continuous such that for each $x \in X$, $y \mapsto f(y, x)$ is G-quasiconvex,*
(iii) *for each $N \in \mathcal{F}(X)$, there exists a nonempty compact G-convex subset L_N of X containing N such that for each $x \in L_N \backslash K$, if $x \notin D$, then $A(x) \cap L_N \neq \emptyset$; if $x \in D$, then there is a $y \in A(x) \cap L_N$ satisfying $f(y, x) < f(x, x)$.*

Then there exists $\hat{x} \in X$ such that for each $i \in I$,

$$\hat{x}_i \in F_i(\hat{x}^i),$$
$$f_i(\hat{x}_i, \hat{x}^i) \leq f_i(y_i, \hat{x}^i), \quad \forall\, y_i \in F_i(\hat{x}^i),$$

i.e., $\hat{x} \in X$ is a equilibrium point of the constrained game Γ.

Proof: Let $X = \prod_{i \in I} X_i$ and $\Gamma = \prod_{i \in I} \Gamma_i$. By Lemma 6.1, (X, Γ) is a G-convex space. Clearly $K = \prod_{i \in I} K_i$ is a nonempty compact subset of X. By the condition (i)–(iii), it is easy to check that A and f satisfy all conditions of Corollary 5.1. Hence there exists $\hat{x} \in X$ such that $\hat{x} \in A(\hat{x})$ and

$$f(\hat{x}, \hat{x}) \leq f(y, \hat{x}), \quad \forall\, y \in A(\hat{x}).$$

By using a similar argument as in the proof of Theorem 6.1, we can show that for each $i \in I$,

$$\hat{x}_i \in F_i\hat{x}^i),$$
$$f_i(\hat{x}_i, \hat{x}^i) \leq f_i(y_i, \hat{x}^i), \quad \forall;\ y_i \in F_i(\hat{x}^i),$$

i.e., $\hat{x} \in X$ is an equilibrium point of the constrained game Γ.

Corollary 6.1: *Let $\Gamma = (X_i, f_i)_{i \in I}$ be a conventional game where each (X_i, Γ_i) is a G-convex space and $X = \prod_{i \in I} X_i$. For each $i \in I$, let K_i be a nonempty compact subset of X_i and $K = \prod_{i \in I} K_i$. Suppose that the function f: $X \times X \to \mathbf{R} \bigcup \{\pm\infty\}$ defined by*

$$f(y, x) = \sum_{i \in I} f_i(y_i, x^i)$$

satisfying the following conditions:

(*i*) *f is continuous such that for each $x \in X$, $y \mapsto f(y, x)$ is G-quasiconvex,*
(*ii*) *for each $N \in \mathcal{F}(X)$, there exists a nonempty compact G-convex subset L_N of X containing N such that for each $x \in L_N \backslash K$, there is a $y \in L_N$ satisfying $f(y, x) < f(x, x)$.*

Then there exists $\hat{x} \in X$ such that for each $i \in I$,

$$f_i(\hat{x}_i, \hat{x}^i) \leq f_i(y_i, \hat{x}^i), \quad \forall\, y_i \in X_i,$$

i.e., $\hat{x} \in X$ is a Nash equilibrium point of the conventional game Γ.

Proof: For each $i \in I$, define the set-valued mapping F_i: $X^i \to 2^{X_i}$ by $F_i(x^i) = X_i$ for each $x^i \in X^i$. Then the conclusion of Corollary 6.1 holds from Theorem 6.2.

Remark 6.2: If for each $i \in I$, (X_i, Γ_i) is a compact G-convex space, then the condition (iii) of Theorem 6.2 and the condition (ii) of Corollary 6.1 are satisfied trivially. Theorem 6.2 is a improving variant of Theorem 8.4.23 of Aubin and Ekeland [15:pp. 350,351]. Theorem 4.1 of Ding [31,32], and Theorem 7.1 of Yuan *et al.* [33] in G-convex spaces without linear structure.

REFERENCES

[1] S. Park and H. Kim (1996). Coincidence theorems for addmissible multifunctions on generalized convex spaces, *J. Math. Anal. Appl.*, **197**, 173–187.

[2] S. Park and H. Kim (1997). Foundations of the KKM theory on generalized convex spaces, *J. Math. Anal. Appl.*, **209**, 551–571.

[3] X.P. Ding (1995). New H-KKM theorems and their applications to geometric property, coincidence theorems, minimax inequality and maximal elements, *Indian J. Pure Appl. Math.*, **26**(1), 1–19.

[4] G. Tian (1992). Generalization of FKKM theorem and the Ky Fan minimax inequality with applications to maximal elements, price equilibrium and complementarity, *J. Math. Anal. Appl.*, **170**, 457–471.

[5] X. Wu and S. Shen (1996). A furhter generalization of Yannelis-Prabhakar's continuous selection theorem and its applications, *J. Math. Anal. Appl.*, **197**, 61–74.

[6] H. Komiya (1986). Coincidence theorem and saddle point theorem, *Proc. Amer. Math. Soc.*, **96**, 599–602.

[7] M. Lassonde (1983). On the use of KKM multifunctions in fixed point theory and related topics, *J. Math. Anal. Appl.*, **97**, 151–201.

[8] C.D. Horvath (1983). Points fixes et coincidences pour les applications multivoques sans convexite, *C.R. Acad. Sci. Paris*, (296), 403–406.

[9] R. Bielawski (1987). Simplicial convexity and its applications, *J. Math. Anal. Appl.*, **127**, 155–171.

[10] C.D. Horvath (1987). Some results on multivalued mappings and inequalities without convexity. In B.L. Lin and S. Simons (Eds), *Nonlinear and Convex Analysis*, New York: Dekker, 99–106.

[11] C.D. Horvath (1991). Contractibility and generalized convexity, *J. Math. Anal. Appl.*, **156**, 341–357.

[12] X.P. Ding (1997). A coincidence theorem involving contractible spaces, *Appl. Math. Lett.*, **10**(3), 53–56.

[13] X.P. Ding (1998). Coincidence theorems involving composites of acyclic mappings in contractible spaces, *Appl. Math. Lett.*, **11**(2), 85–89.

[14] X.P. Ding (1999). Coincidence theorems involving composites of acyclic mappings and applications, *Acta Math. Sci.*, **19**(1), 53–61.

[15] J.P. Aubin and I. Ekeland (1984). *Applied Nonlinear Analysis*, New York: Wiley.

[16] X.P. Ding and E. Tarafdar (1994). Some coincidence theorems and applications, *Bull. Austral. Math. Soc.*, **50**, 73–80.

[17] S. Park and M.P. Chen (1998). A unified approach to variational inequalities on compact convex sets, *Nonlinear Anal.*, **33**, 637–644.

[18] X.Z. Yuan (1996). Extensions of Ky Fan section theorems and minimax inequality theorems, *Acta Math. Hungar.*, **71**(3), 171–182.

[19] C.W. Ha (1987). On a minimax inequality of Ky Fan, *Proc. Amer. Math. Soc.*, **99**, 680–682.

[20] X. Wu (1998). Existence theorems of solutions for generalized quasi-variational inequalities, a minimax theorem, and a section theorem in spaces without linear structure, *J. Math. Anal. Appl.*, **220**, 495–507.

[21] C.W. Ha (1980). Minimax and fixed point theorems, *Math. Ann.*, **248**, 73–77.

[22] K. Fan (1961). A generalization of Tychonoff's fixed point theorem, *Math. Ann.*, **142**, 305–310.

[23] K. Fan (1972). A Minimax inequality and applications, In *Inequality III*, New York: Acad. Press, 103–113.

[24] M.A. Noor and W. Oettli (1994). On general nonlinear complementarity problems and quasi-equilibria, *Le Mathatiche*, **49**(2), 313–331.

[25] P. Cubiotti (1995). Existence of solutions for lower semicontinuus quasi-equilibrium problems, *Computers Math. Applic.*, **30**(12), 11–22.

[26] X.P. Ding (1998). Existence of solutions for equilibrium problems, *J. Sichuan Normal Univ.*, **21**(6), 603–608.

[27] L.J. Lin and S. Park (1998). On some generalized quasi-equilibrium problems, *J. Math. Anal. Appl.*, **224**(2), 167–281.

[28] S.S. Chang, B.S. Lee, X. Wu, Y.J. Cho and G.M. Lee (1996). On generalized quasi-variational inequality problems, *J. Math. Anal. Appl.*, **203**(3), 686–711.

[29] T. Kaczynski and V. Zeidan (1989). An application of Ky Fan fixed point theorem to an optimization problem, *Nonlinear Anal.*, **13**(3), 259–261.

[30] S. Park and J.A. Park (1996). The Idzik type quasivariational inequalities and noncompact optimization problem, *Colloquium Math.*, **71**(2), 287–295.

[31] X.P. Ding (1991). Quasi-variational inequalities and social equilibrium, *Appl. Math. Mech.*, **12**(7), 639–646.

[32] X.P. Ding (1998). Generalized quasivariational inequalities, optimization and equilibrium existence problems, *J. Sichuan Normal Univ.*, **21**(1), 1–5.

[33] X.Z. Yuan, G. Isac, K.K. Tan and J. Yu (1998). The study of minimax inequalities, abstract economics and applications to variational inequalities and Nash equilibria, *Acta Applic. Math.*, **54**(1), 135–166.

[34] K.K. Tan (1997). *G-KKM* theorem, minimax inequalities and saddle points, *Nonlinear Anal.*, **30**(7), 551–571.

SIMAA 4(2002) 175–181

12. Fixed Point Results for Multivalued Contractions on Gauge Spaces*

M. Frigon

Département de Mathématiques et Statistique, Université de Montréal, C.P. 6128, Succ. Centre-ville, Montréal, H3C 3J7, Canada. E-mail: frigon@dms.umontreal.ca

Abstract: In this chapter, we present a fixed point result for set-valued contractions on complete gauge spaces which generalizes fixed point theorems of Nadler, and of Cain and Nashed. Also, we consider contractions $F: X \to E$ defined on an arbitrary closed subset of a gauge space. We show that the property of having a fixed point is invariant by suitable homotopies. From that, we deduce a fixed point theorem of Leray-Schauder type. Then we present an application of those results to first order differential inclusions on the half line.

1. INTRODUCTION AND PRELIMINARIES

In 1969, S. Nadler [14] established a generalization of the well known Banach Contraction Principle for multivalued contractions with closed, bounded, nonempty values, defined on a complete metric space. One year later, Covitz and Nadler [5] proved that the values don't need to be bounded. On the other hand, in 1971, Cain and Nashed [3] extended the notion of singlevalued contraction to Hausdorff locally convex linear spaces. They showed that on sequentially complete subset, the Banach Contraction Principle is still valid.

In this chapter, we present a generalization of those two results to multivalued contractions defined on a complete gauge space E.

In Section 3, we consider multivalued contractions defined on a closed subset of E. We introduce a notion of homotopy for such contractions. Then we show that the property of having a fixed point is invariant by homotopy. Our result generalizes results of Granas and myself [12,13]. It is worthwhile to mention that even in the singlevalued case, our result can not be obtained from the index theory for condensing operators (see [1,9,15]). Indeed, there is no vectorial structure on E; moreover, the set X can have empty interior. The reader is referred to [4,8,10,11] and the references therein for other results on contractions.

Finally, we present an application to first order differential inclusions on the half line.

In what follows, $E = (E, \{d_\alpha\}_{\alpha\in\Lambda})$ denotes a gauge space endowed with a complete gauge structure $\{d_\alpha : \alpha \in \Lambda\}$, see [7] for definitions.

For $r = \{r_\alpha\}_{\alpha\in\Lambda} \in]0,\infty[^\Lambda$, and $x \in E$, we define the *pseudo-ball* centered in x of radius r by

$$B(x,r) = \{y \in E : d_\alpha(x,y) \le r_\alpha \text{ for all } \alpha \in \Lambda\}.$$

* This work was partially supported by CRSNG-Canada.

For all subset X of E, we denote $\bar{X}$ and ∂X respectively the closure and the boundary of X in E. Also, we denote

$$X_{k,M} = \left\{ x \in X : B(x,r) \not\subset X \ \forall r \in]0,\infty[^\Lambda \text{ with } \inf_{\alpha\in\Lambda} \frac{r_\alpha(1-k_\alpha)}{M_\alpha} > 0 \right\},$$

for $k \in [0,1[^\Lambda$ and $M \in [0,\infty[^\Lambda$.

Remark: If $X = \bar{U}$ for U an open subset of E, then $X_{k,M} = \partial U$ for every $k \in [0,1[^\Lambda$ and $M \in [0,\infty[^\Lambda$.

We denote by D_α, the generalized Hausdorff pseudometric induced by d_α; that is, for $X, Y \subset E$,

$$D_\alpha(X,Y) = \inf\{\varepsilon > 0 : \forall x \in X, \forall y \in Y, \exists \hat{x} \in X, \exists \hat{y} \in Y \text{ such that } d_\alpha(x,\hat{y}) < \varepsilon, d_\alpha(\hat{x},y) < \varepsilon\},$$

with the convention that $\inf(\emptyset) = \infty$. In the particular case where E is a complete locally convex space, we say that a subset $X \subset E$ is *bounded* if $D_\alpha(\{0\},X) < \infty$ for every $\alpha \in \Lambda$.

2. FIXED POINT RESULT

In what follows, a multivalued map $F : X \to E$ is a map with closed, nonempty values $F(x) \subset E$.

Definition 2.1: A multivalued map $F : X \to E$ is called an *admissible contraction* with constant $k = \{k_\alpha\}_{\alpha\in\Lambda} \in [0,1[^\Lambda$ if

(i) for every $\alpha \in \Lambda$, $D_\alpha(F(x),F(y)) \leq k_\alpha d_\alpha(x,y)$ for every x, $y \in X$;
(ii) for every $x \in X$ and every $\varepsilon \in]0,\infty[^\Lambda$, there exists $y \in F(x)$ such that $d_\alpha(x,y) \leq d_\alpha(x,F(x)) + \varepsilon_\alpha$ for every $\alpha \in \Lambda$.

Observe that if $\Lambda = \mathbb{N}$, a multivalued map F can be a contraction in the sense of the previous definition without being a contraction in the usual sense when X is endowed with the metric $d(x,y) = \sum_{n\in\mathbb{N}} d_n(x,y)/2^n(1+d_n(x,y))$.

First of all, we establish a fixed point result for a multivalued contractive map defined on a pseudo-ball, and which does not move its center too far away. It generalizes a result of [12].

Proposition 2.2: *Let E be a complete gauge space, $r \in]0,\infty[^\Lambda$, $x_0 \in E$, and $F : B(x_0,r) \to E$ be an admissible multivalued contraction with constant $k \in]0,1[^\Lambda$ such that $d_\alpha(x_0,F(x_0)) < (1-k_\alpha)r_\alpha$ for every $\alpha \in \Lambda$. Then F has a fixed point.*

Proof: By assumptions and Definition 2.1(ii), we can choose $x_1 \in F(x_0)$ such that

$$d_\alpha(x_1,x_0) < (1-k_\alpha)r_\alpha \quad \text{for every } \alpha \in \Lambda.$$

Then, choose $x_2 \in F(x_1)$ such that for every $\alpha \in \Lambda$,

$$\begin{aligned} d_\alpha(x_1, x_2) &< d_\alpha(x_1, F(x_1)) + k_\alpha((1 - k_\alpha)r_\alpha - d_\alpha(x_0, x_1)) \\ &\leq D_\alpha(F(x_0), F(x_1)) + k_\alpha((1 - k_\alpha)r_\alpha - d_\alpha(x_0, x_1)) \\ &\leq k_\alpha d_\alpha(x_0, x_1) + k_\alpha((1 - k_\alpha)r_\alpha - d_\alpha(x_0, x_1)) \\ &= k_\alpha(1 - k_\alpha)r_\alpha. \end{aligned}$$

In repeating this process, we obtain a sequence $\{x_m\}$ such that

$$d_\alpha(x_m, x_{m+1}) < k_\alpha^m(1 - k_\alpha)r_\alpha \quad \text{for every } \alpha \in \Lambda.$$

Therefore, $\{x_m\}$ is a Cauchy sequence and hence converges to $x \in B(x_0, r)$. The continuity of F implies that $x \in F(x)$. □

As a direct consequence of the previous result, we obtain a generalization of Covitz and Nadler's fixed point Theorem [5] and of Cain and Nashed's result [3].

Theorem 2.3: *Let E be a complete gauge space, and let $F: E \to E$ be an admissible multivalued contraction. Then F has a fixed point.*

Proof: Let $k \in [0, 1[^\Lambda$ be a constant of contraction of F. Fix $x_0 \in E$. For every $\alpha \in \Lambda$, choose $r_\alpha > 0$ such that $d_\alpha(x_0, F(x_0)) < (1 - k_\alpha)r_\alpha$. The conclusion follows from the previous proposition.

3. LERAY-SCHAUDER TYPE RESULTS

We introduce the notion of homotopy of contractions.

Definition 3.1: Let X be a closed subset of E. An *homotopy of admissible contractions* is a multivalued map $H: X \times [0, 1] \to E$ such that

(a) there exists $k \in [0, 1[^\Lambda$ such that for every $t \in [0, 1]$, $H(\cdot, t)$ is an admissible contraction with constant k;
(b) there exits $M \in [0, \infty[^\Lambda$ such that $D_\alpha(H(x, t), H(x, s)) \leq M_\alpha|t - s|$ for every s, $t \in [0, 1]$, every $x \in X$, and every $\alpha \in \Lambda$;
(c) $x \notin H(x, t)$ for every $t \in [0, 1]$, and every $x \in X_{k,M}$.

Remark: If $X = \bar{U}$ for U an open subset of E, since $X_{k,M} = \partial U$ for every $k \in [0, 1[^\Lambda$ and $M \in [0, \infty[^\Lambda$, condition (c) becomes: $x \notin H(x, t)$ for every $t \in [0, 1]$, and every $x \in \partial U$. Also, it is easy to see that (b) can be generalized. We stated this condition for sake of simplicity.

Definition 3.2: Let X be a closed subset of E. We say that the multivalued admissible contractions $F, G: X \to E$ are *homotopic* if there exists an homotopy of contractions $H: X \times [0, 1] \to E$ such that $F = H(\cdot, 1)$ and $G = H(\cdot, 0)$.

We obtain the invariance by homotopy of the property of having a fixed point. It is worthwhile to mention that X can have empty interior. The proof is in fact a slight modification of the proof of [12, Th. 4.3]. We present it for sake of completeness.

Theorem 3.3: *Let X be a closed subset of a complete gauge space E, and let $F,G: X \to E$ be two homotopic admissible contractions. Then F has a fixed point if and only if G has a fixed point.*

Proof: Let $H: X \times [01] \to E$ be an homotopy of admissible contractions between F and G. Consider

$$Q = \{(x,t) \in X \times [0,1] : x \in H(x,t)\}.$$

We define on Q the partial order

$$(x,t) \leq (y,s) \quad \text{iff} \quad t \leq s \quad \text{and} \quad d_\alpha(x,y) \leq \frac{2M_\alpha(s-t)}{1-k_\alpha} \quad \text{for every } \alpha \in \Lambda,$$

where k and M are given in the definition of H.

Let $P \subset Q$ be a totally ordered set. Set $t^* = \sup\{t : (x,t) \in P\}$. Let $\{(x_m, t_m)\}$ be a sequence in P such that $(x_m, t_m) \leq (x_{m+1}, t_{m+1})$ and $t_m \to t^*$. The order on Q yealds

$$d_\alpha(x_m, x_l) \leq \frac{2M_\alpha(t_m - t_l)}{1-k_\alpha} \quad \text{for every } m > l \text{ and every } \alpha \in \Lambda.$$

Thus, $\{x_m\}$ is a Cauchy sequence and hence converges to $x^* \in X$. The continuity of H implies that $x^* \in H(x^*, t^*)$, so $(x^*, t^*) \in Q$. It is easy to see the (x^*, t^*) is an upper bound of P.

From Zorn's Lemma, Q has a maximal element $(x_0, t_0) \in Q$, so, $x_0 \in H(x_0, t_0)$.

To conclude, we need to show that $t_0 = 1$. If this is false, since $x_0 \notin X_{k,M}$ by Definition 3.1(c), there exist $r \in [0, \infty[^\Lambda$ and $t_1 \in (t_0, 1]$ such that

$$B(x_0, r) \subset X \quad \text{and} \quad \frac{2M_\alpha(t_1 - t_0)}{1-k_\alpha} = r_\alpha \quad \text{for every } \alpha \in \Lambda.$$

On the other hand, for every $\alpha \in \Lambda$,

$$\begin{aligned} d_\alpha(x_0, H(x_0, t_1)) &\leq d_\alpha(x_0, H(x_0, t_0)) + D_\alpha(H(x_0, t_0), H(x_0, t_1)) \\ &\leq M_\alpha(t_1 - t_0) < (1 - k_\alpha) r_\alpha. \end{aligned}$$

From Proposition 2.2, there exists $x_1 \in B(x_0, r)$ a fixed point of $H(\cdot, t_1)$. So, $(x_1, t_1) \in Q$ and $(x_0, t_0) < (x_1, t_1)$, which is a contradiction. □

In the particular case where E is a complete locally convex space, we deduce the following corollary.

Corollary 3.4 (Nonlinear Alternative): *Let X be a closed subset of a complete locally convex space E, and let $F: X \to E$ be an admissible multivalued contraction with constant k. Assume that F is bounded and $0 \in X \backslash X_{k,M}$, where $M \in [0, \infty[^\Lambda$ with $M_\alpha = D_\alpha(\{0\}, F(X))$. Then one of the following statements holds:*

(1) *F has a fixed point;*
(2) *there exist $t \in]0,1[$ and $x \in X_{k,M}$ such that $x \in tF(x)$.*

This corollary can be stated more simply for contractions defined on the closure of an open set.

Corollary 3.5: *Let U be an open neighborhood of the origin in a complete locally convex space space E, and let $F:\bar{U} \to E$ be an admissible multivalued contraction. Assume that F is bounded. Then one of the following statements holds:*

(1) *F has a fixed point;*
(2) *there exist $t \in]0,1[$ and $x \in \partial U$ such that $x \in tF(x)$.*

4. APPLICATION

Let us consider the first order differential inclusion

$$\begin{aligned} &x'(t) \in f(t, x(t)) \quad \text{a.e. } t \in [0, \infty[, \\ &x(0) = 0 \in H, \end{aligned} \tag{4.1}$$

where H is an Hilbert space, and $f:[0,\infty[\times H \to H$ is a locally Carathéodory multivalued map with closed, bounded, nonempty values; i.e., satisfying:

(i) $t \mapsto f(t,x)$ is measurable for all $x \in H$ ($\{t: f(t,x) \cap C \neq \emptyset\}$ is measurable for every closed subset C);
(ii) $x \mapsto f(t,x)$ is continous for almost every $t \in [0,\infty[$ (here 2^H is endowed with the Hausdorff metric D);
(iii) for all $R > 0$, there exists a function $h_R \in L^1_{\text{loc}}[0,\infty[$ such that for almost every $t \in [0,\infty[$ and for every $x \in H$ with $\|x\| \le R$, we have $D(\{0\}, f(t,x)) \le h_R(t)$.

Theorem 4.1: *Let $(H, \|\cdot\|)$ be an Hilbert space, and $f:[0,\infty[\times H \to H$ a locally Carathéodory multivalued map with closed, bounded, nonempty values. Assume that*

(*a*) *for every $R > 0$, there exists $l_R \in L^1_{\text{loc}}[0,\infty[$ such that for almost every $t \in [0,\infty[$, and every x, $y \in H$ satisfying $\|x\|$, $\|y\| \le R$, we have*

$$D(f(t,x), f(t,y)) \le l_R(t)\|x - y\|;$$

(*b*) *there exist $\theta \in L^1_{\text{loc}}[0,\infty[$ and $\psi:[0,\infty[\to]0,\infty[$ a Borel measurable function such that $D(\{0\}, f(t,x)) \le \theta(t)\psi(\|x\|)$ a.e. $t \in [0,\infty[$, and all $x \in H$, with $1/\psi \in L^1_{\text{loc}}[0,\infty[$; and*

$$\int_0^\infty \frac{dz}{\psi(z)} > \|\theta\|_{L^1[0,r]} \quad \textit{for all } r > 0.$$

Then the problem (4.1) has a solution in $W^{1,1}_{\text{loc}}([0,\infty[, H)$.

Proof: Let $M:[0,\infty[\to [0,\infty[$ be a continuous nondecreasing function such that

$$\int_0^{M(t)} \frac{ds}{\psi(s)} \ge \|\theta\|_{L^1[0,t]}.$$

Let us define $\hat{f}:[0,\infty[\times H \to H$ by

$$\hat{f}(t,x) = \begin{cases} f(t,x), & \text{if } \|x\| \leq M(t), \\ f(t, \frac{M(t)x}{\|x\|}), & \text{if } \|x\| > M(t); \end{cases}$$

and define $F: C([0,\infty[, H) \to C([0,\infty[, H)$ by

$$F(x)(t) = \int_0^t \hat{f}(s, x(s))ds.$$

Using assumption (b) and the definition of M, one shows that if x is a fixed point of F then x is a solution of (4.1).

Set $l(t) = l_{M(n)}(t)$ pour $t \in]n-1, n]$, $n \in \mathbb{N}$, where $l_{M(n)}$ is given in (a). For each $n \in \mathbb{N}$, we define on $C([0,\infty[, H)$ the semi-norm

$$|x|_n = \sup\{e^{-\int_0^t l(s)ds} \|x(t)\| : t \in [0,n]\}.$$

It follows directly from assumption (a) and the theory of multivalued maps (see [2]) that F is an admissible contraction. Theorem 2.3 gives the existence of solution of (4.1).

REFERENCES

[1] R.R. Akhmerov, M.I. Kamenskii, A.S. Potapov, A.E. Rodkina and B.N. Sadovskii (1992). *Measures of noncompactness and condensing operators*. Basel: Birkhäuser.

[2] Y.G. Borisovich, B.D. Gel'man, A.D. Myshkis, and V.V. Obukhovskii (1982). Multivalued mappings, *Itogi Nauki i Tekhniki Ser. Mat. Anal.*, **19**, 127–230 (Russian); English translation: *J. Soviet Math.*, **24**, 719–791.

[3] G.L. Cain, Jr. and M.Z. Nashed (1971). Fixed points and stability for a sum of two operators in locally convex spaces, *Pacific J. Math.*, **39**, 581–592.

[4] J. Carmona Alvárez (1985). Measure of noncompactness and fixed points of nonexpansive condensing mappings in locally convex spaces, *Rev. Real Acad. Cienc. Exact. Fis. Natur. Madrid*, **79**, 53–66.

[5] H. Covitz and S.B. Nadler, Jr. (1970). Multi-valued contraction mappings in generalized metric spaces, *Israel J. Math.*, **8**, 5–11.

[6] R. Datko (1973). On the integration of set-valued mappings in a Banach space, *Fund. Math.*, **78**, 205–208.

[7] J. Dugundji (1966). *Topology*. Boston: Allyn and Bacon.

[8] P.M. Fitzpatrick and W.V. Petryshyn (1975/76). Fixed point theorems and the fixed point index for multivalued mappings in cones, *J. London Math. Soc.*, **12**(2), 75–85.

[9] G. Fournier and M. Martelli (1990). Boundary conditions and vanishing index for α-contractions and condensing maps, Nonlinear functional analysis, Lecture Notes in Pure and Appl. Math., 121, Dekker, New York, 31–48.

[10] M. Frigon (1996). On continuation methods for contractive and nonexpansive mappings. In T. Domnguez Benavides (Ed.), *Recent Advances on Metric Fixed Point Theory*, Sevilla 1995. Sevilla: Universidad de Sevilla, 19–30.

[11] M. Frigon (2000). Fixed point results for generalized contractions in gauge spaces and applications, *Proc. Amer. Math. Soc.*, **128**, 2957–2965

[12] M. Frigon and A. Granas (1994). Résultats de type Leray-Schauder pour des contractions multivoques, *Topol. Methods Nonlinear Anal.*, **4**, 197–208.

[13] M. Frigon and A. Granas (1998). Résultats de type Leray-Schauder pour des contractions sur des espaces de Fréchet, *Ann. Sci. Math. Québec*, **22**, 161–168.

[14] S.B. Nadler, Jr. (1969). Multi-valued contraction mappings, *Pacific J. Math.*, **30**, 415–487.

[15] R. Nussbaum (1971). The fixed point index for local condensing maps, *Ann. Mat. Pura Appl.*, **89**, 217–258.

SIMAA 4(2002) 183–203

13. The Study of Variational Inequalities and Applications to Generalized Complementarity Problems, Fixed Point Theorems of Set-valued Mappings and Minimization Problems

G. Isac[1], E. Tarafdar[2] and George Xian-Zhi Yuan[2,*]

[1]*Department of Mathematics and Computer Science, Royal Military College of Canada, Kingston, K7K 5L0, Canada*
[2]*Department of Mathematics, The University of Queensland, Brisbane, 4072, Australia*

Abstract: As applications of Ky Fan minimax principle and Kneser's minimax theorem, the existence theorems of solutions of variational inequalities for monotone and quasi-monotone set-valued mapping in either topological vector spaces or reflexive Banach spaces have been established. These results are then used to study the existence of solutions for generalized complementarity problems; and some fixed point theorems for set-valued quasi-contractive mappings which are generalizations of pseudo-contractive nonexpansive mappings (which, in turn are generalizations of nonexpansive mappings) are obtained. Finally existence theorems on minimizers of minimization problem for the sum of convex lower semicontinuous functions are established. These results include corresponding results of literatures as special cases.

Keywords and Phrases: Variational inequality, reflexive Banach space, complementarity problem, fixed point theorem, pseudo-contractive, nonexpansive, monotone, semi-monotone, quasi-monotone, Ky Fan minimax inequality

1991 Mathematics Subject Classification: Primary – 49J40, 54C60, 47H10; Secondary – 46C05

1. INTRODUCTION

This chapter consists of two parts. In the first part, by applying Ky Fan minimax principle [12] and Kneser's minimax theorem [17], the existence theorems of solutions of variational inequalities for monotone and quasi-monotone set-valued mappings in either topological vector spaces or reflexive Banach spaces have been established. In the second part, these results are used to study the existence of solutions for generalized complementarity problems, and some fixed point theorems for multi-valued quasi-contractive (see definition below) and multi-valued nonexpansive mappings are obtained. Finally

* Corresponding author. E-mail: xzy@maths.uq.edu.au

existence theorems on minimizers of minimization problems for the sum of convex lower semicontinuous functions are established. These results generalize and unify corresponding results of literature as special cases.

We now recall and introduce some definitions and notations. If X is a non-empty set, we shall denote by 2^X and $\mathcal{F}(X)$ the family of all subsets of X and the family of all non-empty finite subsets of X. In what follows, all vector spaces will be assumed over the real field. If E is a topological vector space, we shall denote by E^* the continuous dual of E and by $\mathrm{Re}\langle w, x\rangle$ the (real part of the) pairing between $w \in E^*$ and $x \in E$.

If X is a non-empty subset of E, a mapping $T: X \to 2^{E^*}$ is said to be

(1) *monotone* on X if for each $x, y \in X$, $u \in T(x)$ and $w \in T(y)$, $\mathrm{Re}\langle w - u, y - x\rangle \geq 0$;

(2) *semi-monotone* (e.g., see [2]) on X if for each $x, y \in X$, $u \in T(x)$ and $w \in T(y)$,

$$\inf_{u \in T(x)} \mathrm{Re}\langle u, y - x\rangle \leq \inf_{w \in T(y)} \mathrm{Re}\langle w, y - x\rangle;$$

(3) *pseudo-monotone* (see Definition 3.4 of [19:p. 263]) if for each $x, y \in X$, $u \in T(x)$ and $v \in T(y)$, $\mathrm{Re}\langle v, x - y\rangle \geq 0$ implies $\mathrm{Re}\langle u, x - y\rangle \geq 0$;

(4) *quasi-monotone* (compare with [6]) if for each $x, y \in K$, $\inf_{v \in T(y)} \mathrm{Re}\langle x - y,\, v\rangle > 0$ implies $\inf_{u \in T(x)} \mathrm{Re}\langle x - y,\, u\rangle > 0$.

Clearly, each monotone mapping is pseudo-monotone and Proposition 3.2 of Karamardian and Schaible [15:p. 39] shows that the definition (4) above is a set-valued generalization of the definition (3) for a single-valued mapping. Also each semi-monotone mapping is quasi-monotone. But the converse is not necessarily true (e.g., see [15,2, Example] and references therein)

Let X be a non-empty convex subset of a topological vector space E and $\psi: X \times X \to [-\infty, +\infty]$ an extended real-valued function. We recall (e.g., see [26]) the following definitions:

(1) ψ is said to be 0-*diagonal quasi-concave in* y if for each $A \in \mathcal{F}(X)$ and $x_0 \in \mathrm{co}(A)$, we have $\min_{y \in A} \psi(x_0, y) \leq 0$;

(2) ψ is said to be 0-*diagonal concave in* y if for each $A := \{y_1, \ldots, y_n\} \in \mathcal{F}(X)$ and $\alpha_i \in (0, 1]$ for $i = 1, \ldots, n$ with $\Sigma_{i=1}^n \alpha_i = 1$ and $x_0 := \Sigma_{i=1}^n \alpha_i x_i$, we have $\Sigma_{i=1}^n \alpha_i \psi(x_0, y_i) \leq 0$.

Let $\psi(x, x) \leq 0$ for each $x \in X$. It is clear that if $\psi(x, y)$ is quasi-concave in y for each $x \in X$, $\psi(x, y)$ is 0-diagonal concave in y. But the converse is not necessarily true (e.g., see Remark 2.2 of [26:p. 215] and references therein). Let X be a topological space and $f: X \to [-\infty, +\infty]$ an extended real valued function. Then f is said to be *compactly lower semicontinuous* if f is lower semicontinuous on each non-empty compact subset of its domain X.

Let E be a vector space and $y \in E$. The set

$$I_X(y) := \{x \in E : x = y + r(u - y) \quad \text{for some} \quad u \in X \text{ and } r > 0\}$$

is called the inward set (e.g., see [11]) of y with respect to X. If E is a topological vector space and $X \subset E$, we denote by either $\overline{X}$ or $cl_E X$ (in short, clX) the closure of X in E. If E is a normed space with norm $\|\cdot\|$, A is a non-empty subset of E and $x \in E$, we denote by $d(x, A) := \inf\{\|x - y\| : y \in A\}$ the distance from x to A. Let $CB(E) := \{A \subset E : A \text{ is non-empty closed and bounded}\}$. Then the Hausdorff metric D on $CB(E)$ induced by the norm is defined by

$$D(A_1, A_2) = \inf\{r > 0 : A_1 \subset B_r(A_2) \quad \text{and} \quad A_2 \subset B_r(A_1)\}$$
$$= \max\{\sup_{x \in A_1} d(x, A_2), \sup_{y \in A_2} d(y, A_1)\}$$

where $B_r(A) := \{x \in E : d(x, A) > r\}$ for any $A \in 2^E$ and $r > 0$.

A mapping T: $X \to CB(E)$ is said to be *nonexpansive* on X if for each $x, y \in X$, we have $D(T(x), T(y)) \leq \|x - y\|$.

A mapping T: $X \to 2^E$ is said to be *pseudo-contractive* (e.g., see [2]) on X if for each $x, y \in X$ and $w \in T(y)$, there exists $u \in T(x)$ such that $\|x - y\| \leq \|(1 + r)(x - y) - r(u - w)\|$ for all $r > 0$. This definition is a set-valued generalization of pseudo-contractive (single-valued) mappings as defined by Browder in [3].

The concept of *quasi-monotone mappings* as given above motivates us to introduce the following definition:

Definition 1.1: Let X be a non-empty subset of a normed space $(E, \|\cdot\|)$ and T: $X \to 2^E$ a set-valued mapping with non-empty values. Then T is said to be *quasi-contractive* if the mapping $I - T$ is a quasi-monotone mapping.

By the definition of a quasi-contractive mapping we shall see that (e.g., Lemma 5.1 below) each pseudo-contractive mapping is quasi-contractive; but the converse is not necessarily true. Thus the quasi-monotone mapping is a generalization of a pseudo-monotone mapping which, in turn, is a generalization of nonexpansive mapping.

2. SOME LEMMAS

In this section, we first state some lemmas which have their own interests and will be used to study existence of solutions of variational inequalities and complementarity problems in the sequel below.

Lemma 2.1: *Let X be a non-empty convex subset of a topological vector space E and T: $X \to 2^{E^*}$ a set-valued quasi-monotone mapping with non-empty values. Then the mapping ψ: $X \times X \to [-\infty, +\infty]$ defined by*

$$\psi(x, y) := \inf_{u \in T(y)} \mathrm{Re}\langle u, x - y\rangle$$

for each $(x, y) \in X \times X$ is 0-diagonal concave in y for each fixed $x \in X$.

Proof: If it were not true, there would exist $n \in \mathbb{N}$, $y_i \in X$ and $\lambda_i \in (0, 1]$ for $i = 1, \ldots, n$ with $\Sigma_{i=1}^n \lambda_i = 1$ such that $\psi(x_0, y_i) > 0$, where $x_0 = \Sigma_{i=1}^n \lambda_i y_i$; i.e., $\inf_{u \in T(y_i)} \mathrm{Re}\langle u, x_0 - y_i\rangle > 0$ for each $i = 1, \ldots, n$. As T is quasi-monotone, it follows that

$$\inf_{u \in T(x_0)} \mathrm{Re}\langle u, x_0 - y_i\rangle > 0 \qquad (*)$$

for each $i = 1, \ldots, n$. Now multiplying $(*)$ both sides by λ_i and summing them for $i = 1, \ldots, n$, we then have

$$0 = \inf_{u \in T(x_0)} \mathrm{Re}\langle u, x_0 - x_0\rangle = \Sigma_{i=1}^n \lambda_i \inf_{u \in T(x_0)} \mathrm{Re}\langle u, x_0 - y_i\rangle > 0$$

which is not true. Thus the conclusion must hold. □

Lemma 2.2: *Let X be a convex subset of a topological vector space E and $T: X \to 2^{E^*}$ a set-valued monotone with non-empty values. Then the mapping $\psi: X \times X \to [-\infty, +\infty]$ defined by*

$$\psi(x, y) := \inf_{u \in T(y)} \mathrm{Re}\langle u, x - y\rangle$$

and

$$\psi(x, y) := \sup_{u \in T(y)} \mathrm{Re}\langle u, x - y\rangle)$$

for each $(x, y) \in X \times X$ is also 0-diagonal concave in y for each fixed $x \in X$.

Proof: The conclusions just follow immediately from definitions of 0-diagonal concave and monotone of mappings. □

Lemma 2.3: *Let X be a non-empty convex subset of a topological vector space E and $T: X \to 2^{E^*}$ a set-valued mapping with non-empty values. Assume the mapping $\psi: X \times X \to [-\infty, +\infty]$ defined by*

$$\psi(x, y) := \inf_{u \in T(y)} \mathrm{Re}\langle u, x - y\rangle$$

for each $(x, y) \in X \times X$ is 0-diagonal concave in y for each fixed $x \in X$. Then we have that for each $x, y \in X$ with $\inf_{u \in T(y)} \mathrm{Re}\langle u, x - y\rangle > 0$, it implies that $\inf_{u \in T(x)} \mathrm{Re}\langle u, y - x\rangle \leq 0$.

Proof: If the conclusion were not true, i.e., there exist $x, y \in X$, $\inf_{u \in T(y)} \mathrm{Re}\langle u, x - y\rangle > 0$, but $\inf_{u \in T(x)} \mathrm{Re}\langle u, y - x\rangle > 0$. Let $x_0 = \frac{1}{2}(x + y)$. Then we have $\psi(x_0, x) > 0$ and $\psi(x_0, y) > 0$, which contradict that $\psi(x, y)$ is 0-diagonal concave in y. Thus the conclusion is true. □

In order to establish the existence of solutions for variational inequalities, we also need the following result which is a generalization of well-known Ky Fan minimax principle [12] which is Theorem 1 of Ding and Tan [8:p. 235].

Lemma 2.4: *Let X be a non-empty convex subset of a Hausdorff topological vector space E and $\psi: X \times X \to [-\infty, +\infty]$ be an extended real-valued functions such that*

(1) *for each fixed $x \in X$, $y \mapsto \psi(x, y)$ is compactly lower semicontinuous;*
(2) *for each $A \in \mathcal{F}(X)$ and $y_0 \in \mathrm{co}(A)$, $\min_{x \in A} \psi(x, y_0) \leq 0$ (or equivalently to say, $\psi(x, y)$ is 0-diagonal concave in x for each fixed $y \in X$);*
(3) *there exist a non-empty compact and convex subset X_0 of X and a non-empty compact subset K of X such that for each $y \in X \backslash K$, there exists $x \in \mathrm{co}(X_0 \cup \{y\})$ with $\psi(x, y) > 0$.*

Then there exists $\hat{y} \in K$ such that $\sup_{x \in X} \psi(x, \hat{y}) \leq 0$.

We shall need the following Kneser's minimax theorem [17].

Lemma 2.5: (*Kneser*) *Let X be a non-empty convex set in a vector space V and Y a non-empty compact convex subset of a Hausdorff topological vector space E. Suppose* *that f is a real-valued function on $X \times Y$ such that for each fixed $x \in X$, $f(x, y)$ is lower semicontinuous and convex on Y, and for each fixed $y \in Y$, $f(x, y)$ is concave on X. Then*

$$\min_{y \in Y} \sup_{x \in X} f(x, y) = \sup_{x \in X} \min_{y \in Y} f(x, y).$$

We also need the following special case of Lemma 3 of Ding and Tan [7]:

Lemma 2.6: *Let X be a non-empty convex subset of a topological vector space E, $h: X \to [-\infty, +\infty]$ a convex function and $T: X \to 2^{E^*}$ lower semicontinuous from the line segments of X to the weak topology on E^*. If $\hat{y} \in X$, then the inequality*

$$\sup_{u \in T(x)} \operatorname{Re}\langle u, \hat{y} - x\rangle \le h(x) - h(\hat{y}) \quad \textit{for all } x \in X$$

implies the inequality

$$\sup_{w \in T(\hat{y})} \operatorname{Re}\langle w, \hat{y} - x\rangle \le h(x) - h(\hat{y}) \quad \textit{for all } x \in X.$$

The same proofs of Lemmas 1 and 2 of Shih and Tan [21] can be modified to obtain the following slight improvement of Lemma 2 [21] and the proof is thus omitted.

Lemma 2.7: *Let X be a non-empty convex subset of a topological vector space E, $h: X \to [-\infty, +\infty]$ a convex and lower semicontinuous function and $T: X \to 2^{E^*}$ a set-valued mapping such that each $T(x)$ is a $\sigma(E^*, E)$ compact subset of E^* and T is upper semicontinuous from the line segments of X to the weak topology of E^*. If $\hat{y} \in X$, then the inequality*

$$\sup_{u \in T(x)} \operatorname{Re}\langle u, \hat{y} - x\rangle \le h(x) - h(\hat{y}) \quad \textit{for all } x \in X$$

implies the inequality

$$\inf_{w \in T(\hat{y})} \operatorname{Re}\langle w, \hat{y} - x\rangle \le h(x) - h(\hat{y}) \quad \textit{for all } x \in X.$$

3. VARIATIONAL INEQUALITIES OF MONOTONE MAPPINGS

As an application of Lemma 2.4, we first have the following existence theorem of solution of variational inequalities of set-valued monotone mappings in topological vector spaces.

Theorem 3.1: *Let X be a non-empty convex subset of a Hausdorff topological vector space E, $h: X \to [-\infty, +\infty]$ a convex and lower semicontinuous function and $T: X \to 2^{E^*}$ set-valued monotone with non-empty values. Suppose there exist a non-empty compact and convex subset X_0 of X and a non-empty compact subset K of X such that for each $y \in X \setminus K$, there exists $x \in \operatorname{co}(X_0 \cup \{y\})$ such that*

$$\sup_{u \in T(x)} \operatorname{Re}\langle u, y - x\rangle + h(y) - h(x) > 0.$$

Then there exists $\hat{y} \in K$ *such that*

$$\sup_{u \in T(x)} \operatorname{Re}\langle u, \hat{y} - x\rangle \leq h(x) - h(\hat{y}) \quad \textit{for all } x \in X.$$

Proof: Define Φ: $X \times X \to [-\infty, +\infty]$ by

$$\Phi(x, y) := \sup_{u \in T(x)} \operatorname{Re}\langle u, y - x\rangle + h(y) - h(x)$$

for each $(x, y) \in X \times X$. Then Φ satisfies all hypotheses of Lemma 2.4. By Lemma 2.4, there exists $\hat{y} \in K$ such that $\Phi(x, \hat{y}) \leq 0$ for all $x \in X$. By the definition of Φ, it follows that

$$\sup_{x \in X} \sup_{u \in T(x)} \operatorname{Re}\langle u, \hat{y} - x\rangle \leq h(x) - h(\hat{y}).$$

Thus the proof is completed. □

For the convenience of our later discussion, we need the following result which is essentially due to Tan and Yuan (Lemma 2 [24]). However, we shall prove it here as an application of Theorem 3.1.

Theorem 3.2: *Let* X *be a non-empty closed convex subset of a reflexive Banach space* $(E, \|\cdot\|)$. *Let* h: $X \to [-\infty, +\infty]$ *be a convex and lower semicontinuous function and* T: $X \to 2^{E^*}$ *set-valued monotone with non-empty values. Assume that the following condition is satisfied*:

(*E*) *For each sequence* $(y_n)_{n=1}^{\infty}$ *in* X *with* $\|y_n\| \to \infty$ *as* $n \to \infty$, *there exists a sequence* $(x_n)_{n=1}^{\infty}$ *in* X *with* $\|x_n\| \leq \|y_n\|$ *for all* $n = 1, 2, \ldots$ *such that*

$$\overline{\lim_{n \to \infty}} \left\{ \sup_{u \in T(x_n)} Re\langle u, y_n - x_n\rangle + h(y_n) - h(x_n) \right\} > 0.$$

Then there exists $\hat{y} \in X$ *such that*

$$\sup_{u \in T(x)} Re\langle u, \hat{y} - x\rangle \leq h(x) - h(\hat{y}) \quad \textit{for all } x \in X.$$

Proof: Define Φ: $X \times X \to [-\infty, +\infty]$ by

$$\Phi(x, y) := \sup_{u \in T(x)} \operatorname{Re}\langle u, y - x\rangle + h(y) - h(x)$$

for each $(x, y) \in X \times X$. Then we have

(1) As T is monotone, $\Phi(x, y)$ is 0-diagonal concave in y for each $x \in X$ by Lemma 2.2;

(2) Since h is convex and lower semicontinuous respect to the norm topology of X, h is lower semicontinuous with respect to the weak topology of X, and hence for each $x \in X$, the function $y \to \Phi(x, y)$ is weakly lower semicontinuous.

Now let $X_N := \{x \in X : \|x\| \leq N\}$ for each $N \in \mathbb{N}$. We may assume that $X_N \neq \emptyset$ for all $N \geq N_0$. Note that for each $N \geq N_0$, X_N is weakly compact and convex since E is reflexive; equip X_N with the weak topology. By applying Theorem 3.1, there exists $\hat{y}_N \in X_N$ such that

$$\Phi(x, \hat{y}_N) \leq 0 \quad \text{for all } x \in X_N. \tag{1}$$

Suppose $\|\hat{y}_N\| \to \infty$ as $N \to \infty$, then by the assumption (E), there exists a sequence $(x_N)_{N \geq N_0}$ in X with $\|x_N\| \leq \|\hat{y}_N\|$ for all $N \geq N_0$ such that

$$\overline{\lim_{N\to\infty}} \sup_{u \in T(x_N)} \mathrm{Re}\langle u, \hat{y}_N - x_N \rangle + h(\hat{y}_N) - h(x_N) > 0. \tag{2}$$

But, for each $N \geq N_0$, $\|x_N\| \leq \|\hat{y}_N\| \leq N$ implies $x_N \in X_N$ so that by (1), $\Phi(x_N, \hat{y}_N) \leq 0$ for all $N \geq N_0$; i.e.,

$$\sup_{u \in T(x_N)} \mathrm{Re}\langle u, \hat{y}_N - x_N \rangle + h(\hat{y}_N) - h(x_N) \leq 0 \quad \text{for all } N \geq N_0$$

which contradicts (2). Therefore we must have $\|\hat{y}_N\| \nrightarrow \infty$ as $N \to \infty$. It follows that there exists a positive integer $M > N_0$ and a subsequence $(\hat{y}_{N(i)})_{i=1}^{\infty}$ of $(\hat{y}_N)_{N \geq N_0}$ such that $\|\hat{y}_{N(i)}\| \leq M$ for all $i = 1, 2, \ldots$. Thus $(\hat{y}_{N(i)})_{i=1}^{\infty}$ is a sequence in the weakly compact set X_M so that by Eberlein-Smulian Theorem (e.g., see [10:p. 430]), there exist another subsequence $(\hat{y}_{N(i(j))})_{j=1}^{\infty}$ of $(\hat{y}_{N(i)})_{i=1}^{\infty}$ and $\hat{y} \in X_M$ such that $(\hat{y}_{N(i(j))})_{j=1}^{\infty}$ converges weakly to $\hat{y}$.

Now let $x \in X$ be given. Choose any positive integer $M' \geq M$ with $x \in X_{M'}$. Take any $j_0 \in \mathbb{N}$ with $N(i(j_0)) \geq M'$; then for all $j \geq j_0$, $x \in X_{M'} \subset X_{N(i(j_0))}$ so that by (1), $\Phi(x, \hat{y}_{N(i(j))}) \leq 0$. Since $(\hat{y}_{N(i(j))})_{j=1}^{\infty}$ converges weakly to $\hat{y}$ and $y \mapsto \Phi(x, y)$ is weakly lower semicontinuous by (2), we must have

$$\Phi(x, \hat{y}) \leq \underline{\lim}_{j\to\infty} \Phi(x, \hat{y}_{N(i(j))}) \leq 0.$$

Hence

$$\sup_{u \in T(x)} \mathrm{Re}\langle u, \hat{y} - x \rangle \leq h(x) - h(\hat{y}) \quad \text{for all} \quad x \in X. \qquad \square$$

Remark 3.1: We note that our coercive condition (E) of Theorem 3.2 is quite general. For instance, our condition (E) includes the following coercive condition which is often used in the literatures.

The Coercive Condition (C): Let X be a non-empty subset of a Banach (more general, normed) space $(E, \|\cdot\|)$ and $T\colon X \to 2^{E^*}$ a set-valued mapping with non-empty values. Suppose there exists $x_0 \in X$ such that

$$\lim_{\|y\|\to\infty, y \in X} \inf_{w \in T(y)} \mathrm{Re}\langle w, y - x_0 \rangle > 0,$$

or more strongly, we have

$$\lim_{\|y\|\to\infty, y \in X} \inf_{w \in T(y)} \mathrm{Re} \frac{\langle w, y - x_0 \rangle}{\|y\|} = \infty.$$

If X is bounded, the condition (E) of Theorem 3.2 is automatically satisfied. For example, we have the following:

Corollary 3.3: *Let X be a non-empty bounded closed convex subset of a reflexive Banach space $(E, \|\cdot\|)$. Let $h: X \to [-\infty, +\infty]$ a convex and lower semicontinuous function and $T: X \to 2^{E^*}$ set-valued monotone with non-empty values. Then there exists $\hat{y} \in X$ such that*

$$\sup_{u \in T(x)} \operatorname{Re}\langle u, \hat{y} - x\rangle \leq h(x) - h(\hat{y}) \quad \textit{for all } x \in X.$$

Proof: As X is bounded, the condition (E) of Theorem 3.2 is satisfied and thus the conclusion follows from Theorem 3.2. □

An another application of Theorem 3.1, we have the following:

Theorem 3.4: *Let X be a non-empty convex subset of a Hausdorff topological vector space E, $h: X \to [-\infty, +\infty]$ be a convex and lower semicontinuous function and $T: X \to 2^{E^*}$ be a set-valued monotone mapping with non-empty values. Suppose there exist a non-empty compact and convex subset X_0 of X and a non-empty compact subset K of X such that for each $y \in X \backslash K$, there exists $x \in \operatorname{co}(X_0 \cup \{y\})$ such that*

$$\sup_{u \in T(x)} \operatorname{Re}\langle u, y - x\rangle + h(y) - h(x) > 0.$$

Then we have the following:

(I) If T is lower semicontinuous from the line segments of X to the weak topology of E^*, then there exists $\hat{y} \in K$ such that

$$\sup_{w \in T(\hat{y})} \operatorname{Re}\langle w, \hat{y} - x\rangle \leq h(x) - h(\hat{y}) \quad \text{for all } x \in X.$$

(II) If T is upper semicontinuous from the line segments of X to the weak topology of E^* and each $T(x)$ is weakly compact, then there exists $\hat{y} \in K$ such that

$$\inf_{w \in T(\hat{y})} \operatorname{Re}\langle w, \hat{y} - x\rangle \leq h(x) - h(\hat{y}) \quad \text{for all } x \in X.$$

If, in addition, $T(\hat{y})$ is also convex, then there exists $\hat{w} \in T(\hat{y})$ such that

$$\operatorname{Re}\langle \hat{w}, \hat{y} - x\rangle \leq h(x) - h(\hat{y}) \quad \text{for all } x \in X.$$

Proof: By Theorem 3.1, there exists $\hat{y} \in K$ such that

$$\sup_{u \in T(x)} \operatorname{Re}\langle u, \hat{y} - x\rangle \leq h(x) - h(\hat{y}) \quad \text{for all } x \in X. \tag{3}$$

(I) If T is lower semicontinuous from the line segments of X to the weak topology of E^*, then by (3) and Lemma 2.6,

$$\sup_{w \in T(\hat{y})} \operatorname{Re}\langle w, \hat{y} - x\rangle \leq h(x) - h(\hat{y}) \quad \text{for all } x \in X.$$

(II) If T is upper semicontinuous from the line segments of X to the weak topology of E^* and each $T(x)$ is weakly compact, by (3) and Lemma 2.7, we have

$$\inf_{w \in T(\hat{y})} \mathrm{Re}\langle w, \hat{y} - x\rangle \leq h(x) - h(\hat{y}) \tag{4}$$

for all $x \in X$.

If, in addition, $T(\hat{y})$ is also convex, define f: $X \times T(\hat{y}) \to [-\infty, +\infty]$ by

$$f(x, w) = \mathrm{Re}\langle w, \hat{y} - x\rangle + h(\hat{y}) - h(x).$$

Note that for each fixed $x \in X$, $w \mapsto f(x, w)$ is weakly lower semicontinuous and affine and for each fixed $w \in T(\hat{y})$, $x \mapsto f(x, w)$ is concave. Thus by Kneser's minimax theorem (i.e., Lemma 2.5)

$$\begin{aligned} &\min_{w \in T(\hat{y})} \{\sup_{x \in X} \mathrm{Re}\langle w, \hat{y} - x\rangle + h(\hat{y}) - h(x)\} \\ &\quad = \sup_{x \in X} \inf_{w \in T(\hat{y})} \mathrm{Re}\langle w, \hat{y} - x\rangle + h(\hat{y}) - h(x)\} \end{aligned} \tag{5}$$

Since $T(\hat{y})$ is weakly compact, there exists $\hat{w} \in T(\hat{y})$ such that by (4) and (5),

$$\begin{aligned} &\sup_{x \in X}\{\mathrm{Re}\langle \hat{w}, \hat{y} - x\rangle + h(\hat{y}) - h(x)\} \\ &= \min_{w \in T(\hat{y})} \sup_{x \in X}\{\mathrm{Re}\langle w, \hat{y} - x\rangle + h(\hat{y}) - h(x)\} \leq 0, \end{aligned}$$

that is, $\mathrm{Re}\langle \hat{w}, \hat{y} - x\rangle \leq h(x) - h(\hat{y})$ for all $x \in X$. □

Remark 3.2: We remark that Theorem 3.4 generalizes Theorem 2 of Yen [25:p. 479–480] in several aspects. By following the exactly the same proof of Theorem 3.3 and as an application of Theorem 3.2 instead of Theorem 3.1, we also have the following non-compact version of variational inequalities in reflexive Banach space under another coercive condition (see also Theorem 1 of [24]).

Theorem 3.4′: *Let X be a non-empty closed convex subset of a reflexive Banach space $(E, \|\cdot\|)$, h: $X \to [-\infty, +\infty]$ a convex and lower semicontinuous function and T: $X \to 2^{E^*}$ a set-valued monotone mapping. Assume that the following condition is satisfied:*

(E) For each sequence $(y_n)_{n=1}^{\infty}$ in X with $\|y_n\| \to \infty$ as $n \to \infty$, there exists a sequence $(x_n)_{n=1}^{\infty}$ in X with $\|x_n\| \leq \|y_n\|$ for all $n = 1, 2, \ldots$ such that

$$\overline{\lim_{n\to\infty}}\{\sup_{u \in T(x_n)} Re\langle u, y_n - x_n\rangle + h(y_n) - h(x_n)\} > 0.$$

(I) If T is lower semicontinuous from the line segments of X to the weak topology of E', then there exists $\hat{y} \in X$ such that

$$\sup_{w \in T(\hat{y})} \mathrm{Re}\langle w, \hat{y} - x\rangle \leq h(x) - h(\hat{y}) \quad \text{for all } x \in X.$$

(II) If T is upper semicontinuous from the line segments of X to the weak topology of E^* and each $T(x)$ is weakly compact, then there exists $\hat{y} \in X$ such that

$$\inf_{w \in T(\hat{y})} \mathrm{Re}\langle w, \hat{y} - x\rangle \leq h(x) - h(\hat{y}) \quad \text{for all } x \in X.$$

If, in addition, $T(\hat{y})$ is also convex, then there exists $\hat{w} \in T(\hat{y})$ such that

$$\mathrm{Re}\langle \hat{w}, \hat{y} - x\rangle \leq h(x) - h(\hat{y}) \quad \text{for all } x \in X.$$

Proof: By following the Proof of Theorem 3.4 and applying Theorem 3.2 instead of Theorem 3.1, we obtain Theorem 3.4′. □

Theorem 3.5: *Let X be a non-empty closed convex subset of a reflexive Banch space $(E, \|\cdot\|)$, $h: X \to [-\infty, +\infty]$ a convex and lower semicontinuous function and $T: X \to 2^{E^*}$ a set-valued monotone mapping with non-empty values. Assume that the following condition is satisfied:*

$(E)_\infty$ *For each sequence $(y_n)_{n=1}^\infty$ in X with $\|y_n\| \to \infty$ as $n \to \infty$, there exists a sequence $(x_n)_{n=1}^\infty$ in X with $\|x_n\| \leq \|y_n\|$ for all $n = 1, 2, \ldots$ such that*

$$\overline{\lim_{n\to\infty}} \{ \sup_{u \in T(x_n)} \mathrm{Re}\langle u, y_n - x_n\rangle + h(y_n) - h(x_n)\}/\|y_n\| = \infty.$$

(I) If T is lower semicontinuous from the line segments in X to the weak topology of E^*, then for each given $w_0 \in E^*$, there exists $\hat{y} \in X$ such that

$$\sup_{w \in T(\hat{y})} \mathrm{Re}\langle w - w_0, \hat{y} - x\rangle \leq h(x) - h(\hat{y}) \quad \text{for all } x \in X.$$

(II) If T is upper semicontinuous from the line segments in X to the weak topology of E^* and each $T(x)$ is weakly compact and convex, then for each given $w_0 \in E^*$, there exist $\hat{y} \in X$ and $\hat{w} \in T(\hat{y})$ such that

$$\mathrm{Re}\langle \hat{w} - w_0, \hat{y} - x\rangle \leq h(x) - h(\hat{y}) \quad \text{for all } x \in X.$$

Proof: Let $w_0 \in E^*$ be given and we define $T^*: X \to 2^{E^*}$ by

$$T^*(y) := T(y) - w_0$$

for each $y \in X$. By the condition $(E)_\infty$, for each sequence $(y_n)_{n=1}^\infty$ in X with $\|y_n\| \to \infty$ as $n \to \infty$, there exists a sequence $(x_n)_{n=1}^\infty$ in X with $\|x_n\| \leq \|y_n\|$ for all $n = 1, 2, \ldots$ such that

$$\begin{aligned}
&\overline{\lim_{n\to\infty}} \{ \sup_{u \in T^*(x_n)} \mathrm{Re}\langle u, y_n - x_n\rangle + h(y_n) - h(x_n)\}/\|y_n\| \\
&= \overline{\lim_{n\to\infty}} \{ \sup_{u \in T(x_n)} \mathrm{Re}\langle u - w_0, y_n - x_n\rangle + h(y_n) - h(x_n)\}/\|y_n\| \\
&\geq \overline{\lim_{n\to\infty}} \{ \sup_{u \in T(x_n)} \mathrm{Re}\langle u, y_n - x_n\rangle + h(y_n) - h(x_n)\}/\|y_n\| - 2\|w_0\| \\
&= \infty
\end{aligned}$$

since $|\mathrm{Re}\langle w_0, y_n - x_n\rangle|/\|y_n\| \leq \|w_0\| + \|w_0\|\|x_n\|/\|y_n\| \leq 2\|w_0\|$.

It follows that

$$\overline{\lim_{n\to\infty}}\{\sup_{u\in T^*(x_n)} \operatorname{Re}\langle w_0, y_n - x_n\rangle + h(y_n) - h(x_n)\} = \infty > 0$$

and hence the conclusion follows from Theorem 3.4′ and we complete the proof. □

We note that the condition (E) in Theorems 3.2 and 3.4′ and the condition that $(E)_\infty$ of Theorem 3.5 are automatically satisfied if X is bounded. Also Theorem 4.1 of Chang and Zhang [5] is a special case of Theorem 3.5 under the case (II). We also note that Theorems 3.4 and 3.4′ are very closely related to but not comparable to Theorems 1 and 2 of Shih and Tan [21] and Theorems 3 and 4 of Shih and Tan [22].

Recall that a subset X of a vector space E is said to be a cone if X is a non-empty convex set such that $\alpha X \subset X$ for all $\alpha \geq 0$. If X is a cone in a topological vector space E, X^* will denote the dual cone of X in E^*, i.e.,

$$X^* = \{y \in E^* : \operatorname{Re}\langle y, x\rangle \geq 0 \quad \text{for all } x \in X\}.$$

The same proof of Lemma 2 of Shih and Tan [20] can be modified to obtain the following results and is thus omitted:

Lemma 3.6: *Let X be a cone in a topological vector space E. Suppose that T: $X \to 2^{E^*}$ is a set-valued mapping with non-empty values and $\hat{y} \in X$. Then the following statements are equivalent*:

(*a*) $\sup_{w\in T(\hat{y})} \operatorname{Re}\langle w, \hat{y} - x\rangle \leq 0$ for all $x \in X$;
(*b*) $\operatorname{Re}\langle w, \hat{y}\rangle = 0$ for all $w \in T(\hat{y})$ and $T(\hat{y}) \subset X^*$.

Lemma 3.7: *Let X be a cone in a topological vector space E, T: $X \to 2^{E^*}, \hat{y} \in X$ and $\hat{w} \in T(\hat{y})$. Then the following statements are equivalent*:

(*a*) $\operatorname{Re}\langle \hat{w}, \hat{y} - x\rangle \leq 0$ *for all* $x \in X$;

(*b*) $\operatorname{Re}\langle \hat{w}, \hat{y}\rangle = 0$ *and* $\hat{w} \in X^*$.

When E is real, Lemma 3.7 was also obtained by S.C. Fang (e.g., see [4:p. 213]).
In view of Lemmas 3.5 and 3.6, by taking $h \equiv 0$ in Theorem 3.4, we have the following theorem on generalized complementarity problem:

Theorem 3.8: *Let X be a closed cone in a reflexive Banach space $(E, \|\cdot\|)$ and T: $X \to 2^{E^*}$ monotone with non-empty values. Assume that the following condition is satisfied*:

$(E)_0$ *For each sequence $(y_n)_{n=1}^\infty$ in X with $\|y_n\| \to \infty$ as $n \to \infty$, there exists a sequence $(x_n)_{n=1}^\infty$ in X with $\|x_n\| \leq \|y_n\|$ for all $n = 1, 2, \ldots$ such that*

$$\overline{\lim_{n\to\infty}} \sup_{u\in T(x_n)} \operatorname{Re}\langle u, y_n - x_n\rangle > 0.$$

(I) If T is lower semicontinuous from the line segments in X to the weak topology of E^*, then there exists $\hat{y} \in X$ such that $\operatorname{Re}\langle w, \hat{y}\rangle = 0$ for all $w \in T(\hat{y})$ and $T(\hat{y}) \subset X^*$.
(II) If T is upper semicontinuous from the line segments in X to the weak topology of E^* and each $T(x)$ is closed bounded and convex, then there exists $\hat{y} \in X$ and $\hat{w} \in T(\hat{y})$ such that $\operatorname{Re}\langle \hat{w}, \hat{y}\rangle = 0$ and $\hat{w} \in X^*$.

Remark 3.3: Before we conclude this section, we would like to remark that if X is bounded, the condition (E) (resp., $(E)_\infty$ or $(E)_0$) of Theorem 3.4 (resp., Theorem 3.5 or Theorem 3.8) is automatically satisfied. Thus our Theorems 3.4, 3.5 and 3.8 are non-compact versions of the corresponding variational inequalities known in the literature.

4. VARIATIONAL INEQUALITIES OF QUASI-MONOTONE MAPPINGS IN REFLEXIVE BANACH SPACES

In this section, we shall first establish the existence theorems of solutions for quasi-monotone mappings in reflexive Banach spaces and then these existence results will be used to derive fixed point theorems for set-valued mappings in Hilbert spaces.

We now state the following simple fact (see, also Lemma 1 of [2]):

Lemma 4.1: *Let E^* be the dual space of a Hausdorff topological vector space E. Let A be a non-empty bounded subset of E and C a non-empty (strongly) compact subset of E^*. Define f: $A \to [-\infty, +\infty]$ by*

$$f(x) = \min_{u \in C} \operatorname{Re}\langle u, x\rangle \quad \text{for all } x \in A.$$

Then f is weakly continuous on A.

Now we have the following result

Lemma 4.2: *Let X be a non-empty convex subset of a Hausdorff topological vector space E, h: $X \to [-\infty, +\infty]$ a convex function and T: $X \to 2^{E^*}$ upper semicontinuous from the line segments of X to the $\sigma(E^*, E)$ topology on E^* such that each $T(x)$ is $\sigma(E^*, E)$ compact. If $\hat{y} \in X$, then the inequality*

$$\inf_{u \in T(x)} \operatorname{Re}\langle u, \hat{y} - x\rangle \leq h(x) - h(\hat{y}) \quad \text{for all } x \in X.$$

implies the inequality

$$\inf_{w \in T(\hat{y})} \operatorname{Re}\langle w, \hat{y} - x\rangle \leq h(x) - h(\hat{y}) \quad \text{for all } x \in X.$$

Proof: Let $x \in X$ be arbitrarily fixed. For each $t \in [0, 1]$, let $z_t = tx + (1 - t)\hat{y} = \hat{y} - t(\hat{y} - x)$. Since X is convex, $z_t \in X$ for all $t \in [0, 1]$. Thus for all $t \in (0, 1]$,

$$t \cdot \inf_{u \in T(z_t)} \operatorname{Re}\langle u, \hat{y} - x\rangle = \inf_{u \in T(z_t)} \operatorname{Re}\langle u, \hat{y} - z_t\rangle$$

$$\leq h(z_t) - h(\hat{y}) \leq t \cdot h(x) + (1 - t) \cdot h(\hat{y}) - h(\hat{y}) = t \cdot (h(x) - h(\hat{y}))$$

so that

$$\inf_{u \in T(z_t)} \operatorname{Re}\langle u, \hat{y} - x\rangle \leq h(x) - h(\hat{y}) \quad \text{for all } t \in (0.1]. \tag{6}$$

If $\inf_{w\in T(\hat{y})} \operatorname{Re}\langle w, \hat{y}-x\rangle > h(x)-h(\hat{y})$, let $G=\{w \in E^*\colon \operatorname{Re}\langle w, \hat{y}-x\rangle + h(\hat{y}) - h(x) > 0\}$; then G is a $\sigma(E^*, E)$-open set in E^* such that $T(\hat{y}) \subset G$. As $z_t \to \hat{y}$ as $t \to 0^+$, by upper semicontinuity of T on $\{z_t\colon t \in [0,1]\}$, there exists $t_0 \in (0,1]$ such that $T(x_t) \subset G$ for all $t \in (0, t_0)$. As $T(z_t)$ is $\sigma(E^*, E)$ compact, $\inf_{u\in T(z_t)} \operatorname{Re}\langle u, \hat{y}-x\rangle + h(\hat{y}) - h(x) > 0$ for all $t \in (0, t_0)$ which contradicts (6). Thus we must have $\inf_{w\in T(\hat{y})} \operatorname{Re}\langle w, \hat{y}-x\rangle \leq h(x)-h(\hat{y})$. □

Theorem 4.3: *Let X be a non-empty closed and convex subset of a reflexive Banach space $(E, \|\cdot\|)$, $h\colon X \to [-\infty, +\infty]$ a convex and lower semicontinuous function and $T\colon X \to 2^{E^*}$ quasi-monotone such that each $T(x)$ is compact in the norm topology on E^*. Assume that the following condition is satisfied:*

(E)* *For each sequence $(y_n)_{n=1}^{\infty}$ in X with $\|y_n\| \to \infty$ as $n \to \infty$, there exists a sequence $(x_n)_{n=1}^{\infty}$ in X with $\|x_n\| \leq \|y_n\|$ for all $n = 1, 2, \dots$ such that*

$$\varlimsup_{n\to\infty} \{ \inf_{u\in T(x_n)} \operatorname{Re}\langle u, y_n - x_n\rangle + h(y_n) - h(x_n)\} > 0.$$

Then there exists $\hat{y} \in X$ such that

$$\inf_{u\in T(x)} \operatorname{Re}\langle u, \hat{y}-x\rangle \leq h(x) - h(\hat{y}) \quad \textit{for all } x \in X.$$

Proof: Define $\Phi\colon X \times X \to [-\infty, +\infty]$ by

$$\Phi(x,y) = \inf_{u\in T(x)} \operatorname{Re}\langle u, y-x\rangle + h(y) - h(x)$$

for each $(x,y) \in X \times X$. Then we have

(1) for each fixed $y \in X$, $\Phi(x,y)$ is 0-diagonal concave in x by Lemma 2.1 as T is quasi-monotone;
(2) since h is convex and lower semicontinuous with respect to the norm topology on X, h is also lower semicontinuous with respect to the weak topology on X. It follows that for each $x \in X$, the function $y \mapsto \Phi(x,y)$ is weakly lower semicontinuous on A for each non-empty (norm-) bounded subset A of X by Lemma 4.1.

Then the same proof of Theorem 3.2 with necessary modifications (all "$\sup_{u\in T(x)}$" and all "$\sup_{u\in T(x_N)}$" being replaced by "$\inf_{u\in T(x)}$" and "$\inf_{u\in T(x_N)}$" respectively), we see that there exists $\hat{y} \in X$ such that

$$\inf_{u\in T(x)} \operatorname{Re}\langle u, \hat{y}-x\rangle \leq h(x) - h(\hat{y}) \quad \text{for all } x \in X.$$ □

Theorem 4.4: *Let X be a non-empty closed and convex subset of a reflexive Banach space $(E, \|\cdot\|)$, $h\colon X \to [-\infty, +\infty]$ be a convex and lower semicontinuous function and $T\colon X \to 2^{E^*}$ be quasi-monotone and upper semicontinuous from the line segments of X to the weak topology on E^* such that each $T(x)$ is compact in the norm topology on E^*. Assume that the following condition is satisfied:*

(E)* *For each sequence $(y_n)_{n=1}^{\infty}$ in X with $\|y_n\| \to \infty$ as $n \to \infty$, there exists a sequence $(x_n)_{n=1}^{\infty}$ in X with $\|x_n\| \leq \|y_n\|$ for all $n = 1, 2, \dots$ such that*

$$\varlimsup_{n\to\infty} \{ \inf_{u\in T(x_n)} \operatorname{Re}\langle u, y_n - x_n\rangle + h(y_n) - h(x_n)\} > 0.$$

Then there exists $\hat{y} \in X$ *such that*

$$\inf_{w \in T(\hat{y})} \mathrm{Re}\langle w, \hat{y} - x\rangle \leq h(x) - h(\hat{y}) \quad \textit{for all } x \in X.$$

If, in addition, $T(\hat{y})$ *is also convex, then there exists* $\hat{w} \in T(\hat{y})$ *such that*

$$\mathrm{Re}\langle \hat{w}, \hat{y} - x\rangle \leq h(x) - h(\hat{y}) \quad \textit{for all } x \in X.$$

Proof: By Theorem 4.3, there exists $\hat{y} \in X$ such that

$$\inf_{u \in T(x)} \mathrm{Re}\langle u, \hat{y} - x\rangle \leq h(x) - h(\hat{y}) \quad \text{for all } x \in X.$$

By Lemma 4.2, we have

$$\inf_{w \in T(\hat{y})} \mathrm{Re}\langle w, \hat{y} - x\rangle \leq h(x) - h(\hat{y}) \quad \text{for all } x \in X.$$

If, in addition, $T(\hat{y})$ is also convex, by Kneser's minimax theorem (i.e., Lemma 2.5) and using the same argument as in the proof of Theorem 3.4′, there exists $\hat{w} \in T(\hat{y})$ such that

$$\mathrm{Re}\langle \hat{w}, \hat{y} - x\rangle \leq h(x) - h(\hat{y}) \quad \text{for all } x \in X. \qquad \square$$

In the proof of Theorem 3.5, if we replace all "$\sup_{u \in T^*(x_n)}$" and all "$\sup_{u \in T(x_n)}$" by "$\inf_{u \in T^*(x_n)}$" and "$\inf_{u \in T(x_n)}$" respectively, we have the following application of Theorem 4.4:

Theorem 4.5: *Let* X *be a non-empty closed and convex subset of a reflexive Banach space* $(E, \|\cdot\|)$, $h\colon X \to [-\infty, +\infty]$ *a convex and lower semicontinuous function and* $T\colon X \to 2^{E^*}$ *quasi-monotone and upper semicontinuous from the line segments of* X *to the weak topology on* E^* *such that each* $T(x)$ *is convex and compact in the norm topology on* E^*. *Assume that the following condition is satisfied:*

$(E)^*_\infty$ *For each sequence* $(y_n)_{n=1}^\infty$ *in* X *with* $\|y_n\| \to \infty$ *as* $n \to \infty$, *there exists a sequence* $(x_n)_{n=1}^\infty$ *in* X *with* $\|x_n\| \leq \|y_n\|$ *for all* $n = 1, 2, \ldots$ *such that*

$$\overline{\lim_{n\to\infty}} \{ \inf_{u \in T(x_n)} \mathrm{Re}\langle u, y_n - x_n\rangle + h(y_n) - h(x_n)\}/\|y_n\| = \infty.$$

Then for each given $w_0 \in E^*$, *there exist* $\hat{y} \in X$ *and* $\hat{w} \in T(\hat{y})$ *such that*

$$\mathrm{Re}\langle \hat{w} - w_0, \hat{y} - x\rangle \leq h(x) - h(\hat{y}) \quad \textit{for all } x \in X.$$

By Lemma 3.7 and by taking $h \equiv 0$ *in Theorem 4.4, we have the following theorem on generalized complementarity problem:*

Theorem 4.6: *Let* X *be a non-empty closed and convex subset of a reflexive Banach space* $(E, \|\cdot\|)$ *and* $T\colon X \to 2^{E^*}$ *quasi-monotone and upper semicontinuous from the line* *segments in X to the weak topology on E^* such that each $T(x)$ is convex and compact in the norm topology on E^*. Assume that the following condition is satisfied:*

$(E)_0^*$ *For each sequence $(y_n)_{n=1}^{\infty}$ in X with $\|y_n\| \to \infty$ as $n \to \infty$, there exists a sequence $(x_n)_{n=1}^{\infty}$ in X with $\|x_n\| \leq \|y_n\|$ for all $n = 1, 2, \ldots$ such that*

$$\overline{\lim_{n\to\infty}}\{\inf_{u\in T(x_n)} \mathrm{Re}\langle u, y_n - x_n\rangle\} > 0.$$

Then there exist $\hat{y} \in X$ and $\hat{w} \in T(\hat{y})$ such that

$$\mathrm{Re}\langle \hat{w}, \hat{y}\rangle = 0 \quad \text{and} \quad \hat{w} \in X^*.$$

5. APPLICATIONS TO FIXED POINT THEOREMS OF QUASI-CONTRACTIVE SET-VALUED MAPPINGS IN HILBERT SPACES

Before we prove fixed point theorem for quasi-contractive set-valued mappings in Hilbert spaces, we first recall the following proposition of Bae *et al.* [2] and we also include its its proof for the sake of completeness.

Lemma 5.1: *Let X be a non-empty subset of a Hilbert space H. Then we have*

(*a*) *If T: $X \to CB(H)$ is nonexpansive and for each $x \in X$, $T(x)$ is weakly compact, then T is pseudo-contractive on X.*
(*b*) *If T: $X \to 2^H$ is pseudo-contractive on X, then $I - T$ is semi-monotone on X where $I(x) := x$ for all $x \in X$.*

Proof: (a) Let $x, y \in X$ and $w \in T(y)$. Since $T(y)$ is weakly compact, there exists $u \in T(x)$ such that $\|w - u\| = d(w, T(x))$. We see that $d(w, T(x)) \leq D(T(x), T(y)) \leq \|x - y\|$, it follows that $\|w - u\| \leq \|x - y\|$. Thus for all $r > 0$,
$\|(1 + r)(x + y) - r(u - w)\| \geq (1 + r)\|x - y\| - r\|u - w\| \geq \|x - y\|$.
Therefore T is pseudo-contractive on X.

(b) Since T is pseudo-contractive on X, for each $x, y \in X$ and for each $w \in T(x)$, there exists $u \in T(x)$ such that for all $r > 0$,

$$\begin{aligned}
&\|x - y\|^2 \leq \|(1 + r)(x - y) - r(u - w)\|^2 \\
&= \|\,(x - y) + r((x - y) - (u - w))\|^2 \\
&= \|x - y\|^2 + 2r \cdot \mathrm{Re}\langle (x - u) - (y - w), x - y\rangle + r^2 \cdot \|(x - y) - (u - w)\|^2.
\end{aligned}$$

Therefore we have

$$2\mathrm{Re}\langle (x - u) - (y - w), x - y)\rangle + r \cdot \|(x - y) - (u - w)\|^2 \geq 0.$$

By taking $r \to 0^+$, we have $\mathrm{Re}\langle (x - u) - (y - w), x - y\rangle \geq 0$ so that

$$\mathrm{Re}\langle y - w, y - x\rangle \geq \mathrm{Re}\langle x - u, y - x\rangle \geq \inf_{u\in T(x)} \mathrm{Re}\langle x - u, y - x\rangle.$$

Thus $\inf_{w\in T(y)} \mathrm{Re}\langle y-w, y-x\rangle \geq \inf_{u\in T(x)} \mathrm{Re}\langle x-u, y-x\rangle$ and $I-T$ is a semi-monotone mapping on X. □

Remark 5.1: Lemma 5.1 shows that the class of quasi-contractive mappings includes the class of pseudo-contractive mappings as a special case.

Now as an application of Theorem 4.4, we have the following fixed point theorem:

Theorem 5.2: *Let X be a non-empty closed convex subset of a Hilbert space H and $T\colon X \to 2^H$ be quasi-contractive and upper semicontinuous from the line segments in X to the weak topology on H such that each $T(x)$ is compact in the norm topology on H. Assume that the following condition is satisfied:*

$(E)^+$ *For each sequence $(y_n)_{n=1}^{\infty}$ in X with $\|y_n\| \to \infty$ as $n \to \infty$, there exists a sequence $(x_n)_{n=1}^{\infty}$ in X with $\|x_n\| \leq \|y_n\|$ for all $n = 1, 2, \ldots$ such that*

$$\overline{\lim_{n\to\infty}}\{\inf_{u\in T(x_n)} \mathrm{Re}\langle x_n - u, y_n - x_n\rangle\} > 0.$$

Then there exists $\hat{y} \in X$ such that

$$\inf_{w\in T(\hat{y})} \mathrm{Re}\langle \hat{y} - w, \hat{y} - x\rangle \leq 0 \quad \textit{for all } x \in X.$$

If $T(\hat{y})$ is also convex, then there exists $\hat{w} \in T(\hat{y})$ such that

$$\mathrm{Re}\langle \hat{y} - \hat{w}, \hat{y} - x\rangle \leq 0 \quad \textit{for all } x \in \overline{I_X(\hat{y})},$$

and if, in addition, either $\hat{y}$ is an interior point of X in H or $P(\hat{y}) \in \overline{I_X(\hat{y})}$, where $P(\hat{y})$ is the projection of $\hat{y}$ on $T(\hat{y})$, then $\hat{y}$ is a fixed point of T, i.e., $\hat{y} \in T(\hat{y})$.

Proof: As T is quasi-monotone, by Theorem 4.4 with $h = 0$, there exists $\hat{y} \in X$ such that

$$\inf_{w\in T^*(\hat{y})} \mathrm{Re}\langle w, \hat{y} - x\rangle \leq 0 \quad \text{for all } x \in X;$$

this is,

$$\inf_{w\in T(\hat{y})} \mathrm{Re}\langle \hat{y} - w, \hat{y} - x\rangle \leq 0 \quad \text{for all } x \in X.$$

If $T(\hat{y})$ is also convex, then by Theorem 4.4 again, there exists $\hat{w} \in T(\hat{y})$ such that

$$\mathrm{Re}\langle \hat{y} - \hat{w}, \hat{y} - x\rangle \leq 0 \quad \text{for all } x \in X. \tag{7}$$

If $x \in I_X(\hat{y})$, then $x = \hat{y} + r(u - \hat{y})$ for some $u \in X$ and $r > 0$. Thus $\hat{y} - x = r(\hat{y} - u)$ so that by (7),

$$\mathrm{Re}\langle \hat{y} - \hat{w}, \hat{y} - x\rangle = r \cdot \mathrm{Re}\langle \hat{y} - \hat{w}, \hat{y} - u\rangle \leq 0.$$

It follows that

$$\mathrm{Re}\langle \hat{y} - \hat{w}, \hat{y} - x\rangle \leq 0 \quad \text{for all } x \in \overline{I_X(\hat{y})}. \tag{8}$$

Now, if $\hat{y}$ is an interior point of X in H, then (8) implies that $\hat{y} = \hat{w} \in T(\hat{y})$. Next suppose $P(\hat{y}) \in \overline{I_X(\hat{y})}$. Since $P(\hat{y})$ is the projection of $\hat{y}$ on $T(\hat{y})$, we must have, by Theorem I.2.3 of Kinderlehrer and Stampacchia [16:p. 9], $P(\hat{y}) \in T(\hat{y})$ and $\text{Re}\langle P(\hat{y}) - \hat{y}, w - P(\hat{y})\rangle \geq 0$ for all $w \in T(\hat{y})$. Since $\hat{w} \in T(\hat{y})$, by (8) we have

$$\begin{aligned} 0 &\leq \text{Re}\langle P(\hat{y}) - \hat{y}, \hat{w} - P(\hat{y})\rangle \\ &= \text{Re}\langle P(\hat{y}) - \hat{y}, \hat{w} - \hat{y} + \hat{y} - P(\hat{y})\rangle \\ &= \text{Re}\langle P(\hat{y}) - \hat{y}, \hat{w} - \hat{y}\rangle - \|\hat{y} - P(\hat{y})\|^2 \\ &= \text{Re}\langle \hat{y} - \hat{w}, \hat{y} - P(\hat{y})\rangle - \|\hat{y} - P(\hat{y})\|^2 \leq -\|P(\hat{y}) - \hat{y}\|^2 \end{aligned}$$

so that $\|P(\hat{y}) - \hat{y}\|^2 \leq 0$ and hence $\hat{y} = P(\hat{y}) \in T(\hat{y})$. □

As each nonexpansive set-valued mapping with weakly compact values is pseudo-contractive, and each pseudo-contractive set-valued mapping is a quasi-contractive mapping, we have the following fixed point theorems in Hilbert spaces:

Theorem 5.3: *Let X be a non-empty closed convex subset of a Hilbert space H and $T\colon X \to 2^H$ be pseudo-contractive and upper semicontinuous (<u>for example</u>, T is non-expansive) such that each $T(x)$ is compact convex and $P(y) \in \overline{I_X(y)}$ for each $y \in \partial X$ where $P(y)$ is the projection of y on $T(y)$ and ∂X is the boundary of X in H. Assume that the following condition is satisfied:*

$(E)^+$ *For each sequence $(y_n)_{n=1}^{\infty}$ in X with $\|y_n\| \to \infty$ as $n \to \infty$, there exists a sequence $(x_n)_{n=1}^{\infty}$ in X with $\|x_n\| \leq \|y_n\|$ for all $n = 1, 2, \ldots$ such that*

$$\overline{\lim_{n\to\infty}} \{ \inf_{u \in T(x_n)} \text{Re}\langle x_n - u, y_n - x_n\rangle\} \; 0.$$

Then T has a fixed point in X.

Except that the set X is required to be closed in H, the above result is a generalization of Theorem 8 of Tan [23:p. 561] to set-valued and non-self maps.

we note that the condition $(E)^*$ in Theorem 4.3 and Theorem 4.4, the condition $(E)^*_\infty$ in Theorem 4.5 and $(E)^+$ of Theorems 5.2 and 5.3 are automatically satisfied if the set X is bounded.

Finally we remark that given any increasing sequence $(N_n)_{n=1}^{\infty}$ of positive integers, the conclusions of Theorems 3.2, 4.3, 3.4–5.3 remain valid if we replace the phrase "... there exists a sequence $(x_n)_{n=1}^{\infty}$ in X with $\|x_n\| \leq \|y_n\|$ for all $n = 1, 2, \ldots$" in the conditions (E), $(E)_\infty$, $(E)_0$, $(E)^*$, $(E)^*_\infty$, $(E)^*_0$ and $(E)^+$ by the phrase "... there exists a sequence $(x_n)_{n=1}^{\infty}$ in X with $\|x_n\| \leq N_n$ for all $n = 1, 2, \ldots$".

We remark that our results in this chapter improve corresponding results of Aubin and Ekeland [1], Bae *et al.* [2], Browder [3], Chan and Pang [4], Chang and Zhang [5], Ding and Tan [6,7], Ding and Tarafdar [9], Shih and Tan [20,22], Tan [23], Tan and Yuan [24], Yen [25] and Zhou and Chen [26]. Finally we also note that as applications of variational inequalities, some existence theorems of fixed points for quasi-contractive set-valued mappings have been established. For an updated art of the subject for fixed point theory in both metric spaces and topological spaces, the interested reader is referred to Goebel and Kirk [13], or a recent book written by Jaworowski *et al.* [14]. For some other study of variational inequalities and topological fixed point theorems as applications of Ky Fan type minimax inequalities in general topological spaces (such as generalized convex spaces and so on), we refer it to Park [18] and Yuan [27].

6. APPLICATIONS TO MINIMIZATION PROBLEMS

In this section, as application of variational inequalities which have been established in Section 3, we shall consider the existence of solutions for the following minimization problem:

$$\inf_{x \in E} f(x) \tag{6.1}$$

where f is the sum of two extended real-valued functions g, h: $E \to (-\infty, +\infty]$, and E is a normed space. Before we prove the existence of solutions for (6.1), we recall the following definition of subdifferential (e.g., see [1:p. 187]):

Definition 6.1: Let X be a non-empty convex subset of a topological vector space E. Suppose f: $X \to (-\infty, +\infty]$ is a function with non-empty domain. If $x_0 \in \mathrm{Dom} f$, the "*subdifferential* $\partial f(x_0)$ *of* f *at* x_0" is the subset (which may be empty) of E^* defined by

$$\partial f(x_0) = \{p \in E^* : f(x_0) - f(x) \leq \langle p, x_0 - x \rangle \text{ for all } x \in X\} \tag{6.2}$$

The element $p \in \partial f(x_0)$ are also called subgradients.

The following simple Proposition shows that the existence of solutions of variational inequalities guarantee sufficiently the existence of the minimizers for the minimization problem (6.1).

Proposition 6.2: *Let X be a non-empty convex subset of a Hausdorff topological vector space E. Suppose $f = g + h$ is the sum of a convex function g and a subdifferential function h defined on a non-empty convex subset X, i.e., g, h: $X \to (-\infty, +\infty]$. Then a point $\hat{x} \in X$ minimizes f if there exists $p \in \partial h(\hat{x})$ such that*

$$\sup_{x \in X}[\langle p, \hat{x} - x \rangle + g(\hat{x}) - g(x)] \leq 0. \tag{6.3}$$

Proof: Suppose there exists $p \in \partial h(\hat{x})$ such that $\langle p, \hat{x} - x \rangle + g(\hat{x}) - g(x) \leq 0$ for all $x \in X$. Then we have that

$$f(\hat{x}) - f(x) = h(\hat{x}) - h(x) + g(\hat{x}) - g(x) \leq \langle p, \hat{x} - x \rangle + g(\hat{x}) - g(x) \leq 0$$

for all $x \in X$. Thus $\hat{x}$ minimizes f and the proof is completed. □

Now we have the following general existence theorem which guarantee the existence of minimizers for the minimization problem (6.1).

Theorem 6.3: *Let X be a non-empty convex subset of a Hausdorff topological vector space E and $f = g + h$ be the sum of two extended-valued functions g, h: $E \to (-\infty, +\infty]$, where g is a convex and lower semicontinuous function and the subdifferential mapping of h which is defined by $T(x) = \partial h(x)$ for each $x \in X$ is upper semicontinuous from the line of X to the weak topology of E^* with non-empty weakly compact and convex values. Suppose there exist a non-empty compact and convex subset X_0 of X and a non-empty compact subset K of X such that for each $y \in X \backslash K$, there exists $x \in \mathrm{co}(X_0 \cup \{y\}$ such that*

$$\sup_{u \in \partial h(x)} \mathrm{Re}\langle u, y - x \rangle + h(y) - h(x) > 0.$$

Then there exists a minimizer x_0 *which is a solution of the problem (6.1), i.e.,* $f(x_0) = \inf_{x \in E} f(x) = \inf_{x \in E} [g(x) + h(x)]$.

Proof: By our hypotheses, g and T satisfy all conditions of Theorem 3.4 (II). By Theorem 3.4 (II), there exists $\hat{x} \in K$ and $p \in T(\hat{x})$ such that

$$\sup_{x \in X} \mathrm{Re}\langle p, \hat{x} - x\rangle \leq g(x) - g(\hat{x}).$$

Then Proposition 6.2 implies that $\hat{x}$ is a solution of the minimization problem (6.1) and thus the proof is completed. □

In what follows, we shall give some sufficient conditions for the function h such that its subdifferential mapping satisfies all conditions posed in Theorem 6.3.

Theorem 6.4: *Let* E *be a normed space and* $f = g + h$ *be the sum of two extended-valued functions* g, h: $E \to (-\infty, +\infty]$, *which are both convex and lower semicontinuous and the domain* Dom$h(:= \{x \in E: -\infty < h(x) < +\infty\}) = E$. *Suppose there exist a non-empty compact and convex subset* X_0 *of* E *and a non-empty compact subset* K *of* E *such that for each* $y \in E\backslash K$, *there exists* $x \in \mathrm{co}(X_0 \cup \{y\}$ *such that*

$$\sup_{u \in \partial h(x)} \mathrm{Re}\langle u, y - x\rangle + h(y) - h(x) > 0.$$

Then there exists a minimizer x_0 *which is a solution of the problem (6.1), i.e.,* $f(x_0) = \inf_{x \in E} f(x) = \inf_{x \in E} [g(x) + h(x)]$.

Proof: By the hypothesis, we have that Dom$h = E$. As h is convex and lower semicontinuous, Theorem 17 of Aubin and Ekeland [1:p. 199] implies that $\partial h(x)$ is a non-empty, bounded closed convex of E^* (and indeed weakly compact) for each $x \in X$. Now define T: $X \to 2^{E^*}$ by $T(x) := \partial h(x)$ for each $x \in X$. By the definition of sub-differential, it is clear to see that T is monotone and upper hemicontinuous on X (e.g., see Theorem 17 of [1:p. 200] again). As $T(x)$ is weakly compact for each $x \in X$, it follows that T is also upper semicontinuous by Theorem 10 of Aubin and Ekeland [1:p. 128]. Thus the conclusion follows from Theorem 6.3. Indeed, by applying the variational inequality Theorem 3.4, there exists $\hat{x} \in X$ and $p \in T(\hat{x})$ such that

$$\sup_{x \in X}[\langle p, \hat{x} - x\rangle + g(\hat{x}) - g(x)] \leq 0.$$

Therefore Proposition 6.2 above shows that $\hat{x}$ is a minimizer of the minimization problem (6.1) and the proof is completed. □

By the same proof of Theorem 6.4 and as an application of Theorem 3.4′ instead of Theorem 3.4, we have the following existence of minimizer for the minimization problem (6.1) and its proof is omitted here.

Theorem 6.5: *Let* E *be a reflexive Banach and* $f = g + h$ *be the sum of two extended-valued functions* g, h: $E \to (-\infty, +\infty]$, *which are both convex and lower semicontinuous and the domain* Dom$h = E$. *Assume that the following condition (E) is satisfied:*

(E) For each sequence $(y_n)_{n=1}^{\infty}$ *in* E *with* $\|y_n\| \to \infty$ *as* $n \to \infty$, *there exists a sequence* $(x_n)_{n=1}^{\infty}$ *in* E *with* $\|x_n\| \leq \|y_n\|$ *for all* $n = 1, 2, \ldots$ *such that*

$$\overline{\lim_{n\to\infty}}\{\sup_{u\in\partial h(x_n)} \mathrm{Re}\langle u, y_n - x_n\rangle + h(y_n) - h(x_n)\} > 0.$$

Then there exists a minimizer x_0 *which is a solution of the Problem (6.1), i.e.,* $f(x_0) = \inf_{x\in E} f(x) = \inf_{x\in E}[g(x) + h(x)]$.

REFERENCES

[1] J.P. Aubin and I. Ekeland (1984). Applied Nonlinear Analysis. New York: Wiley-Interscience.

[2] J.S. Bae, W.K. Kim and K.K. Tan (1993). Another generalization of Ky Fan's minimax inequality and its applications, *Bull. Inst. Math. Acad. Sinica*, **21**, 229–244.

[3] F.E. Browder (1967). Nonlinear mappings of nonexpansive and accretive type in Banach space, *Bull. Amer. Math. Soc.*, **73**, 875–882.

[4] D. Chan and J.S. Pang (1982). The generalized quasi-variational inequality problem, *Mathematics of Operation Research*, **7**, 211–222.

[5] S.S. Chang and Y. Zhang (1991). Generalized KKM Theorem and Variational Inequalities, *J. Math. Anal. Appl.*, **159**, 208–223.

[6] X.P. Ding (1991). Borwder–Hartman–Stampachia type variational inequalities for multi-valued quasi-monotone operator, Journal of Sichuan Normal University **14**(2), 1–8.

[7] X.P. Ding and K.K. Tan (1990). Generalized variational inequalities and generalized quasi-variational inequalities, *J. Math. Anal. Appl.*, **148**, 497–508.

[8] X.P. Ding and K.K. Tan (1992). A minimax inequality with applications to existence of equilibrium point and fixed point theorems, *Colloquium Math.*, **63**, 233–243.

[9] X.P. Ding and E. Tarafdar (1996). Monotone generalized variational inequalities and generalized complementarity problems, *J. Optim. Theory Appl.*, **88**, 107–122.

[10] N. Dunford and J.T. Schwartz (1957). Linear Operators, Part I: General Theory. New York: Interscience Publisher Inc.

[11] B. Halpern and G. Bergman (1968). A fixed point theorem for inward and outward maps, *Trans. Amer. Math. Soc.*, **130**, 353–358.

[12] K. Fan (1972). A minimax inequality and application. In O. Shisha (Ed.), *Inequalities, III*, New York: Academic Press, pp. 103–113.

[13] K. Goebel and W.A. Kirk (1990). Topics in Metric Fixed Point Theory. Cambridge: Cambridge University Press.

[14] J. Jaworowski, W.A. Kirk and S. Park (1995). Antipodal Points and Fixed Points. vol. 28, Lecture Notes Series, Seoul National University, Korea, pp. 21–81.

[15] S. Karamardian and S. Schaible (1990). Seven kinds of monotone maps, *J. Optim. Theory Appl.*, **66**, 37–46.

[16] D. Kinderlehrer and G. Stampacchia (1980). An Introduction to Variational Inequalities and Their Applications. New York: Academic Press.

[17] H. Kneser (1952). Sur un theoreme fondamental de la theorie des jeux. *C.R. Acad. Sci. Paris*, **234**, 2418–1420.

[18] S. Park (1995). Some applications of the KKM theory and fixed point theory for admisible multifunctions. Research report, Department of Mathematics, Seoul National University, Korea.

[19] R. Saigal (1976). Extension of the generalized complementarity problem, *Mathematics of Operations Research*, **1**, 260–266.

[20] M.H. Shih and K.K. Tan (1986). Minimax inequalities and applications, *Contemp. Math.*, **54**, 45–63.

[21] M.H. Shih and K.K. Tan (1988). Browder–Hartman–Stampacchia variational inequalities for multi-valued monotone operators, *J. Math. Anal. Appl.*, **134**, 431–440.

[22] M.H. Shih and K.K. Tan (1988). A minimax inequality and Browder–Hartman–Stampacchia variational inequalities for multi-valued monotone operators. Proceedings of the Fourth Franco – SEAMS Conference Chiang Mai, Thailand.

[23] K.K. Tan (1983). Comparison theorems on minimax inequalities, variational inequalities, and fixed point theorems, *J. London Math. Soc.*, **28**, 555–562.

[24] K.K. Tan and X.Z. Yuan (1994). Variational inequalities on reflexive Banach spaces and applications, J. Natural Geometry, **5**, 43–58.

[25] C.L. Yen (1981). A minimax inequality and its applications to variational inequalities, *Pacific J. Math.*, **97**, 477–481.

[26] J. Zhou and G. Chen (1988). Diagonal convexity conditions for problems in convex analysis and quasi-variational inequalities, *J. Math. Anal. Appl.*, **132**, 213–225.

[27] G.X.Z. Yuan (1999). KKM Theory and Applications in Nonlinear Analysis. New York, Marcel, Dekker.

SIMAA 4(2002) 205–212

14. Remarks on the Existence of Maximal Elements with Respect to a Binary Relation in Non-compact Topological Spaces

George Isac[1] and George Xian-Zhi Yuan[2,*]

[1]*Department of Mathematics and Computer Science, Royal Military College of Canada, Kingston, Ont. K7K 5L0, Canada. E-mail: Isac-G@banyan.rmc.ca*
[2]*Department of Mathematics, The University of Queensland, Brisbane, Australia 4072*

Abstract: In this chapter we prove some non-compact existence theorems for maximal elements with respect to a binary relation in either non-compact topological spaces with MC-structures, K-convex structures, locally convex spaces or metric spaces. Our results include corresponding results in the literature as special cases.

Keywords and Phrases: Maximal element, binary relation, fixed point, MC-structure, K-convex structure, stationary point, dynamical system

1991 Mathematics Subject Classification: Primary – 49J45, 54C08; Secondary – 90C48

1. INTRODUCTION

When a player (resp., agent) is fixed with the problem of choosing a bundle (resp., an action) of products (resp., strategy set), she/he will look for the bundle (resp., strategy) which could maximize her/his preference relation from those she/he can afford. Note that the preference relation can be represented by continuous utility functions which also can be regarded as a special case of binary relation $\mathcal{R}$ defined on a non-empty set X. Thus it seems that the problem of looking for sufficient conditions of which ensure the existence of maximal elements of a binary relation is an important problem in economics (also, in optimization and game) theory.

For a binary relation $\mathcal{R}$ defined on a set X and $x, y \in X$, $x\mathcal{R}y$ means that "*x is preferred to y*". Also for each $x \in X$, the *upper* and *lower* contour sets of an element $x \in X$ are defined as

$$U(x) := \{y \in X : y\mathcal{R}x\} \quad \text{and} \quad U^{-1}(y) := \{x \in X : y\mathcal{R}x\}.$$

An element $x^* \in X$ is said to be a *maximal element* of X if there is no other element which is preferred to it, i.e., $U(x^*) = \emptyset$.

---

* Corresponding author. E-mail: xzy@maths.uq.edu.au

Though the existence theory of maximal elements for the general binary relation $\mathcal{R}$ defined on a set X have been extensively studied in past more than four decades, continuity and/or convexity on upper and lower contour sets or continuity and transitivity conditions on the binary relation $\mathcal{R}$ are usually required; however some of those conditions above are very restrictive such as transitivity of the indifference is an assumption which has been much criticised as being strongly unrealistic (e.g., see [9,11] etc.). Because of this fact, the problem of the existence of maximal elements in non-transitive binary relations is an interesting problem.

Recently, Llinares in [7–8] studied the existence of maximal elements in a non-transitive binary relation (also for acyclic binary relation) defined on a compact set, non-convex in the classical sense but endowed with of generalized convexity and his results also generalize corresponding results given by Sonnenschein [10], Yannelis and Prabhakar [13] and some others.

By the fact that the choice sets (e.g., the set of feasible allocations) generally are not compact in any topology of the choice space even though it is closed and bounded, a typical situation in infinite dimensional linear spaces. This motivates our work in this note to prove some non-compact existence theorems for maximal elements with respect to non-transitive binary relation from viewpoint of fixed point theorems and set-valued dynamical systems.

2. PRELIMINARIES

Let X be a non-empty set. Throughout this chapter, we denote by $|A|$ the cardinality of the set X, 2^X and $\mathcal{F}(X)$ denote the family of all subsets of X and the family of all non-empty finite subsets of X, respectively. A (set-valued) *dynamical system* defined on X is a set-valued mapping Γ: $X \to 2^X$ such that $\Gamma(x) \neq \emptyset$ for each $x \in X$. A point $x \in X$ is said to be a *stationary point* of the set-valued dynamical system Γ if $\Gamma(x) = \{x\}$.

As we will study the existence of maximal elements for a binary relation defined on a non-empty compact set, we need the following definition which was first introduced by Border in [3].

Definition 2.1: Let X be a topological space such that $X = \cup_{n=1}^{\infty} X_n$, where $\{X_n\}_{n=1}^{\infty}$ is an increasing sequence of non-empty sets. A sequence $\{y_n\}_{n=1}^{\infty}$ in X is said to be *escaping* from X (relatively to $\{X_n\}_{n=1}^{\infty}$) if for each $n \in \mathbb{N}$, there exists a positive number $m \in \mathbb{N}$ such that $y_k \notin X_n$ for all $k \geq m$.

In order to study the existence of maximal elements for binary relation $\mathcal{R}$ defined on X in which the lower contour set $U^{-1}(x)$ for each $x \in X$ may not be open, we also recall the following definition (see [14:p. 136] and reference there).

Definition 2.2: Let X be a non-empty-set and Y be a topological space. A mapping F: $X \to 2^Y$ is said to be *transfer open valued* if for each $x \in X$ and $y \in F(x)$, there exist $x' \in Y$ and an open neighbourhood $N(y)$ of y in Y such that $y' \notin F(x')$ for all $y' \in N(y)$

In order to study the existence of maximal elements on a topological space which is non-convex in the classical sense but endowed with the generalized convexity, we also recall the following definitions which are first given by Llinaries in [7–8].

Definition 2.3: A topological space X is said to *have a K-structure* if there exists a so-called *K-convex continuous structure* K which is defined by a continuous mapping K: $X \times X \times [0, 1] \to X$ such that

$$K(x, y, 0) = x \quad \text{and} \quad K(x, y, 1) = y$$

for each $x, y \in X$. Then the pair (X, K) will be said to be a *K-convex space.*

Throughout this paper, a K-convex space X always means that X has a K-convex continuous structure as defined by Definition 2.3 above unless otherwise specified.

Remark 2.1: A non-empty subset A of a K-convex space X is said to be a *K-convex set* if $K(x, y, [0, 1]) \subset A$ for each $x, y \in A$. This means that for each pair x, y in A, the path $K(x, y, [0, 1])$ which joints x and y is contained in A. Moreover it is not difficult to prove that the intersection of any K-convex sets is also a K-convex set.

Definition 2.4: For a given set A in a K-convex space X, the *K-convex hull* $C_K(A)$ of A in X is defined by

$$C_K(A) := \cap\{B : A \subset B \text{ and } K\text{-convex set in } X\}.$$

Definition 2.5: A topological space X is said to have an MC-structure if for any non-empty finite subset $A = \{a_0, a_1, \ldots, a_n\}$ of X, there exists a family of elements $\{b_0, \ldots, b_n\}$ in X and a family of functions G_i^A: $X \times [0, 1] \to X$ for each $i = 0, 1, \ldots, n$ such that

(1) $G_i^A(x, 0) = x$ and $G_i^A(x, 1) = b_i$ for each $x \in X$;
(2) the function G_A: $[0, 1]^n \to X$ defined by

$$G_A(t_0, t_1, \ldots, t_{n-1}) := G_0^A(\cdots(G_{n-1}^A(G_n^A(a_n, 1), t_{n-1}), t_{n-2})\cdots), t_0)$$

for each $(t_0, t_1, \ldots, t_{n-1}) \in [0, 1]^n$ is continuous.

Throughout this note we say that X is an MC-space if and only of X has an MC-structure.

Remark 2.2: By the definition, it is easy to see that each star-shaped set and contractible set are K-convex set; and each convex set in a topological vector space is MC-convex. However by a simple fact that a topological space is K-convex if and only if it is contractible (e.g., see Proposition 1.1 of Llinares [7:p. 9]) and there exists a non-contractible where an MC-structure can be defined (e.g., see Example 1 of Llinaries in [8]), thus both concepts of K-convex and MC-convex structures are independent each other.

More general, if X and Y are two sets and F: $X \to 2^Y$ is a set-valued mapping, then a point $x \in X$ is said to be a *maximal element* of F if $F(x) = \emptyset$.

3. MAXIMAL ELEMENTS IN TOPOLOGICAL SPACES

It is our aim in this note to prove some non-compact version of the existence theorems for maximal elements in topological spaces.

We first state the following result which has been recently established by Llinares in [8] for topological spaces have MC-convex structures.

Lemma 3.1: *Let X be a compact Hausdorff topological MC-space and let U be a binary relation defined on X such that*

(1) *U is transfer open inverse valued;*
(2) *For each $x \in X$ and $A \in \mathcal{F}(X)$ with $A \cap U(x) \neq \emptyset$, we have $x \notin G_{A|U(x)}([0,1]^m)$, where $m = |A \cap U(x)|$.*

Then the set of maximal elements $\{x \in X : U(x) = \emptyset\}$ is non-empty and compact.

Proof: *It is Theorem 3 of Llinares in* [8]. □

By the similar arguments used in the proof of Lemma 3.1, we have also following existence result of maximal elements of topological vector spaces with K-convex structures (e.g., see Theorem II.2 of Llinares [7:p. 11] and thus we omit its proof here).

Lemma 3.2: *Let X be a compact Hausdorff topological space with a K-convex continuous structure. Let U be a binary relation on X satisfying*

(1) *$x \notin C_K(U(x))$ for each $x \in X$;*
(2) *U is transfer open inverse valued.*

Then the set of maximal elements $\{x \in X : U(x) = \emptyset\}$ is non-empty and compact.

Proof: For instance, see Theorem II.2 of Llinares [7:p. 11]). □

We now have the following existence theorems of maximal elements for non-compact topological spaces which have MC-structures.

Theorem 3.3: *Let X be a non-empty subset of a Hausdorff topological space Y such that $X = \cup_{n=1}^{\infty} X_n$ and $Y = \cup_{n=1}^{\infty} Y_n$, where $\{X_n\}_{n=1}^{\infty}$ and $\{Y_n\}_{n=1}^{\infty}$ are increasing sequence of non-empty compacts sets for which $X_n \subset Y_n$ for each $n \in \mathbb{N}$ and Y_n has an MC (resp., K-convex) structure. Suppose F: $Y \to 2^X$ a set-valued mapping such that each $n \in \mathbb{N}$,*

(1) *the mapping F_n: $Y_n \to 2^{X_n}$ is transfer open inverse valued;*
(2) *for each $y \in Y_n$ and $A \in \mathcal{F}(Y_n)$ with $A \cap F_n(y) \neq \emptyset$, then we have $y \notin G_{A|F_n(y)}([0,1]^n)$*
(resp., for each $y \in Y_n$, $y \notin C_K(F_n(y))$);
(3) *For each $\{y_n\}_{n=1}^{\infty}$ in Y with $y_n \in Y_n$ for each $n \in \mathbb{N}$ which is escaping from Y relatively to $\{Y_n\}_{n=1}^{\infty}$, there exists $n_0 \in \mathbb{N}$ and $x_{n_0} \in X_{n_0}$ such that $x_{n_0} \in F(y_{n_0})$.*

Then F has at least one maximal elements and indeed the set of all maximal set $\{x \in Y : F(x) = \emptyset\}$ is non-empty and compact.

Proof: For each $n \in \mathbb{N}$, By Lemma 3.1 (resp., Lemma 3.2), there exists $y_n \in Y_n$ such that

$$F_n(y_n) = F(y_n) \cap X_n = \emptyset. \tag{3.1}$$

Now let $M = \{y_n\}_{n=1}^{\infty}$ be a sequence of elements which satisfy the (3.1). We first claim that there exists $n_1 \in \mathbb{N}$ such that $M \subset Y_{n_1}$. Suppose it was not, i.e., $\{y_n\}_{n=1}^{\infty}$ is escaping from Y relatively to $\{Y_n\}_{n=1}^{\infty}$. By the condition (3), there exists $n_0 \in \mathbb{N}$ and $x_{n_0} \in X_{n_0}$ such that $x_{n_0} \in F(y_{n_0})$ which contradicts that $F_{n_0}(y_{n_0}) = F(y_{n_0}) \cap X_{n_0} = \emptyset$. Thus there exists $n_1 \in \mathbb{N}$ such that $M \subset Y_{n_1}$. Note that Y_{n_1} is compact, without loss of the generality, we may assume that $y_n \to y^* \in Y_{n_1}$. In order to finish the proof, it suffices to show that

$F(y^*) = \emptyset$. Suppose it were not true, i.e., $F(y^*) \neq \emptyset$. Note that $F(Y) \subset X$. There exist $n_2 \in \mathbb{N}$ and $x_{n_2} \in X_{n_2}$ such that

$$x_{n_2} \in F(y^*) \cap X_{n_2} = F_{n_2}(y^*). \tag{3.2}$$

By (3.2), we have that $y^* \in F_{n_2}^{-1}(x_{n_2})$. As $F_{n_2}^{-1}$ is transfer open valued, there exists $x'_{n_2} \in X_{n_2}$ such that $y^* \in \mathrm{int}F_{n_2}^{-1}(x'_{n_2}) \subset F_{n_2}^{-1}(x'_{n_2})$. Note that both $\{X_n\}_{n=1}^{\infty}$ and $\{Y_n\}_{n=1}^{\infty}$ are increasing sequences of sets, we may assume that $n_2 \geq n_1$ and then it follows that $M \subset Y_{n_2}$. Note that $y_n \to y^*$ and $F_{n_2}^{-1}(x'_{n_2})$ is open in Y_{n_2}, there exists $n_3 \in \mathbb{N}$ (and we can also assume that $n_3 \geq n_2$) such that $y_n \in F_{n_2}^{-1}(x'_{n_2})$ for all $n \geq n_3$. It follows that $x'_{n_2} \in F_{n_2}(y_n) \subset F_{n_3}(y_n) \subset F_n(y_n)$ for all $n \geq n_3$. Now let $n_4 := \max\{n_3, n_2\}$. Then we have that $x'_{n_2} \in X_{n_2} \subset X_{n_3} \subset X_{n_4}$ and $x'_{n_2} \in F(y_n) \cap X_n \neq \emptyset$for all $n \geq n_4$. Note that y_n satisfies (3.1) for each $n \in \mathbb{N}$, which is a contradiction. Thus we must have $F(y^*) = \emptyset$ and the proof is completed. □

Remark 3.1: As each topological space with an acyclic binary relation can be given a MC-structure (e.g., see Lemma 2 of Llinares [8]) and there exist topological spaces with non acyclic and non-convex preference (e.g., see Example 2 in [8]) which also have MC-structure, thus Theorem 3.3 include corresponding existence theoremsfor maximal elements given by Bergstrom [2], Border [3], Llinares [7–8], Sonnenschein [10], Yannelis and Prabhakar [13], Walker [12] and others as special case.

In what follows, as applications of set-valued dynamical systems, we shall prove some existence theorems of maximal elements for the binary relation $\mathcal{R}$ defined on a set X which has the irreflexivity in the sense that "$x\mathcal{R}x$ *for each* $x \in X$ (i.e., $x \notin U(x) := \{y \in X : y\mathcal{R}x\}$ for each $x \in X$") in either locally convex spaces or metric spaces (not necessarily having linear structures).

In the paper [4], the following result was proved by the first author, for the convenience we state one of its special case as follows (e.g., see Corollary 1 of Isac [4:p. 308] or Theorem 2 in [6:p. 396]):

Lemma 3.4: Let $(E(\tau, \{P_\alpha\}_{\alpha \in A})$ be a locally convex space, $M \subset E$ a non-empty complete set and Γ: $M \to 2^M$ a (set-valued) dynamical system. We suppose that for each $\alpha \in A$, there exists a lower semicontinuous function Φ_α: $M \to \mathbb{R}_+$ such that for each $x \in M$ and $u \in \Gamma(x)$, we have

$$P_\alpha(x, u) \leq \Phi_\alpha(x) - \Phi(u).$$

Then Γ has a stationary point in M, i.e., there exists $x^* \in M$ such that $\Gamma(x^*) = \{x^*\}$.

Let $S \subset E$ be a bounded complete subset and $\mathcal{R}$ a binary relation defined on S and $\mathcal{R}$ is not necessarily transitive. For each $x \in S$, let $U(x) := \{y \in S : y\mathcal{R}x\}$ and we define the set-valued dynamical system Γ: $S \to 2^S$ defined by

$$\Gamma(x) := U(x) \cup \{x\}$$

for each $x \in S$. Then it is clear that $\Gamma(x) \neq \emptyset$ for each $x \in S$.

Let $\mathbb{K}(\mathcal{R}) := \{u - x : x \in S \text{ and } u \in U(x) \cup \{x\}\}$. Then we have the following definition.

Definition 3.1: We say that the binary relation $\mathcal{R}$ is *nuclear* if for each $\alpha \in A$, there exists $f_\alpha \in E^*$ such that $P_\alpha(u) \leq f_\alpha(u)$ for each $u \in \mathbb{K}(\mathcal{R})$.

Remark 3.2: For the importance of nuclear cone and its application in the study of Pareto optimization, generalized complementarity problems, fixed point theory, the geometry of cones, best approximation theory in locally convex spaces and the study of nuclearity of vector spaces, the interested readers are referred to Isac [5–6] and references wherein.

We now have the following result:

Theorem 3.5: *Let $S \subset E$ be a non-empty bounded complete subset and $\mathcal{R}$ a binary relation (not necessarily transitive) on S. If $\mathcal{R}$ is a nuclear relation, then there exists a maximal elements $x^* \in S$ for the relation $\mathcal{R}$.*

Proof: Since $\mathcal{R}$ is a nuclear, for each $\alpha \in A$, there exists $f_\alpha \in E^*$ such that for each $u \in \mathbb{K}(\mathcal{R})$, we have $P_\alpha(u) \leq f_\alpha(u)$. Let X be any element in S and $u \in \Gamma(x)$. We denote by $m_\alpha := \sup_{v \in S} f_\alpha(v)$. Then we have

$$\begin{aligned} P_\alpha(x-u) = P_\alpha(u-x) \leq f_\alpha(u-x) &= f_\alpha(u) - f_\alpha(x) \\ &= [m_\alpha - f_\alpha(x)] - [m_\alpha - f(u)]. \end{aligned}$$

It we denote by $\Phi_\alpha(v) = m_\alpha - f_\alpha(v)$, it follows that

$$P_\alpha(x-u) \leq \Phi_\alpha(x) - \Phi_\alpha(u)$$

and by Lemma 3.4, Γ has a stationary point $x^* \in S$, i.e., $\Gamma(x^*) = \{x^*\}$. As $x^* \notin U(x^*)$, it follows that $U(x^*) = \emptyset$ and thus x^* is a maximal element for $\mathcal{R}$. □

Remark 3.3: We could have a more general result if we replace the nuclearity by the following condition (*):

(*): *For each $\alpha \in A$, there exists a continuous sub-additive mapping f_α: $E \to \mathbb{R}$ such that $P_\alpha(v) + f_\alpha(v) \leq 0$ for all $v \in \mathbb{K}(\mathcal{R})$.*

Before we conclude this section, we shall study the existence of maximal elements of a binary relation $\mathcal{R}$ when it is not *irreflexive*.

Remark 3.4: When $\mathcal{R}$ does not satisfy the irreflexive condition, we use the following dynamical system Γ: $S \to 2^S$ defined by

$$\Gamma(x) = \begin{cases} U(x) \cup \{x\}, & \text{if } x \notin U(x), \\ U(x) \backslash \{x\}, & \text{if } x \in U(x) \text{ but } U(x) \neq \{x\}, \\ S \backslash \{x\}, & \text{if } U(x) = \{x\} \end{cases}$$

for each $x \in S$. In this case, the set $\mathbb{K}(\mathcal{R})$ in Definition 3.1 is defined as follows:

$$\mathbb{K}(\mathcal{R}) := \{u - x : x \in S \text{ and } u \in \Gamma(x)\}.$$

We then have the following existence theorem of maximal elements for a binary relation $\mathcal{R}$ defined on a metric space X.

Theorem 3.6: *Let (X, d) be a metric space and $\mathcal{R}$ an irreflexive binary relation defined on X. Suppose the mapping U: $X \to 2^X$ defined by*

$$U(x) = \{y \in X : y\mathcal{R}x\}$$

for each $x \in X$ is lower semicontinuous and there exists a function ψ: $X \to (-\infty, +\infty]$ such that for each $x \in X$ and $u \in U(x)$, we have

$$d(x, u) \leq \psi(x) - \psi(u).$$

Then $\mathcal{R}$ has a maximal element, i.e., there exists a point $x^* \in X$ such that $U(x^*) = \emptyset$.

Proof: We define a mapping F: $X \to 2^X$ by $F(x) := U(x) \cup \{x\}$ for each $x \in X$. Then F is lower semicontinuous and $F(x) \neq \emptyset$ for each $x \in X$. Moreover it is clear that for each $x \in X$ and $u \in F(x)$, we also have $d(x, u) \leq \psi(x) - \psi(u)$. Thus F is a dissipative dynamical system (e.g., see Aubin and Ekeland [1:p. 244]). By Corollary 10 of [1:p. 246], it follows that F has a stationary point $x^* \in X$, i.e., $F(x^*) = \{x^*\}$. Note that since $x^* \notin U(x^*)$, we must have that $U(x^*) = \emptyset$, and thus x^* is a maximal element of $\mathcal{R}$. □

Remark 3.5: Note that if the binary relation $\mathcal{R}$ defined on X has open lower contour sets, then the mapping U: $X \to 2^X$ defined by

$$U(x) := \{y \in X : y\mathcal{R}x\}$$

for each $x \in X$ has open inverse values, i.e., $U^{-1}(x) := \{y \in X : x\mathcal{R}y\}$ is open in X for each $x \in X$ and thus the mapping U is lower semicontinuous. Therefore we have the following corollary.

Corollary 3.7: *Let (X, d) be a metric space and $\mathcal{R}$ an irreflexive binary relation defined on X which has open lower property (i.e., $U^{-1}(x) := \{y \in X : x\mathcal{R}y\}$ is open in X for each $x \in X$). Suppose there exists a function ψ: $X \to (-\infty, +\infty]$ such that for each $x \in X$ and $u \in U(x)$, we have*

$$d(x, u) \leq \psi(x) - \psi(u).$$

Then $\mathcal{R}$ has a maximal element, i.e., there exists a point $x^* \in X$ such that $U(x^*) = \emptyset$.

Proof: By Remark 3.4, the mapping U: $X \to 2^X$ defined in Theorem 3.6 is lower semicontinuous. Thus all hypotheses of Theorem 3.6 are satisfied and the proof is completed. □

REFERENCES

[1] J.P. Aubin and I. Ekeland (1984). Applied Nonlinear Analysis. Brisbane, New York: John Wiley and Sons.

[2] T.C. Bergstrom (1975). Maximal elements of acyclic relations on a compact sets, *J. Econom. Theory*, **10**, 403–404.

[3] K.C. Border (1985). Fixed Point Theorems with Applications to Economics and Game Theory. Cambridge Univ. Press

[4] G. Isac (1983). Sur l'existence de l'optimum de Pareto, *Riv. Mat. Univ. Parma*, **9**, 303–325.

[5] G. Isac (1987). Supernormal cones and fixed point theory, *Rocky Mountain J. Math.*, **17**, 219–226.

[6] G. Isac (1994). Pareto optimization in infinite dimensional spaces: the importance of nuclear cones *J. Math. Anal. Appl.*, **182**, 393–404.

[7] J.V. Llinares (1995). Existence of maximal elements in a binary relation relaxing the convexity condition Research Report, WP-AD 95–10, University of Alicante, Spain 1–28.

[8] J.V. Llinares (1998). Unified treatment of the problem of existence of maximal elements in binary relations: a characterization *J. Math. Economics*, **29**, 285–302.

[9] R.D. Luce (1956). Semiorder and a theory of utility discrimination, *Econometrica*, **24**, 178–191.

[10] H. Sonnenschein (1971). Demand theory without transitive preference with applications to the theory of competitive equilibrium. In J.S. Chipman, L. Hurwicz, M.K. Richter and H.F. Sooenschein (Eds), *Preferences Utility and Demand*, New York: Harcourt Brace Javanovich, 213–223.

[11] R.M. Starr (1969). Quasi-equilibria in markets with non-convex preferences, *Econometric*, **37**, 25–38.

[12] M. Walker (1977). On the existence of maximal elements, *J. Econom. Theory*, **16**, 470–474.

[13] N. Yannelis and N. Prabhakar (1983). Existence of maximal elements and equilibria in linear topological spaces, *J. Math. Econom.*, **12**, 233–245.

[14] X.Z. Yuan (1995). The Knaster-Kuratowski-Mazurkiewicz theorem, Ky Fan minimax inequalities and fixed point theorems, *Nonlinear World*, **2**, 131–169.

SIMAA 4(2002) 213–226

15. Periodic Solutions of a Singularly Perturbed System of Differential Inclusions in Banach Spaces*

Mikhail Kamenski[1] and Paolo Nistri[2]

[1]*Department of Mathematics, Voronezh State University, Voronezh, Russia.*
E-mail: mikhail@kam.vsu.ru
[2]*Dip. di Ingegneria dell'Informazione, Università di Siena, 53100 Siena, Italy.*
E-mail: pnistri@dii.unisi.it

Abstract: By means of the Hausdorff measure of noncompactness and the topological degree theory for condensing operators in locally convex spaces we show the existence of periodic solutions of a singularly perturbed system of differential inclusions in infinite dimensional Banach spaces. Moreover, the behaviour of such periodic solutions when the parameter ϵ tends to zero is also investigated.

Keywords: Differential inclusions, condensing operators, periodic solutions, singularly perturbed systems

AMS 1991 Mathematics Subject Classification: Primary – 34A60, 34G20; Secondary – 34C25

INTRODUCTION

The aim of this chapter is to investigate the existence of periodic solutions for a system of differential inclusions in infinite dimensional spaces, depending on a small parameter $\epsilon > 0$, which has the form

$$\begin{aligned} x'(t) &\in Ax(t) + f_1(t, x(t), y(t)), \quad t \geq 0 \\ \epsilon y'(t) &\in By(t) + f_2(t, x(t), y(t)), \end{aligned} \tag{1}$$

where A and B are infinitesimal generators of C^0-semigroups of linear operators e^{At} and $e^{Bt}, t \geq 0, x \in E_1$ and $y \in E_2$ with E_1, E_2 infinite dimensional Banach spaces. The nonlinear multivalued operators $f_i,\ i = 1, 2$, are T-periodic in time with nonempty, convex and compact values and satisfying suitable conditions expressed in terms of the Hausdorff measure of noncompactness. All the assumptions will be precised in the next section.

* The Research of Mikhail Kamenski is supported by R.F.F.I. Grant 96-01-00333 and G.N.A.M.P.A. and the Research of Paolo Nistri is supported by a grant of the University of Siena.

In [2] we provide conditions which guarantee the upper semicontinuity at $\epsilon = 0$ of the solution map $\epsilon \to S_\epsilon$ of the Cauchy problem for (1). The considered topology for the solution pair (x, y) is that of the uniform topology with respect to the x-variable and that of the $L^1(E_2)$-weak convergence with respect to the y-variable. For singularly perturbed systems of differential inclusions in finite dimensional spaces in [10] and [15] the same result is proved by using completely different methods. In this case, if the uniform convergence is also considered for the y-variable then the upper semicontinuity can be obtained for a suitable subset of S_ϵ (see [8,18,19]). In fact, in general, the map $\epsilon \to S_\epsilon$ is not upper semicontinuous at $\epsilon = 0$ (see [9]). Finally, in [11] and [12], see also the references therein, an approach in order to approximate the slow motions of a singularly perturbed control system in finite dimension by a limit differential inclusion was proposed. This approach is based on the averaging method applied to the fast dynamics, as result the uniform convergence of the slow motions to a solution of the limit differential inclusion is obtained. Furthermore, any such solution is the uniform limit of slow motions. Singular perturbation methods for partial differential equations are also intensively studied (see e.g., [16,17]).

In this chapter our attention is devoted to the existence of periodic solutions for small $\epsilon > 0$ and to their behaviour when ϵ tends to zero. It is still convenient for our purposes to consider here the uniform topology for the x-variable and the weak topology for the y-variable. In fact, with this choice we will be able to show the upper semicontinuity at $\epsilon = 0$ of a suitably defined condensing operator, whose fixed points represent the T-periodic solutions of our problem. Roughly speaking, we will show that if the reduced problem at $\epsilon = 0$ admits isolated sets of T-periodic solutions with topological degree different from zero then we provide sufficient conditions to guarantee the existence of T-periodic solutions (x_ϵ, y_ϵ) for small $\epsilon > 0$ and also that for every sequence $\epsilon_n \to 0$ the sequence $(x_{\epsilon_n}, y_{\epsilon_n})$ converge, in the above topology, to a T-periodic solution of the reduced problem. Observe that here the topological degree is that for condensing operators in locally convex spaces.

The methods presented in this chapter are similar to those of [2], but here we use different measures of noncompactness which turn out to be more suitable for the present problem.

The chapter is organized as follows. In Section 1 we state the problem and we formulate the assumptions which permit to solve it. Furthermore, we introduce convenient operators in order to rewrite our problem in terms of a multivalued fixed point problem. In Section 2 we prove in Theorem 1 the relevant properties of the resulting fixed point operator, in particular the condensivity with respect to a suitably introduced measure of noncompactness and the upper semicontinuity in the considered topology. These properties will permit to use the topological degree theory for condensing operators in locally convex spaces in order to prove the existence of fixed points for $\epsilon > 0$ sufficiently small and their behaviour when ϵ tends to zero.

1. STATEMENT OF THE PROBLEM, DEFINITIONS AND ASSUMPTIONS

Through this chapter we consider a system of differential inclusions of the form

$$\begin{aligned} x'(t) &\in Ax(t) + f_1(t, x(t), y(t)), \quad t \geq 0 \\ \epsilon y'(t) &\in By(t) + f_2(t, x(t), y(t)), \end{aligned} \tag{1}$$

where A and B are infinitesimal generators of C^0-semigroups of linear operators e^{At} and e^{Bt}, $t \geq 0$, respectively, acting in the Banach spaces E_1 and E_2 with E_2 satisfying the Radon-Nikodym condition (see [7]), $\epsilon > 0$ is a small parameter and f_i: $\mathbf{R}_+ \times E_1 \times E_2 \rightarrow Kv(E_i), i = 1, 2$, are multivalued operators. Here $Kv(E)$ denotes the set of all the nonempty, convex, compact subsets of the Banach space E.

Statement of the Problem: we want to provide conditions under which system (1) can be rewritten as a fixed point problem in a suitable chosen functional space $\mathcal{F}$ for a multivalued condensing operator F_ϵ. Moreover, these conditions must guarantee that F_ϵ is upper semicontinuous at any $\epsilon \geq 0$ in a prescribed topology of the underlying functional space $\mathcal{F}$.

As a result we can apply the related topological degree theory for multivalued condensing operator defined in $\mathcal{F}$ (see e.g., [1,4]) to derive the following relevant property:

If there exists an open set $U \subset \mathcal{F}$ such that

$$\deg(I - F_0, \overline{U}) \neq 0$$

then for $\epsilon > 0$ sufficiently small the set Σ_ϵ of periodic solutions of (1) is nonempty and the map $\epsilon \rightarrow \Sigma_\epsilon$ is upper semicontinuous in the considered topology of $\mathcal{F}$.

To make precise the setting in which we will solve the above problem we first choose for the functions $t \rightarrow x(t)$ and $t \rightarrow y(t)$ the functional spaces $C_T(E_1)$ and $L^1_T(E_2)$ respectively, and so $\mathcal{F} = C_T(E_1) \times L^1_T(E_2)$. We recall that $C_T(E)$ denotes the space of T-periodic continuous functions x: $[0, T] \rightarrow E$ equipped with the uniform norm: $\max_t \| x(t) \|_E$ and $L^1_T(E)$ is the space of T-periodic strongly measurable functions x: $[0, T] \rightarrow E$ having finite norm $\| x \|_{L^1_T} := \int_0^T \| x(t) \|_E \, dt$. In the sequel by ${}^wE$ we will denote the space E equipped with the weak topology of E, while $Kv\text{-}w(E)$ will denote the set of all the nonempty, convex, weakly compact subsets of E.
We assume that

(S_0) there exist positive constants $\gamma_1, \gamma_2 > 0$ such that

$$\| e^{At} \|_{E_1} \leq e^{-\gamma_1 t} \quad \text{and}$$
$$\| e^{Bt} \|_{E_2} \leq e^{-\gamma_2 t}$$

for any $t \geq 0$. Moreover $D(B^*)$, the domain of the adjoint operator B^*, is dense in E_2^* (see [14]).

(A_0) f_i: $\mathbf{R}_+ \times E_1 \times E_2 \rightarrow Kv(E_i), i = 1, 2$, are T-periodic with respect to time, that is

$$f_i(t + T, x, y) = f_i(t, x, y)$$

for any $t \geq 0$ and $(x, y) \in E_1 \times E_2$. Furthermore, the Nemytskii operators Φ_i: $C_T(E_1) \times L^1_T(E_2) \rightarrow Kv\text{-}w(L^1_T(E_i))$ generated by f_i, $i = 1, 2$, as follows

$$\Phi_i(x, y) = \{g \in L^1_T(E_i) : g(t) \in f_i(t, x(t), y(t)) \\ \text{for almost all (a.a.) } t \in [0, T]\}$$

are well defined.

The following assumptions are formulated in terms of the Nemytskii operators $\Phi_i, i = 1, 2$.

(A_1) For any pair of bounded sets $\Omega_1 \subset C_T(E_1), \Omega_2 \subset Q \subset L^1_T(E_2)$, where Q is the set introduced in (A_3), there exists a function $\varphi \in L^1_T(\mathbf{R})$ such that

$$\| g_i(t) \|_{E_i} \leq \varphi(t)$$

for a.a. $t \in \mathbf{R}$ and any $g_i \in \Phi_i(x, y), i = 1, 2$, whenever $(x, y) \in \Omega_1 \times \Omega_2$.

(A_2) Φ_i are upper semicontinuous multivalued operators from $C_T(E_1) \times {}^wL^1_T(E_2)$ to ${}^wL^1_T(E_i)$.

Remark 1: Explicit conditions on $f_i, i = 1, 2$, which ensure that the related Nemytskii operators are well defined will be given in Section 3. For the finite dimensional case (see [3]).

We also need suitable compactness conditions on $\Phi_i, i = 1, 2$, expressed in terms of the Hausdorff measure of noncompactness. To this aim we give the following definitions.

Definition 1: Let E be a Banach space. Let $\Omega \subset E$ be a bounded set. The Hausdorff measure of noncompactness $\chi_E(\Omega)$ of the set Ω is the infimum of the numbers $\alpha > 0$ such that Ω has a finite α-net in E. For the relevant properties of χ_E we refer to [1].

Definition 2: Let E be a Banach space. Let $\Omega \subset E$ be a bounded set of E. The measure of weak noncompactness $\chi_w(\Omega)$ of the set Ω is the infimum of the number $\alpha > 0$ such that Ω has a weakly compact α-net in E. This measure of weak compactness and its properties have been studied by De Blasi in [6].

Definition 3: Let Ω be a bounded set of $L^1_T(E)$. A function $b \in L^1_T(\mathbf{R})$ is called a weak bound for the Hausdorff measure of noncompactness $\chi_{L^1_T(E)}$ of the set Ω if for every $\delta > 0$ there exist a measurable set $e_\delta \subset [0, T]$ and a compact set $K_\delta \subset E$ such that meas $e_\delta < \delta$ and for every $f \in \Omega$ there exists $g \in L^1_T(E)$ satisfying $g(t) \in K_\delta$ for a.a. $t \in [0, T]$ and

$$\| f(t) - g(t) \|_E \leq b(t) + \delta$$

for a.a. $t \in [0, T] \backslash e_\delta$.

In the sequel the set of all the functions $b \in L^1_T(\mathbf{R})$ with the previous properties will be denoted by $WB(\Omega)$. Observe that we can always assume that $g(t) = 0$ for $t \in e_\delta$.

We introduce now the operators $F_\epsilon, \epsilon \in [0, 1]$, whose fixed points will represent the T-periodic solutions of (1). For this, we need first to define the linear operators Λ_ϵ: $L^1_T(E_1) \times L^1_T(E_2) \to C_T(E_1) \times L^1_T(E_2)$ defined as follows

$$\Lambda_\epsilon \begin{pmatrix} g_1 \\ g_2 \end{pmatrix}(t) = \begin{pmatrix} \Lambda_1 g_1 \\ \Lambda_2(\epsilon) g_2 \end{pmatrix}(t) = \begin{pmatrix} \int_{-\infty}^{t} e^{A(t-s)} g_1(s)\, ds \\ \frac{1}{\epsilon} \int_{-\infty}^{t} e^{(1/\epsilon)B(t-s)} g_2(s)\, ds \end{pmatrix}, \quad \epsilon > 0.$$

While, for $\epsilon = 0$ we set $\Lambda_2(0) = -B^{-1}$. Finally we pose

$$F_\epsilon(x, y) = \left\{ \Lambda_\epsilon \begin{pmatrix} g_1 \\ g_2 \end{pmatrix} : g_1 \in \Phi_1(x, y),\ g_2 \in \Phi_2(x, y) \right\}$$

We formulate now the assumptions under which $F_\epsilon, \epsilon \in [0,1]$, is a well defined condensing operator.

(A_3) There exists a closed, convex set $Q \subset {}^wL^1_T(E_2)$ such that Q is bounded in $L^1(E_2)$ and

$$\Lambda_2(\epsilon)\, \Phi_2 : C_T(E_1) \times Q - \circ Q$$

(A_4) There exist positive constants k_{11} and k_{12} such that for any pair of bounded sets $\Omega_1 \subset C_T(E_1)$ and $\Omega_2 \subset Q$ the constant function

$$K = k_{11} \sup_t \chi_{E_1}(\Omega_1(t)) + k_{12}\chi_w(\Omega_2)$$

belongs to $WB(\Phi_1(\Omega_1 \times \Omega_2)_{[0,T]})$.

(A_5) There exist positive constants k_{21} and k_{22} such that for any pair of bounded sets $\Omega_1 \subset C_T(E_1)$ and $\Omega_2 \subset Q$ one has

$$\chi_w(\Phi_2(\Omega_1 \times \Omega_2)) \leq k_{21} \sup_t \chi_{E_1}(\Omega_1(t)) + k_{22}\chi_w(\Omega_2).$$

Finally, we now formulate the last assumption.

(A_6) The eigenvalues λ of the matrix

$$H = \begin{pmatrix} k_{11}/\gamma_1 & k_{12}/\gamma_1 \\ k_{21}/\gamma_2 & k_{22}/\gamma_2 \end{pmatrix}$$

satisfy $|\lambda| < 1$.

Remark 2: The assumption (A_3) is verified if, for instance, there exist positive constants M and l such that

$$\| f_2(t,x,y) \|_{E_2} \leq M + l \| y \|$$

with $l/\gamma_2 < 1$. In this case, we have $Q = Q_R$, where

$$Q_R := \Big\{ g \in L^1_T(E_2) : \| g(t) \|_{E_2} \leq R, \quad \text{for a.a. } t \in \mathbf{R} \Big\}.$$

and $R > 0$ is sufficiently large.

Definition 4: By a T-periodic solution of system (1) where $\epsilon \geq 0$, we mean a fixed point $(x,y) \in C_T(E_1) \times L^1_T(E_2)$ of the operator F_ϵ.

2. RESULTS

In order to formulate our results we introduce first two measures of noncompactness. The first is defined as follows: given a bounded set $\Omega \subset C_T(E_1) \times L^1_T(E_2)$ we put

$$\overline{\chi}(\Omega) = \begin{pmatrix} \chi_{C_T(E_1)}(P_1(\Omega)) \\ \chi_w(P_2(\Omega)) \end{pmatrix},$$

where P_1 is the projector on the first coordinate of the Cartesian product $C_T(E_1)\times L^1_T(E_2)$, while P_2 is the projector on the second coordinate of the same space. The second measure of noncompactness is defined as

$$\overline{\nu}(\Omega) = \begin{pmatrix} \sup_t \chi_{E_1}(P_1(\Omega(t))) \\ \chi_w(P_2(\Omega)) \end{pmatrix}.$$

We can now prove the following.

Theorem 1: *Assume that the conditions* $(S_0), (A_0) \div (A_5)$ *are satisfied, then the operator* $F_\epsilon\colon C_T(E_1) \times Q \to C_T(E_1) \times Q$ *is upper semicontinuous at any* $\epsilon \in [0,1]$. *Furthermore* F_ϵ *has nonempty, compact, convex values and it is* $(H, \overline{\chi}, \overline{\nu})$*-bounded with respect to all its variables, namely*

$$\overline{\chi}\left(\bigcup_{\epsilon\in[0,1]} F_\epsilon(\Omega) \right) \leq H\overline{\nu}(\Omega) \tag{2}$$

for any bounded set $\Omega \subset C_T(E_1) \times Q$, *where the inequality is understood in the sense of the semiorder induced by the cone* $\mathbf{R}^2_+$.

Proof: Let Ω be a bounded set in $C_T(E_1) \times Q$, then $\Omega \subset P_1\Omega \times P_2\Omega$. By condition (A_4) we have

$$k_{11} \sup_t \chi_{E_1}(P_1\Omega(t)) + k_{12}\chi_w(P_2\Omega) \in WB(\Phi_1(P_1\Omega \times P_2\Omega_2)|_{[0,T]}).$$

Therefore for a given $\delta > 0$ there exist $e_\delta \subset [0, T]$, a compact $K_\delta \subset E_1$ and a set $G \subseteq L^1_T([0,T], E_1)$ such that meas $e_\delta < \delta$ and for all $g \in G$ one has $g(t) \in K_\delta$ for $t \in [0,T]\backslash e_\delta$, $g(t) = 0$ for $t \in e_\delta$. We can extend any $g \in G$ to $\mathbf{R}$ by T-periodicity. For simplicity, we still denote by g such an extension and by G the corresponding set. Observe that $\Lambda_1 G$ is relatively compact in $C_T(E_1)$. Let $f \in \Phi_1(P_1\Omega \times P_2\Omega)$ and evaluate $\| \Lambda_1 f(t) - \Lambda_1 g(t) \|_{E_1}$ as follows

$$\begin{aligned} \| \Lambda_1 f(t) - \Lambda_1 g(t) \|_{E_1} &\leq \int_{-\infty}^t e^{-\gamma_1(t-s)} \| f(s) - g(s) \|_{E_1} ds \\ &\leq \int_{-\infty}^t e^{-\gamma_1(t-s)} (k_{11} \sup_t \chi(P_1\Omega(t)) + k_{12}\chi_w(P_2(\Omega))\, ds \\ &+ \int_{-\infty}^t e^{-\gamma_1(t-s)} \psi_\delta(s) \| f(s) \|_{E_1} ds, \end{aligned}$$

where ψ_δ is the characteristic function of the set $e_\delta + jT, j \in \mathbf{Z}$. Since the function $z(t) = \int_{-\infty}^t e^{-\gamma_1(t-s)}\psi_\delta(s)\| f(s) \|_{E_1} ds$ is T-periodic then it is sufficient to consider z only for $t \in [0,T]$. Let $t \in [0,T]$ and consider

$$\begin{aligned} z(t) = e^{-\gamma_1 t} \sum_{j=0}^{\infty} \int_{-(j+1)T}^{-jT} e^{\gamma_1 s} \psi_\delta(s) \| f(s) \|_{E_1} ds \\ + \int_0^t e^{-\gamma_1(t-s)} \psi_\delta(s) \| f(s) \|_{E_1} ds \end{aligned}$$

$$
\begin{aligned}
&= e^{-\gamma_1 t}\sum_{j=0}^{\infty} e^{-j\gamma_1 T}\int_0^T e^{-\gamma_1(T-s)}\psi_\delta(s)\| f(s)\|_{E_1}\, ds \\
&\quad + \int_0^t e^{-\gamma_1(t-s)}\psi_\delta(s)\| f(s)\|_{E_1}\, ds \\
&= e^{-\gamma_1 t}(1-e^{-\gamma_1 T})^{-1}\int_0^T e^{-\gamma_1(T-s)}\psi_\delta(s)\| f(s)\|_{E_1}\, ds \\
&\quad + \int_0^t e^{-\gamma_1(t-s)}\psi_\delta(s)\| f(s)\|_{E_1}\, ds \\
&\le \frac{2}{1-e^{-\gamma_1 T}}\int_0^T \psi_\delta(s)\| f(s)\|_{E_1}\, ds.
\end{aligned}
$$

Since $\delta > 0$ is arbitrary the last term is zero by (A_1). From (S_0) we obtain

$$
\begin{aligned}
&\chi_{C_T(E_1)}(\Lambda_1\Phi_1(P_1\Omega\times P_2\Omega)) \\
&\le \frac{k_{11}}{\gamma_1}\sup_t \chi_{E_1}(P_1\Omega(t)) + \frac{k_{12}}{\gamma_1}\chi_w(P_2\Omega).
\end{aligned}
$$

By the monotonicity of the Hausdorff measure of noncompactness we get

$$
\chi_{C_T(E_1)}(\Lambda_1\Phi_1(\Omega)) \le \chi_{C_T(E_1)}(\Lambda_1\Phi_1(P_1\Omega\times P_2\Omega))
$$

and so

$$
\chi_{C_T(E_1)}(\Lambda_1\Phi_1(\Omega)) \le \frac{k_{11}}{\gamma_1}\sup_t \chi_{E_1}(P_1\Omega(t)) + \frac{k_{12}}{\gamma_1}\chi_w(P_2\Omega). \tag{3}
$$

Let us now show that the operator F_ϵ: $C_T(E_1)\times Q \to C_T(E_1)\times Q$ is upper semicontinuous at $\epsilon = 0$. Observe that taking into account that the theory of the topological degree for condensing operators is constructed by means of the restriction of the involved operators to a fundamental compact set and the Theorem of Šmulian, we can verify the upper semicontinuity of F_ϵ on the sequences. For this, by assumption (A_2), it is enough to show that the linear operator $\Lambda(\epsilon)$: ${}^wL^1_T(E_2) \to {}^wL^1_T(E_2)$ is sequentially continuous with respect to ϵ. It is clear that it is continuous for $\epsilon > 0$, let us show that it is continuous at $\epsilon = 0$. To this aim, consider $g_n \to g_0$ weakly in $L^1_T(E_2)$ and $\epsilon_n \to 0$. Let v^* be the functional generated by the function

$$
y^*(t) = \sum_{i=0}^{m-1} y_i^*\psi_{[t_i,t_{i+1})}(t), \tag{4}
$$

where $y_i^* \in E_2^*$, $0 = t_0 < t_1 < \cdots < t_m = T$ and $\psi_{[t_i,t_{i+1}]}$ is the characteristic function of the interval $[t_i, t_{i+1}]$, then

$$
\begin{aligned}
&\int_0^T \left\langle y^*(t), \frac{1}{\epsilon_n}\int_{-\infty}^t e^{(1/\epsilon_n)B(t-s)} g_n(s)\, ds \right\rangle dt \\
&= \sum_{i=0}^{m-1}\int_{t_i}^{t_{i+1}} \left\langle y_i^*, \frac{1}{\epsilon_n}\int_{-\infty}^t e^{(1/\epsilon_n)B(t-s)} g_n(s)\, ds \right\rangle dt
\end{aligned}
$$

$$
\begin{aligned}
&= \sum_{i=0}^{m-1} \int_{t_i}^{t_{i+1}} \left\langle y_i^*, \frac{1}{\epsilon_n} \int_{-\infty}^{t_i} e^{(1/\epsilon_n)B(t-s)} g_n(s)\, ds \right\rangle dt \\
&+ \sum_{i=0}^{m-1} \int_{t_i}^{t_{i+1}} \left\langle y_i^*, \frac{1}{\epsilon_n} \int_{t_i}^{t} e^{(1/\epsilon_n)B(t-s)} g_n(s)\, ds \right\rangle dt \\
&= \sum_{i=0}^{m-1} \int_{-\infty}^{t_i} \left\langle \frac{1}{\epsilon_n} \int_{t_i}^{t_{i+1}} e^{(1/\epsilon_n)B^*(t-s)} y_i^*\, dt, g_n(s) \right\rangle ds \\
&+ \sum_{i=0}^{m-1} \int_{t_i}^{t_{i+1}} \left\langle \frac{1}{\epsilon_n} \int_{s}^{t_{i+1}} e^{(1/\epsilon_n)B^*(t-s)} y_i^*\, dt, g_n(s) \right\rangle ds \\
&= \sum_{i=0}^{m-1} \int_{-\infty}^{t_i} \left\langle (B^*)^{-1} \left(e^{(1/\epsilon_n)B^*(t_{i+1}-s)} - e^{(1/\epsilon_n)B^*(t_i-s)} \right) y_i^*, g_n(s) \right\rangle ds \\
&+ \sum_{i=0}^{m-1} \int_{t_i}^{t_{i+1}} \left\langle (B^*)^{-1} \left(e^{(1/\epsilon_n)B^*(t_{i+1}-s)} - I \right) y_i^*, g_n(s) \right\rangle ds \\
&= \sum_{i=0}^{m-1} \int_{-\infty}^{t_{i+1}} \left\langle (B^*)^{-1} e^{(1/\epsilon_n)B^*(t_{i+1}-s)} y_i^*), g_n(s) \right\rangle ds \\
&- \sum_{i=0}^{m-1} \int_{-\infty}^{t_i} \left\langle (B^*)^{-1} e^{(1/\epsilon_n)B^*(t_i-s)} y_i^*, g_n(s) \right\rangle ds \\
&- \int_0^T \left\langle (B^*)^{-1} y^*(s), g_n(s) \right\rangle ds.
\end{aligned}
$$

We now prove that the first two terms tend to zero as $n \to \infty$. For this, note that, since $g_n \to g_0$ weakly in $L^1_T(E_2)$, we have that for any $\delta > 0$ there exists $\mu > 0$ such that for any set $e \subset [0,T]$ with meas$(e) < \mu$ it follows $\int_e \| g_n(s) \|_{E_2} ds < \delta$ for any $n \in \mathbf{N}$, (see [7]). Let $t \in [0,T]$, consider

$$
\begin{aligned}
&\int_{-\infty}^{t} \left\langle (B^*)^{-1} e^{(1/\epsilon_n)B^*(t-s)} y_i^*, g_n(s) \right\rangle ds \\
&= \int_{-\infty}^{0} \left\langle (B^*)^{-1} e^{(1/\epsilon_n)B^*(t-s)} y_i^*, g_n(s) \right\rangle ds \\
&+ \int_{0}^{t} \left\langle (B^*)^{-1} e^{(1/\epsilon_n)B^*(t-s)} y_i^*, g_n(s) \right\rangle ds \\
&= \int_{0}^{T} \left\langle (B^*)^{-1} e^{(1/\epsilon_n)B^*t} (I - e^{(1/\epsilon_n)B^*T})^{-1} e^{(1/\epsilon_n)B^*(T-s)} y_i^*, g_n(s) \right\rangle ds \\
&+ \int_{0}^{t} \left\langle (B^*)^{-1} e^{(1/\epsilon_n)B^*(t-s)} y_i^*, g_n(s) \right\rangle ds.
\end{aligned}
$$

We first write, $\int_0^T = \int_0^{T-\mu} + \int_{T-\mu}^T$ and $\int_0^t = \int_0^{t-\mu} + \int_{t-\mu}^t$. Then we can estimate by means of (S_0) these integrals and taking into account the observation above we can conclude that

$$
\int_{-\infty}^{t} \left\langle (B^*)^{-1} e^{(1/\epsilon_n)B^*(t-s)} y_i^*, g_n(s) \right\rangle ds \to 0
$$

as $n \to \infty$. We leave the details to the reader.

Finally, the last term tends to

$$-\int_0^T \langle y^*(s), B^{-1}g_0(s)\rangle\, ds.$$

In conclusion, we have weak convergence of $\frac{1}{\epsilon_n}\int_{-\infty}^t e^{(1/\epsilon_n)B(t-s)}g_n(s)\,ds$ to $-B^{-1}g_0$ with respect to the functionals v^*. On the other hand E_2 has the Radon-Nikodym property, then by (see Theorem 1 [7:p. 98]) we have that $(L_T^1(E_2))^* = L_T^\infty(E_2^*)$. Now we can approximate in the "almost every" convergence any function $y \in L_T^\infty(E_2^*)$ by a function of the form (4). For this it is possible to consider the continuous function

$$z_h(s) = \frac{1}{h}\int_s^h y(t)\,dt, \quad s \in [0,T]$$

which tends to $y(s)$ as $h \to 0$ for a.a. $s \in [0,T]$ (see Theorem 9 [7:p. 49]) and then approximate $z_h(\,\cdot\,)$ by step functions. Applying now Egorov's Theorem we finally obtain that for any $y \in L_T^\infty(E_2^*)$

$$\int_0^T \left\langle y(t), \frac{1}{\epsilon_n}\int_{-\infty}^t e^{(1/\epsilon_n)B(t-s)}g_n(s)\,ds\right\rangle dt$$

tends to

$$-\int_0^T \langle y(s), B^{-1}g_0(s)\rangle\, ds$$

as $n \to \infty$. Therefore, the operator $\Lambda(\epsilon)\Phi_2(x,y)$ is upper semicontinuous from $[0,1] \times C_T(E_1) \times Q$ to $Kv\text{-}w(Q)$. Observe that if $C \subset L_T^1(E_2)$ is any weakly compact set then $\bigcup_{\epsilon\in[0,1]}\Lambda_2(\epsilon)C$ is weakly compact. To conclude the proof we show that $\Lambda_2(\epsilon)$, $\epsilon \in [0,1]$, is a $\frac{1}{\gamma_2}$-contraction from $L_T^1(E_2)$ to $L_T^1(E_2)$. In fact, let $\epsilon > 0$, and consider

$$\begin{aligned}
&\int_0^T \frac{1}{\epsilon}\left\|\int_{-\infty}^t e^{B(t-s)}g(s)\,ds\right\|_{E_2} dt \le \int_0^T \frac{1}{\epsilon}\int_{-\infty}^t e^{-(1/\epsilon)\gamma_2(t-s)}\|\, g(s)\,\|_{E_2}\, ds\, dt\\
&= \int_0^T \frac{1}{\epsilon}\int_{-\infty}^0 e^{-(1/\epsilon)\gamma_2(t-s)}\|\, g(s)\,\|_{E_2}\, ds\, dt + \int_0^T \frac{1}{\epsilon}\int_0^t e^{-(1/\epsilon)\gamma_2(t-s)}\|\, g(s)\,\|_{E_2}\, ds\, dt\\
&= \int_{-\infty}^0 \frac{1}{\epsilon}\int_0^T e^{-(1/\epsilon)\gamma_2(t-s)}\|\, g(s)\,\|_{E_2}\, ds\, dt + \int_0^T \frac{1}{\epsilon}\int_s^T e^{-(1/\epsilon)\gamma_2(t-s)}\|\, g(s)\,\|_{E_2}\, ds\, dt\\
&= \frac{1}{\gamma_2}\left(\int_{-\infty}^0 e^{(1/\epsilon)\delta s}\|\, g(s)\,\|_{E_2}\, ds - \int_{-\infty}^0 e^{-(1/\epsilon)(T-s)}\|\, g(s)\,\|_{E_2}\, ds\, dt\right.\\
&\quad \left. + \int_0^T (1 - e^{-(1/\epsilon)\gamma_2(T-s)})\|\, g(s)\,\|_{E_2}\, ds\right) = \frac{1}{\gamma_2}\int_0^T \|\, g(s)\,\|_{E_2}\, ds.
\end{aligned}$$

Finally, for $\epsilon = 0$, since $B^{-1} = \int_0^{+\infty} e^{Bt}\,dt$, we have $\|\, B^{-1}\,\| \le \frac{1}{\gamma_2}$. In conclusion

$$\begin{aligned}
&\chi_w\left(\bigcup_{\epsilon\in[0,1]}\Lambda_2(\epsilon)\Phi_2(\Omega)\right) \le \frac{1}{\gamma_2}\chi_w(\Phi_2(\Omega))\\
&\le \frac{k_{21}}{\gamma_2}\sup_t \chi_{E_1}(P_1\Omega(t)) + \frac{k_{22}}{\gamma_2}\chi_w(P_2\Omega).
\end{aligned} \tag{5}$$

In fact, if C is a weakly compact set which constitutes the α-net of $\Phi_2(\Omega)$, then $\bigcup_{\epsilon\in[0,1]} \Lambda_2(\epsilon)C$ is a weakly compact set which represents the $\frac{\alpha}{\gamma_2}$-net of $\bigcup_{\epsilon\in[0,1]} \Lambda_2(\epsilon)\Phi_2(\Omega)$.

From (3) and (5) we have (2). Inequality (2) implies that the values of F_ϵ are compact and as it is easy to see they are also convex and nonempty. This concludes the proof. □

As a consequence of the previous result we have the following.

Corollary 1: *Assume that the assumptions $(S_0), (A_0) \div (A_6)$ are satisfied, then the operator $(\epsilon, x, y) \to F_\epsilon(x, y)$ is $\overline{\chi}$-condensing with respect to all the variables.*

Proof: As it is easy to see $\overline{\nu}(\Omega) \leq \overline{\chi}(\Omega)$. Therefore, if for a bounded set $\Omega \subset C_T(E_1) \times Q$ we have

$$\overline{\chi}(\Omega) \leq \overline{\chi}\left(\bigcup_{\epsilon\in[0,1]} F_\epsilon(\Omega)\right) \tag{6}$$

then by Theorem 1 we obtain

$$\overline{\nu}(\Omega) \leq H\overline{\nu}(\Omega). \tag{7}$$

Applying inequality (7) m times we get

$$\overline{\nu}(\Omega) \leq H^m\overline{\nu}(\Omega). \tag{8}$$

By condition (A_6) we have that $\| H^m \| \to 0$ as $m \to \infty$, hence from (8) we derive $\overline{\nu}(\Omega) = 0$. If we apply Theorem 1 by means of (6) we have $\overline{\chi}(\Omega) \leq H\overline{\nu}(\Omega) = 0$. Therefore $\overline{\chi}(\Omega) = 0$ and so Ω is relatively compact in $C_T(E_1) \times Q$. □

Now, using standard methods of the topological degree theory for multivalued condensing operators in locally convex spaces (see [4]) we can derive the following existence result for system (1).

Theorem 2: *Assume that condition (S_0) and $(A_0) \div (A_6)$ are satisfied and assume that there exists a relative, open, bounded set $U \subset C_T(E_1) \times Q$ for which*

$$\deg(I - F_0, \overline{U}) \neq 0.$$

Then there exists $\epsilon_0 > 0$ such that for all $\epsilon \in [0, \epsilon_0]$ the set Σ_ϵ of the solutions of system (1) belonging to the set U is nonempty and upper semicontinuous with respect to ϵ in the $C_T(E_1) \times {}^w L^1_T(E_2)$ topology.

3. EXAMPLE

In what follows we provide an example illustrating how the assumptions on the Nemytskii operators $\Phi_i, i = 1, 2$, presented in the previous Section can be verified. This will be done by specifying a possible choice and the properties of the nonlinear operators f_i, which generate $\Phi_i, i = 1, 2$. Specifically, we consider the following form for $f_i, i = 1, 2$.

$$f_1(t,x,y) = \psi_1(t,x) + b_{11}(x)y \tag{9}$$

$$f_2(t,x,y) = \psi_2(t,x) + b_{21}(x)y + b_{22}y. \tag{10}$$

We assume the following conditions.

(a_0) The multivalued operators ψ_i: $\mathbf{R} \times E_1 \to Kv(E_i)$ are T-periodic in t, and for any $x \in E_1$ there exists a selection $g(t) \in \psi_i(t,x)$, for a.a. $t \in \mathbf{R}$, belonging to $L^1_T(\mathbf{R})$, $i = 1,2$.
(a_1) For a.a. $t \in \mathbf{R}$ the operators $\psi_i(t,\cdot), i = 1,2$, are upper semicontinuous.
(a_2) There exist positive constants l_{i1} such that

$$\chi_{E_i}(\psi_i(t,\Omega)) \leq l_{i1}\chi_{E_i}(\Omega), \quad i = 1,2.$$

(a_3) There exist positive constants M_i such that

$$\| \psi_i(t,x) \|_{E_i} \leq M_i, \quad i = 1,2.$$

We now formulate the assumptions on the operators

$$b_{i1} : E_1 \to LK(E_2, E_i), \quad i = 1,2, \tag{11}$$

where $LK(E_2, E_i)$ denotes the space of linear compact operators acting from E_2 to E_i.

(a_4) There exist positive constants $m_{i1}, i = 1,2$, such that

$$\| b_{i1}(x) \|_i \leq m_{i1}$$

for any $x \in E_1$. Here $\| \cdot \|_i$ denotes the operator norm in $LK(E_2, E_i)$.
(a_5) The maps $x \to b_{i1}(x), i = 1,2$, are continuous.
(a_6) There exists a positive constant l_{22} such that the bounded linear operator b_{22}: $E_2 \to E_2$ satisfies

$$\| b_{22} \| \leq l_{22}. \tag{12}$$

(a_7) Finally, we assume the following

$$\frac{l_{11}}{\gamma_1} < 1 \quad \text{and} \quad \frac{m_{22} + l_{22}}{\gamma_2} < 1. \tag{13}$$

Let $R > 0$ sufficiently large and let

$$Q_R := \left\{ y \in L^1_T(E_2) : \| y(t) \|_{E_2} \leq R, \quad \text{for a.a.} \quad t \in \mathbf{R} \right\}.$$

Now we prove that by (13) we get: $\Lambda_2(\epsilon)\Phi_2(C_T(E_1) \times Q_R) \subset Q_R$. This was already noticed in Remark 1 (with $Q = Q_R$). For this, let $\epsilon > 0$ and for a.a. $t \in \mathbf{R}$ we have

$$\| \Lambda_2(\epsilon)\Phi_2(x,y)(t) \|_{E_2} \leq \frac{1}{\epsilon}\int_{-\infty}^{t} e^{-(1/\epsilon)\gamma_2(t-s)}[M_2 + (m_{22} + l_{22})R]\, ds$$
$$\leq \frac{M_2 + (m_{22} + l_{22})R}{\gamma_2}.$$

For $\epsilon = 0$, for a.a. $t \in \mathbf{R}$ we have

$$\| \Lambda_2(0)\Phi_2(x,y)(t) \|_{E_2} \leq \frac{M_2 + (m_{22} + l_{22})R}{\gamma_2}.$$

By (13) if $R > \frac{M_2}{\gamma_2 - m_{22} - l_{22}}$ then we get the conclusion.

Let us prove now that the multivalued operators Φ_i are upper semicontinuous from $C_T(E_1) \times {}^wL^1_T(E_2)$ to $Kv\text{-}w(L^1_T(E_i)), i = 1,2$. First observe that under our assumptions on ψ_i we have that the associated Nemytskii operator are upper semicontinuous (see [13]). Therefore, it is sufficient to verify that the operators

$$(x,y) \to b_{i1}(x(\cdot))y(\cdot)$$

are continuous in the topologies which we have introduced in the previous section. From (a_5) we have that if $x_n \to x_0$ in $C_T(E_1)$ and $y_n \to y_0$ in ${}^wL^1_T(E_i)$ then

$$\langle y^*, b_{i1}(x_n)y_n \rangle \to \langle y^*, b_{i1}(x_0)y_0 \rangle. \tag{14}$$

Let us now verify conditions (A_4) and (A_5). From (a_3) we have that for any $\Omega \subset C_T(E_1)$ we have

$$\chi_{E_i}(\psi_i(t, \Omega(t))) \leq l_{i1} \sup_t \chi_{E_1}(\Omega(t)), \quad i = 1,2.$$

Therefore if

$$\Gamma_i(\Omega) := \{g : g \in L^1_T(E_i), g(t) \in \psi_i(t, x(t)), \text{ a.a. } t \in \mathbf{R} \textit{ and } x \in \Omega\},$$

then

$$\chi_{E_i}(\Gamma_i(\Omega)(t)) \leq l_{i1} \sup_t \chi(\Omega(t)).$$

Observe that $\chi_{E_i}(b_{i1}(\Omega_1(t))\Omega_2(t)) = 0$ and so by [5], (11) and (a_5) we have that for $\Omega_1 \subset C_T(E_1)$ and $\Omega_2 \subset Q_R$ we obtain

$$l_{i1} \sup_t \chi_{E_1}(\Omega_1(t)) \in WB(\Gamma_i(\Omega_1) + b_{i1}(\Omega_1)\Omega_2).$$

Consequently,

$$l_{i1} \sup_t \chi_{E_1}(\Omega_1(t)) \in WB(\Phi_1(\Omega_1 \times \Omega_2)).$$

Furthermore, observe that if $\| \Omega(t) \| \leq p(t), p \in L^1_T(\mathbf{R})$ and $\gamma \in WB$ then

$$\chi_w(\Omega) \leq \int_0^T \gamma(t)\, dt$$

and

$$\chi_w(b_{22}\Omega) \leq l_{22}\chi_w(\Omega)$$

we obtain (A_4) and (A_5) with $k_{11} = l_{11}, k_{12} = 0, k_{21} = T\, l_{21}$ and $k_{22} = l_{22}$.

Finally, it is immediate to see that all the possible solutions $(x, y) \in C_T(E_1) \times Q$ of the reduced system at $\epsilon = 0$ are bounded as follows:

$$\| x \|_{C_T(E_1)} \leq M_1 + m_{11}R \quad \text{and} \quad \| y \|_{L_T(E_2)} \leq R.$$

Hence

$$\deg(I - F_0, \overline{U}) \neq 0,$$

where $U = B_{C_T(E_1)}(0, \rho) \times Q_R$ with $\rho > M_1 + m_{11}$. This concludes the example.

REFERENCES

[1] R.R. Akhmerov, M.I. Kamenski, A.S. Potapov, A.E. Rodkina and B.N. Sadovskii (1992). Measure of noncompactness and condensing operators, *Operator Theory Advances and Applications*, **55**, Ed. Birkhäuser Verlag.

[2] A. Andreini, M.I. Kamenski and P. Nistri (2000). A result on the singular perturbation theory for differential inclusions in Banach spaces, *Topol. Methods Nonlinear Anal.*, **15**, 1–15.

[3] J. Appell, E. De Pascale, H.T. Nguyen and P.P. Zabreiko (1995). Multi-valued superpositions, *Dissertationes Mathematicae*, **CCCXLV**, Institute of Mathematics, Polish Academy of Sciences Ed., Warszawa.

[4] Ju.G. Borisovich, B.D. Gelman, A.D. Myshkis and V.V. Obukhovski (1980). Topological methods in the fixed-point theory of multivalued maps, *Russian Math. Surveys*, **35**, 65–143.

[5] J.F. Couchouron and M. Kamenski. A unified topological point of view for integro-differential inclusions. In J. Andres, L. Górniewicz and P. Nistri (Eds), *Differential Inclusions and Optimal Control*, Lecture Notes in Nonlinear Analysis, **2**, J. Schauder Center for Nonlinear Studies, N. Copernicus University.

[6] F.S. De Blasi (1977). On a property of the unit sphere in a Banach space, *Bull. Math. Soc. Sci. Math. R.S. Roumanie (N.S.)*, **21**(69), 259–262.

[7] J. Distel and Jr. Uhl (1977). Vector measures, *Mathematical Surveys* n.15, American Mathematical Society.

[8] A. Dontchev, T.Z. Donchev and I. Slavov (1996). A Tikhonov-type theorem for singularly perturbed differential inclusions, *Nonlinear Analysis TMA*, **26**, 1547–1554.

[9] A. Dontchev and V.M. Veliov (1983). Singular perturbation in Mayer's problem for linear systems, *SIAM J. Control Optim.*, **21**, 566–581.

[10] A. Dontchev and I. Slavov (1991). Upper semicontinuity of singularly perturbed differential inclusions. In H.J. Sebastian and K. Tanner (Eds), *Systems Modeling and Optimization*, Lectures Notes in Control and Information Sciences, **143**, 273–280, New York: Springer.

[11] G. Grammel (1997). Averaging of singularly perturbed systems, *Nonlinear Analysis TMA*, **28**, 1851–1865.

[12] G. Grammel (1996). Singularly perturbed differential inclusions: An averaging approach, *Set-valued Anal.*, **4**, 361–374.

[13] M.I. Kamenski, P. Nistri and P. Zecca (1996). On the periodic solution problem for parabolic inclusions with a large parameter, *Topol. Meth. Nonlin. Anal.*, **8**, 57–77.

[14] A. Pazy (1983). Semigroups of Linear Operators and Applications to Partial Differential Equations, *Applied Mathematical Sciences*, **44**, New York: Springer Verlag, Inc.

[15] M. Quincampoix (1993). Contribution a l'etude des perturbations singuliéres pour les systèmes contrôlés et les inclusions différentielles, *C.R. Acad. Sci. Paris, Série I*, **316**, 133–138.

[16] A.B. Vasil'eva, V.F. Butuzov and L.V. Kalachev (1995). The boundary function method for singular perturbation problems, *SIAM Studies in Applied Mathematics*, **14**, Philadelphia P.A.

[17] A.B. Vasil'eva, and M.G. Dmitriev (1986). Singular perturbations in optimal control problems, *Jour. of Soviet Math.*, **34**, 1579–1629.

[18] V.M. Veliov (1994). Differential inclusions with stable subinclusions, *Nonlinear Analysis TMA*, **23**, 1027–1038.

[19] V. Veliov (1997). A generalization of the Tikhonov theorem for singularly perturbed differential inclusions, *J. Dyn. Contr. Syst.*, **3**, 291–319.

SIMAA 4(2002) 227–249

16. Constrained Differential Inclusions*

Wojciech Kryszewski

Faculty of Mathematics and Computer Science, Nicholas Copernicus University ul. Chopina 12/18, 87-100 Toruń, Poland. E-mail: wkrysz@mat.uni.torun.pl

Keywords: Differential inclusion, structure of solutions, periodic solution, equilibrium

1991 Mathematical Subject Classification: 34A60, 34C25, 47H10

INTRODUCTION

The purpose of the present chapter is to survey and announce some recent results (from [7] and [20]) concerning the existence of solutions and the topological structure of the solution set to the following constrained initial value problem

$$\begin{aligned} &u' \in \varphi(t,u), \quad u(t) \in K,\ t \in J, \\ &u(0) = x_0 \in K, \end{aligned} \tag{IVP}$$

where φ: $J \times K \Rightarrow \mathbb{R}^N, J := [0,T]$ and $T > 0$, is a set-valued mapping (set-valued maps will be denoted by the arrow $\Rightarrow$) and K is a closed subset of the N-dimensional real Euclidean space $\mathbb{R}^N$. As usual, by a *solution* to (IVP) we understand a continuous function u: $J \to \mathbb{R}^N$ such that $u(0) = x_0$, $u(J) \subset K$, for which there is an intergrable function $v \in L^1(J,\mathbb{R}^N)$ such that $u(t) = x_0 + \int_0^t v(s)\,ds$ and $v(t) \in \varphi(t,u(t))$ on J. Clearly u is absolutely continuous, the derivative u' exists almost everywhere (a.e.) and $u'(t) = v(t)$ for almost all (a.a.) $t \in J$.

Along with (IVP) we shall consider the associated periodic problem

$$\begin{aligned} &u' \in \varphi(t,u), \quad u(t) \in K,\ t \in J \\ &u(0) = u(T) \in K. \end{aligned} \tag{PP}$$

Throughout the chapter we assume that the multifunction φ satisfies the following hypotheses:

(H1) For each $t \in J$, $x \in K$, the value $\varphi(t,x) \subset \mathbb{R}^N$ is *nonempty, compact and convex*;
(H2) for a.a. $t \in J$, the map $K \ni x \Rightarrow \varphi(t,x)$ is *upper semicontinuous* and, for all $x \in K$, $\varphi(\cdot,x)$ has a measurable selection, i.e., there is a measurable function w: $J \to \mathbb{R}^N$ such that $w(t) \in \varphi(t,x)$ on J;

* This work was supported by KBN Grant 2 P03A 024 16.

(H3) φ is *(uniformly) bounded*, i.e., there is $c > 0$ such that, for a.a. $t \in J$ and all $x \in K$,

$$\sup_{y \in \varphi(t,x)} |y| \le c.$$

There is no restriction of generality in considering bounded right-hand sides in (IVP) (or (PP)) instead of those having e.g., a *sublinear growth* (i.e., such that there is $c \in L^1(J, \mathbb{R})$ with $\sup_{y \in \varphi(t,x)} |y| \le c(t)(1 + |x|)$ for a.a. $t \in J$ and all $x \in K$). A standard argument involving the Gronwall inequality (comp. [25:p. 52]) implies that we may replace the right-hand side having the sublinear growth and satisfying (H1)–(H2) by a set-valued map satisfying additionally (H3) with $c = 1$ retaining the same set of solutions.

In order to state the most general existence result (sometimes called the viability or weak invariance) recall that the *Bouligand cone tangent to K at* $x \in K$ is defined as

$$T_K(x) := \left\{ u \in \mathbb{R}^N \mid \liminf_{h \to 0^+} \frac{d_K(x + hu)}{h} = 0 \right\},$$

where $d_K = d(\cdot, K)$ is the distance function, i.e., $d_K(x) = d(x, K) := \inf_{y \in K} |x - y|$, and consider the so-called *weak tangency condition*:

(TH) for any $x \in \operatorname{bd} K$, the map $\varphi(\cdot, x) \cap \operatorname{conv} T_K(x)^{\dagger}$ has a measurable selection, i.e., there is a measurable function $w_x \colon J \to \mathbb{R}^N$ such that $w_x(t) \in \varphi(t, x) \cap \operatorname{conv}\, T_K(x)$ for a.a. $t \in J$.

In this setting we have the following fundamental result.

Theorem 1: *Under assumptions (H1)–(H3) and (TH), for any $x_0 \in K$, there exists a solution to problem (IVP). The set*

$$L = \{u \in C(J, \mathbb{R}^N) \mid u'(t) \in \varphi(t, u(t)) \quad a.e. on\ J, u(0) = x_0, u(J) \subset K\}$$

of all solutions to (IVP) is compact in the space $C(J, \mathbb{R}^N)$ of continuous maps $J \to \mathbb{R}^N$.

We note that condition (TH) may be slightly weakened. Namely we may assume that, for any $x \in K$, there is a measurable function $w(\,\cdot\,)$ such that $w(t) \in \varphi(t, x)$ on J and, for any $y \in \mathbb{R}^N$ such that $|y - x| = d_K(y)$, $\langle w(t), y - x \rangle \le 0$ a.e. on J. The set of y satisfying this condition generates the so-called *proximal normal cone* to K at x.

Theorem 1 is, of course, well-known – see e.g., (Cor. 5.1 [25]) and comp. [29]. However, its usual and well-known formulation (e.g. Th. 5.1, 5.2 [25]) relies, instead of (H2) and (TH), on the following assumptions:

- φ is either upper semicontinuous or φ is a Carathéodory map, i.e., $\varphi(\cdot, x)$ measurable for all $x \in K$ and $\varphi(t, \cdot)$ is upper semicontinuous for a.a. $t \in J$;
- φ satisfies the *tangency condition*, i.e.,

$$\varphi(t, x) \cap T_K(x) \ne \emptyset \tag{1}$$

for all $(t, x) \in J \times K$.

It is clear that these assumptions imply both (H2) and (TH). In the upper semicontinuous case, the proof relies on a construction involving transfinite induction, while

† conv stands for the convex hull and bd stands for the boundary of a set.

in the Carathéodory case it is reduced to the upper semicontinuous one via the Scorza-Dragoni type argument. A simple proof using the Euler polygonal approximations of sorts is given in [7] (comp. also [14]).

In case φ is single-valued, the viability was first observed by Nagumo in 1942 and, then, it has been rediscovered several times by several authors. The set-valued (upper semicontinuous) case is due to Haddad [34]. The possibility to replace the tangency condition (1) by: $\varphi(t,x) \cap \operatorname{conv} T_K(x) \neq \emptyset$ on $J \times K$, was noted in [32,33]. Let us note that in the autonomous case (i.e., if φ does not depend on t), condition (1) is also necessary for the existence of solutions to (IVP).

The problem of a topological characterization of the set L of all solutions to (IVP) is much more complicated. In the unconstrained case ($K = \mathbb{R}^N$), the set L of all solutions is characterized in the spirit of the celebrated Aronszajn result [1]: *L is a compact R_δ-set in the space $C(J, \mathbb{R}^N)$* – see e.g., [22,23,35], and comp. (Cor. 5, [3:p. 109]). Recall here that a nonempty compact subset L of a metric space X is called an *R_δ-set* if it is the intersection of a decreasing family of compact contractible sets $L_n \subset X$ (comp. [39]). The situation changes when $K \neq R^N$. Even if φ is defined on $J \times \mathbb{R}^N$ and the *strong tangency condition* (i.e., $\varphi(t,x) \subset T_K(x)$ for all $(t,x) \in J \times K$) is satisfied, then there are solutions "living" in K as well as there may exist solutions starting in K but leaving it.

Example 2: (See Ex. 7.2 [25]) Let $N = 1$, $K = [0, +\infty)$, $f(x) = 2\sqrt{|x|}\operatorname{sgn} x$ and $x_0 = 0$. Then $f(x) \in T_K(x)$ on K but $u(t) = -t^2$ is a solution to $u' = f(u), u(0) = x_0$ such that $u(t) \notin K$ for $t > 0$.

It is clear that the condition:

(CH) for any $x \in \operatorname{bd} K$, the map $\varphi(\cdot, x) \cap C_K(x)$ has a measurable selection,

where $C_K(x)$ is the *Clarke tangent cone to K at x*, i.e.,

$$C_K(x) := \left\{ u \in \mathbb{R}^N \mid \lim_{y \to x,\, y \in K,\, h \to 0^+} \frac{d_K(y + hu)}{h} = 0 \right\},$$

implies (TH) (since $C_K(x) \subset T_K(x)$) and, thus, the existence of solutions to the initial value problem (IVP). But also (CH) does not prevent solutions of (P) to escape from K and does not give means to characterize the set of solutions (the above example applies, too).

Note that the above mentioned strong tangency condition implies that all solutions to (P) stay in K provided $\varphi(t,\cdot)$ is (locally) Lipschitz continuous (comp. Proposition 7.2 [25]). It appears that in this case the strong tangency condition is equivalent to the stronger form of condition (CH), i.e., $\varphi(t,x) \subset C_K(x)$ for $x \in K$. But again nothing interesting can be said about the structure of the set of solutions.

Example 3: (Comp. Ex. 1.1 [8]); see also [3:p. 203] and (Ex. 7.3[25]) Let $K := S_1 \cup S_{-1}$ where $S_k := \{z = (x,y) \in \mathbb{R}^2 \mid (x-k)^2 + y^2 = 1\}$. Let

$$g(x,y) = \begin{cases} (y, 1-x) & \text{for} \quad (x,y) \in S_1 \\ (-y, 1+x) & \text{for} \quad (x,y) \in S_{-1}. \end{cases}$$

For all $z \in K, g(z) \in T_K(z) = C_K(z)$ but the set of all solutions to (IVP) (with φ replaced by g and $x_0 = (0,0)$ is even not connected.

In case K is closed convex, a related result has been proven by Deimling (see, for instance, (Th. 7.2 [25] or Cor. 7.3), see also [27] and [26]). This result was generalized in

[43] where the class of proximate retracts (in the terminology of [30]) is introduced. A compact set K is a *proximate retract* if it admits a metric neighborhood retraction, i.e., for each x from a neighborhood of K, the projection $\pi_K(x) = \{y \in K \mid |y - x| = d_K(x)\}$ is single-valued (comp. Example 10).[†] Plaskacz shows that *if $K \subset \mathbb{R}^N$ is a compact proximate retract, then the set of all solutions to* (IVP) *is an R_δ-set in* $C(J, \mathbb{R}^N)$. This result was extended to the Hilbert space context (see [31]) while Deimling's to a Banach space (see [10]).

Apart from (IVP) and (PP) we shall study the following, intimately related, question. Assume that ψ: $K \Rightarrow \mathbb{R}^N$ is an *upper semicontinuous* set-valued map with closed convex values and consider the existence of equilibria of ψ, i.e., solutions to the problem

$$0 \in \psi(x), \ x \in K. \tag{EQ}$$

It is clear that equilibria do not exist in general unless some additional conditions concerning the set K and the behaviour of ψ with respect to K are not satisfied. Consider the following *tangency* condition:

(T) for any $x \in \operatorname{bd} K, \psi(x) \cap \operatorname{conv} T_K(x) \neq \emptyset$.

The best known result in this direction is

Theorem 4: ([11]) If *K is compact convex, ψ: $K \Rightarrow \mathbb{R}^N$ is upper semicontinuous and has closed convex values, then (EQ) has a solution provided condition (T) is satisfied.*

In view of the observation due to Cornet [17] one may assume that ψ is merely upper hemicontinuous and replace (T) by the so-called *normality condition*:

(N) for all $x \in K$ and all $p \in N_K(x)$, $\inf_{y \in \psi(x)} \langle p, y \rangle \leq 0$,

where $N_K(x)$ is the *normal cone to K at x*, i.e.,

$$N_K(x) = \{p \in \mathbb{R}^N \mid \langle p, x \rangle = \sup_{y \in K} \langle p, y \rangle\}$$

and ψ is *upper hemicontinuous* if, for any $p \in \mathbb{R}^N$, the (real extended) function

$$K \ni x \mapsto \sup_{y \in \psi(x)} \langle p, y \rangle \in \mathbb{R} \cup \{+\infty\}$$

is upper semicontinuous (note that any upper semicontinuous set-valued map is upper hemicontinuous, but not conversely; however an upper hemicontinuous map with compact convex values is upper semicontinuous).

The above result does not hold if convexity is omitted. To see this consider Example 3 with $\psi := g$. Therefore in order to obtain a satisfying existence result one has to impose additional assumptions on K as well as on ψ. The most general result concerning the nonconvex case involves the notion of an $\mathcal{L}$-retract and the so-called Clarke tangency condition. Following [8] we say that $K \subset \mathbb{R}^N$ is an *$\mathcal{L}$-retract* if there is a retraction r: $U \to K$ defined on a neighborhood U of K and a constant $L > 0$ such that

$$|x - r(x)| \leq L d_K(x),$$

and that ψ satisfies the *Clarke tangency* condition if

(C) for any $x \in K, \psi(x) \cap C_K(x) \neq \emptyset$.

[†] In the case of a proximate retract K, $T_K(x) = C_K(x)$ for all $x \in K$.

Theorem 5: ([8]) *Let $K \subset \mathbb{R}^N$ be a compact $\mathcal{L}$-retract with the nontrivial Euler characteristic $\chi(K) \neq 0$.*† *If $\psi\colon K \Rightarrow \mathbb{R}^N$ is an upper semicontinuous set-valued map with closed convex values satisfying (C), then problem (EQ) admits a solution.*

It was recently shown (see [20]), then this result remains valid if we assume that ψ is upper hemicontinuous. Moreover the *Clarke tangency* condition (C) may be slightly relaxed.

Theorem 5 is quite general for the class of $\mathcal{L}$-retracts is rather wide. It contains:

- closed sets bi-Lipschitz homeomorphic to closed convex ones†† and, in particular,
- closed convex sets;
- neighborhood retracts with Lipschitz retractions or compact neighborhood retracts with locally Lipschitz retractions;
- compact sets having the so-called *neighborhood lipschitzian extension property* (defined in [8]) and, in particular,
- the so-called epi-Lipschitz compact sets (defined by Rockafellar [44], see also the next section) in $\mathbb{R}^N$ and compact finite-dimensional C^1 (or Lipschitz) submanifolds of $\mathbb{R}^N$ with (or without) boundaries (or corners).

Observe that the set K considered in Example 3 is a neighborhood retract and $\chi(K) = -1$ but it is not an $\mathcal{L}$-retract.

It is clear that in general condition (C) implies (T). In case of the convex set K we see that $T_K(x) = C_K(x)$ and, thus (T) and (C) are equivalent (more generally these conditions are equivalent if K is *sleek*, i.e., if the cone field $K \ni x \Rightarrow T_K(x)$ is lower semicontinuous (see [5]). Moreover,

Proposition 6: *Tangency conditions (T) and (C) are equivalent if if ψ is single-valued and continuous.*

Proof: Indeed, suppose that ψ satisfies (T). Let $x \in \operatorname{bd} K$ and take a sequence (y_n) such that $y_n \in K$ and $y_n \to x$ as $n \to \infty$. Then

$$\psi(x) = \lim_{n\to\infty} \psi(y_n) \in \operatorname*{Lim\,inf}_{y\to x,\, y\in K} \operatorname{conv} T_K(y) = C_K(x),^{\dagger\dagger\dagger}$$

in view of (Theorem 4.1.10 [5]). Q.E.D.

These conditions are no longer equivalent in case of an $\mathcal{L}$-retract. To this end consider the following example:

Example 7: Let $K \subset \mathbb{R}^3$ be given by

$$K := \{u = (x, y, z) \mid x^2 + y^2 + z^2 \leq 2, \sqrt{x^2 + y^2} \geq z\}$$

† The notion of the Euler characteristic will be briefly recalled below.

†† The Euler characteristic of such sets is automatically nontrivial since they are contractible.

††† Recall the definition of Lim inf: Given a metric space (X, d), a set $A \subset X$ and a set-valued mapping $\Psi\colon A \Rightarrow X$, let $x \in X$ be an accumulation point of A. Then

$$z \in \operatorname*{Lim\,inf}_{y\to x,\, y\in A} \Psi(y) \Longleftrightarrow \forall\, (y_n),\ y_n \to x, y_n \in A\ \ \exists z_n \in \Psi(y_n),\ \ z_n \to z$$

or

$$\operatorname*{Lim\,inf}_{y\to x,\, y\in A} \Psi(y) = \bigcap_{\varepsilon>0} \bigcup_{\eta>0} \bigcap_{y\in B(x,\eta)\cap A} B(\Psi(y), \epsilon),$$

where $B(A, \varepsilon)$ is the open ball of a radius $\varepsilon > 0$ around a set A.

and let $\psi: K \Rightarrow \mathbb{R}^3$ be given by

$$\varphi(u) = \begin{cases} Q & \text{if} \quad u \in K \backslash R \\ \operatorname{conv}(Q \cup \{(-y, x, 0)\}) & \text{if} \quad u \in R, \end{cases}$$

where $R := \{(x, y, 1) \mid x^2 + y^2 = 1\}$ and $Q := \{(x, y, 1) \mid x^2 + y^2 \leq 1\}$. It is clear that ψ is upper semicontinuous, condition (T) holds and K is a contractible $\mathcal{L}$-retract (hence $\chi(K) = 1$). However ψ has no equilibria because $\psi(0) \cap C_K(0) = \emptyset$.

Since epi-Lipschitz sets are $\mathcal{L}$-retracts, Theorem 5 generalizes results from [19]. Observe however that if K is epi-Lipschitz, then it is shown in [19] that the nontriviality of $\chi(K)$ is also a necessary condition for the existence of equilibria.

We shall discuss results leading to a satisfying topological characterization of the set of all solutions to (IVP), the existence of solutions to (PP) as well as some results on the existence of equilibria under less restrictive boundary assumptions on closed sets defined via certain *functional constraints*.

CLASSES OF CONSTRAINT SETS

It seems that the very main assumptions (TH), (T) and/or (CH), (C) of the above stated results may be hard to verify in practice – since the computation of the Bouligand and Clarke tangent cones $T_K(x), C_K(x)(x \in K)$ involves explicitly the geometry of K. In the present chapter we shall be mainly concerned with sets K of the following form. Assume that $f: \mathbb{R}^N \to \mathbb{R}$ is a locally Lipschitz continuous functional and let

$$K = \{x \in \mathbb{R}^n \mid f(x) \leq 0\}. \tag{S}$$

We say that f is a *representing function of K and/or K is represented by f.*

In order to study (IVP), (PP) or (EQ) on the set K of the form (S) one has to compute the respective tangent cones; this may be a difficult task in general. One would prefer to deal with analytic conditions stated in terms of the controlled constraint f, for instance conditions involving the (generalized) gradient of f.

To proceed further we shall need some notions of nonsmooth analysis. Assume that f is as above. For any $x \in \mathbb{R}^N$, the *directional derivative* of f at x in the direction $u \in \mathbb{R}^N$ in the sense of Clarke is defined by the formula

$$Df(x)(u) := \limsup_{y \to x,\, h \to 0^+} \frac{f(y + hu) - f(y)}{h}$$

and the *generalized gradient* of f at x by

$$\partial f(x) := \{p \in \mathbb{R}^N \mid \langle p, u \rangle \leq Df(x)(u) \ \forall\, u \in \mathbb{R}^N\}.$$

It is well-known that the function $\mathbb{R}^N \ni u \mapsto Df(x)(u)$ is Lipschitz, subadditive and positively homogeneous; moreover $Df(x)(\,\cdot\,)$ is the *support function* of $\partial f(x)$, i.e.,

$$Df(x)(u) = \sup_{p \in \partial f(x)} \langle p, u \rangle. \tag{2}$$

Hence, the (negative) polar cone† to $\partial f(x)$,

$$\partial f(x)^- := \{u \in \mathbb{R}^N \mid \langle p, u\rangle \leq 0 \; \forall p \in \partial f(x)\} = \{u \in \mathbb{R}^N \mid Df(x)(u) \leq 0\}.$$

The set $\partial f(x)$ is convex and compact. The (real) function $(x, u) \mapsto Df(x)(u)$ is upper semicontinuous; consequently so is the set-valued map $x \Rightarrow \partial f(x) \subset \mathbb{R}^N$. It is well-known that, given an arbitrary closed set $K \subset \mathbb{R}^N, C_K(x) = \partial d_K(x)^-$.††

It is easy to provide examples showing that in general nothing whatsoever can be said with regard the relations between the cone fields $C_K(\cdot)$ and $\partial f(x)^-$ for the set K of the form (S) save the following

Proposition 8: (Comp. Proposition 16, 7.3 [4]) *If $K \subset \mathbb{R}^N$ is of the form (S), $x_0 \in \operatorname{bd} K$ and $0 \notin \partial f(x_0)$, then $\partial f(x_0)^- \subset C_K(x_0)$ (in general the inclusion is strict).*

Epi-Lipschitz Sets

A nonempty closed subset K of $\mathbb{R}^N$ is said to be *epi-Lipschitz* (or *wedged* in [14]) if, at every $x \in K$, the Clarke normal cone $N_K(x) := C_K(x)^-$ is *pointed* (i.e., $N_K(x) \cap (-N_K(x)) = \{0\}$ or, equivalently, $\operatorname{int} C_K(x) \neq \emptyset$). This important class of sets has been introduced in optimization by Rockafellar [44] and attracted recently a special attention; it includes closed convex subsets of $\mathbb{R}^N$ with nonempty interiors, C^1 submanifolds of $\mathbb{R}^N$ with boundaries (or corners) of full dimension and, more generally, sets in $\mathbb{R}^N$ defined via finite number of C^1 inequality constraints satisfying the so-called constraint qualification assumption.

Given an epi-Lipschitz set $K \subset \mathbb{R}^N$, one considers a Lipschitz continuous function $\Delta_K := d_K - d_{K^c} : \mathbb{R}^N \to \mathbb{R}$ where $K^c = \mathbb{R}^N \backslash K$. Clearly $K = \{x \in \mathbb{R}^N \mid \Delta_K(x) \leq 0\}$, i.e., K is represented by Δ_K. If $x \in \operatorname{bd} K$, then Hiriart-Urruty (see [36,37]) shows that $N_K(x) = cl\left(\bigcup_{\lambda\geq 0} \lambda \partial \Delta_K(x)\right)$ and Cornet, Czarnecki (see [18]) show that $0 \notin \partial \Delta_K(x)$ (hence, actually, $N_K(x) = \bigcup_{\lambda\geq 0} \lambda\partial\Delta_K(x)$). This follows from the inclusion

$$\partial \Delta_K(x) \subset \operatorname{conv}(N_K(x) \cap S^{N-1}), \; x \in \operatorname{bd} K,$$

where S^{N-1} stands for the unit sphere in $\mathbb{R}^N$, derived from facts due to Rockafellar [44]. Hence $C_K(x) = \partial\Delta_K(x)^-$ and the set-valued map $K \ni x \Rightarrow C_K(x)$ is lower semicontinuous.

On the other hand consider a closed set K of the form (S) such that

(E) for any $x \in \operatorname{bd} K, 0 \notin \partial f(x)$.

Then K is epi-Lipschitz since, by Proposition 8, $N_K(x) \subset \partial f(x)^{--}$ for all $x \in \operatorname{bd} K$ and $N_K(x)$ must be pointed. Thus we see that $K \subset \mathbb{R}^N$ is epi-Lipschitz if and only if it is of the form (S) and condition (E) holds.

---

† In general, given a closed $A \subset \mathbb{R}^N$, the *negative polar cone* A^- to A is defined by

$$A^- := \{p \in \mathbb{R}^N \mid \langle p, x\rangle \leq 0 \quad \text{for all } x \in A\}.$$

Then the double polar cone $A^{--} := (A^-)^- = \operatorname{cl}\left(\bigcup_{\lambda\geq 0} \lambda \cdot A\right)$. Therefore, $A^{--} = \bigcup_{\lambda\geq 0} \lambda \cdot A$ provided $0 \notin A$.

†† For the further facts from nonsmooth analysis as well as properties of tangent cones we refer the reader to the monographs [13] and [5].

Regular Sets

We say that the set K of the form (S), represented by a locally Lipschitz function $f\colon \mathbb{R}^N \to \mathbb{R}$, is *regular* if:

(i) there is an open set U such that $K \subset U$ and $0 \notin \partial f(x)$ for $x \in U\setminus K$.

The set K is *strictly regular* if

(ii) for any $x \in \operatorname{bd} K$, $\liminf_{y\to x,\, y\notin K} |||\partial f(y)||| > 0$.

Above, for $x \in \mathbb{R}^N$,

$$|||\partial f(x)||| := \inf_{p\in\partial f(x)} |p| = \inf_{p\in\partial f(x)} \sup_{u\in D(0,1)} \langle p, u\rangle^{\dagger}.$$

By (2), it is easy to see that

$$|||\partial f(x)||| = \sup_{u\in D(0,1)} \inf_{p\in\partial f(x)} \langle p, u\rangle = -\inf_{u\in D(0,1)} Df(x)(u) \tag{3}$$

and $0 \in \partial f(x)$ if and only if $|||\partial f(x)||| = 0$. Moreover, note that the function $\mathbb{R}^N \ni x \mapsto |||\partial f(x)|||$ is lower semicontinuous.

Below we shall provide examples of (strictly) regular sets but, in order to make them more familiar, let us first discuss the definition and establish some of their basic properties.

Remark 9

(1) Observe that the (strict) regularity of the set K does not depend only on K itself: it depends strongly on the behaviour of its representing function. It is not difficult to provide an example of a closed set K being represented by two different functions and such that it is (strictly) regular with respect to one of these functions and failing to have this property with respect to another. For an *a priori* given closed subset $K \subset \mathbb{R}^N$, there is even a C^∞-function representing it. Hence the above condition (i) (or (ii)) is crucial. Moreover, note that our notion of regularity differs from the notion of the so-called *tangential regularity* (see e.g., [5]).

(2) If K is regular, then $\{x \in U \mid 0 \in \partial f(x)\} \subset K$; this means that f has no critical points in $U\setminus K$. We emphasize that above we make assumptions concerning the behaviour of ∂f neither on K nor on $\operatorname{bd} K$.

(3) If K is regular and $C \subset U\setminus K$ is compact, then $\inf_{x\in C} f(x) > 0$ and, since the function $|||\partial f(\cdot)|||$ is lower semicontinuous, we gather that also $\inf_{x\in C} |||\partial f(x)||| > 0$.

(4) Assume that the set K of the form (S) satisfies the following condition:

($*$) *there is a neighborhood U of K such that* $\inf_{y\in U\setminus K} |||\partial f(y)||| > 0$.

It is clear that ($*$) implies (ii).[††] If $\operatorname{bd} K$ is compact, then condition (ii) implies ($*$). To see this observe that the function $K \ni x \mapsto \liminf_{y\to x, y\notin K} |||\partial f(y)|||$ is lower semicontinuous; hence it admits a minimum on $\operatorname{bd} K$. In view of these remarks, one may think of strict regularity in terms of the more friendly looking condition ($*$).

(5) It is clear that a strictly regular set is regular as condition (ii) implies the existence of U from condition (i).

[†] If $A \subset \mathbb{R}^N$, $\varepsilon > 0$, then $B(A,\varepsilon) := \{x \in \mathbb{R}^N | d(x, A) < \varepsilon\}$ and $D(A,\varepsilon) = \{x \in \mathbb{R}^N | d(x, A) \le \varepsilon\}$; in particular $D(0,1) = \{x | |x| \le 1\}$.

[††] Actually condition ($*$) was used in [20] to define strictly regular sets; in [20] the infinite dimensional situation was considered and condition (ii) was too weak for our purposes.

Example 10

(a) The set K from Example 3 is regular when represented by its distance function d_K. There is no way to make it strictly regular.

(b) Any set K satisfying (E) is strictly regular; in particular any epi-Lipschitz set is strictly regular. This is because $|||\partial f(\cdot)|||$ is lower semicontinuous and, hence, for any $x \in \mathrm{bd}\, K$, $\liminf_{y \to x,\, y \notin K} |||\partial f(y)||| \geq |||\partial f(x)||| > 0$.

(c) Let K be a proximate retract (see [43,30]). This means that there is a neighborhood U of K and a (continuous) retraction $r\colon U \to K$ such that $|r(x) - x| = d_K(x)$.† Observe that, for instance, *proximally smooth* sets from [15] are proximate retracts.

Take $x \in U \backslash K$, sequences $y_n \to x, h_n \to 0^+$ and let $u := r(x) - x$. Then

$$d_K(y_n + h_n u) - d_K(y_n) \leq h_n |(r(y_n) - y_n) - u| + d_K(y_n + h_n(r(y_n) - y_n)) - d_K(y_n).$$

It is easy to see that $d_K(y_n + h_n(r(y_n) - y_n)) = (1 - h_n) d_K(y_n)$. Hence

$$\limsup_{n \to \infty} \frac{d_K(y_n + h_n u) - d_K(y_n)}{h_n} \leq -|u|,$$

i.e., $Dd_K(x)(\frac{u}{|u|}) \leq -1$ and, by (3), $|||\partial d_K(x)||| \geq 1$, so condition (*) is satisfied on U. Hence K, represented by d_K, is strictly regular.

(d) In particular any closed convex $K \subset \mathbb{R}^N$, represented by d_K, is strictly regular.

(e) Plaskacz shows that any C^2-manifold (in $\mathbb{R}^N$) is a proximate retract (this is because of the tubular neighborhood theorem). However, he also provides an example of a C^1-manifold without this property. Here let us consider an a C^1-smooth map $g\colon \mathbb{R}^N \to \mathbb{R}^m$ $(m < N)$. Let $c \in \mathbb{R}^m$ be a regular value of g (i.e., $M := g^{-1}(c) \neq \emptyset$ and the derivative $g'(x)$ is surjective for all $x \in M$). Of course M is an $(N - m)$-dimensional C^1-manifold. It easily follows that M is strictly regular. To see this put $f\colon \mathbb{R}^N \to \mathbb{R}$ by $f(y) = |g(y) - c|$. Clearly M is represented by f and f being of C^1-class on $\mathbb{R}^N \backslash M$, i.e., $\partial f(y) = \{\nabla f(y)\}$ (so $|||\partial f(y)||| = |\nabla f(y)|$) for all $y \in \mathbb{R}^N \backslash M$. Finally we find that, for all $x \in M$, $\lim_{y \to x,\, y \notin M} |\nabla f(y)| = \|g'(x)\| > 0$ ($\|\cdot\|$ is the operator norm).

In particular we see that any orientable closed C^1-manifold is strictly regular.

STRUCTURE OF THE SOLUTIONS SET

Now we return to the problem of the topological characterization of the set L of all solutions to problem (IVP). Suppose that K is of the form (S) and consider the following boundary condition (*functional tangency*):

(FH) for any $x \in \mathrm{bd}\, K, \varphi(\cdot, x) \cap \partial f(x)^-$ has a measurable selection.††

Our first result is the following

† Note here that in order to check whether a closed set K is a proximate retract it is enough to show that the map $x \Rightarrow \pi_K(x) = \{y \in K \| x - y| = d_K(x)\}$ is lower semicontinuous on U; this is also equivalent to the single-valuedness of π_K on U.

†† It is clear that this condition is equivalent to: for any $x \in \mathrm{bd}K$, there is a measurable function $w\colon J \to \mathbb{R}^N$ such that $w(t) \in \varphi(t, x)$ on J and $Df(x)(w(t)) \leq 0$ a.e. on J.

Theorem 11: *Suppose that K, represented by f, satisfies condition (E). Let $\varphi\colon J \times K \Rightarrow \mathbb{R}^N$ satisfy conditions (H1)–(H3) and (FH). Then, for any $x_0 \in K$, the set L is an R_δ-set in $C(J, \mathbb{R}^N)$.*

In this situation, $\partial f(x)^- \subset C_K(x) \subset T_K(x)$ (see Proposition 8); thus (FH) implies (CH) and, by Theorem 1, $L \neq \emptyset$.

Proof: Take an arbitrary $x \in \operatorname{bd} K$. There is a measurable function $w_x\colon J \to \mathbb{R}^N$ such that $w_x(t) \in \varphi(t, x)$ on J and, for a.a. $t \in J$ (say on some $J_0 \subset J$ of full measure),

$$Df(x)(w_x(t)) \le 0. \tag{4}$$

Since $0 \notin \partial f(x)$, by (3) we have

$$\inf_{u \in D(0,1)} Df(x)(u) < 0.$$

Choose $\bar{u}_x \in D(0,1), \bar{u}_x \neq 0$ such that $Df(x)(\bar{u}_x) < 0$ and, for any positive integer n, let $u_x^n := \frac{\bar{u}_x}{2n|\bar{u}_x|}$; hence

$$|u_x^n| = \frac{1}{2n} \quad \text{and} \quad Df(x)(u_x^n) < 0. \tag{5}$$

Since w_x is bounded (by c), one easily sees that there is a simple function $\bar{w}_x^n\colon J \to \mathbb{R}^N$ such that $\bar{w}_x^n(J) \subset w_x(J_0)$ and

$$|w_x(t) - \bar{w}_x^n(t)| < \frac{1}{2n} \tag{6}$$

for any $t \in J_0$. Evidently, by (4),

$$Df(x)(\bar{w}_x^n(t)) \le 0 \tag{7}$$

for all $t \in J$. Let

$$v_x^n(t) = \bar{w}_x^n(t) + u_x^n.$$

Clearly the function $v_x^n\colon J \to \mathbb{R}^N$ is simple. Since $Df(x)(\,\cdot\,)$ is subadditive and positively homogeneous, by (7) and (5), for all $t \in J$,

$$Df(x)(v_x^n(t)) \le Df(x)(\bar{w}_x^n(t)) + Df(x)(u_x^n) < 0.$$

Since $v_x^n(\,\cdot\,)$ admits a finite number of values and, for any $w \in \mathbb{R}^N$, the function $Df(\,\cdot\,)(w)$ is upper semicontinuous, there is $\gamma_x^n \in (0, \frac{1}{n})$ such that, for all $y \in B(x, \gamma_x^n)$,

$$Df(y)(v_x^n(t)) < 0, \quad \text{for all } t \in J. \tag{8}$$

If $x \in \operatorname{int} K$, then we choose $\gamma_x^n \in (0, \frac{1}{n})$ such that $B(x, \gamma_x^n) \subset \operatorname{int} K$.

Therefore, for any n, we have constructed an open covering $\{B(x, \gamma_x^n)\}_{x \in K}$ of K; let $\{\lambda_s\}_{s \in S}$ be a locally finite partition of unity (depending on n) consisting of locally Lipschitz functions inscribed into this covering, i.e., for any $s \in S$, there is $x_s \in K$ such that the support $\operatorname{supp} \lambda_s \subset B(x_s, \gamma_{x_s}^n)$. For any $s \in S$, let $v_s^n := v_{x_s}^n$ in case $x_s \in \operatorname{bd} K$ and, in remaining cases let v_s^n be an arbitrary measurable selection of $\varphi(\cdot, x_s)$.

We define a map $g_n\colon J \times K \to \mathbb{R}^N$ by the formula

$$g_n(t, x) = \sum_{s \in S} \lambda_s(x) v_s^n(t), t \in J, x \in K.$$

Clearly, for any $t \in J, g_n(t,\cdot)$ is locally Lipschitz and, for any $x \in K, g_n(\cdot, x)$ is measurable. Hence g_n is a Carathéodory map. For any $x \in K$, let $S(x) = \{s \in S \mid x \in \operatorname{supp} \lambda_s\}$. The set $S(x)$ is finite and, actually,

$$g_n(t,x) = \sum_{s\in S(x)} \lambda_s(x) v_s^n(t).$$

If $x \in \operatorname{bd} K$ and $s \in S(x)$, then the corresponding $x_s \in \operatorname{bd} K$ (otherwise $x \in \operatorname{supp}\lambda_s \subset B(x_s, \gamma_{x_s}^n) \subset \operatorname{int} K$, contradiction). Hence, by (8), $Df(x)(v_s^n(t)) < 0$ on J and, therefore

$$Df(x)(g_n(t,x)) \leq \sum_{s\in S(x)} \lambda_s(x) Df(x)(v_s^n(t)) < 0,$$

i.e., $g_n(t,x) \in \partial f(x)^- \subset T_K(x)$ on J. In view of Theorem 1, this implies that, for any $\tau \in J$ and $z \in K$, there is a unique solution $y(\cdot\,;\tau,z) : [\tau, T] \to \mathbb{R}^N$ to the Cauchy problem

$$\begin{aligned} &u' \in g_n(t,u), \;\; u(t) \in K, \; t \in [\tau, T] \\ &u(\tau) = z \in K. \end{aligned} \qquad (P_n)$$

Take $x \in K$ and observe that, by (5) and (6), for all $s \in S(x), v_s^n(t) \in B(\varphi(t,x_s), \frac{1}{n})$ (if $x \in \operatorname{int} K$, then $v_s^n(t) \in \varphi(t,x_s)$) on J_0 and, since $|x - x_s| < \gamma_{x_s}^n < \frac{1}{n}$, we see that $v_s^n(t) \in B(\varphi(t, B(x,\frac{1}{n})), \frac{1}{n})$ on J_0. Hence, for all $t \in J_0, x \in K$,

$$g_n(t,x) \in \varphi_n(t,x) := \operatorname{cl}\operatorname{conv} D\left(\varphi\left(t, D\left(x, \frac{1}{n}\right)\right), \frac{1}{n}\right).$$

Evidently, for a.a. $t \in J$ (due to the upper semicontinuity of $\varphi(t,\cdot)$),

$$\varphi(t,x) = \bigcap_{n=1}^{\infty} \varphi_n(t,x). \qquad (9)$$

Clearly, for each n, φ_n satisfies assumptions of Theorem 1 (it is bounded; has compact convex values; for a.a. $t \in J, \varphi_n(t,\cdot)$ is upper semicontinuous and, for any $x \in K$, $g_n(\cdot, x)$ is a measurable selection of $\varphi_n(t,x) \cap T_K(x)$). Hence the set L_n of all solutions to (P) (with φ replaced by φ_n) is nonempty and compact in $C(J, \mathbb{R}^N)$.

We shall now show that L_n is contractible. To this end consider a mapping h: $L_n \times [0,1] \to C(J, \mathbb{R}^N)$ given by the formula: for $u \in L_n$ and $\alpha \in [0,1]$,

$$h(u,\alpha)(t) = \begin{cases} u(t) & \text{if} \quad t \in [0, \alpha T] \\ y(t; \alpha T, u(\alpha T)) & \text{if} \quad t \in [\alpha T, T]. \end{cases}$$

It is clear that $h(u,\alpha) \in L_n$ (i.e., h: $L_n \times [0,1] \to L_n$) and $h(u,0) = y(\cdot\,;0,x_0)$, $h(u,1) = u$. Moreover h is continuous since $y(\cdot\,;\tau,u_0)$ depends continuously on τ and u_0.

Since by (9), $L = \bigcap_{n=1}^{\infty} L_n$, we infer that L is an R_δ-set by definition. Q.E.D.

In the language of epi-Lipschitz sets we have

Corollary 12: *Assume that K is an epi-Lipschitz subset of $\mathbb{R}^N$. If φ: $J \times K \Rightarrow \mathbb{R}^N$ satisfies conditions (H1)–(H3) and (CH), then, for any $x_0 \in K$, the set of all solutions to (IVP) is an R_δ-set in $C(J, \mathbb{R}^N)$.*

In particular, if φ is a single-valued Carathéodory map, then the same holds if condition (CH) is replaced by (TH), i.e.,

$$\varphi(t,x) \in \operatorname{conv} T_K(x) \quad \textit{for all } x \in \operatorname{bd} K \quad \textit{and a.a.} \quad t \in J.$$

Proof. It is enough to put $f = \Delta_K$ in Theorem 11 and recall that $\partial\Delta_K(x)^- = C_K(x)(x \in \operatorname{bd} K)$. The last part follows by the argument used to show Proposition 6. Q.E.D.

It is obvious that epi-Lipschitz sets have nonempty interiors; thus e.g., a convex closed set is not epi-Lipschitz in general (it is epi-Lipschitz when considered as a subset of the affine subspace spanned by itself; hence Theorem 11 may serve as a generalization of the above mentioned Deimling result). Therefore Theorem 11 (or Corollary 12) is not fully satisfying (for instance note that proximate retracts are, in general, "thin", i.e., have empty interiors) and there is a need to establish a result generalizing Theorem 11 and extending the result of Plaskacz. This is the main reason for introducing the class of regular sets.

To state the analogous characterization for a regular set K we need to employ the following limit form of boundary condition (FH)

(LFH) for any $x \in \operatorname{bd} K$, the map $\varphi(\cdot, x) \cap \operatorname{Lim\,inf}_{y \to x,\, y \notin K} \partial f(y)^-$ has a measurable selection.

In the situation of the next result, assumption (LFH) cannot be replaced by assumption (FH). Example 3 provides a suitable counter example. This condition is by all means very restrictive, but is the best we could achieve under rather weak assumptions concerning the set K. Namely

Theorem 13: *Suppose that K, represented by f, is regular. Let φ satisfy conditions (H1)–(H3) and (LFH). Then, for any $x_0 \in K$, the set L of all solutions to (IVP) is an R_δ-set in $C(J, \mathbb{R}^N)$.*

Proof: For simplicity let us assume additionally that K is compact (this, in fact does not restrict the generality since when $x_0 \in K$ is fixed, then there is a compact set $K' \subset K$ such that $u(J) \subset K'$ for any $u \in L$) – the full argument may be found in [7].

Recall the definition of a regular set (condition (i)) and take an open Ω such that $\operatorname{cl}\Omega$ is compact and $K \subset \Omega \subset \operatorname{cl}\Omega \subset U$.

Step 1

For any $\varepsilon > 0$, there is $\eta > 0$ such that $B(K,\eta) \subset \Omega$ and, for any $x \in \operatorname{bd} K$ and $y \in B(x,\eta) \setminus K$, there is a measurable function $u = u_{x,y}\colon J \to \mathbb{R}^N$ such that

$$u(t) \in B(\varphi(\{t\} \times B_K(x,\varepsilon)), \varepsilon) \cap \partial f(y)^-$$

for a.a $t \in J$ (where $B_K(x,\varepsilon) := B(x,\varepsilon) \cap K$).

To see this let $\varepsilon > 0$, fix $x \in \operatorname{bd} K$ and take a measurable $w_x\colon J \to \mathbb{R}^N$ such that $w_x(t) \in \varphi(t,x) \cap \operatorname{Lim\,inf}_{y \to x, y \notin K} \partial f(y)^-$ on $J_0 \subset J$ of full measure. Since w_x is bounded, there is a simple function $\bar{w}_x\colon J \to \mathbb{R}^N$ such that $\bar{w}_x(J) \subset w_x(J_0)$ and

$$|\bar{w}_x(t) - w_x(t)| < \varepsilon/2 \quad \text{on } J_0. \tag{10}$$

It is clear that

$$\bar{w}_x(t) \in \operatorname*{Lim\,inf}_{y \to x, y \notin K} \partial f(y)^-$$

for all $t \in J_0$. Since $\bar{w}_x(\cdot)$ has finite number of values, by the definition of Lim inf, there is $0 < \eta(x) < 2\varepsilon$ such that

$$B(\bar{w}_x(t), \varepsilon/2) \cap \partial f(y)^- \neq \emptyset \tag{11}$$

for all $t \in J_0$ provided $y \in \mathbb{R}^N, y \notin K$ and $|y - x| < \eta(x)$.

Choose $x_1, \ldots, x_n \in \operatorname{bd} K$ such that $\bigcup_{i=1}^n B(x_i, \eta(x_i)/2) \supset \operatorname{bd} K$ and define

$$\eta = \min_{i=1,\ldots,n} \frac{\eta(x_i)}{2}.$$

We diminish η if necessary so that $B(K, \eta) \subset \Omega$.

Take any $x \in \operatorname{bd} K$. There is $1 \leq i_0 \leq n$ such that $x \in B(x_0, \eta(x_0)/2)$ where $x_0 := x_{i_0}$. Let $w_0 := \bar{w}_{x_0}$ and take $y \in B(x, \eta) \backslash K$. Then $y \in B(x_0, \eta(x_0))$ and, by (11), for all $t \in J$,

$$B(w_0(t), \varepsilon/2) \cap \partial f(y)^- \neq \emptyset. \tag{12}$$

There exists a measurable $u = u_{x,y}\colon J \to \mathbb{R}^N$ such that $u(t) \in B(w_0(t), \varepsilon/2) \cap \partial f(y)^-$ on J. Observe now that, by (10) and (12),

$$u(t) \in B(w_0(t), \varepsilon/2) \subset B(\varphi(t, x_0), \varepsilon) \subset B(\varphi(\{t\} \times B_K(x, \varepsilon)), \varepsilon)$$

on J_0, so u satisfies our requirements.

Fix $\varepsilon > 0$ and define $\psi = \psi_\varepsilon\colon J \times K \to \mathbb{R}^N$ by the formula

$$\psi(t, x) := \operatorname{cl} \operatorname{conv} D(\varphi(\{t\} \times D_K(x, \varepsilon)), \varepsilon)$$

for $t \in J, x \in K$. Evidently, $\varphi(t, x) \subset \psi(t, x)$ on $J \times K$ and, for a.a. $t \in J, \psi(t, \cdot)$ is upper semicontinuous, ψ is bounded (by $c + \varepsilon$) and has compact convex values. It is clear that, for any $x \in K, \psi(\cdot, x)$ has a measurable selector.

Step 2

There is an extension $\Phi = \Phi_\varepsilon\colon J \times \mathbb{R}^N \Rightarrow \mathbb{R}^N$ of ψ_ε.

Let $\{(U_s, a_s)\}_{s \in S}$ be the so-called *Dugundji system* for $\mathbb{R}^N \backslash K$ (see [9]), i.e., for all $s \in S$,

- $U_s \subset \mathbb{R}^N \backslash K, a_s \in \operatorname{bd} K$;
- if $x \in U_s$, then $|x - a_s| \leq 2d_K(x)$;
- $\mathcal{U} = \{U_s\}_{s \in S}$ is a locally finite open covering of $\mathbb{R}^N \backslash K$.

One defines

$$\Phi(t, x) = \Phi_\varepsilon(t, x) = \begin{cases} \psi(t, x) & \text{for } x \in K, t \in J \\ \sum_{s \in S} \lambda_s(x) \psi(t, a_s) & \text{for } x \notin K, t \in J, \end{cases}$$

where $\{\lambda_s\}_{s\in S}$ is a locally finite partition of unity subordinated to $\mathcal{U}$. It is routine to check that $\Phi\colon J\times\mathbb{R}^N \Rightarrow \mathbb{R}^N$ is well-defined and, for a.a. $t\in J$, the map $\Phi(t,\cdot)$ is upper semicontinuous, Φ is bounded (again by $c+\varepsilon$) and has compact convex values. Moreover, for any $x\in\mathbb{R}^N$, $\Phi(\cdot,x)$ has a measurable selection.

Now choose $\eta\in(0,\varepsilon)$ as in Step 1.

Step 3

For all $x\in B(K,\eta/2)\backslash K$, there is a measurable $w=w_x\colon J\to\mathbb{R}^N$ such that

$$w(t)\in\Phi(t,x)\cap\partial f(x)^-$$

a.e. on J.

Indeed: let $x\in B(K,\eta/2)\backslash K$, i.e., $d_K(x)<\eta/2$ and $x\notin K$. Let $S(x)=\{s\in S\mid x\in U_s\}$. It is clear that the set $S(x)$ is finite and

$$\Phi(t,x)=\sum_{s\in S(x)}\lambda_s(x)\psi(t,a_s).$$

Let $s\in S(x)$; then $|x-a_s|\le 2d_K(x)<\eta$. Hence $x\in B(a_s,\eta)\backslash K$. By Step 1, there exists a measurable $u_s=u_{a_s,x}\colon J\to\mathbb{R}^N$ such that $u_s(t)\in\psi(t,a_s)\cap\partial f(x)^-$ for a.a. $t\in J$. Define $w=w_x\colon J\to\mathbb{R}^N$ by the formula

$$w(t)=\sum_{s\in S(x)}\lambda_s(x)u_s(t)\quad\text{for}\quad t\in J.$$

Obviously w is measurable, $w(t)\in\partial f(x)^-$ (since the latter set is convex) and $w(t)\in\Phi(t,x)$ by the very definition of Φ.

For any $a\in\mathbb{R}$, let

$$f^a:=\{x\in\operatorname{cl}\Omega\mid f(x)\le a\}.$$

It is clear that f^a is closed and is of the form (S): it is represented by the locally Lipschitz function $\mathbb{R}^N\ni x\mapsto g(x)=\max\{\Delta(x),f(x)-a\}$ where $\Delta(x):=d(x,\operatorname{cl}\Omega)-d(x,\mathbb{R}^N\backslash\operatorname{cl}\Omega)$. It is clear that there is $a=a(\varepsilon)>0$ such that $f^a\subset B(K,\eta/2)$. Moreover, for any $x\in\operatorname{bd}f^a$, we have $x\in\Omega$; hence for all y from a (sufficiently small) neighborhood of x, $\Delta(y)<f(y)-a$ and thus $g(y)=f(y)-a$. It follows that $x\in\Omega\backslash K\subset U\backslash K$ and $0\notin\partial f(x)=\partial g(x)$.

Step 4

For sufficiently small $\varepsilon>0$, the set f^a ($a=a(\varepsilon)$) and $\Phi=\Phi_\varepsilon$ satisfy assumptions of Theorem 11. Therefore the set

$$L_\varepsilon:=\{u\in C(J,\mathbb{R}^N)\mid u'(t)\in\Phi_\varepsilon(t,u(t))\quad\text{for a.a}\quad t\in J, u(0)=x_0, u(J)\subset f^{a(\varepsilon)}\}$$

is an R_δ-subset of $C(J,\mathbb{R}^N)$.

Indeed, Φ satisfies conditions (H1)–(H3) on $\mathbb{R}^N$, for any $x\in f^a$, $\Phi(\cdot,x)$ has a measurable selection and, if $x\in\operatorname{bd}f^a$, then, for sufficiently small ε, in view of Step 4, $0\notin\partial g(x)$; moreover $x\in B(K,\eta/2)\backslash K$. Hence by Step 3, there is a measurable $w(\cdot)$ such that

$$w(t)\in\Phi(t,x)\cap\partial f(x)^-=\Phi(t,x)\cap\partial g(x)^-$$

on J.

It follows from the construction that if $0 < \varepsilon_1 \le \varepsilon_2$, then $\psi_{\varepsilon_1} \subset \psi_{\varepsilon_2}$ and $\Phi_{\varepsilon_1} \subset \Phi_{\varepsilon_2}$. Moreover, without loss of generality we may assume that also $a(\varepsilon_1) \le a(\varepsilon_2)$.

For a positive integer n, put

$$R_n := L_{1/n}.$$

For all n, R_n is an R_δ-set (in $C(J, \mathbb{R}^N)$) and the family $\{R_n\}_{n=1}^{\infty}$ is decreasing, i.e.,

$$R_{n+1} \subset R_n.$$

Step 5

$L = \bigcap_{n=1}^{\infty} R_n$.

Let $u \in L$. Then $u(0) = x_0$ and $u(t) \in K \subset f^{a(1/n)}$ for all $t \in J$. Since

$$u'(t) \in \varphi(t, u(t)) \subset \psi_{1/n}(t, u(t)) = \Phi_{1/n}(t, u(t))$$

for a.a. $t \in J$, we gather that $u \in R_n$.

If $u \in R_n$ for any n, then clearly $u(t) \in K$ on J; hence $u'(t) \in \Phi_{1/n}(t, u(t)) = \psi_{1/n}(t, u(t))$. This, together with the upper semicontinuity of $\varphi(t, \cdot)$ for a.a. $t \in J$, implies that $u'(t) \in \varphi(t, u(t))$ a.e. on J, i.e., $u \in L$.

Clearly L, as the intersection of a decreasing family of compact R_δ-sets, is again an R_δ-set (comp. [39]). Q.E.D.

Corollary 14: *Suppose that K of the form (S) is strictly regular. If φ satisfies conditions (H1)–(H3) and (FH), then the set of all solutions to (IVP) is a nonempty compact R_δ-set in $C(J, \mathbb{R}^N)$. In particular, if $f = d_K$, then (FH) is equivalent to (CH); in this situation if φ is a single-valued Carathéodory map, then (CH) may be replaced by (TH).*

Proof: We shall show that, for any $x \in \operatorname{bd} K$,

$$\partial f(x)^- \subset \operatorname*{Lim\,inf}_{y \to x,\, y \notin K} \partial f(y)^-;$$

hence (FH) entails (LFH) and our result follows from Theorem 13. To this end take $x \in \operatorname{bd} K, u \in \partial f(x)^-$ and suppose to the contrary that $u \notin \operatorname{Lim\,inf}_{y \to x,\, y \notin K} \partial f(y)^-$. Then there are $\varepsilon > 0$ and a sequence $(y_n)_{n=1}^{\infty}, y_n \notin K$ and $y_n \to x$ such that $B(u, \varepsilon) \cap \partial f(y_n)^- = \emptyset$. The separation theorem implies the existence of a sequence $(q_n)_{n=1}^{\infty}$ in $\mathbb{R}^N$ such that $|q_n| = 1$ and, for each n,

$$\sup_{z \in \partial f(y_n)^-} \langle q_n, z \rangle \le \langle q_n, u \rangle - \varepsilon |q_n| = \langle q_n, u \rangle - \varepsilon. \tag{13}$$

Since $\partial f(y_n)^-$ is a cone,

$$\sup_{z \in \partial f(y_n)^-} \langle q_n, z \rangle = 0, \tag{14}$$

i.e., $q_n \in \partial f(y_n)^{--}$ for all n.

We have $\partial f(y_n)^{--} = \operatorname{cl} \bigcup_{\lambda \ge 0} \lambda \partial f(y_n)$, but since $|||\partial f(y_n)||| \ge m$ for some $m > 0$, we gather that $\partial f(y_n)^{--} = \bigcup_{\lambda \ge 0} \lambda \partial f(y_n)$, i.e., for each $n, q_n = \lambda_n p_n$ where $\lambda_n \ge 0$ and

$p_n \in \partial f(y_n)$. Since $\partial f(\,\cdot\,)$ is upper semicontinuous, passing to a subsequence if necessary, we may assume that $p_n \to p \in \mathbb{R}^N$ and $p \in \partial f(x)$. Since $|p_n| \geq m$ for almost all $n, p \neq 0$; hence (again for a subsequence) $\lambda_n \to \lambda > 0$ and $q_n \to q = \lambda p \in \partial f(x)^{--}$.

Therefore, by (13) and (14), $0 < \varepsilon \leq \langle q, u \rangle \leq 0$, a contradiction. Q.E.D.

PERIODIC SOLUTIONS AND EQUILIBRIA

Assume now that K, represented by a locally Lipschitz function $f\colon \mathbb{R}^N \to \mathbb{R}$, is a compact set of the form (S). Remark 9 (4) imply that K is (resp. strictly) regular if there is an open set Ω with compact closure such that $K \subset \Omega \subset$ and, for any $x \in \operatorname{cl}\Omega \setminus K, 0 \notin \partial f(x)$ (resp. $\inf_{x \in \operatorname{cl}\Omega\setminus K} |||\partial f(x)||| \geq m > 0$).

Therefore, $M := \inf_{x \in \operatorname{bd}\Omega} f(x) > 0$ and, for any $a \in (0, M), f^a := \{x \in \operatorname{cl}\Omega \mid f(x) \leq a\} \subset \Omega$. Moreover, $\inf\{|||\partial f(x)||| \mid x \in \operatorname{cl}\Omega, f(x) \geq a\} > 0$. This shows that f^a is a compact strictly regular set (represented by $\Omega \ni x \mapsto g(x) := f(x) - a$). In fact $0 \notin \partial g(x) = \partial f(x)$ when $x \in \operatorname{bd} f^a$. After [20], let us recall

Proposition 15: *(i) Assume that K is regular. Then, for any $0 < a < M$, the set f^a is a strong deformation retract of $\Omega_M := \{x \in \operatorname{cl}\Omega \mid f(x) < M\}$.*† *In particular, if $b \in (0, M)$ and $0 < a \leq b$, then f^a is a strong deformation retract of f^b.*

(ii) If K is strictly regular, then K is a strong deformation retract of Ω_M and of f^a for all $a \in [0, M)$.

We provide a rough sketch of the proof. First observe that, according to above remarks, part (i) follows from (ii). To show (ii), one first constructs a continuous function $\mu\colon \operatorname{Dom}(f) \to \mathbb{R}$ such that, for any $x \in \operatorname{cl}\Omega_M, 0 < m \leq \mu(x) \leq |||\partial f(x)|||$. For instance, we may define μ by the formula

$$\mu(x) := \inf_{y \in \operatorname{cl}\Omega_M \setminus K} (|||\partial f(y)||| + |f(x) - f(y)|), \; x \in \mathbb{R}^N.$$

Next we construct a locally Lipschitz vector field $V\colon \mathbb{R}^N \setminus K \to \mathbb{R}^N$ such that $|V(x)| \leq 2$ and

$$\forall x \in \operatorname{cl}\Omega_M \quad \inf_{p \in \partial f(x)} \langle p, V(x) \rangle > \frac{1}{2}\mu(x).$$

The construction follows the steps of the standard construction of the so-called pseudo-gradient field well-known in the critical point theory – see e.g., [46].

For any $x \in \Omega_M \setminus K$, the Cauchy initial value problem

$$\frac{d}{dt}\sigma(t, x) = -V(\sigma(t, x))$$
$$\sigma(0, x) = x$$

---

†Recall that a closed set A of a (topological) space X is a *strong deformation retract* of X if there is a retraction $r\colon X \to A$ such that $i \circ r\colon X \to X$, where $i\colon A \to X$ is the inclusion, is homotopic relative to A with the identity id_X on X – see e.g., [45]. If A is a strong deformation retract in X, then, for any homology theory H_* (resp. cohomology H^*) the inclusion i induces an isomorphism $H_*(A) \to H_*(X)$ (resp. $H^*(X) \to H^*(A)$).

has a unique solution $\sigma(\cdot, x)$ in $\Omega_M \backslash K$ defined on the (right) maximal interval of existence $[0, T(x))$; moreover σ is continuous on the set $\{(t, x) \in [0, +\infty) \times \Omega_M \backslash K \mid 0 \le t < T(x)\}$. It is also well-known that the function T: $\Omega_M \backslash K \to [0, +\infty]$ is lower semicontinuous.

One proves that, for any $x \in \Omega_M, T(x) \le \frac{2f(x)}{m}, \sigma^*(x) := \lim_{t \to T(x)^-} \sigma(t, x)$ exists, $\sigma^*(x) \in \Omega_M$ and, actually, $f(\sigma^*(x)) = 0$. Moreover, it is shown that the function T is continuous. Hence, if we put η: $\Omega_M \times [0, 1] \to \Omega_M$ by the formula

$$\eta(x, t) = \begin{cases} x & \text{for} \quad (x, t) \in K \times [0, 1] \\ \sigma(tT(x), x) & \text{for} \quad t \in [0, 1), x \in \Omega_M \backslash K \\ \sigma^*(x) & \text{for} \quad t = 1, x \in \Omega_M \backslash K, \end{cases}$$

then η is continuous and provides a homotopy (relative to K) joining the identity (on Ω_M) to $i \circ r$ where i: $K \to \Omega_M$ is the inclusion and r: $\Omega_M \to K$ is the retraction given by the formula

$$r(x) = \begin{cases} x & \text{for} \quad x \in K \\ \sigma^*(x) & \text{for} \quad x \in \Omega_M \backslash K. \end{cases}$$

Remark 16: It appears (see [20]) that a compact strictly regular set K together with the retraction r defined above is an $\mathcal{L}$-retract. The constant L from the definition may be computed via condition $(*)$ from Remark 9 (4).

Denote by $\check{H}^*(\cdot) = \{\check{H}^q(\cdot)\}_{q \ge 0}$ the Čech cohomology with rational coefficients. If X is a topological space, then for each nonnegative integer $q \ge 0$, $\check{H}^q(X)$ is a vector space over rationals; it is well-known that if X is a compact absolute neighborhood retract (e.g., a compact neighborhood retract in a metric space), then X is of *finite-type*, i.e., for all $q \ge 0$, the space $\check{H}^q(X)$ is finite dimensional and, for almost all q, $\check{H}^q(X) = 0$. Thus, the *Euler characteristic* $\chi(X) = \sum_{q \ge 0} \dim \check{H}^q(X)$ is a well-defined integer.

Proposition 15 implies that a compact strictly regular set K is a neighborhood retract in $\mathbb{R}^N$ and, hence, is of finite-type and its Euler characteristic $\chi(K)$ is defined. The same holds for compact regular sets.

Proposition 17: *If a compact set $K \subset \mathbb{R}^N$ is regular, then the Euler characteristic $\chi(K)$ is well-defined.*

Proof: Using the notation introduced at the beginning of this section, we see that for sufficiently large positive integer n, say $n \ge n_0, K_n := f^{1/n} \subset \Omega_M$. Moreover, for $n \ge n_0, K_{n+1} \subset K_n$ and, by Proposition 15, K_{n+1} is a strong deformation retract of K_n. Therefore there is an isomorphism

$$\check{H}^q(K_n) \cong \check{H}^q(K_{n+1}) \tag{15}$$

for all $q \ge 0$; this isomorphism is induced by the inclusion $K_{n+1} \hookrightarrow K_n$. The Čech cohomology has the property of *continuity*: it means, in particular, that for all $q \ge 0$,

$$\check{H}^q(K) \cong \lim_{\substack{\longrightarrow \\ n \to \infty}} \check{H}^q(K_n),$$

i.e., $\check{H}^q(K)$ is isomorphic to the direct limit of the direct system $\{\check{H}^q(K_n); i_{mn}\}_{n \ge m \ge n_0}$ where i_{mn}: $\check{H}^q(K_m) \to \check{H}^q(K_n)$ is induced by the inclusion $K_n \hookrightarrow K_m$ $(m \le n)$. Since, by

(15), i_{mn} is an isomorphism, we easily gather that $\check{H}^q(K) \cong \check{H}^q(K_n)$ for all $q \geq 0$ and $n \geq n_0$. Recalling that K_n is a compact strictly regular set (and, therefore, is of finite-type) we complete the proof. Q.E.D.

The above stated results concerning the structure of solutions to problem (IVP) enable us to obtain certain criteria of the existence of solutions to problem (PP). First we need some additional facts from the fixed point theory of set-valued maps. Let X be a metric space. A set-valued map $\ell: X \times [0,1] \Rightarrow X$ is *decomposable* if there is a Banach space Y and two maps $S: X \times [0,1] \Rightarrow Y$, $g: Y \to X$ where g is (single-valued) continuous and S is upper semicontinuous with values being R_δ-sets and $\ell = g \circ S$. Decomposable maps have been considered in e.g., [6]; they fall into the class of *admissible* maps studied in e.g., [30] (the rich bibliography is given therein). The following result holds true (see [6,43] and comp. [30] for even more general result).

Proposition 18: *If X is a compact absolute neighborhood retract, $\chi(X) \neq 0$, $\ell: X \times [0,1] \Rightarrow X$ is decomposable, $\ell(x,0) = \{x\}$ on X, then $\ell(\cdot,1)$ has a fixed point, i.e., there is $x \in X$ such that $x \in \ell(x,1)$.*

The proof in [6] uses the so-called graph approximations (for the most general results in this direction—see [41]) while in order to get this result for the class of admissible maps one employs the methods of algebraic topology.

Theorem 19: *Let K, represented by a locally Lipschitz function $f: \mathbb{R}^N \to \mathbb{R}$, be a compact regular set and suppose that φ satisfies conditions (H1)–(H3) and (LFH). Then there is a solution to the two point boundary problem (PP) provided $\chi(K) \neq 0$.*

Proof: Let $\varepsilon > 0$. We shall use the notation from the proof of Theorem 13. There is $0 < a = a(\varepsilon) < \varepsilon$ such that $f^a \subset \Omega$ and $\Phi = \Phi_\varepsilon$ satisfies the assumptions of Theorem 11 on f^a. In view of Propositions 15 and 17, f^a is a neighborhood retract with the nontrivial Euler characteristic.

Fix $x \in f^a$ and let $L_\varepsilon(x)$ be the set of all solutions to problem $(P)_\varepsilon$ (that is problem (P) with K, φ and x_0 replaced by f^a, Φ and x). Evidently $L_\varepsilon(x)$ is an R_δ-set in $C(J, \mathbb{R}^N)$.

For any $\tau \in [0,T]$, let $e_\tau: C(J,\mathbb{R}^N) \to \mathbb{R}^N$ be given by the formula $e_\tau(u) = u(\tau)$. It is clear that a set-valued map $\ell: [0,T] \times f^a \Rightarrow f^a$ (given by $\ell(\tau,x) = e_\tau(L_\varepsilon(x)), \tau \in [0,T], x \in f^a$) is decomposable. Since, of course $\ell(0,x) = \{x\}$ on K, Proposition 18 implies that there is $x_\varepsilon \in f^a$ such that $x_\varepsilon \in \ell(T, x_\varepsilon)$, i.e., there is a solution u_ε to problem $(P)_\varepsilon$ such that $u_\varepsilon(0) = x_\varepsilon = u_\varepsilon(T)$. The compactness of $\operatorname{cl}\Omega$ implies that letting $\varepsilon \to 0$ (and passing to a subsequence if necessary), $x_\varepsilon \to x \in K$. Clearly the corresponding solutions u_ε are equicontinuous; hence (passing to a subsequence if necessary) $u_\varepsilon \to u \in C(J,\mathbb{R}^N)$ uniformly as $\varepsilon \to 0$, u is absolutely continuous and $u'_\varepsilon \to u'$ weakly in $L^1(J,\mathbb{R}^N)$. By the Convergence Theorem (see Th. 3.6 [4]), u is the required solution to problem (PP). Q.E.D.

Corollary 20: *Let K be a compact strictly regular set of the form (S). If $\chi(K) \neq 0, \varphi$ satisfies (H1)–(H3) and (FH), then (PP) admits a solution.*

Corollary 21: *If K is a compact epi-Lipschitz set with $\chi(K) \neq 0, \varphi$ satisfies (H1)–(H3) and (CH), then (PP) has a solution. The same holds if φ is a single-valued Carathéodory map and (CH) is replaced by (TH).*

Let us now pass to problem (EQ).

Corollary 22: *If K is a compact regular set with the nontrivial Euler characteristic and $\psi\colon K \Rightarrow \mathbb{R}^N$ is an upper semicontinuous set-valued map with nonempty compact convex values such that*

$$\forall x \in \operatorname{bd} K \quad \psi(x) \cap \operatorname*{Lim\,inf}_{y \to x,\, y \notin K} \partial f(y)^- \neq \emptyset,$$

then ψ has and equilibrium, i.e., *there is $x_0 \in K$ such that $0 \in \psi(x_0)$.*

The same holds if K is strictly regular (and, in particular, K satisfies (E) or K is epi-Lipschitz) and

$$\forall x \in \operatorname{bd} K \quad \psi(x) \cap \partial f(x)^- \neq \emptyset.$$

Proof: It is enough to consider a sequence $(u_n(0))$ where u_n is a solution to the problem $u_n'(t) \in \psi(u_n(t))$ a.e. on $[0, 1/n]$ such that $u_n(0) = u_n(1/n)$ existing in view of Theorem 19. This sequence has an accumulation point $x_0 \in K$. It is clear that $0 \in \psi(x_0)$. Q.E.D.

The second part of Corollary 22 recovers the main result from [19].

Corollary 22 may be easily generalized to the case of upper hemicontinuous ψ.

Theorem 23: *Assume that $K \subset \mathbb{R}^N$ is a compact regular set with the nontrivial Euler characteristic and let $\psi\colon K \Rightarrow \mathbb{R}^N$ be an upper hemicontinuous map with nonempty closed convex values such that*

$$\forall x \in \operatorname{bd} K \quad \psi(x) \cap \operatorname*{Lim\,inf}_{y \to x,\, y \notin K} \partial f(y)^- \neq \emptyset.$$

Then ψ has an equilibrium, i.e., there is $x_0 \in K$ such that $0 \in \psi(x_0)$.

The same holds if K is strictly regular and the above condition is replaced by

$$\forall x \in \operatorname{bd} K \quad \psi(x) \cap \partial f(x)^- \neq \emptyset. \tag{16}$$

Proof: The second statement follows from the first one since, in case K is strictly regular, for any $x \in \operatorname{bd} K$,

$$\partial f(x)^- \subset \operatorname*{Lim\,inf}_{y \to x,\, y \notin K} \partial f(y)^-.$$

In order to proceed, observe that, for any $\varepsilon > 0$, both the map

$$K \ni x \Rightarrow \operatorname{cl} \operatorname{conv} D(\psi(D_K(x, \varepsilon)), \varepsilon)$$

and its extension Ψ (defined similarly as Φ_ε in Step 2 of the proof of Theorem 13) are upper hemicontinuous. Therefore, using arguments similar to those of the proof of Theorem 19, without loss of generality we may assume that K satisfies condition (E) and that ψ satisfies condition (16) (being an analogue of (FH) stated in terms of ψ).

Now suppose to the contrary that ψ has no zeros in K. The separation theorem implies that, for each $x \in K$, there is $p_x \in \mathbb{R}^N$ with $|p_x| = 1$ such that $\inf_{y \in \psi(x)} \langle p_x, y \rangle > 0$. The upper hemicontinuity of ψ implies that $U(x) := \{z \in K \mid \inf_{y \in \psi(z)} \langle p_x, y \rangle > 0\}$ is open (in K) and the collection $\{U(x)\}_{x \in K}$ forms an open covering of K. Let $\{\lambda_s\}_{s \in S}$ be a locally finite partition of unity inscribed into this cover, i.e., for any $s \in S$, there is $x_s \in K$ such that $\operatorname{supp} \lambda_s \subset U(x_s)$. Let, for $x \in K, p(x) = \sum_{s\in S} \lambda_s(x) p_s$ where $p_s := p_{x_s}$. Clearly $p\colon K \to \mathbb{R}^N$ is continuous and

$$\forall x \in K \quad \inf_{y\in\psi(x)} \langle p(x), y\rangle > 0. \tag{17}$$

We claim that there is $x_0 \in \operatorname{bd} K$ such that

$$\sup_{q\in\partial f(x_0)^-} \langle p(x_0), q\rangle \le 0. \tag{18}$$

Indeed, otherwise, for any $x \in \operatorname{bd} K$, there is $q_x \in \partial f(x)^-$ such that $\langle p(x), q_x\rangle > 0$. Using arguments similar to those used at the beginning of the proof of Theorem 11, we easily construct a continuous map $q\colon K \to \mathbb{R}^N$ such that $\langle p(x), q(x)\rangle > 0$ on K and $q(x) \in \partial f(x)^-$ on $\operatorname{bd} K$. But q fulfills assumptions of Corollary 22 and, hence, has a zero. This is a contradiction.

But we see that, in view of (16), (18) contradicts (17) and our result is proved.

Q.E.D.

By inspection of the proof we easily see that assumption (16) may be slightly weakened. Namely it is enough to assume that

$$\forall x \in \operatorname{bd} K \quad \inf_{y\in\psi(x)} Df(x)(y) \le 0. \tag{19}$$

Clearly (19) is equivalent to (16) in case ψ has compact values.

In view of Remark 16, Corollary 21, Theorem 22 and the above observation constitute an important contribution to Theorem 5. Namely it is worthwhile to note that, in case of a compact strictly regular set one may replace the Clarke tangency (C) by condition (19) stated in terms of the functional representing K being much easier to verify in practice.

To get immediate applications of the previous results let us consider the following situation. Suppose that $V\colon \mathbb{R}^N \to \mathbb{R}$ is a continuously differentiable *coercive* functional, i.e., such that

$$\lim_{|x|\to+\infty} V(x) = +\infty.$$

Hence, for any $\alpha \in \mathbb{R}$, the sublevel set $V^\alpha := \{x \in \mathbb{R}^N \mid V(x) \le \alpha\}$ is compact. Further suppose that there is $r_0 > 0$ such that the gradient $\nabla V(x) \neq 0$ for $x \in \mathbb{R}^N$ with $|x| \ge r_0$. Therefore, for any $r \ge r_0$, the Brouwer degree $\deg(\nabla V, B(0,r), 0)$ of ∇V on the ball $B(0,r)$ with respect to 0 is defined and, in fact, does not depend on r. Let us put

$$\operatorname{ind} V := \deg(\nabla V, B(0,r), 0) \quad r \ge r_0.$$

The coercivity of V implies that $\operatorname{ind} V = 1$ (see Th. 12.9[40]). For an arbitrary $M > \sup_{|x|\le r_0} V(x)$, we have $D(0, r_0) \subset V^M$ and, for $x \in \operatorname{bd} V^M = \{x \in \mathbb{R}^N \mid V(x) = M\}, \nabla V(x) \neq 0$. Hence V^M is a C^1-smooth manifold with boundary $\operatorname{bd} V^M$ and, for $x \in \operatorname{bd} V^M, T_{V^M}(x) = \{y \in \mathbb{R}^N \mid \langle y, \nabla V(x)\rangle \le 0\} = \{\nabla V(x)\}^-$.

The celebrated Poincaré-Hopf formula (see e.g., Th. 3.1 [12] or Ch. 5[38]) asserts that the Euler charactersitic $\chi(V^M) = \operatorname{ind} V = 1$ since one may (in view of the Sard theorem) assume that critical points of V (i.e., zeros of ∇V) are isolated.

Theorem 24: *Let φ: $J \times \mathbb{R}^N \Rightarrow \mathbb{R}^N$ satisfy (H1)–(H3) and assume that V is a guiding potential for φ (a Lyapunov fuction) in the sense that there is $r_1 > 0$ such that, for any $x \in \mathbb{R}^N$ with $|x| \geq r_1$, there is a measurable function w:$J \to \mathbb{R}^N$ such that $w(t) \in \varphi(t, x)$ on J and $\langle \nabla V(x), w(t) \rangle \leq 0$ a.e. on J.*† *Then (PP) admits a solution.*

Proof: Taking large M (such that $M > \sup_{|x| \leq r} V(x)$ where $r = \max\{r_0, r_1\}$), we see that the set $K := V^M$ is a compact set of the form (S) (it is represented by $f(x) = V(x) - M$) satisfying (E); so in particular K is a strictly regular set with $\chi(K) \neq 0$. Our assumption implies that condition (FH) is satisfied. Hence the assertion follows from Corollary 20 (or 21). Q.E.D.

In the same spirit we obtain

Theorem 25: *Suppose that ψ: $\mathbb{R}^N \Rightarrow \mathbb{R}^N$ is upper hemicontinuous and has closed convex values. If*

$$\inf_{y \in \psi(x)} \langle \nabla V(x), y \rangle \leq 0 \quad \text{for} \quad |x| \geq r_1,$$

then ψ has an equilibrium.

One may easily replace the differentiability of V by the assumption that V is locally Lipschitz and consider its generalized gradient instead of ∇V in order to obtain direct generalizations of Theorems 23 and 24 (the presented proofs remain unchanged). In this case the only (serious) difficulty is to show that $\chi(V^M) = \operatorname{ind} V$ (a version of the Poincaré-Hopf formula). It is done by means of the smoothing technique due to Ćwiszewski and the present author which will appear in [21].

FINAL REMARKS

The detailed proofs of the main results of this chapter, i.e., Theorems 1, 11, 13, 17, 23 and Corollaries 12, 14, 22 will appear in [7]. Auxiliary results: Propositions 15 and 17 are presented in [20].

Theorem 4 [11] and its improvement due to Cornet [17] are true in an arbitrary Banach space, so does Theorem 5 [8] and its generalization due to Ćwiszewski and Kryszewski ([20] where the infinite-dimensional counterpart of the notion of a regular set was introduced). In [20] the possibility to relax the compactness assumption is thoroughly studied.

The infinite-dimensional versions of results concerning the R_δ-characterization of the solution sets to (IVP) are known in the unconstrained case (see[28]) and when K is convex ([10]); when the constraint set K in nonconvex some results are also true – see the forthcoming chapter *On the solution sets of differential inclusions and the periodic problem in Banach spaces* by R. Bader and the present author.

A version of Theorem 24 appears in [30] (see also the rich bibliography there). The possibility to consider guiding potentials being merely locally Lipschitz is discussed in [24] in a different setting.

† This holds, in particular, if φ is a Carathéodory map and, for any $|x| \geq r_1$, $\inf_{y \in \varphi(t,x)} \langle \nabla V(x), y \rangle \leq 0$ for a.a. $t \in J$.

REFERENCES

[1] N. Aronszajn (1942). Le correspondant topologique de l'unicite dans la théorie des équations differentielles, *Ann. of Math.*, **43**, 730–738.

[2] J.-P. Aubin (1991). *Viability Theory.* Birkhäuser.

[3] J.-P. Aubin and A. Cellina (1984). *Differential inclusions.* Berlin: Springer.

[4] J.-P. Aubin and I. Ekeland (1987). *Applied Nonlinear Analysis.* New York: Wiley- Interscience.

[5] J.-P. Aubin and H. Frankowska (1991). *Set-valued Analysis.* Boston: Birkhäuser.

[6] R. Bader and W. Kryszewski (1994). Fixed point index for compositions of set-valued maps with proximally ∞-connected values on arbitarry ANRs, *Set Valued Anal.*, **2**, 459–480.

[7] R. Bader and W. Kryszewski. On the solution sets of differential inclusions and the periodic problem, (to appear) in *Set Valued Analysis.*

[8] H. Ben-El-Mechaiekh and W. Kryszewski (1997). Equilibria of set-valued maps on nonconvex domains, *Trans. Amer. Math. Soc.*, **349**, 4159–4179.

[9] C. Bessaga and A. Pełczyński (1975). *Selected topics in infinite-dimensional topology.* Warszawa: PWN.

[10] D. Bothe (1992). Multivalued differential equations on graphs and applications, Ph.D. Thesis, Paderborn Univ.

[11] F. Browder (1968). The fixed point theory of multivalued mappings in topological vector spaces, *Math. Ann.*, **117**, 283–301.

[12] K.-C. Chang (1993). *Infinite Dimensional Morse Theory and Multiple Solution Problems.* Boston: Birkhäuser.

[13] F.H. Clarke (1983). *Optimization and nonsmooth analysis.* New York: Wiley-Interscience.

[14] F.H. Clarke, Yu. S. Ledyaev, R.J. Stern and P.R. Wolenski (1995). Qualitative properties of trajectories of control systems: a survey, *J. Dynam. Control Systems*, **1**, 1–48.

[15] F.H. Clarke, Yu. S. Ledyaev and R.J. Stern (1997). Complements, approximations, smoothings and invariance properties, *J. Convex Anal.*, **4**, 189–219.

[16] F.H. Clarke, Yu. S. Ledyaev and R.J. Stern (1995). Fixed points and equilibria in nonconvex sets, *Nonlinear Anal.*, **25**, 145–161.

[17] B. Cornet (1975). Paris avec handicaps et théorèmes de surjectivité de correspondences, *C.R. Acad. Sci. Paris Sér.*, **A281**, 479–482.

[18] B. Cornet and M.-O. Czarnecki (1997). Représentations lisses de sous-ensembles épi-lipschitziens de $\mathbb{R}^n$, *C.R. Acad. Sci. Paris Sér.*, **I325**, 475–480.

[19] B. Cornet and M.-O. Czarnecki. Necessary and sufficient conditions for the existence of (generalized) equilibria on compact epi-lipschitzian domain, submitted to *Comm. Appl. Nonlinear Anal.*

[20] A. Ćwiszewski and W. Kryszewski. Equilibria of set-valued maps: a variational approach, (to appear) in Nonlin. Anal. TMA.

[21] A. Ćwiszewski and W. Kryszewski. Approximate smoothings of locally Lipschitz functionals, (to appear) in Boll. Un. Mat. Ital.

[22] J.L. Davy (1972). Properties of the solution set of a generalized differential equation, *Bull. Austr. Math. Soc.*, **6**, 379–398.

[23] F. De Blasi and J. Myjak (1985). On the solution sets for differential inclusions, *Bull. Acad. Polon. Sci.*, **33**, 17–23.

[24] F. De Blasi, L. Górniewicz and G. Pianigiani (1999). Topological degree and periodic solutions of differential inclusions, *Nonlinear Anal. TMA*, **37**, 217–245.

[25] K. Deimling (1992). *Multivalued differential equations.* Berlin: Walter de Gruyter.

[26] K. Deimling (1989). Extremal solutions of multivalued differential equations II, *Results in Math.*, **15**, 197–201.

[27] K. Deimling and M.R. Rao (1988). On solution sets of multivalued differential equations, *Appl. Anal.*, **30**, 129–135.

[28] R. Dragoni, J.W. Macki, P. Nistri and P. Zecca (1996). *Solution Sets of differential equations in abstract spaces.* Longman.

[29] H. Frankowska, S. Plaskacz and T. Rzeżuchowski (1995). Measurable viability theorems and Hamilton-Jacobi-Bellman equation, *J. Diff. Eq.*, **116**, 265–305.

[30] L. Górniewicz (1995). Topological approach to differential inclusions. In A. Granas and M. Frigon (Eds), *Topological Methods in Differential Equations and Inclusions*, NATO ASI Series, Kluwer Acad. Publ., 129–190.

[31] L. Górniewicz, P. Nistri and V. Obukhovski (1997). Differential inclusions on proximate retracts of Hilbert spaces, *Int. J. Nonlinear Diff. Eq. TMA*, **3**, 13–26.

[32] H.G. Guseinov, A.I. Subbotin and V.N. Ushakov (1985). Derivatives for multivalued mappings with applications to game-theoretic problems of control, *Problems of Control Information Theory*, **14**, 155–167.

[33] H.G. Guseinov and V.N. Ushakov (1990). Strongly and weakly invariant sets with respect to differential inclusions, *Diff. Ur.*, **26**, 1888–1894.

[34] G. Haddad (1981). Monotone trajectories of differential inclusions and functional differentail inclusions with memory, *Israel J. Math.* **39**, 83–100.

[35] C. Himmelberg and F. Van Vleck (1980). On the topological triviality of solution sets, *Rocky Mountain J. Math.*, **10**, 247–252.

[36] J.-B. Hiriart-Urruty (1979). New concepts in nondifferentiable programming, Journées d'Analyse Non Convexe (Mai 1977 Pau), *Bull. Soc. math. France*, **60**, 57–85.

[37] J.-B. Hiriart-Urruty (1979). Tangent cones, generalized gradients and mathematical programming, *Math. Oper. Res.*, **4**, 79–97.

[38] M.W. Hirsch (1976). *Differential Topology*. Heidelberg: Springer-Verlag.

[39] D.M. Hyman (1969). On decreasing sequences of compact absolute retracts, *Fund. Math.*, **64**, 91–97.

[40] M.A. Krasnoselskii and P.P. Zabreiko (1975). *Geometrical Methods in Nonlinear Analysis*. Moscow: Nauka (Russian).

[41] W. Kryszewski (1998). Graph-approximation of set-valued maps on noncompact domains, *Topology and Appl.*, **83**, 1–21.

[42] S. Plaskacz (1992). On the solution sets for differential inclusions, *Boll. Un. Mat. Ital.*, **6A**, 387–394.

[43] S. Plaskacz (1990). Periodic solutions of differential inclusions on compact subsets of $\mathbb{R}^n$, *J. Math. Anal. Appl.*, **148**, 202–212.

[44] R.T. Rockafellar (1979). Clarke's tangent cones and boundaries of closed sets in $\mathbb{R}^n$, *Nonlinear Anal.*, **3**, 145–154.

[45] E. Spanier (1972). *Algebraic Topology*. New York: McGraw-Hill; Glenview Illinois.

[46] M. Willem (1996). *Minimax Theorems*. Boston: Birkhäuser.

SIMAA 4(2002) 251–286

17. Nonlinear Boundary Value Problems with Multivalued Terms

Nikolaos Matzakos and Nikolaos S. Papageorgiou*

National Technical University, Department of Mathematics, Zografou Campus, Athens 157 80, Greece

Abstract: In this chapter we consider nonlinear differential inclusions, with a maximal monotone term and nonlinear set-valued boundary value terms. We develop the existence theory for this problem for the cases where the maximal monotone set-valued term is defined on all of $\mathbb{R}^N$ and is defined on a proper subset of $\mathbb{R}^N$. This is done for both the convex and nonconvex problems. The proofs are based on the theory of maximal monotone operators and on a set-valued version of the Leray-Schauder alternative theorem. Then in Section 4, for various special cases of the original boundary value problem, we examine in detail the topological structure of the solution set. In Section 5, for the Dirichlet problem, we prove the existence of extremal solutions and we also present a strong relaxation theorem. Finally in Section 6, we present several cases of interest, which fit in our general framework and illustrate the unifying power of our work. The special cases incorporate the classical Dirichlet, Neumann and periodic problems.

Keywords and Phrases: Multifunction, maximal monotone operator, coercive operator, Yosida approximation, measurable selection, compact R_δ set, strong relaxation, Dirichlet, Neumann and periodic problems

1991 AMS Subject Classification: 34B15

1. INTRODUCTION

In the last few years many researchers, motivated by various physical applications such as to non-Newtonian fluids, diffusion of flows in porous media, nonlinear elasticity and others, started studying equations in which the differential operator is not linear, but instead is nonlinear homogeneous. The homogeneity property is helpful in the derivation of a spectral theory for the corresponding nonlinear eigenvalue problem, which in turn leads to useful nonresonance conditions. The case of the so-called one dimensional p-Laplacian, which is the differential operator $x \longrightarrow (\|x'\|^{p-2}x')'$, has been investigated by number of authors. Initially they studied the Dirichlet problem, which presents fewer technical difficulties, since the differential operator is invertible and degree theoretic techniques are possible. We refer to the works of Boccardo *et al.* [4], Del Pino *et al.* [7] and Drabek [9]. More recently, people started looking at the periodic problem. In this case the lack of invertibility of the nonlinear differential operator, is a source of

* Corresponding author. E-mail: npapag@math.ntua.gr

additional technical difficulties which require new techniques. We refer to the works of Dang–Oppenheimer [6], Del Pino *et al.* [8], Fabry–Fayyad [13], Guo [15] and Manasevich–Mawhin [21]. In the works of Dang–Oppenheimer and Manasevich–Mawhin, the differential operator is more general, of the form $x \to (\varphi(x'))'$, with $\varphi(\cdot)$ a strictly monotone function. Also all the previously mentioned works on the periodic problem (with the exception of Manasevich–Mawhin [20]), deal with the scalar equation.

In this chapter, we study nonlinear vector differential equations, with a differential operator which is more general than the p-Laplacian, not necessarily homogeneous and with nonlinear boundary conditions. The perturbation term in the right hand side is multivalued and is the sum of a maximal monotone map which depends only on x and of another set-valued map, which depends on both x and x'. The maximal monotone term is in general unbounded and it need not be defined everywhere. This way our formulation incorporates problems with unilateral constraints. In this respect our work seems to be the first for quasilinear problems. Our work here extends the recent semilinear works of Kandilakis–Papageorgiou [19] and Halidias–Papageorgiou [16], where $A = 0$. Other semilinear works with $A = 0$ are those by Erbe–Krawcewicz [11], Frigon [14] and Kravvaritis–Papageorgiou [20].

The structure of the chapter is as follows. In Section 2, we highlight the background material from analysis, which is needed in order to follow the arguments of the chapter. In Section 3 we pass to the results of this chapter and develop the existence theory for both the convex and nonconvex versions of the inclusion. In Section 4, we determine the topological structure of the solution set of various special cases of the original problem. In Section 5 we study the existence of the extremal solutions (i.e., the multivalued perturbation term $F(t, x, y)$ is replaces by $\text{ext}F(t, x, y) =$ extreme points of $F(t, x, y)$). Also for the particular version of the problem, where the maximal monotone operator is not present and the boundary conditions are homogeneous Dirichlet, we prove a strong relaxation theorem. Finally is Section 6, we present several special cases of the problem, which include the classical Dirichlet, Neumann and periodic problems. This illustrates the unifying character of our work.

2. PRELIMINARIES

Our approach will be based on notions and results from multivalued analysis and nonlinear analysis. So in this section we recall some basic definitions and facts from above areas of analysis that will be used in what follows. Our basic references are the books of Hu-Papageorgiou [18] and Zeidler [24].

Let (Ω, Σ) be a measurable space and X a separable Banach space. We introduce the following notations: $P_{f(c)}(X) = \{A \subseteq X\colon \text{nonempty, closed (and convex)}\}$ and $P_{(w)k(c)}(X) = \{A \subseteq X\colon \text{nonempty, (weakly-) compact (and convex)}\}$. A multifunction $F\colon \Omega \to P_f(X)$ is said to be measurable, if for all $x \in X, \omega \to d(x, F(\omega))$ is measurable. A multifunction $F\colon \Omega \to 2^X \backslash \{\emptyset\}$ is said to be graph measurable, if $\text{GrF} = \{(\omega, x) \in \Omega \times X : x \in F(\omega)\} \in \Sigma \times B(X)$, with $B(X)$ being the Borel σ-field of X. For $P_f(X)$-valued maps, measurability implies graph measurability and the converse is true if Σ is complete (i.e., $\Sigma = \hat{\Sigma} =$ the universal σ-field). Let μ be a finite measure on (Ω, Σ). For a multifunction $F\colon \Omega \to 2^X \backslash \{\emptyset\}$ and $1 \le p \le \infty$, we define $S_F^p = \{f \in L^p(\Omega, X)\colon f(\omega) \in F(\omega)\mu - \text{a.e.}\}$. This set may be empty. For a graph measurable multifunction it is nonempty if and only if $\inf\{\|x\| : x \in F(\omega)\} \le \varphi(\omega)$ μ–a.e. on Ω, with $\varphi \in L^p(\Omega)$.

Let Y, Z be Hausdorff topological spaces. A multifunction $G: Y \to 2^Z \backslash \{\emptyset\}$ is said to be lower semicontinuous (lsc) (resp., upper semicontinuous (usc)), if for all $C \subseteq Z$ closed, the set $G^+(C) = \{y \in Y : G(y) \subseteq C\}$(resp., $G^-(C) = \{y \in Y : G(y) \cap C \neq \emptyset\}$) is closed. An usc multifunction has closed graph in $Y \times Z$, while the converse is true if the multifunction is locally compact. A multifunction which is both usc and lsc, is said to be continuous (or sometimes Vietoris continuous). If Z is a metric space, on $P_f(Z)$ we can define a generalized metric known as the Hausdorff metric by $h(C, E) = \max[\sup_{c \in C} d(c, E), \sup_{e \in E} d(e, C)], C, E \in P_f(Z)$. A multifunction $G: Y \to P_f(Z)$ which is continuous into the metric space $(P_f(Z), h)$ is said to be h-continuous. For $P_k(Z)$-valued multifunctions continuity and h-continuity are equivalent notions. Finally a multifunction $G: Y \to 2^Z \backslash \{\emptyset\}$ is compact, if it is usc and maps bounded sets into relatively compact sets.

Let X be a reflexive Banach space and X^* its dual. A map $A: D \subseteq X \to 2^{X^*}$ is said to be monotone, if for all $(x, x^*), (y, y^*) \in GrA$, we have $(x^* - y^*, x - y) \geq 0$. If $(x^* - y^*, x - y) = 0$ implies $x = y$ then we say that A is strictly monotone. The map A is said to be maximal monotone, if $(x^* - y^*, x - y) \geq 0$ for all $(x, x^*) \in GrA$, imply $(y, y^*) \in GrA$, i.e., the graph of A is maximal with respect to inclusion among the graphs of monotone maps. It is easy to see that the graph of a maximal monotone A is sequentially closed in $X \times X^*_w$ and $X_w \times X^*$ (here by X_w and X^*_w we denote the spaces X and X^* furnished with the weak topology). A map $A: X \to X^*$ which is single valued and defined everywhere, is said to be demicontinuous, if $x_n \to x$ in X, implies $A(x_n) \xrightarrow{w} A(x)$ in X^*. A monotone, demicontinuous map, is maximal monotone. A map $A: D \subseteq X \to 2^{X^*}$ is said to be coercive, if D is bounded or if D is unbounded and $\inf\{\|x^*\| : x^* \in A(x)\} \to +\infty$ as $\|x\| \to \infty$. A maximal monotone coercive map is surjective.

Let Y, Z be Banach spaces and $K: Y \to Z$. We say: (a) K is completely continuous, if $y_n \xrightarrow{w} y$ in Y, implies $K(y_n) \to K(y)$ in Z; (b) K is compact, if it is continuous and maps bounded sets into relatively compact sets. In general these two notions are not comparable. However, if Y is reflexive, then complete continuity of K implies compactness. Moreover, if Y is reflexive and K is linear, then the two notions are equivalent.

We will need the following multivalued generalization of the Leray–Schauder alternative theorem, due to Bader [2]. Let X, Y be Banach spaces, $G: X \to P_{wkc}(Y)$ an usc multifunction from X into Y_w and $K: Y \to X$ a completely continuous map. Set $\Phi = K \circ G$.

Proposition 1: If X, Y *and* Φ *are as above and* Φ *is compact, then either the set* $S = \{x \in X : x \in \lambda\Phi(x)$ for some $0 < \lambda < 1\}$ *is unbounded or* Φ *has a fixed point.*

3. EXISTENCE THEORY

Let $T = [0, b]$ the problem under consideration is the following:

$$\begin{aligned} &(\varphi(x'(t)))' \in A(x(t)) + F(t, x(t), x'(t) \text{ a.e. on } T \\ &(\varphi(x'(0)), -\varphi(x'(b))) \in \xi(x(0), x(b)). \end{aligned} \tag{1}$$

Here $\varphi: \mathbb{R}^N \to \mathbb{R}^N$ is defined by $\varphi(y) = a(\|y\|^p)\|y\|^{p-2}y$, $2 \leq p < \infty$. We introduce the following set of hypotheses for the data of (1)

$H(a)$: $a \in C(\mathbb{R}_+, \mathbb{R}_+), r \to a(r^p)r^{p-1}$ is strictly increasing and $cr^p \le a(r^p)r^p \le c_1 r^p$ for some $0 < c < c_1$ and all $r \ge 0$.

Remark: When $a(r) \equiv 1$, then the differential operator is the one-dimensional p-Laplacian. Other possibilities are the following: $a(r) = 1 + \frac{1}{(1+r)^p}$ and $a(r) = r^{\frac{1-p}{p}}\beta(r)$ with $\beta \in C(\mathbb{R}_+, \mathbb{R}_+)$ strictly increasing such that $ct \le \beta(t) \le c_1 t$

$H(A)_1$: A: $\mathbb{R}^N \to 2^{\mathbb{R}^N}$ is a maximal monotone with $\operatorname{dom} A = \mathbb{R}^N$ and $0 \in A(0)$

Remark: The hypothesis $0 \in A(0)$ is not an essential restriction, since we can always satisfy it by appropriately translating things. Also since $\operatorname{dom} A = \mathbb{R}^N$, the minimal section $A^0(x) = \inf\{\|u\| : u \in A(x)\}$ is bounded on compact sets (see Definition III.2.27 [18:p. 325]).

If hypothesis $H(A)_1$ is in force, then we impose the following conditions on the multivalued perturbation $F(t, x, y)$.

$H(F)_1$: $F{:}T \times \mathbb{R}^N \times \mathbb{R}^N \to P_{kc}(\mathbb{R}^N)$ is a mutlifunction such that:

(i) for all $x, y \in \mathbb{R}^N, t \to F(t, x, y)$ is measurable;
(ii) for almost all $t \in T, (x, y) \to F(t, x, y)$ has closed graph;
(iii) for almost all $t \in T$, all $x, y \in \mathbb{R}^N$ and all $u \in F(t, x, y)$

$$(u, x)_{\mathbb{R}^N} \ge -c_2\|x\|^p - \gamma_1\|x\|^r\|y\|^{p-r} - \eta(t)\|x\|^s,$$

with $c_2, \gamma_1 > 0, \eta \in L^1(T)_+$ and $1 \le r, s < p$;

(iv) there exists $M > 0$ such that if $\|x_0\| > M$ and $(x_0, y_0)_{\mathbb{R}^N} = 0$, then we can find $\delta > 0$ and $\theta > 0$ such that for almost all $t \in T$

$$\inf[(u, x)_{\mathbb{R}^N} + c\|y\|^p : \|x - x_0\| + \|y - y_0\| < \delta, u \in F(t, x, y)] \ge \theta > 0;$$

(v) for almost all $t \in T$, all $x, y \in \mathbb{R}^N$ and all $u \in F(t, x, y)$

$$\|u\| \le \gamma_2(t, \|x\|) + \gamma_3(t, \|x\|)\|y\|^{p-1}$$

with $\sup_{0 \le r \le k} \gamma_2(t, r) \le \eta_{1,k}(t)$ a.e. on $T, \eta_{1,k} \in L^q(T)$ $(\frac{1}{p} + \frac{1}{q} = 1)$ and

$$\sup_{0 \le r \le k} \gamma_3(t, r) \le \eta_{2,k}(t) \quad \text{a.e. on } T, \eta_{2,k} \in L^\infty(T).$$

Remark: Hypothesis $H(F)_1(iv)$ is a slight generalization of the well-known Nagumo–Hartman condition (see [17:p. 433]).

We can drop the requirement that $\operatorname{dom} A = \mathbb{R}^N$, provided we strengthen a little the growth hypothesis on $F(t, x, y)$.

$H(A)_2$: A: $\mathbb{R}^N \to 2^{\mathbb{R}^N}$ is maximal monotone and $0 \in A(0)$.

Remark: Again the condition $0 \in A(0)$ can be weakened to $0 \in \operatorname{dom} A$. Then the requirement $0 \in A(0)$ can be achieved by appropriate translation.

$H(F)_2$: $F: T \times \mathbb{R}^N \times \mathbb{R}^N \to P_{kc}(\mathbb{R}^N)$ is a multifunction such that hypotheses $H(F)_1$(i) $\to$ (iv) hold and

(v′) for almost all $t \in T$, all $x, y \in \mathbb{R}^N$ and all $u \in F(t, x, y)$

$$\|u\| \leq \gamma_2(t, \|x\|) + \gamma_3(t, \|x\|)\|y\|$$

with

$$\sup_{0 \leq r \leq k} \gamma_2(t, r) \leq \eta_{1,k}(t) \quad \text{a.e on } T, \quad \eta_{1,k} \in L^2(T) \text{ and}$$

$$\sup_{0 \leq r \leq k} \gamma_3(t, r) \leq \eta_{2,k}(t) \quad \text{a.e on } T \ \eta_{2,k} \in L^{\frac{2p}{p-2}}(T).$$

The hypotheses on the boundary multifunction are the following:

$H(\xi)$: $\xi: \mathbb{R}^N \times \mathbb{R}^N \to 2^{\mathbb{R}^N \times \mathbb{R}^N}$ is a maximal monotone map such that $(0, 0) \in \xi(0, 0)$ and one of the following conditions holds:

(i) for every $(k', d') \in \xi(k, d)$, we have $(k', k)_{\mathbb{R}^N} \geq 0$ and $(d', d)_{\mathbb{R}^N} \geq 0$; or
(ii) $\text{dom}\xi = \{(k, d) \in \mathbb{R}^N \times \mathbb{R}^N : k = d\}$.

H_0: for all $[(k, d), (k', d')] \in Gr\xi$, we have $(A_\lambda(k), k')_{\mathbb{R}^N} + (A_\lambda(d), d')_{\mathbb{R}^N} \geq 0$ for all $\lambda > 0$ (here $A_\lambda(\cdot)$ denotes the Yosida approximation of A (see Remark III, 2.30 [18:p. 327]).

Remark: If $\xi = \partial\vartheta$ with $\vartheta: \mathbb{R}^N \times \mathbb{R}^N \to \mathbb{R}$ continuous convex (hence locally Lipschitz), then if by $\partial_i\vartheta, i = 1, 2$, we denote the partial subdifferential of $\vartheta(k, d)$ with respect to k and ϑ respectively, then $\partial\vartheta(k, d) \subseteq \partial_1\vartheta(k, d) \times \partial_2\vartheta(k, d)$. In this setting hypothesis H_0 follows from the inequalities $\varphi(J_\lambda(k), d) \leq \varphi(k, d)$ and $\varphi(k, J_\lambda(d)) \leq \varphi(k, d)$, with $J_\lambda(x)$ being the resolvent map corresponding to A, i.e., $J_\lambda(x) = (I + \lambda A)^{-1}x$ (see Theorem III.8.5 [18:p. 414]). In general hypothesis H_0 links the monotone operator A of the inclusion, with the monotone map ξ of the boundary conditions.

It is easy to verify that $y \to \varphi(y) = a(\|y\|^p)\|y\|^{p-2}y$ is strictly monotone and continuous. Therefore $\psi = \varphi^{-1} : \mathbb{R}^N \to \mathbb{R}^N$ is well-defined map.

Lemma 2: $\psi: \mathbb{R}^N \to \mathbb{R}^N$ *is continuous*

Proof: Let $u_n \to u$ as $n \to \infty$. Since φ is maximal monotone (being monotone and continuous), so is ψ. Thus ψ is locally bounded (see Theorem III.1.21 [18:p. 306]). So if $x_n = \psi(u_n), n \geq 1$, by passing to a subsequence if necessary we may assume that $x_n \to x$. Since $\varphi(x_n) = u_n$, we have $\varphi(x) = u$, thus $x = \psi(u)$ which proves the continuity of ψ. □

We start by examining the following auxiliary problem:

$$\begin{gathered} -(\varphi(x'(t)))' + \|x(t)\|^{p-2}x(t) = h(t) \quad \text{a.e. on } T \\ (\varphi(x'(0)), -\varphi(x'(b))) \in \xi(x(0), x(b)), h \in L^q(T, \mathbb{R}^N). \end{gathered} \tag{2}$$

Proposition 3: *If* $\xi:\mathbb{R}^N \times \mathbb{R}^N \to 2^{\mathbb{R}^N \times \mathbb{R}^N}$ *is a maximal monotone with* $(0, 0) \in \xi(0, 0)$ *and* $h \in L^q(T, \mathbb{R}^N)$*, then problem (2) has a unique solution* $x \in C^1(T, \mathbb{R}^N)$.

Proof: For given $v, w \in \mathbb{R}^N$, we consider the following two point boundary value problem:

$$\begin{aligned} &-(\varphi(x'(t)))' + \|x(t)\|^{p-2}x(t) = h(t) \quad \text{a.e. on } T \\ &x(0) = v, x(b) = w. \end{aligned} \tag{3}$$

Let $\mu(t) = (1 - \frac{t}{b})v + \frac{t}{b}w$, so that $\mu(0) = v$, $\mu(b) = w$. Let $y(t) = x(t) - \mu(t)$ and rewrite problem (3) as a homogeneous Dirichlet problem in terms of y:

$$\begin{aligned} &(\varphi((y+\mu))'(t))' + \|(y+\mu)(t)\|^{p-2}(y+\mu)(t) = h(t) \quad \text{a.e. on } T \\ &y(0) = y(b) = 0. \end{aligned} \tag{4}$$

We will solve (4). Then $x = y + \mu$ will be the solution of (3). To this end let $a_1 \colon W_0^{1,p}(T, \mathbb{R}^N) \to W^{-1,q}(T, \mathbb{R}^N)$ be defined by

$$\langle a_1(x), y \rangle = \int_0^b a(\|(x+\mu)'(t)\|^p)\|(x+\mu)'(t)\|^{p-2}((x+\mu)'(t), y'(t))_{\mathbb{R}^N} dt$$

for all $x, y \in W_0^{1,p}(T, \mathbb{R}^N)$. Here and in what follows by $\langle \cdot, \cdot \rangle$ denote the duality brackets for the pair $(W_0^{1,p}(Z), W^{-1,q}(Z))$. Using the strict monotonicity of $r \to a(r^p)r^{p-1}$ it is straightforward to check that $\langle a_1(x) - a_1(y), x - y \rangle \geq 0$ for all $x, y \in W_0^{1,p}(T, \mathbb{R}^N)$ (i.e., a_1 is monotone) and the inequality is strict, unless $\|(x+\mu)'(t)\| = \|(y+\mu)'(t)\|$ a.e on T. Also let $J_1 \colon W_0^{1,p}(T, \mathbb{R}^N) \to L^q(T, \mathbb{R}^N)$ be defined by

$$\langle J_1(x), y \rangle = \int_0^b \|(x+\mu)(t)\|^{p-2}(x(t) + \mu(t), y(t))_{\mathbb{R}^N} dt, \quad x, y \in W_0^{1,p}(T, \mathbb{R}^N).$$

Again we can easily check that for all $x, y \in W_0^{1,p}(T, \mathbb{R}^N)$, we have $\langle J_1(x) - J_1(y), x - y \rangle \geq 0$ and the inequality is strict, unless $\|(x+\mu)(t)\| = \|(y+\mu)(t)\|$ for all $t \in T$. Set $V_1(x) = a_1(x) + J_1(x)$

Then $\langle V_1(x) - V_1(y), x - y \rangle \geq 0$ and the inequality is strict unless $\ell_1(t) = \|(x+\mu)'(t)\| = \|(y+\mu)'(t)\|$ a.e. on T and $\ell_2(t) = \|(x+\mu)(t)\| = \|(y+\mu)(t)\|$ for all $t \in T$.

Then if $\langle V_1(x) - V_1(y), x - y \rangle = 0$, we have

$$\int_0^b a(\ell_1(t)^p)\ell_1^{p-2}(t)\|(x' - y')(t)\|^2 dt + \int_0^b \ell_2(t)^{p-2}\|(x-y)(t)\|^p dt = 0$$

$$\Rightarrow x'(t) = y'(t) \quad \text{a.e. on } T \; x(t) = y(t) \quad \text{for all } t \in T$$

$$\Rightarrow V_1(\cdot) \text{ is strictly monotone.}$$

Moreover, using the extended dominated convergence theorem (see for example Appendix, Theorem A.2.54 [18:p. 907]), we can easily check that V_1 is demicontinuous. Hence V_1 is maximal monotone. Finally using hypothesis $H(a)$, for every $x \in W_0^{1,p}(T, \mathbb{R}^N)$ we have

$$\begin{aligned} \langle V_1(x), x \rangle \geq{}& \beta_1(\|(x+\mu)'\|_p^p - 1) + \|x+\mu\|_p^p \\ &- \beta_2(\|(x+\mu)'\|_p^{p-1} + 1) - \|x+\mu\|_p^{p-1}\|\mu\|_p \end{aligned}$$

for some $\beta_1, \beta_2 > 0$. Thus V_1 is coercive and so we can apply Theorem III. 2.17, of Hu-Papageorgiou [18:p. 322] and obtain $y \in W_0^{1,p}(T, \mathbb{R}^N)$ such that $V_1(y) = h$. The strict monotonicity of V_1 implies y is unique. Evidently y solves (4) and so $x = y + \mu \in W^{1,p}(T, \mathbb{R}^N)$ solves (3). Note that $a(\|x'(\cdot)\|^p)\|x'(\cdot)\|^{p-2}x'(\cdot) = \varphi(x'(\cdot)) \in W^{1,q}(T, \mathbb{R}^N) \subseteq C(T, \mathbb{R}^N)$, hence $\psi(\varphi(x'(\cdot))) = x'(\cdot) \in C(T, \mathbb{R}^N)$ (see Lemma 2 and so $x \in C^1(T, \mathbb{R}^N)$. Let $\vartheta\colon \mathbb{R}^N \times \mathbb{R}^N \to C^1(T, \mathbb{R}^N)$, be the map which to each pair $(v, w) \in \mathbb{R}^N \times \mathbb{R}^N$ assigns the unique solution $\vartheta(v, w) \in C^1(T, \mathbb{R}^N)$ of (3) established above.

Let $\rho\colon \mathbb{R}^N \times \mathbb{R}^N \to \mathbb{R}^N \times \mathbb{R}^N$ be defined by

$$\rho(k, d) = (-\varphi(\vartheta(k, d)'(0)), \varphi(\vartheta(k, d)'(b))).$$

Claim 1: *ρ is monotone.*
Let $x = \vartheta(k, d)$ and $y = \vartheta(k_1, d_1)$. Using Green's identity, we have

$$\begin{aligned}
&(\rho(k,d) - \rho(k_1,d_1), \begin{pmatrix} k - k_1 \\ d - d_1 \end{pmatrix})_{\mathbb{R}^{2N}} \\
&= (-\varphi(x'(0)) + \varphi(x_1'(0)), k - k_1)_{\mathbb{R}^N} + (\varphi(x'(b)) - \varphi(x_1'(b)), d - d_1)_{\mathbb{R}^N} \\
&= \int_0^b (a(\|x'(t)\|^p)\|x'(t)\|^{p-2}(x'(t), x'(t) - x_1'(t))_{\mathbb{R}^N} dt \\
&\quad - \int_0^b (a(\|x_1'(t)\|^p)\|x_1'(t)\|^{p-2}(x_1'(t), x'(t) - x_1'(t))_{\mathbb{R}^N} dt \\
&\quad + \int_0^b ((a(\|x'(t)\|^p)\|x'(t)\|^{p-2}x'(t)) - a(\|x_1'(t)\|^p)\|x_1'(t)\|^{p-2}x'(t))', x'(t) \\
&\quad - x_1'(t))_{\mathbb{R}^N} dt
\end{aligned}$$

Note that since $r \to a(r^p)r^{p-1}$ is strictly increasing, we have

$$\begin{aligned}
&\int_0^b (a(\|x'(t)\|^p)\|x'(t)\|^{p-2}(x'(t), x'(t) - x_1'(t))_{\mathbb{R}^N} dt \\
&- \int_0^b (a(\|x_1'(t)\|^p)\|x_1'(t)\|^{p-2}(x_1'(t), x'(t) - x_1'(t))_{\mathbb{R}^N} dt \\
&\geq \int_0^b (a(\|x'(t)\|^p)\|x'(t)\|^{p-1} - a(\|x_1'(t)\|^p)\|x_1'(t)\|^{p-1}(\|x'(t)\| - \|x_1'(t)\|) dt \geq 0
\end{aligned}$$

Also we have

$$\begin{aligned}
&\int_0^b ((a(\|x'(t)\|^p)\|x'(t)\|^{p-2}(x'(t) - a(\|x_1'(t)\|^p)\|x_1'(t)\|^{p-2}x_1'(t))', x'(t) - x_1'(t))_{\mathbb{R}^N} dt \\
&= \int_0^b (\|x(t)\|^{p-2}x(t) - \|x_1(t)\|^{p-2}x_1(t), x(t) - x_1(t))_{\mathbb{R}^N} dt \\
&\geq 2^{2-p}\|x - x_1\|_p^p \geq 0,
\end{aligned}$$

$\Rightarrow \rho$ is monotone.

Claim 2: *ρ is continuous.*
Let $k_n \to k$ and $d_n \to d$ in $\mathbb{R}^N$. Let $x_n = \vartheta(k_n, d_n), n \geq 1$, and $x = \vartheta(k, d)$. As before we introduce the maps $\mu_n(t) = (1-\frac{t}{b})k_n + \frac{t}{b}d_n, n \geq 1$, and $\mu(t) = (1-\frac{t}{b})k + \frac{t}{b}d$. Let $y_n = x_n - \mu_n$. We have:

$$-\int_0^b ((a(\|(y_n+\mu_n)'(t)\|^p)\|(y+\mu)'(t)\|^{p-2}(y_n+\mu_n)'(t))', y_n(t))_{\mathbb{R}^N}dt$$

$$+\int_0^b (\|(y_n+\mu_n)(t)\|^{p-2})(y_n+\mu_n)(t), y_n(t)_{\mathbb{R}^N}dt = \int_0^b (h(t), y_n(t))_{\mathbb{R}^N}dt$$

$$\Rightarrow \int_0^b (a(\|(y_n+\mu_n)'(t)\|^p)\|(y_n+\mu_n)'(t)\|^{p-2}((y_n+\mu_n)'(t), y_n'(t))_{\mathbb{R}^N}dt$$

$$+\int_0^b \|(y_n+\mu_n)(t)\|^{p-2}((y_n+\mu_n)(t), y(t))_{\mathbb{R}^N}dt = \int_0^b (h(t), y_n(t))_{\mathbb{R}^N}dt$$

(by Green's identity)

$$\Rightarrow \beta_3(\|(y_n+\mu_n)'\|_{p-1}^p - 1) - \beta_4(\|(y_n+\mu_n)'\|_p^{p-1} - 1) - \|y_n+\mu_n\|_p^p - \|y_n+\mu_n\|_p^{p-1}\|\mu_n\|_p$$

$$\leq \|h\|_q\|y_n+\mu_n\|_p + \beta_5 \text{ for some } \beta_3, \beta_4, \beta_5 > 0,$$

$$\Rightarrow \{x_n = y_n + \mu_n\}_{n\geq 1} \subseteq W^{1,p}(T, \mathbb{R}^N) \text{ is bounded}$$

Then using hypothesis $H(a)$, we have that $\{a(\|x_n'\|^p)\|x_n'\|^{p-2}x_n'\}_{n\geq 1} \subseteq L^q(T, \mathbb{R}^N)$ is bounded and so $\{\|x_n\|^{p-2}x_n\}_{n\geq 1} \subseteq L^q(T, \mathbb{R}^N)$. Moreover, directly from equation (3), it follows that $\{(a(\|x_n'\|^p)\|x_n'\|^{p-2}x_n')'\}_{n\geq 1} \subseteq L^q(T, \mathbb{R}^N)$ is bounded. Thus by passing to an appropriate subsequence if necessary, we may assume that

$$x_n \xrightarrow{w} z \text{ in } W^{1,p}(T, \mathbb{R}^N) \quad \text{and} \quad a(\|x_n'\|^p)\|x_n'\|^{p-2}x_n' \xrightarrow{w} v \text{ in } W^{1,p}(T, \mathbb{R}^N)$$

$$\Rightarrow x_n \to z \text{ in } C(T, \mathbb{R}^N) \quad \text{and} \quad (a(\|x_n'\|^p)\|x_n'\|^{p-2}x_n')' \xrightarrow{w} v' \text{ in } L^q(T, \mathbb{R}^N).$$

Hence by passing to the limit as $n \to \infty$

$$-v(t) + \|z(t)\|^{p-2}z(t) = h(t) \text{ a.e. on } T$$

$$z(0) = k, z(b) = d$$

Also $a(\|x_n'(t)\|^p)\|x_n'(t)\|^{p-2}x_n'(t) \to^w v(t)$ in $C(T, \mathbb{R}^N)$, hence $\psi(a(\|x_n'\|^p)\|x_n'\|^{p-2}x_n') = x_n'(t) \to \psi(v(t))$ for all $t \in T$ and so $x_n' \to \psi(v)$ in $L^q(T, \mathbb{R}^N)$ (in fact in $C(T, \mathbb{R}^N)$). Therefore $\psi(v) = z' \Rightarrow v = \phi(z') = a(\|z'\|^p)\|z'\|^{p-2}z'$ and so it follows that $z = \varphi(k, d)$, i.e., ϑ is continuous.

From the continuity of ϑ we obtain at once the continuity of $\rho(\cdot, \cdot)$.

Claim 3: *ϱ is coercive.*

Let $x = \vartheta(k, d)$. Using Green's identity and hypothesis H(a), we have

$$\frac{\left(\rho(k,d), \begin{pmatrix} k \\ d \end{pmatrix}\right)_{\mathbb{R}^{2N}}}{\left\|\begin{pmatrix} k \\ d \end{pmatrix}\right\|} = \frac{(-\varphi(x'(0)), k)_{\mathbb{R}^N} + (\varphi(x'(b)), d)_{\mathbb{R}^N}}{\left\|\begin{pmatrix} k \\ d \end{pmatrix}\right\|}$$

$$= \frac{\int_0^b a(\|x'(t)\|^p)\|x'(t)\|^p dt + \int_0^b ((a(\|x'(t)\|^p)\|x'(t)\|^{p-2}x'(t))', x'(t))_{\mathbb{R}^N} dt}{\left\|\begin{pmatrix} k \\ d \end{pmatrix}\right\|}$$

$$\geq \frac{\beta_6(\|x'\|_p^p - 1) + \|x\|_p^p - \|h\|_q \|x\|_p}{\left\|\begin{pmatrix} k \\ d \end{pmatrix}\right\|} \quad \text{for some} \quad \beta_6 > 0$$

$$\geq \frac{\beta_7\|x\|^p - \beta_8 - \|h\|_q\|x\|}{\left\|\begin{pmatrix} k \\ d \end{pmatrix}\right\|} \quad \text{for some} \quad \beta_7, \beta_8 > 0$$

$$\geq \frac{\beta_9\|x\|_\infty^p - \beta_{10}\|x\|_\infty - \beta_8}{\left\|\begin{pmatrix} k \\ d \end{pmatrix}\right\|} \quad \text{for some} \quad \beta_9, \beta_{10} > 0$$

If $\left\|\begin{pmatrix} k \\ d \end{pmatrix}\right\| \to \infty$ then $\|x\|_\infty \to 0$ and from the above inequality if follows, then ρ is coercive.

From claims 1,2,3, we see that ρ is maximal monotone, coercive. Hence so is $\sigma = \rho + \xi$ and we can find $k, d \in \mathbb{R}^N$ such that $0 \in \sigma(k, d) \Rightarrow (\vartheta(k, d)'(0), -\vartheta(k, d)'(b)) \in \xi(k, d) \Rightarrow x = \vartheta(k, d)$ is the unique solution of (2). ☐

Now let

$$D = \{x \in C^1(T, \mathbb{R}^N) : a(\|x'(\cdot)\|^p)\|x'(\cdot)\|^{p-2}x'(\cdot) \in W^{1,q}(T, \mathbb{R}^N),$$
$$(\varphi(x'(0)), -\varphi(x'(b))) \in \xi(x(0), x(b))\}$$

and let V: $D \subseteq L^p(T, \mathbb{R}^N) \to L^q(T, \mathbb{R}^N)$ be defined by

$$V(x)(\cdot) = -(a(\|x'(\cdot)\|^p)\|x'(\cdot)\|^{p-2}x'(\cdot))', x \in D.$$

Proposition 4: *If ξ: $\mathbb{R}^N \times \mathbb{R}^N \to 2^{\mathbb{R}^N \times \mathbb{R}^N}$ is maximal monotone and $(0,0) \in \xi(0,0)$, then V is maximal monotone.*

Proof: It is easy to see that V is monotone For the maximality of V, let J: $L^p(T, \mathbb{R}^N) \to L^q(T, \mathbb{R}^N)$ be defined by $J(x)(\cdot) = \|x(\cdot)\|^{p-2}x(\cdot)$. From Proposition (3), we know that $R(V + J) = L^q(T, \mathbb{R}^N)$. We will show that this surjectivity property implies the maximality of V. Indeed suppose that for some $y \in L^p(T, \mathbb{R}^N)$ and $v \in L^q(T, \mathbb{R}^N)$, we have

$$(V(x) - v, x - y)_{pq} \geq 0 \text{ for all } x \in D$$

Let $x_1 \in D$ such that $v + J(y) = V(x_1) + J(x_1)$. We have:

$$(V(x) - V(x_1) - J(x_1) + J(y), x - x_1)_{pq} \geq 0$$
$$\Rightarrow (J(x) - J(x_1), x_1 - y)_{pq} \geq 0$$

i.e., $x_1 = y$ from the strict monotonicity of J.
Also $v = V(x_1)$ and so $(y, v) \in GrV$, which proves the maximality of V. □

Now let $\lambda > 0$, let A_λ be the Yosida approximation of A (i.e., $A_\lambda = \frac{1}{\lambda}(I - J_\lambda)$ with $J_\lambda = (I + \lambda A)^{-1}$) and consider the following approximation of problem (1)

$$\begin{cases} (\varphi(x'(t))' \in A_\lambda(x(t)) + F(t, x(t), x'(t) \text{ a.e. on } T \\ (\varphi(x'(0)), -\varphi(x'(b)) \in \xi(x(0), x(b)) \end{cases} \tag{5}$$

Proposition 5: *If hypothesis $H(a)$, $H(A)_2, H(F)_1$ or $H(F)_2$ and $H(\xi)$ hold, then problem (5) has a solution $x \in C(T, \mathbb{R}^N)$*

proof: Let $\hat{A}_\lambda$: $L^p(T, \mathbb{R}^N) \to L^q(T, \mathbb{R}^N)$ be the Nemitsky operator corresponding to A, i.e., $\hat{A}_\lambda(x)(\cdot) = A_\lambda(x(\cdot))$. Let $K_\lambda = V + \hat{A}_\lambda + J : D \subseteq L^p(T, \mathbb{R}^N) \to L^q(T, \mathbb{R}^N)$. From proposition 4 and Theorem III.3.3 of Hu-Papageorgiou [18:p. 334], we have that K_λ is maximal monotone. Also because $A_\lambda(0) = 0$, we have

$$(K_\lambda(x), x)_{pq} \geq (V(x), x)_{pq} + (J(x), x)_{pq}$$

Note that $(J(x), x)_{pq} = \|x\|_p^p$ and

$$\begin{aligned}(V(x), x)_{pq} &= -\int_0^b ((a(\|x'(t)\|^p)\|x'(t)\|^{p-2}x'(t))', x(t))_{\mathbb{R}^N} dt \\ &= -(\varphi(x'(b)), x(b))_{\mathbb{R}^N} + (\varphi(x'(0)), x(0))_{\mathbb{R}^N} + \int_0^b a(\|x'(t)\|^p)\|x'(t)\|^p dt \\ &\geq c(\|x'\|_p^p - 1),\end{aligned}$$

the last inequality being a consequence of the fact that $x \in D$ satisfies the boundary conditions and of hypothesis $H(a)$. So finally we have

$$(K_\lambda(x), x)_{pq} \geq c\|x'\|_p^p + \|x\|_p^p - c$$
$$\Rightarrow K_\lambda \text{ is coercive.}$$

But a maximal monotone coercive operator is surjective. Moreover, by virtue of the strict monotonicity of J, K_λ is one-to-one. Therefore, K_λ^{-1}: $L^q(T, \mathbb{R}^N) \to D \subseteq W^{1,p}(T, \mathbb{R}^N)$ is well-defined.

Claim 1: K_λ^{-1}: $L^q(T, \mathbb{R}^N) \to D \subseteq W^{1,p}(T, \mathbb{R}^N)$ is compact.
Let $u_n \to^w u$ in $L^q(T, \mathbb{R}^N)$ and let $x_n = K_\lambda^{-1}(u_n), n \geq 1$, and $x = K_\lambda^{-1}(u)$. We have

$$(V(x_n), x_n)_{pq} + (\hat{A}_\lambda(x_n), x_n)_{pq} + (J(x_n), x_n)_{pq} = (u_n, x_n)_{pq}$$
$$\Rightarrow c(\|x_n'\|_{p-1}^p - 1) + \|x_n\|_p^p \leq \|u_n\|_q \|x_n\|_p$$
$$\Rightarrow \{x_n\}_{n \geq 1} \subseteq W^{1,p}(T, \mathbb{R}^N) \text{ is bounded.}$$

We may assume that $x_n \xrightarrow{w} y$ in $W^{1,p}(T,\mathbb{R}^N)$ and $x_n \to y$ in $L^p(T,\mathbb{R}^N)$. Since $(x_n, u_n) \in GrK_\lambda, n \geq 1$, and GrK_λ is demiclosed (see Section 2), we have that $(y,u) \in GrK_\lambda$. So $u = K_\lambda(y) \Rightarrow u = V(y) + \hat{A}_\lambda(y) + J(y)$, i.e., $y = x$. Also we have

$$\lim(V(x_n) + \hat{A}_\lambda(x_n) + J(x_n), x_n - x)_{pq} = \lim(u_n, x_n - x)_{pq} = 0$$
$$\Rightarrow \lim(V(x_n), x_n - x) = 0$$

(since $\hat{A}_\lambda(x_n) \to \hat{A}_\lambda(x)$ and $J(x_n) \to J(x)$ in $L^q(T,\mathbb{R}^N)$) But V being maximal monotone (Proposition 4)), it is generalized pseudomonotone (see: Remark III.6.3 [18: p. 365]. So we have that

$$V(x_n) \xrightarrow{w} V(x) \text{ in } L^q(T,\mathbb{R}^N) \text{ and } (V(x_n), x_n)_{pq} \to (V(x), x)_{pq}$$

Note that $\left\{a(\|x_n'(\cdot)\|^p)\|x_n'(\cdot)\|^{p-2}x_n'(\cdot)\right\}_{n\geq 1} \subseteq W^{1,q}(T,\mathbb{R}^N)$ and

$$a(\|x_n'(\cdot)\|^p)\|x_n'(\cdot)\|^{p-2}x_n'(\cdot) \xrightarrow{w} a(\|x'(\cdot)\|^p)\|x'(\cdot)\|^{p-2}x'(\cdot) \text{ in } W^{1,q}(T,\mathbb{R}^N)$$
$$\Rightarrow a(\|x_n'(\cdot)\|^p)\|x_n'(\cdot)\|^{p-2}x_n'(\cdot) \to a(\|x'(\cdot)\|^p)\|x'(\cdot)\|^{p-2}x'(\cdot)$$

in $C(T,\mathbb{R}^N)$
Acting with ψ which is continuous, we obtain

$$x_n' \to x' \text{ in } L^p(T,\mathbb{R}^N)$$
$$\Rightarrow x_n \to x \text{ in } W^{1,p}(T,\mathbb{R}^N)$$
$$\Rightarrow K_\lambda^{-1} \text{ is compact as claimed.}$$

Next, let $N_1\colon W^{1,p}(T,\mathbb{R}^N) \to 2^{L^q(T,\mathbb{R}^N)}$ be defined by $N_1(x) = S^q_{F(\cdot,x(\cdot),x'(\cdot))}$

$$= \left\{f \in L^q(T,\mathbb{R}^N) : f(t) \in F(t,x(t),x'(t)) \text{ a.e. on } T\right\}.$$

From Halidias–Papageorgiou [16] (proof of Theorem 3, see also [14]), we know that $N_1(\cdot)$ has nonempty, w-compact and convex values and is usc for $W^{1,p}(T,\mathbb{R}^N)$ into $L^q(T,\mathbb{R}^N)$ furnished with the weak topology. Let $N = -N_1 + J : W^{1,p}(T,\mathbb{R}^N) \to P_{wkc}(L^q(T,\mathbb{R}^N))$. Using Proposition 1, we will solve the following abstract fixed point problem: $x \in K_\lambda^{-1}N(x)$.

Claim 2: $S = \left\{x \in W^{1,p}(T,\mathbb{R}^N) : x \in \beta K_\lambda^{-1}N(x), 0 < \beta < 1\right\}$ is bounded. Let $x \in S$. We have

$$V\left(\frac{1}{\beta}x\right) + \hat{A}_\lambda\left(\frac{1}{\beta}x\right) + J\left(\frac{1}{\beta}x\right) \in -N_1(x) + J(x)$$

$$\Rightarrow \left(V\left(\frac{1}{\beta}x\right), x\right)_{pq} + \left(\hat{A}_\lambda\left(\frac{1}{\beta}x\right), x\right)_{pq} + \frac{1}{\beta^{p-1}}\|x\|_p^p = -(f,x)_{pq} + \|x\|_p^p \text{ with } f \in N_1(x).$$

Since $0 = \hat{A}(0)$, we have $\left(\hat{A}_\lambda\left(\frac{1}{\beta}x\right), x\right)_{pq} \geq 0$. Also using the boundary conditions, we have

$$\begin{aligned}
\left(V\left(\frac{1}{\beta}x\right), x\right)_{pq} &= \int_0^b \left(-\left(a\left(\frac{1}{\beta^p}\|x'(t)\|^p\right)\frac{1}{\beta^{p-1}}\|x'(t)\|^{p-2}x'(t)\right)', x(t)\right)_{\mathbb{R}^N} dt \\
&\geq \int_0^b a\left(\frac{1}{\beta^p}\|x'(t)\|^p\right)\frac{1}{\beta^{p-1}}\|x'(t)\|^{p-2}(x'(t), x'(t))_{\mathbb{R}^N} dt \\
&\geq \int_0^b \beta c \frac{1}{\beta^p}\|x'(t)\|^p \text{ (hypothesis } H(a)) \\
&= \int_0^b c\frac{1}{\beta^{p-1}}\|x'(t)\|^p
\end{aligned}$$

So we obtain

$$\begin{aligned}
&\frac{c}{\beta^{p-1}}\|x'\|_p^p - c + \frac{1}{\beta^{p-1}}\|x\|_p^p \leq -(f, x)_{pq} + \|x\|_p^p \\
&\Rightarrow c\|x'\|_p^p \leq -\beta^{p-1}(f, x)_{pq} + (\beta^{p-1} - 1)\|x\|_p^p + c\beta^{p-1} \\
&\leq -\beta^{p-1}(f, x)_{pq} + c\beta^{p-1} \text{ (because } 0 < \beta < 1).
\end{aligned} \tag{6}$$

We will do the proof assuming that hypotheses $H(F)_1$ are in effect. The proof similar if $H(F)_2$ holds. Using hypothesis $H(F)_1(iii)$ we have

$$\begin{aligned}
&-\beta^{p-1}(f, x)_{pq} = \beta^{p-1}\int_0^b -(f(t), x(t))_{\mathbb{R}^N} dt \\
&\leq \beta^{p-1}c_2\|x\|_p^p + \beta^{p-1}\gamma_1\int_0^b \|x(t)\|^r\|x'(t)\|^{p-r}dt + \beta^{p-1}\|\eta\|_1\|x\|_\infty^s.
\end{aligned}$$

Let $\tau = p - r$ and set $\omega = \frac{p}{r}, \omega' = \frac{p}{\tau}$ $\left(\frac{1}{\omega} + \frac{1}{\omega'} = 1\right)$. Apply Holder's inequality we obtain

$$\begin{aligned}
&\int_0^b \|x(t)\|^r\|x'(t)\|^{p-r}dt \leq \left(\int_0^b \|x(t)\|^{r\omega}dt\right)^{\frac{1}{\omega}}\left(\int_0^b \|x'(t)\|^{\tau\omega'}dt\right)^{\frac{1}{\omega'}} \leq \|x\|_p^r\|x'\|_p^\tau \\
&\Rightarrow -\beta^{p-1}(f, x)_{pq} \leq \beta^{p-1}c_2\|x\|_p^p + \beta^{p-1}\gamma_1\|x\|_p^r\|x'\|_p^\tau + \beta^{p-1}\|\eta\|_1\|x\|_\infty^s.
\end{aligned} \tag{7}$$

We will show that for all $x \in S$, $\|x\|_\infty \leq M$. To this end let $e(t) = \|x(t)\|^p$ and let $t_0 \in T$ be the point where $e(\cdot)$ attains its maximum. Suppose $e(t_0) > M^p$ and first assume that $0 < t_0 < b$. Then $0 = r'(t_0) = p\|x(t_0)\|^{p-2}(x'(t_0), x(t_0))_{\mathbb{R}^N} \Rightarrow (x'(t_0), x(t_0))_{\mathbb{R}^N} = 0$. By virtue of hypothesis $H(F)_1(iv)$, we know that there exist $\eta, \delta > 0$ such that

$$\inf[(u, x)_{\mathbb{R}^N} + c\|y\|^p : u \in F(t, x, y), \|x - x(t_0)\| + \|y - x'(t_0)\| < \delta] \geq \theta > 0$$

Because $x \in S$, we have that $x \in D$ and so $\varphi(x'(\cdot)) \in W^{1,q}(T, \mathbb{R}^N) \Rightarrow \psi(\varphi(x'(\cdot))) = x'(\cdot) \in C(T, \mathbb{R}^N)$.

Thus we can find $\delta_1 > 0$ such that if $t_0 < t \leq t_0 + \delta_1$ we have $\|x(t) - x(t_0)\| + \|x'(t) - x'(t_0)\| < \delta$. Hence we can write that

$$(f(t), x(t))_{\mathbb{R}^N} + c\|x'(t)\|^p \geq \theta \quad \text{a.e. on } (t_0, t_0 + \delta_1] \tag{8}$$

From the equation $V(\frac{1}{\beta}x) + \hat{A}_\lambda(\frac{1}{\beta}x) + J(\frac{1}{\beta}x) = -f + J(x),\ f \in N(x)$, we have

$$-\left(a\left(\frac{1}{\beta^p}\|x'(t)\|^p\right)\frac{1}{\beta^{p-1}}\|x'(t)\|\right)' + \hat{A}_\lambda\left(\frac{1}{\beta}x(t)\right) = -f(t) + \left(1 - \frac{1}{\beta^{p-1}}\right)\|x(t)\|^{p-2}x(t)$$

a.e. on T.

Using this equation in (8) and since $0 < \beta < 1$, we obtain for almost all $t \in (t_0, t_0 + \delta_1]$

$$\begin{aligned}
&\left(\left(a\left(\frac{1}{\beta^p}\|x'(t)\|^p\right)\frac{1}{\beta^{p-1}}\|x'(t)\|^{p-2}x'(t)\right)', x(t)\right)_{\mathbb{R}^N} \\
&\quad - \left(A_\lambda\left(\frac{1}{\beta}x(t)\right), x(t)\right)_{\mathbb{R}^N} + c\|x'(t)\|^p \geq \theta \\
\Rightarrow &\int_{t_0}^{t}\left(\left(a\left(\frac{1}{\beta^p}\|x'(s)\|^p\right)\frac{1}{\beta^{p-1}}\|x'(s)\|^{p-2}x'(s)\right)', x(s)\right)_{\mathbb{R}^N} ds \\
&\quad - \int_{t_0}^{t}\left(A_\lambda\left(\frac{1}{\beta}x(s)\right), x(s)\right)_{\mathbb{R}^N} ds + c\int_{t_0}^{t}\|x'(s)\|^p ds \geq \vartheta(t - t_0), t \in (t_0, t_0 + \delta_1)
\end{aligned}$$

By Green's identity and hypothesis $H(a)$, we have

$$\begin{aligned}
&\int_{t_0}^{t}\left(\left(a\left(\frac{1}{\beta^p}\|x'(s)\|^p\right)\frac{1}{\beta^{p-1}}\|x'(s)\|^{p-2}x'(s)\right)', x(s)\right)_{\mathbb{R}^N} ds \\
&= \left(\varphi\left(\frac{1}{\beta}x'(t)\right), x(t)\right)_{\mathbb{R}^N} - \left(\varphi\left(\frac{1}{\beta}x'(t_0)\right), x(t_0)\right)_{\mathbb{R}^N} \\
&\quad - \int_{t_0}^{t} a\left(\frac{1}{\beta^p}\|x'(s)^p\|\right)\frac{1}{\beta^{p-1}}\|x'(s)\|^p ds \\
&\leq \left(\varphi\left(\frac{1}{\beta}x'(t)\right), x(t)\right)_{\mathbb{R}^N} - \beta\int_{t_0}^{t} c\frac{1}{\beta^p}\|x'(s)\|^p ds.
\end{aligned}$$

Therefore finally we can write that

$$\begin{aligned}
&\left(\varphi\left(\frac{1}{\beta}x'(t)\right), x(t)\right)_{\mathbb{R}^N} - \left(\frac{c}{\beta^{p-1}} - c\right)\int_{t_0}^{t} c\|x'(s)\|^p ds \geq \vartheta(t - t_0) \\
&\Rightarrow a\left(\frac{1}{\beta^p}\|x'(t)\|^p\right)\|x'(t)\|^{p-2}(x'(t), x(t))_{\mathbb{R}^N} \geq \vartheta(t - t_0) > 0 \quad (\text{since } 0 < \beta < 1) \\
&\Rightarrow (x'(t), x(t))_{\mathbb{R}^N} \geq 0 \text{ i.e., } r'(t) > 0 \text{ for all } t \in (t_0, t_0 + \delta_1]
\end{aligned}$$

which contradicts the choice of t_0.

If $t_0 = 0$, then $r'(0) \leq 0$ and so $(x'(0), x(0))_{\mathbb{R}^N} \leq 0$. If hypothesis $H(\xi)(i)$ holds, then $a(\|x'(0)\|)\|x'(0)\|^{p-2}(x'(0), x(0))_{\mathbb{R}^N} \geq 0 \Rightarrow (x'(0), x(0))_{\mathbb{R}^N} \geq 0 \Rightarrow (x'(0), x(0))_{\mathbb{R}^N} = 0$ and we continue as above. If hypothesis $H(\xi)(ii)$ holds, then $x(0) = x(b) \Rightarrow r(0) = r(b)$ and so $r'(0) \leq 0 \leq r'(b) \Rightarrow (x'(0), x(0))_{\mathbb{R}^N} \leq 0 \leq (x'(b), x(b))_{\mathbb{R}^N}$. Also from the boundary condition have

$$\begin{aligned}
0 \geq\ &(\varphi(x'(0)), x(0))_{\mathbb{R}^N} \geq (\varphi(x'(b)), x(b))_{\mathbb{R}^N} \geq 0 \\
\Rightarrow\ &(x'(0), x(0))_{\mathbb{R}^N} = (x'(b), x(b))_{\mathbb{R}^N} = 0 \\
\Rightarrow\ &r'(0) = r'(b) = 0
\end{aligned}$$

and we proceed as above.

The argument is similar if $t_0 = b$. So finally we have that for all $x \in S$, $\|x\|_\infty \leq M$.

Using this fact in inequality (7), we obtain

$$\beta^{p-1}(-f, x) \leq \beta_{11} + \beta_{12}\|x'\|_p^r$$

for some $\beta_{11}, \beta_{12} > 0$. Using this inequality in (6), it follows that $S \subseteq W^{1,p}(T, \mathbb{R}^N)$ is bounded. We can apply Proposition 1 to obtain $x \in D$ such that $k \in K_\lambda^{-1} N_1(x)$. Evidently $x \in W_0^{1,p}(T, \mathbb{R}^N)$ is a solution of (5).

Now we are ready to prove our first existence theorem for the problem, when $\operatorname{dom} A = \mathbb{R}^N$ (i.e., under hypothesis $H(A)_1$)

Theorem 6: *If hypotheses* H (a), $H(A)_1, H(F)_1$ *and* $H(\xi)$ *hold, then problem (1) has a solution* $x \in C^1(T, \mathbb{R}^N)$.

Proof: Let $\lambda_n \downarrow 0$ as $n \to \infty$ and let $x_n \in C^1(T, \mathbb{R}^N)$ be solutions of the auxiliary problem (5). As in the proof of Proposition 5, from the Nagumo–Hartman condition (hypothesis $H(F)_1(iv)$) we have that $\|x_n(t)\| \leq M$ for all $n \geq$ and all $t \in T$. We have

$$\begin{aligned}
&V(x_n) + \hat{A}_{\lambda_n}(x_n) = -f_n \text{ with } f_n \in N(x_n), n \geq 1, \\
&\Rightarrow (V(x_n), x_n)_{pq} + (\hat{A}_{\lambda_n}(x_n), x_n)_{pq} = -(f_n, x_n)_{pq} \\
&\Rightarrow c\|x_n'\|_p^p \leq \|f_n\|_q \|x_n\|_p \leq \beta_{13}\|f_n\|_q \text{ for some } \beta_{13} > 0.
\end{aligned}$$

By virtue hypothesis $H(F)_1(v)$ we have

$$\|f_n\|_q \leq \beta_{14} + \beta_{15}\|x_n'\|_p^{p-1} \text{ for some } \beta_{14}, \beta_{15} > 0.$$

Hence we obtain

$$\begin{aligned}
&c\|x_n'\|_p^p \leq \beta_{16} + \beta_{17}\|x_n'\|_p^{p-1} \text{ for some } \beta_{16}, \beta_{17} > 0 \\
&\Rightarrow \{x_n'\}_{n\geq 1} \subseteq L^p(T, \mathbb{R}^N)) \text{ is bounded,} \\
&\Rightarrow \{x_n\}_{n\geq 1} \subseteq W^{1,p}(T, \mathbb{R}^N) \text{ is bounded.}
\end{aligned}$$

Thus we may assume $x_n \xrightarrow{w} x$ in $W^{1,p}(T, \mathbb{R}^N)$ and $x_n \to x$ in $L^p(T, \mathbb{R}^N)$. Since $\operatorname{dom} A = \mathbb{R}^N$ (hypothesis $H(A)_1$), we have that

$$\|A_{\lambda_n}(x_n(t))\| \leq \|A^0(x_n(t))\| \leq M_1 \text{ for all } n \geq 1 \text{ and all } t \in T$$

(see Proposition III.2.29 [18:p. 325]). Thus we may assume that $\hat{A}_{\lambda_n}(x_n) \xrightarrow{w} u$ in $L^q(T, \mathbb{R}^N)$. Also $\{f_n\}_{n\geq 1} \subseteq L^q(T, \mathbb{R}^N)$ is bounded and so we may assume that $f_n \xrightarrow{w} f$ in $L^q(T\mathbb{R}^N)$. We have

$$\begin{aligned}
&\lim[(V(x_n)x_n - x)_{pq} + (\hat{A}_{\lambda_n}(x_n), x_n - x)_{pq}] = \lim(-f_n, x_n - x)_{pq} = 0 \\
&\Rightarrow \lim(V(x_n), x_n - x)_{pq} = 0 \\
&\Rightarrow V(x_n) \xrightarrow{w} V(x) \text{ in } L^p(T, \mathbb{R}^N)
\end{aligned}$$

(since V is maximal monotone, (Proposition 4)).

Thus in the limit we have

$$V(x) + u = -f$$
$$\Rightarrow (a(\|x'(t)\|^p)\|x'(t)\|^{p-2}x'(t))' + u(t) = -f(t) \text{ a.e. on } T.$$

Hence $\varphi(x'(\cdot)) \in W^{1,q}(T, \mathbb{R}^N) \Rightarrow \psi(\varphi(x'(\cdot))) = x'(\cdot) \in C(T, \mathbb{R}^N) \Rightarrow x \in C^1(T, \mathbb{R}^N)$. Also

$$(\varphi(x_n'(0)), -\varphi(x_n'(b))) \in \xi(x_n(0)), x_n(b)), n \geq 1.$$

Note that $\{\varphi(x_n'(\cdot))\}_{n\geq 1} \subseteq W^{1,q}(T, \mathbb{R}^N)$ is bounded and we already know that $\varphi(x_n'(\cdot))' \to \varphi(x'(\cdot))'$ in $L^q(T, \mathbb{R}^N)$. So $\varphi(x_n'(\cdot)) \to \varphi(x'(\cdot))$ in $W^{1,q}(T, \mathbb{R}^N)$ $\varphi(x_n'(t)) \to \varphi(x'(\cdot))$ in $C(T, \mathbb{R}^N) \Rightarrow \psi(\varphi(x_n'(t))) = x_n'(t) \to x(t) = \psi(\varphi(x'(t)))$ for all $t \in T$. Because $Gr\xi \subseteq \mathbb{R}^N \times \mathbb{R}^N$ is closed (by maximal monotonicity of ξ), we have that $(\varphi(x'(0)), -\varphi(x'(b))) \in \xi(x(0), x(b))$. If we can show that $u(t) \in A(x(t))$ a.e. on T, we will be done. To this end let $\hat{A} : \hat{D} \subseteq L^p(T, \mathbb{R}^N) \to L^q(T, \mathbb{R}^N)$ be defined by

$$\hat{A}(x) = \{g \in L^q(T, \mathbb{R}^N) : g(t) \in A(x(t)) \text{ a.e. on } T\}$$

for all

$$x \in \hat{D} = \{x \in L^p(T, \mathbb{R}^N) : \quad \text{there is} \quad g \in L^q(T, \mathbb{R}^N) \text{ such that } g(t) \in A(x(t)) \text{ a.e. on } T\}$$

the realization of A on the dual pair $(L^p(T, \mathbb{R}^N), L^q(T, \mathbb{R}^N))$. We will show that $\hat{A}$ is maximal monotone. To this end, as before, it suffices to show that $R(\hat{A} + J) = L^q(T, \mathbb{R}^N)$. So let $h \in L^p(T, \mathbb{R}^N)$. Set $R(t) = \|h(t)\|^{\frac{1}{p-1}} + 1$. Evidently $R \in L^p(T)_+$. Consider the set

$$Q(t) = \{(x, a) \in \mathbb{R}^N \times \mathbb{R}^N : a + j_p(x), a \in A(x), \|x\| \leq R(t)\}$$

with $j_p \colon \mathbb{R}^N \to \mathbb{R}^N$ being defined by $j_p(x) = \|x\|^{p-2}x$. From Theorem III. 6.28 of Hu–Papageorgiou [17:p. 371], we have that $S(t) \neq \emptyset$ for all $t \in T$. Also easy to see that $GrQ \in \mathcal{L} \times B(\mathbb{R}^N) \times B(\mathbb{R}^N)$ (with $\mathcal{L}$ being the Lebesque σ-field of T and $B(\mathbb{R}^N)$ the Borel σ-field of $\mathbb{R}^N$). Thus we can apply the Yankov-von Neumann–Aumann selection theorem (see Theorem II.2.14 [17:p. 158]), to obtain $x, a \colon T \to \mathbb{R}^N$ measurable maps such that $(x(t), a(t)) \in Q(t)$ a.e on T. Hence $x \in L^p(T, \mathbb{R}^N)$, $a(t) + j_p(x(t))$ a.e. on T and $a \in L^q(T, \mathbb{R}^N)$. Therefore $R(\hat{A} + J) = L^q(T, \mathbb{R}^N)$ and implies that $\hat{A} \colon \hat{D} \subseteq L^p(T, \mathbb{R}^N) \to L^q(T, \mathbb{R}^N)$ is maximal monotone.

Recall that $A_{\lambda_n}(x_n(t)) \in A(J_{\lambda_n}(x_n(t))) \in A(J_{\lambda_n}(x_n(t)))$ a.e. on T (see Proposition III.2.29 [18:p. 325]). So $(J_{\lambda_n}(x_n(\cdot)), A_{\lambda_n}(x_n(\cdot))) \in Gr\hat{A}$. Moreover, we have

$$\|J_{\lambda_n}(x_n(t)) - x(t)\| \leq \|J_{\lambda_n}(x_n(t)) - J_{\lambda_n}(x(t))\| + \|J_{\lambda_n}(x(t)) - x(t)\|$$
$$\leq \|x_n(t) - x(t)\| + \|J_{\lambda_n}(x(t)) - x(t)\| \to 0 \text{ as } n \to \infty$$
$$\Rightarrow J_{\lambda_n}(x_n(\cdot)) \to x(\cdot) \text{ in } L^p(T, \mathbb{R}^N).$$

Since $\hat{A}$ is maximal monotone, $Gr\hat{A}$ is demiclosed (i.e., sequentially closed in $L^p(T, \mathbb{R}^N) \times L^q(T, \mathbb{R}^N)_w$). So in the limit we have $(u, x) \in Gr\hat{A} \Rightarrow u(t) \in A(x(t))$ a.e. on T. Therefore $x \in C^1(T, \mathbb{R}^N)$ is a solution of (1). □

The next existence theorem, takes care of the case when dom $A \neq \mathbb{R}^N$.

Theorem 7: *If hypothesis H(a), $H(A)_2, H(F)_2$ and $H(\xi)$ and H_0 hold, then problem (1) has a solution $x \in C^1(T, \mathbb{R}^N)$*

Proof: Let $\lambda_n \downarrow 0$ and let $x_n \in C^1(T, \mathbb{R}^N)$ be solution of the corresponding auxiliary problem (5). As before we can check that $\{x_n\}_{n\geq 1} \subseteq W^{1,p}(T, \mathbb{R}^N)$ is bounded and so we may assume that $x_n \xrightarrow{w} x$ in $W_{1,p}(T, \mathbb{R}^N)$. For every $n \geq 1$, $A_{\lambda_n}(x_n(\cdot))) \in C(T, \mathbb{R}^N)$ and we have

$$V(x_n) + \hat{A}_{\lambda_n}(x_n) = -f_n \text{ with } f_n \in N_1(x_n)$$

$$\Rightarrow (V(x_n), \hat{A}_{\lambda_n}(x_n))_{pq} + \left\|\hat{A}_{\lambda_n}(x_n)\right\|_2^2 = -(f_n, \hat{A}_{\lambda_n}(x_n))_{pq}$$

We have:

$$\begin{aligned}(V(x_n), \hat{A}_{\lambda_n}(x_n)_{pq} &= -\int_0^b ((a(\|x_n'(t)\|^p)\|x_n'(t)\|^{p-2}x_n'(t))', A_{\lambda_n}(x_n(t)))_{\mathbb{R}^N}dt, t \in T \\ &= -(\varphi(x_n'(b)), A_{\lambda_n}(x_n(b)))_{\mathbb{R}^N} + (\varphi(x_n'(0)), A_{\lambda_n}(x_n(0)))_{\mathbb{R}^N} \\ &+ \int_0^b a((\|x_n'(t)\|^p)\|x_n'(t)\|^{p-2}(x_n'(t), \frac{d}{dt}A_{\lambda_n}(x_n(t)))_{\mathbb{R}^N}dt \\ &\text{(by Green's identity)} \\ &\geq \int_0^b a((\|x_n'(t)\|^p)\|x_n'(t)\|^{p-2}(x_n'(t), \frac{d}{dt}A_{\lambda_n}(x_n(t)))_{\mathbb{R}^N}dt \\ &\text{(hypothesis} H_0).\end{aligned}$$

Recall that $A_{\lambda_n}(\cdot)$ is Lipschitz continuous, thus differentiable almost everywhere on $\mathbb{R}^N$ (Rademacher's theorem, see Theorem 2 [12:p. 81]). Also $A_{\lambda_n}(\cdot)$ is maximal monotone. So for every $x \in \mathbb{R}^N$ differentiability point of $A_{\lambda_n}(\cdot)$, we have

$$\left(y, \frac{A_{\lambda_n}(x+sy) - A_{\lambda_n}(x)}{s}\right)_{\mathbb{R}^N} \geq 0 \text{ for all } s \in \mathbb{R} \text{ and all } y \in \mathbb{R}^N$$

$$\Rightarrow (y, A'_{\lambda_n}(x)y)_{\mathbb{R}^N} \geq 0 \text{ for all } y \in \mathbb{R}^N.$$

From Marcus–Mizel [21], we know that for almost all $t \in T$

$$\frac{d}{dt}A_{\lambda_n}(x_n(t)) = A'_{\lambda_n}(x_n(t))x_n'(t).$$

Thus we can write that

$$(V(x_n), \hat{A}_{\lambda_n}(x_n))_{pq} \geq \int_0^b a(\|x_n'(t)\|^p)\|x_n'(t)\|^{p-2}(x_n'(t), A'_{\lambda_n}(x_n(t))x_n'(t))_{\mathbb{R}^N} \geq 0$$

There finally we obtain

$$\left\|\hat{A}_{\lambda_n}(x_n)\right\|_2^2 \leq \|f_n\|_2\left\|\hat{A}_{\lambda_n}(x_n)\right\|_2 \text{ hypothesis } H(F)_1(v))$$

$$\Rightarrow \left\|\hat{A}_{\lambda_n}(x_n)\right\|_2 \leq \beta_{18} \quad \text{for some } \beta_{18} > 0 \text{ and all } x \geq 1.$$

Since by hypothesis $p \geq 2$, we have $q \leq 2$ and so we may assume that $\hat{A}_{\lambda_n}(x_n) \xrightarrow{w} u$ in $L^q(T, \mathbb{R}^N)$. Arguing as in the proof of Theorem 6, in the limit we have

$$(a(\|x'(t)\|^p)\|x'(t)\|^{p-2}x'(t))' = u(t) + f(t) \quad \text{a.e. on } T$$
$$(\phi(x'(0)), -\phi(x'(b)) \in \xi(x(0), x(b))$$

and $x_n \to x$ in $W^{1,p}(T, \mathbb{R}^N)$ and $f_n \xrightarrow{w} f$ in $L^q(T, \mathbb{R}^N)$. Using Proposition VII.3.9 of Hu-Papageorgiou [18:p. 694], we have that $f(t) \in F(t, x(t), x'(t))$ a.e. on T. We need to show that $u(t) \in A(x(t))$ a.e. on T. Since for every $\lambda > 0$, $J_\lambda(\cdot)$ is nonexpansive using the chain rule of Marcus–Mizel [21], we have that $J_{\lambda_n}(x_n(\cdot)) \in W^{1,p}(T, \mathbb{R}^N)$ and $\frac{d}{dt} J'_{\lambda_n}(x_n(t))x'_n(t)$ a.e. on T, with $\|J'_{\lambda_n}(x_n(t))\| \leq 1$ a.e. on T. So $\{J_{\lambda_n}(x_n(\cdot))\}_{n\geq 1} \subseteq W^{1,p}(T, \mathbb{R}^N)$ is bounded and we may assume that $\hat{J}_{\lambda_n}(x_n) \xrightarrow{w} z$ in $W^{1,p}(T, \mathbb{R}^N) \Rightarrow \hat{J}_{\lambda_n}(x_n) \to z$ in $C(T, \mathbb{R}^N)$. From the definition of the Yosida approximation, we have that

$$J_{\lambda_n}(x_n(t))_{\lambda_n} + \lambda_n A_{\lambda_n}(x_n(t)) = x_n(t)$$

for all $t \in T$ and $n \geq 1$,, hence

$$\hat{J}_{\lambda_n}(x_n) + \lambda_n \hat{A}_{\lambda_n}(x_n) = x_n.$$

Passing to the limit we obtain $z = x$, hence $\hat{J}_{\lambda_n}(x_n) \to x$ in $C(T, \mathbb{R}^N)$.

Let $\Gamma = \{t \in T$: there exist $y \in \mathbb{R}^N$ and $v \in A(y)$ such that $(u(t) - v, x(t) - y)_{\mathbb{R}^N} < 0\}$. If we can show that Γ is Lebesgue-null, by virtue of the maximal monotonicity of A, we will have that $u(t) \in A(x(t))$ a.e. on T, which is what we want. So let

$$E(t) = \{(y, v) \in \mathbb{R}^N \times \mathbb{R}^N : v \in A(y), (u(t) - v, x(t) - y)_{\mathbb{R}^N} < 0\}$$
$$\Rightarrow \Gamma = \operatorname{dom} E.$$

Note that $GrE \in \mathcal{L} \times B(\mathbb{R}^N) \times B(\mathbb{R}^N)$ and so from the Yankov-von Neumann–Aumann projection theorem (see Theorem II.1.33 [18:p. 149]), we have that $\operatorname{proj}_T GrE = \operatorname{dom}E \in \mathcal{L}$. Suppose $|\Gamma| > 0$ ($|\cdot|$ denoting the Lebesgue measure on T). We apply the Yankov-von Neumann–Aumann selection theorem, to obtain $y: \Gamma \to \mathbb{R}^N$ and $v: \Gamma \to \mathbb{R}^N$ measurable maps such that $(y(t), v(t)) \in E(t)$ for all $t \in T$. From Lusin's theorem, we know that there exist $\Gamma_1 \subseteq \Gamma$ closed, $0 < |\Gamma_1| \leq |\Gamma|$ such that $y_{|\Gamma_1}, v_{|\Gamma_1}$ are both continuous. Recall that $A_{\lambda_n}(x_n(t)) \in A(J_{\lambda_n}(x_n(t)))$ for all $t \in T$ and so

$$(A_{\lambda_n}(x_n(t)) - v(t)), J_{\lambda_n}(x_n(t)) - y(t))_{\mathbb{R}^N} \geq 0 \text{ for all } t \in \Gamma_1 \text{ and all } n \geq 1$$
$$\int_{\Gamma_1} (A_{\lambda_n}(x_n(t)) - v(t)), J_{\lambda_n}(x_n(t) - y(t))_{\mathbb{R}^N} dt \geq 0.$$

Passing to the limit, we obtain

$$\int_{\Gamma_1} (u(t) - v(t), x(t) - y(t))_{\mathbb{R}^N} dt \geq 0.$$

On the other hand since $|\Gamma_1| > 0$ and $(y(t), v(t)) \in E(t)$ for all $t \in \Gamma_1$, we have

$$\int_{\Gamma_1} (u(t) - v(t), x(t) - y(t))_{\mathbb{R}^N} dt < 0,$$

a contradiction. Therefore $|\Gamma| = 0$ and so $u(t) \in A(x(t))$ a.e. on T, which implies that $x \in C^1(T, \mathbb{R}^N)$ is a solution of problem (1).

We can have "nonconvex" variants of Theorems 6 and 7. The hypotheses on $F(t,x,y)$ are now the following:

$H(F)_3$: $F: T \times \mathbb{R}^N \times \mathbb{R}^N \to P_k(\mathbb{R}^N)$ is a multifunction such that

(i) $(t,x,y) \to F(t,x,y)$ is graph measurable;
(ii) for almost all $t \in T, (x,y) \to F(t,x,y)$ is lsc;
(iii) for almost all $t \in T$, all $x, y \in \mathbb{R}^N$ and all $u \in F(t,x,y)$

$$(u,x)_{\mathbb{R}^N} \geq -\, c_2\|x\|^p - \gamma_1\|x\|^r\|y\|^{p-r} - \eta(t)\|x\|^s$$

with $c_2, \gamma_1 > 0, \eta \in L^1(T)_+$ and $1 \leq r, s < p$;

(iv) there exists $M > 0$ such that if $\|x_0\| > M$ and $(x_0, y_0)_{\mathbb{R}^N} = 0$, then we can find $\delta > 0, \theta > 0$ such that for almost all $t \in T$

$$\inf[(u,x)_{\mathbb{R}^N} + \|y\|^p: \|x - x_0\| - \|y - y_0\| < \delta, u \in F(t,x,y)] \geq \theta > 0;$$

(v) for almost all $\in T$, all $x, y \in \mathbb{R}^N$ and all $u \in F(t,x,y)$

$$\|u\| \leq \gamma_2(t, \|x\|) + \gamma_3(t, \|x\|)\|y\|^{p-1}$$

with

$$\sup_{0 \leq r \leq k} \gamma_2(t,r) \leq \eta_{1,k}(t) \text{ a.e on } T,\ \eta_{1,k} \in L^q(T)\ \left(\frac{1}{p} + \frac{1}{q} = 1\right)$$

and

$$\sup_{0 \leq r \leq k} \gamma_3(t,r) \leq \eta_{2,k}(t) \text{ a.e. on } T,\ \eta_{2,k} \in L^\infty(T).$$

When $\operatorname{dom} A \neq \mathbb{R}^N$, our hypothesis on $F(t,x,y)$ are the following:

$H(F)_4$: $F: T \times \mathbb{R}^N \times \mathbb{R}^N \to P_k(\mathbb{R}^N)$ is a multifunction satisfying hypothesis $H(F)_3$(i) $\to$ (iv) and

(v′) for almost all $t \in T$, all $x, y \in \mathbb{R}^N$ and all $u \in F(t,x,y)$

$$\|u\| \leq \gamma_2(t, \|x\|) + \gamma_3(t, \|x\|)\|y\|$$

with

$$\sup_{0 \leq r \leq k} \gamma_2(t,r) \leq \eta_{1,k}(t) \quad \text{a.e. on } T,\ \eta_{1,k} \in L^2(T)$$

and

$$\sup_{0 \leq r \leq k} \gamma_2(t,r) \leq \eta_{2,k}(t) \text{ a.e. on } T,\ \eta_{2,k} \in L^{\frac{2p}{p-2}}$$

Theorem 8: *If hypothesis $H(a), H(A)_1, H(F)_3$ and $H(\xi)$ hold, then problem (1) has a solution $v \in C^1(T, \mathbb{R}^N)$*

Proof: Let $N_1: W^{1,p}(T, \mathbb{R}^N) \to P_f(L^q(T, \mathbb{R}^N))$ be the multivalued Nemitsky operator corresponding to F, i.e., $N_1(x) = S^q_{F(\cdot, x(\cdot), x'(\cdot))}$. From Halidias–Papageorgiou [16] (see also [14]), we know that N_1 is lsc and clearly has decomposable values (i.e., if $u_1, u_2 \in N_1(x)$ and $C \in \mathcal{L}$, then $\chi_C u_1 + \chi_C u_2 \in N_1(x)$). We can apply Theorem II.8.7, of Hu–Papageorgiou [18:p. 245], and obtain w: $W^{1,p}(T, \mathbb{R}^N) \to L^q(T, \mathbb{R}^N)$ a continuous map such that $w(x) \in N_1(x)$ for all $x \in W^{1,p}(T, \mathbb{R}^N)$. Then arguing as in the proof of theorem (6), we solve the operator inclusion

$$V(x) + \hat{A}(x) + w(x) \ni 0, x \in D,$$

by passing to the limit as $\lambda \downarrow 0$ in the auxiliary problem

$$V(x) + \hat{A}_\lambda(x) + w(x) = 0.$$

□

Similarly, we can have the following "nonconvex" counterpart of theorem 7

Theorem 9: *If hypothesis $H(a), H(A)_2, H(F)_4$ and $H(\xi)$ and H_0 hold, then problem (1) has a solution $x \in C^1(T, \mathbb{R}^N)$.*

4. STRUCTURE OF THE SOLUTION SET

In this section we investigate the topological structure of the solution set of various special versions of the problem (1)

Theorem 10: *If hypothesis $H(a), H(A)_1, H(F)_1$ and $H(\xi)$ hold, then the solution set $S \subseteq C^1(T, \mathbb{R}^N)$ of problem (1) is nonempty, compact.*

Proof: From Theorem 6 and its proof, we know that $S \subseteq W^{1,p}(T, \mathbb{R}^N)$ is nonempty and bounded. Hence $S \subseteq C(T, \mathbb{R}^N)$ is relatively compact (from the compact embedding of $W^{1,p}(T, \mathbb{R}^N)$ in $C(T, \mathbb{R}^N)$). Since $\operatorname{dom} A = \mathbb{R}^N$ (hypothesis $H(A)_1$), from Theorem III.1.28, p. 308 and Corollary I.2.20, p. 42 of Hu-Papageorgiou [18], we have that $\sup[\|A(x(t))\| : x \in S, t \in T] \leq M_1$. For every $x \in S$, we have

$$(a(\|x'(t)\|^p)\|x'(t)\|^{p-2}x'(t))' \in A(x(t)) + f_x(t) \text{ a.e. on } T$$
$$(\varphi(x'(0)), -\varphi(x'(b))) \in \xi(x(0), x(b)),$$

with $f_x \in S^q_{F(\cdot, x(\cdot), x'(\cdot))}$. By virtue of hypothesis $H(F)_1(v)$, we have that $\{f_x\}_{x\in S} \subseteq L^q(T, \mathbb{R}^N)$ is bounded. So from the inclusion above it follows that $\{\varphi(x')\}_{x\in S} \subseteq W^{1,q}(T, \mathbb{R}^N)$ is bounded and so $\{\varphi(x')\}_{x\in S} \subseteq C(T, \mathbb{R}^N)$ is relatively compact, hence $\sup[\|\varphi(x'(t)) : x \in S, t \in T\|] \leq M_2$. Then $\psi|_{\bar{B}_{M_2}(0)}$ is uniformly continuous. So given $\varepsilon > 0$,we can find $\delta_1 = \delta_1(\varepsilon) > 0$ such that if $\|r - u\| < \delta_1$, then $\|\psi(r) - \psi(u)\| < \varepsilon$. From the relative compactness of $\{\varphi(x')\}_{x\in S} \subseteq C(T, \mathbb{R}^N)$ we know that there exists a $\delta > 0$ such that if $|t - s| < \delta$, then $\|\varphi(x'(t)) - \varphi(x'(s))\| < \delta_1$ for all $x \in S$. Since $\varphi(x'(t)) \in \bar{B}_{M_2}(0)$ for all $x \in S, t \in T$, we have $\|\psi(\varphi(x'(t))) - \psi(\varphi(x'(s)))\| < \varepsilon$, hence $\|x'(t) - x'(s)\| < \varepsilon$ for all $x \in S$. Moreover, we claim that $\sup[\|x'(t)\| : x \in S, t \in T] \leq M_3$. Indeed if this is not the case, we can find $x_n \in S, t_n \in T$ such that $\|x'_n(t_n)\| \to \infty$. Then we have that $M_2 \geq \|\varphi(x'_n(t_n))\| = a(\|x'_n(t_n)\|^p)\|x'_n(t_n)\|^{p-1} \to \infty$ a contradiction. So $\{x'\}_{x\in S}$ is equicontinuous and equibounded, thus by the Arzela–Ascoli theorem, we have that $\{x'\}_{x\in S} \subseteq C(T,\mathbb{R}^N)$ is relatively compact. So we infer that $S \subseteq C^1(T,\mathbb{R}^N)$ is relatively compact. To finish the proof we need to show that $S \subseteq C^1(T,\mathbb{R}^N)$ is closed. To this end let $\{x_n\}_{n\geq 1} \subseteq S$ and assume that $x_n \to x$ in $C(T,\mathbb{R}^N)$. We have

$$a(\|x_n'(t)\|^p)\|x_n'(t)\|^{p-2}x_n'(t) \in A(x_n(t)) + f_n(t) \quad \text{a.e. on } T$$
$$(\varphi(x_n'(0)), -\varphi(x_n'(b))) \in \xi(x_n(0), x_n(b)),$$

with $f_n \in S^q_{F(\cdot,x_n(\cdot),x_n'(\cdot))}$. Again we may assume that $f_n \xrightarrow{w} f$ in $L^q(T,\mathbb{R}^N)$ and using Proposition VII.3.9, of Hu-Papageorgiou [18:p. 694], we have that

$$f(t) \in \overline{\text{conv}}\, \lim\sup F(t,x_n(t),x_n'(t)) \subseteq F(t,x(t),x'(t)) \quad \text{a.e. on } T,$$

the last inclusion being a consequence of fact that F is $P_{kc}(\mathbb{R}^N)$-valued and since $GrF(t,\cdot,\cdot)$ is closed. Also we have $\varphi(x_n') \xrightarrow{w} u$ in $W^{1,q}(T,\mathbb{R}^N)$ and for every $n \geq 1$

$$V(x_n) + v_n + f_n = 0$$

with $v_n \in \hat{A}(x_n)$. We may assume that $v_n \xrightarrow{w} v$ in $L^q(T,\mathbb{R}^N)$ and because $A(\cdot)$ is usc with values in $P_{kc}(\mathbb{R}^N)$ (see Theorem III.1.28, p. 308, Theorem III.1.21, p. 306 and Proposition III.1.14, p. 304 [18]), we have

$$v(t) \in \overline{\text{conv}}\, \lim\sup A(x_n(t) \subseteq A(x(t)) \quad \text{a.e. on } T.$$

Thus, we have

$$\lim\sup(V(x_n), x_n - x) = 0$$
$$\Rightarrow V(x_n) \xrightarrow{w} V(x) \in L^q(T,\mathbb{R}^N)$$

(since V is maximal monotone, see proposition 4). Hence $u = \varphi(x')$ and so $\varphi(x_n') \xrightarrow{w} \varphi(x')$ in $W^{1,q}(T,\mathbb{R}^N)$. Therefore in the limit we have

$$\varphi(x'(t))' = v(t) + f(t) \quad \text{a.e. on } T,$$

with $v(t) \in A(x(t))$ and $f(t) \in F(t,x(t),x'(t))$ a.e. on T. Moreover, because $Gr\xi$ is closed we also have $(\varphi(x'(0)), -\varphi(x'(b))) \in \xi(x(0),x(b))$. So $x \in S$, which proves the compactness of S in $C^1(T,\mathbb{R}^N)$ □

When $F(t,x,y) = f(t,x,y)$ is a single valued, we can strengthen the conclusion of Theorem 10. First let us recall the following notion:

Definition: A metric space is said to be an absolute retract (AR for short), if it can replace $\mathbb{R}$ in Tietze's extension theorem, i.e. for every metric space Z and nonempty closed set $A \subseteq Z$, every continuous map $f\colon A \to X$ has a continuous extension $\hat{f}\colon Z \to X$. A compact metric space X, is a compact R_δ-set, if there exists a decreasing sequence $\{A_n\}_{n\geq 1}$ of compact absolute retracts such that $X = \cap_{n\geq 1} A_n$.

Remark: By Dugundji's extension theorem (see Theorem I.2.88 [18:p. 70]) we know that a closed, convex subset of a normed space is an AR. Compact R_δ sets and AR's are acyclic. In the context of normed spaces, the reader should think of acyclic sets as lying between the smaller class of convex sets and larger class of connected sets.

Our hypotheses on the single valued map $F(t,x,y) = f(t,x,y)$ are the following

$H(f)_1$: $f\colon T \times \mathbb{R}^N \times \mathbb{R}^N \to \mathbb{R}^N$ is a function such that

(i) for all $x, y \in \mathbb{R}^N$, $t \to f(t,x,y)$ is measurable;

(ii) for almost all $t \in T, (x,y) \to f(t,x,y)$ is continuous and $x \to f(t,x,y)$ is monotone;

(iii) for almost all $t \in T$ and all $x, y \in \mathbb{R}^N$ we have

$$(f(t,x,y), x)_{\mathbb{R}^N} \geq -c_2\|x\|^p - \gamma_1\|x\|^r\|y\|^{p-r} - \eta(t)\|x\|^s$$

with $c_2, \gamma_1 > 0, \eta \in L^1(T)$ and $1 \leq r, s < p$;

(iv) there exists $M > 0$ such that if $\|x_0\| > M$ and $(x_0, y_0)_{\mathbb{R}^N} = 0$, then we can find $\delta > 0$ and $\theta > 0$ such that

$$\inf[(f(t,x,y),x)_{\mathbb{R}^N} + \|x\|^p\colon \|x - x_0\| + \|y - y_0\| < \delta] \geq \theta > 0;$$

(v) for almost all $t \in T$ and all $x, y \in \mathbb{R}^N$ $\|f(t,x,y)\| \leq \gamma_2(t, \|x\|) + \gamma_3(t, \|x\|)\|y\|^{p-1}$ with

$$\sup_{0 \leq r \leq k}\gamma_2(t,r) \leq \eta_{1,k}(t) \quad \text{a.e. on } T,\ \eta_{1,k} \in L^q(T)\left(\frac{1}{p} + \frac{1}{q} = 1\right)$$
$$\text{and } \sup_{0 \leq r \leq k}\gamma_3(t,r) \leq \eta_{2,k}(t) \quad \text{a.e. on } T\ \eta_{2,k} \in L^\infty(T).$$

The problem under consideration is the following:

$$\begin{aligned} &(a(\|x'(t))\|^p)\|x'(t)\|^{p-2}x'(t))' \in A(x(t)) + f(t,x(t),x'(t)) \quad \text{a.e. on } T \\ &(\varphi(x'(0)), -\varphi(x'(b))) \in \xi(x(0), x(b)). \end{aligned} \tag{9}$$

For problem (9) we have the following structural result:

Theorem 11: *If hypothesis $H(a), H(A)_1, H(f)_1$ and $H(\xi)$ hold, then the solution set $S \subseteq W^{1,p}(T,\mathbb{R}^N)$ of (9) is nonempty and compact R_δ.*

Proof: Let $N_1(x) = -N_f(x) + J(x)$ where $N_f(x)(\cdot) = f(\cdot, x(\cdot), x'(\cdot))$ is the Nemitsky operator corresponding to f. So $N_1\colon W^{1,p}(T,\mathbb{R}^N) \to L^q(T,\mathbb{R}^N)$ and is continuous and bounded. Also let $K\colon D \subseteq L^p(T,\mathbb{R}^N) \to 2^{L^q(T,\mathbb{R}^N)}$ be defined by $K(x) = V(x) + \hat{A}(x) + J(x)$. Since $\hat{A}\colon L^p(T,\mathbb{R}^N) \to 2^{L^q(T,\mathbb{R}^N)}\backslash\{\emptyset\}$ is maximal monotone (see the proof of Theorem 6), we see that K is maximal monotone, strictly monotone and coercive, thus surjective. So $K^{-1}\colon L^q(T,\mathbb{R}^N) \to D \subseteq W^{1,p}(T,\mathbb{R}^N)$ is well defined and as in the proof of Proposition 5 (Claim 1), we can check that K^{-1} is completely continuous. Let $N_{1n} = \frac{n-1}{n}N_1$ and set $Q_n = K^{-1}N_{1n}, Q = K^{-1}N_1, n \geq 1$.

From the proof of Theorem 6, we know that the solution set $S \subseteq W^{1,p}(T,\mathbb{R}^N)$ of (9) is bounded. So there exists $M_1 > M > 0$ such that $\|x(t)\| \leq M_1$ for all $x \in S$ and all $t \in T$. We will show that

$$\sup[\|Q_n(x) - Q(x)\| : \|x\| \leq M_1] \to 0, \text{ as } n \to \infty$$

To this end let $\|x\| \leq M_1$ and

$$y_n = K^{-1}N_{1,n}(x), y = K^{-1}N_1(x), n \geq 1$$

We have:

$$\begin{aligned} &K(y_n) = N_{1n}(x), K(y) = N_1(x) \\ &\Rightarrow V(y_n) + \hat{A}(y_n) + J(y_n) \ni N_{1n}(x), n \geq 1 \\ &\text{and } V(y) + \hat{A}(y) + J(y) \ni N_1(x). \end{aligned}$$

Subtracting the last two inclusions, taking duality brackets in $(L^p(T, \mathbb{R}^N), L^q(T, \mathbb{R}^N))$ with $y_n - y$ and exploiting the monotonicity of $\hat{A}$, we have

$$\begin{aligned} &(V(y_n) - V(y), y_n - y)_{pq} + (J(y_n) - J(y), y_n - y)_{pq} \leq (N_{1n}(x) - N_1(x), y_n - y)_{pq} \\ &\Rightarrow c\|y'_n - y'\|_p^p + \|y_n - y\|_p^p \leq (1 - \frac{n-1}{n})\|N_1(x)\|_q \|y_n - y\|_p \\ &\Rightarrow \hat{c}\|y_n - y\|^p \leq \left(1 - \frac{n-1}{n}\right) M_2 \|y_n - y\| \end{aligned}$$

with $\hat{c} = \min\{c, 1\}$ and $M_2 > 0$ independent of x (see hypothesis $H(f)_1(v)$ and recall that $S \subseteq W^{1,p}(T, \mathbb{R}^N)$ is bounded). So have that

$$\begin{aligned} &\hat{c}\,\|y_n - y\|^{p-1} \leq (1 - \frac{n-1}{n})M_2 \to 0 \\ &\Rightarrow \sup\{\|Q_n(x) - Q(x)\| : \|x\| \leq M_1 \to 0 \text{ as } n \to \infty\}. \end{aligned}$$

Also by choosing $M_2 > M > 0$ large enough, so that

$$\{x \in W^{1,p}(T, \mathbb{R}^N) : x \in \beta K^{-1} N_1(x), 0 < \beta < 1\} \subseteq B_{M_2}(0),$$

from the homotopy invariance of the Leray–Schauder degree (see Hu–Papageorgiou Theorem IV.1.1 [18:p. 450]), we have

$$D_{LS}(I - Q, B_{M_2}(0), 0) \neq \emptyset,$$

$$(\text{recall } B_{M_2}(0) = \{x \in W^{1,p}(T, \mathbb{R}^N) : \|x\| < M_2\}).$$

Finally let

$$\|y\| \leq r_n = \sup[\|Q_n(x) - Q(x)\| : \|x\| \leq M_2]$$

and suppose

$$\begin{aligned} &x_1 = Q_n(x_1) + y,\ x_2 = Q_n(x_2) + y \\ &\Rightarrow K(x_1 - y) = N_{1n}(x_1),\ K(x_2 - y) = N_{1n}(x_2). \end{aligned}$$

Subtracting, taking the duality brackets in $(L^p(T,\mathbb{R}^N), L^q(T,\mathbb{R}^N))$ with $x_1 - x_2$ and using the monotonicity of $\hat{A}$ and $f(t,\cdot,y)$ (hypothesis $H(f)_1$(iii)), we obtain

$$(K(x_1-y)-K(x_2-y), x_1-x_2)_{pq} = -\frac{n-1}{n}(N_f(x_1)-N_f(x_2), x_1-x_2)_{pq}$$

$$+\frac{n-1}{n}(J(x_1)-J(x_2), x_1-x_2)_{pq}$$

$$\Rightarrow c\|x_1'-x_2'\|_p^p + \frac{1}{n}\|x_1-x_2\|_p^p \leq 0$$

$$\Rightarrow \|x_1-x_2\| = 0, \text{ i.e. } x_1 = x_2.$$

Thus we can apply Lemma 2 of Bebernes–Martelli [3] (see also [10,1]) and deduce that

$$S_0 = \{x \in W^{1,p}(T,\mathbb{R}^N) : x \in K^{-1}N_1(x)\}$$

is nonempty, compact R_δ. But clearly $S_0 = S$. □

Remark: In particular $S \subseteq W^{1,p}(T,\mathbb{R}^N)$ is acyclic and path-connected. Theorem 11 improves significantly Theorem 3 of Nieto [23], who studied the semilinear (i.e., $p=2$, $a(r) \equiv 1$), scalar (i.e., $N=1$) problem, with $A \equiv 0$, f independent of x' and with Dirichlet boundary conditions.

Recall that if $p=q=2$, then $\hat{A}$: $D_0 \subseteq L^2(T,\mathbb{R}^N) \to 2^{L^2(T,\mathbb{R}^N)}$ is maximal monotone, even if dom$A \neq \mathbb{R}^N$ (see Example III.2.33, [18:p. 328]). Then a careful reading of the proof of Theorem 10, reveals that when $p=q=2$, we can drop the requirement that dom$A = \mathbb{R}^N$, i.e., replace hypothesis $H(A)_1$ by $H(A)_2$ and still have a compactness result for S. So we can state the following theorem.

Theorem 12: *If hypothesis $H(a)$, $H(A)_2$, $H(F)_2$, $H(\xi)$ and H_0 hold, then the solution set $S \subseteq W_0^{1,2}(T,\mathbb{R}^N)$ is nonempty, compact.*

In the rest of this section we will prove structural results for the solution set of some particular cases of the general boundary value problem (1). We start with the semilinear problem (i.e., $p=q=2$, $a(r) \equiv 1$), with $A \equiv 0$ and Dirichlet boundary conditions. So the multivalued boundary value problem under consideration is the following:

$$\begin{aligned} &x''(t) \in F(t, x(t), x'(t)) \quad \text{a.e. on } T \\ &x(0) = x(b) = 0 \end{aligned} \tag{10}$$

The hypothesis on the data of (10) are the following:

$H(F)_5$: F: $T \times \mathbb{R}^N \times \mathbb{R}^N \to P_{kc}(\mathbb{R}^N)$ is a multifunction such that

(i) for every $x, y \in \mathbb{R}^N$, $t \to F(t,x,y)$ is measurable;
(ii) for almost all $t \in T$ and all $x, y, x_1, y_1 \in \mathbb{R}^N$ we have

$$h(F(t,x,y), F(t,x_1,y_1)) \leq k(t)(\|x-x_1\| + \|y-y_1\|)$$

with $k \in L^1(T)$, $\|k\|_1 < \frac{1}{2}$;

(iii) there exists $M > 0$ such that if $\|x_0\| > M$ and $(x_0, y_0)_{\mathbb{R}^N} = 0$, then we can find $\delta > 0$ and $\vartheta > 0$ such that for almost all $t \in T$

$$\inf[(u, x)_{\mathbb{R}^N} + \|y\|^p \colon \|x - x_0\| + \|y - y_0\| < \delta, u \in F(t, x, y)] \geq \vartheta > 0;$$

(iv) for every $r > 0$, there exists $\gamma_r \in L^2(T)$ such that for almost all $t \in T$, all $\|x\| \leq r$ and all $u \in F(t, x, y)$, we have

$$\|u\| \leq \gamma_r(t)\|y\|.$$

Theorem 13: *If hypotheses $H(F)_5$ hold, then the solution set $S \subseteq C^1(T, \mathbb{R}^N)$ of (10) is nonempty and compact, absolute retract.*

Proof: Arguing as in the proof of proposition 5, using hypothesis $H(F)_5$(iii), we can show that for every $x \in S$, $\|x\|_\infty \leq M$. Then from the inclusion we have

$$\|x'\|_2^2 \leq M\|\gamma_M\|_2\|x'\|_2$$
$$\Rightarrow \|x'\|_2 \leq M\|\gamma_M\|_2 = M_1 \text{ for all } x \in S.$$

So $\|x''\|_2 \leq \|\gamma_M\|_2 M_1 = M_2$ for all $x \in S$. We have proved that $S \subseteq W^{2,2}(T, \mathbb{R}^N)$ is bounded, thus $S \subseteq C^1(T, \mathbb{R}^N)$ is relatively compact. In fact as before we can check that $S \subseteq C^1(T, \mathbb{R}^N)$ is closed, thus compact.

It remains to show that $S \subseteq C^1(T, \mathbb{R}^N)$ is a nonempty, absolute retract. To this end let $G(t, s)$ be the Green's function corresponding to the differential operator $L(x) = -x''$ for all $x \in W^{2,1}(T, \mathbb{R}^N) \cap W_0^{1,1}(T, \mathbb{R}^N)$. So for every $u \in L^1(T, \mathbb{R}^N)$, the solution of the problem

$$-x''(t) = u(t) \text{ a.e. on } T$$
$$x(0) = x(b) = 0$$

is given by $x(t) = \int_0^b G(t, s)v(s)ds,\ t \in T, x \in C^1(T, \mathbb{R}^N)$. It is well-known that

$$G(t, s) = \begin{cases} s(b - t) & \text{if } 0 \leq s \leq t \leq b \\ t(b - s) & \text{if } 0 \leq t \leq s \leq b. \end{cases}$$

Let $K \colon C^1(T, \mathbb{R}^N) \to 2^{C^1(T, \mathbb{R}^N)}$ be defined by

$$K(x) = \left\{ y \in C^1(T, \mathbb{R}^N) \colon y(t) = \int_0^b G(t, s)u(s)ds \right\}.$$

Clearly K has nonempty, closed and convex values. Let $x_1, x_2 \in C^1(T, \mathbb{R}^N)$ and $y_1 \in K(x_1)$. We have $y_1(t) = \int_0^b G(t, s)v_1(s)ds, v_1 \in S^2_{F(\cdot, x(\cdot), x'(\cdot))}$. For every $t \in T$, let

$$\Gamma(t) = \{u \in F(t, x_2(t), x_2'(t)) : \|u_1(t) - u\| = d(u_1(t), F(t, x_2(t), x_2'(t)))\}.$$

Note that hypotheses $H(F)_5(i)$ and (ii), imply that $(t,x,y) \to F(t,x,y)$ is measurable (see Proposition II.7.9 [18:p. 229]) and $t \to F(t, x_2(t), x_2'(t))$ is measurable. Hence $Gr\Gamma \in \mathcal{L} \times B(\mathbb{R}^N)$ and so via the Yankov-von Neuamann–Aumann selection theorem, we can produce a measurable function $u_2\colon T \to \mathbb{R}^N$ such that $u_2(t) \in \Gamma(t)$ for all $t \in T$. Then we have $u_2 \in S^2_{F(\cdot, x_2(\cdot), x_2'(\cdot))}$ and $\|u_1(t) - u_2(t)\| = d(u_1(t), F(t, x_1(t), x_1'(t)))$ a.e. on T.

Let $y_2 \in K(x_2)$ be defined by $y_2(t) = \int_0^b G(t,s)u_2(s)ds, t \in T$. We have:

$$d(y_1, K(x_1)) \le \|y_1 - y_2\|_{C(T,\mathbb{R}^N)}.$$

For every $t \in T$, we have

$$\begin{aligned}
\|y_1(t) - y_2(t)\|_{C(T,\mathbb{R}^N)} &\le \int_0^b \|G\|_\infty \|u_1(s) - u_2(s)\| ds \\
&\le \int_0^b d(u_1(s), F(s, x_2(s), x_2'(s)))ds \\
&\le \int_0^b k(s)(\|x_1(s) - x_2(s)\| + \|x_1'(s) - x_2'(s)\| ds \\
&\le \|k\|_1 \|x_1 - x_2\|_{C^1(T,\mathbb{R}^N)}.
\end{aligned}$$

Similarly, since $\|G_t\| \le 1$, we have $\|y_1'(t) - y_2'(t)\| \le \|k\|_1 \|x_1 - x_2\|_{C^1(T,\mathbb{R}^N)}$.

Therefore finally we have

$$\begin{aligned}
&\|y_1 - y_2\|_{C^1(T,\mathbb{R}^N)} \le 2\|k\|_1 \|x_1 - x_2\|_{C^1(T,\mathbb{R}^N} \\
&\Rightarrow h(K(x_1), K(x_2)) \le 2\|k\|_1 \|x_1 - x_2\|_{C^1(T,\mathbb{R}^N}
\end{aligned}$$

Since $\|k\|_1 < \frac{1}{2}$ (hypothesis $H(F)_5(ii)$), it follows that K is an h-contraction. Invoking theorem IV.5.5, of Hu–Papageorgiou [18:p. 562], we have that $S_0 = \{x \in C^1(T, \mathbb{R}^N\colon x \in K(x)\}$ is a nonempty, absolute retract of $C^1(T,\mathbb{R}^N)$. Since $S_0 = S$, the proof is finished. □

Since a compact absolute retract, is compact R_δ (hence acyclic), we obtain:

Corollary 14: *If hypotheses $H(F)_5$ hold, then the solution set $S \subseteq C^1(T,\mathbb{R}^N)$ of 10 is nonempty, compact R_δ, thus acyclic and path connected.*

Next we prove a structural result for the "nonconvex" problem. In this case our hypotheses on $F(t,x,y)$ are the following:

$H(F)_6$: $F\colon T \times \mathbb{R}^N \times \mathbb{R}^N \to P_k(\mathbb{R}^N)$ is a multifunction such that

(i) for every $x, y \in \mathbb{R}^N, t \to F(t,x,y)$ is measurable;

(ii) for almost all $t \in T$ and all $x, y, x_1, y_1 \in \mathbb{R}^N$ we have

$$h(F(t,x,y), F(t,x_1,y_1)) \le k(t)(\|x - x_1\| + \|y - y_1\|)$$

with $k \in L^1(T)_+, 2\|k\|_1 < 1$;

(iii) for every $r > 0$, there exist $\gamma_r \in L^1(T)$ such that for almost all $t \in T$, all $\|x\| \le r$ and all $u \in F(t,x,y)$, we have $\|u\| \le \gamma_r(t)\|y\|$.

Theorem 15: *If hypotheses $H(F)_6$ hold, then the solution set $S \subseteq C^1(T,\mathbb{R}^N)$ of (10) is a nonempty retract.*

Proof: Let η: $L^1(T,\mathbb{R}^N) \to C^1(T,\mathbb{R}^N)$ be the map which to each $u \in L^1(T,\mathbb{R}^N)$ assigns the unique solution of the Dirichlet problem $-x''(t) = u(t)$ a.e. on T, $x(0) = x(b) = 0$. We know that $\eta(u)(t) = \int_0^b G(t,s)u(s)ds, t \in T$. Also let N: $L^1(T,\mathbb{R}^N) \to P_f(L^1(T,\mathbb{R}^N))$ be the multivalued Nemitsky operator corresponding to F, i.e., $N(u) = S^1_{F(\cdot,\eta(\cdot),\eta'(\cdot))}$. Let $u_1, u_2 \in L^1(T,\mathbb{R}^N)$ and $f_1 \in N(u_1)$. As before, via the Yankov-von Neumann–Aumann selection theorem, we can produce $f_2 \in N(u_1)$ such that

$$d(f_1(t), F(t,\eta(u_2)(t),\eta(u_2)'(t))) = \|f_1(t) - f_2(t)\| \text{ a.e. on } T.$$

Then we have

$$d(f_1, N(u_2)) \leq \|f_1 - f_2\|_1 \leq \int_0^b k(t)\|\eta(u_1) - \eta(u_2)\|_{C^1(T,\mathbb{R}^N)}dt \quad (\text{hypothesis} H(F)_6(ii))$$

$$= \|k\|_1\|\eta(u_1) - \eta(u_2)\|_{C^1(T,\mathbb{R}^N)}$$

$$\leq 2\|k\|_1\|u_1 - u_2\|_1$$

$$\Rightarrow h(N(u_1), N(u_2)) \leq 2\|k\|_1\|u_1 - u_2\|_1$$

$\Rightarrow N$ is an h-contraction (since $2\|k\|_1 < 1$, see hypothesis $H(F)_6(ii)$).

Form Bressan *et al.* [5], we have that

$$\text{Fix}N = \{u \in L^1(T,\mathbb{R}^N) : u \in N(u)\}$$

is a nonempty retract of $L^1(T,\mathbb{R}^N)$. Let r: $L^1(T,\mathbb{R}^N) \to \text{Fix}N$ be the corresponding retraction. Define γ: $W^{2,1}(T,\mathbb{R}^N) \to S$ by $\gamma(x)(t) = \int_0^b G(t,s)r(x'')(s)ds$. It is easy to see that this is a retraction. Hence S is a retract of $W^{2,1}(T,\mathbb{R}^N)$. □

An immediate consequence of theorem 15 is the following result:

Corollary 16: *If hypotheses $H(F)_6$ hold, then the solution set $S \subseteq C^1(T,\mathbb{R}^N)$ of (10) is an absolute retract and so is path connected.*

Proof: By Dugundji's theorem, $W^{2,1}(T,\mathbb{R}^N)$ is an absolute retract. A retract of an absolute retract, is an absolute retract itself. So $S \subseteq W^{2,1}(T,\mathbb{R}^N)$ is a nonempty, absolute retract, thus is path-connected. □

5. EXTREMAL SOLUTION AND STRONG RELAXATION

In this section we establish the existence of extremal solutions for problem (1) when the boundary conditions are Dirichlet. Subsequently for the special case where $A \equiv 0$ and $p = 2$, we prove a strong relaxation theorem, namely we show that the extremal solutions are dense in the solution set of the convexified problem for the $W_0^{1,2}(T,\mathbb{R}^N)$-norm topology. Evidently such a result is of special interest in control theory, in connection with the "bang–bang principle"

The problem under consideration is the following:

$$\begin{aligned} &(a(\|x'(t)\|^p)\|x'(t)\|^{p-2}x'(t))' \in A(x(t)) + \text{ext}F(t, x(t), x'(t)) \text{ a.e. on } T \\ &x(0) = x(b) = 0. \end{aligned} \tag{11}$$

By $\text{ext}F(t, x, y)$ we denote the extreme points of the set $F(t, x, y)$. Since $\text{ext}F(t, x, y)$ is not in general closed and $(x, y) \to \text{ext}F(t, x, y)$ is not general regular, even if $F(t, \cdot, \cdot)$ has good continuity properties, we see that the existence of solutions for problem (11) can not be deduced from the existence theorems of Section 3. A new approach is needed. For this new approach to work, we need the following hypotheses on the data of (11):

$H(F)_7$: $F: T \times \mathbb{R}^N \times \mathbb{R}^N \to P_{kc}(\mathbb{R}^N)$ is a multifunction such that:

(i) for all $x, y \in \mathbb{R}^N, t \to F(t, x, y)$ is measurable;
(ii) for almost all $t \in T, (x, y) \to F(t, x, y)$ is h-continuous;
(iii) for almost all $t \in T$, all $x, y \in \mathbb{R}^N$ and all $u \in F(t, x, y)$

$$\|u\| \leq \gamma(t) + c_2(\|x\|^{p-1} + \|y\|^{p-1})$$

with $\gamma \in L^q(T)$ $(\frac{1}{p} + \frac{1}{q} = 1), c_2 > 0$ and $c > c_2(\beta + \beta^{\frac{1}{p}})$, where $c > 0$ is as in hypothesis H(a) and $\beta > 0$ is such that $\|x\|_p^p \leq \beta\|x'\|_p^p$ for all $x \in W_0^{1,p}(T, \mathbb{R}^N)$ ($\beta > 0$ exists by Poincare's inequality and if $a \equiv 1$, then $\beta = \lambda_1^{-1}$, with λ_1 being the first eigenvalue of the one dimensional p-Laplacian with Dirichlet boundary conditions).

Theorem 17: *If hypotheses H(a), $H(A_1)$ and $H(F)_7$ hold, then the solution set $S_e \subseteq C^1(T, \mathbb{R}^N)$ of 11 is nonempty.*

Proof: First we will derive some *a priori* bounds for the elements of S_e. So let $x \in S_e$. We have $x \in W_0^{1,p}(T, \mathbb{R}^N)$ and

$$(a(\|x'(t)\|^p)\|x'(t)\|^{p-2}x'(t))' \in A(x(t)) + f(t) \text{ a.e. on } T, f \in S^q_{\text{ext}F(\cdot, x(\cdot), x'(\cdot))}$$

$$\Rightarrow c\|x'\|_p^p \leq \|f\|_q\|x\|_p \leq \beta^{\frac{1}{p}}\|x'\|_p \quad (\text{hypothesis } H(a))$$

$$\Rightarrow \frac{c}{\beta^{\frac{1}{p}}\|x'\|_p^{p-1}} \leq \|\gamma\|_q c_2(\|x\|_p^{p-1} \leq + \|x'\|_p^{p-1}) \quad (\text{hypothesis } H(F)_7(iii))$$

$$\leq \|\gamma\|_q + c_2(\beta^{\frac{1}{q}} + 1)\|x'\|_p^{p-1})$$

$$\Rightarrow \left(\frac{c}{\beta^{\frac{1}{p}}} - c_2(\beta^{\frac{1}{q}} + 1)\right)\|x'\|_p^{p-1} \leq \|\gamma\|_q.$$

By virtue of hypothesis $H(F)_7(iii)$ $\frac{c}{\beta^{\frac{1}{q}}} > c_2\left(\beta^{\frac{1}{p}} + 1\right)$ and so from the last inequality we infer that

$$\begin{aligned} &\|x'\|_p \leq M_1 \quad \text{for all } x \in S_e, M_1 > 0 \\ \Rightarrow &\|x\|_\infty \leq M_2 \quad \text{for all } x \in S_e, M_2 > 0 \end{aligned}$$

Since $\operatorname{dom} A = \mathbb{R}^N$, A is usc (see Theorem III.1.28 [18:p. 308]) and so $\sup[\|A(x(t))\| : x \in S_e, t \in T] \leq M_3$, with $M_3 > 0$. So directly from the inclusion we infer that $\left\{a(\|x'(\cdot)\|^p)\|x'(\cdot)\|^{p-2}x'(\cdot)\right\}_{x \in S_e} \subseteq W^{1,q}(T, \mathbb{R}^N)$ is bounded, hence relatively compact in $C(T, \mathbb{R}^N)$. Thus by virtue of hypothesis $H(a)$, we can find $M_4 > 0$ such that $\|x'(t)\| \leq M_4$ for all $x \in S_e$ and all $t \in T$.

Let p_{M_i}: $\mathbb{R}^N \to \mathbb{R}^N$ be the M_i-radial retraction, i = 2, 4; i.e.,

$$p_{M_i}(x) = \begin{cases} x & \text{if } \|x\| \leq M_i \\ \frac{M_i x}{\|x\|} & \text{if } \|x\| > M_i. \end{cases}$$

Recall that $p_{M_i}(\cdot)$ is nonexpansive. Set $\hat{F}(t, x, y) = F(t, p_{M_2}(x), p_{M_4}(y))$. Evidently $t \to \hat{F}(t, x, y)$ is measurable and $(x, y) \to \hat{F}(t, x, y)$ is h-Lipschitz. Moreover, for almost all $t \in T$, all $x, y \in \mathbb{R}^N$ and all $u \in F(t, x, y)$ we have $\|u\| \leq k(t)$ with $k \in L^q_+$. Set $W = \left\{u \in L^q(T, \mathbb{R}^N) : \|u(t)\| \leq k(t) \text{ a.e. on } T\right\}$. Given $u \in W$, we can find a unique $x \in C^1(T, \mathbb{R}^N)$ such that

$$(a(\|x'(t)\|^p)\|x'(t)\|^{p-2}x'(t))' \in A(x(t)) + u(t) \text{ a.e. on } T$$
$$x(0) = x(b) = 0.$$

Scalarly multiply with $x(t)$ and then integrate over T. Using Green's identity (integration by parts), we obtain

$$c\|x'\|_p^{p-1} \leq \beta\|k\|_q$$
$$\Rightarrow \|x'\|_p^{p-1} \leq \left(\frac{\beta}{c}\|k\|_q\right)^{\frac{q}{p}} = M_5.$$

Let $\mu = \max\left[b^{\frac{1}{p}}(M_2 + M_4), \beta M_5\right]$ and set $K = \bar{B}(0, \mu) \subseteq W_0^{1,p}(T, \mathbb{R}^N)$ (i.e., $\bar{B}(0, \mu) = \left\{x \in W_0^{1,p}(T, \mathbb{R}^N) : \|x\| \leq \mu\right\}$). The set is w-compact, convex, thus it is compact, convex when viewed as a subset of $L^p(T, \mathbb{R}^N)$ (from the compact embedding of $W_0^{1,p}(T, \mathbb{R}^N)$ into $L^p(T, \mathbb{R}^N)$). Let G_1: $K \to P_{fc}(L^1(T, \mathbb{R}^N))$ be defined by $G(x) = -S^1_{\hat{F}(\cdot, x(\cdot), x'(\cdot))}$.

Using Theorem II.8.31 of Hu-Papageorgiou [18:p. 260], we obtain g_1:$K \subseteq L^p(T, \mathbb{R}^N) \to L^1_w(T, \mathbb{R}^N)$ a continuous map such that $g_1(x) \in \operatorname{ext}G_1(x)$. Here by $L^1_w(T, \mathbb{R}^N)$ we denote the space $L^1(T, \mathbb{R}^N)$ furnished with weak norm $\|h\|_w = \sup[\|\int_s^t h(r)dr\| : 0 \leq s \leq t \leq b]$. From theorem II.4.6, of Hu-Papageorgiou [18:p. 192] we know that $\operatorname{ext}G_1(x) = -S^1_{\operatorname{ext}\hat{F}(\cdot, x(\cdot), x'(\cdot))}$. Let m_k: $L^p(T, \mathbb{R}^N) \to K$ be the metric projection map. Since K is compact, convex in the uniformly convex space $L^p(T, \mathbb{R}^N)$, it is single-valued and continuous. Let $\hat{g}_1 = g_1 \circ m_K$. Then $\hat{g}_1$: $L^p(T, \mathbb{R}^N) \to L^1_w(T, \mathbb{R}^N)$ is continuous and $\hat{g}_1(x)(t) \in -\operatorname{ext}\hat{F}(t, m_K(x)(t), m_K(x)'(t))$ a.e. on T. Let R: $V_1 + \hat{A}$: $W_0^{1,p}(T, \mathbb{R}^N) \to 2^{W^{-1,q}(T, \mathbb{R}^N)} \backslash \{\emptyset\}$, where V_1: $W_0^{1,p}(T, \mathbb{R}^N) \to W^{-1,q}(T, \mathbb{R}^N)$ is defined by $\langle V_1(x), y\rangle = \int_0^b a(\|x'(t)\|^p)\|x'(t)\|^{p-2}(x'(t), y'(t))_{\mathbb{R}^N} dt$ and $\hat{A}$ is the realization of $A(\cdot)$ on $(L^p(T, \mathbb{R}^N), L^q(T, \mathbb{R}^N))$. We know that $V_1(\cdot)$ is maximal monotone, strictly monotone, coercive and $\hat{A}$ is maximal monotone (see the proof of Theorem 6). Hence because $0 \in \hat{A}(0)$, we have that $V_1 + \hat{A}$ is maximal monotone, strictly monotone and coercive. Therefore R^{-1}: $W^{-1,q}(T, \mathbb{R}^N) \to W_0^{1,p}(T, \mathbb{R}^N)$ is well-defined, single-valued and of course maximal monotone. Thus R^{-1} is demicontinuous (recall that a maximal monotone map has a demiclosed graph). The map $\hat{g}_1 \circ R^{-1}$: $W^{-1,q}(T, \mathbb{R}^N) \to$

$W^{-1,q}(T,\mathbb{R}^N)$ is continuous. Indeed if $u_n \to u$ in $W^{-1,q}(T,\mathbb{R}^N)$, we have just seen that $R^{-1}(u_n) \xrightarrow{w} R^{-1}(u)$ in $W_0^{1,p}(T,\mathbb{R}^N)$, hence $R^{-1}(u_n) \to R^{-1}(u)$ in $L^p(T,\mathbb{R}^N)$ and so $\mu_K(R^{-1}(u_n)) \to \mu_K(R^{-1}(u))$ in $L^p(T,\mathbb{R}^N)$. Therefore $g_1(\mu_K(R^{-1}(u_n))) \xrightarrow{\|\cdot\|_w} g_1(\mu_K(R^{-1}(u)))$. From the lemma of Kravvaritis–Papageorgiou [20], we have

$$\begin{aligned}
&\hat{g}_1(R^{-1}(u_n)) \xrightarrow{w} \hat{g}(R^{-1}(u)) \quad \text{in} \quad L^q(T,\mathbb{R}^N)\\
\Rightarrow &\hat{g}_1(R^{-1}(u_n)) \to \hat{g}(R^{-1}(u)) \quad \text{in} \quad W^{-1,q}(T,\mathbb{R}^N),\\
\Rightarrow &\hat{g}_1 \circ R^{-1} : W^{-1,q}(T,\mathbb{R}^N) \to W^{-1,q}(T,\mathbb{R}^N) \text{ is continuous.}
\end{aligned}$$

Also $\hat{g}_1 \circ R^{-1}(W^{-1,q}(T,\mathbb{R}^N)) = \hat{g}_1(W_0^{1,p}(T,\mathbb{R}^N)) = g_1(K)$ and from the properties of g_1, we have that $g_1(K)$ is compact in $L^1_w(T,\mathbb{R}^N)$ and as above, via the lemma of Kravvaritis–Papageorgiou [20], we have that $g_1(K) \subseteq L^q(T,\mathbb{R}^N)$ is w-compact, hence $g_1(K)$ is compact in $W^{-1,q}(T,\mathbb{R}^N)$. Thus we can apply Schauder's fixed point theorem and obtain $u \in W^{-1,q}(T,\mathbb{R}^N)$ such that $u = \hat{g}_1(R^{-1}(u))$. Let $x = R^{-1}(u) \in W^{1,p}(T,\mathbb{R}^N)$. We have $u \in R(x)$ and so

$$V_1(x) + \hat{A}(x) \ni \hat{g}_1(x),$$

with $\hat{g}_1(x)(t) = g_1(m_K(x))(t) \in -\text{ext}\,\hat{F}(t, m_K(x)(t), m_K(x)'(t))$ a.e. on T. We have

$$\|x\| \leq \beta M_5$$

$$\Rightarrow x \in K \quad \text{and so} \quad m_K(x) = x.$$

Thus we have

$$\hat{g}_1(x)(t) = g_1(x)(t) \in -\text{ext}\hat{F}(t, x(t), x'(t)) \text{ a.e. on } T$$

and $(a(\|x'(t)\|^p)\|x'(t)\|^{p-2}x'(t))' \in A(x(t)) + g_1(x)(t)$ a.e. on T.

Because $|\hat{F}(t,x,y)| = \sup[\|u\| : u \in \hat{F}(t,x,y)] \leq \gamma(t) + c_2(\|x\|^{p-1} + \|y\|^{p-1})$ (recall that $\|p_{M_i}(x)\| \leq \|x\|, i = 2, 4$), arguing as in the first part of the proof, we obtain $\|x\|_\infty \leq M_2$ and $\|x'\|_\infty \leq M_4$. So $\hat{F}(t, x(t), x'(t)) = F(x, x(t), x'(t))$ for all $t \in T$ and this proves that $x \in C^1(T,\mathbb{R}^N)$ is a solution of problem (11), i.e., $x \in S_e$. □

Next we will show that for the particular version of problem (11), where $p = 2$ and $A \equiv 0$, the solutions of the convexified problem can obtain as limits in $W_0^{1,2}(T,\mathbb{R}^N)$ of a sequence of extremal solutions (strong relaxation theorem). So we consider the following two problems:

$$\begin{aligned}
&(a(\|x'(t)\|^2)\|x'(t)\|^{p-2}x'(t))' \in F(t, x(t), x'(t)) \text{ a.e. on } T\\
&x(0) = x(b) = 0.
\end{aligned} \tag{12}$$

$$\begin{aligned}
&(a(\|x'(t)\|^2)\|x'(t)\|^{p-2}x'(t))' \in \text{ext}F(t, x(t), x'(t)) \text{ a.e. on } T\\
&x(0) = x(b) = 0.
\end{aligned} \tag{13}$$

We denote the solution set of (12) by $S_c \subseteq C^1(T, \mathbb{R}^N)$ and the solution set of (13) by $S_e \subseteq C^1(T, \mathbb{R}^N)$. We will show that S_e is dense in S_c for the $W_0^{1,2}(T, \mathbb{R}^N)$-norm topology. For this we will need the following hypotheses on the multifunction $F(t, x, y)$:

$H(F)_8$: $F: T \times \mathbb{R}^N \times \mathbb{R}^N \to P_{kc}(\mathbb{R}^N)$ is a multifunction such that

(i) for all $x, y \in \mathbb{R}^N, t \to F(t, x, y)$ is measurable;

(ii) for almost all $t \in T, x, y, x_1, y_1 \in \mathbb{R}^N$

$$h(F(t, x, y), F(t, x_1, y_1)) \leq k(t)[\|x - x_1\| + \|y - y_1\|]$$

with $k \in L^\infty(T)$ and $\|k\|_\infty(\beta + \beta^{\frac{1}{2}}) < c$, where $c > 0$ is as in hypothesis $H(a)$ and $\beta > 0$ is such that $\|x\|_2^2 \leq \beta\|x'\|_2^2$ for all $x \in W_0^{1,2}(T, \mathbb{R}^N)$; again if $a \equiv 1$, then $\beta^{-1} = \lambda_1$ the first eigenvalue of the one dimensional p-Laplacian with Dirichlet boundary conditions.

(iii) for almost all $t \in T$, all $x, y \in \mathbb{R}^N$ and all $u \in F(t, x, y)$

$$\|u\| \leq \gamma(t) + c_2(\|x\| + \|y\|)$$

with $\gamma \in L^2(T), c_2 > 0$ and $c_2(\eta - 1)^2 < c$.

Theorem 18: *If hypotheses $H(a)$ (with $p = 2$) and $H(F)_8$ hold, then* $\bar{S}_e^{W_0^{1,2}(T, \mathbb{R}^N)} = S_c$

Proof: From the proof of Theorem 17, we know that we can replace $F(t, x, y)$ by $\hat{F}(t, x, y) = F(t, p_{M_2}(x), p_{M_4}(y))$. Let $K = \bar{B}(0, \mu) \subseteq W_0^{1,2}(T, \mathbb{R}^N)$ be as in the proof of Theorem 17.

Let $x \in S_c$. Then by definition, we have $x \in C^1(T, \mathbb{R}^N)$ and

$$(a(\|x'(t)\|^p)x'(t))' = f(t) \text{ a.e. on } T, f \in S^2_{F(\cdot, x(\cdot), x'(\cdot))}.$$

Given $y \in K$ and $\varepsilon > 0$ and let $\Gamma_\varepsilon \to 2^{\mathbb{R}^N} \setminus \{\emptyset\}$ be defined by

$$\Gamma_\varepsilon(t) = \{u \in \mathbb{R}^N : \|f(t) - u\| < \varepsilon + d(f(t), F(t, y(t), y'(t))), u \in F(t, y(t), y'(t))\}.$$

Note that by virtue of hypotheses $H(F)_8(i)$ and (ii), $(t, x, y) \to F(t, x, y)$ is measurable (see Proposition II.7.9 [18:p. 229]). Hence $t \to d(f(t), F(t, y(t), y'(t)))$ and $(t, u) \to d(u, F(t, y(t), y'(t)))$ is Caratheodory, thus jointly measurable (see Proposition II.1.6 [18:p. 142]). Therefore we have

$$\begin{aligned} Gr\Gamma_\varepsilon = \{(t, u) \in T \times \mathbb{R}^N &: \|f(t) - u\| - d(f(t), F(t, y(t), y'(t))) < \varepsilon \\ &d(u, F(t, y(t), y'(t))) = 0\} \in \mathcal{L} \times B(\mathbb{R}^N). \end{aligned}$$

So we can apply the Yankov-von Neumann-Aumann selection theorem and obtain $u: T \to \mathbb{R}^N$ a measurable function such that $u(t) \in \Gamma_\varepsilon(t)$ for all $t \in T$. Hence if we define $L_\varepsilon: K \to 2^{L^1(T, \mathbb{R}^N)}$ by

$$L_\varepsilon(y) = \left\{u \in S^1_{F(\cdot, x(\cdot), x'(\cdot))} : \|f(t) - u(t)\| + d(f(t), F(t, y(t), y'(t))) \text{ a.e. on } T\right\}$$

from what was done above, we see that for all $y \in K$, $L_\epsilon(y) \neq \emptyset$. Moreover, from Lemma II.8.3 of Hu-Papageorgiou [18:p. 239], we have that L_ε is lsc. Thus so is $y \to \bar{L}_\varepsilon(y)$ (see Proposition I.2.38 [18]). Apply theorem II.8.7 of Hu-Papageorgiou [18:p. 245], to obtain $u_\varepsilon: K \to L^1(T, \mathbb{R}^N)$ a continuous map such that $u_\varepsilon(y) \in \bar{L}_\varepsilon(y)$ for all $y \in K$. Hence we have

$$\begin{aligned}\|f(t) - u_\varepsilon(y)(t)\| &\leq \varepsilon + d(f(t), F(t, y(t), y'(t)))\\ &\leq \varepsilon + k(t)(\|x(t) - y(t)\| + \|x'(t) - y'(t)\|) \text{ a.e. on } T.\end{aligned}$$

In addition, by virtue of theorem II.8.31, of Hu-Papageorgiou [18:p. 260], we can find $v_\varepsilon: K \subseteq L^2(T, \mathbb{R}^N) \to L^1_w(T, \mathbb{R}^N)$ a continuous map such that $v_\varepsilon(y) \in \text{ext} S^1_{F(\cdot, y(\cdot), y'(\cdot))} = S^1_{F(\cdot, y(\cdot), y'(\cdot))}$ and

$$\|u_\epsilon(y) - v_\epsilon(y)\|_w < \epsilon \text{ for all } y \in K.$$

Let $\epsilon_n \downarrow 0$ and set $u_n = u_{\varepsilon_n}$, $v_n = v_{\varepsilon_n}$. Arguing as in the proof of theorem 17, we can generate $x_n \in K$ such that

$$V_1(x_n) + v_n(x_n) = 0, \; n \geq 1. \tag{14}$$

Recall that $\{x_n\}_{n\geq 1} \subseteq W_0^{1,2}(T, \mathbb{R}^N)$ is bounded and so we may assume that $x_n \xrightarrow{w} y$ in $W_0^{1,2}(T, \mathbb{R}^N)$. Moreover, directly from (14), we see that $\{\varphi(x_n')\}_{n\geq 1} \subseteq W^{1,2}(T, \mathbb{R}^N)$ is bounded, hence relatively compact in $C(T, \mathbb{R}^N)$. Thus we may assume that $\varphi(x_n') \to w$ in $C(T, \mathbb{R}^N)$ and so $x_n'(t) \to \psi(w(t))$ for all $t \in T$, which by the dominated convergence theorem implies that $x_n' \to \psi(w)$ in $L^2(T, \mathbb{R}^N)$. Since we already know that $x_n \xrightarrow{w} y$ in $W_0^{1,p}(T, \mathbb{R}^N)$, it follows that $y = \psi(w)$. Therefore we have proved that $x_n \to y$ in $W_0^{1,2}(T, \mathbb{R}^N)$. We have:

$$\begin{aligned}&\langle V_1(x_n) - V_1(x), x_n - x\rangle + \langle v_n(x_n) - f, x_n - x\rangle = 0\\ &\Rightarrow c\|x_n' - x'\|_2^2 \leq \int_0^b (f(t) - u_n(x_n)(t), x_n(t) - x(t))_{\mathbb{R}^N}\\ &+ \int_0^b (u_n(x_n)(t) - v_n(x_n)(t), x_n(t) - x(t))_{\mathbb{R}^N} dt.\end{aligned}$$

Note that

$$\begin{aligned}&\int_0^b (f(t) - u_n(x_n)(t), x_n(t) - x(t))_{\mathbb{R}^N} dt \leq\\ &\frac{1}{n} b^{\frac{1}{2}} \|x_n - x\|_2 + \|k\|_\infty (\|x_n - x\|_2 + \|x_n' - x'\|_2) \|x_n - x\|_2\end{aligned}$$

Also since by construction $\|u_n(x_n) - v_n(x_n)\|_w \leq \frac{1}{n} \to 0$, we have $u_n(x_n) - v_n(x_n) \xrightarrow{w} 0$ in $L^2(T, \mathbb{R}^N)$ (see [20]). Therefore

$$\int_0^b (u_n(x_n)(t) - v_n(x_n)(t), x_n(t) - x(t))_{\mathbb{R}^N} dt \to 0$$

Thus passing to the limit as $n \to \infty$, we obtain

$$c\|y' - x'\|_2^2 \leq \|k\|_\infty \|y - x\|_2^2 + \|k\|_\infty \|y' - x'\|_2 \|y - x\|_2$$

$$\leq \|k\|_\infty (\beta + \beta^{\frac{1}{2}}) \|y' - x'\|_2^2$$

$$\Rightarrow y' = x' \quad \text{(hypothesis } H(F)_8(ii)\text{)}.$$

Because $x, y \in W_0^{1,p}(T, \mathbb{R}^N)$, we conclude that $x = y$. Thus we have proved that $S_c \subseteq \hat{S}^{W_0^{1,2}(T,\mathbb{R}^N)}$. But from previous considerations we know that S_c closed in $W_0^{1,p}(T, \mathbb{R}^N)$. So finally we conclude that $S_c = \hat{S}_e^{W_0^{1,2}(T,\mathbb{R}^N)}$ (strong relaxation). □

Remark: Theorem 18 extends Theorem 25 of Halidias-Papageorgiou [16], where F is independent of y and $a(r^2) \equiv 1$.

6. SPECIAL CASES

In this section we present some special cases of interest, which illustrate the unifying character of our work in this chapter. In particular, we produce as special cases of problem (1), the classical Dirichlet, Neumann and periodic problems, which have been studied extensively in the literature, with $p = 2$ and $a(r) = 1$ (semilinear problem).

(a) Let $K_1, K_2 \subseteq \mathbb{R}^N$ be nonempty, closed and convex sets with $0 \in K_1 \cap K_2$. Let $\delta_{K_1 \times K_2}$ be the indicator of the set $K_1 \times K_2$, i.e.,

$$\delta_{K_1 \times K_2}(x) = \begin{cases} 0 & \text{if } x \in K_1 \times K_2 \\ +\infty & \text{otherwise.} \end{cases}$$

It is well-known that $\delta_{K_1 \times K_2}(x)$ is proper, convex and lower semicontinuous. Let $\partial\delta_{K_1 \times K_2}$ be its subdifferential in the sense of convex analysis. We know that $\partial\delta_{K_1 \times K_2} = N_{K_1 \times K_2} = N_{K_1} \times N_{K_2}$ (see Proposition VII.3.8 [18:p. 636]), where for any $K \in P_{fc}(\mathbb{R}^N)$ by $N_K(x)$ we denote the normal cone to K at x. Set $\xi = \partial\delta_{K_1 \times K_2} = N_{K_1 \times K_2} = N_{K_1} \times N_{K_2}$. Also let $a(r^p) = 1 + \frac{1}{(1+r^p)^p}$. Then we have the following special case of problem (1):

$$\begin{aligned} &((1 + \frac{1}{(1 + \|x'(t)\|^p)^p}) \|x'(t)\|^{p-2} x'(t))' \in A(x(t)) + F(t, x(t), x'(t)) \text{ a.e. on } T \\ &x(0) \in K_1, x(b) \in K_2, \\ &(x'(0), x(0))_{\mathbb{R}^N} = \sigma(x'(0), K_1), (x'(b), x(b))_{\mathbb{R}^N} = \sigma(-x'(b), K_2). \end{aligned} \tag{15}$$

Then we can apply to this problem the existence results of Section 3. We state only the "convex" result:

Corollary 19: *If hypotheses $H(A)_1, H(F)_1$ or $H(A)_2, H(F)_2, H_0$ hold, then problem (15) has a solution $x \in C^1(T, \mathbb{R}^N)$.*

Remark: Note that in this case hypothesis $H(\xi)$ is automatically satisfied, since if $(k', d') \in \xi(k, d) = N_{K_1}(k) \times N_{K_2}(d)$, then from the definition of the normal cone, we have $(k', k)_{\mathbb{R}^N} \geq 0$ and $(d', d)_{\mathbb{R}^N} \geq 0$ (see Definition VI.5.33 [18:pp. 634–635]).

In particular let $K_1, K_2 \subseteq \mathbb{R}_+^N$ be as before and let $A = \partial\psi$ with $\psi = \delta_{\mathbb{R}_+^N}$. Then for all $x \in \mathbb{R}_+^N$, we have

$$A(x) = N_{\mathbb{R}_+^N}(x) = \begin{cases} \{0\} & \text{if } x_k > 0 \text{ for all } k \in \{1, \dots, N\} \\ -\mathbb{R}_+^N & \text{if } x_k = 0 \text{ for some } k \in \{1, \dots, N\}. \end{cases}$$

By direct calculation, we can check that $A_\lambda(x) = \frac{1}{\lambda}(x - m_{\mathbb{R}_+^N}(x))$ (as before by $m_{\mathbb{R}_+^N}(\cdot)$ we denote the metric projection on $\mathbb{R}^N$). Then for $x \in K_1$ or $x \in K_2$ we have that $m_{\mathbb{R}_+^N}(x) = x$ and so $A_\lambda(x) = 0$, which means that hypotheses H_0 holds. Provided that hypothesis $H(F)_2$ hold (note that in the present situation $\mathrm{dom}A \neq \mathbb{R}^N$), we can apply Corollary 19 and obtain a solution for the following variational inequality problem (note that the problem can be viewed as a viability problem):

$$\begin{aligned}
&((1 + \frac{1}{(1 + \|x'(t)\|^p)^p})\|x'(t)\|^{p-2}x'(t))' \in F(t, x(t), x'(t)) \text{ a.e. on} \\
&T_+ = \{t \in T : x_k(t) > 0 \text{ for all } k \in \{1, \dots N\}\} \\
&((1 + \frac{1}{(1 + \|x'(t)\|^p)^p})\|x'(t)\|^{p-2}x'(t))' \in F(t, x(t), x'(t)) - \mathbb{R}_+^N \text{ a.e. on} \\
&T_0 = \{t \in T : x_k(t) = 0 \text{ for some } k \in \{1, \dots N\}\} \\
&x(t) \geq 0 \text{ for all}, t \in T,\ x(0) \in K_1, x(b) \in K_2 \\
&(x'(0), x(0))_{\mathbb{R}^N} = \sigma(x'(0), K_1), (-x'(b), x(b))_{\mathbb{R}^N} = \sigma(-x'(b), K_2).
\end{aligned}$$

Remark: To our knowledge this is the first existence result for variational inequalities of quasilinear ordinary equations.

(b) In case (a), let $K_1 = K_2 = \{0\}$, then $N_{K_1}(x) = \mathbb{R}^N$ for all $x \in K_1$ and $N_{K_2}(y) = \mathbb{R}^N$ for all $y \in K_2$. Thus there are no restrictions on the derivative at $t = 0$ and $t = b$ and so problem (1) becomes the Dirichlet problem

$$\begin{aligned}
&((1 + \frac{1}{(1 + \|x'(t)\|^p)^p})\|x'(t)\|^{p-2}x'(t))' \in A(x(t)) + F(t, x(t), x'(t)) \text{ a.e. on } T \\
&x(0) = x(b) = 0.
\end{aligned} \tag{16}$$

Note that in this situation hypothesis H_0 is automatically satisfied because $A_\lambda(0) = 0$. So we can state the following existence result:

Corollary 20: *If hypotheses $H(A)_1, H(F)_1$ or $H(A)_2, H(F)_2$ hold, then problem (16) has a solution $x \in C^1(T, \mathbb{R}^N)$.*

(c) In case (a) let $K_1 = K_2 = \mathbb{R}^N$. Then in this case $N_{K_1}(x) = N_{K_2}(x) = \{0\}$ for all $x \in \mathbb{R}^N$. So now there are no restrictions on the values of the values of the function at $t = 0$ and $t = b$. The resulting boundary value problem, is the Neumann problem:

$$\begin{aligned}
&((1 + \frac{1}{(1 + \|x'(t)\|^p)^p})\|x'(t)\|^{p-2}x'(t))' \in A(x(t)) + F(t, x(t), x'(t)) \text{ a.e. on } T \\
&x'(0) = x'(b) = 0.
\end{aligned} \tag{17}$$

Since $\xi(k,d) = \partial\delta_{K_1\times K_2}(k,d) = N_{K_1}(k)\times N_{K_2}(d) = \{(0,0)\}$, it is clear that hypothesis H_0 holds trivially. So we can state the following existence result:

Corollary 21: *If hypotheses $H(A)_1, H(F)_1$ or $H(A)_2, H(F)_2$ hold, then problem (17) has a solution $x \in C^1(T,\mathbb{R}^N)$.*

(d) Now let $K = \{(x,y)\in\mathbb{R}^N\times\mathbb{R}^N : x = y\}$ and set $\xi = \partial\delta_K = K^{\perp} = \{(v,w) \in\mathbb{R}^N\times\mathbb{R}^N : v = -w\}$ In this case the resulting boundary value problem is the periodic problem.

$$\begin{aligned}&((1+\frac{1}{(1+\|x'(t)\|^p)^{p}})\|x'(t)\|^{p-2}x'(t))' \in A(x(t)) + F(t,x(t),x'(t)) \text{ a.e. on } T\\ &x(0) = x(b), x'(0) = x'(b).\end{aligned} \tag{18}$$

From the definition of K and ξ, it is clear that hypothesis H_0 holds. So we can state the following existence result for the problem (18):

Corollary 22: *If hypotheses $H(A)_1, H(F)_1$ or $H(A)_2, H(F)_2$ hold, then problem* (18) *has a solution $x \in C^1(T,\mathbb{R}^N)$.*

(e) For the next boundary value problem, let ξ: $\mathbb{R}^N\times\mathbb{R}^N \to \mathbb{R}^N\times\mathbb{R}^N$ be defined by

$$\xi(x,y) = \left(\frac{1}{\vartheta^{p-1}}\varphi(x), \frac{1}{\eta^{p-1}}\varphi(y)\right), \vartheta,\eta > 0.$$

Evidently ξ is monotone, continuous, thus maximal monotone and $\xi(0,0) = (0,0)$. With this choice of ξ, problem (1) gives us the following Sturm-Liouville type problem:

$$\begin{aligned}&\left(\left(1+\frac{1}{(1+\|x'(t)\|^p)^{p}}\right)\|x'(t)\|^{p-2}x'(t)\right)' \in A(x(t)) + F(t,x(t),x'(t)) \text{ a.e. on } T\\ &x'(0)) = \vartheta x'(0), x(b) = \eta x'(b).\end{aligned} \tag{19}$$

If $(k',d') = \xi(k,d)$, then $(k',k)_{\mathbb{R}^N} = \frac{1}{\vartheta^{p-1}}\|k\|^d \geq 0$ and $(d',d)_{\mathbb{R}^N} = \frac{1}{\eta^{p-1}}\|d\|^d \geq 0$, which imply that hypothesis $H(\xi)(i)$ holds. Also we have

$$(A_\lambda(k), k')_{\mathbb{R}^N} = (A_\lambda(k), \frac{1}{\vartheta^{p-1}}\varphi(k))_{\mathbb{R}^N} = \frac{1}{\vartheta^{p-1}}\|k\|^{p-2}(A_\lambda(k),k)_{\mathbb{R}^N} \geq 0$$

because $A_\lambda(\cdot)$ is monotone and $A_\lambda(0) = 0$. Similarly we show that $(A_\lambda(d),d)_{\mathbb{R}^N} \geq 0$. Thus we have satisfied hypothesis H_0 and so we are in a position to apply the existence theory of Section 3. We can state the following existence result for the problem (19):

Corollary 23: *If hypotheses $H(A)_1, H(F)_1$ or $H(A)_2, H(F)_2$ hold, then problem (19) has a solution $x \in C^1(T,\mathbb{R}^N)$.*

(f) Let ξ: $\mathbb{R}^N\times\mathbb{R}^N \to \mathbb{R}^N\times\mathbb{R}^N$ be defined by

$$\xi(x,y) = (x - h_1(x), y - h_2(y))$$

with $h_1, h_2: \mathbb{R}^N \to \mathbb{R}^N$ nonexpansive maps with $h_1(0) = h_2(0) = 0$. Then ξ is monotone, continuous, thus maximal monotone (see Example III.1.11 [18:p. 304]). Also $\xi(0,0) = (0,0)$. Assume that $(h_i(x), x)_{\mathbb{R}^N} \leq c\|x\|^2$ for all $x \in \mathbb{R}^N$ and for $i = 1, 2$, with $0 < c < 1$. Moreover, suppose that for all $x \in \mathbb{R}^N$ and all $\lambda > 0$, $(A_\lambda(x), b_i(x))_{\mathbb{R}^N} \leq (A_\lambda(x), x)_{\mathbb{R}^N}, i = 1, 2$. Then with theses assumptions, hypotheses $H(\xi)$ and H_0 are satisfied. The boundary value problem under consideration is the following:

$$\begin{aligned} &((1 + \frac{1}{(1 + \|x'(t)\|^p)^p} \|x'(t)\|^{p-2} x'(t))' \in A(x(t)) + F(t, x(t), x'(t)) \text{ a.e. on } T \\ &\varphi(x'(0)) = x(0) - h_1(x(0)) \\ &- \varphi(x'(b)) = x(b) - h_2(x(b)). \end{aligned} \tag{20}$$

Corollary 24: *If hypotheses $H(A)_1, H(F)_1$ or $H(A)_2, H(F)_2$ hold and $\xi(x, y)$ is as above, then problem (20) has a solution $x \in C^1(T, \mathbb{R}^N)$.*

Remark: In all the above corollaries, if hypotheses $H(A)_1, H(F)_1$ are in effect, then the solution set of the corresponding problem is compact in $C^1(T, \mathbb{R}^N)$. Moreover, if $F(t, x, y) = f(t, x, y)$ is single valued and monotone in x and hypotheses $H(A)_1, H(F)_1$ hold, then the solution set of all these problems is a compact R_δ in $W^{1,p}(T, \mathbb{R}^N)$. Finally in all these special cases, we can also have $a(r) \equiv 1$ (the cases of the one-dimensional p-Laplacian) and $a(r) = r^{\frac{1-p}{p}} \beta(r)$ with $\beta \in C(\mathbb{R}_+, \mathbb{R}_+)$ strictly increasing and $cr \leq \beta(r) \leq c_1 r$ for all $r \geq 0$.

REFERENCES

[1] N. Aronszajn (1942). Le correspodent topologique de l'unicite dans la theorie des equations differentielles, *Annals of Math.*, **43**, 730–738.

[2] R. Bader. A fixed point index theory for a class of set-valued maps. University of Munchen, Germany, (preprint).

[3] J. Bebernes and M. Martelli (1980). On the structure of the solution set for periodic boundary value problems, *Nonlin. Anal.*, **4**, 821–830.

[4] L. Boccardo, P. Drabek, D. Giachetti and M. Kucera (1986). Generalization of Fredholm alternative for nonlinear diffrential operators, *Nonlin. Anal.*, **10**, 1083–1103.

[5] A. Bressan, A. Cellina and A. Fryszkowski (1991). On a class of absolute retracts in spaces of integrable functions, *Proc. AMS*, **111**, 413–418.

[6] H. Dang and S.F. Oppenheimer (1996). Existence and uniqueness results for some nonlinear boundary value problems, *J. Math. Anal. Appl.*, **198**, 35–48.

[7] M. Del Pino, M. Elgueta and R. Manasevich (1989). A homotopic deformation along p of a Leray-Schauder degree result and existence for $(|u'|^{p-2}u')' + f(t, u) = 0$, $u(0) = u(T) = 0$, *J. Diff. Eqns.*, **80**, 1–13.

[8] M. Del Pino, R. Manasevich, and A. Murua (1992). Existence and multiplicity of solutions with prescribed period for a second order quasilinear o.d.e., *Nonlinear Anal.*, **18**, 79–92.

[9] P. Drabek (1986). Solvability of boundary value problems with homogeneous ordinary differetial operator, *Rend. Ist. Mat. Univ. Trieste*, **8**, 105–124.

[10] R. Dragoni, J. Macki, P. Nistri and P. Zecca (1996). Solution Sets of Differential Equations in Abstarct Spaces. Essex, UK: Longman.

[11] L. Erbe and W. Krawcewicz (1992). Boundary value problems for differential inclusions $y'' \in F(t, y, y')$, *Ann. Polon. Math.*, **56**, 195–226.

[12] L. Evans and R. Gariepy (1992). Measure Theory and Fine Properties of Functions, Boca Raton, Florida: CRC Press.

[13] C. Fabry and D. Fayyad (1992). Periodic solutions of second order differential equations with a p-Laplacian and asymmetric nonlinearities, *Rend. Istit. Mat. Univ. Trieste*, **24**, 207–227.

[14] M. Frigon (1995). Theoremes d'existence des solutions d'inclusions differentielle Topological Methods in Diferential Equations and Inclusions. NATO ASI Series, section C, Vol. **472**. Kluwer, Dordrecht, The Netherlands, 51–87.

[15] Z. Guo (1993). Boundary value problems of a class of quasilinear differential equation, *Diff. Intergral Eqns.*, **6**, 705–719.

[16] N. Halidias and N.S. Papageorgiou (1998). Existence and relaxation results for nonlinear second order multivalued boundary value problems in $\mathbb{R}^N$, *J. Diff. Eqns.*, **147**, 123–154.

[17] P. Hartman (1964). Ordinary Differential Equations. New York: J. Wiley.

[18] S. Hu and N.S. Papageorgiou (1997). Handbook of Multivalued Analysis. Volume I: Theory. Kluwer, Dordrecht, The Netherlands.

[19] D. Kandilakis and N.S. Papageorgiou (1996). Existense theorems for nonlinear boundary value problem for econd order differential incluions, *J. Diff. Eqns.*, **132**, 107–125.

[20] D. Kravvaritis and N.S. Papageorgiou (1994). Boundary value problems for nonconvex differential inclusions, *J. Math. Anal. Appl.*, **185**, 146–160.

[21] R. Manasevich and J. Mawhin (1998). Periodic solutions for nonlinear systems with p-Laplacian-like operators, *J. Diff. Eqns.*, **145**, 367–393.

[22] M. Marcus and V. Mizel (1972). Absolute continuity on tracks and mappings of Sobolev spaces, *Arch. Rational Mech. Anal.*, **45**, 294–320.

[23] J. Nieto (1987). Hukuhara-Kneser property for nonlinear Dirichlet problem, *J. Math. Anal. Appl.*, **128**, 57–63.

[24] E. Zeidler (1990). Nonlinear Functional Analysis and its Applications II. New York: Springer-Verlag.

SIMAA 4(2002) 287–310

18. Optimal Control of a Class of Nonlinear Parabolic Problems with Multivalued Terms

Nikolaos Matzakos, Nikolaos S. Papageorgiou* and Nikolaos Yannakakis

National Technical University, Department of Mathematics, Zografou Campus, Athens 157 80, Greece

Abstract: In this chapter we examine nonlinear parabolic optimal control problems involving multivalued terms. First we consider a Lagrange optimal control problem with the differential operator being in divergence form and nonmonotone. Then we turn our attention to parametric problems with a nonlinear, multivalued, maximal monotone differential operator. In this case the relevant problem is a minimax problem in which we minimize the maximum cost over all parameters. We prove the existence of an optimal control using the notion of G-convergence of operators.

Keywords and Phrases: Evolution triple, compact embedding, upper and lower semicontinuous multifunctions, measurable multifunctions, measurable selection, monotone operator, pseudomonotone operator, L-generalized pseudomonotone operator, G-convergence, τ-topology, optimal control, minimax problem

1990 AMS Subject Classification: 49J35, 49J27

1. INTRODUCTION

In this chapter we study nonlinear parabolic optimal control problems, with multivalued terms. First, under very general hypotheses on the data, we establish the existence of optimal state-control pairs. This is done by employing the well known "direct method". However, the result on the nonemptiness and compactness of the set of admissible states is new and of independent interest. Subsequently we turn our attention to a parametric version of the problem. Since it is not known which parameter value is in force, the system analyst first evaluates the maximum cost over all possible parameter values and then minimizes this maximum cost over all feasible controls. Thus the relevant optimization problem is a minimax problem. Under reasonable hypotheses on the data, we show that there exists an optimal control solving the minimax problem. A basic tool in our analysis is the notion of G-convergence of nonlinear multivalued elliptic operators, as this was defined and studied by Chiado Piat-Dal Maso-Defranceschi [3].

* Corresponding author. E-mail: npapag@math.ntua.gr

2. PRELIMINARIES

In this section we recall some basic definitions and facts from Multivalued Analysis and from Nonlinear Analysis, which we will need in the sequel. Our basic references are the books of Hu-Papageorgiou [6] and Zeidler [12].

Let (Ω, Σ) be a measurable space and X a separable Banach space. Throughout this work we will be using the following notations:

$$P_{f(c)}(X) = \{C \subseteq X : C \text{ is nonempty, closed (and convex)}\}$$
$$P_{(w)k(c)}(X) = \{C \subseteq X : C \text{ is nonempty, (weakly-) compact (and convex)}\}.$$

A multifunction

$$F : \Omega \to P_f(X)$$

is said to be measurable, if for all $x \in X$,

$$\omega \to d(x, F(\omega)) = \inf\{\|x - v\| : v \in F(\omega)\}$$

is measurable. Let $\mu(\cdot)$ be a finite measure on (Ω, Σ). For $1 \leq p \leq \infty$ and $F: \Omega \to 2^X \backslash \{\emptyset\}$ a multifunction, we define the set

$$S_F^p = \{f \in L^p(\Omega, X) : f(\omega) \in F(\omega) \ \ \mu - \text{a.e.}\}.$$

In general this set may be empty. If

$$GrF = \{(\omega, x) \in \Omega \times X : x \in F(\omega)\} \in \Sigma \times \mathcal{B}(X)$$

(i.e., if F is graph measurable), then

$$S_F^p \neq \emptyset$$

if and only if

$$\omega \to \inf\{\|x\| : x \in F(\omega)\} \in L^p(\Omega).$$

For $P_f(X)$-valued multifunctions, measurability implies graph measurability, while the converse is true if Σ is μ-complete.

Let Y, Z be Hausdorff topological spaces and

$$G: Y \to 2^Z \backslash \{\emptyset\}$$

a multifunction. We say that $G(\cdot)$ is upper semicontinuous (usc) (resp., lower semicontinuous (lsc)), if for every $C \subseteq Z$ closed the set

$$G^-(C) = \{y \in Y : G(y) \cap C \neq \emptyset\} \ \text{(resp., } G^+(C) = \{y \in Y : G(y) \subseteq C\})$$

is closed. An upper semicontinuous multifunction has closed graph, while the converse is true provided the multifunction is locally compact.

Let H be a separable Hilbert space and let X be a separable reflexive Banach space, which is embedded continuously and densely in H. Identifying H with its dual (pivot space), we have

$$X \subseteq H \subseteq X^*$$

with all injections being continuous and dense. The triple (X, H, X^*) is called "evolution triple". By $\|\cdot\|$ (resp., $\|\cdot\|$, $\|\cdot\|_*$) we denote the norm of H (resp., of X, X^*) and by $\langle\cdot,\cdot\rangle$ (resp., $(\cdot,\cdot)$) the duality brackets for the pair (X, X^*) (resp., the inner product of H). The two are compatible in the sense that

$$\langle\cdot,\cdot\rangle|_{X\times H} = (\cdot,\cdot).$$

Let $T = [0, b]$ and $1 < p, q < \infty$, $\frac{1}{p} + \frac{1}{q} = 1$. We define:

$$W_{pq}(T) = \{x \in L^p(T, X) : \dot{x} \in L^q(T, X^*)\}.$$

The time derivative in this definition, is understood in the sense of vector-valued distributions. Furnished with the norm

$$\|x\|_{pq} = \left\{\|x\|_p^2 + \|\dot{x}\|_q^2\right\}^{\frac{1}{2}}$$

becomes a separable reflexive Banach space. It is well known that $W_{pq}(T)$ is embedded continuously in the Banach space $C(T, H)$. Moreover, if X is embedded compactly in H (hence so does H into X^*), then $W_{pq}(T)$ is embedded compactly in $L^p(T, H)$.

Let Y be a reflexive Banach space and

$$L \colon D \subseteq Y \to Y^*$$

a linear, maximal monotone operator. A map

$$V \colon Y \to P_{wkc}(Y^*)$$

is said to be L-generalized pseudomonotone, if

$$\begin{aligned}
&y_n \xrightarrow{w} y \text{ in } Y,\\
&Ly_n \xrightarrow{w} Ly \text{ in } Y^*,\\
&y_n^* \in V(y_n),\ n \geq 1\\
&\text{and}\\
&\overline{\lim}(y_n^*, y_n - y)_{Y^*,Y} \leq 0,
\end{aligned}$$

imply that

$$y^* \in V(y) \quad \text{and} \quad (y_n^*, y_n)_{Y^*,Y} \to (y^*, y)_{Y^*,Y}.$$

The following surjectivity result for L-generalized pseudomonotone operators will be used in Section 3. For V single-valued, is due to Lions [7]. The multivalued version used here was proved by Papageorgiou-Papalini-Renzacci [10].

Proposition 1: *If Y is a reflexive Banach space, Y and Y^* are both strictly convex,*

$$L: D \subseteq Y \to Y^*$$

is a linear maximal monotone operator and

$$V: Y \to P_{wkc}(Y^*)$$

is L-generalized pseudomonotone, bounded and coercive, then $L + V$ is surjective (i.e., $R(L+V) = Y^*$). □

Next let $Z \subseteq \mathbb{R}^N$ be a bounded domain. By $M_Z(\mathbb{R}^N)$ we denote the class of all multifunctions

$$a: Z \times \mathbb{R}^N \to P_{kc}(\mathbb{R}^N) \text{ such that}$$

(i) $(z, y) \to a(z, y)$ is graph measurable;
(ii) for almost all $z \in Z$, $a(z, \cdot)$ is maximal monotone;
(iii) for almost all $z \in Z$, all $y \in \mathbb{R}^N$ and all $v \in a(z, y)$ we have

$$\|v\| \leq c_1(z) + c_2\|y\|^{p-1} \quad \text{and} \quad (v, y)_{\mathbb{R}^N} \geq c_3\|y\|^p - c_4(z)$$

with c_2, $c_3 > 0$, $c_1 \in L^p(Z)$ $(\frac{1}{p} + \frac{1}{q} = 1)$ and $c_4 \in L^1(Z)$.

Let

$$\alpha: W_0^{1,p}(Z) \to P_{wkc}(L^q(Z, \mathbb{R}^N))\text{-}$$

be defined by

$$\alpha(x) = S^q_{a(\cdot, Dy(\cdot))}.$$

Let σ_1 be the weak topology on $L^q(Z, \mathbb{R}^N)$ and by σ_2 the topology on $L^q(Z, \mathbb{R}^N)$ generated by the pseudometric

$$d(g_1, g_2) = \|\text{div}\, g_1 - \text{div} g_2\|_{W^{-1,q}(Z)}.$$

Let $\sigma = \sigma_1 \vee \sigma_2$ (i.e., the weakest topology on $L^q(Z, \mathbb{R}^N)$, which is stronger than both σ_1, σ_2). So

$$g_n \xrightarrow{\sigma} g$$

if and only if

$$g_n \xrightarrow{w} g \text{ in } L^q(Z, \mathbb{R}^N) \quad \text{and} \quad \text{div}\, g_n \to \text{div}\, g \text{ in } W^{-1,q}(Z).$$

For a sequence $\{a_n\}_{n \geq 1} \subseteq M_Z(\mathbb{R}^N)$, we say that

$$a_n \xrightarrow{G} a \text{ if and only if } K_{seq}(w \times \sigma) - \overline{\lim}\, Gr\alpha_n \subseteq Gr\alpha.$$

Recall (see for example Definition VII.1.3 [6:p. 660]) that

$$K_{seq}(w \times \sigma) - \overline{\lim}\, Gra_n = \left\{ \begin{array}{l} (x, g) \in W_0^{1,p}(Z) \times L^q(Z, \mathbb{R}^N) : \exists\, (x_{n_k}, g_{n_k}) \in Gra_{n_k},\ x_{n_k} \xrightarrow{w} x \\ \text{in } W_0^{1,p}(Z),\ g_{n_k} \xrightarrow{\sigma} g \text{ in } L^q(Z, \mathbb{R}^N) \text{ and } n_1 < n_2 < \ldots < n_k \ldots \end{array} \right\}.$$

3. EXISTENCE OF OPTIMAL PAIRS

In this section we prove the existence of optimal state-control pairs for a broad class of nonlinear parabolic optimal control problems with multivalued dynamics.

So let $T = [0, b]$ and let $Z \subseteq \mathbb{R}^N$ be a bounded domain with a C^1-boundary Γ. In this section we deal with the following optimal control problem:

$$\begin{aligned} & J(x, u) = \int_0^b \int_Z L(t, z, x(t, z), u(t, z))dzdt \to \inf = m \\ & \frac{\partial x}{\partial t} - \operatorname{div} a(t, z, x, Dx) \ni f(t, z, x(t, z)) + k(t, z)u(t, z) \text{ on } T \times Z \\ & x(0, z) = x_0(z) \text{ a.e. on } Z,\quad x|_{T\times\Gamma} = 0,\quad u(t, \cdot) \in U(t) \text{ a.e. on } T. \end{aligned} \tag{1}$$

Our hypotheses on the data of (1) are the following:

H(a): $a: T \times Z \times \mathbb{R} \times \mathbb{R}^N \to P_{kc}(\mathbb{R}^N)$ is a multifunction such that

(i) $(t, z, x, y) \to a(t, z, x, y)$ is graph measurable;

(ii) for all $(t, z) \in T \times Z$ and all $x \in \mathbb{R}$, $y \to a(t, z, x, y)$ is maximal monotone and strictly monotone;

(iii) for all $(t, z) \in T \times Z$, $(x, y) \to a(t, z, x, y)$ has closed graph and for all $(t, z, y) \in T \times Z \times \mathbb{R}^N$, $x \to a(t, z, x, y)$ is lsc;

(iv) for almost all $(t, z) \in T \times Z$, all $x \in \mathbb{R}$, all $y \in \mathbb{R}^N$ and all $v \in a(t, z, x, y)$

$$\|v\| \le \gamma(t, z) + c(|x|^{p-1} + \|y\|^{p-1})$$

with $\gamma \in L^q(T \times Z)$, $2 \le p < \infty$, $\frac{1}{p} + \frac{1}{q} = 1$, $c > 0$;

(v) for almost all $(t, z) \in T \times Z$, all $x \in \mathbb{R}$, all $y \in \mathbb{R}^N$ and all $v \in a(t, z, x, y)$, we have

$$(v, y)_{\mathbb{R}^N} \ge c_1\|y\|^p - \gamma_1(t, z)$$

with $\gamma_1 \in L^1(T \times Z)$, $c_1 > 0$.

H(f): $f: T \times Z \times \mathbb{R} \to \mathbb{R}$ is a function such that

(i) for all $x \in \mathbb{R}$, $(t, z) \to f(t, z, x)$ is measurable;

(ii) for all $(t, z) \in T \times Z$, $x \to f(t, z, x)$ is continuous;

(iii) for almost all $(t, z) \in T \times Z$ and all $x \in \mathbb{R}$, we have

$$|f(t, z, x)| \le \gamma_2(t, z) + c_2|x|^{p-1}$$

with $\gamma_2 \in L^q(T \times Z)$, $c_2 > 0$;

(iv) for almost all $(t,z) \in T \times Z$ and all $|x| > M(t,z)$ we have

$$f(t,z,x)x \leq 0$$

with $M \in L^p(T \times Z)_+$.

$H(U)$: $U\colon T \to P_{fc}(L^q(Z))$ is a measurable multifunction such that for almost all $t \in T$ and all $u \in U(t)$

$$\|u\|_q \leq M.$$

$H(L)$: $L\colon T \times Z \times \mathbb{R} \times \mathbb{R} \to \overline{\mathbb{R}} = \mathbb{R} \cup \{\infty\}$ is an integrand such that

(i) $(t,z,x,u) \to L(t,z,x,u)$ is measurable;
(ii) for all $(t,z) \in T \times Z$, $(x,u) \to L(t,z,x,u)$ is lower semicontinuous and for all $(t,z,x) \in T \times Z \times \mathbb{R}$, $u \to L(t,z,x,u)$ is convex;
(iii) for almost all $(t,z) \in T \times Z$ and all $(x,u) \in \mathbb{R} \times \mathbb{R}$ we have

$$\gamma_4(t,z) - c_4(|x| + |u|) \leq L(t,z,x,u)$$

with $\gamma_4 \in L^1(T \times Z)$, $c_4 > 0$.

$H(k)$: $k \in L^q(T \times Z)$.

By a solution of the multivalued nonlinear parabolic equation in (1), we mean a function $x \in L^p(T, W_0^{1,p}(Z)) \cap C(T, L^2(Z))$ such that $\frac{\partial x}{\partial t} \in L^q(T, W^{-1,q}(Z))$ and for all $\varphi \in C_0^\infty(Z)$ we have for almost all $t \in T$

$$\left\langle \frac{\partial x}{\partial t}(t,z), \varphi \right\rangle + \int_Z (v(t,z), D\varphi(z))_{\mathbb{R}^N} dz = \int_Z f(t,z,x(t,z))\varphi(z)dz + \int_Z k(t,z)u(t,z)\varphi(z)dz$$

with $v \in L^q(T \times Z)$, $v(t,z) \in a(t,z,x(t,z),Dx(t,z))$ a.e. on $T \times Z$ and $\langle \cdot,\cdot \rangle$ denoting the duality brackets of the pair $(W_0^{1,p}(Z), W^{-1,q}(Z))$. Also $x(0,z) = x_0(z)$ a.e. on Z with $x_0 \in L^2(Z)$. In what follows we set $X = W_0^{1,p}(Z)$, $H = L^2(Z)$, $X^* = W^{-1,q}(Z)$. Then (X, H, X^*) is an evolution triple with the embedding of X into H being compact (Sobolev embedding theorem). So from the above definition, we see that if $x(t,z)$ is a solution of the parabolic inclusion, then $x \in W_{pq}(T)$.

Let $\alpha\colon T \times X \to 2^{X^*}$ be defined by

$$\alpha(t,x) = S^q_{a(t,\cdot,x(\cdot),Dx(\cdot))}$$

Also let $A\colon T \times X \to 2^{X^*}$ be defined by

$$A(t,x) = \{-\mathrm{div} g : g \in \alpha(t,x)\}.$$

We will show that for every $t \in T$, $A(t,\cdot)$ is pseudomonotone. To this end we need an auxiliary result. Fix $t \in$ and $y \in W_0^{1,p}(Z)$ and consider the map

$$E_y(t)\colon \to 2^{X^*}$$

defined by

$$E_y(t)(x) = \left\{-\mathrm{div}\, u : u \in S^q_{a(t,\cdot,y(\cdot),Dx(\cdot))}\right\}.$$

Lemma 1: *If hypotheses $H(a)$ hold, then for all $t \in T$ and $y \in W_0^{1,p}(T)$, $x \to E_y(t)(x)$ is maximal monotone.*

Proof: By virtue of hypothesis $H(a)$(i), $z \to a(t, z, y(z), Dx(z))$ is graph measurable and so we can apply the Yankov-von Neumann-Aumann selection theorem (see Theorem II.2.14 [6:p. 158] and infer that

$$S^q_{a(t,\cdot,y(\cdot),Dx(\cdot))} \neq \emptyset, \text{ (see hypothesis } H(a)(\text{iv})).$$

Therefore $E_y(t)$ has values in $P_{wkc}(X^*)$ and it is immediate from hypothesis $H(a)(ii)$ that $E_y(t)(\cdot)$ is monotone. According to Theorem III.1.33; of Hu-Papageorgiou [6:p. 309], in order to prove the maximality of $E_y(t)(\cdot)$, it suffices to show that for every $x, w \in X$, the multifunction

$$\lambda \to E_y(t)(x + \lambda w)$$

is usc from [0,1] into X^*_w (here X^*_w denotes the space $X^* = W^{-1,q}(Z)$ equipped with the weak topology). By virtue of hypothesis $H(a)$(iv), $E_y(t)(\cdot)$ is locally weakly compact. So in order to prove the desired upper semicontinuity, it suffices to show that the graph of the multifunction $\lambda \to E_y(t)(x + \lambda w)$ is sequentially closed in $X \times X^*$ (see Proposition I.2.23 [6:p. 43]). So let $\lambda_n \to \lambda$ in [0,1], $v_n \xrightarrow{w} v$ in X^* and $v_n \in E_y(t)(x + \lambda_n w)$, $n \geq 1$. By definition we have $v_n = -\text{div} h_n$, $h_n \in S^q_{a(t,\cdot,y(\cdot),D(x+\lambda_n w)(\cdot))}$. We may assume that $h_n \xrightarrow{w} h$ in $L^q(Z, \mathbb{R}^N)$ (hypothesis $H(a)$(iv)) and so $v = -\text{div} h$. Also invoking Proposition VII.3.9, of Hu-Papageorgiou [6:p. 694], we have that

$$\begin{aligned} h(z) &\in \overline{\text{conv}}\, w - \overline{\lim}\, a(t, z, y(z), D(x + \lambda_n w)(z)) \\ &\subseteq a(t, z, y(z), D(x + \lambda w)(z)) \quad \text{a.e. on } Z. \end{aligned}$$

This last inclusion follows from the fact that $D(x + \lambda_n w)(z) \to D(x + \lambda w)(z)$ a.e. on Z in $\mathbb{R}^N$ and because $(x, y) \to a(t, z, x, y)$ has closed graph and values in $P_{kc}(\mathbb{R}^N)$. So

$$h \in S^q_{a(t,\cdot,y(\cdot),D(x+\lambda w)(\cdot))}$$

and thus $v \in E_y(t)(x + \lambda w)$. This proves the desired upper semicontinuity of $\lambda \to E_y(t)(x + \lambda w)$ and so we have the maximal monotonicity of $E_y(t)(\cdot)$. □

Using this lemma, we can prove the next proposition which is crucial in our subsequent considerations.

Proposition 2: *If hypotheses $H(a)$ hold, then for every $t \in T$, $A(t, \cdot): X \to 2^{X^*} \setminus \{\emptyset\}$ is bounded and pseudomonotone.*

Proof: It is clear that for every $t \in T$, $A(t, \cdot)$ is $P_{wkc}(X^*)$-valued and bounded (see hypothesis $H(a)$(iv)). So according to Proposition III.6.11, of Hu-Papageorgiou [6:p. 366], it suffices to show that

$$\text{If } x_n \xrightarrow{w} x \text{ in } X^* \quad \text{and} \quad \overline{\lim} \langle v_n, x_n - x \rangle \leq 0,$$

$$\text{then} \quad v \in A(t, x) \quad \text{and} \quad \langle v_n, x_n \rangle \to \langle v, x \rangle \quad \text{as} \quad n \to \infty$$

By definition we have $v_n = -\text{div} g_n$, with $g_n \in a(t, x_n)$. Because of $H(a)$(iv),

$$\{g_n\}_{n\geq 1} \subseteq L^q(Z, \mathbb{R}^N)$$

is bounded and so we may assume that $g_n \xrightarrow{w} g$ in $L^q(Z, \mathbb{R}^N)$. Hence $v = -\text{div} g$. Let $y \in X = W_0^{1,p}(Z)$ be arbitrary and for $t \in T$ consider the multifunction

$$\hat{a}_t \colon Z \to P_{kc}(\mathbb{R}^N)$$

defined by $\hat{a}_t(z) = a(t, z, x(z), Dy(z))$. Evidently $a_t(\cdot)$ is graph measurable and so by the Yankov-von Neumann-Aumann selection theorem (see Theorem II.2.14 [6:p. 158]), we can find

$$w \colon Z \to \mathbb{R}^N$$

a measurable map such that $w(z) \in \hat{a}_t(z)$ a.e. on Z. Another straightforward application of the Yankov-von Neumann-Aumann selection theorem gives measurable functions

$$w_n \colon Z \to \mathbb{R}^N, \quad n \geq 1,$$

such that

$$w_n(z) \in a(t, z, x_n(z), Dy(z)) \quad \text{a.e. on } Z \text{ and}$$
$$\|w(z) - w_n(z)\| = d(w(z), a(t, z, x_n(z), Dy(z))) \quad \text{a.e. on} \quad Z$$

So $w_n(z) \in S^q_{a(t,\cdot,x_n(\cdot),Dy(\cdot))}$ and we have

$$\|w(z) - w_n(z)\| \leq h^*(a(t, z, x(z), Dy(z)), a(t, z, x_n(z), Dy(z))) \quad \text{a.e. on } Z,$$

with h^* being the Hausdorff semidistance (see Definition I.1.1 [6:p. 5]). Because $x_n \xrightarrow{w} x$ in $X = W_0^{1,p}(Z)$ and the latter is embedded compactly in $L^p(Z)$ we have $x_n \to x$ in $L^p(Z)$ and by passing to a subsequence if necessary, we may assume that $x_n(z) \to x(z)$ a.e. on Z. Since by hypothesis $H(a)$(iii) $a(t, z, \cdot, y)$ is lsc and $P_{kc}(\mathbb{R}^N)$-valued, it is h-lsc (see Theorem I.2.68 [6:p. 62]) and so

$$h^*(a(t, z, x(z), Dy(z)), a(t, z, x_n(z), Dy(z))) \to 0 \quad \text{a.e. on } Z,$$
$$\Rightarrow \|w(z) - w_n(z)\| \to 0 \quad \text{a.e. on } Z$$
$$\Rightarrow w_n \to w \text{ in } L^q(Z, \mathbb{R}^N) \text{ (by the generalized dominated convergence theorem)}$$
$$\Rightarrow \text{div} w_n \to \text{div} w \text{ in } X^* = W^{-1,q}(Z).$$

Exploiting the monotonicity of $a(t, z, x_n(z), \cdot)$ (hypothesis $H(a)(iii)$), we have

$$\begin{aligned}
0 &\leq \int_Z (g_n(z) - w_n(z), Dx_n(z) - Dy(z))_{\mathbb{R}^N} dz \\
&= \int_Z (g_n(z), Dx_n(z) - Dx(z))_{\mathbb{R}^N} dz + \int_Z (g_n(z), Dx(z) - Dy(z))_{\mathbb{R}^N} \\
&+ \int_Z (w_n(z), Dy(z) - Dx_n(z))_{\mathbb{R}^N} dz \\
&= \langle -\text{div}\, g_n, x_n - x\rangle + \langle -\text{div}\, g_n, x - y\rangle + \langle -\text{div}\, w_n, y - x_n\rangle
\end{aligned}$$

$$\Rightarrow 0 \leq \langle -\text{div}\, g + \text{div}\, w, x - y\rangle \tag{2}$$

(recall the choice of the sequences $\{x_n\}_{n\geq 1} \subseteq X$ and $\{v_n = -\operatorname{div} g_n\}_{n\geq 1} \subseteq X^*$). Since the pair $[y, -\operatorname{div} w] \in GrE_x(t)$ was arbitrary and from lemma 2 $E_x(t)(\cdot)$ is maximal monotone, from (2) we infer that $-\operatorname{div} g \in E_x(t)(x) = A(t, x)$. Also recall that

$$0 \leq \langle -\operatorname{div} g_n, x_n - x\rangle + \langle -\operatorname{div} g_n, x - y\rangle + \langle -\operatorname{div} w_n, y - y_n\rangle,\ n \geq 1.$$

Let $y = x \in X$. Then we obtain

$$0 \leq \underline{\lim}\langle -\operatorname{div} g_n, x_n - x\rangle = \underline{\lim}\langle v_n, x_n - x\rangle.$$

From the choice of $\{x_n\}_{n\geq 1} \subseteq X$ and of $\{v_n = -\operatorname{div} g_n\}_{n\geq 1} \subseteq X^*$, we infer that

$$\langle v_n, x_n\rangle \to \langle v, x\rangle \text{ as } n \to \infty$$

which completes the proof of the proposition. □

We consider the dynamical equation of (1), namely the problem

$$\begin{aligned} &\frac{\partial x}{\partial t} - \operatorname{div} a(t, z, x, Dx) \ni f(t, z, x(t, z)) + k(t, z)u(t, z) \text{ on } T \times Z \\ &x(0, z) = x_0(z) \quad \text{a.e. on } Z, \quad u(t, \cdot) \in U(t) \quad \text{a.e. on } T. \end{aligned} \tag{3}$$

Fix $u \in S_U^\infty$, an admissible control, and let $S(u) \subseteq W_{pq}(T)$ be the solution set of (3) for the given control function.

Proposition 3: *If hypotheses $H(a), H(k)$ hold and $x_0 \in L^2(Z)$, then for every $u \in S_U^\infty$, $S(u) \subseteq W_{pq}(T)$ is nonempty and $\cup[S(u)\colon u \in S_U^\infty] \subseteq C(T, H)$ is compact.*

Proof: First assume that $x_0 \in W_0^{1,p}(Z)$. Let

$$F\colon T \times L^p(Z) \to L^q(Z)$$

be defined by $F(t, x)(\cdot) = f(t, \cdot, x(\cdot))$ (the Nemitsky operator corresponding to the function f) and $K(t) \in \mathcal{L}(L^\infty(Z), L^q(Z))$ be defined by $K(t)u(\cdot) = k(t, \cdot)u(\cdot)$. We rewrite (3) as the following abstract evolution equation, defined in the framework of the evolution triple

$$\begin{gathered} (X = W_0^{1,p},\ H = L^2(Z),\ X^* = W^{-1,q}(Z)), \\ \dot{x}(t) + A(t, x(t)) \ni F(t, x(t)) + K(t)u(t) \quad \text{a.e. on } T \\ x(0) = x_0. \end{gathered} \tag{4}$$

We solve (4) and in this way we obtain a solution for (3) too. To this end let

$$W_{pq}^0(T) = \{x \in L^p(T, X) : \dot{x} \in L^q(T, X^*),\ x(0) = 0\}.$$

Since $W_{pq}(T) \subseteq C(T, H)$ the evaluation at $t = 0$ makes sense. Let

$$L\colon W_{pq}^0(T) \subseteq L^p(T, X) \to L^q(T, X^*)$$

be defined by $Ly = \dot{y}$. It is well-known (see for example Proposition III.9.3 [6:p. 419]), that L is linear, maximal monotone. Also let

$$V_u\colon L^p(T, X) \to 2^{L^q(T, X^*)}$$

be defined by

$$V_u(y) = \{v \in L^q(T, X^*) : v(t) = h(t) - F(t, y(t) + x_0) \\ - K(t)u(t),\ h(t) \in A(t, y(t) + x_0) \quad \text{a.e. on } T\}.$$

Claim: V is L-generalized pseudomonotone.
First we show that $t \to A(t, y(t) + x_0)$ is measurable. For every $x \in X$

$$\begin{aligned}\sigma(x, A(t, y(t) + x_0)) &= \sup[\langle h, x\rangle : h \in A(t, y(t) + x_0)] \\ &= \sup[(g, Dx)_{pq} : g \in \alpha(t, y(t) + x_0)] \\ &= \sup\left[\int_Z (g(z), Dx(z))_{\mathbb{R}^N} dz : g \in S^q_{a(t,\cdot,y(t,\cdot)+x_0(\cdot),Dy(t,\cdot)+Dx_0(\cdot))}\right] \\ &= \int_Z \sigma(Dx(z), a(t, z, y(t, z) + x_0(z), Dy(t, z) + Dx_0(z)))dz\end{aligned}$$

(see Theorem II.3.24 [6:p. 183] and by $(\cdot,\cdot)_{pq}$ we denote the duality brackets of $(L^p(Z, \mathbb{R}^N),\ L^q(Z, \mathbb{R}^N)))$.

By virtue of hypothesis $H(a)(i)$ and Fubini's theorem,

$$t \to \int_Z \sigma(Dx(z), a(t, y(t, z) + x_0(z), Dy(t, z) + Dx_0(z)))dz$$

is measurable. Hence for every $x \in X$, $t \to \sigma(x, A(t, y(t) + x_0))$ is measurable and because $A(t, y(t) + x_0) \in P_{wkc}(X^*)$, from Proposition II.2.39, of Hu-Papageorgiou [6:p. 166], we have that $t \to A(t, y(t) + x_0)$ is measurable. From this it follows that $S^q_{A(\cdot, y(\cdot)+x_0)}$ is nonempty, bounded, closed and convex. Therefore V_u is $P_{wkc}(L^q(T, X^*))$-valued.

Next let $y_n \xrightarrow{w} y$ in $L^p(T, X)$, $\dot{y}_n \xrightarrow{w} \dot{y} \in L^q(T, X^*)$, $h_n \in L^q(T, X^*)$, $h_n(t) \in A(t, y_n(t) + x_0)$ a.e. on T, $h_n \xrightarrow{w} h$ in $L^q(T, X^*)$ and $\overline{\lim}((h_n, y_n - y)) \leq 0$ (by $((\cdot,\cdot))$ we denote the duality brackets for the pair $(L^p(T, X), L^q(T, X^*))$, i.e., $((h, y)) = \int_0^b < h(t), y(t) > dt$ for $h \in L^q(T, X^*)$, $y \in L^p(T, X)$). Let

$$\xi_n(t) = \langle h_n(t), y_n(t) - y(t)\rangle.$$

By definition we have $h_n(t) = -\text{div}\, g_n(t)$, with $g_n \in L^q(T \times Z)$,

$$g_n(t, z) \in a(t, z, y_n(t, z) + x_0(z), Dy_n(t, z) + Dx_0(z)) > \text{a.e. on } T \times Z.$$

Using hypotheses $H(a)(iv)$ and (v), we have

$$\begin{aligned}\langle h_n(t), y_n(t) - y(t)\rangle &= \langle -\text{div} g_n(t), y_n(t) - y(t)\rangle \\ &= (g_n(t), Dy_n(t) - Dy(t))_{pq} \\ &= \int_Z (g_n(t, z), Dy_n(t, z) - Dy(t, z))_{\mathbb{R}^N} dz \\ &\geq c_1\|Dy_n(t)\|^p_{L^p(Z)} - \|\gamma_1(t, \cdot)\|_{L^1(Z)} \\ &\quad - (c_5(t) + c_6\|Dy_n(t, \cdot)\|^{p-1}_{L^p(Z)})\|Dy(t, \cdot)\|_{L^p(Z)}\end{aligned}$$

for some $c_5 \in L^p(T)$ and $c_6 > 0$. By virtue of Poincare's inequality

$$\xi_n(t) \geq c_1\|y_n(t)\|^p - \|\gamma_1(t,\cdot)\|_{L^1(Z)} - (c_5(t) + c_6\|y_n(t)\|^{p-1})\|y(t)\| \quad \text{a.e. on } T. \tag{5}$$

Let $S = \{t \in T\colon \underline{\lim}\xi_n(t) < 0\}$ which is Lebesgue measurable. Suppose $\lambda(S) > 0$ (by $\lambda(\cdot)$ we denote the Lebesgue measure on T). From (5), it follows that if $N \subseteq T$ is the Lebesgue-null set of all $t \in S \cap (T \backslash N) \neq 0$, then $\{y_n(t,\cdot)\}_{n\geq 1} \subseteq X = W_0^{1,p}(Z)$ is bounded. So by passing to a subsequence (depending of course on $t \in S \cap (T\backslash N)$), we may say that $y_n(t,\cdot) \xrightarrow{w} \hat{y}(t,\cdot)$ in X. On the other hand from the choice of the sequence $\{y_n\}_{n\geq 1}$, we have $y_n \xrightarrow{w} y$ in $W_{pq}(T)$. Since the latter is embedded continuously in $C(T, L^2(Z))$, for all $t \in T$ we have $y_n(t,\cdot) \xrightarrow{w} y(t,\cdot)$ in $L^2(Z)$. Also because X is embedded compactly in $L^2(Z)$, we also have $y_n(t,\cdot) \to \hat{y}(t,\cdot)$ in $L^2(Z)$. Therefore finally we can say that for all $t \in S \cap (T\backslash N)$ and for the whole sequence $\{y_n(t,\cdot)\}_{n\geq 1}$, we have $y_n(t,\cdot) \xrightarrow{w} y(t,\cdot)$ in X. Fix $t \in S \cap (T\backslash N)$ and choose a suitable subsequence $\{n_k\}$ of $\{n\}$ (depending in general on t), such that $\underline{\lim}\xi_n(t) = \lim \xi_{n_k}(t)$. Given that $A(t,\cdot)$ is pseudomonotone (Proposition 3), we have that

$$\langle h_{n_k}(t), y_{n_k}(t) - y(t)\rangle \to 0$$

which contradicts the fact that $t \in S$. This means that $\lambda(S) = 0$ and so we have

$$0 \leq \underline{\lim}\ \xi_n(t) \quad \text{a.e. on } T.$$

Using Fatou's lemma, we have

$$0 \leq \int_0^b \underline{\lim}\xi_n(t)dt \leq \underline{\lim} \int_0^b \xi_n(t)dt \leq \overline{\lim} \int_0^b \xi_n(t)dt = \overline{\lim}((h_n, y_n - y)) \leq 0$$
$$\Rightarrow \int_0^b \xi_n(t)dt \to 0.$$

Let $|\xi_n(t)| = \xi_n^+(t) + \xi_n^-(t) = \xi_n(t) + 2\xi_n^-(t)$. Note that $\xi_n^-(t) \to 0$ a.e. on T. Also from (5) we see that $\eta_n(t) \leq \xi_n(t)$ a.e. on T, with $\{\eta_n\}_{n\geq 1} \subseteq L^1(T)$ uniformly integrable. We have

$$0 \leq \xi_n^-(t) \leq \eta_n^-(t)$$

a.e. on T, hence $\{\xi_n^-\}_{n\geq 1} \subseteq L^1(T)$ is uniformly integrable and we can apply the extended dominated convergence theorem (see for example Appendix, Theorem A.2.54, [6:p. 907]) and obtain $\int_0^b \xi_n^-(t)dt \to 0$. It follows that

$$\int_0^b |\xi_n(t)|dt = \int_0^b \xi_n(t)dt + 2\int_0^b \xi_n^-(t)dt \to 0$$

i.e., $\xi \to 0$ in $L^1(T)$. Thus by passing to a subsequence if necessary, we may assume that $\xi_n(t) \to 0$ a.e. on T. Hence $\langle h_n(t), y_n(t) - y(t)\rangle \to 0$ a.e. on T and so

$$((h_n, y_n - y)) \to 0.$$

Let $r \in L^p(T, X)$ and set $\theta(t) = \underline{\lim}\langle h_n(t), y_n(t) - r(t)\rangle$ and

$$\Gamma(t) = \{w \in A(t, y(t) + x_0) : \langle w, y(t) - r(t)\rangle \leq \theta(t)\}.$$

Since $A(t, \cdot)$ is pseudomonotone (Proposition 2), we see that $\Gamma(t) \neq \emptyset$ for all $t \in T \backslash N$. Also we know that

$$t \to A(t, y(t) + x_0)$$

is measurable and so it follows easily that $Gr\Gamma \in \mathcal{L} \times \mathcal{B}(X^*)$, with $\mathcal{L}$ being the Lebesgue σ-field of T and $\mathcal{B}(X^*)$ the Borel σ-field of X^*. Apply the Yankov-von Neumann-Aumann selection theorem, to obtain

$$w \colon T \to X^*$$

a measurable map such that $w(t) \in \Gamma(t)$ a.e. on T. Evidently $w \in L^q(T, X^*)$. So we have proved that for every $r \in L^p(T, X)$, we can find

$$w \in S^q_{A(\cdot, y(\cdot)+x_0)}$$

such that

$$((w, y - r)) \leq \underline{\lim}((h_n, y_n - r)).$$

But recall that $((h_n, y_n)) \to ((h, y))$ and so

$$((w, y - r)) \leq ((h, y - r)). \tag{6}$$

We claim that $h \in S^q_{A(\cdot, y(\cdot)+x_0)}$. Suppose not. Then by the strong separation theorem, we can find $s \in L^p(T, X)$ such that

$$((h, s)) < \inf\left[((w, s)) : w \in S^q_{A(\cdot, y(\cdot)+r_0)}\right] \tag{7}$$

In (6), let $r = y - s$. Then $((w, s)) \leq ((h, s))$, which contradicts (7). Moreover, since

$$((h_n, y_n - y)) \to 0,$$

we have $((h_n, y_n)) \to ((h, y))$. This proves the L-generalized pseudomonotonicity of the multivalued map

$$y \to S^q_{A(\cdot, y(\cdot)+x_0)}.$$

Let

$$G \colon L^p(T \times Z) \to L^q(T \times Z)$$

be defined by

$$G(y)(\cdot) = F(\cdot, y(\cdot) + x_0) + Ku.$$

If $y_n \to y$ in $L^p(T, X)$, $Ly_n = \dot{y}_n \xrightarrow{w} Ly = \dot{y}$ in $L^q(T, X^*)$, we have that $y_n \xrightarrow{w} y$ in $W_{pq}(T)$ and since $W_{pq}(T)$ is embedded compactly in $L^p(T \times Z)$ we have $y_n \to y$ in $L^p(T \times Z)$. Therefore it follows at once that $G(y_n) \to G(y)$ and $((G(y_n), y_n)) \to ((G(y), y))$. Combining this with the already established L-generalized pseudomonotonicity of $y \to S^q_{A(\cdot, y(\cdot)+x_0)}$, we conclude that V_u is L-generalized pseudomonotone. This proves the claim.

Now we rewrite (4) as the following equivalent abstract operator equation

$$Ly + V_u(y) \ni 0. \tag{8}$$

By virtue of Proposition 1, inclusion (8) has a solution $y \in W^0_{pq}(T)$. Then $x = y + x_0$ is a solution of (4), i.e., $x \in S(u) \neq \emptyset$. Therefore we have proved the nonemptiness of the solution set $S(u)$ when the initial condition $x_0 \in X = W^{1,p}_0(Z)$ (regular initial condition). Now assume that $x_0 \in H$. Then we can find $x^n_0 \in X$ such that $x^n_0 \to x_0$ in $H = L^2(Z)$. Let $\{x_n\}_{n\geq 1} \subseteq W_{pq}(T)$ be solutions of (8) (hence of (4) too). By standard *a priori* estimation (see [11]), we can show that $\{x_n\}_{n\geq 1} \subseteq W_{pq}(T)$ is bounded and so we may assume that $x_n \xrightarrow{w} x$ in $W_{pq}(T)$. For every $n \geq 1$ we have

$$\dot{x}_n + V_u(x_n - x_0) \ni 0.$$

Using the integration by parts formula for functions in $W_{pq}(T)$ (see for example Proposition 23.23 [12:pp. 422–423]), we have

$$((\dot{x}_n, x_n - x)) + ((v_n, x_n - x)) = 0,\ v_n \in V_u(x_n - x_0) \text{ we may assume } v_n \xrightarrow{w} v \text{ in } L^q(T, X^*))$$

$$\Rightarrow -\frac{1}{2}\|x^n_0 - x_0|^2 - ((\dot{x}, x_n - x)) + ((v_n, x_n - x)) \leq 0$$

$$\Rightarrow \overline{\lim}((v_n, x_n - x)) \leq 0$$

But from the claim, we have that V_u is L-generalized pseudomonotone. So $v \in V_u(x - x_0)$ and we have proved that $x \in W_{pq}(T)$ is a solution of (4) when $x_0 \in H = L^2(Z)$, i.e., $S(u) \neq \emptyset$.

Next we will establish the compactness in $C(T, H)$ of $\hat{S} = \bigcup[S(u) : u \in S^\infty_U]$. To this end let $\{x_n\}_{n\geq 1} \subseteq \hat{S}$. First we will derive some *a priori* bounds for this sequence. We may assume that $u_n \xrightarrow{w^*} u$ in $L^\infty(T \times Z)$ and $u \in S^\infty_U$. We have

$$\begin{aligned} \dot{x}_n(t) + A(t, x_n(t)) &\ni F(t, x_n(t)) + K(t)u_n(t) \quad \text{a.e. on } T \\ x_n(0) &= x_0. \end{aligned} \tag{9}$$

Take the duality brackets with $x_n(t)$ and then integrate over $T_t = [0, t]$. We obtain

$$\frac{1}{2}|x_n(t)|^2 - \frac{1}{2}|x_0|^2 + \int_0^t \langle -\operatorname{div} g_n(s), x_n(s)\rangle ds = \int_0^t (F(s, x_n(s)), x_n(s))_{pq} ds + \int_0^t (K(s)u_n(s), x_n(s))_{pq} ds, \quad (g_n \in S^q_{a(t,\cdot,x_n(\cdot),Dx_n(\cdot))}).$$

$$\Rightarrow \frac{1}{2}|x_n(t)|^2 - \frac{1}{2}|x_0|^2 + \int_0^t (g_n(s), Dx_n(s))_{pq} ds = \int_0^t (F(s, x_n(s)), x_n(s))_{pq} ds + \int_0^t (K(s)u_n(s), x_n(s))_{pq} ds$$

$$\Rightarrow \frac{1}{2}|x_n(t)|^2 + c_1 \int_0^b \|Dx_n(s)\|^p ds \leq c_7 + \frac{\varepsilon^p}{p} \int_0^t \|x_n(s)\|^p_{L^p(Z)} ds + \frac{1}{\varepsilon^q q}\|Ku_n\|_{L^q(T\times Z)}, \tag{10}$$

for some c_7, and $\varepsilon > 0$ arbitrary. To obtain the last inequality, we have used hypothesis $H(a)$(v) on the term $\int_0^t (g_n(s), Dx_n(s))_{pq} ds$, hypothesis $H(f)$(iv) on the term $\int_0^b (F(s, x_n(s)), x_n(s))_{pq} ds$ and Young's inequality with $\varepsilon > 0$ on the term $\int_0^t (K(s)u_n(s), x_n(s))_{pq} ds$. From Poincare's inequality and choosing $\varepsilon > 0$ small enough, from (10) we have that $\{x_n\}_{n\geq 1} \subseteq L^p(T, X)$ is bounded. Then directly from (9), it follows that $\{\dot{x}_n\}_{n\geq 1} \subseteq L^q(T, X^*)$ is bounded. So $\{x_n\}_{n\geq 1} \subseteq W_{pq}(T)$ is bounded and by passing to a subsequence if necessary, we may assume that $x_n \xrightarrow{w} x$ in $W_{pq}(T)$, $x_n \to x$ in $L^p(T \times Z)$ (from the compact embedding of $W_{pq}(T)$ into $L^p(T \times Z)$) and $x_n(t) \to x(t)$ for all $t \in T \backslash N$, $\lambda(N) = 0$. Also

$$\begin{aligned} G_{u_n}(x_n)(\cdot) &= F(\cdot, x_n(\cdot)) + K(\cdot)u(\cdot) \to G_u(x)(\cdot) \\ &= F(\cdot, x(\cdot)) + K(\cdot)u(\cdot) \text{ in } L^q(T \times Z). \end{aligned}$$

Note that the sequence $\{\langle \dot{x}_n(\cdot), x_n(\cdot) - x(\cdot)\rangle\}_{n\geq 1}$ is uniformly integrable. So given $\varepsilon > 0$, we can find $t \in T \backslash N$ such that

$$\int_t^b |\langle \dot{x}_n(s), x_n(s) - x(s)\rangle| ds < \varepsilon.$$

Let $((\cdot, \cdot))_t$ denote the duality brackets of the pair $(L^p([0, t], X), L^q([0, t], X^*))$. Using the integration by parts formula for functions in $W_{pq}(T)$, we have

$$((\dot{x}_n, x_n - x))_t = \frac{1}{2}|x_n(t) - x(t)|^2 + ((\dot{x}, x_n - x))_t.$$

Note that $\frac{1}{2}|x_n(t) - x(t)|^2 \to 0$ as $n \to \infty$, since $t \in T \backslash N$. Moreover, because $x_n \xrightarrow{w} x$ in $L^p([0, t], X)$ we have $((\dot{x}, x_n - x))_t \to 0$ as $n \to \infty$. Thus it follows that $((\dot{x}_n, x_n - x))_t \to 0$ as $n \to \infty$. We have

$$((\dot{x}_n, x_n - x)) = ((\dot{x}_n, x_n - x))_t + \int_t^b \langle \dot{x}_n(s), x_n(s) - x(s)\rangle ds$$

$$\Rightarrow ((\dot{x}_n, x_n - x)) \geq ((\dot{x}_n, x_n - x))_t - \varepsilon$$
$$\Rightarrow \underline{\lim}((\dot{x}_n, x_n - x)) \geq -\varepsilon.$$

Let $\varepsilon \downarrow 0$ to conclude that

$$\underline{\lim}((\dot{x}_n, x_n - x)) \geq 0.$$

In a similar fashion we can show that

$$\overline{\lim}((\dot{x}_n, x_n - x)) \leq 0,$$

$$\Rightarrow ((\dot{x}_n, x_n - x)) \to 0 \text{ as } n \to \infty.$$

Then for $h_n \in L^q(T, X^*)$, $h_n(t) \in A(t, x_n(t))$ a.e. on T, we have that

$$((\dot{x}_n, x_n - x)) + ((h_n, x_n - x)) = (G_{u_n}(x_n), x_n - x)_{pq}.$$

We may assume that $h_n \xrightarrow{w} h$ in $L^q(T, X^*)$. Since $((\dot{x}_n, x_n - x))$, $(G_{u_n}(x_n), x_n - x)_{pq} \to 0$, we have

$$\lim((h_n, x_n - x)) = 0.$$

By virtue of the L-generalized pseudomonotonicity of $x \to \hat{A}(x) = S^q_{A(\cdot, x(\cdot))}$ established earlier, we have that $h \in \hat{A}(x)$ and $((h_n, x_n)) \to ((h, x))$. Using once more the integration by parts formula for functions in $W_{pq}(T)$, we have that for all $t \in T$

$$\frac{1}{2}|x_n(t) - x(t)|^2 = ((G_{u_n}(x_n) - G_u(x), x_n - x))_t + ((h_n - h, x_n - x))_t$$

$$\Rightarrow \frac{1}{2}|x_n(t) - x(t)|^2 \le \int_0^b |(G_{u_n}(x_n)(s) - G_u(x)(s), x_n(s) - x(s))_{pq}|ds$$

$$+ \int_0^b |\langle h_n(s), x_n(s) - x(s)\rangle|ds + ((h, x_n - x))_t.$$

Note that $\int_0^b |(G_{u_n}(x_n)(s) - G_u(x)(s), x_n(s) - x(s))_{pq}|ds \to 0$. Also let

$$\xi_n(t) = \langle h_n(t), x_n(t) - x(t)\rangle.$$

From the first part of the proof we know that $\xi_n \to 0$ in $L^1(T)$. So we have

$$\int_0^b |\langle h_n(s), x_n(s) - x(s)\rangle|ds \to 0 \text{ as } n \to \infty.$$

Set $\rho_n(t) = \int_0^t \langle h(s), x_n(s) - x(s)\rangle ds$ and choose $t_n \in T$ such that $\rho_n(t_n) = \max_T \rho_n$. We may assume that $t_n \to t$. We have

$$\begin{aligned}\rho_n(t_n) &= \int_0^{t_n} \langle h(s), x_n(s) - x(s)\rangle ds \\ &= \int_0^b \langle \chi_{[0,t_n]}(s)h(s), x_n(s) - x(s)\rangle ds \\ &= ((\chi_{[0,t_n]}h, x_n - x))\end{aligned}$$

Observe that

$$\int_0^b \|\chi_{[0,t_n]}(s)h(s) - \chi_{[0,t]}(s)h(s)\|_*^q ds = \int_{t_n \wedge t}^{t_n \vee t} \|h(s)\|_*^q ds \to 0 \text{ as } n \to \infty,$$

$$\Rightarrow \chi_{[0,t_n]}h \to \chi_{[0,t]}h \text{ in } L^q(T, X^*)$$

$$\Rightarrow ((\chi_{[0,t_n]}h, x_n - x)) \to 0, \text{ i.e. } \rho_n(t_n) \to 0.$$

Therefore finally we conclude that

$$\max_T |x_n(t) - x(t)| = \|x_n - x\|_{C(T,H)} \to 0$$

Since $\hat{x} + h = G_u(x)$, with $h \in \hat{A}(x)$, $u \in S_U^\infty$, we conclude that $x \in S(u)$ and so $\hat{S} \subseteq C(T, H)$ is compact. □

Using this proposition, we can now prove the existence of an optimal state-control pair for problem (1).

Theorem 1: *If hypotheses $H(a), H(U), H(k), H(L)$ hold and $x_0 \in L^2(Z)$, then problem* (1) *has a solution $(x^*, u^*) \in W_{pq}(T) \times S_U^\infty$.*

Proof: Let J: $L^p(T \times Z) \times L^p(T \times Z) \to \overline{\mathbb{R}} = \mathbb{R} \cup \{\infty\}$ be defined by

$$J(x, u) = \int_0^b \int_Z L(t, z, x(t, z), u(t, z))dzdt.$$

Let $\{(x_n, u_n)\}_{n \geq 1} \subseteq W_{pq}(T) \times S_U^\infty$ be a minimizing sequence for problem (1). By virtue of Proposition 3, we may assume that $x_n \to x^*$ in $C(T, H)$ and also $u_n \xrightarrow{w} u^*$ in $L^p(T \times Z)$, with $x^* \in S(u)$ (see the proof of Proposition 3). Using Theorem 1 of Berkovitz [2], we have that

$$\begin{aligned} m &= \underline{\lim}\, J(x_n, u_n) \geq J(x^*, u^*) \\ &\Rightarrow J(x^*, u^*) = m, \text{ since } x^* \in S(u^*) \end{aligned}$$ □

Also we can solve a time-optimal control problem. Let $\Sigma(t) \subseteq L^2(Z)$ be the moving target set. The goal is to reach $\Sigma(t)$ in minimum time, i.e.,

$$\begin{aligned} &\inf\{t \in T\colon x(t, \cdot) \in \Sigma(t)\} \\ &\frac{\partial x}{\partial t} - \operatorname{div} a(t, z, x(t, z), Dx(t, z)) \ni f(t, z, x(t, z)) + k(t, z)u(t, z) \text{ on } T \times Z \qquad (11) \\ &x|_{T \times \Gamma} = 0, \ \ x(0, z) = x_0(z) \ \text{ a.e. on } Z, \ \ u(t, \cdot) \in U(t) \ \text{ a.e. on } T. \end{aligned}$$

We make two additional hypotheses concerning the target set $\Sigma(t)$. The first is a controllability hypothesis and the second concerns the motion of the target.

H_0: $T_0 = \{t \in T\colon x(t, \cdot) \in \Sigma(t),\ \ x \text{ is a solution of } (3)\} \neq \emptyset$

$H(\Sigma)$: $\Sigma\colon T \to P_f(L^2(Z))$ has a closed graph.

Theorem 2: *If hypotheses $H(a), H(U), H(k), H_0, H(\Sigma)$ hold and $x_0 \in H = L^2(Z)$ then problem* (11) *has an optimal control pair $(x^*, u^*) \in W_{pq}(T) \times S_U^\infty$.*

Proof: Let $\{t_n\}_{n \geq 1} \subseteq T_0$ be a minimizing sequence for problem (11). We may assume that $t_n \downarrow t^*$ in T. Let $(x_n, u_n) \in W_{pq}(T) \times S_U^\infty$, $n \geq 1$, be admissible state-control pairs such that $x_n(t_n) \in \Sigma(t_n)$. Using Proposition 3 we may say that $x_n \to x^*$ in $C(T, H)$, also $u_n \xrightarrow{w^*} u^*$ in $L^\infty(T)$, $u^* \in S_U^\infty$ and $x^* S(u^*)$ (see the proof of Proposition 3). Then $x_n(t_n) \to x^*(t^*)$ in $H = L^2(Z)$ and because of hypothesis $H(\Sigma)$, $x^*(t^*) \in \Sigma(t^*)$. This proves that $(x^*, u^*) \in W_{pq}(T) \times S_U^\infty$ is the time-optimal state-control pair. □

4. A MINIMAX PROBLEM

In this section, we deal with parametric control systems. In this case the relevant optimization problem, is a minimax problem. So let E be a complete metric space (the parameter space).

We consider the following nonlinear distributed parameter control system:

$$\begin{aligned} &\frac{\partial x}{\partial t} - \operatorname{div} a(z, Dx(t,z), \lambda) \ni f(t,z,x(t,z),\lambda) + (K(\lambda)u)(t,z) \text{ on } T \times Z \\ &x|_{T\times\Gamma} = 0, \;\; x(0,z) = x_0(z,\lambda) \quad \text{a.e. on } Z, \;\; u(t,\cdot) \in U(t) \quad \text{a.e. on } \; T. \end{aligned} \tag{12}$$

Here for every $\lambda \in E$, $K(\lambda)$: $L^q(T \times Z) \to L^q(T \times Z)$ is a compact, linear operator. Note that in (12), the nonlinear differential operator in divergence force is time-invariant and independent of x. Hence the corresponding abstract multivalued map A: $X \to 2^{X^*}$ is maximal monotone. So by an appropriate Lipschitz hypothesis on $f(t, z, \cdot, \lambda)$ (see $H(f)_1$ below), we can guarantee that given any control-parameter pair $(u, \lambda) \in S^q_u \times E$, problem (12) has a unique solution (state) $x(u, \lambda) \in W_{pq}(T)$ (see Lemma 2 below). For the triple $(u, \lambda, x(u, \lambda))$ the performance of the system is measured by the integral cost functional

$$J(u,\lambda) = \int_0^b \int_Z L(t,z,x(u,\lambda)(t,z),u(t,z),\lambda)dzdt.$$

Since it is not known *a priori* which disturbance element $\lambda \in E$ is in force, the best a system analyst can do, is to minimize the maximum cost. So he is confronted with the following minimax optimization problem

$$\inf_{u\in S^q_U} \sup_{\lambda\in E} J(u,\lambda) = \theta. \tag{13}$$

For a given feasible control $u \in S^q_U$, let

$$m(u) = \sup_{\lambda\in E} J(u,\lambda)$$

which is the maximum cost corresponding to the feasible control $u \in S^q_U$. Our goal is to find a feasible control $u^* \in S^q_U$ such that

$$\theta = m(u^*)$$

Such a control will be called "optimal".

Let us introduce the hypotheses on the data of the minimax control problem.

$H(a)_1$: a: $Z \times \mathbb{R}^N \times E \to P_{kc}(\mathbb{R}^N)$ is a multifunction such that

(i) for every $\lambda \in E$, $(z, y) \to a(z, y, \lambda)$ is graph measurable;
(ii) for all $z \in Z$ and all $\lambda \in E$, $y \to a(z, y, \lambda)$ is maximal monotone;
(iii) for almost all $z \in Z$, all $y \in \mathbb{R}^N$ and all $\lambda \in K \subseteq E$ compact, we have

$$|a(z,y,\lambda)| \le \gamma_K(z) + c_K\|y\|^{p-1}$$

with $\gamma_K \in L^q(Z)$, $2 \le p < \infty$, $\frac{1}{p} + \frac{1}{q} = 1$, $c_K > 0$;
(iv) for almost all $z \in Z$, all $y \in \mathbb{R}^{N^p}$, all $\lambda \in K \subseteq E$ compact and all $v \in a(z, y, \lambda)$ we have

$$(v,y)^N_{\mathbf{R}} \ge c_{1K}\|y\|^p - \gamma_{1K}(z)$$

with $\gamma_{1K} \in L^1(Z)$, $c_{1K} > 0$;
(v) if $\lambda_n \to \lambda$ in E, then $a(\cdot,\cdot,\lambda_n) \xrightarrow{G} a(\cdot,\cdot,\lambda)$.

$H(f)_1$: $f\colon T \times Z \times \mathbb{R} \times E \to \mathbb{R}$ is a function such that

(i) for every $(x, \lambda) \in \mathbb{R} \times E$, $(t, z) \to f(t, z, x, \lambda)$ is measurable;
(ii) for all $(t, z, x) \in T \times Z \times \mathbb{R}$, $\lambda \to f(t, z, x, \lambda)$ is continuous;
(iii) for almost all $(t, z) \in T \times Z$, all $\lambda \in K \subseteq E$ compact and all $x, y \in \mathbb{R}$

$$|f(t, z, x, \lambda) - f(t, z, y, \lambda)| \leq \eta_K(t, z)|x - y|$$

with $\eta_K \in L^1(T, L^\infty(Z))$;
(iv) for almost all $(t, z) \in T \times Z$, all $x \in \mathbb{R}$ and all $\lambda \in K \subseteq E$ compact, we have

$$|f(t, z, x, \lambda)| \leq \gamma_{2K}(t, z) + c_{2K}|x|^{p-1}$$

with $\gamma_{2K} \in L^q(T \times Z)$, $c_{2K} > 0$;
(v) for almost all $(t, z) \in T \times Z$, all $\lambda \in K \subseteq E$ compact and all $|x| > M_K(t, z)$ we have

$$f(t, z, x, \lambda)x \leq 0$$

with $M_K \in L^p(T \times Z)_+$.

$H(U)_1$: $U\colon T \to P_{fc}(L^q(Z))$ is a measurable multifunction s.t.

$$t \to |U(t)| = \sup\left\{\|u\|_q : u \in U(t)\right\} \in L^q(T).$$

We start with an auxiliary result, which is crucial in what follows. This result is a parabolic compensated compactness lemma (see Lemma 3.4 [3] for an analogous result in the elliptic case). First a definition:

Definition 1: *Let $\{(x_n, g_n)\}_{n\geq 1} \subseteq W_{pq}(T) \times L^q(T \times Z, \mathbb{R}^N)$. We say that the sequence τ-converges to (x, g) if*

$$x_n \xrightarrow{w} x \text{ in } W_{pq}(T),\ g_n \xrightarrow{w} g \text{ in } L^q(T \times Z, \mathbb{R}^N)$$
$$\text{and } \dot{x}_n - \operatorname{div} g_n \to \dot{x} - \operatorname{div} g \text{ in } L^q(T, X^*).$$

Lemma 2: *If $(x_n, g_n) \xrightarrow{\tau} (x, g)$ in $W_{pq}(T) \times L^q(T, L^p(Z, \mathbb{R}^N))$, then for all $\varphi \in C_0^\infty(T \times Z)$,*

$$((g_n, \varphi D x_n))_{p,q} \to ((g, \varphi D x))_{p,q}$$

(by $((\cdot,\cdot))_{p,q}$ we denote the duality brackets for the pair $(L^p(T \times Z, \mathbb{R}^N)$, $L^q(T \times Z, \mathbb{R}^N))$).

Proof: Let $h_n = \dot{x}_n - \operatorname{div} g_n$ and $h = \dot{x} - \operatorname{div} g$. Then from the definition of τ-convergence, we have

$$((h_n, x_n\varphi)) = ((\dot{x}_n, x_n\varphi)) + ((-\operatorname{div} g_n, x_n\varphi)).$$

Using the integration by parts formula for functions in $W_{pq}(T)$, we have

$$((\dot{x}_n, x_n\varphi)) = -\frac{1}{2}((x_n, x_n\varphi')).$$

So we have

$$\begin{aligned}((h_n, x_n\varphi)) &= -\frac{1}{2}((x_n, x_n\varphi')) + ((-\operatorname{div} g_n, x_n\varphi)) \\ &= -\frac{1}{2}((x_n, x_n\varphi'))_{p,q} + ((g_n, \varphi Dx_n))_{p,q} + ((g_n, x_n D\varphi))_{p,q}.\end{aligned}$$

Recall that since $x_n \xrightarrow{w} x$ in $W_{pq}(T)$, we have that $x_n \to x$ in $L^p(T \times Z)$ (from the compact embedding of $W_{pq}(T)$ into $L^p(T \times Z)$). Thus by passing to the limit as $n \to \infty$, we have

$$\lim((g_n, \varphi Dx_n))_{p,q} = ((h, x\varphi)) + \frac{1}{2}((x, x\varphi'))_{p,q} - ((g, xD\varphi))_{p,q}$$

$$\begin{aligned}\Rightarrow \lim((g_n, \varphi Dx_n))_{p,q} &= ((\dot{x}, x\varphi)) - ((\operatorname{div} g, x\varphi)) + \frac{1}{2}((x, x\varphi'))_{p,q} \\ &\quad - ((g, xD\varphi))_{p,q} \\ &= -\frac{1}{2}((x, x\varphi'))_{p,q} + ((g, \varphi Dx))_{p,q} \\ &\quad + ((g, xD\varphi))_{p,q} + \frac{1}{2}((x, x\varphi'))_{p,q} - ((g, xD\varphi))_{p,q} \\ &= ((g, \varphi Dx))_{p,q}.\end{aligned}$$

□

Next we establish the existence of a unique state $x(u, \lambda) \in W_{pq}(T)$ for every given control-parameter pair $(u, \lambda) \in S_U^q \times E$.

Lemma 3: *If hypotheses $H(a)_1, H(f)_1, H(U)$ hold and $x_0(\cdot, \lambda) \in L^2(Z)$, then for every*

$$(u, \lambda) \in S_U^q \times E,$$

a state $x(u, \lambda) \in W_{pq}(T)$ of the system (12) exists and in fact is unique.

Proof: As before let

$$\alpha_1: X \times E \to P_{wkc}(L^q(Z, \mathbb{R}^N)) \quad \text{and} \quad A_1: X \times E \to P_{wkc}(X^*)$$

be defined by

$$\alpha_1(x, \lambda) = S^q_{a(\cdot, Dx(\cdot), \lambda)} \text{ and } A_1(x, \lambda) = \{-\operatorname{div} g: g \in \alpha_1(x, \lambda)\}.$$

In this case by virtue of hypotheses $H(a)_1$, we can check that for all $\lambda \in E$, $\alpha_1(\cdot, \lambda)$ is monotone, demicontinuous hence maximal monotone. Also let

$$F_1: T \times L^p(Z) \times E \to L^q(Z)$$

be defined by

$$F_1(t, x, \lambda)(\cdot) = f(t, \cdot, x(\cdot), \lambda) \text{ (Nemitsky map).}$$

The system (12) can be equivalently rewritten as the following abstract evolution inclusion

$$\begin{aligned} \dot{x}(t) + A_1(x(t),\lambda) \ni F_1(t,x(t),\lambda) + (K(\lambda)u)(t) \quad \text{a.e. on } T \\ x(0) = x_0(\lambda), \quad u(t) \in U(t) \quad \text{a.e. on } T. \end{aligned} \tag{14}$$

We know (see [11] or Proposition 3) that given $(u,\lambda) \in S_U^q \times E$ problem (14) has a solution. We will show that this solution is unique. Suppose $x, y \in W_{pq}(T)$ were two solutions of (14) corresponding to the pair $(u,\lambda) \in S_U^q \times E$. We have

$$\dot{x}(t) - \dot{y}(t) + v_1(t) - v_2(t) = F_1(t,x_1(t),\lambda) - F_1(t,y(t),\lambda) \quad \text{a.e. on } T,$$
$$x(0) = y(0) = x_0(\lambda),$$

where $v_1(t) \in A_1(x(t),\lambda)$, $v_2(t) \in A_1(y(t),\lambda)$ a.e. on T, $v_1, v_2 \in L^q(T,X^*)$. Take the duality brackets of the above equation with $x(t) - y(t)$ and then integrate over $T_t = [0,t]$. Using the integration by parts formula for functions in $W_{pq}(T)$, we obtain

$$\begin{aligned} \frac{1}{2}|x(t)-y(t)|^2 &+ \int_0^b \langle v_1(s) - v_2(s), x(s) - y(s)\rangle ds \\ &= \int_0^t (F_1(s,x(s)) - F_1(s,y(s)), x(s) - y(s))ds \end{aligned}$$

$$\begin{aligned} \Rightarrow \frac{1}{2}|x(t)-y(t)|^2 &\le \int_0^t \int_Z (f(s,z,x(s,z),\lambda) - f(s,z,y(s,z),\lambda))(x(s,z) - y(s,z))dzds \\ &\quad (\text{see } H(a)_1(ii)) \\ &\le \int_0^t \int_Z \eta_{\{\lambda\}}(s,z)|x(s,z) - y(s,z)|^2 dzds \\ &\le \int_0^t \|\eta_{\{\lambda\}}(s,\cdot)\|_\infty |x(s) - y(s)|^2 ds. \end{aligned}$$

From Gronwall's inequality, it follows that $x(t) = y(t)$ for all $t \in T$. □

By virtue of Lemma 3, we can define the map

$$\theta\colon S_U^q \times E \to L^p(T \times Z)$$

by

$$\theta(u,\lambda) = x(u,\lambda)$$

(= the unique state of (12) corresponding to the pair (u,λ)).

On $S_U^q \subseteq L^\infty(Z)$ we consider the relative w-topology. This way S_U^q becomes a compact metrizable space. We introduce the following hypothesis concerning $K(\lambda)$.

$H(K)_1$: $K\colon E \to \mathcal{L}(L^p(T \times Z))$ is continuous and for every $\lambda \in E$, $K(\lambda)$ is compact.

Proposition 4: *If hypotheses $H(a)_1, H(f)_1, H(k)_1$ hold and $\lambda \to x_0(\cdot,\lambda)$ is continuous from E into $H = L^2(Z)$, then $\theta\colon S_U^q \times E \to L^p(T \times Z)$ is continuous.*

Proof: Let $(u_n, \lambda_n) \to (u, \lambda)$ in $S_U^q \times E$. Set $x_n = \theta(u_n, \lambda_n)$, $x = \theta(u, \lambda)$. We have

$$\dot{x}_n(t) + v_n(t) = F_1(t, x_n(t), \lambda_n) + (K(\lambda_n)u_n)(t) \quad \text{a.e. on } T.$$
$$x_n(0) = x_0(\lambda_n), \quad v_n(t) \in A(x_n(t), \lambda_n) \quad \text{a.e. on } T.$$

As in the proof of proposition 3 (see also [11]), we can check that $\{x_n\}_{n\geq 1} \subseteq W_{pq}(T)$ and $\{v_n\}_{n\geq 1} \subseteq L^q(T, X^*)$ are bounded sequences. By definition $v_n = -\text{div} g_n$, with $g_n \in \alpha_1(x_n, \lambda_n)$, $n \geq 1$ and so $\{g_n\}_{n\geq 1} \subseteq L^q(T \times Z, \mathbb{R}^N)$ is bounded. Thus by passing to a subsequence if necessary, we may assume that

$$x_n \xrightarrow{w} y \text{ in } W_{pq}(T) \quad \text{and} \quad g_n \xrightarrow{w} g \text{ in } L^q(T \times Z, \mathbb{R}^N).$$

Then $F_1(t, x_n(t), \lambda_n) \to F_1(t, y(t), \lambda)$ and $(K(\lambda_n)u_n)(t) \to (K(\lambda)u)(t)$ a.e. on T in $L^q(Z)$ (see hypotheses $H(f)_1$ and $H(k)_1$). Thus if $v = -\text{div}\, g \in L^q(T, X^*)$, in the limit as $n \to \infty$, we have

$$\dot{x}(t) + v(t) = F_1(t, x(t), \lambda) + (K(\lambda)u)(t) \quad \text{a.e. on } T,\ x(0) = x_0(\lambda).$$

We need to show that $g(t, z) \in a(z, Dy(t, z), \lambda)$ a.e. on $T \times Z$, hence $v(t) \in A_1(y(t), \lambda)$ a.e. on T and so $y = \theta(u, \lambda) = x$ and we will be done. To this end let $h \in W^{-1,q}(Z)$ and let w_n be (weak) solutions of the following stationary (elliptic) problems

$$\begin{aligned} -\text{div} a(z, Dw_n(z), \lambda_n) &\ni h(z) \text{ on } Z \\ w_n|_\Gamma &= 0 \end{aligned} \tag{15}$$

In (15) the multivalued nonlinear operator $(A_1(y, \lambda))$ is maximal monotone, coercive, thus surjective. So problem (15), has solutions $\{w_n\}_{n\geq 1} \subseteq W_0^{1,p}(Z)$, which of course are not in general unique. From hypothesis $H(a)_1(v)$ and remark 3.10 Chiado Piat-Dal Maso-Defranceschi [3], by passing to a subsequence if necessary, we may assume that

$$w_n \xrightarrow{w} w \text{ in } W_0^{1,p}(Z) \quad \text{and} \quad r_n \xrightarrow{w} r \text{ in } L^q(Z, \mathbb{R}^N)$$

with $r_n(z) \in a(z, Dw_n(z), \lambda_n)$ a.e. on Z, $-\text{div} r_n = h_n$, $-\text{div} r = h$, $r(z) \in a(z, Dw(z), \lambda)$ a.e. on Z.

Let $\varphi \in C_0^\infty(T \times Z)$, $\varphi \geq 0$. Exploiting the monotonicity of $a(z, \cdot, \lambda_n)$, we have

$$0 \leq \int_0^b \int_Z (r_n(z) - g_n(t, z), Dw_n(z) - Dx_n(t, z))_{\mathbb{R}^N} \varphi(t, z)\, dz\, dt$$

Note that $[w_n - x_n, r_n - g_n] \xrightarrow{\tau} [y - x, r - g]$ (recall the definition of τ-convergence). Using Lemma 2, in the limit as $n \to \infty$, we obtain

$$0 \leq \int_0^b \int_Z (r(z) - g(t, z), Dw(z) - Dy(t, z))_{\mathbb{R}^N} \varphi(t, z)\, dz\, dt$$

for all $\varphi \in C_0^\infty(T \times Z)$

$$\Rightarrow 0 \leq (r(z) - g(t, z), Dw(z) - Dy(t, z))_{\mathbb{R}^N} \quad \text{a.e. on } T \times Z. \tag{16}$$

Fix $\xi \in \mathbb{R}^N$ and let $Z_1 \subset\subset Z$, $\psi \in C_0^\infty$ such that $\psi|_{Z_1} \equiv 1$. Evidently, we can find $h \in W^{-1,q}(Z)$ such that for $w(z) = \psi(z)(\xi, z)_{\mathbb{R}^N} \in C_0^\infty$, we have

$$-\text{div} a(z, Dw(z), \lambda) \ni h(z) \text{ on } Z.$$

Using this in (16), we obtain

$$0 \leq (r(z) - g(t,z), \xi - Dy(t,z))_{\mathbb{R}^N} \quad \text{a.e. on } T \times Z, \quad \text{for all } \xi \in \mathbb{R}^N.$$

Let $\xi_1 \in \mathbb{R}^N$ and set $\rho_n(t,z) = \frac{1}{n}\xi_1 + (1 - \frac{1}{n})Dy(t,z)$. Then for every $r_n \in S^q_{a(\cdot, D\rho_n(t,\cdot),\lambda)}$ we have

$$0 \leq (r_n(z) - g(t,z), \frac{1}{n}(\xi_1 - Dy(t,z)))_{\mathbb{R}^N} \quad \text{a.e. on } T \times Z.$$

We may assume that $r_n \xrightarrow{w} \hat{r}$ in $L^q(Z, \mathbb{R}^N)$. Invoking Proposition VII.3.9, p. 694, of Hu-Papageorgiou [6], we infer that

$$\hat{r}(z) \in \overline{\text{conv}}\, \overline{\lim_{n\to\infty}}\, a(z, \frac{1}{n}\xi + (1 - \frac{1}{n})Dy(t,z), \lambda) \subseteq a(z, Dy(t,z), \lambda) \quad \text{a.e. on } T \times Z_1.$$

In addition, we have for all $\varphi \in C_0^\infty(T \times Z)$

$$\begin{aligned} 0 &\leq \int_0^b \int_{Z_1} (r_n(z) - g(t,z), \xi_1 - Dy(t,z))_{\mathbb{R}^N} \varphi(t,z)\,dz\,dt \\ \Rightarrow 0 &\leq \int_0^b \int_{Z_1} (\hat{r}(z) - g(t,z), \xi_1 - Dy(t,z))_{\mathbb{R}^N} \varphi(t,z)\,dz\,dt \\ \Rightarrow 0 &\leq (\hat{r}(z) - g(t,z), \xi_1 - Dy(t,z))_{\mathbb{R}^N} \quad \text{a.e. on } T \times Z_1. \end{aligned}$$

Since $Z_1 \subset\subset Z$ was arbitrary, we infer that

$$0 \leq (\hat{r}(z) - g(t,z), \xi_1 - Dy(t,z))_{\mathbb{R}^N} \quad \text{a.e. on } T \times Z, \quad \text{for all } \xi_1 \in \mathbb{R}^N.$$

Set $\xi_1 = Dy(t,z) \pm \eta$, $\eta \in \mathbb{R}^N$. It follows that $\hat{r}(z) = g(t,z)$ a.e. on $T \times Z$, hence we have $g(t,z) \in a(z, Dy(t,z), \lambda)$ a.e. on $T \times Z$. As we already indicated, this implies that

$$y = \theta(u, \lambda) = x$$

and so we have proved the proposition. □

Now we introduce our hypotheses on the cost integrand.

$H(L)_1$: $L: T \times Z \times \mathbb{R} \times \mathbb{R} \times E \to \overline{\mathbb{R}} = \mathbb{R} \cup \{+\infty\}$ is an integrand such that

(i) $(t, z, x, u, \lambda) \to L(t, z, x, u, \lambda)$ is measurable;
(ii) for all $(t,z) \in T \times Z$, $(x, u, \lambda) \to L(t, z, x, u, \lambda)$ is lower semicontinuous;
(iii) for all $(t, z, x, \lambda) \in T \times Z \times \mathbb{R} \times E$, $u \to L(t, z, x, u, \lambda)$ is convex;
(iv) for almost all $(t,z) \in T \times Z$, all $x, u \in \mathbb{R}$ and all $\lambda \in K \subseteq E$ compact, we have

$$\hat{c}_K(|x| + |u|) - \hat{\gamma}_K(t,z) \leq L(t, z, x, u, \lambda)$$

with $\hat{c}_K > 0$, $\hat{\gamma}_K \in L^1(T \times Z)$.

Proposition 5: *If hypotheses $H(a)_1, H(f)_1, H(k)_1, H(L)_1$ hold and $\lambda \to x_0(\cdot, \lambda)$ is continuous from E into $L^2(Z)$, then $(u, \lambda) \to J(u, \lambda)$ is lower semicontinuous on $S_U^q \times E$.*

Proof: We need to show that for every $\mu \in \mathbb{R}$, the sublevel set

$$\Gamma_\mu = \{(u, \lambda) \in S_U^q \times E : J(u, \lambda) \le \mu\}$$

is closed. To this end let $\{(u_n, \lambda_n)\}_{n\ge1} \subseteq \Gamma_\mu$ and assume that $(u_n, \lambda_n) \to (u, \lambda)$ in $S_U^q \times E$. Since E is a complete metric space, it is isometrically isomorphic to a closed subspace of a Banach space Y (see [9]). Let $K = \{\lambda_n, \lambda\}_{n\ge1} \subseteq E$ compact and let Y_1 be the closed subspace generated by the isometric image of the compact K. Then Y_1 being compactly generated is separable and so we can use Theorem 2.1 of Balder 1 and have that

$$(x, u, \lambda) \to I(x, u, \lambda) = \int_0^b \hat{L}(t, x(t), u(t), \lambda)dt = \int_0^b \int_Z L(t, z, x(t, z), u(t, z), \lambda)dzdt$$

is lower semicontinuous on $L^p(Z) \times L^p(Z) \times E$. Thus we have

$$\int_0^b \int_Z L(t, z, x(t, z), u(t, z), \lambda)dzdt \le \underline{\lim} \int_0^b \int_Z L(t, z, x_n(t, z), u_n(t, z), \lambda)dzdt$$

$$\Rightarrow J(u, \lambda) \le \underline{\lim} J(u_n, \lambda_n) \le \mu$$
$$\Rightarrow (u, \lambda) \in \Gamma_\mu.$$

This proves the desired lower semicontinuity of the cost functional. □

Now we have all the necessary tools to solve the minimax problem (13).

Theorem 3: *If hypotheses $H(a)_1, H(f)_1, H(k)_1, H(L)_1$ hold and $\lambda \to x_0(\cdot, \lambda)$ is continuous from E into $L^2(Z)$, then there exists $u^* \in S_U^q$ s.t. $m(u^*) = \theta$ (i.e., u^* is a solution of the minimax problem).*

Proof: By definition

$$m(u) = \sup_{\lambda \in E} J(u, \lambda).$$

Using Proposition 5 and Theorem I.3.1, of Hu-Papageorgiou [6:p. 82], we have that

$$u \to m(u)$$

is lower semicontinuous on $L^q(T \times Z)_w$. Consider the minimization problem

$$\inf[m(u) : u \in S_U^q] = \theta.$$

If $\{u_n\}_{n\ge1} \subseteq S_U^q$ is a minimizing sequence, then we may assume that

$$u_n \xrightarrow{w} u^* \text{ in } L^p(T \times Z)$$

and so

$$\theta = \underline{\lim}\, m(u_n) \geq m(u^*) \Rightarrow m(u^*) = \theta$$

(because $u^* \in S_U^q$). □

Remark: Let $a(z, y, \lambda) = \beta(z, \lambda)\|y\|^{p-2}y$, with $\beta(\cdot, \lambda) \in L^\infty(Z)_+$ and assume that if $\lambda_n \to \lambda$ in E, then $\beta(z, \lambda_n) \to \beta(z, \lambda)$ a.e. on Z. Note that $a(z, y, \lambda) = \beta(z, \lambda)\partial\eta(y)$ where $\eta(y) = \frac{1}{p}\|y\|^p$ and $\partial\eta$ is the convex subdifferential. Then combining Theorem 3.2 of Defranceschi [5] with Theorem 5.14 of Dal Maso [4], we have that

$$a(\cdot, \cdot, \lambda_n) \xrightarrow{G} a(\cdot, \cdot, \lambda).$$

If $N = 1$ and $a(z, y, \lambda) = \beta(z, \lambda)|y|^{p-2}y$, $z \in \mathbb{R}$, then employing this time the result of Marcellini-Sbordone [8], we have that

$$a(\cdot, \cdot, \lambda_n) \xrightarrow{G} a(\cdot, \cdot, \lambda)$$

if and only if

$$\frac{1}{\beta(\cdot, \lambda_n)} \xrightarrow{w^*} \frac{1}{\beta(\cdot, \lambda)}$$

in $L^\infty(Z)$, provided $0 < \theta_{1K} \leq \beta(z, \lambda) \leq \theta_{2K}$ a.e. on $Z \subseteq \mathbb{R}$, with $\lambda \in K \subseteq E$ compact.

REFERENCES

[1] E. Balder (1987). Necessary and sufficient conditions for L_1-strong-weak lower semicontinuity of integral functionals, *Nonlin. Anal.*, **11**, 1399–1404.

[2] L. Berkovitz (1974). Lower semicontinuity of integral functionals, *Trans. AMS*, **192**, 51–57.

[3] V. Chiado Piat-G, dal Maso-A and Defranceschi (1990). G-convergence of monotone operators, *Ann. Inst. H. PoincareAnalyse Non Lineaire*, **7**, 123–160.

[4] G. Dal Maso (1993). An Introduction to Γ-Convergence, Boston: Birkhauser.

[5] A. Defranceschi (1989). *G*-convergence of cyclically monotone operators, *Asymptotic Anal.*, **2**, 21–37.

[6] S. Hu and N.S. Papageorgiou (1997). Handbook of Multivalued Analysis. Volume I: Theory. The Netherlands: Kluwer, Dordrecht.

[7] J.-L. Lions (1969). Quelques Methodes de Resolution des Problemes aux Limites Non Lineaires. Dunod, Paris.

[8] P. Marcellini-C and Sbordone (1977). Dualita e perturbazione di funzionali integrali, *Ricerche Mat.*, **26**, 383–421.

[9] E. Michael (1964). A short proof of the Arens-Eells embedding theorem, *Proc. AMS*, **15**, 415–416.

[10] N.S. Papageorgiou, F. Papalini and F. Renzacci (1999). Existence of solutions and periodic solutions for nonlinear evolution inclusions, *Rend. Circolo Mat. Palermo*, **48**, 341–364.

[11] N.S. Papageorgiou and N. Shahzad (1997). Properties of the solution set of nonlinear evolution inclusions, *Appl. Math. Optim.*, **36**, 1–20.

[12] E. Zeidler (1990). Nonlinear Functional Analysis and its Applications II. New York: Springer Verlag.

SIMAA 4(2002) 311–359

19. Continuation Theory for A-Proper Mappings and their Uniform Limits and Nonlinear Perturbations of Fredholm Mappings

P.S. Milojević

Department of Mathematics and CAMS, New Jersey Institute of Technology, Newark, NJ, USA 07102-1982. E-mail: pemilo@m.njit.edu

1. INTRODUCTION

Since the classical continuation theorem of Leray and Shauder, which has yielded a large number of existence results in the theory of nonlinear integral and differential equations, various new continuation results have been established for diverse classes of nonlinear operators. We refer to [63–68,53] for a detailed discussion and references.

This chapter presents the continuation theory for multivalued A-proper and their uniform limits began in [62–68] and to apply it to equations involving nonlinear perturbations of Fredholm mappings of index zero. We present a fixed point theory for single valued and multivalued approximable mappings and discuss its applications. Solvability of strongly nonlinear equations is studied in great detail.

Applications of the abstract theory to the (approximation) solvability of equations involving nonlinear perturbations of Fredholm mappings of index zero are given. Constructive solvability and the error estimates of approximate solutions are also presented.

Throughout the work we discuss possible applications of the abstract results to various special classes of A-proper mappings and their uniform limits involving, for example mappings of type (KS), ball-condensing, a-stable, of monotone type, etc. as well as such perturbations of Fredholm mappings. Some applications of our theory to boundary value problems for differential equations are also indicated.

2. BASIC DEFINITIONS AND EXAMPLES OF A-PROPER MAPPINGS

Let $\{X_n\}$ and $\{Y_n\}$ be finite dimensional subspaces of Banach spaces X and Y respectively such that $\dim X_n = \dim Y_n$ for each n and $\mathrm{dist}(x, X_n) \to 0$ as $n \to \infty$ for each $x \in X$. Let P_n: $X \to Y_n$ and Q_n: $Y \to Y_n$ be linear projections onto X_n and Y_n respectively such that $P_n x \to x$ for each $x \in X$ and $\delta = \max \|Q_n\| < \infty$. Then $\Gamma = \{X_n, P_n; Y_n, Q_n\}$ is a projection scheme for (X, Y). In some cases we may need that X_n are infinite dimensional (see Section 10).

Definition 2.1: A map $T: D \subset X \to Y$ is said to be *approximation-proper* (*A-proper* for short) with respect to Γ if: (i) $Q_nT: D \cap X_n \to Y_n$ is demicontinuous for each n; and (ii) whenever $\{x_{n_k} \in D \cap X_{n_k}\}$ is bounded and $\|Q_{n_k}Tx_{n_k} - Q_{n_k}f\| \to 0$ for some $f \in Y$, then a subsequence $x_{n_{k(i)}} \to x$ and $Tx = f$. T is said to be *pseudo A-proper* w.r.t. Γ if in (ii) above we do not require that a subsequence of $\{x_{n_k}\}$ converges to x for which $f \in Tx$. If f is given in advance, we say that T is (pseudo) A-proper at f.

There is an extensive literature on the study of (pseudo) A-proper maps and we refer to [91], [54–85] and the references therein.

To demonstrate the generality and the unifying nature of the solvability theory developed in this chapter, we state now a number of examples of A-proper and pseudo A-proper maps. Many other ones can be found later on.

To state some results for ϕ-condensing maps, we recall that the *set measure of noncompactness* of a bounded set $D \subset X$ is defined as $\gamma(D) = \inf\{d > 0 : D$ has a finite covering by sets of diameter less than $d\}$. The *ball-measure of noncompactness* of D is defined as $\chi(D) = \inf\{r > 0 | D \subset \cup_{i=1}^n B(x_i, r),\ x \in X,\ n \in N\}$. Let ϕ denote either the set or the ball-measure of noncompactness. Then a map $N: D \subset X \to X$ is said to be $k - \phi$ *contractive* (ϕ-*condensing*) if $\phi(N(Q)) \leq k\phi(Q)$ (respectively $\phi(N(Q)) < \phi(Q)$) whenever $Q \subset D$ (with $\phi(Q) \neq 0$).

Recall that $N: X \to Y$ is K-monotone for some $K: X \to Y^*$ if $(Nx - Ny, K(x - y)) \geq 0$ for all $x, y \in X$. It is said to be generalized pseudo-K-monotone (of type (KM)) if whenever $x_n \rightharpoonup x$ and $\limsup (Nx_n, K(x_n - x)) \leq 0$ then $(Nx_n, K(x_n - x)) \to 0$ and $Nx_n \rightharpoonup Nx$ (then $Nx_n \rightharpoonup Nx$). Recall that N is said to be of type (KS_+) if $x_n \rightharpoonup x$ and $\limsup (Nx_n, K(x_n - x)) \leq 0$ imply that $x_n \to x$. If $x_n \rightharpoonup x$ implies that $\limsup (Nx_n, K(x_n - x)) \geq 0$, N is said to be of type (KP). If $Y = X^*$ and K is the identity map, then these maps are called monotone, generalized pseudo monotone, of type (M) and (S_+) respectively. If $Y = X$ and $K = J$ the duality map, then J-monotone maps are called accretive. It is known that bounded monotone maps are of type (M). We say that N is demicontinuous if $x_n \to x$ in X implies that $Nx_n \rightharpoonup Nx$. It is well known that $I - N$ is A-proper if N is ball-condensing and that K-monotone like maps are pseudo A-proper under some conditions on N and K (cf. [91], [65–67]). We shall now see that suitable sums of these maps and their perturbations by Fredholm or hyperbolic like maps are A-proper or pseudo A-proper.

Theorem 2.1: [57,63]: *Let $D \subset X$ be closed, $\Gamma = \{X_n, P_n, Y_n, Q_n\}$ be a projectional scheme for (X, Y) with $\delta = \max \|Q_n\|$, $T: D \to Y$ be continuous such that for some $c > 0$*

$$\chi(\{Q_nTx_n\}) \geq c\chi(\{x_n\}) \quad \textit{for each bounded sequence } \{x_n \mid x_n \in D \cap X_n\}$$

and $F: D \to Y$ be continuous. Then $T + F: D \to Y$ is A-proper if F is k-ball contractive with $k\delta < c$ or it is ball condensing if $\delta = c = 1$.

The following consequence states that ball-condensing perturbations of stable A-proper maps are also A-proper.

Theorem 2.2: [57]: *Let $D \subset X$ be closed, $T: X \to Y$ be continuous and A-proper w.r.t. a projection scheme Γ and a-stable, i.e., for some $c > 0$ and n_0*

$$\|Q_nTx - Q_nTy\| \geq c\|x - y\| \quad \textit{for } x, y \in X_n \text{ and } n \geq n_0$$

and $F: D \to Y$ *be continuous. Then* $T + F: D \to Y$ *is A-proper w.r.t.* Γ *if F is k-ball contrctive with* $k\delta < c$, *or it is ball-condensing if* $\delta = c = 1$.

Remark 2.1: The A-properness of T in Theorem 2.2 is equivalent to T being surjective. In particular, as T we can take a c-strongly K-monotone map for a suitable $K: X \to Y^*$, i.e., $(Tx - Ty, K(x - y)) \geq c\|x - y\|^2$ for all $x, y \in X$. When $X = Y = H$ a Hilbert space and F is k-ball contractive, this is due to Toland and Webb [100].

The next result shows that the a-stability can be weaken to a Garding like ineqaulity.

Theorem 2.3: [63,66]: *Let X be densely and compactly embedded in* $(X_0, \|\cdot\|_0)$ *and* $T: V \subset X \to X_0$ *be such that* $Q_nT: X_n \to Y_n$ *is continuous and*

$$(Tx - Ty, K(x - y)) \geq c\|x - y\|^2 - c_0\|x - y\|_0^2 \quad \textit{for all } x, y \in V$$

for some $K: X \to X_0$ *with* $(x, Kx) \geq \|x\|_0^2$ *on X and* $F: V \to Y$ *be continuous. Then* $T + F: V \to Y$ *is A-proper w.r.t.* Γ *for* (V, X_0) *with* $Q_n^*Kx = Kx$ *on* X_n *if F is k-ball contractive with* $k\delta < c$, *or it is ball-condensing if* $\delta = c = 1$.

The a-stability of T above can be replaced by the semi a-stability. Namely, let us look at more general classes of A-proper maps, which are ball-condensing perturbations of semi a-stable maps and, in particular, of strongly semimonotone and strongly semiaccretive maps. Such classes of maps have been introduced and studied by Milojević [62–64] (cf. also [65–67]).

Let $C_b(D, Y)$ denote the normed linear space with the supremum norm of all continuous bounded functions from the topological space D into the normed linear space Y.

Definition 2.2: [63]: A map $T: X \to Y$ is said to be semi a-stable w.r.t. a projectionally complete scheme $\Gamma = \{X_n, P_n, Y_n, Q_n\}$ for (X, Y) if there is a map $U: X \times X \to Y$ such that $Tx = U(x, x)$ for $x \in X$, and

(i) The map $x \to U(x, \cdot)|\bar{D}$ is compact from $\bar{D}$ into $C_b(\bar{D}, Y)$ for each bounded subset D of X;
(ii) For each $x \in X$, $U(x, \cdot)$ is a-stable w.r.t. Γ, i.e., for some $c > 0$ independent on x and each large n

$$\|Q_nU(x, u) - Q_n(x, v)\| \geq c\|u - v\| \quad u, v \in X_n.$$

In particular, $T: X \to Y$ is c-strongly semi K-monotone if there exist maps $U: X \times X \to Y$ and $K: X \to 2^{Y^*}$ such that $Tx = U(x, x)$, (i) of Definition 2.2 holds and, for each $x \in X$, $U(x, \cdot)$ is c-strongly K-monotone. We also require that $\|y\| \leq m\|x\|$ for each $y \in Kx$, $x \in X$ and some $m > 0$, and $Q_n^*Kx \subset Kx$, $x \in X_n$. If $Y = X^*$ (resp. $Y = X$) and $K = I$ (resp. $K = J$, the duality map), such maps are called c-strongly semimonotone (resp. semiaccretive). When $c = 0$, such maps are called semi K-monotone.

For such maps, the following result was proved by Milojević [63] (cf. [67]).

Theorem 2.4: *Let* $D \subset X$ *be closed,* $T: X \to Y$ *be continuous, surjective and semi a-stable w.r.t. a projection scheme* Γ *and* $F: D \to Y$ *be continuous and either k-ball contractive with* $k\delta < c$, *or it is ball-condensing if* $\delta = c = 1$, *where* $\delta = \max \|Q_n\|$. *Then* $T + F: D \to Y$ *is A-proper w.r.t.* Γ.

The proof of Theorem 2.3 follows from Theorem 2.1 since T satisfies condition (2.1) as shown in ([63]). Since c-strongly accretive maps are surjective, we have the following special case [63].

Corollary 2.1: *Let X be a π_1 Banach space, $D \subset X$ be closed, T: $X \to X$ be continuous and c-strongly semiaccretive and F: $D \to X$ be either k-ball contractive with $k < c$, or it is ball-condensing if $c = 1$. Then $T + F$: $D \to X$ is A-proper w.r.t. Γ.*

3. FIXED POINT THEORY

We shall discuss here fixed point theories for essential A-proper maps acting from a subset of a convex to itself and for weakly inward essential maps.

3.1. Fixed Points for Essential A-proper Maps

In this section we present a fixed point theory, developed by the author in [69/71], for maps T with $I - T$ (pseudo) A-proper based only on the notion of approximation-essential (A-essential) maps. The study of such maps has been initiated by the author in [63] and continued in [67,71]. Let $B = \bar{B} \subset D \subset X$ and denote by $F_B(D, Y)$ the class of maps T: $D \to Y$ such that Q_nT: $B \cap X_n \to Y_n \backslash \{0\}$ is continuous and compact for each n sufficiently large. If $\Gamma = \{X_n, P_n\}$ is a scheme for X, $C \subset X$ is convex with $P_nC \subset C$ and $B = \bar{B} \subset D \subset C$, we denote by $F_B(D, C)$ the class of maps of the form $T = I - F$ such that F: $D \to C$ and P_nT: $B \cap X_n \to X_n\backslash\{0\}$ for n large.

Definition 3.1: [63]: We say that $T \in F_B(D, Y)$ is *A-essential* w.r.t. Γ_i, $\dim X_n - \dim Y_n = i \geq 0$, if, for each n large, all continuous compact maps $G \in F_{B \cap X_n}(D \cap X_n, Y_n)$ such that $G_n|B \cap X_n = Q_nT|B \cap X_n$ has a zero in $D \cap X_n$; if not then we say that T is A-inessential relative to Γ_i. If $Y = X$, the A-essentiality of $T = I - F \in F_B(D, C)$ is defined similarly.

For many examples of A-essential maps we refer to [63,67,71].

Theorem 3.1: [69/71] (Nonlinear Alternative): *Let $C \subset X$ be convex, U be open in C with $0 \in U$ and F: $\bar{U} \to C$ such that $I - tF$, $0 \leq t \leq 1$, is A-proper at 0 w.r.t. $\Gamma = \{X_n, P_n\}$ with $P_nC \subset C$. Then*

(*i*) $0 \in (I - F)(\bar{U})$ *and, if* $0 \notin (I - F)(\partial U)$, *the equation* $Fx = x$ *is approximation solvable, and/or*

(*ii*) *there is an* $x \in \partial U$ *such that* $tFx = x$ *for some* $t \in (0, 1)$.

The proof of this result is based on using the homotopy $H(t, x) = x - tFx$ and the corresponding topological transversality theorem in [63,71]. A number of fixed point results can be deduced from Theorem 3.1 by imposing conditions on T that prevent alternative (ii) to happen, such as, for example, Leray-Schauder, Rothe and Altman (see [21,91,61]). In particular, we have

Theorem 3.2: [71]: (Nonlinear Alternative): *Let $C \subset X$ be convex, $0 \in C$, F: $C \to C$ such that $F(C \cap U)$ is bounded for each neighborhood of 0 and $I - tF$, $0 \leq t \leq 1$, is*

A-proper at 0 w.r.t. Γ *with* $P_nC \subset C$. *If* $S = \{x \in C \mid x = tFx \text{ for some } t \in (0,1)\}$, *then either S is unbounded or the equation* $Fx = x$ *is approximation solvable.*

For the existence of nontrivial fixed points, we have

Theorem 3.3: [71]: *Let* $C \subset X$, *U and V be two open sets in C with* $0 \in U \subset V$ *and* $F\colon \bar{V} \to C$ *such that* $I - tF$, $0 \le t \le 1$, *is A-proper w.r.t.* $\Gamma = \{X_n, P_n\}$ *with* $P_nC \subset C$. *Suppose that* $(I - F)(\bar{U})$ *is p-bounded for some continuous seminorm p and*

(1) $x \neq Fx + tw$ *for* $t > 0$, $x \in \partial U$ *and some* $w \in C$ *with* $p(w) \neq 0$;
(2) $Fx \neq \lambda x$ *for all* $\lambda > 1$ *and* $x \in \partial V$.

Then the equation $Fx = x$ *is approximation solvable in* $V \setminus \bar{U}$.

In view of numerous examples of A-proper maps, many specific fixed point results can be derived from the above theorems. We limit ourselves here to ball condensing perturbations of k-strongly pseudo cotractive maps, i.e. maps $T\colon X \to X$ such that $(Tx - Ty, x - y)_- = \inf\{(Tx - Ty, u) \mid u \in J(x - y)\} \le k\|x - y\|^2$ for all $x, y \in X$, where $J\colon X \to 2^{X^*}$ is the duality map.

Theorem 3.4: [61]: (*a*) *Let* $D \subset X$ *be open and bounded,* $0 \in D$, *C be a cone in X,* $T\colon X \to X$ *be continuous k-strongly pseudo contractive with* $T(C) \subset C$ *and* $F\colon \bar{D} \cap C \to C$ *be continuous and* k_1*-ball contractive with* $k_1 < 1 - k$ *or ball-condensing if* $k = 0$. *Let* $C_n = C \cap X_n$ *and*

(1) $Tx + Fx \neq \lambda x$ *for all* $x \in \partial_C(D \cap C)$, *all* $\lambda \ge \mu \ge 1$, *all large n and some* μ;
(2) *there is* $c_0 > 0$ *such that if* $P_nTx + P_nFx = \lambda x$ *for some* $x \in \partial_{C_n} D \cap C_n$ *and any n, then* $\lambda \le c_0$.

Then the equation $Tx + Fx = \mu x$ *is approximation solvable.*

(*b*) If $k_1 = 1 - k$ *with* $k \le 1$ *in* (*a*) *and also* $(\mu I - T - F)(\bar{D} \cap C)$ *is closed, then the equation* $Tx + Fx = \mu x$ *is solvable.*
(*c*) *In particular, if* $T = I - A$ *in* (*a*) (*or* (*b*)) *with* $A\colon X \to X$ *c-strongly accretive,* $k < c$ *or F is ball condensing if* $c = 1$ (*or* $k = c \ge 0$), $(I - A - F)(C) \subset C$ *and* $I - A - F$ *satisfies (1)–(2) with* $\mu = 1$ (*and* $(A + F)(\bar{D} \cap C)$ *is closed*), *then* $Ax + Fx = 0$ *is approximation solvable (solvable, respectively).*

3.2. Fixed Points for Essential Weakly Inward A-proper Maps

In this section we shall assume that C is a closed and convex subset of X and present a fixed point theory for weakly inward maps $F\colon D \subset C \to X$ using again the A-essentialness approach. Weakly inward maps have been first studied by Halperin and Bergman [30] and then by many other authors (see [18]). For $x \in C$, define the inward set of x relative to C by $I_C(x) = \{x + c(z - x) : z \in C, c \ge 0\}$. It is a convex set containing C. Denote its closure by $\bar{I}_C(x)$. A map $F\colon D \subset C \to X$ is said to be weakly inward if $Fx \in \bar{I}_C(x)$ for each $x \in D$.

Let $\Gamma = \{X_n, P_n\}$ be a scheme for X, $C \subset X$ be convex with $P_nC \subset C$ and $B = \bar{B} \subset D \subset C$. We denote by $F_{B,w}(D, X)$ the class of maps of the form $T = I - F$ such that $F\colon D \to X$, $P_nF\colon D \cap X_n \to X_n$ is weakly inward on $D \cap X_n$ relative to $C \cap X_n$ and $P_nT\colon B \cap X_n \to X_n \setminus \{0\}$ is compact for n large.

Modifying slighltly Definition 3.1, we define A-essential weakly inward maps.

Definition 3.2: We say that $T \in F_{B,w}(D, X)$ is *A-essential* w.r.t. Γ, if, for each n large, all continuous compact maps $G \in F_{B\cap X_n, w}(D \cap X_n, X_n)$ such that $G_n|B \cap X_n = Q_nT|B \cap X_n$ has a fixed point in $D \cap X_n$; if not then we say that T is A-inessential relative to Γ.

For these maps, we have the following version of Theorem 3.1.

Theorem 3.5: (Nonlinear Alternative): *Let $C \subset X$ be closed and convex, U be open in C with $0 \in U$ and $F: \bar{U} \to X$ such that $F(\bar{U})$ is bounded, $I - tF$, $0 \le t \le 1$, is A-proper at 0 w.r.t. $\Gamma = \{X_n, P_n\}$ with $P_nC \subset C$ and $P_nF: \bar{U} \cap X_n \to X_n$ is weakly inward on $\bar{U} \cap X_n$ relative to $C \cap X_n$ for each large n. Then*

(*i*) $0 \in (I - F)(\bar{U})$ *and, if* $0 \notin (I - F)(\partial U)$, *the equation* $Fx = x$ *is approximation solvable, and/or*
(*ii*) *there is an* $x \in \partial U$ *such that* $tFx = x$ *for some* $t \in (0, 1)$.

The proof of this result is based on using the homotopy $H(t, x) = x - tFx$ and the corresponding topological transversality theorem for weakly inward maps. Again, as above, a number of fixed point results can be deduced from Theorem 3.5 by imposing conditions on T that prevent alternative (ii) to happen, such as, for example, Leray-Schauder, Rothe and Altman. In particular, we have the following version of Theorem 3.2.

Theorem 3.6: (Nonlinear Alternative): *Let $C \subset X$ be closed and convex, $0 \in C$, $F: C \to X$ such that $P_nF: C \cap X_n \to X$ is weakly inward relative to $C \cap X_n$ for each large n, $F(C \cap U)$ is bounded for each bounded neighborhood of 0 and $I - tF$, $0 \le t \le 1$, is A-proper at* 0 w.r.t. Γ *with $P_nC \subset C$. If* $S = \{x \in C \mid x = tFx$ for some $t \in (0, 1)\}$, *then either S is unbounded or the equation $Fx = x$ is approximation solvable.*

For the existence of nontrivial fixed points, we have

Theorem 3.7: *Let $C \subset X$ closed and convex, U and V be two open sets in C with $0 \in U \subset V$ and $F: \bar{V} \to X$ such that $I - tF$, $0 \le t \le 1$, is A-proper w.r.t. $\Gamma = \{X_n, P_n\}$ with $P_nC \subset C$. Suppose that $(I - F)(\bar{U})$ is bounded and, for some $w \in C$, $w \neq 0$, $H(t, x) = Fx + \lambda tx$ is weakly inward on $\bar{V}$ for each $t \in [0, 1]$ and $\lambda > 0$,*

(1) $x \neq Fx + tw$ *for* $t > 0$, $x \in \partial U$;
(2) $Fx \neq \lambda x$ *for all* $\lambda > 1$ *and* $x \in \partial V$.

Then the equation $Fx = x$ is approximation solvable in $V \backslash \bar{U}$.

4. MULTIVALUED A-PROPER AND PSEUDO A-PROPER MAPPINGS

In this section we shall define multivalued (pseudo) A-proper maps and present some basic continuation theory for such maps and their uniform limits. The study of such multivalued maps has began in [54] and was continued in [54–69]. As we shall see below, these maps include multivalued monotone and generalized pseudo monotone maps, maps of type (M), multivalued condensing perturbations of the identity. It turns out that multivalued mappings play an important role in the theory of evolution equations, variational inequalities, contingent ordinary, partial differential and integral equations, optimal control theory, etc.

We begin by defining general approximation schemes. Let X and Y be two normed linear spaces, $\{E_n\}$ and $\{F_n\}$ two sequences of oriented finite dimensional spaces and V_n and W_n, continuous linear mappings of E_n into X and of Y onto F_n respectively.

Definition 4.1: A quadruple of sequences $\Gamma = \{E_n, V_n;\ F_n, W_n\}$ is said to be an admissible scheme for $(X,\ Y)$ if $\dim E_n = \dim F_n$ for each n, V_n is injective, $\text{dist}(x, V_n(E_n)) \to 0$ as $n \to \infty$ for each x in X and $\delta = \sup \|W_n\| < \infty$.

To state some examples of such schemes, assume that $\{X_n\}$ is a sequence of oriented finite dimensional subspaces of X such that $\text{dist}(x, X_n) \to 0$ as $n \to \infty$ for each x in X and let V_n be the linear injection of X_n into X.

(a) Let $\{Y_n\}$ be a sequence of finite dimensional oriented subspaces of Y with $\dim X_n = \dim Y_n$ for each n and Q_n be continuous linear mappings of Y onto Y_n with $\delta = \sup \|Q_n\| < \infty$. Then $\Gamma = \{X_n, V_n;\ Y_n, Q_n\}$ is admissible for (X, Y).
(b) If $Y = X$, $Y_n = X_n$ and $W_n = P_n$, where P_n's are projections of X onto X_n with $P_n(x) \to x$ for each x in X and $\delta = \sup \|P_n\| < \infty$, then $\Gamma = \{X_n, V_n;\ X_n, P_n\}$ is an admissible projection scheme for (X, X).
(c) If $Y = X^*$, $Y_n = R(P_n^*)$, $V_n = P_n|X_n = I_n$ and $W_n = P_n^*$, then $\Gamma = \{X_n, P_n;\ Y_n, P_n^*\}$ is an admissible projection scheme (X, X^*).
(d) If $Y = X^*$, $Y_n = X_n^*$ and $W_n = V_n^*$, then $\Gamma = \{X_n, V_n;\ X_n *, V_n^*\}$ is an admissible injection scheme for $(X, X^*\}$.

If P_n is as in (b) and Q_n is a projection of Y onto Y_n with $Q_n(y) \to y$ for all y in Y, then $\Gamma_0 = \{X_n, P_n;\ Y_n, Q_n\}$ is called a projectionally complete scheme for (X, Y).

Let D be a subset of X and V be a subspace of X. Set $T_n = W_n T V_n$.

Definition 4.2: A mapping T: $D \cap V \to 2^Y$ is approximation-proper (A-proper for short) w.r.t. Γ if, whenever $\{V_n(u_{n_k}) \in D \cap V\}$ is bounded and $\|y_n - W_n(f)\| \to 0$ for some $y_{n_k} \in T_{n_k}(u_{n_k})$ and f in Y, then some subsequence $V_{n_{k(i)}}(u_{n_{k(i)}}) \to 0$.

T is weakly A-proper if we require the weak convergence $V_{n_k}(u_{n_k}) \rightharpoonup x$ in Definition 4.2.

We say that H: $[0,1] \times (D \cap V) \to 2^Y$ is an A-proper homotopy w.r.t. Γ if, whenever $\{V_{n_k}(u_{n_k}) \in D \cap V\}$ is bounded and $\|W_{n_k}(y_k) - W_{n_k}(f)\| \to 0$ for some $y_{n_k} \in H(t_k, V_{n_k}(u_{n_k}))$, $t_k \in [0,1]$ with $t_k \to t_0$ and f in Y, then some subsequence $V_{n_{k(i)}}(u_{n_{k(i)}}) \to x$ and $f \in H(t_0, x)$.

Consider the operator equation

$$f \in T(x) \quad (x \in D \cap V,\ f \in Y) \tag{4.1}$$

and a sequence of finite dimensional equations induced by Γ

$$W_n(f) \in T_n(u) \quad (u \in E_n, n = 1, 2, \ldots). \tag{4.2}$$

Definition 4.3: We say that equation (4.1) is feebly approximation-solvable (w.r.t. Γ) if there exists a solution u_n of equation (4.2) for infinitely many n and some further subsequence $V_{n_k}(u_{n_k}) \to x_0$ in X with $f \in T(x_0)$.

It is clear that if T is A-proper and for some set of solutions $\{u_{n_k}\}$ of equation (4.2), $\{V_{n_k}(u_{n_k})\}$ is bounded, then equation (4.1) is feebly approximation-solvable. We note that if T is injective, i.e., $T(x) \cap T(y) = \emptyset$ whenever $x \neq y$, and $\{V_n(u_n)\}$ is bounded,

$\{u_n\}$ being solutions of equation (4.2), then $V_n(u_n) \to x_0$ and $f \in T(x_0)$. In this case we say that equation (4.1) is strongly approximation-solvable.

Now we shall state some basic examples of A-proper and mappings needed in the sequel. Let $C(X)$, $K(X)$ and $BK(X)$ denote the family of all nonempty compact, closed convex, and bounded closed convex subsets of X, respectively. We begin with the following

Definition 4.4: (a) Let X and Y be real Banach spaces, $D \subset X$ a closed subset of X and $K: X \to 2^{X^*}$. Then $T: D \to BK(Y)$ is said to be of type (KS) if,

(i) for each finite-dimensional subspace F of X, T is upper semicontinuous (u.s.c for short) from $D \cap F$ into the weak topology of Y;
(ii) for each $\{x_n\} \subset D$ with $x_n \rightharpoonup x$ in X and $(u_n, f_n) \to 0$ for some $u_n \in Tx_n$ and $f_n \in K(x_n - x)$, we have that $x_n \to x$ in X.

(b) $T: D \to BK(Y)$ is said to be of type (KS_+) if (i) holds and

(iii) for each $\{x_n\} \subset D$ with $x_n \rightharpoonup x$ in X and $\limsup\,(u_n, g_n) \leq 0$ for some $u_n \in Tx_n$, $f_n \in K(x_n - x)$, we have that $x_n \to x$ in X.

Example 4.1: Let X and Y be reflexive Banach spaces, $D \subset X$ closed, $\Gamma = \{X_n, V_n, Y_n, Q_n\}$ an admissible scheme for (X, Y) and $T: D \to BK(Y)$. Let $K: X \to Y^*$ be bounded and such that

(i) $Kx = 0$ implies $x = 0$, K is positively homogeneous of order $a > 0$ and the range $R(K)$ is dense in Y^*;
(ii) for each $x \in X_n$ and $y \in Y$, we have $(Q_n y, Kx) = (y, Kx)$;
(iii) K is weakly continuous at 0 and is uniformly continuous on closed balls in X.

Then, if T is K-quasibounded, demiclosed and of type (KS), it is A-proper w.r.t. Γ_a.

Regarding the intertwined representation of mappings of type (KS_+), we refer to [63,67].

It turns out that monotone mappings can be treated via the A-proper mapping theory due to the fact that they are uniform limits of A-proper mappings or are pseudo A-proper. To see this, we recall some basic facts about them.

Definition 4.5: Let $K: X \to 2^{Y^*}$. A mapping $T: X \to BK(Y)$ is said to be quasi-K-monotone if

(i) T is u.s.c. from each finite-dimensional subspace F of X to the weak topology of Y.
(ii) $x_n \rightharpoonup x$ in X implies that for each $u_n \in T(x_n)$ and $f_n \in K(x_n - x)$, $\limsup\,(u_n, f_n) \geq 0$.

Singlevalued quasi-monotone mappings from X into X^* were introduced and studied by Hess [32] and Calvert and Webb [15] and then studied by many authors (cf. [63,67]. Such mappings are uniform limits of A-proper mappings as shown below.

We say that a map $T: X \to BK(Y)$ is K-quasibounded if for any bounded sequence $\{x_n\} \subset X$, $y_n \in T(x_n)$ and $f_n \in K(x_n)$ with $(y_n, f_n) \leq c\|x_n\|$ for each n and some constant $c > 0$, the sequence $\{y_n\}$ is bounded in Y.

Example 4.2: Let X and Y be reflexive Banach spaces with an admissible scheme $\Gamma_a = \{X_n, V_n, Y_n, Q_n\}$ and $K: X \to Y^*$ be a bounded mapping for which conditions (i), (ii) and (iii) of Example 4.1 hold. Let $T: X \to BK(Y)$ be K-quasibounded, demiclosed and quasi-K-monotone and $G: X \to BK(Y)$ be bounded, demiclosed and of type (KS_+). Then $T + \alpha G$ is of type (KS_+) and therefore is A-proper w.r.t. Γ_a for each $\alpha > 0$.

The next two classes of mappings are given by

Definition 4.5: (a): A mapping $T: X \to BK(Y)$ is said to be pseudo-K-monotons if

(i) of Definition 4.4 holds; and
(ii) if $x_n \rightharpoonup x$ in X and if $u_n \in T(x_n)$ and $f_n \in K(x_n - x)$ are such that $\limsup (u_n, f_n) \le 0$, then for each element $v \in X$ there exist $u(v) \in T(x)$, $g \in K(x - v)$ and $g_n \in K(x_n - v)$ such that $\liminf (u_n, g_n) \ge (u(v), g)$.

(b) $T: X \to BK(Y)$ is said to be generalized pseudo K-monotone if (i) of Definition 4.4 holds; and

(iii) if $x_n \rightharpoonup x$ in X and $u_n \in T(x_n)$, $f_n \in K(x_n - x)$ with $u_n \rightharpoonup u$ in Y and $\limsup (u_n, f_n) \le 0$ imply that $u \in T(x)$ and $(u_n, f_n) \to 0$.

In the singlevalued case, the first class of pseudo-monotone mappings was studied by Leray and Lions [45] and explicitely they were introduced (in a somewhat different way) and studied by Brezis [8] (here, $K = I$, $Y = X^*$). Later on they have been studied by many other authors (cf. Lions [46] and Browder [11]. Generalized pseudo-monotone mappings were introduced by Browder and Hess [14]. These two classes of mappings are also uniform limits of A-proper mappings. More precisely, we have (cf. [63,67]).

Example 4.3: Let X and Y be reflexive, $T: X \to BK(Y)$ be K-quasibounded, $\Gamma_a = \{X_n, V_n, Y_n^*, Q_n\}$ an admissible scheme for (X, Y) and K a bounded mapping for which conditions (i), (ii) and (iii) of Example 4.1 hold. Let $G: X \to BK(Y)$ be bounded demicontinuous and of type (KS_+).

(a) If T is demicontinuous and K-monotone with K weakly continuous, then $T + \alpha G$ is A-proper w.r.t. Γ_a for each $\alpha > 0$, $T(\bar{B}(0, r))$ is closed for each $r > 0$ and T is weakly A-proper w.r.t. Γ_a.
(b) If T is pseudo K-monotone and either domiclosed or $R(K) = Y^*$ and if K is weakly continuous with $K(0) = 0$, then $T + \alpha G$ is A-proper w.r.t. Γ_a for $\alpha > 0$, $T(\bar{B}(0, r))$ is closed for each $r > 0$ and T is weakly A-proper w.r.t. Γ_a.
(c) If T is generalized pseudo K-monotone, then $T + \alpha G$ is A-proper w.r.t. Γ_a for each $\alpha > 0$ and $T(\bar{B}(0, r))$ is closed for each $r > 0$. Moreover, if in addition, K is continuous, then T is weakly A-proper w.r.t. Γ_a.

The next general class of nonlinear mappings is that of type (KM).

Definition 4.6: A mapping $T: X \to BK(Y)$ is said to be of type (KM) if (i) of Definition 4.4 holds and if $x_n \rightharpoonup x$ in X and $u_n \in T(x_n)$, $f_n \in K(x_n - x)$ with $u_n \rightharpoonup u$ in Y and $\limsup (u_n, f_n) \le 0$ imply $u \in T(x)$.

This type of mappings was introduced by Brezis [8] when $Y = X^*$, $K = I$ and later studied by many authors (cf. Lions [46]). They are also of weakly A-proper type. Namely, we have proven in [63] the following

Example 4.4: Let X and Y be reflexive, $K: X \to Y^*$ bounded and as in Example 4.1 and $T: X \to BK(Y)$ be K-quasibounded and of type (KM). Then T is weakly A-proper w.r.t. Γ_a.

We note also that weakly closed and weakly continuous mappings are also pseudo A-proper (cf. [63]).

5. CONTINUATION THEOREMS FOR A-PROPER MAPPINGS

In this section we shall prove a number of continuation results for A-proper like mappings. Most of the results have been first announced in [63] and proven in [67–69]. Applications of the results of this section to equations involving nonlinear perturbations of Fredholm linear mappings can be found in Section 9. Their applications to other classes of nonlinear mappings like, for example, ball-condensing perturbations of (strongly) accretive, or K-monotone mappings, etc., were given in [67–69].

Throughout this Section 5 shall denote a subspace of X and $\Gamma = \{E_n, V_n, F_n, W_n\}$ an admissible scheme for (V, Y).

To facilitate the statements of our results, we separate the following condition on $H(t,x)$ defined in $[0,1] \times (B_n \cap V)$:

(5.1) If for some $t_k \in (0,1)$ with $t_k \to 1$ and a bounded sequence $\{V_{n_k}(u_{n_k}) \mid u_{n_k} \in V_{n_k}^{-1}(\bar{D} \cap V)\}$ we have that $W_{n_k}(y_k) - W_{n_k}(f) \to 0$ for some $y_k \in H(t_k, V_{n_k}(u_{n_k}))$ and $f \in Y$, then there are $z_k \in H(1, V_{n_k}(u_k))$ such that $W_{n_k}(z_k) - W_{n_k}(f) \to 0$ as $k \to \infty$.

We say that $H(t,x)$ is α-continuous at 1 uniformly for $x \in \bar{D} \cap V$ if $\alpha(H(t_n, x), H(1,x)) = \sup\{d(y, H(1,x)) \mid y \in H(t_n, x)\} \to 0$ as $t_n \to 1$ uniformly for $x \in \bar{D} \cap V$. It is easy to show that if $H(t,x)$ is α-continuous at 1 uniformly for $x \in \bar{D} \cap V$ and $W_n H(t,x) \in C(F_n)$ for $x \in \bar{D} \cap V_n(E_n)$, then (5.1) holds.

We say that a mapping $T: \bar{D} \cap V \to K(Y)$ satisfies condition $(*)$ if, whenever $\{x_n\}$ is bounded in X and $d(f, Tx_n) \to 0$ for some f in Y, then there is some $x \in \bar{D} \cap V$ such that $f \in T(x)$.

Our results are based on the degree theory for upper demicontinuous (u.d.c. for short) mappings $T: \bar{G} \subset E_n \to K(F_n)$ as developed in [47], [43].

Our first result has been obtained in [63,67,68]. In what follows, we let V be dense subspace of a Banach space X, $D \subset X$ be an open and bounded subset and Y be a Banach space.

Theorem 5.1: *Let a homotopy $H: [0,1] \times (\bar{D} \cap V) \to K(Y)$ be such that*

(*i*) *H is an A-proper homotopy w.r.t. Γ on $[0,\epsilon] \times (\partial D \cap V)$ for each $\epsilon \in (0,1)$ and H_1 is pseudo A-proper w.r.t. Γ;*

(*ii*) *$f \notin H(t,x)$ and $tf \notin H(0,x)$ for $t \in [0,1]$, $x \in \partial D \cap V$;*

(*iii*) $\deg(W_n H_0 V_n, V_n^{-1}(D \cap X_n), 0) \neq 0$ *for all large n.*

(*a*) *Let condition (5.1) hold, and, in particular, $H(t,x)$ be continuous at 1 uniformly for $x \in \partial D \cap V$, with $W_n H(t,x) \in C(F_n)$ on $\bar{D} \cap V_n(E_n)$.*

Then the equation $f \in H(1,x)$ is solvable in $\bar{D} \cap V$. It is approximation solvable if H_1 is A-proper.

(*b*) *Let H_1 satisfy condition $(*)$, $H(t,x)$ be α-continuous at 1 uniformly for $x \in \bar{D} \cap V$ and H_t be pseudo A-proper w.r.t. Γ for each $t \in (t_0, 1)$ for some t_0. Then the equation $f \in H(1,x)$ is solvable.*

When $H(t,x)$ is not an A-proper homotopy, the following extension of Theorem 5.1 is valid.

Theorem 5.2: *Let H: $[0,1]\times(\bar{D}\cap V)\to K(Y)$ be such that for a given f in Y there exists an $n_f\geq 1$ with*

(*i*) $W_n(f)\notin W_nH(t,V_n(u))$ *for* $u\in\partial D_n$, $t\in[0,1)$, $n\geq n_f$;
(*ii*) $tW_n(f)\notin W_nH(0,V_n(u))$ *for* $u\in\partial D_n$, $t\in[0,1]$, $n\geq n_f$
(*iii*) $\deg(W_nH_0V_n, V_n^{-1}(D\cap X_n),0)\neq 0$ *for all large* n.

Suppose that H_1 is pseudo A-proper w.r.t. Γ. Then the equation $f\in H(1,\ x)$ is solvable.

Definition 5.1: A mapping H: $[0,1]\times V\to K(Y)$ satisfies condition (+) if $\{x_n\}\subset V$ is bounded whenever $y_n\to f$ for some $y_n\in H(t_n,x_n)$ with $t_n\in[0,1]$.

We continue our exposition with a second type of continuation results for A-proper mappings.

Theorem 5.3: *Let a homotopy H: $[0,1]\times(\bar{D}\cap V)\to K(Y)$ be such that*

(*i*) *H is an A-proper homotopy at 0 w.r.t. Γ on $[0,\epsilon]\times(\partial D\cap V)$ for each $\epsilon\in(0,1)$ and H_1 is pseudo A-proper w.r.t. Γ;*
(*ii*) $tf\notin H(1,x), 0\notin H(t,x)$ *for* $t\in[0,1]$, $x\in\partial D\cap V$;
(*iii*) $0\notin H(t,x)$ *for* $t\in[0,1]$, $x\in\partial D\cap V$;
(*iv*) $\deg(W_nH_0V_n, V_n^{-1}(D\cap X_n),0)\neq 0$ *for all large* n.

(*a*) *Let condition (5.1) hold, and, in particular, $H(t,x)$ be continuous at 1 uniformly for $x\in\partial D\cap V$, with $W_nH(t,x)\in C(F_n)$ on $\bar{D}\cap V_n(E_n)$.*

Then the equation $f\in H(1,x)$ is solvable in $\bar{D}\cap V$. It is approximation solvable if H_1 is A-proper.

(*b*) *Let H_1 satisfy condition (*), $H(t,x)$ be α-continuous at 1 uniformly for $x\in\bar{D}\cap V$ and H_t be pseudo A-proper w.r.t. Γ for each $t\in(t_0,1)$ for some t_0. Then the equation $f\in H(1,x)$ is solvable without assuming (ii).*

When $H(t,x)$ is not an A-proper homotopy, the following extension of Theorem 5.3 is valid.

Theorem 5.4: *Let H: $[0,1]\times(\bar{D}\cap V)\to K(Y)$ be such that for a given f in Y there exists an $n_f\geq 1$ with*

(*i*) $tW_n(f)\notin W_nH(1,V_n(u))$ *for* $u\in\partial D_n$, $t\in[0,1)$, $n\geq n_f$;
(*ii*) $0\notin W_nH(1,V_n(u))$ *for* $u\in\partial D_n$, $t\in[0,1]$, $n\geq n_f$;
(*iii*) $\deg(W_nH_0V_n, V_n^{-1}(D\cap X_n),0)\neq 0$ *for all large* n.

Suppose that H_1 is pseudo A-proper w.r.t. Γ. Then the equation $f\in H(1,x)$ is solvable without assuming (ii).

Finally, let us consider now a third general continuation type result.

Theorem 5.5: *Let a homotopy H: $[0,1]\times(\bar{D}\cap V)\to K(Y)$ be such that*

(*i*) *H is an A-proper homotopy w.r.t. Γ on $[0,\epsilon]\times(\partial D\cap V)$ for each $\epsilon\in(0,1)$ and H_1 is pseudo A-proper w.r.t. Γ;*

(*ii*) $tf \notin H(t,x)$ *for* $t \in [0,1]$, $x \in \partial D \cap V$;
(*iii*) $\deg(W_n H_0 V_n, V_n^{-1}(D \cap X_n), 0) \neq 0$ *for all large* n.

(*a*) *Let condition (5.1) hold, and, in particular,* $H(t,x)$ *be continuous at 1 uniformly for* $x \in \partial D \cap V$, *with* $W_n H(t,x) \in C(F_n)$ *on* $\bar{D} \cap V_n(E_n)$.

Then the equation $f \in H(1,x)$ *is solvable in* $\bar{D} \cap V$. *It is approximation solvable if* H_1 *is A-proper.*

(*b*) *Let* H_1 *satisfy condition* $(*)$, $H(t,x)$ *be* α*-continuous at 1 uniformly for* $x \in \bar{D} \cap V$ *and* H_t *be pseudo A-proper w.r.t.* Γ *for each* $t \in (t_0, 1)$ *for some* t_0. *Then the equation* $f \in H(1,x)$ *is solvable.*

When $H(t,x)$ is not an A-proper homotopy, the following extension of Theorem 5.5 is valid.

Theorem 5.6: *Let* H: $[0,1] \times (\bar{D} \cap V) \to K(Y)$ *be such that for a given* f *in* Y *there exists an* $n_f \geq 1$ *with*

(*i*) $tW_n(f) \notin W_n H(t, V_n(u))$ for $u \in \partial D_n$, $t \in [0,1)$, $n \geq n_f$;
(*iii*) $\deg(W_n H_0 V_n, V_n^{-1}(D \cap X_n), 0) \neq 0$ *for all large* n.

Suppose that H_1 *is pseudo A-proper w.r.t.* Γ. *Then the equation* $f \in H(1,x)$ *is solvable.*

6. CONTINUATION THEOREMS FOR UNIFORM LIMITS OF A-PROPER MAPPINGS

In this section we shall extend nonconstructively the continuation theorems from Section 5 to a larger class of mappings that includes uniform limits of A-proper mappings. Applications of these results to nonlinear perturbations of Fredholm mappings will be given in Section 7, while their applications to monotone like and other types of mappings will be given elsewhere.

We need the following condition

(6.1) If for $\mu \in (0, \mu_0)$ fixed, we have that $W_{n_k}(y_k) + \mu W_{n_k}(z_k) - W_{n_k}(f) \to 0$ for some $y_k \in H(t_k, V_{n_k}(u_k))$, $z_k \in GV_{n_k}(u_k)$, $u_k \in \bar{D}_{n_k}$, $f \in Y$ and $t_k \to 1$, then $W_{n_{k(i)}}(v_i) + \mu W_{n_{k(i)}}(w_i) - W_{n_{k(i)}}(f) \to 0$ for some $v_i \in H(1, V_{n_{k(i)}}(u_{k(i)}))$ and $w_i \in GV_{n_{k(i)}}(u_{n(i)})$.

Our first basic continuation theorem has been proved in [63,67].

Theorem 6.1: *Let* G: $\bar{D} \cap V \to BK(Y)$ *be bounded,* $f \in Y$, $\mu_0 > 0$ *and a homotopy* H: $[0,1] \times (\bar{D} \cap V) \to K(Y)$ *be such that for each fixed* $\mu \in (0, \mu_0)$.

(*i*) $H + \mu G$ *is an A-proper homotopy w.r.t.* Γ *on* $[0,\epsilon] \times (\partial D \cap V)$ *for each* $\epsilon \in (0,1)$
(*ii*) $f \notin H(t,x) + \mu Gx$ *for* $t \in [0,1]$, $x \in \partial D \cap V$;
(*iii*) $tf \notin H(0,x) + \mu Gx$ *for* $t \in [0,1]$, $x \in \partial D \cap V$;
(*iv*) $\deg(W_n H_0 V_n + \mu W_n G V_n, V_n^{-1}(D \cap X_n), 0) \neq 0$ *for all large* n.

Let condition (6.1) hold. Then, if $H_1 + \mu G$ *is pseudo A-proper w.r.t.* Γ *for each* $\mu \in (0, \mu_0)$ *and* H_1 *satisfies condition* $(*)$, *the equation* $f \in H(1,x)$ *is solvable.*

When $H_t + \mu G$ is not an A-proper homotopy, the following extension of Theorem 6.1 is valid.

Theorem 6.2: *Let G: $\bar{D} \cap V \to BK(Y)$ be bounded, $f \in Y$, $\mu_0 > 0$ and a homotopy H: $[0,1] \times (\bar{D} \cap V) \to K(Y)$ be such that there is an $n_0 = n(n, \mu)$ for each $\mu \in (0, \mu_0)$ with*

(*i*) $W_n(f) \notin W_n H(t, V_n(u)) + \mu W_n G V_n(u)$ *for* $u \in \partial D_n$, $t \in [0,1)$, $n \geq n_0$;
(*ii*) $tW_n(f) \notin W_n H(0, V_n(u)) + \mu W_n G V_n(u)$ *for* $u \in \partial D_n$, $t \in [0,1]$, $n \geq n_0$;
(*iii*) $\deg(W_n H_0 V_n + \mu W_n G V_n, V_n^{-1}(D \cap X_n), 0) \neq 0$ *for all large* n.

Suppose that H_1 *satisfies condition* $(*)$. *Then the equation* $f \in H(1, x)$ *is solvable.*

We continue our exposition by looking at an analogue of Theorem 5.3 for uniform limits of A-proper mappings.

Theorem 6.3: *Let G: $\bar{D} \cap V \to BK(Y)$ be bounded, $f \in Y$, $\mu_0 > 0$ and a homotopy H: $[0,1] \times (\bar{D} \cap V) \to K(Y)$ be such that for each $\mu \in (0, \mu_0)$*

(*i*) $H + \mu G$ *is an A-proper homotopy w.r.t.* Γ *on* $[0, \epsilon] \times (\partial D \cap V)$ *for each* $\epsilon \in (0,1)$;
(*ii*) $tf \notin H(1, x) + \mu G x$ *for* $t \in [0,1]$, $x \in \partial D \cap V$;
(*iii*) $0 \notin H(t, x) + \mu G x$ *for* $t \in [0,1]$, $x \in \partial D \cap V$;
(*iv*) $\deg(W_n H_0 V_n + \mu W_n G V_n, V_n^{-1}(D \cap X_n), 0) \neq 0$ *for all large* n.

Let condition (6.1) hold and H_1 *satisfy condition* $(*)$.

(*a*) *If* $H_1 + \mu G$ *is A-proper w.r.t.* Γ *for each* $\mu \in (0, \mu_0)$, *then the equation* $f \in H(1, x)$ *is solvable.*
(*b*) *If* $H_1 + \mu G$ *is pseudo A-proper w.r.t.* Γ *for each* $\mu \in (0, \mu_0)$, *then the equation* $0 \in H(1, x)$ *is solvable without assuming (ii).*

When $H_t + \mu G$ is not an A-proper homotopy, the following extension of Theorem 6.3 is valid.

Theorem 6.4: *Let G: $\bar{D} \cap V \to BK(Y)$ be bounded, $f \in Y$, $\mu_0 > 0$ and a homotopy H: $[0,1] \times (\bar{D} \cap V) \to K(Y)$ be such that there is an $n_0 = n(n, \mu)$ for each $\mu \in (0, \mu_0)$ with*

(*i*) $tW_n(f) \notin W_n H(1, V_n(u)) + \mu W_n G V_n(u)$ *for* $u \in \partial D_n$, $t \in [0,1]$, $n \geq n_0$;
(*ii*) $0 \notin W_n H(t, V_n(u)) + \mu W_n G V_n(u)$ *for* $u \in \partial D_n$, $t \in [0,1]$, $n \geq n_0$;
(*iii*) $\deg(W_n H_0 V_n + \mu W_n G V_n, V_n^{-1}(D \cap X_n), 0) \neq 0$ *for all large* n.

Suppose that H_1 *satisfies condition* $(*)$ *and* $H_1 + \mu G$ *is pseudo A-proper w.r.t.* Γ. *Then the equation* $f \in H(1, x)$ *is solvable.*

Next, we shall extend Theorem 5.5 to uniform limits of A-proper mappings.

Theorem 6.5: *Let G: $\bar{D} \cap V \to BK(Y)$ be bounded, $f \in Y$, $\mu_0 > 0$ and a homotopy H: $[0,1] \times (\bar{D} \cap V) \to K(Y)$ be such that for each fixed $\mu \in (0, \mu_0)$*

(*i*) $H + \mu G$ *is an A-proper homotopy w.r.t.* Γ *on* $[0, \epsilon] \times (\partial D \cap V)$ *for each* $\epsilon \in (0,1)$;
(*ii*) $tf \notin H(t, x) + \mu G x$ *for* $t \in [0,1]$, $x \in \partial D \cap V$;
(*iii*) $\deg(W_n H_0 V_n + \mu W_n G V_n, V_n^{-1}(D \cap X_n), 0) \neq 0$ *for all large* n.

Let condition (6.1) hold, H_1 *satisfy condition* $(*)$ *and* $H_1 + \mu G$ *be pseudo A-proper w.r.t.* Γ *for each* $\mu \in (0, \mu_0)$. *Then the equation* $f \in H(1, x)$ *is solvable.*

When $H_t + \mu G$ is not an A-proper homotopy, the following extension of Theorem 6.5 is valid.

Theorem 6.6: *Let* G: $\bar{D} \cap V \to BK(Y)$ *be bounded,* $f \in Y$, $\mu_0 > 0$ *and a homotopy* H: $[0, 1] \times (\bar{D} \cap V) \to K(Y)$ *be such that there is an* $n_0 = n(n, \mu)$ *for each* $\mu \in (0, \mu_0)$ *with*

(i) $tW_n(f) \notin W_nH(t, V_n(u)) + \mu W_nGV_n(u)$ *for* $u \in \partial D_n$, $t \in [0, 1]$, $n \geq n_0$;
(iii) *For each* $\mu \in (0, \mu_0)$, $\deg(W_nH_0V_n + \mu W_nGV_n, V_n^{-1}(D \cap X_n), 0) \neq 0$ *for all large* n.

Suppose that H_1 *satisfies condition* $(*)$ *and* $H_1 + \mu G$ *is pseudo A-proper w.r.t.* Γ. *Then the equation* $f \in H(1, x)$ *is solvable.*

7. FIXED POINT THEORY FOR MULTIVALUED APPROXIMABLE MAPPINGS

In this section we study several new classes of multivalued approximable mappings acting in a cone. Such mappings appear naturally in studying differential equations using a finite difference method (see [44]). We follow [61].

Let X be a normed linear space and $\{X_n\}$ a sequence of oriented finite dimensional Banach spaces (not necessarily subspaces of X) such that dim $X_n \to \infty$ as $n \to \infty$. Suppose that for each n there exists a linear mapping V_n from X_n into X such that for some constant $K > 0$, $\|V_nx\| \leq K\|x\|$ for all x in X_n and all n. The pair $\{X_n, V_n\}$ is also referred to as an approximation scheme for X.

Definition 7.1: A multivalued mapping T: $D \subset X \to 2^X$ is said to be *approximately compact* (A_γ-compact for short) at some $f \in X$ with respect to the scheme $\{X_n, V_n\}$) if

(1) there exists a sequence of upper semicontinuous mappings $\{T_n\}$ from $D_n \subset X_n$ into $CK(X_n) =$ the family of all nonempty compact and convex subsets of X_n;
(2) there exists $\gamma > 0$ such that, if for some positive $\mu \geq \gamma$ and a sequence $\{x_{n_k} \mid x_{n_k} \in D_{n_k}\}$ with $\|x_{n_k}\| \leq M$ for all k and some $M > 0$, we have that $V_{n_k}(y_{n_k}) - \mu x_{n_k} \to f$ in X as $k \to \infty$ for some $y_{n_k} \in T_{n_k}(x_{n_k})$, then there exist a subsequence $\{x_{n_{k(i)}}\}$ and $x \in X$ such that $V_{n_{k(i)}}(x_{n_{k(i)}}) \to x_0$ in X and $f \in T(x_0) - \mu x_0$.

A slight variant of the definition of A_1-compact mappings at 0 in the single-valued case was given in Lees–Schultz [44].

By a cone $K \subset X$ we mean a closed subset of X such that $\alpha x + \beta y \in K$ whenever $x, y \in K$ and $\alpha > 0$, $\beta > 0$. Let $B(0, r)$ and $B_n(0, r)$ be the open balls in X and X_n, respectively centered at the origin and of radius r and $\partial(B \cap K)$ denotes the boundary of B relative to K.

Our first fixed point result for positive A_γ-compact mappings is given by

Theorem 7.1: *Let* X *be a Banach space,* K *and* K_n *cones in* X *and* X_n *respectively,* $D \subset X$ *and* $D_n \subset X_n$ *open and bounded subsets containing the origin of* X *and* X_n, *respectively and* T: $\bar{D} \cap K \to 2^K$ A_γ*-compact at 0 with* T_n: $D_n \cap K_n \to CK(K_n)$. *Suppose that* $\{D_n\}$ *is uniformly bounded, i.e., there exists an* $M > 0$ *such that* $\|x\| \leq M$ *for all* $x \in D_n$ *and all* n, *and that* T_n *satisfies the Leray-Schauder condition on* $\partial_{K_n}(D_n \cap K_n)$, *i.e., for all large* n

(μ) $\alpha x \notin T_n(x)$ *for all* $x \in \partial_{K_n}(D_n \cap K_n)$ *and* $\alpha \geq \mu$ *for some positive* μ.

Then the equation $\mu x \in T(x)$ is feebly approximation solvable. If the equation $\mu x \in T(x)$ is uniquely solvable, then it is strongly approximation solvable in the sense that the entire sequence of approximate solutions $V_n(x_n) \to x$ and $\mu x \in T(x)$.

Some conditions that imply condition (μ) are given next.

Theorem 7.2: *Let X, K and K_n be as in Theorem 7.1 and T: $K \to 2^K$ A_γ-compact at 0 with T_n: $K_n \to CK(K_n)$ such that:*

(i) *there exists a $C_1 > 0$ such that $\|V_n(x)\| \geq C_1\|x\|$ for all n and $x \in X_n$;*
(ii) *there exists an $r_0 > 0$ such that T satisfies the Leray-Schauder condition (μ) on $\partial_K(B(0,r) \cap K)$ for all $r \in [r_0, Cr_0/C_1]$ and some positive $\gamma \geq \mu$;*
(iii) *there exists a $C_2 > 0$ such that if $\lambda x \in T_n(x)$ for some $x \in \partial_{K_n}(B_n(0,r_0/C_1) \cap K_n)$ and all n, then $\lambda \leq C_2$.*

Then the equation $\mu \in T(x)$ is feebly approximation solvable in $B(0,r_0/C_2) \cap K$ with approximate solutions lying in $B_n(0,r_0/C_2) \cap K_n$.

When $K = X$, $K_n = X_n$, $\mu = 1$ D and D_n are balls and T and T_n are singlevalued, Theorems 7.1–7.2 were proved by Lees–Scultz for their A_1-compact mappings.

Corollary 7.1: *Let X be a Banach space with a projectionally complete scheme $\Gamma = \{X_n, P_n\}$, $K \subset X$ a cone, $K_n = K \cap X_n$ and $D \subset X$ open and bounded with $0 \in D$. Let T: $\bar{D} \cap K \to 2^K$ be such that $I - tT$ is A-proper at 0 for all $t \geq \gamma$ with P_n: $\bar{D} \cap K_n \to CK(X_n)$ u.s.c. and $P_n(K) \subset K$ for all large n. Suppose that:*

(i) *T satisfies the Leray-Schauder condition on $\partial_K(D \cap K)$;*
(ii) *$T_n = P_nT$ satisfies condition (ii) of Theorem 7.2 on $\partial_{K_n}(D_n \cap K_n)$ for all large n. Then the equation $\mu x \in T(x)$ is feebly approximation solvable in $D \cap K$.*

As consequences of this result we just mention here the following ones (many others can be found in [61]).

Corollary 7.2: *Let X be a reflexive, π_1-Banach space, K, D and Γ as in Corollary 7.1. Suppose that T: $K \to CK(K)$ is a generalized contraction (i.e., for each $x \in K$ there exists $\alpha(x) \in (0,1)$ such that $\delta(Tx,Ty) \leq \alpha(x)\|x-y\|$ for all $x, y \in K$, where δ is the Hausdorff distance induced by the norm of X) and C: $\bar{D} \to CK(K)$ demiclosed and compact. Then, if $T + C$ satifies condition (μ) on $\partial_K(D \cap K)$ for some $\mu \geq 1$, the equation $\mu x \in T(x) + C(x)$ is feebly approximation solvable.*

Corollary 7.3: *Let X be a reflexive π_1-Banach space and T: $X \to CK(X)$ a generalized contraction. Then for each $f \in X$ the equation $f \in x - Tx$ is feebly approximation solvable in $B(0, r_f)$ with $r_f = \max\{\|v + f\| : v \in T(0)\}/(1 - \alpha(0))$.*

For the singlevalued case with $K = X$ and T: $X \to X$ see [23] and the references therein. Moreover if T is k-strict contractive, $k < 1$, Corollaries 7.2–7.3 hold without the reflexivity of X.

8. SEMILINEAR EQUATIONS WITHOUT RESONANCE

In this section we study semilinear equations without resonance of the form

$$Ax + Nx = f \quad (x \in D(A),\ f \in Y) \tag{8.1}$$

with asymptotically linear or positively homogeneous perturbations $N: X \to Y$ of Fredholm maps of nonnegative index and of linear maps with infinite dimensional null space $A: D(A) \subset X \to Y$.

We begin with the following generalized first Fredholm theorem for nonlinear maps that are asymptotically close to a suitable map.

Theorem 8.1: [54,55,59]: *Let A, T: $X \to Y$ be nonlinear maps and for some function $c: R \to R$ with $c(r) \to \infty$ as $r \to \infty$*

$$\|Q_n Ax\| \geq c(\|x\|) \quad \text{for } x \in X_n,\ n \geq n_0 \geq 1;$$
$$|T - A| = \limsup_{\|x\| \to \infty} \|Tx - Ax\|/c(\|x\|) < 1/\delta.$$

Let T be pseudo A-proper w.r.t. Γ and either A is odd or $T = A + N$ and $\deg(Q_n A, B(0, r) \cap X_n, 0) \neq 0$ for each r large and each $n \geq n_0(r)$. Then $T(X) = Y$.

Theorem 8.2: [54,55,59]: *Let T: $X \to Y$ be asymptotically positively k-homogeneous, i.e., $T = A + N$ with the quasinorm $|N|$ of N, defined as above with $c(r) = r^k$, sufficienly small and A is positively k-homogeneous, $k > 0$, A-proper w.r.t. Γ and $x = 0$ if $Ax = 0$. Let T be pseudo A-proper w.r.t. Γ and $\deg(Q_n A, B(0, r) \cap X_n, 0) \neq 0$ for each large r and each $n \geq n_0(r)$. Then $Ax + Nx = f$ is solvable for each $f \in Y$.*

Corollary 8.1: [55,59,62]: *Let $T = I - N$: $X \to X$ be pseudo A-proper w.r.t. $\Gamma = \{X_n, P_n\}$, $\delta = \max \|P_n\|$ and the quasinorm $\delta|N| < 1$. Then $(I - N)(X) = X$.*

We refer to these works for applications to BVP's for nonlinear elliptic partial differential equations. The following special case is suitable for studying semilinear equations without resonance at infinity.

Theorem 8.3: [63,68]: *Let A: $D(A) \subset X \to Y$ and N_∞: $X \to Y$ be linear maps, $V = (D(A), \|\cdot\|_0)$ be a Banach space densely and continuously embedded in X and N: $V \to Y$ be a nonlinear map such that $A + N$: $V \to Y$ is pseudo A-proper w.r.t. $\Gamma = \{X_n, Y_n, Q_n\}$ and $Q_n(A + N_\infty)x = (A + N_\infty)x$ on X_n. Suppose that for some positive constants a, b, c and r, with $a < c$,*

$$\|Nx - N_\infty x\| \leq a\|x\|_0 + b \quad \textit{for } \|x\|_0 \geq r \tag{8.2}$$

and either $\|(A - N_\infty)^{-1} y\|_0 \leq c\|y\|$ on Y, or $X = Y = H$ is a Hilbert space, A and N_∞ are selfadjoint and

$$0 < a < \min\{|\mu| \mid \mu \in \sigma(A - N_\infty)\}. \tag{8.3}$$

Then equation (8.1) is solvable for each $f \in Y$.

Next, we state a full extension of the Fredholm Alternative for equation (8.1).

Theorem 8.4: [72]: (Fredholm Alternative): *Let A: $X \to Y$ be a linear continuous Fredholm map of index zero and N: $Y \to Y$ be continuous and have a sufficiently small quasinorm $|N|$. Let A and $T = A + N$ be A-proper w.r.t. Γ with $\ker A \subset X_n$. Then (i) either $\ker(A) = \{0\}$ and equation (8.1) is approximation solvable for each $f \in Y$, or (ii) $\ker(A) \neq \{0\}$ and if $\operatorname{codim} R(A) = m > 0$ and $R(N) \subset R(A)$, then for each $f \in R(A)$ $(= N(A^*)^\perp)$, and only such ones, there is a connected closed subset K of $T^{-1}(f)$ whose dimension at each point is at least m.*

The proof of (i) follows from Theorem 8.1, while (ii) is based on the covering dimension result in [24]. Its applications to nonlinear Hammerstein integral equations can be found in [62].

9. SEMILINEAR EQUATIONS AT RESONANCE INVOLVING FREDHOLM MAPS

In this section, we shall study equation (8.1) in resonance, i.e., when there is an interaction between N and the spectrum of A. We shall look at both cases when A is a Fredholm map of index zero and of positive index. There is an extensive literature on this type of problems (see the references), where condensing and monotone operator theories have been used. We shall study these problems using the (pseudo) A-proper mapping approach as developed in [63–69].

9.1. Nonlinear Perturbations of Fredholm Maps of Index Zero

Using the continuation theorems from Sections 5–6, we begin by discussing some basic continuation theory for semilinear operator equations of the form

$$Ax + F(t, x) = f \quad (x \in D(A) \subset X, \; f \in Y),$$

where A: $D(A) \subset X \to Y$ is a linear Fredholm map of index $i(A) = \dim \ker A - \operatorname{codim} R(A) = 0$ and F: $[0,1] \times \bar{D} \to Y$ is a suitable nonlinear homotopy. Here, X and Y are Banach spaces, $D \subset X$ is an open subset and $\ker A$ and $R(A)$ are the kernel and the range of A. Applications to equation (8.1) are also given.

Througout this section we assume that A: $D(A) \subset X \to Y$ is a Fredholm map of index zero, $X_0 = \ker A$, $\tilde{Y} = R(A)$ and $\tilde{X}$ and Y_0 are closed subspaces of X and Y, respectively, such that $X = X_0 \oplus \tilde{X}$ and $Y = Y_0 \oplus \tilde{Y}$ with $\dim X_0 = \dim Y_0$. Let P: $X \to X_0$ and Q: $Y \to Y_0$ be linear projections onto X_0 and Y_0 respectively. We assume that $\Gamma = \{X_n, Y_n, Q_n\}$ is a projectionally complete scheme for (X, Y) with $X_0 \subset X_n \subset D(A)$, $Y_0 \subset Y_n$, $Q_n Ax = Ax$ for $x \in X_n$, and $Q_n y \to y$ for each $y \in Y$. For each $x \in X$, we have $x = x_0 + x_1$, $x_0 \in X_0$, $x_1 \in \tilde{X}$.

One of the basic assumptions in our results is nonvanishing of the Brouwer degree of a suitable nonlinear map acting on $D \cap X_0$. This can be assured under various conditions on the nonlinearity and, in particular, if it possesses some symmetry property.

We begin with the following continuation result for general A-proper homotopies. Recall that H: $[0,1] \times D \subset X \to Y$ is an A-proper homotopy relative to Γ if whenever $\{x_{n_k} \mid x_{n_k} \in D \cap X_{nk}\}$ is bounded and $t_k \in [0,1]$ with $t_k \to t$ are such that $Q_{n_k} H(t_k, x_{n_k}) \to f$, then a subsequence $x_{n_{k_i}} \to x$ and $H(t, x) = f$.

We assume that D is an open and bounded subset of X. For the sake of simplicity of exposition, we look at a singlevalued N now.

Theorem 9.1: [63,68,76]: *Let A: $D(A) \subset X \to Y$ be Fredholm of index zero, and F: $[0,1] \times X \to Y$ be nonlinear such that $H(t,x) = Ax + F(t,x)$ is an A-proper homotopy on $[0,1] \times (\bar{D} \cap D(A))$ w.r.t. Γ. Suppose that*

(1) $A(x) + F(t, x) \neq f$ *for* $x \in \partial D \cap D(A)$, $t \in [0, 1)$;

(2) $F(0, \cdot)(\bar{D}) \subset Y_0$;

(3) $F(0,x) \neq tf_0$ *for* $x \in \partial D \cap X_0,\ \ t \in [0,1]$;
(4) $\deg((F(0,\cdot)|_{\bar{D}_0 \cap X_0}, D \cap X_0, 0) \neq 0$.

Then the equation $Ax + F(1,x) = f$ *is feebly approximation-solvable in* $D(A) \cap \bar{D}$.

The proof is based on Theorem 5.1.

Remark 9.1: If $f = 0$, then it is enough to assume that (1) in Theorem 9.1 holds for $t \in (0,1)$ in view of (2).

An easy application of Theorem 9.1 yields the following result for equation (8.1).

Corollary 9.1: [59,63,68]: *Let* $A + N$: $D(A) \cap \bar{D} \subset X \to Y$ *be A-proper w.r.t.* Γ *with* N *bounded and nonlinear. Let* $f = f_0 + f_1 \in Y_0 \oplus \tilde{Y}$ *and suppose that*

(1) $A(x) + tNx \neq f$ *and either* $Ax \neq f_1$ *or* $QNx \neq f_0$ *for* $x \in \partial D \cap D(A),\ t \in [0,1)$;
(2) $QNx \neq tf_0$ *for* $x \in \partial D \cap X_0,\ \ t \in [0,1]$;
(3) $\deg((QN, D \cap X_0, 0) \neq 0$.

Then equation (8.1) is feebly approximation solvable.

Proof: Let $F(t,x) = tNx + (1-t)QNx$. Then, it is easy to see that the conditions of Theorem 9.1 are satisfied. □

Remark 9.2: If $f = 0$, then Corollary 9.1 is still valid if (1) is replaced by

(1′) $Ax + tNx \neq 0$ for $x \in \partial D \cap D(A),\ \ t \in (0,1)$.

For pseudo A-proper maps, we have the following extension of Theorem 9.1.

Theorem 9.2: *Let* A: $D(A) \subset X \to Y$ *be Fredholm of index zero and* F: $[0,1] \times X \to Y$ *be such that conditions (2)–(4) of Theorem 9.1 hold and*

(1_n) $Ax + Q_nF(t,x) \neq Q_nf$ *for* $x \in \partial D \cap X_n$, $t \in [0,1)$ *and all large* n.

Then, if $A + F_1$ *is pseudo A-proper w.r.t.* Γ, *the equation* $Ax + F(1,x) = f$ *is solvable.*

Corollary 9.2 [63,64,68]: *Let* A: $V = D(A) \subset X \to Y$ *be Fredholm of index zero and A-proper w.r.t.* Γ *and* N: $V \to Y$ *be nonlinear such that*

(1) *For a given* $f \in Y$, *there are constants* $M_f > 0$ *and* $N_f > 0$ *such that* $QN(x_0 + x_1) \neq tQf$ *for* $\|x_1\| \leq r,\ r > N_f,\ \|x_0\| \geq rM_f,\ t \in [0,1]$.
(2) $M = (I - N)$ *is quasibounded, i.e.,* $|M| = \limsup_{\|x\|_1 \to \infty} \|Mx\|/\|x\|_1 < \infty$ *and* $\|Q_n\||M| \max\{1, M_f\} < c$, *where* $\|x = (x_0, x_1)\|_1 = \max\{\|x_0\|, \|x_1\|\}$ *and* $\|Q_n Ax_1\| \geq c\,\|x_1\|$ *on* X.

Let $\deg(QN, B(0,r) \cap X_0, 0) \neq 0$ *for* $r > N_f$ *large and* $A + N$ *be pseudo A-proper w.r.t.* Γ. *Then the equation* $Ax + Nx = f$ *is solvable.*

Remark 9.3: [63,64,68]: If $\|Nx\| \leq \alpha + \beta\|x\|$ on X for some $\alpha \geq 0$, $\beta \geq 0$ and $\gamma \in [0,1]$, then condition (1) in Corollary 9.2 can be relaxed to

(1′) $QN(\rho x_0 + \rho^{\gamma} x_1) \neq tQf$ for all $x_0 \in \partial B(0,1) \cap X_0$, $\|x_1\| \leq r$, $r > N_f$, $\rho \geq r\rho(r)$ for some $\rho(r) \geq M_f$, $t \in [0,1]$.

Another result suitable for applications is

Theorem 9.3: [64,68]: *Let A: $V = D(A) \subset X \to Y$ be Fredholm of index zero and A-proper w.r.t. Γ, N: $V \to Y$ be nonlinear and the quasinorm $|(I-Q)N|$ be sufficiently small. Let K: $X_0 \to Y_0^*$ be such that $(QQ_ny, Kx_0) = (Qy, Kx_0)$ on X_0 and $(Lx_0, Kx_0) > 0$ for $x_0 \in X_0 \backslash \{0\}$ and some linear isomorphism L: $X_0 \to Y_0$. Assume*

(1) *For a given $f \in Y$, there are $M_f > 0$ and $N_f > 0$ such that for each $r > N_f$, $\|x_1\| \leq r$, $x_0 \in \partial B(0,1) \cap X_0$ and $\rho \geq rM_f$ either (i) $(QN(\rho x_0 + x_1), Kx_0) < (Qf, Kx_0)$ or (ii) $(QN(\rho x_0 + x_1), Kx_0) > (Qf, Kx_0)$.*

Then, if $A + N$: $V \to Y$ is pseudo A-proper w.r.t. Γ, the equation $Ax + Nx = f$ is solvable.

When A and $A + N$ are of type (S) in a Hilbert space and N is uniformly bounded the result is due to Jarusek–Necas [37].

Other continuation theorems can be found in [53,62–71]. A similar continuation theorem for $Ax + F(t,x) = f$ is valid if $F(0,\cdot)$ has some symmetry properties, i.e. it is equivariant under some compact Lie group (see [71,75] for a detailed study of such maps).

9.2. Nonlinear Perturbations of Fredholm Maps of Positive Index

The results of part A can be extended to Fredholm maps A of positive index $i(A) = m$. As before, we have the decompositions $X = X_0 \oplus \tilde{X}$ and $Y = Y_0 \oplus \tilde{Y}$ with $X_0 = \ker A$, $\tilde{Y} = R(A)$ and $i(A) = \dim X_0 - \dim Y_0 = m$. Let $\Gamma_m = \{X_n, Y_n, Q_n\}$ be a projection scheme for (X, Y) with $X_0 \subset X_n$, $Q_n(\tilde{Y}) \subset \tilde{Y}$ and $\dim X_n - \dim Y_n = m$ for each n. We also assume that $Q_nAx = Ax$ for $x \in X_n$. Decompose X_0 as $X_0 = W \oplus Z$ with $\dim W = m$ and $\dim Z = \dim Y_0$, and let Q: $Y \to Y_0$ be a linear projection onto Y_0. If we try to establish continuation theorems of the type discussed in Part A, we find that

$$\deg(QN, D \cap X_0, 0) = 0$$

since QN: $\bar{D} \cap X_0 \to Y_0$ and $\dim Y_0 < \dim X_0$. To overcome this difficulty, we have presented in [63,72,76] two approaches to the study of such equations based on the notions of complementing and essential maps. A third approach, based on G-equivariant maps, is given in [72]. Using the complementing mapping approach and some results in [24], as well as some index theories, we have developed in [72,76] a theory on the dimension and the index of the solution set of equation (8.1) and gave some applications to semilinear elliptic BVP's. Due to the lack of space, we omit any detailed discussion of this theory and refer the reader to the above works.

9.3. Semilinear Equations and Essential Maps

Another approach to semilinear equations with $i(A) = m > 0$ is to use stable homotopies. For compact uniformly bounded nonlinearities, this approach was used first by Nirenberg [88] and subsequently by many other authors for compact or A-compact nonlinearities (cf. [53] and the references in there). For a much broader class of nonlinearities N, such that $A + N$ is A-proper, this approach was developed by the author [63,66,71,76].

Below we just give a sample of such results in the setting of Theorem 9.1.

Theorem 9.4: [66,76]: *Let $i(A) = m > 0$, F: $[0,1] \times B(0, R) \to Y$ be nonlinear such that $H(t, x) = Ax + F(t, x)$ is an A-proper homotopy w.r.t. $\Gamma_m = \{X_n, Y_n, Q_n\}$ on $[0,1] \times (\bar{B} \cap D(A))$. Suppose that*

(1) $Ax + F(t, x) \neq f$ *for* $x \in \partial B \cap D(A)$, $t \in [0, 1)$;
(2) $F(0, \cdot)(\bar{B}) \subset Y_0$;
(3) $F(0, x) \neq tf_0$ *for* $x \in \partial B \cap X_0$, $t \in [0, 1]$;
(4) $F(0, \cdot)$: $\partial B(0, R) \cap X_0 \to Y_0 \backslash \{0\}$ *has nontrivial stable homotopy.*

Then the equation $Ax + F(1, x) = f$ is solvable.

Corollary 9.3: [66]: *Let $i(A) = m > 0$ and $A + N$: $\bar{B}(0, R) \cap D(A) \to Y$ be A-proper w.r.t. Γ_m with N nonlinear and bounded. Suppose that*

(1) $A(x) + tNx \neq f$ *and either* $Ax \neq f_1$ *or* $QNx \neq f_0$ *for* $x \in \partial B \cap D(A)$, $t \in (0, 1)$;
(2) $QNx \neq tf_0$ *for* $x \in \partial B \cap X_0$, $t \in [0, 1]$;
(3) QN: $\partial B(0, R) \cap X_0 \to Y_0 \backslash \{0\}$ *has a nontrivial stable homotopy.*

Then equation (8.1) is solvable.

10. EXISTENCE RESULTS FOR NONLINEAR PERTURBATIONS OF FREDHOLM MAPS INVOLVING UNIFORM LIMITS OF A-PROPER MAPS

In this section we shall discuss various applications of the abstract results developed in Section 6 to the solvability of equation (8.1) and consider briefly possible applications of these results to monotone type mappings.

In view of Theorems 6.1 and 6.4, we have the following existence result for equation (8.1).

Theorem 10.1: *Let $H(t, x) + \mu Gx = Ax + F(t, x) + \mu Gx$ with $F(t, x) = tNx + (1 - t)QNx$. Suppose that for a given f in Y_1 and each $\mu \in (0, \mu_0)$*

(*i*) $H + \mu G$ *is an A-proper homotopy w.r.t.* Γ *on* $[0, \epsilon] \times (\partial D \cap V)$ *for each* $\epsilon \in (0, 1)$;
(*ii*) $f \notin Ax + t(I - Q)Nx + \mu(I - Q)Gx$ *for* $x \in (V \backslash X_0) \cap \partial D$, $t \in [0, 1]$;
(*iii*) $0 \notin QNx + \mu QGx$ *for* $x \in X_0 \cap \partial D$;
(*iv*) $\deg(W_n H_0 V_n + \mu W_n G V_n, V_n^{-1}(D \cap X_n), 0) \neq 0$ *for all large n.*

Let condition (6.1) hold. Then, if $H_1 + \mu G$ is pseudo A-proper w.r.t. Γ for each $\mu \in (0, \mu_0)$ and H_1 satisfies condition (*), the equation $f \in H(1, x)$ is solvable.

Remark 10.1: If A and $A + tN + \mu G$ are A-proper w.r.t. Γ for each $t \in [0, 1)$ and $\mu \in (0, \mu_0)$ and N is also bounded, then conditions (6.1) and (i) in the above result are satisfied.

Using a homotopy of the form $H_\mu(t, x) = H(t, x) + \mu Gx = Ax + F(t, x) + \mu Gx$ with $F(t, x) = tNx + (1 - t)Cx$ with some linear and compact C with $A + C$: $V \to Y$ bijective, we deduce from Theorems 6.2 and 6.4 the following

Theorem 10.2: *Suppose also that G is odd and that for a given f in Y the homotopy $H_\mu(t, x) = Ax + F(t, x) + \mu Gx$ with $F(t, x) = tNx + (1 - t)Cx$ for some linear and compact C with $A + C$: $V \to Y$ bijective, satisfies either conditions (i)–(iii) or (iv) below for each $\mu \in (0, \mu_0)$*

(*i*) *Condition (6.1) holds and* $A + F(t, x) + \mu G$ *is an A-proper homotopy w.r.t.* Γ *on* $[0, \epsilon] \times (\partial D \cap V)$ *for each* $\epsilon \in (0, 1)$;
(*ii*) $f \notin A + F(t, x) + \mu Gx$ *for* $t \in [0, 1]$, $x \in \partial D \cap V$;
(*iii*) $tf \notin Ax + Cx + \mu Gx$ *for* $t \in [0, 1]$, $x \in \partial D \cap V$;
(*iv*) *For each large* n, $tQ_n f \notin Q_n(Ax + F(t, x) + \mu Gx)$ *for* $x \in \partial D \cap X_n$, $t \in [0, 1]$.

Then, if $A + N + \mu G$ *is pseudo A-proper w.r.t.* Γ *for each* $\mu \in (0, \mu_0)$ *and* $A + N$ *satisfies condition* $(*)$, *equation (8.1) is solvable.*

Theorem 10.2 is useful in applications if we know some easily verifyable conditions on N that imply (i)–(iv). This will be considered next.

We shall discuss some special cases under the following conditions on the nonlinear map considered first in this form by Fitzpatrick [23]. We suppose that there is a continuous bilinear form $[\cdot, \cdot]$ on $Y \times X$ such that $y \in Y_1$ if and only if $[y, x] = 0$ for all $x \in X_0$. Let $\{\phi_1, \ldots, \phi_m\}$ be a basis in X_0 and J: $Y_0 \to X_0$ be given by $Jy = \Sigma_{i=1}^m [y, \phi_i]\phi_i$.

(10.1) $\|Nx\|/\|x\| \to 0$ and either (i) $\liminf [Nx_n, y] \langle [f, y]$ or (ii) $\limsup [Nx_n, y] \rangle [f, y]$ whenever $\{x_n\} \subset V$ is such that $\|x_n\| \to \infty$ and $x_n\|x_n\|^{-1} \to y \in X_0$.

(10.2) N has a sublinear growth, i.e., $\|Nx\| \le \alpha + \beta\|x\|^\gamma$ for each $x \in X$ and some $\alpha \ge 0$, $\beta \ge 0$ and $\gamma \in [0, 1)$, and either (i) $\liminf [N(\rho_n u_n + \rho_n^\gamma v_n), x_0] \langle [f, x_0]$ or (ii) $\limsup [N(\rho_n u_n + \rho_n^\gamma v_n), x_0] \rangle [f, x_0]$ whenever $x_0 \in \partial B(0, 1) \cap X_0$, $\rho_n \to \infty$, $\{v_n\} \subset \tilde{X}$ is bounded in X and $\{u_n\} \subset X_0$ with $u_n \to x_0$.

(10.3) N has a sublinear growth and either (i) $\liminf [Nu_n, v_n] \langle [f, y]$ or $\limsup [Nu_n, v_n] \rangle [f, y]$ whenever $\{u_n\} \subset V$ is such that $\|Pu_n\| \to 0$ and $v_n = Pu_n/\|Pu_n\| \to y \in X_0$

We need that a scheme $\Gamma = \{X_n, P_n, Y_n, Q_n\}$ satisfies

Property P: (i) $Y_0 \subset Y_n$, (ii) $Q_n(A + C)x = (A + C)x$ on X_n and (iii) $[Q_n y, x_0] = [y, x_0]$ for $y \in Y$ and $x_0 \in X_0$.

Property (P) holds in many practically important cases (see [63,64,68]). Using Theorem 10.2, we have

Theorem 10.4: [59,62,63,68]: (*a*) *Let* $H_t = A + tN + (1 - t)C$ *with* $A + C$ *injective. Suppose that either*

(*i*) H_t *is an A-proper homotopy on* $[0, 1) \times (\bar{D} \cap V)$ *w.r.t.* Γ *for* (V, Y) *and* $H(t, x) \neq tf$ *for* $x \in \partial D \cap V$, $t \in [0, 1]$, *or*
(*ii*) *either one of conditions (10.1)–(10.3) holds and* Γ *has Property (P).*

Then equation (8.1) is solvable for this f *if* $A + N$ *is pseudo A-proper (approximation solvable if* $A + N$ *is A-proper)*

(*b*) *Suppose that* $H_{t\mu} = H_t + \mu G$, G *bounded, and either*

  (*i*) $H_{t\mu}$ *is an A-proper homotopy on* $[0, 1) \times (\bar{D} \cap V)$ *w.r.t.* Γ *for* (V, Y) *and* $H_\mu(t, x) \neq tf$ *for* $x \in \partial D \cap V$, *and all* $\mu > 0$ *small, or*
  (*ii*) *condition (ii) in (a) holds.*

Then equation (8.1) is solvable for this f *if* $A + N + \mu G$ *is pseudo A-proper w.r.t.* Γ *for each* $\mu > 0$ *small and* $A + N$ *satisfies condition (++), i.e., if* $\{x_n\}$ *is bounded and* $(A + N)x_n \to f$, *then* $Ax + Nx = f$ *for some* x.

As our examples of (pseudo) A-proper maps below show, Theorem 10.4 is applicable to ball-condensing, K-(generalized pseudo) monotone and of type (KM) maps. We refer to [63,64,68] for detailes and applications to BVP's for nonlinear elliptic equations. For k-set contractive perturbations, we refer to Fitzpatrick [23]. There is an extensive literature on these problems (see [7 or 8], [26], etc. for surveys).

Next, we shall discuss specific (pseudo) A-properness of perturbations of linear Fredholm maps and of general linear maps by nonlinear maps of the above types to which our general results apply. Such maps appear naturally in studying BVP's for semilinear elliptic differential equations and periodic-BVP's for semilinear hyperbolic and parabolic equations.

Theorem 10.5: [63–68]: *(a) Let A: $D(A) = V \subset X \to Y$ be Fredholm of index $i \geq 0$ ($i = 0$, resp.), $D \subset X$ be open and bounded and N: $\bar{D} \to Y$ be continuous and either k-ball contractive with $k\delta < c$ or ball-condensing if $\delta = c = 1$, where $\delta = \max \|Q_n\|$ and $\|Ax_1\| \geq c\|x_1\|$ for $x_1 \in X_1$, the complement of the* $\ker A$*($\|(A+C)x\| \geq c\|x\|$ on x for some compact map C, resp.). Then $A + N$: $\bar{D} \cap V \to Y$ is A-proper w.r.t. a suitable scheme Γ induced by A.*

(b) *If either N or $(A|X_1)^{-1}$ is compact or $N(A+C)^{-1}$ is ball-condensing with $\delta = 1$, then $A + N$ is A-proper w.r.t. Γ.*

(c) *If A: $X \to X^*$ is of type (S) (and so Fredholm with $i = 0$) and N: $X \to X^*$ is demicontinuous, quasibounded with $A + N$ of type (S), then $A + N$ is A-proper w.r.t. a suitable scheme Γ.*

The proof of Theorem 10.5 is based on showing that $\chi(\{Ax_n\}) \geq c\chi(\{x_n\})$ for each bounded sequence $\{x_n \in X_n\}$ and some $c > 0$ and then using similar arguments as in Theorem 2.1. Hence, a version of Theorem 10.5 holds with the ball-condenseness replaced by by the a-stability of N.

Now we discuss the pseudo A-properness of $A + N$ for monotone like perturbations. However, as has been noticed by the author in [61,63], in most of the examples of pseudo A-proper maps, and in particular the ones stated below, we have also a weak convergence of $\{x_{n_k}\}$ to x in Definition 2.1 (ii). In such cases one calls such maps weakly A-proper.

We recall that N is K-quasibounded if whenever $\{x_n\} \subset X$ is bounded and $(Nx_n, Kx_n) \leq c\|x_n\|$ for each n and some $c > 0$, then $\{Nx_n\}$ is bounded.

Theorem 10.6: [63–68]: *(a) Let X and Y be reflexive Banach spaces, K: $X \to Y^*$ be a linear homeomorphism and A: $X \to Y$ be linear and continuous such that*

$$(Ax, Kx) \geq c(\|x\|) - \phi(x) \quad \textit{for } x \in X$$

for some continuous function c: $R^+ \to R^+$ such that $r \to 0$ if $c(r) \to 0$ and a weakly upper semicontinuous at 0 function ϕ: $X \to R$. If N: $X \to Y$ is either generalized pseudo K-monotone or of type (KM) and $A + N$ is K-quasibounded, then $A + N$: $X \to Y$ is pseudo A-proper w.r.t. a suitable scheme Γ.

(b) *Let X, Y and K be as in (a), A: $X \to Y$ be demicontinuous and of type (KS_+) and N: $X \to Y$ be pseudo K-monotone and $A + N$ be K-quasibounded, then $A + N$: $X \to Y$ is pseudo A-proper w.r.t. Γ.*

(c) *If Y is a Hilbert space, $i(A) = 0$ and N: $X \to Y$ is B-quasibounded and of type (BM) with $B = A + C$ for some compact C, then $A + N$ is pseudo A-proper w.r.t. Γ.*

(d) Let X, Y, K, A and N be as in (a) or (b). If G: $X \to Y$ is bounded, demiclosed and of type (KS_+), then $A + tN + \mu G$ is pseudo A-proper w.r.t. Γ if N is of type (KM) and is A-proper in other cases for each $\mu > 0$, $t > 0$.

Now we look at perturbations of general linear maps with infinite dimensional null space.

Theorem 10.7: [69,70,72,75,78]: *Let A: $D(A) \subset H \to H$ be a closed linear densely defined map with $R(A) = \ker A^{\perp}$ and N: $H \to H$ be bounded and of type (M). Suppose that either one of the following conditions holds:*

(i) A^{-1}: $R(A) \to R(A)$ is compact,
(ii) $A = A^*$, $0 \in \sigma(A)$ and $\sigma(A) \cap (0,\infty) \neq \emptyset$ and consists of isolated eigenvalues having finite multiplicities.

Then $\pm A + N$ is pseudo A-proper w.r.t. $\Gamma = \{H_n, P_n\}$ for H with $P_n Ax = Ax$ on H.

For a smaller class of nonlinearities we shall now prove the A-properness of some homotopies needed later.

Theorem 10.8: [72]: *Let A: $D(A) \subset H \to H$ be a closed linear densely defined map with $R(A) = \ker A^{\perp}$ and either (i) or (ii) in Theorem 10.7 holds. Let N: $H \to H$ be bounded and generalized pseudo monotone and G: $H \to H$ be bounded and of type (S_+) and (P). Then, for each $t \in [0,1)$, $H_t = \pm A + (1-t)G + tN$ is A-proper w.r.t. $\Gamma = \{H_n, P_n\}$ for H with $P_n Ax = Ax$ on H_n and $A + N$ is pseudo A-proper w.r.t. Γ.*

Remark 10.2: As G we can take P or rI, $r > 0$, or a strongly monotone map. Using the same type of arguments we can show that $\pm A + N + \alpha G$ is also A-*proper* w.r.t. Γ for each $\alpha > 0$ under the conditions of Theorem 10.6 (cf. [72]). As in [72,78] Theorem 10.6 is valid also for g.p.m. (A), $P(A)$ and $S(A)$ type of maps when (i) or (ii) holds.

In the next few results we do not assume that A^{-1} is compact, which is the case when studying semilinear hyperbolic equations in more than one space variable.

Theorem 10.9: [81]: *Let A: $D(A) \subset H \to H$ be a linear selfadjoint map, N: $H \to H$ be a continuous gradient map and $\alpha \leq \beta$ be real numbers not in $\sigma(A)$ such that $[\alpha,\beta] \cap \sigma(A)$ consists of a (positive) finite number of eigenvalues of A having finite multiplicity and*

$$\alpha\|x-y\|^2 \leq (Nx - Ny, x-y) \leq \beta\|x-y\|^2 \quad for\ x, y \in H, \tag{10.1}$$

then $A - N$: $D(A) \subset H \to H$ is A-proper w.r.t. $\Gamma = \{H_n, P_n\}$ with $P_n Ax = Ax$ on H_n.

The next result does not require that N is of gradient type.

Theorem 10.10: *Let A: $D(A) \subset H \to H$ be a linear densely defined map and N: $H \to H$ be Lipschitz continuous, i.e., there are constants k and r_0 such that*

$$\|Nx - Ny\| \leq k\|x-y\| \quad for\ all\ x, y \in \bar{B}(0, r_0).$$

Suppose that $[-k+\mu, k+\mu] \cap \sigma(A)$ contains at most a finite number of spectral points all of which are eigenvalues of finite multiplicity. Then, there is $\epsilon > 0$ such that for each $\lambda \in [\mu-\epsilon, \mu+\epsilon]$, the map $A + N + \lambda I$: $D(A) \cap \bar{B}(0, r_0) \subset H \to H$ is A-proper w.r.t. $\Gamma = \{H_n, P_n\}$ with $P_n Ax = Ax$ on H_n.

For our next result, we need some preliminaries. Recall that a closed subspace $X \subset H$ is said to reduce A if A commutes with the orthogonal projection P of H onto X. Let $P^{\pm}$ be the orthogonal projections of H onto its closed subspaces $H^{\pm}$. Regarding A and H, we require that the following conditions hold for a scheme $\Gamma = \{H_n, P_n\}$ (cf. [1]):

(10.2) (i) $H_1 \subset H_2 \subset \ldots$ with each H_n closed in H, $\dim H_n = \infty$ and $\cup H_n$ is dense in H, (ii) each H_n reduces A, (iii) the orthogonal projections P_n: $H \to H_n$ commute with $P^{\pm}$, (iv) $H_n = H_n^- \oplus H_n^+$, where $H_n^{\pm} = H^{\pm} \cap H_n$, (v) $Q^{\pm}(D(A) \cap H_n) \subset D(A)$ for each n, where $Q^{\pm}$: $H^- \oplus H^+ \to H^{\pm}$ are projections.

If H has a complete system of orthonormal eigenvectors $\{e_n\}$ of A and $H = H^- + H^+$, such a scheme can be constructed by setting $H_n = \text{lin sp}\{e_1, \ldots, e_n\}$. A scheme Γ consisting of a family $\{H_\alpha | \alpha \in \Lambda\}$ of closed vector subspaces of H, directed by inclusion, that satisfies (10.2) was introduced by Amann [1]. Under some mild assumptions on the spectrum of A, such a scheme was constructed in [1].

For such maps we have the following class of (pseudo) A-proper maps [Mi-32].

Theorem 10.11: *Let A: $D(A) \subset H \to H$ be selfadjoint, $H^{\pm}$ be closed subspaces of H with $H = H^- + H^+$ and $H^- \cap H^+ = \{0\}$ and $\Gamma = \{H_n, P_n\}$ be a scheme for H that satisfies (10.2). Suppose that N: $H \to H$ is a gradient map such that there are self-adjoint maps $B^{\pm} \in L(H)$ and $\gamma \geq 0$ such that*

(i) $N - B^-$ and $B^+ - N$ are monotone maps; and
(ii) $((A - B^-)x, x) \leq -\gamma \|x\|^2$ for $x \in D(A) \cap H^-$ and $((A - B^+)x, x) \geq \gamma \|x\|^2$ for $x \in D(A) \cap H^+$.

Then $\pm A + N$: $D(A) \subset H \to H$ is (pseudo) A-proper w.r.t. Γ if $\gamma > 0$ ($\gamma = 0$, respectively).

Remark 10.3: If $H = H^- + H^+$ with $H^- \cap H^+ = \{0\}$ and (10.2) is not required, then the above proof shows that $(A - N)K$ is a $\gamma/2$-strongly monotone (monotone if $\gamma = 0$) map on H and is therefore (pseudo) A-proper relative to a general scheme $\Gamma = \{H_n, P_n\}$. For the case when $H \neq H^- + H^-$, we refer to [1] and [80].

For our next class of A-proper maps we assume that A: $D(A) \subset X \to Y$ is a linear closed and densely defined map with $X_0 = \ker A$ infinite dimensional and closed range $\tilde{Y} = R(A)$. Suppose that there are closed subspaces $\tilde{X}$ and Y_0 of X and Y such that $X = X_0 \oplus \tilde{X}$ and $Y = Y_0 \oplus \tilde{Y}$. Then we have a partial inverse of A: A^{-1}: $\tilde{Y} \to \tilde{X}$. Let P: $X \to X_0$ and Q: $Y \to Y_0$ be linear projections onto X_0 and Y_0. Let $X_{i1} \subset X_{i2} \subset \ldots$ and $Y_{i1} \subset Y_{i2} \subset \ldots$, $i = 0, 1$ be finite dimensional subspaces with $\cup X_{0n}$, $\cup X_{1n}$, $\cup Y_{0n}$ and $\cup Y_{1n}$ dense in X_0, $\tilde{X}$, Y_0 and $\tilde{Y}$ respectively. Let Q_{0n}: $Y_0 \to Y_{0n}$ and Q_{1n}: $\tilde{Y} \to Y_{1n}$ be projections onto Y_{0n} and Y_{1n}. For $(y_0, y_1) \in Y_0 \oplus \tilde{Y}$, define $Q_n(y_0, y_1) = (Q_{0n}, Q_{1n})(y_0, y_1) = (Q_{0n} y_0, Q_{1n} y_1)$. Then $\Gamma = \{X_n = X_{0n} \oplus X_{1n}, Y_n = Y_{0n} \oplus Y_{1n}, Q_n\}$ is a projection scheme for (X, Y). Let $\delta_i = \max \|Q_{in}\|$, $i = 0, 1$.

Theorem 10.12: [63,72,78]: *Let A^{-1}: $\tilde{Y} \to \tilde{X}$ be compact, N: $X \to Y$ be bounded and such that for each fixed sequences $\{x_{in} \in X_{in}\}$, $i = 0, 1$ and some $c > 0$*

$$\chi(\{Q_{0n} Q N(x_{0n} + x_{1n})\}) \geq c\chi(\{x_{0n}\}). \tag{10.3}$$

Let F: $X \to Y$ be continuous and k-ball-contractive with $k\delta_0 \|Q\| < c$, or ball-condensing if $c = \delta_0 \|Q_0\| = 1$ and either N is continous or Y is reflexive and N is demicontinous. Then $\pm A + N + F$: $D(A) \subset X \to Y$ is A-proper w.r.t. Γ with $Q_{1n} Ax = Ax$ on X_{1n}.

Lemma 10.1: [72]: *Suppose that $N: X \to Y$ is demicontinuous and either*

(*i*) *N is strongly K-monotone with $K: X \to Y^*$ such that $\|Kx\| \le c_1\|x\|$ and $Q^*Q_{0n}^*Kx = Kx$ for $x \in X_{0n}$; or*
(*ii*) *For each $x_{0n}, z_{0n} \in X_{0n}$ and each $x_{1n} \in X_{1n}$, and some $c > 0$*

$$\|Q_{0n}QN(x_{0n} + x_{1n}) - Q_{0n}QN(z_{0n} + x_{1n})\| \ge c\|x_{0n} - z_{0n}\| \quad n \ge 1.$$

Then condition (10.3) holds.

Finally, if A^{-1} is just continuous, then we need to impose stronger conditions on N.

Theorem 10.13: [78]: *Let $A^{-1}: \tilde{Y} \to \tilde{X}$ be continuous and $N: X \to Y$ be continuous, k_1-ball contractive and satisfy condition (10.3). Suppose that $F: X \to Y$ is continuous and k_2-ball contractive with $(k_1 + k_2)\delta_1\|I - Q\|\ \|A^{-1}\| + k_2\delta_0\|Q\|c^{-1} < 1$. Then $\pm A + N + F: D(A) \subset X \to Y$ is A-proper w.r.t. Γ with $Q_{1n}Ax = Ax$ on X_{1n}.*

11. SEMILINEAR EQUATIONS WITH GENERAL LINEAR MAPS

In this section, we shall study equation (8.1) when $\dim\ker A = \infty$ and the nonlinearity N interacts in some sense with the spectrum of A. We discuss an extension of the basic result of Brezis–Nirenberg [9] to pseudo A-proper maps $A + N$. As discussed above, $A + N$ is pseudo A-proper for some classes of maps N even if A^{-1} is not compact. Hence, the obtained results are suitable also for applications to resonance problems for systems of semilinear hyperbolic equations in more than one space variables.

Assume that $A: D(A) \subset H \to H$ satisfies:

(11.1) There are positive constants $a_\pm$ and a_0 such that

(*i*) $-a_+^{-1}\|Ax\|^2 \le (Ax, x) \le a_-^{-1}\|Ax\|^2$ for $x \in D(A)$;
(*ii*) $\|x\| \le a_0\|Ax\|$ for $x \in \tilde{H} = \ker A^\perp$.

Let Int(D) denote the interior of D and conv D be the convex hull of D. We have

Theorem 11.1: [71,72,79,80]: *Let a linear closed map $A: D(A) \subset H \to H$ satisfy (11.1) and $N: H \to H$ be such that $\pm A + N: D(A) \subset H \to H$ is pseudo A-proper w.r.t. $\Gamma = \{H_n, P_n\}$ with $P_nAx = Ax$ on H_n. Suppose that*

(11.2) *There are $\gamma < a_\pm$ and $\tau < a_0^{-2}(\gamma^{-1} - a_\pm^{-1})$ such that for every $y \in H$ and every $\delta > 0$ there exist $c_i(y), i = 1, 2$, and $k(\delta)$ such that for each $x \in H$*

$$(Nx - Ny, x) \ge \gamma^{-1}\|Nx\|^2 - c_1(y)\|Nx\| - \tau\|x_1\| - c_2(y)\|(\delta\|x_0\| + k(\delta)).$$

Then Int($R(A)$ + conv $R(N)$) $\subset R(\pm A + N)$. *Moreover, if N is onto, so is $\pm A + N$.*

Now, condition (11.2) implies that $\limsup_{\|x\|\to\infty} \|Nx\|/\|x\| \le \gamma$, i.e. N is a quasibounded map with the quasinorm $|N| \le \gamma$. As pointed out in [9], if N is potential and quasibounded, then a condition of type (11.2) holds. Moreover, if a is the smallest positive constant such that $(Ax, x) \ge -a^{-1}\|Ax\|^2$ on $D(A)$ and A is self-adjoint, then $Ax = \lambda x$ implies that $\lambda \ge -a$ and $-a$ is an eigenvalue of A. Hence, roughly speaking, the condition $0 < \gamma < a$ and $\limsup_{\|x\|\to\infty} \|Nx\|/\|x\| \le \gamma$ mean that the nonlinearity N asymptotically stays away from the nonzero eigenvalues of A.

Theorem 11.1 has been proved by Brezis–Nirenberg [9] when $N: H \to H$ is a monotone map and the partial inverse A^{-1} is compact using the Leray-Shauder and monotone operator theories. In view of the various examples of pseudo A-proper maps $A + N$ discussed before, this result holds also for many other classes of maps A and N, even when A^{-1} is not compact. For example, we have the following corollary.

Corollary 11.1: *Let A and N satisfy conditions (11.1)–(11.2) and $N = N_1 + N_2$ be such that N_1 is c-strongly monotone, k_1-ball contractive and N_2 is k_2 ball contractive, with k_1, k_2 sufficiently small, and continuous. Then $A + N$ is surjective.*

In view of Theorem 10.11, we have the following special case of Theorem 11.1.

Corollary 11.2: *Let A: $D(A) \subset H \to H$ and N: $H \to H$ be as in Theorem 10.11 and satisfy conditions (11.1)–(11.2). Then* $\text{Int}\,(R(A) + \text{conv}\, R(A)) \subset R(\pm A + N)$ *and $\pm A + N$ is surjective if such is N.*

12. UNIQUE APPROXIMATION SOLVABILITY OF NONLINEAR EQUATIONS AND ERROR ESTIMATES

In this section we shall discuss various unique approximation solvability and homeomorphism results for

$$Tx = f. \tag{12.1}$$

Error results will also be considered.

12.1. Unique Approximation Solvability and Homeomorphism Results

In this section, we shall give some unique approximation solvability and homeomorphism theorems for nonlinear maps.

We say that T satisfies condition (+) if whenever $Tx_n \to f$ in Y then $\{x_n\}$ is bounded in X.

Theorem 12.1: [81]: *Let T: $X \to Y$ be continuous, A-proper w.r.t. Γ, satisfy condition (+) and be locally invertible on X. Then T is a homeomorphism onto Y and the equation $Tx = f$ is uniquely approximation solvable for each $f \in Y$.*

Regarding approximation and error estimates for approximate solutions of $Tx = f$ involving differentiable A-proper maps we have the following basic result, which extends the result of Krasnoselski [38] and Vainikko [101] for compact maps.

Theorem 12.2: [54]: *Let T: $\bar{U} \subset X \to Y$ be A-proper w.r.t. Γ and x_0 be a solution of equation (12.1). If T is Frechet differentiable at x_0 and $T'(x_0)$ is A-proper w.r.t. Γ and injective, then equation (12.1) is strongly approximation solvable in $B_r(x_0)$ for some $r > 0$ (i.e. $Q_nTx_n = Q_nf$ for some $x_n \in \bar{B}_r(x_0) \cap X_n$ and all large n and $x_n \to x_0$) with x_0 being the isolated solution in $\bar{B}$. Moreover, if $c_0 = \|(T'(x_0))^{-1}\|$, then for any $\epsilon \in (0, c_0)$ there is an $n_0 \geq 1$ such that the approximate solutions of equation (12.1) $x_n \in \bar{B}_r(x_0) \cap X_n$ satisfy*

$$\|x_n - x_0\| \le (c_0 - \epsilon)^{-1}\|Tx_n - f\| \quad \textit{for } n \ge n_0. \tag{12.2}$$

If T is also continuously Frechet differentiable at x_0, then there is an $n_1 \ge n_0$ such that equation (12.1) is uniquely approximation solvable in $\bar{B}$ and the unique approximate solutions $x_n \in B_r(x_0) \cap X_n$ of $Q_nTx = Q_nf$, $n \ge n_1$, satisfy

$$\|x_n - x_0\| \le k\|P_nx_0 - x_0\| \le c\,\mathrm{dist}(x_0, X_n), \tag{12.3}$$

for some k depending on c_0, $\|T(x_0)\|$, ϵ, $\delta = \sup \|Q_n\|$, where $c = 2k\delta_1$, $\delta_1 = \sup \|P_n\|$.

We have the following approximation and error result.

Theorem 12.3: [59]: *(a) Let T: $X \to Y$ be Fréchet differentiable, A-proper w.r.t. Γ and have the closed range in Y. Then, if $T'(x)$ is injective and A-proper w.r.t. Γ for each $x \in X$, equation (12.1) is strongly approximation solvable in a neighborhood $B_r(x_0)$ of each of its solution x_0 for each $f \in Y$ and (12.2) holds.*

(b) If, in addition, T is either locally injective or continuously Fréchet differentiable in X, then T is a homeomorphism and, in the latter case, equation (12.1) is uniquely approximation solvable for each $f \in Y$ and (12.3) holds.

Remark 12.1: If $T = I - C, C$ compact and continuously Fréchet differentiable, the homeomorphism assertion only is due to Krasnoselskii–Zabreiko [42].

Next, we shall establish a constructive solvability and error estimates for the approximate solutions of nonlinear equations (12.1) involving A-proper maps which have a multivalued derivative at a solution as defined below.

Definition 12.2: Let U be open in X and T: $\bar{U} \to Y$. A positively homogeneous map A: $X \to 2^Y$, with $A(x)$ convex and closed for each $x \in X$, is said to be a *multivalued derivative* of T at $x_0 \in U$ if there exists a map $R = R(x_0)$: $\bar{U} - x_0 \to 2^Y$ such that $\|y\|/\|x - x_0\| \to 0$ as $x \to x_0$ for each $y \in R(x - x_0)$ and

$$Tx - Tx_0 \in A(x - x_0) + R(x - x_0) \quad \textit{for } x \textit{ near } x_0.$$

Let T: $D \subset X \to 2^Y$ be a multivalued map. We recall [54].

A multivalued map A: $X \to 2^Y$ is said to be m-bounded if there is a positive constant m such that $\|y\| \le m\|x\|$ for all $x \in X$, $y \in Ax$. It is c-coercive if $\|u\| \ge c\|x\|$ for $x \in X$ and $u \in Ax$.

Our basic result, announced in [70,78], is proved in [85].

Theorem 12.4: *Let T: $\bar{U} \subset X \to Y$ be A-proper w.r.t. Γ and x_0 be a solution of equation (12.1). Suppose that A is an odd multivalued derivative of T at x_0 and there exist constants $c_0 > 0$ and $n_0 \ge 1$ such that*

$$\|Q_nu\| \ge c_0\|x\| \quad \textit{for } x \in X_n, \ u \in Ax, \ n \ge n_0. \tag{12.4}$$

(a) *If x_0 is an isolated solution, then equation (12.1) is strongly approximation solvable in $B_r(x_0)$ for some $r > 0$.*

(b) *If, in addition, A is c_1-coercive for some $c_1 > 0$, then x_0 is an isolated solution, the conclusion of (a) holds and, for $\epsilon \in (0, c_0)$, approximate solutions x_n satisfy*

$$\|x_n - x_0\| \le (c_0 - \epsilon)^{-1}\|Tx_n - f\| \quad \textit{for } n \ge n_1 \ge n_0. \tag{12.5}$$

(c) *If x_0 is an isolated solution in $B_r(x_0)$, A is c_2-bounded for some c_2 and*

$$Tx - Ty \in A(x-y) + R(x-y) \quad whenever \quad x - y \in B_r \tag{12.6}$$

and $z/\|x-y\| \to 0$ as $x \to x_0$ and $y \to x_0$ for each $z \in R(x-y)$, then equation (12.1) is uniquely approximation solvable in $B_r(x_0)$ and the unique solutions $x_n \in B_r(x_0) \cap X_n$ of $Q_nTx = Q_nf$ satisfy

$$\|x_n - x_0\| \leq k\|P_nx_0 - x_0\| \leq c\mathrm{dist}(x_0, X_n), \tag{12.7}$$

where k depends on c_0, c_2, ϵ and δ and $c = 2k\delta_1$, $\delta_1 = \sup\|P_n\|$.

Now, we extend Theorem 12.3 to nondifferntiable maps.

Theorem 12.5: (cf. [59]): *Let T: $X \to Y$ be continuous, locally injective, A-proper w.r.t Γ for (X, Y) and have the closed range $R(T)$. Suppose that T has an odd multivalued derivative $A(x_0)$: $X \to 2^Y$ at each $x_0 \in X$ and there exist an $n_0 = n_0(x_0) \geq 1$ and $c_0 = c_0(x_0) > 0$ such that*

$$\|Q_nu\| \geq c_0\|x\| \quad \text{for } x \in X_n, \quad u \in A(x_0)x, \quad n \geq n_0. \tag{12.8}$$

Then, T is a homeomorphism onto Y and equation (12.1) is strongly approximation solvable for each $f \in Y$. If, in addition, $A(x_0)$ is c_1-coercive for some $c_1 = c_1(x_0) > 0$, then the estimate (12.5) holds. Moreover, if T and each $A(x_0)$ satisfy (12.6) and $A(x_0)$ is $c_2(x_0)$-bounded, for some $c_2(x_0) > 0$, then, for each $f \in Y$, equation (12.1) is uniquely approximation solvable and (12.7) holds.

12.2. Applications to Semilinear Equations

In this section we shall consider semilinear maps of the form

$$Ax - Nx = f \tag{12.9}$$

with A linear and not necessarily continuous. We have the following constructive inverse function theorem. Using Theorem 12.4, we obtain the following constructive solvability result.

Theorem 12.6: [78]: *Let A: $D(A) \subset X \to Y$ be a closed linear densely defined map and C: $X \to Y$ be linear and such that $A - C$: $D(A) \subset X \to Y$ is a bijection and $d = \|(A-C)^{-1}\|^{-1}$. Suppose that N: $X \to Y$ is nonlinear and continuous.*

(a) *Let, for some $k \in (0, d)$*

$$\|(N-C)x - (N-C)y\| \leq k\|x-y\| \quad for\ all\ x, y \in X. \tag{12.10}$$

Then equation (12.9) is uniquely solvable for each $f \in Y$ and the solution is the limit of the iterative process

$$Ax_n - Cx_n = Nx_{n-1} - Cx_{n-1} + f. \tag{12.11}$$

(b) *Equation (12.9) is uniquely approximation solvable w.r.t. $\Gamma = \{X_n, P_n, Y_n, Q_n\}$ with $Q_n(A-C)x = (A-C)x$ on Y_n and $\delta = \max\|Q_n\| = 1$ for each $f \in Y$ and the approximate solutions $\{x_n \in X_n\}$ satisfy*

$$\|x_n - x\| \leq c\|(A+N)x_n - f\| \quad \text{for some } c \text{ and all large } n. \tag{12.12}$$

If A is defined on all of X, then the approximate solutions satisfy also (12.7).

(c) *If $k = d$, X is uniformly convex, $\delta = 1$ and*

$$\|Nx - Cx\| \leq a\|x\| + b \quad \text{for some } a < k, b > 0. \tag{12.13}$$

then equation (12.9) is solvable for each $f \in H$.

Let us now discuss some special cases in the Hilbert space H setting. For $c \in \sigma(A) \cap (-\infty, 0]$, define $d_c^- = \text{dist}(c, \sigma(A) \cap (-\infty, c))$. The following result with $c = 0$ was proved by the author (Proposition 2.7 [78]). In this form it is given in [85,86].

Theorem 12.7: *Let A: $D(A) \subset H \to H$ be a selfadjoint map and N: $H \to H$ satisfy*

(i) $(Nx - Ny, x - y) \geq \alpha\|x - y\|^2$ *for all* $x, y \in H$;
(ii) $\|Nx - Ny\| \leq \beta\|x - y\|$ *for all* $x, y \in H$.

(a) *If (i)–(ii) hold and* $\beta^2 < \alpha d_c^- + c(d_c^- - c - 2\alpha)$ *for some* $c \leq 0$, *then equation (12.9) is uniquely solvable and the solution is the limit of the iterative process (12.11). Moreover, equation (12.9) is uniquely approximation solvable w.r.t.* $\Gamma = \{H_n, P_n\}$ *with* $\delta = \max\|P_n\| = 1$ *for each* $f \in H$ *and (12.12) holds. If A is defined on all of H, then the approximate solutions satisfy also (12.7).*

(b) *If* $\beta^2 \leq \alpha d_c^- + c(d_c^- - c - 2\alpha)$ *and, for some* $a < \lambda = c - d_c^-/2$ *and* $b > 0$,

$$\|Nx - \lambda x\| \leq a\|x\| + b \quad \text{for all } x \in H,$$

then equation (12.9) is solvable for each $f \in H$.

Remark 12.2: Since $\alpha \leq \beta$, the conditions imposed on α and β require that they belong to $(|c|, |c| + d_c^-)$. Hence, $c \leq 0$ is chosen so that this fact holds.

Remark 12.3: Theorem 12.7 extends a result of Smiley [97] in various ways, whose proof is based on the Liapunov–Schmidt alternative method, and the obtained error estimate is of a different type.

Theorem 12.8: *Let A: $D(A) \subset H \to H$ be selfadjoint, N: $H \to H$ be a gradient map and* $C, B^\pm$: $H \to H$ *be selfadjoint maps such that*

(i) $(B^-(x - y), x - y) \leq (Nx - Ny, x - y) \leq (B^+(x - y), x - y)$ *for all* $x, y \in H$;
(ii) $\|B^\pm - C\| \leq d = \min\{|\lambda| \mid \lambda \in \sigma(A - C)\}$.

(a) *If the inequality is strict in (ii), then equation (12.9) is uniquely solvable and the solution is the limit of the iterative process (12.11). Moreover, equation (12.9) is uniquely approximation solvable w.r.t.* $\Gamma = \{H_n, P_n\}$ *with* $\max\|P_n\| = 1$ *for H for each* $f \in H$ *and the approximate solutions satisfy (12.12). If A is defined on all of H, then the approximate solutions satisfy also (12.7).*

(b) *If, in addition, there are* $0 < a < d$ *and* $b \geq 0$ *such that*

$$\|Nx - Cx\| \leq a\|x\| + b \quad \text{for all } x \in H$$

then equation (12.9) is solvable for each $f \in H$.

Remark 12.4: If $B^- = \alpha I$ and $B^+ = \beta I$ and $[\alpha, \beta]$ is contained in the resolvent set $\rho(A)$ of A, then we can take $C = \lambda I$ for some $\lambda \in (\alpha, \beta)$ in part a) of Theorem 12.8 (cf. [78]). In this case the unique solvability of equation (12.9) was proved by Amann [1] and a different proof of it was given in Mawhin [50]. Part (b) allows a bigger value of β as the following result shows. For $c \in \sigma(A) \cap (0, \infty)$, define $d_c^+ = \text{dist}(c, \sigma(A) \cap (c, \infty))$. We have [85,86]

Theorem 12.9: *Let A: $D(A) \subset H \to H$ be selfadjoint, N: $H \to H$ be a gradient map and α, $\beta \in R$ be such that*

$$\alpha\|x - y\|^2 \leq (Nx - Ny, x - y) \leq \beta\|x - y\|^2 \quad \text{for } x, y \in H.$$

(*a*) *If either $c \in \sigma(A) \cap (-\infty, 0]$ and $-c < \alpha \leq \beta < -c + d_c^-$, or $c \in \sigma(A) \cap (0, \infty)$ and $-c - d_c^+ < \alpha \leq \beta < -c$, then equation (12.9) is uniquely solvable and the solution is the limit of the iterative process (12.11). Moreover, equation (12.9) is uniquely approximation solvable w.r.t. $\Gamma = \{H_n, P_n\}$ with* max $\|P_n\| = 1$ *for each $f \in H$ and (12.12) holds. If A is defined on all of H, then the approximate solutions satisfy also (12.7).*

(*b*) *If the conditions in (a) hold with each "$<$" sign replaced by "$\leq$" and, for some $a < \lambda$ with $\lambda = c - d_c^-/2$ if $c \leq 0$ and $\lambda = c + d_c^+/2$ if $c > 0$, and $b > 0$,*

$$\|Nx - \lambda x\| \leq a\|x\| + b \quad \text{for all } x \in H,$$

then equation (12.9) is solvable for each $f \in H$.

When N is also Gateaux differentiable, Theorem 12.9 was proved in [78]. Without the constructive solvability assertions and the error estimates, it is due to Ben-Naoum–Mawhin [5,6] when $c = 0$.

When $[\alpha, \beta] \cap \sigma(A)$ consists of a (positive) finite number of eigenvalues of A having finite multiplicity, we have the following constructive and much simpler proof of a result of Amann–Zehnder [2]. For another simple proof of it with $N_\infty = \lambda I$, we refer to Mawhin [50].

Theorem 12.10: *Let A: $D(A) \subset H \to H$ be a linear selfadjoint map, N: $H \to H$ be a continuous gradient map and $\alpha \leq \beta$ be real numbers not in $\sigma(A)$ such that $[\alpha, \beta] \cap \sigma(A)$ consists of a (positive) finite number of eigenvalues of A having finite multiplicity and*

$$\alpha\|x - y\|^2 \leq (Nx - Ny, x - y) \leq \beta\|x - y\|^2 \quad \text{for } x, y \in H.$$

Suppose that for some selfadjoint map N_∞ and positive constants a, b, c and r, with $a < c$,

$$\|Nx - N_\infty x\| \leq a\|x\| + b \quad \text{for } \|x\| \geq r,$$
$$0 < a < \min\{|\mu| \mid \mu \in \sigma(A - N_\infty)\}.$$

Then equation (12.9) is approximation solvable for each $f \in H$.

The next result deals with conditions which imply the contractivity property of the nonlinear map in a suitable reformulation of equation (12.9).

Theorem 12.11: *Let $A: D(A) \subset H \to H$ be selfadjoint, $N: H \to H$ be a gradient map and $C, B^{\pm}: H \to H$ be selfadjoint maps such that*

(*i*) $N - B^-$ *and* $B^+ - N$ *are monotone;*
and either one of the following conditions holds:
(*ii*) $H = H^- \oplus H^+$ *for some closed subspaces* $H^{\pm}$ *and the projections* $P^{\pm}: H \to H^{\pm}$ *are such that* $P^{\pm}(D(A)) \subset D(A)$ *and for some* $\gamma > 0$

$$((A - B^-)x, x) \le -\gamma\|x\|^2 \quad x \in D(A) \cap H^-; \tag{12.14}$$

$$((A - B^+)x, x) \ge \gamma\|x\|^2 \quad x \in D(A) \cap H^+; \tag{12.15}$$

(*iii*) $((A - B^-)x, x) < 0$ *for* $x \in D(A) \cap H^-$ *and* $((A - B^+)x, x) > 0$ *for* $x \in D(A) \cap H^+$;

and either $A - (1-t)B^- - tB^+$ *has a closed range or* A *has a compact resolvent*

(*iv*) $A - (1-t)B^- - tB^+$ *has a bounded inverse for each* $t \in [0, 1]$.

Then, for each $f \in H$, *equation (12.9) is uniquely solvable, (12.10) holds and, if (ii) holds, it is uniquely approximation solvable w.r.t.* $\Gamma = \{P_n, H_n\}$ *with* $P_nAx = Ax$ *on* H_n *and the approximate solutions satisfy (12.11). If* A *is defined on all of* H, *then the approximate solutions satisfy also (12.7).*

Remark 12.5: Theorem 12.11 (ii) gives a constructive proof of a result of Tersian [99] and part (iv) is due to Fonda–Mawhin [24]. Our proof of the unique solvability in (ii) is new. This result extends also many other earlier ones (Amann [1], Dancer [17], etc.). For various applications to ordinary, elliptic and hyperbolic equations we refer to the above cited works.

Corollary 12.1: *Let $A: D(A) \subset H \to H$ be selfadjoint, $N: H \to H$ be a gradient map such that for some selfadjoint maps $B^{\pm}: H \to H$*

(*i*) $N - B^-$ *and* $B^+ - N$ *are monotone;*
(*ii*) $B^{\pm} = \sum_{i=1}^{m} \lambda_i^{\pm} P_i^{\pm}$ *commute with* A, *where* $P_i^{\pm}: H \to \ker(B^{\pm} - \lambda_i^{\pm})$ *are orthogonal projections with* $P_i^- = P_i^+$ *for* $1 \le i \le m$, $\lambda_1^{\pm} \le \cdots \le \lambda_m^{\pm}$ *and* $\lambda_i^{\pm}$ *are pairwise distinct;*
(*iii*) $\bigcup_{i=1}^{m} [\lambda_i^-, \lambda_i^+] \subset \rho(A)$, *the resolvent set of* A.

Then equation (12.9) is uniquely approximation solvable w.r.t. Γ *for each* $f \in H$ *and (12.10)–(12.11) hold. If* A *is defined on all of* H, *then the approximate solutions satisfy also (12.7).*

13. SOLVABILITY OF NONLINEAR EQUATIONS

In this section we shall discuss a basic solvability result for (pseudo) A-proper maps. It has been first announced by Milojević in [56] and details have been given in [57] and in [65], which was submitted for publication in 1978 (cf. also [69]).

Theorem 13.1: [56,57,59,65]: *Let* $T: X \to Y$ *satisfy condition* (+), $G: X \to Y$ *be bounded and* $K: X \to Y^*$ *be such that*

(1) $H_\mu(t, x) = tTx + \mu Gx$ *is an A-proper homotopy at 0 on* $[0, 1] \times X \backslash B(0, R)$ *for some large* R *and all* $\mu > 0$ *small;*

(2) $Tx \neq \lambda Gx$ *on* $X \backslash B(0,r)$ *for some* r *and all* $\lambda < 0$ *(i.e.,* $H_\mu(t,x) \neq 0$ *there);*
(3) T *is either pseudo A-proper w.r.t.* Γ *or* $T(\bar{B}(0,r))$ *is closed for each* $r > 0$;
(4) $(Gx, Kx) = \|Gx\| \, \|Kx\| > 0$ *for* $x \neq 0$ *and* G *is either A-proper w.r.t.* Γ *and odd, or* $\deg(\mu Q_n G, B_n(0,r), 0) \neq 0$ *for all* $r > 0$, $\mu > 0$ *small, and all large* n.

Then $Tx = f$ *is solvable (approximation solvable if* T *is A-proper) for each* $f \in Y$.

In particular, we have the following basic solvability result for (pseudo) A-proper maps.

Theorem 13.2: [56,57,59,63,65,69]: *Let* T *satisfy condition* $(+)$, *conditions (2)–(4) of Theorem 13.1 hold and either one of the following coditions holds*

($1'$) T *is bounded and* G *and* $T + \mu G$ *are A-proper w.r.t.* Γ *for all* $\mu > 0$ *small;*
($1''$) *there is* $K_n \colon X_n \to Y_n^*$ *such that* $(Q_n g, K_n x) = (g, Kx)$ *for all* $x \in X_n$, $g \in Y$ *and each* n, *and either* T *is* K*-quasibounded or* $(Tx, Kx) \geq -c\|Kx\|$ *for all* $x \in X$ *and some* $c > 0$.

Then $Tx = f$ *is solvable (approximation solvable if* T *is A-proper) for each* $f \in Y$.

We note that some additional assumptions on K, K_n used in [56,57,59,65] have been removed in [63] (cf. [69]). When $Y = X$, then we can take $G = J$, the duality map, and $K = I$ and when $Y = X^*$, take $G = I$ and $K = I$. Hence, we get

Corollary 13.1: *Let* $T \colon X \to X^*$ *be demicontinuous, satisfy condition* $(+)$ *and either*

(*i*) T *is monotone, or*
(*ii*) T *is pseudo monotone,* $Tx \neq \lambda Jx$ *for* $x \notin B(0,r)$ *for some* r *and all* $\lambda < 0$ *and either* T *is quasibounded or* $(Tx, x) \geq -c\|x\|$ *for* $x \in X$.

Then $T(X) = X^*$.

Corollary 13.2: *Let* $A \colon X \to X$ *be continuous and accretive,* $F \colon X \to X$ *be ball condensing and* $I + A - F$ *satisfy condition* $(+)$. *Suppose that* $(F - A)x \neq \lambda x$ *for* $x \notin B(0,r)$ *and* $\lambda > 1$. *Then the equation* $Fx - Ax - x = f$ *is solvable for each* $f \in X$.

14. SOLVABILITY OF STRONGLY NONLINEAR OPERATOR EQUATIONS AND APPLICATIONS

In this section, we discuss two methods of solvability of strongly nonlinear operator equations introduced in [74/83], prove a number of solvability results and give some applications to strongly nonlinear differential equations.

Strongly nonlinear equations have been studied earlier by Gossez [27], Browder [11,12], Kato [40] using topological methods. Recently, Pohozaev [94,95] introduced a new powerfull method – the fibering method, which is also suitable for studying strongly nonlinear problems (see [92], [95]).

Let $\{X, X^+\}$ and $\{Y, Y^+\}$ be two compatible pairs of Banach spaces in duality and $T \colon Y \subset D(T) \subset X \to Y^+$ be a nonlinear map. We shall study the solvability of fully nonlinear equations of the form

$$Tx = f \quad (x \in D(T), f \in Y^+), \tag{14.1}$$

for a new class of nonlinear maps using an approach suitable for the study of strongly nonlinear maps. Let H be a Hilbert space such that $X \subset H$. In the next section, we give some applications to the solvability of semilinear equations with strongly nonlinear perturbation of the form

$$Ax + Nx = f, \tag{14.2}$$

where $A: D(A) \subset H \to H$ is a linear densely defined map and $N: Y \subset D(N) \subset X \to Y^+$ is a nonlinear map. A second approach to the study of equation (14.1), based on a double approximation, is discussed in Section 16. Section 17 is devoted to applications of the abstract results to BVP's for nonlinear elliptic and periodic semilinear hyperbolic equations with strong nonlinearities.

We continue by describing a setting for studying strongly nonlinear equations. A pair of Banach spaces $\{X, X^+\}$ is called a dual pair if there exists a nondegenerate continuous bilinear form $\langle,\rangle_X$ on $X \times X^+$, i.e., if $\langle x, y\rangle_X = 0$ for all $x \in X$, then $y = 0$, and if $\langle x, y\rangle_X = 0$ for all $y \in X^+$, then $x = 0$. Here, X^+ need not be the dual space X^* of X.

Suppose that there are two pairs $\{X, X^+\}$ and $\{Y, Y^+\}$ of Banach spaces in duality with $Y \subset X$ and $X^+ \subset Y^+$ and the injections are continuous and dense. Suppose that X and Y is separable and the two dual pairs are compatible, i.e., $\langle x, y\rangle_X = \langle x, y\rangle_Y$ for all $x \in X^+$ and $y \in Y$. Let $Z \subset Y^+$ be a Banach space. We now define a new class of maps introduced in [74,83].

Definition 14.1: A map $T: Y \subset D(T) \subset X \to Y^+$, with $T(Y) \subset Z$, is said to be *(pseudo) A-proper* relative to (Y, X) w.r.t. $\Gamma = \{Y_n, Z_n, R_n\}$ for (Y, Z) if the map $T_n = R_n T: D \cap Y_n \to Z_n$ is continuous for each n and whenever $\{x_{n_k} \in D(T) \cap Y_{n_k}\}$ is bounded in X and $T_{n_k} x_{n_k} \to f$ in Y^+ for some $f \in X^+$, then a subsequence of $\{x_{n_k}\}$ converges to $x \in D(T)$ in X (respectively, there is an $x \in D(T)$) such that $Tx = f$.

If $Y^+ = Y^* = Z$ and $Q_n: Y \to Y_n$ are linear projections onto Y_n, then the above definition of the (pseudo) A-properness coincides with the usual one w.r.t. $\Gamma = \{Y_n, Q_n, Y_n^*, Q_n^*\}$.

Recall that $T: D(T) \to Y^+$ is quasibounded if, whenever $\{x_n\} \subset Y$ is bounded in X and $\langle Tx_n, x_n\rangle \leq \text{const.}\|x_n\|_X$, then $\{Tx_n\}$ is bounded in Y^+. It satisfies condition $(+)$ if, whenever $\{x_n\} \subset Y$ and $Tx_n \to f$ in Y^+, then $\{x_n\}$ is bounded in X.

Our first result is for odd strongly nonlinear maps. Such equations have been first studied by Pohozaev [93].

Theorem 14.1: *Let $T: Y \subset D(T) \subset X \to Y^+$ be odd, satisfy condition $(+)$ and either T is A-proper relative to (Y, X) w.r.t. $\Gamma = \{Y_n, Z_n, R_n\}$ for (Y, Z) or T is pseudo A-proper and $T + \mu G$ is A-proper w.r.t. Γ for some odd G and all $\mu > 0$ small. Then the equation $Tx = f$ is solvable in $D(T)$ for each $f \in X^+$.*

Theorem 14.2: *Let $T: Y \subset D(T) \subset X \to Y^+$ be odd, k-positively homogeneous and A-proper relative to (Y, X) w.r.t. $\Gamma = \{Y_n, Z_n, R_n\}$ for (Y, Z) and $x = 0$ if $Tx = 0$. Then the equation $Tx = f$ is solvable for each $f \in X^+$.*

Remark 14.1: Theorems 14.1–14.2 are due to Pohozaev [93] for maps $T: X \to X^*$ and a for general $T: X \to Y$ see [91].

Extending the arguments of the proof of Theorem 8.2 to this setting, we have the following result.

Theorem 14.3: *Let $T: Y \subset D(T) \subset X \to Y^+$ be pseudo A-proper relative to (Y, X) w.r.t. Γ for (Y, Z), asymptotically positively k-homogeneous, i.e., $T = A + N$ with the quasinorm of N, $|N| = \limsup_{\|x\| \to \infty} \|Nx\|/\|x\|^k$ sufficienly small and A is positively k-homogeneous, $k > 0$. Let A be A-proper relative to (Y, X) w.r.t. Γ for (Y, Z), $x = 0$ if $Ax = 0$ and $\deg(Q_n A, B(0, r) \cap Y_n, 0) \neq 0$ for each large r and each $n \geq n_0(r)$. Then $Ax + Nx = f$ is solvable for each $f \in X^+$.*

Next, we extend Theorem 13.1 to strongly nonlinear maps [83].

Theorem 14.4: *Let D be an open bounded subset of X containig the origin 0, $T: Y \subset D(T) \subset X \to Y^+$ be pseudo A-proper relative to (Y, X) w.r.t. $\Gamma = \{Y_n, Z_n, R_n\}$ for (Y, Z), $B: X \to Y^+$ be a bounded map with $\deg(B_n, D \cap Y_n, 0) \neq 0$ for each large n, $V = \cup_n Y_n$ and $f \in X^+$ be given. Suppose that either*

(1) *There is a positive γ such that*

$$\|Tx - tf\|_{Y^+} \geq \gamma/2 \quad \text{for all } x \in \partial D \cap Y,\ t \in [0, 1];$$

and

(2) *T is quasibounded, $T + \mu B$ is A-proper w.r.t. Γ for all $\mu \in (0, \mu_0)$, $\langle Bx, x \rangle_Y = \|Bx\|_{Y^+} \|x\|_X$ for $x \in V$ and*

$$Tx \neq \lambda Bx \quad \text{for all } x \in \partial D \cap Y,\ \lambda < 0;$$

or

(3) *There is an $n_0 \geq 1$ such that*

$$R_n Tx + R_n f \neq \lambda R_n Bx \quad \text{for } \lambda < 0,\ x \in \partial D \cap Y_n,\ \ n \geq n_0;$$

or

(4) *Let $\langle B_n x, y \rangle = \langle Bx, y \rangle$ for $x \in Y_n$ and $y \in X^+$, $T(V) \subset H$, $G(V) \subset H$ and*

$$\langle Tx - f, Bx \rangle_Y > 0 \quad \text{for all } x \in \partial D \cap V.$$

Then the equations $Tx = f$ is solvable.

Theorem 14.5: *Let X be reflexive, $A, C, N: Y \subset D \subset X \to Y^+$ be such that A is demicontinuous, $A + N - C$ is A-proper relative to (Y, X) w.r.t. $\Gamma = \{Y_n, Z_n, R_n\}$ and $B: X \to Y^+$ and $K: X \to Y$ be bounded and such that $\|Kx\| \leq M\|x\|$, $\langle Bx, Kx \rangle \geq \|x\|^2$ and*

(1) *$\langle Ax, Kx \rangle \geq 0$ on X and, for $p > 1$, $A(tx) = t^q Ax$ for all $x \in D$, $t \geq 0$ and some $q \in [1, p)$ or $q \in (0, p)$ provided also $\langle Ax, Kx \rangle \geq m\|x\|^2$ on D for some $m > 0$.*
(2) *C is compact, positively homogeneous and quasibounded.*
(3) *$N = N_1 + N_2$ such that N_1 is pseudo A-proper relative to (Y, X) w.r.t. Γ at 0, $\langle N_i x, Kx \rangle \geq 0$, $i = 1, 2$, $x = 0$ if $N_1 x = 0$, $N_1(tx) = t^p N_1 x$ for $x \in D$, $t \geq 0$ and $\|N_2 x\| \leq a\|x\|^{p_0} + b$ for $x \in X$, and some $p_0 < p$ and positive a and b.*

Then the equation $Ax + Nx - Cx = f$ is approximation solvable for each $f \in X^+$.

In particular, we have the following partial extension of Pohozaev's result [93] in another direction.

Corollary 14.1: *Let X be reflexive, $A, C, N\colon X \to X^*$ be such that A is demicontinuous, $A + N - C$ is A-proper w.r.t. $\Gamma = \{X_n, X_n^*, Q_n^*\}$ and $B\colon X \to X^*$ is bounded and such that $\langle Bx, x\rangle \geq \|x\|^2$ and*

(1) *$\langle Ax, x\rangle \geq 0$ on X and, for $p > 1$, $A(tx) = t^q Ax$ for all $x \in X$, $t \geq 0$ and some $q \in [1, p)$.*
(2) *C is compact, positively homogeneous and quasibounded.*
(3) *$N = N_1 + N_2$ such that N_1 is pseudo A-proper w.r.t. Γ at 0, $\langle N_i x, x\rangle \geq 0$, $i = 1, 2$, $x = 0$ if $N_1 x = 0$, $N_1(tx) = t^p N_1 x$ for $x \in X$, $t \geq 0$ and $\|N_2 x\| \leq a\|x\|^{p_0} + b$ for $x \in X$, and some $p_0 < p$ and positive a and b.*

Then the equation $Ax + Nx - Cx = f$ is approximation solvable for each $f \in X^*$.

Next, we give some examples of (pseudo) A-proper maps w.r.t. Γ relative to (Y, X).

Definition 14.2: Let $Y \subset D(T) \subset X$ and $T\colon D(T) \to Y^+$.

(a) Then T is said to be of type (M) relative to (Y, X) if
  (i) T is continuous from each finite dimensional subspace of Y into the weak topology of Y^+, and
  (ii) if $\{x_n\} \subset Y$, $x_n \rightharpoonup x$ in X, $Tx_n \rightharpoonup y$ in Y^+ with $y \in X^+$ and $\limsup\langle Tx_n, x_n\rangle \leq \langle y, x\rangle$, then $x \in D(T)$ and $Tx = y$.

If y in (ii) is given in advance, we say that T is of type (M) at y relative to (Y, X).

(b) T is said to be pseudo monotone relative to (Y, X) if (i) holds and whenever $\{x_n\}$ is as in (ii), then $x \in D(T)$, $\langle Tx_n, x_n\rangle \to \langle y, x\rangle$ and $Tx = y$.
(c) T is said to be of type (S_+) if (i) holds and whenever $\{x_n\}$ is as in (ii), then $x_n \to x$ in X.

Proposition 14.1: [84]: *Let $T\colon Y \subset D(T) \subset X \to Y^+$ be of type (M) relative to (X, Y), then $T\colon Y \subset D(T) \subset X \to Y^+$ is pseudo A-proper relative to (Y, X) w.r.t. $\Gamma = \{H_n, P_n\}$ for (Y, H) in the sense that whenever $\{x_{n_k} \in H_{n_k}\}$ is bounded in X, $\{Tx_{n_k}\}$ is bounded in Y^+ and $P_{n_k}Tx_{n_k} \to f$ in Y^+ for $f \in X^+$, then there is an $x \in D(T)$ such that $Tx = f$.*

Proposition 14.2: [84]: *Let $T\colon Y \subset D(T) \subset X \to Y^+$ be quasibounded and pseudo monotone relative to (X, Y) with $T(Y) \subset H$ and $B\colon X \to X^+$ be of type (S_+). Then $T + B$ is A-proper relative to (Y, X) w.r.t. $\Gamma = \{H_n, P_n\}$.*

In the above examples one can look at the (pseudo) A-properness w.r.t. other schemes. For example, we have the following version of Proposition 14.2.

Proposition 14.3: [84]: *Let $T\colon Y \subset D(T) \subset X \to Y^*$ be quasibounded and pseudo monotone relative to (X, Y) and $B\colon Y \subset D(B) \subset X \to Y^*$ be of type (S_+) relative to (Y, X). Then $T + B$ is A-proper relative to (Y, X) w.r.t. $\Gamma = \{Y_n, P_n^*\}$ for (Y, Y^*) with $\cup_{n\geq 1} Y_n$ dense in X and $P_n\colon Y \to Y_n$ a linear projection.*

Proposition 14.4: [84]: *Let $T\colon Y \subset D(T) \subset X \to Y^*$ be of type (S_+) relative to (X, Y). Then T is A-proper relative to (Y, X) w.r.t. $\Gamma = \{Y_n, P_n^*\}$ for (Y, Y^*) with $\cup_{n\geq 1} Y_n$ dense in X and $P_n\colon Y \to Y_n$ a linear projection.*

In a general situation, one can use a scheme $\Gamma = \{Y_n, Q_n^*\}$ for (Y, Y^+) *if the corresponding bilinear form* $\langle , \rangle$ is such that $\langle Q_n^* f, y\rangle = \langle f, Q_n y\rangle$ for all $f \in Y^+$, $y \in Y$.

For our next example, we introduce the following new class of maps.

Definition 14.3: A map $T: Y \subset D(T) \subset X \to Y^+$ is said to be of *type (F) relative to* (X, Y) if (i) of Definition 14.2 holds and whenever $\{x_n\} \subset Y$ is such that $x_n \rightharpoonup x$ in X and $Tx_n \rightharpoonup u$, then $x \in D(T)$, $\langle Tx, x\rangle \leq \liminf \langle Tx_n, x_n\rangle$ and $Tx = u$.

Proposition 14.5: *Let $T: Y \subset D(T) \subset X \to Y^+$ be quasibounded and of type (F) relative to (X, Y) with $T(Y) \subset H$. Then*

(a) *If $B: X \to X^+$ is of type (S_+) relative to (X, Y), then $T + B$ is A-proper relative to (Y, X) w.r.t. $\Gamma = \{H_n, P_n\}$.*
(b) *If B is pseudo monotone relative to (Y, X) and G is of type (S_+) relative to (Y, X), then $T + B + \mu G$ is A-proper relative to (Y, X) w.r.t. Γ for (Y, X) for each $\mu > 0$.*
(c) *If B is of type (M) relative to (X, Y), then $T + B$ is pseudo A-proper relative to (Y, X) w.r.t. Γ for (Y, X).*

We have the following special case of Proposition 14.5, useful in applications.

Proposition 14.6: [84]: *Let $T: Y \subset D(T) \subset X \to Y^+$, $T(Y) \subset H$, be quasibounded, weakly continuous and such that for each $r > 0$, some constant $q(r) > 0$ and x, y with $\|x - y\|_X \geq r$,*

$$\langle Tx - Ty, x - y\rangle \geq q(r)c(\|x - y\|_X) - \phi(x - y),$$

where $\phi: X \to R$ is weakly upper semicontinuous at 0 with $\phi(0) = 0$, and $c: R^+ \to R^+$ is continuous, $c(0) = 0$ and $r \to 0$ whenever $c(r) \to 0$. Then $T + B$ is A-proper (pseudo A-proper) relative to (Y, X) w.r.t. $\Gamma = \{H_n, P_n\}$ if $B: X \to X^+$ is of type (S_+) (respectively, of type (M)) relative to (X, Y).

We note that as T in Proposition 14.6 we can take a continuous linear map satisfying a Garding like inequality.

Using the above propositions, one can state many particular case of Theorems 14.1–14.5. We just state the following one.

Corollary 14.2: *Let $T: Y \subset D(T) \subset X \to Y^+$ be of type (M) relative to (X, Y) with $T(Y) \subset H$ and conditions (1)–(4) of Theorem 6 hold with $B = I$, the identity map, except the A-properness of $T + \mu B$. Then the equation $Tx = f$ is solvable.*

In view of Proposition 14.4, we have

Corollary 14.3: *Let $T: Y \subset D(T) \subset X \to Y^*$ be of type (S_+) relative to (X, Y) be such that for a bounded open set D in X*

$$Tx \neq 0 \quad \text{for } x \in \partial D \cap D(T).$$

If $\deg(P_n^* T, D \cap Y_n, 0) \neq 0$ for all large n, then the equation $Tx = 0$ is approximation solvable.

In [Mi-1991] we have developed a perturbation method for studying strongly nonlinear maps. Using that method, we have

Theorem 14.6: [84 (1992)]: *Let X be a reflexive Banach space, $A: Y \subset D(A) \subset X \to 2^{Y^+}$ be coercive with $0 \in A(0)$, $B: X \to Y^+$, $K: X \to Y$ be bounded and such that $\|Kx\| \leq M\|x\|$ and $\langle Bx, Kx\rangle \geq \|x\|^2$, $C: X \to Y^+$ be compact, positively homogeneous and quasibounded and $N: Y \subset D(N) \subset X \to Y^+$ be nonlinear and $\langle Nx, Kx\rangle \geq c(x/\|x\|)\ \|x\|^{p+1}$ for all $x \in D(A)\backslash 0$ and some $p > 0$, where a function $c: X \to R^+$ is such that $x \rightharpoonup 0$ if $c(x) \to 0$ and $x = 0$ if $c(x) = 0$. Suppose that $A + N$ satisfies condition $(*)$ and, for each $m > 0$, $R_m = (A + 1/mB)^{-1}: Y^+ \to X$ exists and $T_m = I - (A + 1/mB)^{-1}N: Y \subset D(N) \subset X \to X$ is A-proper relative to (Y, X) w.r.t. $\Gamma = \{Y_n, P_n\}$ for (Y, X). Then the equation $f \in Ax - Cx + Nx$ is solvable for each $f \in X^+$.*

Theorems 14.6 extend some results of Guan [29] involving compact perturbations of monotone like maps.

15. SEMILINEAR EQUATIONS WITH STRONG NONLINEARITIES

One could prove a number of solvability results for equation (14.2) using the results from Section 14. To that end, we need to discuss first the pseudo A-properness of $A + N$ relative to two Banach spaces under various conditions on A and N.

Proposition 15.1: [84]: *Let $A: D(A) \subset H \to H$ be a densely defined linear map with closed range and compact partial inverse. Then*

(*a*) *If $N: Y \subset D(N) \subset X \to Y^+$ is of type (M) relative to (X, Y) with $N(Y) \subset H$, then $A + N: Y \subset D(A) \cap D(N) \subset X \to Y^+$ is pseudo A-proper relative to (Y, X) w.r.t. $\Gamma = \{H_n, P_n\}$ for (Y, H) with $P_n Ax = Ax$ on H_n in the sense that whenever $\{x_{n_k} \in H_{n_k}\}$ is bounded in X, $\{Nx_{n_k}\}$ is bounded in Y^+ and $Ax_{n_k} + P_{n_k}Nx_{n_k} \to f$ in Y^+ for $f \in X^+$, then there is an $x \in D(A) \cap D(N)$ such that $Ax + Nx = f$.*

(*b*) *If $N : Y \subset D(N) \subset X \to Y^+$ is completely continuous (i.e., if $x_n \rightharpoonup x$ in X then $Nx_n \to Nx$ in Y^+) and $M: Y \subset D(M) \subset X \to Y^+$ is of type (M) relative to (X, Y) with $(N + M)(Y) \subset H$, then $A + N + M: Y \subset D(A) \cap D(N) \cap D(M) \subset X \to Y^+$ is pseudo A-proper relative to (Y, X) w.r.t. $\Gamma = \{H_n, P_n\}$ for (Y, H) with $P_n Ax = Ax$ on H_n in the same sense as in (a).*

Now, using the results of Section 14, one can derive the corresponding solvability results for equation (14.2). We just note that as an immediate consequence of Proposition 14.1 and Theorem 14.4, we have [83]

Theorem 15.1: *Let $A: D(A) \subset H \to H$ be a densely defined linear map with closed range and compact partial inverse, $N: Y \subset D(N) \subset X \to Y^+$ be of type (M) relative to (X, Y) with $N(Y) \subset H$. Suppose that for a given $f \in X^+$ and a linear map $B: X \to H$ with $P_n Bx = Bx$ on H_n there is an open and bounded subset D of X such that $D \cap Y \subset D(N)$ and*

$$\langle Ax + Nx - f, Bx\rangle_{Y^+} > 0 \quad \text{for } x \in \partial D \cap \cup_{n \geq 1} H_n.$$

Then the equation $Ax + Nx = f$ is solvable in X.

16. DOUBLE APPROXIMATION AND STRONGLY NONLINEAR EQUATIONS

In this section, we shall introduce another approach to strongly nonlinear equations based on a double approximation [84]. Namely, we shall first approximate $T: D(T) \subset X \to Y$ by a suitable sequence of pseudo A-proper maps $\{T_n\}$ and use the above results for each T_n, which are also based on approximations of each T_n.

Definition 16.1: A map $T: D(T) \subset X \to Y$ is approximable by $\{T_n: D(T_n) \subset X \to Y\}$ if each T_n is pseudo A-proper w.r.t. $\Gamma_n = \{X_{n,k}, Y_{n,k}, Q_{n,k}\}$ for (X, Y) and whenever $\{x_{n_i} \in D(T_{n_i})\}$ is bounded in X and $T_{n_i}x_{n_i} \to f$ in Y, then $Tx = f$ for some $x \in D(T)$.

Proposition 16.2: *Let X be reflexive, $A: D(A) \subset X \to X^*$ be demicontinuous of type (S_+), $N: D(N) \subset X \to X^*$ and $N_n: D(N_n) \subset X \to X^*$ be approximations for N such that each $A + N_n: D(A) \cap D(N_n) \subset X \to X^*$ is pseudo A-proper w.r.t. $\Gamma = \{X_k, Y_k, Q_k\}$ for (X, X^*) and*

(C) *If $\{x_n \in D(N_n)\}$ is such that $x_n \rightharpoonup x$ in X and $N_n x_n \rightharpoonup z$ in X^*, then $x \in D(N)$, $Nx = z$ and $\langle Nx, x\rangle \leq \liminf\langle N_n x_n, x_n\rangle$.*

Then $A + N$ is approximable by $A + N_n: D(A) \cap D(N_n) \subset X \to X^*$.

Proposition 16.3: *Let X, N and N_n be as in Proposition 16.2 and $A: D(A) \subset X \to X^*$ be bounded and pseudo monotone. Then $A + N: D(A) \cap D(N) \subset X \to X^*$ is approximable by $A + N_n: D(A) \cap D(N_n) \subset X \to X^*$.*

Proposition 16.4: *Let H be a Hilbert space, $A: D(A) \subset H \to H$ be linear densely defined with closed range $R(A)$ and compact partial inverse and the null space $N(A) = N(A^*)$. Let $N: H \to H$ be bounded and of type (M). Let $V_n \subset \ker(A)$ be finite dimensional such that $\cup_{n \geq 1} V_n$ is dense in $\ker A$, $W_n = V_n \oplus R(A)$ and $P_n: H \to W_n$ be orthogonal projections. Then $A + N: D(A) \subset H \to H$ is approximable by $A|W_n \cap D(A) + P_n N: D(A) \cap W_n \to W_n$.*

Using versions of Theorems 14.2 and 14.4 and the properties of T, we have [83]

Theorem 16.1: *Let $D \subset X$ be open and bounded with $0 \in D$, $T: \bar{D} \to X^*$ be approximable by $T_n: \bar{D} \to X^*$ and, for a given $f \in X^*$ and each n, either*

(a) *T_n is odd on ∂D, or*
(b) *$\langle T_n x, x\rangle \geq \langle f, x\rangle$ for all $x \in \partial D$, or*
(c) *$T_n x - f \neq \lambda Jx$ for all $x \in \partial D$, $\lambda < 0$ and X is uniformly convex, where J is the normalized duality map.*

Then the equation $Tx = f$ is solvable in $\bar{D}$.

Corollary 16.1: *Let $A: \bar{D} \subset X \to X^*$ be either of type (S_+) or pseudo monotone, $N: \bar{D} \to X^*$ be nonlinear and $N_n: \bar{D} \to X^*$ be such that $A + N$ is approximable by $A + N_n$ and condition (C) in Proposition 16.2 holds. Then, if $T = A + N$ satisfies conditions (a)–(c) of Theorem 16.1, the equation $Ax + Nx = f$ is solvable.*

17. APPLICATIONS TO SEMILINEAR DIFFERENTIAL EQUATIONS

In this section, we shall give applications of the above abstract results to BVP's for nonlinear elliptic PDE's in divergence form and to periodic-boundary value problems for semilinear hyperbolic equations.

17.1. Weak Solvability of Nonlinear Elliptic Problems

Consider the following formal differential equation

$$\sum_{|\alpha|\le m}(-1)^{|\alpha|}D^{\alpha}A_{\alpha}(x,u,\ldots,D^{m}u)+\sum_{|\alpha|<m}(-1)^{|\alpha}D^{\alpha}B_{\alpha}(x,u,\ldots,D^{m}u)=f \text{ in } Q, \tag{17.1}$$

where $Q\subset R^n$ is a bounded domain, $u\in V$, $f\in V^*$, where V is a closed subspace of W_p^m such that $\dot{W}_p^m\subset V$. Suppose

(A1) For each $|\alpha|\le m$, let $A_\alpha(x,\xi)=A_\alpha^{(0)}(x,\xi)+A_\alpha^{(1)}(x,\xi)$ be such that each $A_\alpha^{(i)}\colon Q\times R^{s_m}\to R$ satisfies the Caratheodory conditions and

$$|A_\alpha^{(i)}(x,\xi)|\le c_i(1+|\xi|)^{p_i-1}$$

for some constants c_0 and c_1, $p_0=p-1$, $p_1<p-1$.

(A2) For each $|\alpha|\le m$, $A_\alpha^{(0)}(x,\xi)$ is odd and $p-1$-homogeneous in ξ.

(A3) Let $c_1\colon R^+\to R$ be continuous and $r\to 0$ if $c_1(r)\to 0$, $c: R^+\times R^+\to R^+$ be such that $c(R,\cdot)$ is weakly uppersemicontinuous at 0 and $c(R,0)=0$ for each $R>0$ and, for $w\in V$ and $u,v\in \bar{B}(0,r)\subset V$, we have for some $k\le m-1$, $i=0,1$

$$\sum_{|\alpha|=m}\int_Q[A_\alpha^{(i)}(x,D^\gamma w,D^m u)-A_\alpha^{(i)}(x,D^\gamma w,D^m v)]D^\alpha(u-v)$$

$$\ge c_1(\|u-v\|_{p,m})-c(R,\|u-v\|_{p,k}).$$

(B1) For each $|\alpha|\le m$, $B_\alpha\colon Q\times R^{s_m}\to R$ satisfies the Caratheodory conditions and for some $c_2>0$

$$|B_\alpha(x,\xi)|\le c_2(1+|\xi|)^{p-1}.$$

(B2) Let $c\colon R^+\times R^+\to R^+$ be continuous and $c(R,tr)/t\to 0$ as $t\to 0^+$ for each r and R and suppose that for each $R>0$ and $u,v\in\bar{B}(0,R)\subset V$ the integral inequality in (A3) holds for the B_α's with $c_1\equiv 0$.

Then the weak solvability of BVP (17.1) is equivalent to the solvability of the equation

$$A_0u+A_1u+Nu=f,$$

where A_0, A_1 and N are the corresponding operators induced by the functions $A_\alpha^{(0)}$, $A_\alpha^{(1)}$ and B_α respectively in the weak formulation of BVP (17.1).

Using Theorem 8.2, we obtain the following result.

Theorem 17.1: [84]: *Let (A1)–(A3) and (B1)–(B2) hold with $c_1(t) = c(t) \equiv 0$ for $i = 1$ in (A3), c_1 and c_2 sufficiently small and $u = 0$ if $A_0 u = 0$. Then BVP (17.1) has a weak solution for each $f \in V^*$.*

Theorem 13.2 yields the following extensions of results of Pohozaev [93] and Dubinski [20].

Theorem 17.2: [84]: *Let (A1), (A3) and (B1)–(B2) hold, N satisfy condition (+) and either*

(a) $A_\alpha(x, -\xi) = -A_\alpha(x, \xi)$ *for $x \in Q$ a.e., and all $|\xi|$ large and $|\alpha| \le m$, or*

(b) *For each $h \in V^*$ there exists an $r_h > 0$ such that $Nu \ne \lambda Ju$ for $u \in \partial B(0, r_h)$, $\lambda < 0$, where $J\colon V \to V^*$ is the normalized duality map.*

Then the generalized BVP for (17.1) is approximation solvable in V for each $f \in L_q$. It is solvable if (A3) holds with $c_1 \equiv 0$.

17.2. Strongly Nonlinear Perturbations of Nonlinear Elliptic Operators

We consider now

$$Au + F(x, u) = f(x) \quad u \in \dot{W}_p^m, \tag{17.2}$$

where $Au = \sum_{|\alpha| \le m} (-1)^{|\alpha|} D^\alpha A_\alpha(x, u(x), \ldots, D^m u)$ is an elliptic operator in divergence form that satisfies the polynomial growth and the ellipticity conditions that make it an operator of type (S_+) or pseudo monotone from $X = \dot{W}_p^m(Q)$ into $X^* = W_q^{-m}(Q)$, $1/p + 1/q = 1$, and the perturbation $F(x, u)$ satisfies no *a priori* growth condition but only a sign condition of the form (F1).

Assume that each $A_\alpha\colon Q \times R^{s_m} \to R$ is a Caratheodory function and

(A1) For some $p > 1$, $C_1 > 0$ and $k_1 \in L_q(Q)$,

$$|A_\alpha(x, \xi)| \le C_1(k_1(x) + |\xi|^{p-1}) \quad \text{for } |\alpha| \le m, x \in Q, \xi \in R^{s_m}.$$

Then the map $A\colon V \to X^*$, for any subspace V with $X \subset V \subset W_p^*$, defined by

$$(Au, v) = \int_Q \sum_{|\alpha| \le m} A_\alpha(x, \xi(u)) D^\alpha v \quad u, v \in V$$

is bounded and continuous. It is well known that A is pseudo monotone if it satisfies the Leray–Lions condition:

(A2) For each $x \in Q$ and $\eta \in R^{s_{m-1}}$ and any pair of distinct vectors $\zeta, \zeta^* \in R^{s'_m}$, corresponding to the derivatives of order $|\alpha| = m$,

$$\sum_{|\alpha| = m} (A_\alpha(x, \eta, \zeta) - A_\alpha(x, \eta, \zeta^*))(\zeta_\alpha - \zeta_\alpha^*) > 0.$$

A is of type (S_+) if also the following coercivity condition holds.

(A3) $\sum_{|\alpha|\leq m} A_\alpha(x,\xi)\xi_\alpha \geq c_2|\zeta|^p - k_2(x)$ for some $c_2 > 0$, $k_2 \in L_1(Q)$, all $x \in Q$ and $\xi = (\eta,\zeta) \in R^{s_m}$.

Regarding F we assume the following sign condition:

(F1) Let $F\colon Q \times R \to R$ be a Caratheodory function and for each $s \geq 0$, there is a function $h_s \in L^1_{\text{loc}}$ such that

$$\sup_{|y|\leq s} |F(x,y)| \leq h_s(x) \quad \text{and} \quad F(x,y)y \geq 0 \quad \text{for a.e. } x \in Q, y \in R.$$

Define $N\colon D(N) \subset X = \dot{W}^m_p(Q) \to X^* = W^{-q}_p(Q)$ by $Nu = F(x,u(x))$ for $u \in D(N) = \{u \mid u \in \dot{W}^m_p(Q), Nu \in W^{-m}_q(Q) \cap L^1_{\text{loc}}(Q)\}$.

Next, as in Browder [11–13] we define approximations N_k of N. Let $\{Q_k\}$ be a sequence of relatively compact open subsets of Q with $Q_k \subset Q_{k+1}$ for each k and $Q = \cup_{k\geq 1} Q_k$. Let ξ_k be the characteristic function of Q_k and define the trancation of F at the level k by $F^{(k)}(x,y) = F(x,y)$ if $|F(x,y)| \leq k$ and $F^{(k)}(x,y) = k\ \text{sign}(F(x,y))$ if $|F(x,y)| \geq k$. Define $N_k : D(N_k) \subset X \to X^*$ by $N_k u(x) = \xi_k(x)F^{(k)}(x,u(x))$. Each N_k is a compact map of X into $X^* \cap L^\infty_c(Q)$ and therefore $A + N_k$ is of type (S_+) if such is A. Moreover, condition (C) holds as can be seen from Browder [11–13].

Using the above approximations, Browder [11–13] has defined a topological degree for maps defined by equation (17.2). However, using only the Brouwer degree and a double approximation method of Section 15, we can study equation (17.2) in a simpler fashion. For example, Propositions 16.2–16.3 and Theorem 16.1 yield the following solvability result.

Theorem 17.3: *Let (A1), (F1) and (A2) hold and $A + N$ satisfy condition $(+)$. Suppose that either $F(x,\cdot)$ and A are odd, i.e., for each $|\alpha| \leq m$, $A_\alpha(x,-\xi) = -A_\alpha(x,\xi)$ for each $x \in Q$, $\xi \in R^{s_m}$, or (A3) holds and $Au + Nu - f \neq \lambda Ju$ for $\|u\| = r$, all $\lambda < 0$ and some $r > 0$. Then equation (17.2) is solvable in $\dot{W}^m_p$.*

17.3. Periodic Solutions of Semilinear Hyperbolic Equations

In this part we shall show how the abstract theory can be applied to the existence of weak T-periodic solutions of some hyperbolic equations with strong nonlinearities without resonance. We refer to [83] for a detailed study of these equations using the (pseudo) A-proper mapping approach.

Let $Q \subset R^n$ be a bounded domain with smooth boundary, $\Omega = (0,T) \times Q$ and let $H = L_2(\Omega, R^m)$ with the inner product defined by

$$(u,v) = \int_0^T \int_Q (u(t,x), v(t,x))\,dx\,dt,$$

where $(u(t,x), v(t,x))$, $(t,x) \in \Omega$, is the inner product in R^m. Let L_1 be a linear self-adjoint elliptic operator in space variables $x \in R^n$ with coefficients independent of t such that the induced bilinear form $a(u,v)$ on the Sobolev space $W^1_2(Q, R^m)$ is continuous and symmetric. Let $V = W_2^1$ be a closed subspace of $W_2^1(Q, R^m)$, containing the test functions, such that $a(u, v)$ is semi-coercive on V, i.e. there are constants $a_1 > 0$ and $a_2 \geq 0$ such that

$$a(u, v) \geq a_1\|u\|_{2,1}^2 - a_2\|u\|_{L_2}^2 \quad \text{for all } u \in V.$$

Define a linear map $L_0 \colon D(L_0) \subset L_2(Q, R^m) \to L_2(Q, R^m)$ by

$$(L_0 u, v) = a(u, v) \quad \text{for each } v \in V,$$

where

$$D(L_0) = \{u \in V \mid a(u, \cdot) \text{ is continuous on V in the } L_2 - \text{norm}\}.$$

It is well known that L_0 is selfadjoint and has a compact resolvent, since W_2^1 is compactly embedded in L_2. Next, define a selfadjoint map with compact resolvent $L \colon D(L) \subset H_1 = L_2(Q, R^m) \to H_1$ by:

$$D(L) = [D(L_0)]^m \quad \text{and} \quad L = \text{diag}\,(L_0, \ldots, L_0).$$

Since L has a compact resolvent, there is an orthonormal basis $\{\psi_j \mid j \in J\}$ in H_1 and a sequence of its eigenvalues $\{\mu_j \mid j \in J\}$ such that $|\mu_j| \to \infty$.
We shall give an application to the system of hyperbolic equations with strong nonlinearities.

We begin with a perturbed system of the form

$$\begin{cases} u_{tt} - L_1 u + g(x, u)u_t + F(x, u) + H(t, x, u) = 0 & (t, x) \in (0, T) \times Q \\ u^{2\alpha_i}(t, \cdot) = 0 \quad \text{for } (t, x) \in [0, T] \times \partial Q, & \\ \alpha_i = k \quad \text{for } 0 \leq k \leq m - 1, 1 \leq i \leq n, & \end{cases} \tag{17.3}$$

where $\alpha = (\alpha_1, \ldots, \alpha_n)$. Let $\sigma(L)$ be the spectrum of the eigenvalue problem $L_1 u = \lambda u$ subject to the boundary conditions in (17.3). Assume that

(H1) $g \colon Q \times R^m \to R^m$ is continuos and for some positive constants a_i, $i = 0, 1, 2$, and q

$$a_0 \leq |g(x, y)| \leq a_1|y|^q + a_2 \quad \text{for } (x, y) \in Q \times R^m.$$

(H2) $F \colon Q \times R^m \to R^m$ is continuous and for some positive constants b_i, $i = 0, 1, 2$, c and r

$$|F(x, y)| \leq b_1|y|^r + b_2 \quad \text{and} \quad yF(x, y) \geq -(1 - b_0)|y|^2 - c \quad \text{for } (x, y) \in Q \times R^m.$$

(H3) $H \colon \Omega \times R^m \to R^m$ is a Caratheodory function such that for some $h \in L_2(\Omega, R)$

$$|H(t, x, y)| \leq h(t, x) \quad \text{for } (t, x, y) \in \Omega \times R^m.$$

Let $H = L_2(\Omega, R^m)$ and let X_i, $i = 1, 2$, denote the closure of all T-periodic in t infinitely differential functions from $\Omega \to R^m$ satisfying the boundary conditions in (17.3) with respect to the norm

$$\|u\|_i = \int_0^T \int_Q \left(\sum_{k=0}^{i} (D_t^k u)^2 + \sum_{k=0}^{im} (D_x^k u)^2 \right) dx dt.$$

Let $A_1u = u_{tt} + L_1u$ for $u \in X_2$ and set

$$Au = \sum_{j,k} u_{j,k}(\mu_j - \tau^2k^2)\psi_j(x)e^{i\tau kt}, \quad u_{j,-k} = \bar{u}_{j,k} \quad \text{for}$$

$$u \in D(A) = \left\{ u = \sum_{j,k} u_{j,k}\psi_j(x)e^{i\tau kt} \,\middle|\, \sum_{j,k} |u_{jk}(\mu_j - \tau^2k^2)|^2 < \infty \right\}.$$

Then $A\colon D(A) \subset H \to H$ is a variational extension of A_1 and $(Au, v) = \langle A_1u, v\rangle$ for $u \in X_2$ and $v \in X_1$. Define $Gu = g(x, u)u_t$ for $u \in X_1$ and $Nu = F(x, u) + H(t, x, u)$ for $u \in D(N) = \{u \in X_1 \mid F(x, u), F(x, u)u \text{ are in } L_2(\Omega, R^m)\}$.

We say that a weak solution of (17.3) is a function $u \in X_1$ such that

$$(u, A_1v) + (Gu, v) + (Nu, v) = 0 \quad \text{for all } v \in X_2.$$

Hence, finding weak solutions of (17.3) reduces to solving the following operator equation

$$Au + Gu + Nu = 0 \quad u \in D(A).$$

Theorem 17.4: [83]: *Suppose that assumptions (H1)–(H3) hold and let* $\sigma(L) = \{\lambda_j > 0, j = 1, 2, \ldots\}$. *Then problem (17.3) has a weak solution* $u \in X_1$.

Regarding the solvability of (17.3) with F monotone and strongly nonlinear and $n = 1$, we refer to Plotnikov [92].

18. DOUBLE APPROXIMATION METHOD FOR SEMILINEAR EQUATIONS

Next, we shall prove a general solvability result for equation (12.8) using a double approximation of $A - N$. Assume that $A\colon D(A) \subset X \to Y$ is linear closed and densely defined map with $X_0 = \ker(A)$ infinite dimensional and the range $\tilde{Y} = R(A)$ closed. Suppose there are closed subspaces $\tilde{X}$ and Y_0 of X and Y such that $X = X_0 \oplus \tilde{X}$ and $Y = Y_0 \oplus \tilde{Y}$. Let $X_{0n} \subset X_0$ and $Y_{0n} \subset Y_0$ be finite dimensional subspaces such that $\bigcup_{n\geq 1} X_{0n}$ and $\bigcup_{n\geq 1} Y_{0n}$ are dense in X_0 and Y_0, respectively. Set $X_n = X_{0n} \oplus \tilde{X}$, $Y_n = Y_{0n} \oplus \tilde{Y}$ and let $Q_n\colon Y \to Y_n$ be a projection onto Y_n. Then $\Gamma = \{X_n, Y_n, Q_n\}$ is an approximation scheme for (X, Y) and $A_n - N_n = A \mid X_n - Q_nN\colon D(A) \cap X_n \to Y_n$ is an approximation of $A - N$, where A_n is a Fredholm map. In what follows, we shall assume that $A - N$ is pseudo A-proper w.r.t. Γ and that, for each n, $A_n - N_n\colon D(A) \cap X_n \to Y_n$ is (pseudo) A-proper w.r.t. a scheme $\Gamma_n = \{X_{n,k}, Y_{n,k}, Q_{n,k}\}$ for (X_n, Y_n). In this setting, we shall first solve the approximate equations

$$A_nx - N_nx = Ax - Q_nNx = Q_nf \quad (x \in X_n,\ f \in Y)$$

and then use the pseudo A-properness of $A - N$ to solve equation (12.9).

Our basic homotopy result in this setting is based on the following continuation result [63,64,81].

Theorem 18.1: *Let* $A\colon D(A) \subset X \to Y$ *be a linear densely defined map with closed range* $R(A)$, $N\colon X \to Y$ *be nonlinear and such that* $A - N$ *is pseudo A-proper w.r.t.*

$\Gamma = \{X_n, Y_n, Q_n\}$. *Let* $D \subset X$ *be a bounded and open subset and* $C: X \to Y$ *be a nonlinear map such that* $C(\bar{D})$ *is bounded and* $H(t,x) = Ax - (1-t)Cx - tNx$ *is such that for each* n *large,*

(i) $H_n(t,\cdot) = A - (1-t)Q_nC - tQ_nN$ *is an A-proper homotopy w.r.t.* $\Gamma_n = \{X_{n,k}, Y_{n,k}, Q_{n,k}\}$ *on* $[0,\epsilon] \times \partial D \cap D(A) \cap X_n$ *for each* $\epsilon \in (0,1)$ *and* $H_n(1,\cdot)$ *is pseudo A-proper w.r.t.* Γ_n;

(ii) $H_n(t,x)$ *is continuous at 1 uniformly for* $x \in \partial D \cap D(A) \cap X_n$;

(iii) $H_n(t,x) \neq Q_nf$ *and* $H_n(0,x) \neq tQ_nf$ *for* $t \in [0,1]$, $x \in \partial D \cap D(A) \cap X_n$;

(iv) $\deg(Q_{n,k}H_n(0,\cdot), D \cap X_{n,k}, 0) \neq 0$ *for all k.*

Then equation (12.9) is solvable in $\bar{D} \cap D(A)$.

The next results give some classes of maps C and N to which Theorem 18.1 is applicable.

Theorem 18.2: *Let H be a Hilbert space,* $A: D(A) \subset H \to H$ *be linear densely defined with closed range* $R(A)$ *and compact partial inverse and the null space* $N(A) = N(A^*)$. *Let* $N: H \to H$ *be bounded and of type (M). Let* $H_{0n} \subset \ker(A)$ *be finite dimensional such that* $\cup_{n\geq 1} H_{0n}$ *is dense in* $\ker A$, $H_n = H_{0n} \oplus R(A)$ *and* $P_n: H \to H_n$ *be orthogonal projections. Let* $D \subset H$ *be an open and bounded subset and, for all large* n,

$$P_nAx - P_nNx = P_nf \quad \text{for some } x \in \bar{D} \cap H_n \cap D(A).$$

Then equation (12.9) is solvable in $\bar{D}$.

Theorem 18.3: *Let* A: $D(A) \subset H \to H$ *be a closed linear densely defined map with* $R(A) = \ker A^{\perp}$ *and either* A^{-1}: $R(A) \to R(A)$ *is compact or* $A = A^*$, $0 \in \sigma(A)$ *and* $\sigma(A) \cap (0,\infty) \neq \emptyset$ *and consists of isolated eigenvalues having finite multiplicities. Let* N: $H \to H$ *be bounded and generalized pseudo monotone,* C: $H \to H$ *be bounded and of type* (S_+) *and (P) and* $D \subset H$ *be a bounded and open subset. Let* H_{0n} *be a finite dimensional subspace of* $H_0 = \ker(A)$, $H_n = H_{0n} \oplus R(A)$, P_n: $H \to H_n$ *be the orthogonal projection,* $\Gamma = \{H_n, P_n\}$ *and* $\Gamma_n = \{H_{n,k}, P_{n,k}\}$ *be a scheme for* H_n *with* $P_{n,k}Ax = Ax$ *on* $H_{n,k}$. *Suppose that for some* $f \in H$

(i) $H(t,x) = Ax - (1-t)Cx - tNx \neq f$ *and* $H(0,x) \neq tf$ *for* $t \in [0,1]$ $x \in \partial D \cap D(A)$

(ii) *For each large* n, $\deg(P_{n,k}(A - P_nC), D \cap H_{n,k}, 0) \neq 0$ *for all k.*

Then equation (12.9) is solvable in $\bar{D} \cap D(A)$.

REFERENCES

[1] H. Amann (1982). On the unique solvability of semilinear operator equations in Hilbert spaces, *J. Math. Pures Appl.*, **61**, 149–175.

[2] H. Amann and E. Zehnder (1980). Nontrivial solutions for a class of nonresonance problems and applications to nonlinear differential equations, *Annali Scuola Norm. sup. Pisa, Ser. IV*, **7**, 539–603.

[3] J.P. Aubin and I. Ekeland (1984). *Applied Nonlinear Analysis*. New York: J. Wiley and Sons.

[4] J.P. Aubin and H. Frankowska (1987). On the inverse function theorems for set valued maps, *J. Math. pures et appl.*, **66**, 71–89.

[5] A.K. Ben-Naoum and J. Mawhin (1992). The periodic-Dirichlet problem for some semilinear wave equations, *J. Differential equations*, **96**, 340–354.

[6] A.K. Ben-Naoum and J. Mawhin (1993). Periodic solutions of some semilinear wave equations on balls and on spheres, *Topological Methods in Nonlinear Analysis*, **1**, 113–137.

[7] M.S. Berger (1977). *Nonlinearity and Functional Analysis*, New York: Academic Press.

[8] H. Brezis (1968). Equationes et inequation es non-lineaires dans las espaces vetorieles en dualits, *Ann. Inst. Fourier* (Grenoble), **18**, 115–175.

[9] H. Brezis and L. Nirenberg (1978). Characterizations of the ranges of some nonlinear operators and applications to boundary value problems, *Ann. Scuola Norm. Sup. Pisa*, **59**, 225–326.

[10] H. Brezis and L. Nirenberg (1978). Forced vibrations for a nonlinear wave equation, *Comm. Pure Appl. Math.*, **31**, 1–30.

[11] F.E. Browder (1973). Existence theory for boundary value problems for quasilinear elliptic systems with strongly nonlinear lower order terms, in *Proc. Symp. Pure Math.*, **23**, 269–286, *Amer. Math. Soc.*, Providence, R.I.

[12] F. Browder (1976). Nonlinear operators and nonlinear equations of evolution in Banach spaces, *Proc. Symp. Pure Math.* **18**(2), AMS.

[13] F.E. Browder (1986). Degree theory for nonlinear mappings, *Proc. Symposia in Pure Mathematics*, AMS, **45**(1), 203–226.

[14] F.E. Browder and P. Hess (1972). Nonlinear mappings of monotone type in Banach spaces, *J. Funct. Anal.*, **11**, 251–294.

[15] B. Calvert and J.R.L. Webb (1971). An existence theorem for quasimonotone operators, *Rend. Accad. Naz. Lincei*, **8**, 362–368.

[16] L. Cesari (1976). Functional analysis, nonlinear differential equations and the alternative method, in Lecture Notes in *Pure and Appl. Math.*, **19**, 1–197 N.Y.: M. Dekker.

[17] E.N. Dancer (1984). Order intervals of self-adjoint linear operators and nonlinear homeomorphisms, *Pacific J. Math.*, **115**, 57–72.

[18] K. Deimling (1985). *Nonlinear Functional Analysis*, Berlin: Springer.

[19] C. Dolph (1949). Nonlinear integral equations of Hammerstein type, *Trans. Amer. Math. Soc.*, **66**, 289–307.

[20] Yu.A. Dubinski (1976). Nonlinear elliptic and parabolic equations, *Itogi nauki i texniki. VINITI*, **9**, 5–130.

[21] J. Dugundji and A. Granas (1982). *Fixed Point Theory*, Warsaw.

[22] P.M. Fitzpatrick (1978). Existence results for equations involving noncompact perturbations of Fredholm mappings with applications to differential equations, *J. Math. Anal. Appl.*, **61**, 157–165.

[23] P.M. Fitzpatrick (1974). On the structure of the set of solutions of equations involving A-proper mappings, *Trans. Amer. Math. Soc.*, **189**, 107–131.

[24] A. Fonda and J. Mawhin (1992). Iterative and variational methods for the solvability of some semilinear equations in Hilbert spaces, *J. Diff. Eq.*

[25] S. Fucik (1978). *Ranges of nonlinear operators*, Lecture Notes, 5 volumes, Univ. Carolinae Prangensis, Praha.

[26] R.S. Gaines and J.L. Mawhin (1977). *Coincidence Degree and Nonlinear Differential Equations*, Lecture Notes 568, Berlin: Springer-Verlag.

[27] G.M. Goncarov (1970). *On some existence theorems for the solutions of a class of nonlinear operator equations*, Math. Notes, **7**, 137–141.

[28] J.R. Gossez (1974). Nonlinear elliptic boundary value problems for equations with rapidly (or slowly) increasing coefficients, *Trans. Amer. Math. Soc.*, **190**, 163–206.

[29] Z. Guan (1994). Solvability of operator equations with compact perturbations of operators of monotone type, *Proc. AMS*, **121**, 93–102.

[30] B. Halpern and G. Bergman (1968). A fixed point theorem for inward and outward maps, *Trans. AMS*, **130**, 353–358.

[31] J.D. Hamilton (1972). Noncompact mappings and cones in Banach spaces, *Archives Rational Mech. Anal.*, **48**, 153–162.

[32] P. Hess (1972a). On the Fradholm alternative for nonlinear functional equations in Banach spaces, *Proc. Amer. Math. Soc.*, **33**, 55–62.

[33] P. Hess (1972b). On nonlinear mappings of monotone type homotopic to odd operators, *J. Funct. Anal.*, **11**, 138–167.

[34] P. Hess (1973). On nonlinear mappings of monotone type with respect to two Banach spaces, *J. Math. Pures Appl.*, **52**, 13–26.

[35] G. Hetzer (1980). A continuation theorem and the variational solvability of quasilinear elliptic boundary value problems at resonance, *J. Nonlinear Analysis, TMA*, **4**(4), 773–780.

[36] J. Jarusek (1979). Ranges of non-linear operators in Banach spaces, *Math. Nachr.*, **92**, 203–210.

[37] J. Jarusek and J. Necas (1977). Sur les domaines des valeurs des operateurs non-lineaires, *Casopis pro Pestovani Mat.*, **102**, 61–72.

[38] R.I. Kachurovskii (1971). Generalizations of the Fredholm theorems and of theorems on linear operators with closed range to some classes of nonlinear operators, *Soviet Math. Dokl.*, **12**, 487–491.

[39] S. Kaniel (1965). Quasi-compact non-linear operators in Banach spaces and applications, *Arch. Ration. Mech. Anal.*, **20**, 259–278.

[40] T. Kato (1984). Locally coercive nonlinear equations with applications to some periodic solutions, *Duke Math. J.*, **51**, 923–936.

[41] M.A. Krasnosel'skii (1964). *Topological Methods in the Theory of Nonlinear Integral equations*, N.Y.: MacMillan.

[42] M.A. Krasnoselskii and P.O. Zabreiko (1984). *Geometrical methods of Nonlinear Analysis*, Springer-Verlag.

[43] J.M. Lasry and R. Robert (1974–75). *Degre at theorems de point fixe pour les functions multivoques; applications, Sominaire Goulaovic–Lions–Schwartz*, Ecole Polytechnique, Paris, Cedex 05.

[44] M. Lees and M.H. Shultz (1966). A Leray-Schauder principle for A-compact mappings and the numerical solution of non-linear two-posnt boundary value problems, Numerical Solutions of Nonlinear Differential Equations, *Proc. Adv. Sympos., Madison, Wis.*, 1966, New York: Wiley, pp. 167–179.

[45] J. Leray and J.L. Lions (1965). Quelques resultats de Visik sur les problemes elliptiques non lineaires par les methodes de Minty-Browder, *Bull. Sec. Math. France*, **93**, 97–107.

[46] J.L. Lions (1969). *Quelques methodes de resolution des problemes aux limites non lineaires, Dunod*; Paris: Gauthier-Villars.

[47] T.W. Ma (1972). Topological degrees for set valued compact vector fields in locally convex spaces, *Dissartationes Math.*, **92**, 1–43.

[48] J. Mawhin (1979). *Topological degree methods in nonlinear boundary value problems*, Regional Conference Series in Math., **40**, AMS, Providence, R.I.

[49] J. Mawhin (1981). Conservative systems of semi-linear wave equations with periodic-Dirichlet boundary conditions, *J. Diff. Equations*, **42**, 116–128.

[50] J. Mawhin (1982). Nonlinear functional analysis and periodic solutions of semilinear wave equations, in Lakshmikantham (Ed.), *Nonlinear Phenomena in Math. Sci.*, Arlington, NY: Acad. Press, 671–681.

[51] J. Mawhin (1981). Semilinear equations of gradient type in Hilbert spaces and applications to differential equations, in *Nonlinear Differential Equations: Invariance, Stability and Bifurcation*, NY: Acad. Press, 269–282.

[52] J. Mawhin (1981). *Compacticité, monotonie et concexité dans l'etude des problmes aux limites semi-linéaires*, Séminaire d'analyse moderne, no. 19, Univ. de Sherbrooke.

[53] J. Mawhin and K.F. Rybakovski (1987). Continuation theorems for semi-linear equations in Banach spaces: a survey. In Th.M. Rassias (Ed.), *Nonlinear Analysis*, Singapore: World Scientific Publ. Co., 367–405.

[54] P.S. Milojević (1975). Multivalued maps of A-proper and condensing type and boundary value problems, Ph.D. Thesis, Rutgers University, 1–212.

[55] P.S. Milojević (1976). Some generalizations of the first Fredholm theorem to multivalued condensing and A-proper mappings, *Boll. Unione Mat. Ital.*, **13–B**, 619–633.

[56] P.S. Milojević (1977a). *Surjectivity results for A-proper, their uniform limits and pseudo A-proper maps with applications*, Notices AMS (January), 77T–B27.

[57] P.S. Milojević (1977b). A generalization of the Leray-Schauder theorem and surjectivity results for multivalued A-proper and pseudo A-proper mappings, *Nonlinear Anal., TMA*, **1**(3), 263–276.

[58] P.S. Milojević (1977c). On the solvability and continuation type results for nonlinear equations with applications, *Proc. 3rd Intern. Symposium on Topology and Its Applications*, Belgrade, 468–485.

[59] P.S. Milojević (1978). Some generalizations of the first Fredholm theorem to multivalued A-proper mappings with applications to nonlinear elliptic equations, *J. Math. Anal. Appl.*, **65**(2), 468–502.

[60] P.S. Milojević (1979a). The solvability of operator equations with asymptotic quasibounded nonlinearity, *Proc. AMS*, **76**, 293–298.

[61] P.S. Milojević (1979b). Fixed point theorems for multivalued approximable mappings, *Proc. AMS*, **73**, 65–72.

[62] P.S. Milojević (1980a). Fredholm alternatives and surjectivity results for multivalued A-proper and condensing mappings with applications to nonlinear integral and differential equations, *Czech. Math. J.*, **30**(105), 387–417.

[63] P.S. Milojević (1980b). Teoria de Aplicacoes A-proprias e pseudo A-proprias, Habilitation Memoir, Univer. Federal de Minas Gerais, Belo Horizonte, Brasil, 1–208.

[64] P.S. Milojević (1980c). Continuation theorems and solvability of equations involving nonlinear noncompact perturbations of Fredholm mappings, *Proc. 12th Seminario Brasileiro de Analize*, ITA, Sao Jose dos Campos, 163–189.

[65] P.S. Milojević (1982a). On the solvability and continuation type results for nonlinear equations with applications, II, *Canadian Math. Bull.*, **25**, 98–109.

[66] P.S. Milojević (1982b). Solvability of operator equations involving nonlinear perturbations of Fredholm mappings of nonnegative index, *Differential Equations*, Springer Verlag Lecture Notes in Math., **957**, 212–228.

[67] P.S. Milojević (1982c). Continuation theory for A-proper and strongly A-closed mappings and their uniform limits and nonlinear perturbations of Fredholm mappings, in A. Barroso (Ed.), *Proc. Int. Seminar in Funct. Anal., Holomorphy and Approx. Theory* (Rio de Janeiro, August 1980), Math. Studies, **71**, North-Holland, 299–372.

[68] P.S. Milojević (1983a). Approximation-solvability results for equations involving nonlinear perturbations of Fredholm mappings with applications to differential equations, in G. Zapata (Ed.), *Proc. Int. Sem. Funct. Anal. Holom. Approx. Theory* (Rio de Janeiro, August 1979), Lecture notes in Pure and Appl. Math., **83**, NY: M. Dekker, 305–358.

[69] P.S. Milojević (1984). Approximation solvability of some noncoercive nonlinear equations and semilinear problems at resonance, in G.I. Zapata (Ed.), *Functional Anal. Holomorphy and Approx. Theory, II*, North-Holland: Elsevier Science Pub., 259–295.

[70] Mi $(1984)_2$

[71] P.S. Milojević (1986a). Quelques resultats de point fixe et de surjectivite pour des applications A-propres, C. R. Acad. Sc. Paris, t. 303, Serie 1, no. 2, 49–52. Abstracts AMS 32, 1984, p. 278.

[72] P.S. Milojević (1986b). On the index and the covering dimension of the solution set of semilinear equations, *Proc. Symp. Pure Math.*, **45**(2), *Amer. Math. Soc.*, 183–205. Abstracts AMS, **31**, 1984, p. 243.

[73] P.S. Milojević (1987a). Fredholm theory and semilinear equations without resonance involving noncompact perturbations, I, II, Applications, *Publications de l'Institut Math.*, **42**, 71–82 and 83–95.

[74] P.S. Milojević (1987b). Solvability of some semilinear equations with strong nonlinearities and applications to elliptic problems, *Applicable Analysis*, **25**(3), 181–196.

[75] P.S. Milojević (1988). Solvability of nonlinear operator equations with applictions to hyperbolic equations, in B. Stankovic *et al.* (Eds), *Proc. Generalized Funct. Convergence Structures and their Applic.*, NY: Plenum Press, 245–250.

[76] P.S. Milojević (1989a). Continuation theory for semilinear operator equations, Facta Universitatis, *Ser. Math. Inform.*, **4**, 63–71.

[77] P.S. Milojević (1989b). Solvability of semilinear hyperbolic equations at resonance, in A.R. Aftabizadeh (Ed.), *Proc. Diff. equation and Applic.* Ohio University Press, pp. 216–221.

[78] P.S. Milojević (1989c). Solvability of semilinear operator equations and applications to semilinear hyperbolic equations, in P.S. Milojević (Ed.), *Nonlinear Functoinal Analysis*, Marcel Dekker, **121**, 95–178.

[79] P.S. Milojević (1990a). Solvability of operator equations and periodic solutions of semilinear hyperbolic equations, *Publication de l'Institut de Mathematique*, 133–168.

[80] P.S. Milojević (1990b). Periodic solutions of semilinear hyperbolic equations at resonance, *J. Math, Anal. Appl.*, **146**, 546–569. Abstracts AMS, January 1986, T825–47–106.

[81] P.S. Milojević (1992). Nonlinear Fredholm theory and applications, in J. Wiener and J. Hale (Eds), *Partial Differential Equations*, Pitman Research Notes in Math., **273**, 133–152.

[82] P.S. Milojević (1994a). Approximation-solvability of nonlinear equations and applications, in W. Bray, P.S. Milojević and C.V. Stanojević (Eds.), *Fourier Analysis*, Lecture Notes in Pure and Applied Mathematics, **157**, 311–373, NY: Marcel Dekker Inc.

[83] P.S. Milojević (1995b). Solvability of strongly nonlinear operator equations and applications, *Differential Equations*, **31**(3), 502–516 (in Russian). English translation: *Diff. Equations*, **31**(3), 466–479.

[84] P.S. Milojević (1995). On the dimension and the index of the solution set of nonlinear equations, *Transactions Amer. Math. Soc.*, **347**(3), 835–856.

[85] P.S. Milojević (1996). Approximation-solvability of semilinear equations and applications. In A.G. Kartsatos (Ed.), *Theory and Applications of Nonlinear Operators of Accretive and Monotone Type*, Lecture Notes in Pure and Applied Math., **178**, 149–208, NY: M. Dekker.

[86] P.S. Milojević (1997). Implicit function theorems, approximation solvability of nonlinear equations, and error estimates, *J. Math. Anal. Applic.*, **211**, 424–459. Abstracts AMS, **13**, (January 1992), pp. 94–95.

[87] J. Necas (1972). Fredholm alternative for nonlinear operators and applications to partial differential equations and integral equations, *Cas. Pest. Mat.*, **9**, 15–21.

[88] L. Nirenberg (1971). An application of generalized degree to a class of nonlinear problems, *Troisieme Colloque SBRM d'analyse fonctionelle*, Vander, Louvain, 57–74.

[89] L. Nirenberg (1974). *Topics in nonlinear functional analysis*, Courant Institute Lecture Notes.

[90] R.D. Nussbaum (1969). The fixed point index and fixed point theorems for k-set contractions, Ph.D. Dissartaion, Univ. of Chicago, Chicago, **111**.

[91] W.V. Petryshyn (1975). On the approximation-solvability of equations involving A-proper and pseudo A-proper mappings, *Bull. Amer. Math. Soc.*, **81**, 223–312.

[92] P.I. Plotnikov (1988). Existence of a countable set of periodic solutions of the problem of forced oscilations for a weakly nonlinear wave equation, *Matem. Sbornik*, **136**, 543–556.

[93] S.I. Pohozaev (1967). On the solvability of nonlinear equations with odd operators, *Functional. Anal. Appl.*, **1**, 66–73.

[94] S.I. Pohozaev (1979). On an approach to nonlinear equations, *Dokl. Akad. Nauk. SSSR*, **247**, 1327–1331.

[95] S.I. Pohozaev (1990). *On the method of fibering a solution in nonlinear boundary value problems*, Trudi Matem. Inst. Steklova, **192**.

[96] K. Schmitt (1976). Approximate solutions of boundary value problems for systems of nonlinear differential equations, *Proc. S. Lefschetz Conf.*, Mexico City, 1975, Math. Notes and Symposia, **2**, 345–354.

[97] M.W. Smiley (1987). Eigenfunction methods and nonlinear hyperbolic boundary value problems at resonance, *J. Math. Anal. Appl.*, **122**, 129–151.

[98] M.W. Smiley (1987). Time periodic solutions of nonlinear wave equations in balls, in Oscillations, Bifurcation and Chaos (Toronto 1986), *Canad. Math. Soc. Confer. Proc.*, 287–297.

[99] S.A. Tersian (1986). A minimax theorem and applications to nonresonance problems for semilinear equations, *Nonl. Anal. TMA*, **10**, 651–668.

[100] J.F. Toland (1977). Global bifurcation theory via Galerkin method, *Nonlinear Anal.*, **1**, 305–317.

[101] G. Vainikko (1866). On the convergence of the collocation method for nonlinear differential equations, *Z. Vychisl. Mat. Math. Fiz.*, **6**, 35–42.

SIMAA 4(2002) 361–368

20. Existence Theorems for Strongly Accretive Operators in Banach Spaces

Claudio H. Morales

Department of Mathematics, University of Alabama in Huntsville, Huntsville, AL 35899, USA

1. INTRODUCTION

Let X be a (real) Banach space and let α: $[0, \infty) \to [0, \infty)$ be a function for which $\alpha(0) = 0$ and the $\liminf_{r \to r_0} \alpha(r) > 0$ for every $r_0 > 0$. An operator A: $D(A) \subset X \to 2^X$ is called *α-strongly accretive* if for each $x, y \in D(A)$ there exists $j \in J(x - y)$ such that

$$\langle u - v, j \rangle \geq \alpha(\| x - y \|) \| x - y \| \tag{1}$$

for $u \in A(x)$ and $v \in A(y)$, where J: $X \to 2^{X^*}$ is normalized duality mapping which is defined by

$$J(x) = \{j \in X^*: \langle x, j \rangle = \| x \|^2, \| j \| = \| x \| \}.$$

For $\alpha(r) = kr$, with $0 < k < 1$, the mapping A is known to be called, simply, *strongly accretive*, and while α is the zero function, A is called *accretive*. In the latter case, if in addition, the range of $A + \lambda I$ is precisely X for all $\lambda > 0$, then A is said to be *m-accretive*. On the other hand, if A is strongly accretive (accretive), then $I - A$ is known to be called *strongly pseudo-contractive* (*pseudo-contractive*). Also, a mapping A is said to be *locally α-strongly accretive* if for each $x \in D(A)$, there exists a neighborhood $N(x)$ where A is globally α-strongly accretive.

In 1967, Browder [2] proved that every Lipschitz and accretive mapping A from X into X is m-accretive. Later in 1970, Martin [6], extended this result to general continuous mappings. However, much later in 1985, this writer (see [7]) extended the results to multi-valued mappings. Most recently, Chen [4] proved a multi-valued version of this very result under the additional assumption that $A(x)$ must be weakly compact, via using differential equations argument. It is our objective here to present some of the results obtained in [7] and [8], which appear to be unknown for a wider audience. Nevertheless, we intend to preserve an elementary argument approach. However, in this case we add simpler and shorter proofs of various results which are intimately related to the above-mentioned Browder's Theorem.

We begin with the definition of the Hausdorff metric H as it was stated in Assad and Kirk [1]. Let $\epsilon > 0$ and K be a bounded subset of X. Then

$$N_\epsilon(K) = \{x \in X : \text{dist}(x, K) < \epsilon\}.$$

denotes an ϵ-neighborhood of K. Suppose K and L are bounded subsets of X, we then define the Hausdorff metric H by

$$H(K,L) = \inf\{\epsilon : K \subset N_\epsilon(L) \quad \text{and} \quad L \subset N_\epsilon(K)\}.$$

Using this notion, we say that a mapping T from D into X is said to be *continuous at* x_0 if for every $\epsilon > 0$ there exits $\delta > 0$ such that

$$H(T(x), T(x_0)) < \epsilon \quad \text{whenever} \quad \| x - x_0 \| < \delta.$$

Additionally, T is said to be *closed* if maps closed set onto closed sets.

Throughout this chapter we use $\bar{K}$ and ∂K to denote, respectively, the closure and the boundary of K. For $u, v, \in X$, we use seg$[u, v]$ to denote the segment $\{tu + (1-t)v : t \in [0,1]\}$. Also, we use $|K|$ to denote $\inf\{\| x \| : x \in K\}$.

2. PRELIMINARIES

We assume that X is a real Banach space and $B(X)$ denotes the metric space of all nonempty bounded and closed subsets of X provided with the Hausdorff metric H. Before we state our main results, we need some basic facts which will be used in the coming proofs. We begin with a result due to Kirk [5].

Theorem K: (Theorem 3.1 of [5]). *Let X be a Banach space and let D be an open subset of X. Suppose A: $D \to B(X)$ is continuous (relative to the Hausdorff metric) and strongly accretive. Then $A(D)$ is open in X.*

Lemma 1: *If $K, L \in B(X)$ and $x \in K$, then for each positive number ϵ, there exists $y \in L$ such that*

$$\| x - y \| \le H(K, L) + \epsilon.$$

Lemma 2: *Let D be a subset of a Banach space X with $0 \in D$, and let A: $D \to B(X)$ be an α-strongly accretive mapping (with $\liminf_{r\to\infty} \alpha(r) > |A(0)|$). Then:*

(*i*) *the set $E = \{x \in D : tx \in A(x)$ for some $t < 0\}$ is bounded;*

(*ii*) *if $\{x_n - u_n\}$ is a bounded sequence in X for $u_n \in A(x_n), t_n \to t$ with $t_n \in [0,1]$, and $z_n = (1-t_n)x_n + t_n u_n \to z$, then $\{x_n\}$ is a Cauchy sequence.*

Proof: (i) Let $tx \in A(x)$ for some $t < 0$. Select $u \in A(x)$ such that $u = tx$. Then there exists $j \in J(X)$ for which

$$\alpha(\| x \|) \| x \| \le \langle tx - v, j \rangle$$

for all $v \in A(0)$. Therefore $\alpha(\| x \|) \le \| v \|$ for all $v \in A(0)$, which implies that E is bounded.

(ii) Choose $j \in J(x_n - x_m)$. Then by (1) we have

$$\begin{aligned}
\langle z_n - z_m, j \rangle &= \langle (1-t_n)x_n - (1-t_m)x_m, j \rangle + \langle t_n u_n - t_m u_m, j \rangle \\
&= (1-t_n)\langle x_n - x_m, j \rangle + (t_m - t_n)\langle x_m - u_m, j \rangle + t_n \langle u_n - u_m, j \rangle \\
&\ge (1-t_n)\langle x_n - x_m, j \rangle + (t_m - t_n)\langle x_m - u_m, j \rangle \\
&\quad + t_n \alpha(\| x_n - x_m \|) \| x_n - x_m \|,
\end{aligned}$$

which implies that

$$(1-t_n)\,\|x_n - x_m\| + t_n\alpha(\|x_n - x_m\|) \leq |t_m - t_n|\,\|x_m - u_m\| + \|z_n - z_m\|\,.$$

Therefore $\{x_n\}$ is a Cauchy sequence.

Lemma 3: *Let K be a closed subset of a Banach space X and let T: $K \to B(X)$ be continuous. Suppose $h_t(x) = (1-t)x + tT(x)$ for $t \in (0,1]$ and $z_n \in h_{t_n}(x_n)$ where $z_n \to z, t_n \to t_0$ and $x_n \to x_0$. Then $z \in h_{t_0}(x_0)$.*

Proof: Since T is continuous, for a given $\epsilon > 0$ there exists $N \in \mathbb{N}$ such that

$$H(T(x_n), T(x_0)) < \epsilon/2 \quad \text{for all } n \geq N. \tag{2}$$

Since $z_n \in h_{t_n}(x_n)$, we may choose $u_n \in T(x_n)$ so that $z_n = (1-t_n)x_n + t_n u_n$. Moreover, by Lemma 1, we may select $v_n \in T(x_0)$ satisfying

$$\|u_n - v_n\| \leq H(T(x_n), T(x_0)) + \epsilon/2. \tag{3}$$

Let $w_n = (1-t_0)x_0 + t_0 v_n$ for each n, then

$$\begin{aligned}\|z_n - w_n\| &= \|(1-t_n)x_n + t_n u_n - [(1-t_0)x_0 + t_0 v_n]\| \\ &\leq |1-t_n|\,\|x_n - x_0\| + |t_0 - t_n|\,\|x_0 - u_n\| + t_0\,\|u_n - v_n\|\,.\end{aligned}$$

By making use of (2) and (3), we get

$$\|z_n - w_n\| \leq |1-t_n|\,\|x_n - x_0\| + |t_0 - t_n|\,\|x_0 - u_n\| + \epsilon \tag{4}$$

for all $n \geq N$. Due to (2), the sequence $\{u_n\}$ is bounded, and hence by letting $n \to \infty$ in (4) we conclude

$$\limsup_{n\to\infty} \|w_n - z\| \leq \epsilon.$$

Since ϵ is arbitrary and $w_n \in h_{t_0}(x_0)$ for all n, the sequence $\{w_n\}$ converges to z, and hence $z \in h_{t_0}(x_0)$.

3. MAIN RESULTS

We are now ready to state one of our main results.

Theorem 1: *Let X be a (real) Banach space and let D be an open subset of X. Suppose A: $\bar{D} \to B(X)$ is a continuous (relative to the Hausdorff metric) and α-strongly accretive mapping, (with $\liminf_{r\to\infty} \alpha(r) > |A(0)|$) which satisfies for some $z \in D$*

$$t(x-z) \notin A(x) \quad \text{for } x \in \partial D \quad \text{and} \quad t < 0. \tag{5}$$

Then there exists a unique $x \in \bar{D}$ such that $0 \in A(x)$.

Proof: By replacing $A(x)$ with $A(x+z)$ and D by $D-z$, one may select $z=0$ in (5). Since the set E (defined in Lemma 2) is bounded, there is no loss of generality in assuming D is bounded.

Let h_t: $\bar{D} \to B(X)$ be defined by $h_t(x) = (1-t)x + tA(x)$ for each $t \in [0,1]$, and let

$$M = \{t \in [0,1] : 0 \in h_t(x) \text{ for some } x \in D\}.$$

Clearly, $M \neq \phi$ since $0 \in M$. Our goal is to show that $1 \in M$. To see this, let $\{t_n\}$ be a sequence in M with $t_n \to t$ as $n \to \infty$. Then, for each n, there exists $x_n \in D$ so that $0 \in h_{t_n}(x_n)$. This means, we may select $u_n \in A(x_n)$ for which $(1-t_n)x_n + t_n u_n = 0$. Hence, by Lemma 2(ii), we derive that $\{x_n\}$ is a Cauchy sequence in D. Therefore $x_n \to x \in \bar{D}$, and this combined with Lemma 3 implies that $0 \in (1-t)x + tA(x)$. However, by assumption (5) we obtain that $x \in D$. Therefore, M is closed in $[0,1]$.

On the other hand, suppose M is not open. Then there exits $t \in M$ $(t < 1)$ and a sequence $\{t_n\}$ in $[0,1)$ for which $t_n \notin M$ and $t_n \to t$. Then $0 \in h_t(x_0)$ for some $x_0 \in D$ and by selecting $u_0 \in A(x_0)$ we have $(1-t)x_0 + tu_0 = 0$. Now due to the continuity of A we may choose a closed ball B centered at x_0 and contained in D such that

$$H(A(x), A(x_0)) < 1 \quad \text{for all } x \in B. \tag{6}$$

Define $y_n = (1-t_n)x_0 + t_n u_0$ for each $n \in \mathbb{N}$. Then

$$y_n \in h_{t_n}(x_0) \subset h_{t_n}(B),$$

while $0 \notin h_{t_n}(B)$. This implies that there exists $u_n \in [0, y_n] \cap \partial h_{t_n}(B)$. Since h_{t_n} is strongly accretive for each $n \in \mathbb{N}$, Theorem K implies that $h_{t_n}(\text{int}(B))$ is open, while by (1) $h_{t_n}(B)$ is closed. Hence, we may derive that $\partial h_{t_n}(B) \subset h_{t_n}(\partial B)$, which yields to the existence of a point $x_n \in \partial B$ so that $u_n \in h_{t_n}(x_n)$. Since $y_n \to 0$, so does $\{u_n\}$, and thus Lemma 2(ii) combined with (6) imply that $\{x_n\}$ is a Cauchy sequence which must converge to some $\bar{x} \in \partial B$. Therefore, by Lemma 3, $0 \in h_t(\bar{x})$. This is a contradiction! Therefore, M is open and the proof is complete.

As a consequence of Theorem 1, we can derive the following corollaries.

Corollary 1: *Let X be a (real) Banach space and let D be a bounded open subset of X. Suppose A: $\bar{D} \to B(X)$ is a continuous (relative to the Hausdorff metric) and accretive mapping which satisfies for some $z \in D$.*

$$t(x-z) \notin A(x) \quad \textit{for } x \in \partial D \quad \textit{and} \quad t < 0. \tag{7}$$

Then $\inf\{|A(x)|: x \in \bar{D}\} = 0$.

Proof: Without loss of generality, we may assume that $z=0$ in (7). For $\lambda > 0$, the mapping $A_\lambda(x) = \lambda x + A(x)$ is strongly accretive and satisfies (7). Hence, by Theorem 1, $0 \in A_\lambda(x)$ for some $x \in \bar{D}$. Select $\lambda_n \to 0^+$ as $n \to \infty$. Then there exists $x_n \in \bar{D}$ such that $0 \in \lambda_n x_n + A(x_n)$. Since D is bounded, $|A(x_n)| \to 0$, which completes the proof.

Corollary 2: *Let X be a (real) Banach space and let D be a bounded open subset of X. Suppose A: $\bar{D} \to B(X)$ is a continuous (relative to the Hausdorff metric) and accretive mapping which satisfies for some $z \in D$*

$$|A(z)| < |A(x)| \quad \textit{for } x \in \partial D. \tag{8}$$

then $\inf\{|A(x)| : x \in \bar{D}\} = 0$.

Proof: We shall show that, indeed, the assumption (7) can be derived from (8). To this end, suppose there exists $x \in \partial D$ so that $t(x-z) \in A(x)$ for some $t < 0$. Then there exists $j \in J(x-z)$ such that

$$\langle t(x-z) - v, j\rangle \geq 0 \quad \text{for } v \in A(z).$$

This implies that $-t \parallel x - z \parallel \leq \parallel v \parallel$ for all $v \in A(z)$, and thus $|A(x)| \leq |A(z)|$, which is a contradiction. Therefore (7) holds, and the proof is complete.

Corollary 3: *Let X be a (real) Banach space and let K be a closed ball in X. Let T: $K \to B(K)$ be a continuous and strongly pseudo-contractive mapping. Then there exists $x_0 \in K$ such that $x_0 \in T(x_0)$.*

We are now ready to prove our result concerning m-accretive operators, while the operator A takes, simply, closed and bounded values on X.

Theorem 2: *Let X be a (real) Banach space and let A: $X \to B(X)$ be a continuous (relative to the Hausdorff metric) and accretive mapping. Then A is m-accretive.*

Proof: Let $\lambda > 0$ and let z be an arbitrary element in X. Define the mapping A_z: $X \to B(X)$ by $A_z(x) = \lambda x + A(x) - z$. Then A_z is strongly accretive on X. We now show that the set

$$E(z) = \{x \in X : tx \in A_z(x) \quad \text{for some } t < 0\}$$

is bounded. To see this, let $tx \in A_z(x)$ for some $t < 0$ and select $y \in A(x)$ such that $tx = \lambda x + y - z$. Then

$$\langle y - v, j\rangle = \langle (t - \lambda)x + z - v, j\rangle \geq 0$$

for some $j \in J(x)$ and for all $v \in A(0)$. This implies that

$$(\lambda - t) \parallel x \parallel^2 \leq |A(0)| + \parallel z \parallel .$$

Therefore $E(z)$ is bounded. Due to this latter fact, we may select $r > 0$ large enough such that the closure of $E(z)$ is contained in the open ball $B(0; r)$. This means the mapping A_z satisfies the following boundary condition:

$$t(x-z) \notin A(x) \quad \text{for } x \in \partial B(0;\, r) \quad \text{and} \quad t < 0.$$

Therefore, by Theorem 1, there exists $x \in \bar{B}(0; r)$ such that $0 \in A_z(x)$, i.e., $z \in \lambda x + A(x)$.

Theorem 3: *Let X be a (real) Banach space and let D be an open subset of X. Suppose A: $D \to B(X)$ is a continuous (with respect to the Hausdorff metric), locally closed, locally one-to-one and locally accretive mapping. Then $A(D)$ is open in X.*

Proof: Let $x_0 \in D$ and let $y_0 \in A(x_0)$. Then due to the local assumptions on A, we may select a closed ball $\bar{B}(x_0; r)$ contained in D such that A is globally accretive, closed and one-to-one on $\bar{B}(x_0; r)$. Then the number

$$\delta = \inf\{|A(x) - y_0| : \| x - x_0 \| = r\} > 0.$$

We now choose $\eta > 0$ so that $\eta(1 + r) < \delta$. For $\lambda \in (0, \eta)$, define the mapping h_t: $\bar{B}(x_0; r) \to B(X)$ by $h_t(x) = \lambda(x - x_0) + A(x) - [ty + (1 - t)y_0]$ for $t \in [0, 1]$ and $y \in \bar{B}(y_0; \eta)$. In addition, we can easily see that for $x, y \in \bar{B}(x_0; r)$, there exists $j \in J(x - y)$ such that

$$\langle u - v, j \rangle \geq \lambda \| x - y \|^2 \quad \text{for } u \in h_t(x) \quad \text{and} \quad v \in h_t(y),$$

which means that h_t is strongly accretive on B. We now define the set,

$$M_\lambda = \{t \in [0, 1] : 0 \in h_t(x) \quad \text{for some } x \in B(x_0; r)\}.$$

Then for each $\lambda > 0, M_\lambda$ is nonempty (since $1 \in M_\lambda$). Our goal is to demonstrate that $M_\lambda = 1$. To accomplish this, we first show that M_λ is a closed subset of $[0, 1]$. Let $\{t_n\}$ be a sequence in M_λ for which $t_n \to t \in [0, 1]$ as $n \to \infty$. Then for each $n \in \mathbb{N}$ there exists $x_n \in B(x_0; r)$ such that $0 \in h_{t_n}(x_n)$. This means, we may select $u_n \in A(x_n)$ such that

$$\lambda(x_n - x_0) + u_n - [t_n y + (1 - t_n)y_0] = 0. \tag{9}$$

Then we may derive that

$$\begin{aligned} \lambda \| x_n - x_m \| &\leq |h_t(x_n) - h_t(x_m)| \\ &\leq \| \lambda(x_n - x_m) + u_n - u_m \| \\ &\leq |t_n - t_m| \, \| y_0 - y \| . \end{aligned}$$

Therefore, $\{x_n\}$ converges, say to $\bar{x} \in \bar{B}(x_0; r)$, and hence $u_n \to u$ for some $u \in X$. Since A is continuous at $\bar{x}$, for a given $\epsilon > 0$ there exists $N \in \mathbb{N}$ such that

$$H(A(x_n), A(\bar{x})) < \epsilon/2 \quad \text{for } n \geq N.$$

Moreover, by Lemma 1, we may select $v_n \in A(\bar{x})$ satisfying

$$\| u_n - v_n \| \leq H(A(x_n), A(\bar{x})) + \epsilon/2$$

for $n \geq N$. Therefore, $u \in A(\bar{x})$ and hence by (9), $0 \in h_t(\bar{x})$ and

$$\begin{aligned} \| u - y_0 \| &\leq t \, \| y_0 - y \| + \lambda \, \| \bar{x} - x_0 \| \\ &\leq \eta + \lambda r \\ &< \delta. \end{aligned}$$

This implies that $\| \bar{x} - x_0 \| < r$, and thus $t \in M_\lambda$. Therefore, M_λ is closed.

On the other hand, suppose M_λ is not open. Then there exists $t \in M_\lambda$ $(t < 1)$ and a sequence $\{t_n\}$ in $[0, 1)$ for which $t_n \notin M_\lambda$ and $t_n \to t$. Then $0 \in h_t(z)$ for some

$z \in B(x_0; r)$, and by selecting $u_1 \in A(z)$ we have $\lambda(z - x_0) + u_1 - [ty + (1-t)y_0] = 0$. Now, due to the continuity of A, we may choose a closed ball $\bar{B}(z; \nu)$ contained in $B(x_0; r)$ such that

$$H(A(x), A(z)) < 1 \quad \text{for all } x \in \overline{B}(z; \nu). \tag{9}$$

Define $y_n = \lambda(z - x_0) + u_1 - [t_n y + (1 - t_n)y_0]$ for each $n \in \mathbb{N}$. Then

$$y_n \in h_{t_n}(z) \subset h_{t_n}(\bar{B}(z; \nu)),$$

while $0 \notin h_{t_n}(\bar{B}(z; \nu))$. Now we may follow the argument given in the proof of Theorem 1 to conclude that there exist $w \in \partial B(z; \nu)$ for which $0 \in h_t(w)$. However, this contradicts the one-to-oness of h_t on $\bar{B}(x_0; r)$. Therefore, M_λ must be also open in $[0, 1]$, and thus $1 \in M_\lambda$. This means,

$$0 \in \lambda(x - x_0) + A(x) - y \quad \text{for each } \lambda \in (0, \eta).$$

If we choose $\lambda_n \to 0$ as $n \to \infty$, then there exist $x_n \in \bar{B}(x_0; r)$ and $u_n \in A(x_n)$ such that $\lambda_n(x_n - x_0) + u_n = y$. Therefore, $u_n \to y$ as $n \to \infty$. It follows from the closedness of A on $\bar{B}(x_0; r)$ that $y \in A(x)$ for some $x \in \bar{B}(x_0; r)$, implying that $B(y_0; \eta) \subset A(D)$. Hence, $A(D)$ is open in X.

Corollary 4: *Let X and D be as in Theorem 3, and let $A: D \to B(X)$ be locally α-strongly accretive with $\liminf_{r \to \infty} \alpha(r) > 0$. Then $A(D)$ is open in X.*

Proof: Let $y_0 \in A(D)$. Then there exists $x_0 \in D$ such that $y_0 \in A(x_0)$. Also there exists a closed ball $\bar{B}(x_0; r) \subset D$ such that A is α-strongly accretive, and hence one-to-one, on $\bar{B}(x_0; r)$. Due to Theorem 3, it just remains to show that $A(\bar{B}(x_0; r))$ is closed. To see this, let $y_n \in A(\bar{B}(x_0; r))$ such that $y_n \to y \in X$. Then there exists $x_n \in \bar{B}(x_0; r)$ such that $y_n \in A(x_n)$. However, by combining Lemmas 2 and 3, we obtain that $y \in A(\bar{B}(x_0; r))$. This completes the proof.

As a final result, we discuss another Invariance of Domain results for a family of operators slightly different than the α-strongly accretive mentioned above. These former operators appear to be first introduced by Browder [3].

Theorem 4: *Let X be a (real) Banach space, A a continuous mapping from X into $B(X)$ and c: $[0, \infty) \to [0, \infty)$ be a continuous function having $c(t) > 0$ for all $t \in [0, \infty)$. Suppose for each $z \in X$ there exists a neighborhood $N(z)$ of z and a $j \in J(x - y)$ such that*

$$\langle u - v, j\rangle \geq c(\max(\| x \|, \| y \|)\ \| x - y \|^2 \tag{10}$$

for all $x, y \in N(z)$ and, all $u \in A(x)$ and $v \in A(y)$. Then A is an open mapping on X.

Proof: Let $w \in A(X)$. Then there exists $z \in X$ such that $w \in A(z)$. Now we choose an open ball B centered at z such that (10) holds for all $x, y \in B$. Then due to the assumptions on c we derive that

$$k = \inf\{c(\| x \|) : x \in B\} > 0.$$

This means, for $x, y \in B$ we obtain

$$\langle u - v, j \rangle \geq k \parallel x - y \parallel^2 \quad \text{for } u \in A(x) \quad \text{and} \quad v \in A(y).$$

Therefore, by Corollary 4, we conclude that $A(B)$ is open, and hence A is an open mapping.

We should observe that Ray and Walker [9] proved Theorem 4 under the standard additional assumptions on the function c such as being non-increasing with $\int^\infty c(t)dt = \infty$.

REFERENCES

[1] N.A. Assad and W.A. Kirk (1972). Fixed point theorems for set-valued mappings of contractive type, *Pacific J. of Math.*, **43**, 553–562.

[2] F.E. Browder (1967). Nonlinear mappings of nonexpansive and accretive type in Banach spaces, *Bull. Amer. Math. Soc.*, **73**, 875–881.

[3] F.E. Browder (1976). Nonlinear operators and nonlinear equations of evolution in Banach spaces, *Proc. Symp. Pure Math.*, **18**, Pt. 2, *Amer. Math. Soc.*, Providence, R.I.

[4] Y.Q. Chen (1999). On the range of nonlinear set-valued accretive operators in Banach spaces, *J. Math. Anal. Appl.*, **233**, 827–842.

[5] W.A. Kirk (1983). Local expansions and accretive mappings, *Internat. J. Math. & Math. Sci.*, **6**, 419–429.

[6] R.H. Martin (1970). A global existence theorem for autonomous differential equations in a Banach space, *Proc. Amer. Math. Soc.*, **26**, 307–314.

[7] C.H. Morales (1985). Surjectivity theorems for multi-valued mappings of accretive type, *Comment. Math. Univ. Carol.*, **26**, 397–413.

[8] C.H. Morales (1986). Zeros for strongly accretive set-valued mappings, *Comment. Math. Univ. Carol.*, **27**, 455–469.

[9] W.O. Ray and A.M. Walker (1982). Mapping theorems for Gateaux differentiable and accretive operators, *J. Nonlinear Anal.*, **6**, 423–433.

SIMAA 4(2002) 369–381

21. A Kneser Type Property for the Solution Set of a Semilinear Differential Inclusion with Lower Semicontinuous Nonlinearity

Valeri Obukhovskii[1,*] and Pietro Zecca[2,**]

[1]*Department of Mathematics, Voronezh University, 394693 Voronezh, Russia.*
E-mail: valeri@ob.vsu.ru
[2]*Dipartimento di Energetica "S. Stecco" Università di Firenze, 50139 Firenze, Italy.*
E-mail: pzecca@ingfi1.ing.unifi.it

1. INTRODUCTION

The present work is devoted to the study of the classical Kneser type property for a semilinear differential inclusion with lower semicontinuous multivalued nonlinearity in a separable Banach space. For nonlinear lower semicontinuous differential inclusions in a finite dimensional space this problem was investigated by A. Bressan [3,4] and K. Deimling [8]. Conditions under which the solution set for differential equations in an infinite dimensional space is a continuum are presented in the book of R. Dragoni, J.W. Macki, P. Nistri and P. Zecca [9]. At last, in the chapter of A. Bressan and V. Staicu [6] the connectedness of the set of solutions is proved for a differential inclusion whose right hand side is a bounded lower semicontinuous perturbation of a maximal monotone operator generating a compact semigroup of contractions.

In our approach we suppose that the linear part of the inclusion is the infinitesimal generator of an arbitrary C_0-semigroup which is not assumed to be neither compact or contractive, and the almost lower semicontinuous nonlinearity satisfies a general boundedness condition. Moreover, we do not pose any additional assumption on the geometry of the Banach space and do not assume reflexivity.

First, following the method of directionally continuous selections (see e.g., [4–6]) we construct a semilinear differential inclusion with upper semicontinuous nonlinearity associated with the given one. Then, using the topological properties of the solution set of the associated problem (see [12]), we prove the main result (Theorem 8).

2. PRELIMINARIES

Let (X, ρ_X) and (Y, ρ_Y) be metric spaces. $K(Y)$ denotes the collection of all nonempty compact subsets of Y.

* The work of V. Obukhovskii is partially supported by C.N.R. (Italy) and the Russian Foundation for Basic Research, Grant 99-01-00333.

** The work of Pietro Zecca is partially supported by a national grant ex 40% MURST.

Definition 1: A multivalued map (multimap)

$$\mathcal{F}: X \to K(Y)$$

is said to be *lower (upper) semicontinuous at a point* $x \in X$ if for every $\varepsilon > 0$ there exists a $\delta > 0$ such that

$$\mathcal{F}(x) \subset W_\varepsilon \mathcal{F}(x'), \ (\mathcal{F}(x') \subset W_\varepsilon \mathcal{F}(x))$$

for all $x' \in X$, $\rho_X(x, x') < \delta$, where W_ε denotes the ε-neighborhood of a set.

If this property holds at each point $x \in X, \mathcal{F}$ is called *lower semicontinuous* (l.s.c.) (*upper semicontinuous* (u.s.c.)) respectively.

A multimap both l.s.c. and u.s.c. is called *continuous* (see e.g., [2,10,12] for further details).

We recall also the following notions.

Definition 2: Let E be a Banach space, 2^E denote the collection of all subsets of E and let $(\mathbf{A}, \succeq)$ be a partially ordered set. A map

$$\beta: 2^E \to \mathbf{A}$$

is called a *measure of noncompactness* (MNC) in E if

$$\beta(\overline{\mathrm{co}}\,\Omega) = \beta(\Omega)$$

for every $\Omega \in 2^E$.
A MNC is called:

(i) *monotone,* if $\Omega_0, \ \Omega_1 \in 2^E, \ \Omega_0 \subseteq \Omega_1$ implies $\beta(\Omega_0) \leq \beta(\Omega_1)$;
(ii) *nonsingular,* if $\beta(\{a\} \cup \Omega) = \beta(\Omega)$ for every $a \in E, \ \Omega \in 2^E$;
(iii) *regular,* if $\beta(\Omega) = 0$ is equivalent to the relative compactness of Ω.
(iv) *algebraically semiadditive,* if $\beta(\Omega_1 + \Omega_2) \leq \beta(\Omega_1) + \beta(\Omega_2)$.

As an example of MNC possessing all these properties we can consider the *Hausdorff MNC*

$$\chi(\Omega) = \inf\{\varepsilon > 0 : \Omega \text{ has a finite } \varepsilon\text{-net}\}.$$

In the sequel we will need the following property of the Hausdorff MNC that can be easily verified.

Proposition 3: (See [1]) *If L: $E \to E$ is a bounded linear operator then, for every bounded set $\Omega \subset E$, we have*

$$\chi(L\,\Omega) \leq \|L\| \cdot \chi(\Omega).$$

Denote by $P(E)$ the collection of all nonempty subsets of E, we recall the following:

Definition 4: A multifunction $\mathcal{G}$: $[0, a] \to P(E)$ is said to be:

(i) *integrable* if it admits a summable selection $g \in L^1([0, a]; E)$, i.e., $g(t) \in \mathcal{G}(t)$ for a.a. $t \in [0, a]$;

(ii) *integrably bounded* if there exists a summable function $\nu \in L^1_+([0,a])$ such that

$$\|\mathcal{G}(t)\| := \sup\{\|g\|, g \in \mathcal{G}(t) \leq \nu(t) \ \text{ for a.a. } t \in [0,a]\}.$$

The set of all summable selections of the multifunction $\mathcal{G}$:$[0,a] \to P(E)$ will be denoted as $S^1_{\mathcal{G}}$. If the multifunction $\mathcal{G}$: $[0,a] \to P(E)$ is integrable then its integral can be defined as

$$\int_0^a \mathcal{G}(s)ds := \left\{\int_0^a g(s)ds : g \in S^1_{\mathcal{G}}\right\}.$$

The multifunction $\mathcal{G}$: $[0,a] \to K(E)$ is said to be *measurable* if $\mathcal{G}^{-1}(V) = \{t \in [0,a] : \mathcal{G}(t) \subset V\}$ is Lebesgue measurable for every open set $V \subseteq E$ (see e.g., [2, 10, 12] for equivalent definitions and details).

Lemma 5: ([13], *see also* [12]). *Let the space E be separable and the multifunction* $\mathcal{G}$: $[0,a] \to P(E)$ *be integrable, integrably bounded and* $\chi(\mathcal{G}(t)) \leq q(t)$ *for a.a.* $t \in [0,a]$ *where* $q(\cdot) \in L^1_+[0,a]$. *Then*

$$\chi\left(\int_0^\tau \mathcal{G}(s)ds\right) \leq \int_0^\tau q(s)ds, \quad \text{for all } \tau \in [0,\mathrm{a}].$$

In particular, if the multifunction $\mathcal{G}$: $[0,a] \to K(E)$ *is measurable and integrably bounded then the function* $\chi(\mathcal{G}(\cdot))$ *is integrable and*

$$\chi\left(\int_0^\tau \mathcal{G}(s)ds\right) \leq \int_0^\tau \chi(\mathcal{G}(s))ds, \quad \text{for all } \tau \in [0,\mathrm{a}].$$

In the sequel we will suppose that:

(A) A: $D(A) \subseteq E \to E$ is a densely defined linear operator generating a C_0-semigroup $\exp\{At\}$.

We will say that the map S: $L^1([0,a];\ E) \to C([0,a];\ E)$ defined as

$$S(g)(t) = \int_0^t \exp\{A(t-s) \cdot g(s)ds\}$$

is the *Cauchy operator*.

We need the following property.

Lemma 6: ([7, 12]). *If the sequence* $\{g_n\} \subset L^1([0,a];\ E)$ *is semicompact, i.e., it is integrably bounded and the set* $\{g_n(t)\}$ *is relatively compact in E for a.a.* $t \in [0,a]$, *then the sequence* $\{S\,g_n\} \subset C([0,a;\ E])$ *is relatively compact.*

3. RESULTS

Let us consider the Cauchy problem for a semilinear differential inclusion in a separable Banach space E

$$x'(t) \in Ax(t) + F(t, x(t)) \tag{1}$$

$$x(0) = x_0 \tag{2}$$

under assumption (A) on the linear part $A\colon D(A) \subseteq E \to E$ and the following hypotheses on the multivalued nonlinearity $F\colon [0,d] \times E \to K(E)$:

(F1) F is *almost lower semicontinuous* (a.l.s.c.) in the sense that there exists a sequence of disjoint compact sets $\{I_n\}$, $I_n \subseteq [0,d]$ such that meas $([0,d] \setminus \cup_n I_n) = 0$, and the restriction of F on each set $J_n = I_n \times E$ is l.s.c. (see [6, 8]).

(F2) there exists a constant $K > 0$ such that for all $x \in E$

$$\|F(t,x)\| \leq K\,(1 + \|x\|), \quad \text{for a.a. t} \in [0,\text{d}];$$

(F3) there exists a function $k(\cdot) \in L^1_+([0,d])$ such that for every bounded set $\Omega \subset E$ we have that

$$\lim_{\tau \to +0} \chi(F(Q_{t,\tau} \times \Omega)) \leq k(t) \cdot \chi(\Omega), \quad \text{for a.a. t} \in [0,\text{d}],$$

where $Q_{t,\tau} = [t-\tau, t+\tau] \cap [0,d]$.

It is easy to see (cfr. e.g., [2,10,12]) that for every function $x(\cdot) \in C([0,d]; E)$, the multifunction $F(t, x(t))$ is l.s.c. on $\cup_n I_n$ and integrable by (F2), so we can define the *superposition multioperator* $\mathcal{P}_F\colon C([0,d]; E) \to P(L^1([0,d]; E))$ as:

$$\mathcal{P}_F(x) = S^1_{F(\cdot, x(\cdot))}.$$

and the integral multioperator $\Gamma_F\colon C([0,d]; E) \to P(C([0,d]; E))$ as

$$\Gamma_F(x) = \left\{ y\colon y(t) = \exp\{At\}x_0 + \int_0^t \exp\{A(t-s)\}\, f(s)ds,\ f \in \mathcal{P}_F(x) \right\}.$$

We will say that the function $x(\cdot) \in C([0,d]; E)$, is a *mild solution* of problem (1), (2) on the interval $[0,h]$ if $x(\cdot)$ has the following representation

$$x(t) = \exp\{At\}x_0 + \int_0^t \exp\{A(t-s)\}\, f(s)ds,\ f \in \mathcal{P}_F(x)\ .$$

It is clear that every mild solution $x(\cdot)$ of (1)–(2) is a fixed point $x \in \Gamma_F(x)$ for the integral multioperator Γ_F.

Following [11] we can deduce the following existence result.

Theorem 7: *Under conditions (A), (F1)–(F3) there exists a mild solution $x(\cdot) \in C([0,d]; E)$ of problem (1)–(2).*

Let us denote the set of all mild solutions of (1)–(2) by the symbol $\Sigma^F_{x_0}$. Our main result is the following statement concerning the topological structure of the set $\Sigma^F_{x_0}$.

Theorem 8: *Under conditions (A), (F1)–(F3) the solution set $\Sigma^F_{x_0}$ is connected. In particular each set $\Sigma^F_{x_0}(t) = \{x(t)\colon x \in \Sigma^F_{x_0}\}$, $t \in [0,d]$ is also connected.*

The proof will be divided into three main steps and several lemmas.

Proof:

Step 1

Denote by $Cv(E)[Kv(E)]$ the collection of all nonempty, convex closed [compact] subsets of E. Let us consider the multimap $\hat{F}\colon [0,d] \times E \to Cv(E)$ defined as $\hat{F}(t,x) = \bigcap_{\varepsilon > 0} F^\varepsilon(t,x)$ where $F^\varepsilon(t,x) = \overline{\text{co}}\,\{F(s,y)\colon |s-t| < \varepsilon,\ \|y - x\| < \varepsilon\}$.

For any bounded set $\Omega \subset E$ the multimap F satisfies the following χ-regularity condition:

$$\chi(\hat{F}(t,\Omega)) \leq k(t) \cdot \chi(\Omega) \quad \text{for a.a. } t \in [0,d]. \tag{3}$$

In fact, for any $t \in [0,d]$ for which the estimate (F3) holds, take an arbitrary $\delta > 0$ and choose τ, $0 < \tau \leq \delta$ such that

$$\chi(F(Q_{t,\tau} \times W_\delta(\Omega))) \leq k(t) \cdot \chi(W_\delta(\Omega)) + \delta \leq k(t) \cdot (\chi(\Omega) + \delta) + \delta.$$

Then,

$$\chi(\hat{F}(t,\Omega)) \leq \chi(F^\tau(t,\Omega)) \leq \chi(F(Q_{t,\tau} \times W_\delta(\Omega))) \leq k(t) \cdot (\chi(\Omega) + \delta) + \delta$$

and estimate (3) follows from the arbitrariness of δ.

We construct now a nonempty compact convex subset $X \subset C([0,d];\, E) = C$ with the following properties:

$$x(0) = 0 \quad \text{for all } x \in X \tag{4}$$

$$\exp\{At\}x_0 + \int_0^t \exp\{A(t-s)\}\, \overline{\text{co}}\, \hat{F}(s, X(s))\, ds \subseteq X(t) \quad \text{for } t \in [0,d] \tag{5}$$

$$\Sigma_{x_0}^F \subset X \tag{6}$$

Note that from condition (F2), using a Gronwall type inequality, the *a priori* boundedness for the set $\Sigma_{x_0}^F$ can be proved. Therefore we can assume, without loss of generality, that the multimap F and hence $\hat{F}$ are bounded:

$$\|\hat{F}(t,x)\| \leq K \quad \text{for a.a. } t \in [0,d] \quad \text{and} \quad x \in E.$$

We construct a decreasing sequence of closed convex sets $\{X_i\}_{i=1}^\infty \subset C$ by inductive process as follows:

$$X^0 = \{x \in C : x(0) = x_0,\ \|x\|_C \leq r\},$$

where $r = M(\|x_0\| + Kd)$, $M = \max_{t \in [0,d]} \|\exp\{A\,t\}\|$;

$$Y^1 = \left\{ y \in C : y(t) = \exp\{At\}x_0 + \int_0^t \exp\{A(t-s)\} f(s) ds,\ f \in S^1_{\overline{\text{co}}\, F(\cdot, X^0(\cdot))} \right\};$$

$X^1 = \overline{Y^1}$.

If X^{n-1} is constructed then

$$Y^n = \left\{ y \in C : y(t) = \exp\{A\,t\}x_0 + \int_0^t \exp\{A(t-s)\} f(s) ds,\ f \in S^1_{\overline{\text{co}} F(\cdot, X^{n-1}(\cdot))} \right\}$$

and $X^n = \overline{Y^n}$.

First of all let us note that by construction we have that

$$F(t,x) \subseteq \hat{F}(t,x) \quad \text{for all } (t,x) \in [0,d] \times E. \tag{7}$$

Hence the multifunctions $\overline{\text{co}}\, F(\cdot\,, X^{n-1}(\,\cdot\,)), i \geq 1$ are integrable and all the sets X^n are nonempty since $\sum_{x_0}^{F} \subset X^n$ for all $n \geq 1$.

Consider, in the space $C([0,d]; E)$, the measure of noncompactness

$$\psi(\Omega) = \sup_{t\in[0,d]} \left\{ \chi(\Omega(t)) \cdot \exp\left\{ -R \int_0^t k(s)ds \right\} \right\},$$

where χ is the Hausdorff MNC in E, $R > M$ is an arbitrary number and $k(\,\cdot\,)$ is the function from condition (F3). It is easy to see that the MNC ψ is monotone and non-singular. Using condition (3) we have the following estimate

$$\begin{aligned}
&\chi\big(\exp\{A(t-s)\}\,\overline{\text{co}}\,\hat{F}(s, X^{n-1}(s))\big)\\
&\leq \|\exp\{A(t-s)\|\chi\big(\overline{\text{co}}\,\hat{F}(s, X^{n-1}(s))\big)\\
&\leq M\,\chi\big(\hat{F}(s, X^{n-1}(s))\big)\\
&\leq M\,k(s)\chi(X^{n-1})\\
&\leq M\,k(s)\psi(X^{n-1})\exp\left\{ R\int_0^s k(\theta)d\theta \right\} \quad \text{for all } s \in [0,t].
\end{aligned}$$

Now, for every $t \in [0,d]$ the multifunction

$$s \multimap \exp\{A(t-s)\}\,\overline{co}\hat{F}(s, X^{n-1}(s)), \;\; s \in [0,t]$$

is bounded and hence, using Lemma 5 we have that for every $t \in [0,d]$

$$\begin{aligned}
\chi(Y^n) &= \chi\left(\exp\{At\}x_0 + \int_0^t \exp\{A(t-s)\}\,\overline{\text{co}}\,\hat{F}(s, X^{n-1}(s))\,ds \right)\\
&= \chi\left(\int_0^t \exp\{A(t-s)\}\,\overline{\text{co}}\,\hat{F}(s, X^{n-1}(s))\,ds \right)\\
&\leq M\psi(X^{n-1}) \cdot \int_0^t k(s) \cdot \exp\left\{ R\int_0^s k(\theta)d\theta \right\} ds\\
&\leq M/R\psi(X^{n-1}) \cdot \exp\left\{ R\int_0^t k(s)ds \right\}.
\end{aligned}$$

Therefore,

$$\chi(Y^n(t)) \cdot \exp\left\{ -R\int_0^t k(s)ds \right\} \leq M/R\,\psi(X^{n-1})$$

and

$$\psi(Y^n) \leq M/R\,\psi(X^{n-1}).$$

Finally, we have $\psi(X^n) \le (M/R)\,\psi(X^{n-1})$, $n \ge 1$ and therefore

$$\psi(X^n) \underset{n\to\infty}{\to} 0.$$

If we consider the nonempty set $\tilde{X} = \cap_{n\ge1} X^n$, using the monotonicity property of the MNC ψ we obtain that $\psi(\tilde{X}) = 0$ and hence $\chi(\tilde{X}) = 0$.

Moreover, condition (3) implies

$$\chi\big(\overline{\mathrm{co}}\,\hat{F}(t, \tilde{X}(t))\big) = 0 \quad \text{for a.a. } t \in [0, d].$$

Then, from the property of the Cauchy operator described in Lemma 6, it follows that the set $X \subseteq \tilde{X}$ given by

$$X = \overline{Y};$$
$$Y = \left\{ y \in C : y(t) = \exp\{At\}x_0 + \int_0^t \exp\{A(t-s)\}\, f(s)ds\,,\ f \in S^1_{\overline{\mathrm{co}}\,\hat{F}(s,\tilde{X}(s))} \right\}$$

is compact. The set X is the one we were looking for. □

Step 2 An associated differential inclusion with u.s.c. nonlinearity.

We consider the compact set $D \subset [0, d] \times E$ given by

$$D = \{(s, y) : s \in [0, d],\ y = x(s),\ x \in X\}.$$

For $(t, x) \in D, \varepsilon > 0$ and $N > \mu = M \cdot K$ consider the set

$$V(t, x, \varepsilon) = \{(s,\ y) \in D : t \le s < t + \varepsilon,$$
$$\|\, y - \exp\{A\tau\}x \,\| \le N\,(s - t) \quad \text{for some } \tau \in [0, \mathrm{s} - \mathrm{t}]\}$$

The following properties are known from [6] (see also [10]).

Lemma 9: *The family of sets $\{V(t, x, \varepsilon) : (t, x) \in D,\ \varepsilon > 0\}$ form a base of closed open neighborhoods for a topology $\mathcal{I}^+$ on D, stronger than the usual metric topology.*

Lemma 10: *The multifunction $F\colon D \to K(E)$ admits an almost $\mathcal{I}^+$-continuous selection $\gamma\colon D \to E$ in the sense that γ is $\mathcal{I}^+$-continuous on every set $D_n = J_n \cap D$.*

Now, for $(t, x) \in D_n$ we set

$$G_n(t, x) = \ \cap_{\varepsilon>0} G^{\varepsilon}_n(t, x)$$
$$= \cap_{\varepsilon>0}\overline{\mathrm{co}}\{\gamma(s, y) : (s, y) \in D_j, |s - t| < \varepsilon, \|y - x\| < \varepsilon\}.$$

and, for $(t, x) \in D$ define

$$G(t, x) = \begin{cases} G_n(t, x) & \text{if} \quad (t, x) \in D_n, \\ \gamma(t, x) & \text{if} \quad (t, x) \notin \cup_n D_n. \end{cases}$$

Note that from the construction it follows

$$\gamma(t,x) \in G(t,x), \quad \text{for all } (t,x) \in D \tag{8}$$

and

$$G(t,x) \subseteq \hat{F}(t,x), \quad \text{for all } (t,x) \in D \tag{9}$$

We also observe that from condition (F3) and the compactness of the sets D_n it follows that for every sequence $\varepsilon_k \to 0+$ the corresponding sequence $\chi(F_n^{\varepsilon_k}(t,x))$ tends to zero. This implies that for all $(t,x) \in D$ the nonempty convex sets $G(t,x)$ are compact. Moreover the following Lemma holds:

Lemma 11: *The multimap G is u.s.c. on $\cup_n D_n$.*

Proof: Suppose that the map is not u.s.c. Then there exists a point $(t,x) \in D_n$, a number $\delta > 0$ and a sequence $(t_k,x_k) \in D_n$, $(t_k,x_k) \to (t,x)$, $y_k \in G_n(t_k,x_k)$ such that $\text{dist}(y_k, G_n(t,x)) > \delta$.
Without loss of generality we can assume that the sequence

$$\varepsilon_k = 2\max\{|\,t_k - t\,|, |\,x_k - x\,|\}$$

is nonincreasing. Then we obtain that $y_k \in G_n^{\varepsilon_k}(t,x)$ and we can consider the decreasing sequence of closed sets

$$L_k = G_n^{\varepsilon_k}(t,x) \cap (E \setminus W_\delta(G_n(t,x))).$$

Each set L_k is nonempty since it contains y_k and $\chi(L_k) \to 0$ by the observation above. The nonsingularity property of the Hausdorff MNC implies $\chi(\{y_k\}) = 0$ so the sequence $\{y_k\}$ is relatively compact and admits limit point y.
On one hand $y \in G_n(t,x)$ and from the other one dist $(y, G_n(t,x)) \geq \delta$ giving a contradiction. □

Now, consider the problem:

$$x'(t) \in Ax(t) + G(t, x(t)), t \in [0,d] \tag{10}$$

$$x(0) = x_0 \tag{11}$$

on the compact set D.
For any function $x \in X$ the multifunction $G(.\times(.))$ is a.e. u.s.c. and hence measurable. Since the multimap G is bounded, the integral multioperator Γ_G is defined and u.s.c. on X.
Let us study its action on X in some details.
Choose an arbitrary function $x \in X$. From the construction of X (see step 1) it is easy to see that for any $\eta > 0$, x can be uniformly η-approximated by the function

$$\tilde{x}(t) = \exp\{A\,t\}x_0 + \int_0^t \exp\{A\,(t-s)\} \cdot v(s)ds,$$

where $v(\,\cdot\,) \in L^1([0,d], E)$ and $\|v(t)\| \leq K$ for a.a. $t \in [0,d]$.

Lemma 12: *The function x satisfies the following condition:*

$$\{(s,y) \in D : t \leq s < t + \varepsilon;\ y \in \overline{B}_{(s-t)(N-\mu)}(x(s))\} \subset V(t,\ x(t), \varepsilon).$$

The statement of the Lemma means that for points s close enough to t the set $V(t, x(t), \varepsilon)$ is a metric neighborhood of the point $(s, x(s))$ (in the relative topology of the space D).

Proof: Let $(s, y) \in D$ and $y \in \overline{B}_{(s-t)(N-\mu)}(x(s))$, then

$$\begin{aligned}
&\| y - \exp\{A(s-t)\} \cdot x(t)\| \\
&\leq \| y - \exp\{A(s-t)\} \cdot \tilde{x}(t)\| \\
&+ \| \exp\{A(s-t)\} \cdot \tilde{x}(t) - \exp\{A(s-t)\} \cdot x(t)\| \\
&\leq \| y - \tilde{x}(s)\| + \| \tilde{x}(s) - \exp\{A(s-t)\} \cdot \tilde{x}(t)\| + M\eta \\
&\leq (s-t)(N-\mu) + \eta + \| \exp\{A(s-t)\} \cdot \tilde{x}(t) \\
&+ \int_t^s \exp\{A(s-\theta)\} \cdot v(\theta)\, d\theta - \exp\{A(s-t)\} \cdot \tilde{x}(t)\| + M\eta \\
&\leq (s-t)(N-\mu) + \int_t^s \| \exp\{A(s-\theta)\}\| \cdot \|v(\theta)\|\, d\theta \\
&\leq (s-t)(N-\mu) + (s-t)\,\mu + \eta\,(1+M) \\
&= N\,(s-t) + \eta\,(1+M).
\end{aligned}$$

As η is arbitrary we get

$$\| y - \exp\{A(s-t)\}\, x(t)\| \leq N\,(s-t). \qquad \square$$

Now, if we take $z \in \Gamma_G$ we have that z can be represented as

$$z(t) = \exp\{A(t)\}\, x_0 + \int_0^t \exp\{A(t-s)\}\, g(s) ds, \text{ where } g \in \mathcal{P}_G(x).$$

Lemma 13:

$$g(t) = \gamma(t, x(t))$$

for almost all $t \in (0, d)$.

Proof: For each value of the index n we consider the set $I_n^* \subseteq I_n$ consisting of all $t \in I_n$ with properties:

(i) $g(t) \in G_n(t, x(t))$;
(ii) there exists a sequence $\{t_k\} \subset I_n$ strictly decreasing to t such that $g(t_k) \in G_n(t_k, x(t_k))$ for all k and $g(t_k) \to g(t)$.

It is known (see Lemma 2.3 in [4]) that $\text{meas}(I_n \setminus I_n^*) = 0$. Following [6] we will show that $g(t) = \gamma(t, x(t))$ for all $t \in I_n^*$.

Suppose, to the contrary, that there exists a point $t \in I_n^*$ and $\varepsilon > 0$ such that

$$\|g(t) - \gamma(t, x(t))\| = \varepsilon\,. \tag{12}$$

Since γ is $\mathcal{I}^+$-continuous on D_n there exists $\delta > 0$ such that

$$\|\gamma(s, y) - \gamma(t, x(t))\| < \varepsilon/2$$

for all $(s, y) \in V(t, x(t), \delta)$, $s \in I_n$.

Now let $\{t_k\} \subset I_n$ be a sequence with the properties described in (ii). Choose k_0 such that for $k \geq k_0$,

$$0 < t_k - t < \delta$$

and

$$\|g(t_k) - g(t)\| < \varepsilon/2. \tag{13}$$

From Lemma 12 we know that the set $V(t, x(t), \delta)$ is a neighborhood of $(t_k, x(t_k))$ in the usual relative metric topology of D for all $k \geq k_0$. So let us suppose that the ω-neighborhood W_ω of $(t_k, x(t_k))$ in D is contained in $V(t, x(t), \delta)$. Then, for $k \geq k_0$ we have

$$\begin{aligned} g(t_k) \in G_n(t_k, x(t_k)) &= \overline{\text{co}}\{\gamma(s,y) : (s,y) \in W_\omega, s \in I_n\} \\ &\subset \overline{\text{co}}\{\gamma(s,y) : (s,y) \in V(t, x(t), \delta),\ s \in I_n\} \subset B_{\varepsilon/2}(\gamma(t, x(t))), \end{aligned}$$

i.e.,

$$\|\gamma(t, x(t)) - g(t_k)\| < \varepsilon/2. \tag{14}$$

It is clear that (13) and (14) give the contradiction to (12) proving the Lemma. □

From the above Lemma it follows that the integral multioperator Γ_G is single-valued on X and moreover, since it has the form

$$\Gamma_G(x) = \exp\{A\,(t)\}\,x_0 + \int_0^t \exp\{A\,(t-s)\}\,\gamma(s, x(s))ds,$$

it is a continuous selection of the integral multioperator Γ_F. Consequently, Γ_G maps the set X into itself since Γ_F has the same property (see (5) and (7)). Furthermore, every mild solution of the problem (10), (11) is a mild solution of (1), (2).

We want now to consider the Cauchy problem for a differential inclusion whose nonlinearity is defined on the whole $[0, d] \times E$. To this aim take the metric projection P: $[0, d] \times E \multimap Kv(E)$

$$P(t, x) = \{y \in X(t) : \|x - y\| = \text{dist}(x, X(t))\}$$

and the multimap $\tilde{G}$: $[0, d] \times E \multimap Kv(E)$ defined by

$$\tilde{G}(t, x) = \overline{\text{co}}\, G(t, P(t, x))$$

Lemma 14: *The multimap $\tilde{G}$ satisfies the following conditions*

(*i*) *the multifunction $\tilde{G}(\cdot, x)$: $[0, d] \multimap Kv(E)$ admits a measurable selection for every $x \in E$;*

(*ii*) *the multimap $\tilde{G}(t, \cdot)$:$E \multimap Kv(E)$ is u.s.c. for a.a. $t \in [0, d]$;*

(*iii*) *$\|\tilde{G}(t, x)\| \leq K$ for a.a. $t \in [0, d]$ and $x \in E$;*

(*iv*) *the multimap $\tilde{G}(t, \cdot)$: $E \multimap Kv(E)$ is compact for a.a. $t \in [0, d]$.*

Proof: First of all let us note that the multimap $X(\cdot)$: $[0, d] \multimap Kv(E)$, $t \multimap X(t)$ is continuous since X is compact (see e.g., [12]) and hence the multimap P has a closed graph. In fact, let $t_n \to t_0 \in [0, d]$; $x_n \to x_0 \in E$; $y_n \in P(t_n, x_n)$, $y_n \to y_0 \in E$.

From the equality

$$|\operatorname{dist}(x_n, X(t_n)) - \operatorname{dist}(x_0, X(t_0))| \le \|x_n - x_0\| + h(X(t_n), X(t_0)),$$

where h is the Hausdorff metric, it follows

$$\lim_{n\to\infty} \operatorname{dist}(x_n, X(t_n)) = (x_0, X(t_0)).$$

On the other hand,

$$\lim_{n\to\infty} \operatorname{dist}(x_n, X(t_n)) = \lim_{n\to\infty} \| x_n - y_n\| = \|x_0 - y_0\|,$$

so $y_0 \in P(t_0, x_0)$.

The range of the multimap P is contained in the compact set $X([0, d])$, so the multimap P is u.s.c. (see e.g., [2,10,12]). Applying continuity properties of multimaps we obtain that the multimap $\tilde{G}$ satisfies condition (ii).

Furthermore, for fixed $x \in E$ the multifunction $P(\cdot, x) : [0, d] \to Kv(E)$ is u.s.c. hence measurable. Therefore it admits a measurable selection $p(\cdot)$. From Castaing theorem (see [2, 12]), it follows that the multifunction $G(t, p(t))$ has a measurable selection implying that the multimap $\tilde{G}$ satisfies condition (i). Condition (iii) follows from (9) and the boundedness of multifunction $\hat{F}$.

To prove (iv) is sufficient to note that, under the action of a compact-valued u.s.c. multimap, the image of a compact set is compact (see the same sources). □

We can now consider the Cauchy problem defined on $[0, d] \times E$

$$x'(t) \in Ax(t) + \tilde{G}(t, x(t)), \quad t \in [0, d] \tag{15}$$

$$x(0) = x_0 \tag{16}$$

From Lemma 14 and corresponding existence result for semilinear differential inclusions with upper semicontinuous nonlinearities (see Theorem 5.2.2 in [12]) we conclude that the set $\Sigma_{x_0}^{\tilde{G}}$ of all mild solutions of (15), (16) is a nonempty compact subset of $\mathcal{C}$. Moreover, from the statement on the topological structure of the solution set (see Corollary 5.3.1 in [12]) it follows that the set $\Sigma_{x_0}^{\tilde{G}}$ is connected.

Lemma 15: $\Sigma_{x_0}^{\tilde{G}} = \Sigma_{x_0}^{G}$

Proof: Let $x \in \Sigma_{x_0}^{\tilde{G}}([0, d])$. Then

$$\begin{aligned}
x(t) &\in \exp\{At\}\, x_0 + \int_0^t \exp\{A(t-s)\}\, \tilde{G}(s, x(s))ds \\
&= \exp\{At\}\, x_0 + \int_0^t \exp\{A(t-s)\}\, \overline{\text{co}}\, G(s, P(s, x(s)))ds \\
&\subseteq \exp\{At\}\, x_0 + \int_0^t \exp\{A(t-s)\}\, \overline{\text{co}}\, G(s, X(s))ds \\
&\subseteq \exp\{At\}\, x_0 + \int_0^t \exp\{A(t-s)\}\, \overline{\text{co}}\, \hat{F}(s, X(s))ds \subseteq X(t)
\end{aligned}$$

and hence $P(t, x(t)) = \{x(t)\}$. Then

$$x(t) \in \exp\{At\}x_0 + \int_0^t \exp\{A(t-s)\}\, g(s)ds,$$

where $g \in S^1_{\tilde{G}(\cdot,x(\cdot))} = S^1_{G(\cdot,x(\cdot))}$ and so $x \in \Sigma^G_{x_0}$.
The inclusion $\Sigma^G_{x_0} \subseteq \Sigma^{\tilde{G}}_{x_0}$ follows easily from (6). □

We will say that problem (15), (16) with u.s.c. part is associated to the initial Cauchy problem (1), (2).

Step 3 Connectedness of the integral funnel of (1), (2).
We are now in the position to proof the connectedness of the set $\Sigma^F_{x_0}$.
Let $x^1, x^2 \in \Sigma^F_{x_0}$ solutions of (1), (2). They have the form

$$x^i = \exp\{At\}x_0 + \int_0^t \exp\{A(t-s)\}\, f^i(s)ds$$

where $f^i \in \mathcal{P}_F(x^i)$, $i = 1, 2$.
Consider the multimaps $F^i(t, x) \subseteq F(t, x)$, $i = 1, 2$ defined as

$$F^i(t,x) = \begin{cases} \{f^i(t)\} & \text{if} \quad x = x^i(t) \\ F(t,x) & \text{if} \quad x \neq x^i(t). \end{cases}$$

Since the functions $f^i, i = 1, 2$ are measurable, there exists a sequence of disjoint compact sets $\{I_k\}$, $I_k \subset [0, d]$ such that meas$([0, d]/\cup_k I_k)$ and the restriction of f^i on each I_k is continuous. Hence each multimap F^i, $i = 1, 2$ is a.l.s.c. and satisfies properties (*F*2), (*F*3). Then, in accordance to Step 1, it is possible to construct a nonempty convex subset $X \subset C$ containing the mild solutions x^1, x^2, invariant with respect to the action of Γ_F and consequently of Γ_{F^i}, $i = 1, 2$. In accordance with Step 2 to each semilinear inclusion with nonlinearity F^i we can associate a differential inclusion with u.s.c. nonlinear part $\tilde{G}^i$, $i = 1, 2$. Note that by construction (remind (8)) it follows that

$$f^i(t) = F^i(t, x^i(t)) = \gamma^i(t, x^i(t)) \in \tilde{G}^i(t, x^i(t)) \quad \text{for a.a. } t \in [0, d],\ i = 1, 2$$

hence x^i can be considered as a solution of the problem with nonlinearity $\tilde{G}^i$.
Consider now the following parametrized family of semilinear differential inclusions

$$x'(t) \in Ax(t) + \tilde{G}_\lambda(t, x(t)), \quad t \in [0, d], \quad \lambda \in [0, 1] \tag{17}$$

$$x(0) = 0, \tag{18}$$

where the one-parameter family $\tilde{G}_\lambda$ is defined as

$$\tilde{G}_\lambda(t,x) = \begin{cases} \tilde{G}^1(t,x) & \text{if} \quad t \in [0, \lambda d) \\ \overline{\text{co}}\,(\tilde{G}^1(t,x) \cup \tilde{G}^2(t,x)) & \text{if} \quad t = \lambda d \\ \tilde{G}^2(t,x) & \text{if} \quad t \in (\lambda d, d]. \end{cases}$$

From the results on the existence and topological structure of the solution sets for semilinear differential inclusions with u.s.c. nonlinearities (see [12]) it follows that for every $\lambda \in [0,1]$ the set $\Sigma_{x_0}^{\tilde{G}_\lambda}$ of all mild solutions of (17), (18) is a nonempty compact connected subset of C and it is easy to see that $\Sigma_{x_0}^{\tilde{G}_\lambda} \subset \Sigma_{x_0}^{F}$ for all $\lambda \in [0,d]$. Moreover, the family (17), (18) satisfies the conditions of the theorem on continuous dependence of the solution set on a parameter (see Theorem 5.2.5 of [12]). Hence the multimap

$$\lambda \multimap \Sigma_{x_0}^{\tilde{G}_\lambda}, \ \lambda \in [0,d]$$

is u.s.c. and then it is easy to see that the set

$$\bigcup_{\lambda \in [0,d]} \Sigma_{x_0}^{\tilde{G}_\lambda}$$

is connected. It remains to observe only that $x^1 \in \Sigma_{x_0}^{\tilde{G}_1}$; $x^2 \in \Sigma_{x_0}^{\tilde{G}_0}$. □

REFERENCES

[1] R.R. Akmerov, M.I. Kamenskii, A.S. Potapov, A.E. Rodkina and B.N. Sadovskii (1992). *Measures of Noncompactness and Condensing Operators*. OT, Berlin: Birkhauser Verlag, **55**.

[2] Yu.G. Borisovich, B.D. Gelman, A.D. Myshkis and V.V. Obukhovskii (1986). *Introduction to the Theory of Multivalued Maps*. Voronezh: Voronezh Gos Univ., (in Russian).

[3] A. Bressan (1989). On the Qualitative Theory of Lower Semicontinuous Differential Inclusions, *J. of Diff. Equations*, **77**, 379–391.

[4] A. Bressan (1990). Upper and Lower Semicontinuous Differential Inclusions: A Unified Approach. In H. Sussmann (Ed.), *Nonlinear Controllability and Optimal Control*, M. Dekker, 21–31.

[5] A. Bressan and A. Cortesi (1989). Directionally Continuous Selections in Banach Spaces, *Nonlinear Analysis: Th. Meth. and Appl.*, **13**, 987–992.

[6] A. Bressan and V. Staicu (1994). On Nonconvex Perturbations of Maximal Monotone Differential Inclusions, *Set Valued Analysis*, **2**, 415–437.

[7] J.F. Couchouron and M. Kamenskii (1998). A Unified Topological Point of View for Integro-Differential Inclusions. In J. Andres, L. Gorniewicz and P. Nistri (Eds), *Differential Inclusions and Optimal Control*, Lect. Notes in Nonlinear Analysis, **2**, 123–137.

[8] K. Deimling (1992). *Multivalued Differential Equations*. Berlin-New York: W. De Gruyter.

[9] R. Dragoni, J.W. Macki, P. Nistri and P. Zecca (1996). *Solution Sets of Differential Equations in Abstract Space*, Pitman Research Notes in Math., **342**, Longman.

[10] S. Hu and N.S. Papageorgiou (1997). *Handbook of Multivalued Analysis, Vol. I: Theory*. Dordrecht-Boston-London: Kluwer Acad. Publ.

[11] M. Kamenskii, V. Obukhovskii and P. Zecca (2000). On Semilinear Differential Inclusions with Lower Semicontinuous Nonlinearities, *Annali di Matematica Pura e Appl.*, **CLXXV** (ser. IV), (to appear), 235–244.

[12] M. Kamenskii, V. Obukhovskii and P. Zecca (2001). Condensing Multivalued Maps and Semilinear Differential Inclusions in Banach Spaces, Manuscript, De Gruyter Series in Nonlinear Analysis and Application, (to appear).

[13] V.V. Obukhovskii (1991). Semilinear Functional-Differential Inclusions in a Banach Space and Controlled Parabolic Systems, *Soviet J. Automat. Inform. Sci.*, **24**, 71–79.

SIMAA 4(2002) 383–402

22. A Nonlinear Multivalued Problem with Nonlinear Boundary Conditions

Michela Palmucci[1] and Francesca Papalini[2,*]

[1]*Department of Mathematics, University of Perugia, Via Vanvitelli, 1 Perugia, 06123, Italy*
[2]*Department of Mathematics, University of Ancona, Via Brecce Bianche, Ancona 60131, Italy*

Abstract: In this chapter we study multivalued problem with nonlinear boundary conditions in which the usual differential operator $x \mapsto x''$ is substitued by the operator $x \mapsto (\phi(x'))'$. We prove the existence of solutions and extremal solutions in the order interval delimited by the lower and upper solution. The boundary conditions considered by us contains, as a particular case, the Dirichlet, periodic, Nuemann and Sturm-Liouville conditions.

Keywords and Phrases: Upper solution, lower solution, order interval, extremal solutions, nonlinear boundary conditions

1991 AMS Subject Classification: 34B15

1. INTRODUCTION

In the last years several Authors have studied differential equations of the form

$$(\phi(x'(t)))' = f(t, x(t), x'(t)), \quad \text{a.e. on } I = [a, b] \tag{1}$$

in which the usual linear differential operator $x \mapsto x''$ is substitued by the operator $x \mapsto (\phi(x'))'$, where ϕ: $\mathbb{R} \to \mathbb{R}$ is a suitable monotone function. This is the case, for instance, of the so called one-dimensional p-Laplacian operator obtained by putting

$$\phi(s) = \begin{cases} |s|^{p-2}s & \text{if } s \neq 0 \\ 0 & \text{if } s = 0, \end{cases}$$

where $p > 1$ is a fixed real number. This kind of equations find a lot of applications to non-Newtonian fluid theory, diffusion of flows in porus media, nonlinear elasticity and theory of capillary surfaces (see for istance [3,12,19]).

Different kind of equations, containing this type of operator, with different boundary conditions, are extensively studied in order to obtain existence and multiplicity of

* Corresponding author. E-mail: papalini@dipmat.unian.it

solutions. See for example [7,9,22,24], for the Dirichlet problem, and [10,13,17,25] for the periodic and other boundary conditions. We want also mention the chapters of M. Zang [30] and Manásevich and Mawhin [21] in which the system case is examined.

Recently Wang *et al.* [29] have studied the equation (1) in which $\phi\colon \mathbb{R} \to \mathbb{R}$ is a continuous, increasing function with $\phi(\mathbb{R}) = \mathbb{R}$ and, using the method of lower and upper solutions, they proved the existence of solutions satisfying Dirichlet and mixed boundary conditions and assuming the continuity of the function f.

Later Wang and Gao [28] have generalized these results to the case in which f is a Caratheodory function. Subsequently in [4], Cabada and Pouso have obtained the existence of solutions for the periodic and Neumann problem supposing, also in this case, that f is a Caratheodory function and in presence of lower and upper solutions. All these results for linear boundary conditions have been generalized in [5] by Cabada and Pouso to a case of nonlinear boundary conditions which contain, in particular the Dirichlet, periodic, Nuemann and mixed boundary conditions. Similar nonlinear conditions were been considered in [14] for the second order equation $x'' = f(t, x, x')$.

In this chapter we study the scalar multivalued boundary value problem

$$
\begin{aligned}
&(\phi(x'(t)))' \in F(t, x(t), x'(t)), \quad \text{a.e. on } I \\
&g(x(a), x(b), x'(a), x'(b)) = 0 \\
&x(b) = h(x(a)),
\end{aligned}
$$

where $F\colon [a,b] \times \mathbb{R} \times \mathbb{R} \to 2^{\mathbb{R}}$ is a Caratheodory multifunction, $g\colon \mathbb{R}^4 \to \mathbb{R}$ is a continuous function nondecreasing in the third variable and increasing in the fourth one, while $h\colon \mathbb{R} \to \mathbb{R}$ is a continuous and nonincreasing function. This problem contains, as a particular case, the periodic problem and for it we prove the existence of solutions (cf. Theorem 2) making use of the method of lower and upper solutions. In the context of multivalued boundary value problems this method has been used recently in [15,16,26], in which semilinear differential inclusions with Dirichlet, periodic or Sturm-Liouville conditions are studied. Moreover we succeed in obtaining the existence of extremal solutions for the previous problem in the order interval delimited by lower and upper solution (cf. Theorem 3). These theorems contain, as a particular case, the analogous result of Cabada and Pouso (cf. Theorem 2.1 [5]).

In the second part of this chapter we also study these two problems

$$
\begin{aligned}
&(\phi(x'(t)))' \in F(t, x(t), x'(t)), \quad \text{a.e. on } I \\
&p(x(a), x'(a)) = 0 \\
&q(x(b), x'(b)) = 0,
\end{aligned}
$$

$$
\begin{aligned}
&(\phi(x'(t)))' \in F(t, x(t), x'(t)), \quad \text{a.e. on } I \\
&q(x(a), x'(a)) = 0 \\
&p(x(b), x'(b)) = 0,
\end{aligned}
$$

where $p\colon \mathbb{R}^2 \to \mathbb{R}$ is a continuous function nonincreasing in the second variable and $q\colon \mathbb{R}^2 \to \mathbb{R}$ is a continuous function nondecreasing in the second variable and we obtain for them the existence of extremal solutions. In this way we also cover the case of Sturm-Liouville and so Dirichlet problem, which are contained in the first problem, and the case of Nuemann and mixed boundary problem, which are contained in the second one.

Finally we observe that the results of this chapter also improve the corresponding results obtained for semilinear differential inclusions by [16,26].

2. MATHEMATICAL PRELIMINARIES

Let X, Y be Hausdorff topological vector spaces and $F\colon X \to 2^Y$ be a multifunction. We denote by

$$R(F) = \bigcup_{x \in X} F(x) \quad \text{and} \quad GrF = \{(x,y) : y \in F(x)\}$$

the range and the graph of F respectively; moreover, for every subset A of Y, we put $F^{-1}(A) = \{x \in X : F(x) \cap A \neq \emptyset\}$ and $F^{+}(A) = \{x \in X : F(x) \subset A\}$.

F is called *upper semicontinuous* (u.s.c.) on X if $F^{-1}(A)$ is closed, for every closed subset A of X (or, equivalently, if $F^{+}(A)$ is open, for every open subset A of X). It is known another definition of upper semicontinuity called metric upper semicontinuity: F is said to be *metric upper semicontinuous* (u.s.c.)$_m$ on X if $\forall x_0 \in X$ and for every neighbourhood U of zero in Y there exists a neighbourhood $I(x_0)$ of x_0, with the property

$$F(x) \subset F(x_0) + U, \quad \forall x \in I(x_0).$$

In general every (u.s.c) multifunction is also (u.s.c.)$_m$ and the two definitions are equivalent, for instance, for compact-valued multifunctions.

The multifunction F is said to have *closed graph* if the set GrF is closed in $X \times Y$. Now we suppose that X and Y are a Banach spaces; F is called *weakly upper semicontinuous* (w-u.s.c.) if for every sequence $\{x_n\}_n$, $x_n \to x$ in X and for every sequence $\{y_n\}_n$, $y_n \in F(x_n)$, $\forall n \in \mathbb{N}$, $y_n \rightarrow^w y$ in Y, we have that $y \in F(x)$.

Let (T, Σ, μ) be a measure space; a closed-valued multifunction $F\colon T \to 2^X$ ($F\colon T \to \{A \subset X\colon$ A is closed$\}$) is said to be *measurable* (*weakly measurable*) if $F^{-1}(B) \in \Sigma$ for every closed (open) subset B of X. If some values of F are empty subsets of X, then F is measurable if $T_0 = \{t \in T : F(t) = \emptyset\}$ belongs to Σ and on $T - T_0$ F is weakly measurable.

If X is a Banach space and I is the closed interval $[a, b]$, we denote by $W^{m,p}(I, X)$ the space of the functions $u \in L^p(I, X)$ which have the distributional derivatives $u^{(k)}$, $k = 1, \ldots, m$, which belong to the space $L^p(I, X)$. It is known (cf. [1]) that the Sobolev space $W^{m,p}(I, X)$ is a Banach space with the norm defined by

$$\|u\|_{m,p} = \|u\|_p + \sum_{k=1}^{m} \|u^{(k)}\|_p;$$

moreover, if X is reflexive then $W^{m,p}(I, X)$ can be identified with the space of absolutely continuous functions which have strong derivatives $u^{(k)}$, $k = 1, \ldots, m$, with the property that $u^{(k)}$, $k = 1, \ldots, m-1$, is absolutely continuous and $u^{(m)} \in L^p(I, X)$.

3. EXISTENCE RESULTS

Let $I = [a, b]$. We start by the following second order differential inclusion with Dirichlet boundary conditions:

$$\begin{aligned} &(\phi(x'(t)))' \in F(t, x(t), x'(t)), \quad \text{a.e. on } I \\ &x(a) = \nu_0, \quad x(b) = \nu_1 \end{aligned} \tag{I}$$

where $F: I \times \mathbb{R} \times \mathbb{R} \to 2^{\mathbb{R}}$ is a multifunction with nonempty, compact and convex values and $\nu_0, \nu_1 \in \mathbb{R}$.

For the problem (*I*) we give the definition of lower and upper solutions: a function $\alpha \in C^1(I, \mathbb{R})$ is said to be a *lower solution* for problem (I) if $\phi \circ \alpha' \in W^{1,1}(I, \mathbb{R})$ and

$$F(t, \alpha(t), \alpha'(t)) \cap]-\infty, (\phi(\alpha'(t)))'] \neq \emptyset, \quad \text{a.e. on } I$$
$$\alpha(a) \leq \nu_0, \quad \alpha(b) \leq \nu_1.$$

Since F has nonempty, compact and convex values in $\mathbb{R}$, we can represent F in this way:

$$F(t, x, y) = \left[\underline{f}(t, x, y), \bar{f}(t, x, y)\right], \quad \text{for all } (t, x, y) \in I \times \mathbb{R} \times \mathbb{R}, \tag{2}$$

where $\underline{f}, \bar{f}: I \times \mathbb{R} \times \mathbb{R} \to \mathbb{R}$ are suitable functions.

So we can say that $\alpha \in C^1(I, \mathbb{R})$ is a lower solution of problem (*I*) if $\phi \circ \alpha' \in W^{1,1}(I, \mathbb{R})$ and

$$(\phi(\alpha'(t)))' \geq \underline{f}(t, \alpha(t), \alpha'(t)), \quad \text{a.e. on } I$$
$$\alpha(a) \leq \nu_0, \quad \alpha(b) \leq \nu_1.$$

A function $\beta \in C^1(I, \mathbb{R})$ is said to be an *upper solution* for problem (I) if $\phi \circ \beta' \in W^{1,1}(I, \mathbb{R})$ and

$$F(t, \beta(t), \beta'(t)) \cap [(\phi(\beta'(t)))', +\infty[\neq \emptyset, \quad \text{a.e. on } I$$
$$\beta(a) \geq \nu_0, \quad \beta(b) \geq \nu_1.$$

So, using the representation (2) of F, we say that $\beta \in C^1(I, \mathbb{R})$ is an upper solution of problem (I) if $\phi \circ \beta' \in W^{1,1}(I, \mathbb{R})$ and

$$(\phi(\beta'(t)))' \leq \bar{f}(t, \beta(t), \beta'(t)), \quad \text{a.e. on } I$$
$$\beta(a) \geq \nu_0, \quad \beta(b) \geq \nu_1.$$

A function $x \in C^1(I, \mathbb{R})$ is a *solution* of problem (I) if $\phi \circ x' \in W^{1,1}(I, \mathbb{R})$ and

$$(\phi(x'(t)))' \in F(t, x(t), x'(t)), \quad \text{a.e. on } I$$
$$x(a) = \nu_0, \quad x(b) = \nu_1.$$

We introduce the following hypotheses on the multifunction F:

$H(F)_1$: $F: I \times \mathbb{R} \times \mathbb{R} \to 2^{\mathbb{R}}$ is a multifunction with nonempty, compact and convex values such that

(i) $\forall x, y \in \mathbb{R}$, $t \mapsto F(t, x, y)$ is measurable;
(ii) for a.e. on I, $(x, y) \mapsto F(t, x, y)$ is (u.s.c);
(iii) $\forall r > 0, \exists \gamma_r \in L^1(I, \mathbb{R})$ such that $\|F(t, x, y)\| \leq \gamma_r(t)$ a.e. on I and $\forall x, y \in \mathbb{R}$ with $|x|, |y| \leq r$.

$H(F)_2$: F satisfies a Nagumo condition relative to the pair α and β, with $\alpha, \beta \in C(I, \mathbb{R}), \alpha \leq \beta$ in I, i.e., there exists a function $k \in L^p(I, \mathbb{R})$, $1 \leq p \leq \infty$, and $\theta: [0, \infty[\to]0, \infty[$ continuous, such that

$$\|F(t, x, y)\| \leq k(t)\theta(|y|) \quad \text{a.e. on } (t, x, y) \in \Omega,$$

where $\Omega = \{(t, x, y) \in I \times \mathbb{R} \times \mathbb{R} : \alpha(t) \leq x \leq \beta(t)\}$, and moreover

$$\int_{\phi(v)}^{+\infty} \frac{|\phi^{-1}(x)|^{(p-1)/p}}{\theta(|\phi^{-1}(x)|)} dx, \int_{-\infty}^{\phi(-v)} \frac{|\phi^{-1}(x)|^{(p-1)/p}}{\theta(|\phi^{-1}(x)|)} dx > \mu^{(p-1)/p} \|k\|_p,$$

with

$$\mu = \max_{t \in I} \beta(t) - \min_{t \in I} \alpha(t),$$

$$v = \frac{\max\{|\alpha(a) - \beta(b)|, |\alpha(b) - \beta(a)|\}}{b - a}$$

and

$$\|k\|_p = \begin{cases} \sup_{t \in I} |k(t)| & \text{if } p = \infty \\ \left\{ \int_a^b |k(t)|^p dt \right\}^{1/p} & \text{if } 1 \leq p < \infty, \end{cases}$$

where $(p - 1)/p \equiv 1$ for $p = \infty$.

H_0: ϕ: $\mathbb{R} \to \mathbb{R}$ is continuous, increasing and $\phi(\mathbb{R}) = \mathbb{R}$.
It holds the following existence result for problem (I).

Theorem 1: *Let α and β be lower and upper solutions, respectively, for problem (I) such that $\alpha \leq \beta$ in I. Assume that hypotheses H_0, $H(F)_1$ are satisfied and the hypotheses $H(F)_2$ hold with respect to α and β.*

Then problem (I) has at least one solution in the order interval $[\alpha, \beta] = \{x \in C^1(I, \mathbb{R}) : \alpha(t) \leq x(t) \leq \beta(t)\}$.

Proof: The Nagumo condition implies the existence of two real number $M_- < 0 < M_+$, such that

$$M_- < -v \leq v < M_+, \quad M_- < \alpha'(t), \quad M_+ > \beta'(t) \quad \text{for all } t \in I \quad \text{and}$$

$$\int_{\phi(v)}^{\phi(M_+)} \frac{|\phi^{-1}(x)|^{(p-1)/p}}{\theta(|\phi^{-1}(x)|)} dx, \int_{\phi(M_-)}^{\phi(-v)} \frac{|\phi^{-1}(x)|^{(p-1)/p}}{\theta(|\phi^{-1}(x)|)} dx > \mu^{(p-1)/p} \|k\|_p.$$

Let τ: $W^{1,1}(I, \mathbb{R}) \to W^{1,1}(I, \mathbb{R})$ the truncation operator defined by

$$\tau(x)(t) = \begin{cases} \beta(t) & \text{if } x(t) \geq \beta(t) \\ x(t) & \text{if } \alpha(t) \leq x(t) \leq \beta(t) \\ \alpha(t) & \text{if } x(t) \leq \alpha(t), \end{cases}$$

for every $x \in W^{1,1}(I, \mathbb{R})$ and for all $t \in I$; by Lemma 2.7 of [18] we get that $\tau(x) \in W^{1,1}(I, \mathbb{R})$ and

$$\tau(x)'(t) = \begin{cases} \beta'(t) & \text{if } x(t) \geq \beta(t) \\ x'(t) & \text{if } \alpha(t) \leq x(t) \leq \beta(t) \\ \alpha'(t) & \text{if } x(t) \leq \alpha(t), \end{cases}$$

for each $x \in W^{1,1}(I, \mathbb{R})$ and a.e. on I. Put $N = \max\{-M_-, M_+\}$, we denote by $q_N: \mathbb{R} \to \mathbb{R}$ and $u: I \times \mathbb{R} \to \mathbb{R}$ the truncation function and the penalty function respectively, which are defined by

$$q_N(x) = \begin{cases} N & \text{if } x \geq N \\ x & \text{if } -N \leq x \leq N \\ -N & \text{if } x \leq -N, \end{cases}$$

for all $x \in \mathbb{R}$ and

$$u(t,x) = \begin{cases} x - \beta(t) & \text{if } x \geq \beta(t) \\ 0 & \text{if } \alpha(t) \leq x \leq \beta(t) \\ x - \alpha(t) & \text{if } x \leq \alpha(t), \end{cases}$$

for every $(t, x) \in I \times \mathbb{R}$.

Now let $\tilde{F}: I \times \mathbb{R} \times \mathbb{R} \to 2^{\mathbb{R}} - \{\emptyset\}$ the multifunction defined by

$$\tilde{F}(t,x,y) = \overline{\text{co}}\left(\bigcup_{i=1}^{3} F_i(t,x,y)\right), \quad \forall (t,x,y) \in I \times \mathbb{R} \times \mathbb{R},$$

where $F_i: I \times \mathbb{R} \times \mathbb{R} \to 2^{\mathbb{R}}$, $i = 1, 2, 3$, are the following multifunctions

$$F_1(t,x,y) = \begin{cases} F(t,x,y) & \text{if } \alpha(t) \leq x \leq \beta(t) \\ \emptyset & \text{otherwise,} \end{cases}$$

$$F_2(t,x,y) = \begin{cases} F(t,\alpha(t),y) \cap]-\infty, (\phi(\alpha'(t)))'] & \text{if } x \leq \alpha(t) \text{ and } F(t,\alpha(t),y) \cap]-\infty, (\phi(\alpha'(t)))'] \neq \emptyset \\ \{(\phi(\alpha'(t)))'\} & \text{if } x \leq \alpha(t) \text{ and } F(t,\alpha(t),y) \cap]-\infty, (\phi(\alpha'(t)))'] = \emptyset \\ \emptyset & \text{otherwise,} \end{cases}$$

$$F_3(t,x,y) = \begin{cases} F(t,\beta(t),y) \cap [(\phi(\beta'(t)))', +\infty[& \text{if } x \geq \beta(t) \text{ and } F(t,\beta(t),y) \cap [(\phi(\beta'(t)))', +\infty[\neq \emptyset \\ \{(\phi(\beta'(t)))'\} & \text{if } x \geq \beta(t) \text{ and } F(t,\beta(t),y) \cap [(\phi(\beta'(t)))', +\infty[= \emptyset \\ \emptyset & \text{otherwise,} \end{cases}$$

for all $(t, x, y) \in I \times \mathbb{R} \times \mathbb{R}$.

It is obvious that $\tilde{F}$ has nonempty, closed and convex values, moreover, using the properties of measurable applications (cf. [20]), from measurability of the functions $t \mapsto F_i(t,x,y)$, $i = 1, 2, 3$, we deduce the maesurability of the multifunction $t \mapsto \tilde{F}(t,x,y)$. In a similar way, using the properties of (u.s.c) multifunctions (cf. [2]) and the fact that, a.e. on I, the applications $(x,y) \mapsto F_i(t,x,y)$, $i = 1, 2, 3$, are (u.s.c.), we obtain the upper semicontinuity for the multifunction $(x,y) \mapsto \tilde{F}(t,x,y)$. Finally, from $H(F)_1$ (iii), it follows that $\tilde{F}$ is integrably bounded on the bounded subsets of $L^1(I, \mathbb{R})$ (i.e., $\forall r > 0, \exists \tilde{\gamma}_r \in L^1(I, \mathbb{R})$ such that $\|\tilde{F}(t,x,y)\| \leq \tilde{\gamma}_r(t)$ a.e. on I and $\forall x, y \in \mathbb{R}$ with $|x|, |y| \leq r$).

We consider now the following modified problem

$$\begin{aligned}&(\phi(x'(t)))' \in \tilde{F}(t, x(t), q_N(\tau(x)'(t))) + \arctan(u(t, x(t))), \quad \text{a.e. on } I\\&x(a) = \nu_0, \quad x(b) = \nu_1.\end{aligned} \tag{3}$$

Claim 1: *If x is a solution of problem* (3), *then $x \in [\alpha, \beta]$*
Let x a solution of problem (3), we shall only see that $\alpha(t) \leq x(t)$, $\forall t \in T$, in a similar way it can be proved that $x(t) \leq \beta(t)$, $\forall t \in T$.

By boundary conditions we have $\alpha(a) \leq x(a) \leq \beta(a)$ and $\alpha(b) \leq x(b) \leq \beta(b)$. If there exists $t_0 \in (a, b)$ such that

$$x(t_0) - \alpha(t_0) = \min_{t \in T}\{(x - \alpha)(t)\} < 0,$$

then, since $x - \alpha \in C^1(I, \mathbb{R})$, we have $(x - \alpha)'(t_0) = 0$. Moreover, there exist $a \leq t_1 < t_0 < t_2 \leq b$ such that $x(t) < \alpha(t)$ in (t_1, t_2) and $(x - \alpha)(t_1) = (x - \alpha)(t_2) = 0$. Thus

$$\begin{aligned}&(\phi(x'(t)))' \in \big(F(t, \alpha(t), \alpha'(t)) \cap]-\infty, (\phi(\alpha'(t)))']\big) + \arctan((x - \alpha)(t)),\\&\text{a.e. on } (t_1, t_2)\end{aligned}$$

and so

$$(\phi(x'(t)))' - (\phi(\alpha'(t)))' \leq \arctan((x - \alpha)(t)) < 0, \text{ a.e. on } (t_1, t_2).$$

Then we deduce that

$$\phi(x'(t)) - \phi(\alpha'(t)) < \phi(x'(t_0)) - \phi(\alpha'(t_0)), \quad \text{for all } t \in (t_0, t_2).$$

Now, by monotonicity and injectivity of ϕ, we obtain $x'(t) < \alpha'(t)$, for all $t \in (t_0, t_2)$, hence we conclude

$$(x - \alpha)(t_2) < (x - \alpha)(t_0) < 0$$

which is a contradiction with the choice of t_2. Therefore $x \in [\alpha, \beta]$ as claimed.

Claim 2: *If x is a solution of problem* (3), *then $|x'(t)| \leq N$ for all $t \in I$.*
Let x be a solution of problem (3), we shall prove that $M_- \leq x'(t) \leq M_+$, for all $t \in I$.

By the mean-value theorem, there exists $t_0 \in (a, b)$ such that

$$x'(t_0) = \frac{x(b) - x(a)}{b - a},$$

then, by definition of v and since $M_- < -v \leq v < M_+$, we have

$$M_- < -v \leq \frac{\alpha(b) - \beta(a)}{b - a} \leq x'(t_0) \leq \frac{\beta(b) - \alpha(a)}{b - a} \leq v < M_+.$$

Let $v_0 = |x'(t_0)|$. If there exists a point $\tilde{t} \in I$ for which $x'(\tilde{t}) > M_+$ or $x'(\tilde{t}) < M_-$, by continuity of x' we can choose $t_1 \in I$ verifying one of the following situations:

(i) $x'(t_0) = v_0$, $x'(t_1) = M_+$ and $v_0 \leq x'(t) \leq M_+$ for all $t \in (t_0, t_1)$;
(ii) $x'(t_1) = M_+$, $x'(t_0) = v_0$ and $v_0 \leq x'(t) \leq M_+$ for all $t \in (t_1, t_0)$;

(iii) $x'(t_1) = -v_0$, $x'(t_0) = M_-$ and $M_- \le x'(t) \le -v_0$ for all $t \in (t_0, t_1)$;
(iv) $x'(t_1) = M_-$, $x'(t_0) = -v_0$ and $M_- \le x'(t) \le -v_0$ for all $t \in (t_1, t_0)$.

Assume that the situation (i) holds (in the other cases we proceed in a similar way).

Since $-N \le M_- \le v_0 \le x'(t) \le M_+ \le N$ for all $t \in (t_0, t_1)$, from Banach Lemma (cf. [23]) we have

$$(\phi(x'(t)))' \in \tilde{F}(t, x(t), q_N(x'(t))) = F(t, x(t), x'(t)), \quad \text{a.e on } (t_0, t_1)$$

so, by the Nagumo condition

$$|(\phi(x'(t)))'| \le k(t)\theta(|x'(t)|), \quad \text{a.e on } (t_0, t_1).$$

Note that $\phi^{-1}(s) \ge 0$ for $s \in [\phi(v_0), \phi(M_+)]$ and $\phi(v_0) \le \phi(v)$, therefore we have

$$\int_{\phi(v_0)}^{\phi(M_+)} \frac{|\phi^{-1}(s)|^{(p-1)/p}}{\theta(|\phi^{-1}(s)|)} ds \ge \int_{\phi(v)}^{\phi(M_+)} \frac{|\phi^{-1}(s)|^{(p-1)/p}}{\theta(|\phi^{-1}(s)|)} ds > \mu^{(p-1)/p}\|k\|_p. \tag{4}$$

Consider now the function φ: $[t_0, t_1] \to [\phi(v_0), \phi(M_+)]$ defined by

$$\varphi(r) = \phi(x'(r)).$$

By definition of solution $\varphi \in W^{1,1}(I, \mathbb{R})$ and using the continuity of the function under the integral we obtain

$$\int_{\phi(v_0)}^{\phi(M_+)} \frac{|\phi^{-1}(s)|^{(p-1)/p}}{\theta(|\phi^{-1}(s)|)} ds = \int_{t_0}^{t_1} \frac{\varphi'(t))[x'(t)]^{(p-1)/p}}{\theta(x'(t))} dt = \int_{t_0}^{t_1} \frac{|(\phi(x'(t)))'|[x'(t)]^{(p-1)/p}}{\theta(x'(t))} dt,$$

then we deduce

$$\int_{\phi(v_0)}^{\phi(M_+)} \frac{|\phi^{-1}(s)|^{(p-1)/p}}{\theta(|\phi^{-1}(s)|)} ds \le \int_{t_0}^{t_1} k(t)[x'(t)]^{(p-1)/p} dt$$

and by Hölder's inequality it follows

$$\int_{\phi(v_0)}^{\phi(M_+)} \frac{|\phi^{-1}(s)|^{(p-1)/p}}{\theta(|\phi^{-1}(s)|)} ds \le \|k\|_p \mu^{(p-1)/p}$$

hence, taking into account (4) we get a contradiction. Therefore $|x'(t)| \le N$, for all $t \in I$, as claimed.

Claim 3: *The problem* (3) *has at least a solution.*
Put $\bar{F}(t, x, y) = \tilde{F}(t, x, y) + \arctan(u(t, x))$, for every $(t, x, y) \in I \times \mathbb{R} \times \mathbb{R}$; we denote by N: $C^1(I, \mathbb{R}) \to 2^{L^1(I, \mathbb{R})}$ the Nemytskii operator associated to $\bar{F}$, which is

$$N(x) = \{y \in L^1(I, \mathbb{R}) : y(t) \in \bar{F}(t, x(t), q_N(\tau(x)'(t))) \quad \text{a.e. on } I\}, \quad \forall x \in C^1(I, \mathbb{R}).$$

Since $\bar{F}$ is a Caratheodory multifunction with nonempty, convex and compact values, it follows easily that the values of N are nonempty, closed and convex, moreover N is weakly upper semicontinuous (the proof of this is the same, with minor modifications, of that of Proposition 1.4 [27:p. 960]).

Let A: $C^1(I, \mathbb{R}) \to 2^{C(I,\mathbb{R})}$ the Caratheodory operator associated to $\bar{F}$ defined by

$$A(x) = \left\{ y \in C(I, \mathbb{R}) : \exists f \in N(x) \text{ with } y(t) = \int_a^t f(s)ds, \forall t \in I \right\}, \quad \forall x \in C^1(I, \mathbb{R}).$$

Observe that for every $x \in C^1(I, \mathbb{R})$ the set $A(x)$ is nonempty, convex and compact: convexity and nonemptyness are simple consequences of convexity and nonemptyness of $N(x)$; so we shall only prove the compactness. To this end, let $\{y_n\}_n \subset A(x)$ be a sequence. For every $n \in \mathbb{N}$ there is a function $f_n \in N(x)$ such that $y_n(t) = \int_a^t f_n(s)ds$, $\forall t \in I$. Recalling that $\bar{F}$ is integrably bounded on the bounded subset of $L^1(I, \mathbb{R})$, we infer that

$$|f_n(t)| \leq \psi(t), \quad \text{a.e. on } I \tag{5}$$

where

$$\psi(t) = \tilde{\gamma}_r(t) + \frac{\pi}{2}, \tag{6}$$

and $\tilde{\gamma}_r$ is the function determined in correspondence of $r = \max\{N, \|\alpha\|_\infty, \|\beta\|_\infty\}$.

From (5) and using the uniform continuity of the function $t \mapsto \int_a^t \psi(s)ds$ on I, by passing to a subsequence if necessary, we deduce the existence of $y \in C(I, \mathbb{R})$ such that $y_n \to y$ in $C^1(I, \mathbb{R})$ and invoking Dunford–Pettis theorem, it is possible to find $f \in N(x)$ such that $y(t) = \int_a^t f(s)ds$, hence we conclude that $A(x)$ is compact in $C(I, \mathbb{R})$.

Proceeding as above, we establish also that GrF is closed in $C^1(I, \mathbb{R}) \times C(I, \mathbb{R})$ and $A(C^1(I, \mathbb{R}))$ is a relative compact subset of $C(I, \mathbb{R})$, which implies the upper semicontinuity of the operator A.

Now, put $D = \{y \in C(I, \mathbb{R}) : \|y\|_\infty \leq \|\psi\|\}$, where ψ is the function introduced by (6). We consider the following operator g: $D \subset C(I, \mathbb{R}) \to C^1(I, \mathbb{R})$ defined, for each $y \in C(I, \mathbb{R})$ and every $t \in I$, by

$$g(y)(t) = \nu_0 + \int_a^t \phi^{-1}(\tau_y + y(r))dr,$$

where, for all $y \in C(I, \mathbb{R})$, $\tau_y \in \mathbb{R}$ denotes the unique element satisfying

$$\int_a^b \phi^{-1}(\tau_y + y(r))dr = \nu_1 - \nu_0. \tag{7}$$

We start showing the existence and uniqueness of such τ_y.
For every $y \in D$, we define φ_y: $\mathbb{R} \to \mathbb{R}$ as

$$\varphi_y(\xi) = \int_a^b \phi^{-1}(\xi + y(r))dr \quad \text{for all } \xi \in \mathbb{R}.$$

Clearly, φ_y is continuous and increasing, moreover, by definition of D, we have

$$\xi - \|\psi\|_1 \leq \xi + y(r) \leq \xi + \|\psi\|_1 \quad \text{for each } r \in I, \xi \in \mathbb{R} \text{ and } y \in D$$

then, from hypothesis H_0, it follows that

$$f_-(\xi) \equiv (b-a)\phi^{-1}(\xi - \|\psi\|_1) \le \varphi_y(\xi) \le (b-a)\phi^{-1}(\xi + \|\psi\|_1) \equiv f_+(\xi) \qquad (8)$$

for every $\xi \in \mathbb{R}$ and for each $y \in D$.
The functions f_- and f_+ are continuous, increasing and surjective; moreover, by (8) we obtain $\varphi_y(\mathbb{R}) = \mathbb{R}$, for all $y \in D$, so the existence of τ_y is proved. Uniqueness follows immediately by injectivity of φ_y.

Note that, since τ_y satisfies (7), by mean-value theorem, for every $y \in D$, we have

$$\nu_1 - \nu_0 = \int_a^b \phi^{-1}(\tau_y + y(r))dr = (b-a)\phi^{-1}(\tau_y + y(\bar{r}))$$

with suitable $\bar{r} \in (a, b)$ and so, recalling definition of D, there exists a costant $L > 0$ such that

$$|\tau_y| \le L \quad \text{for all } y \in D. \qquad (9)$$

Now we prove two properties on the operator g that we shall use in the followings: g is uniformly continuous on co($R(A)$) and g(co($R(A)$)) is relative compact in $C^1(I, \mathbb{R})$. To this end let y_1, y_2 be elements of co($R(A)$) and we denote by τ_1, τ_2 the points τ_{y_1} and τ_{y_2} (cf. (7)). Using mean-value theorem, there exists $\hat{r} \in (a, b)$ such that

$$(b-a)[\phi^{-1}(\tau_1 + y_1(\hat{r})) - \phi^{-1}(\tau_2 + y_2(\hat{r}))] = 0$$

so, by monotonicity of ϕ^{-1}, we have

$$|\tau_1 - \tau_2| \le \|y_1 - y_2\|_\infty,$$

therefore, since

$$|\tau_1 + y_1(r) - \tau_2 - y_2(r)| \le |\tau_1 - \tau_2| + \|y_1 - y_2\|_\infty, \quad \text{for all } r \in I$$

from the uniform continuity of ϕ^{-1} on $[-M, M]$, with $M = L + \|\psi\|_1$, we conclude that g is uniformly continuous on co($R(A)$).

For proving that the set g(co($R(A)$)) is relative compact in $C^1(I, \mathbb{R})$, we fix a sequence $\{y_n\}_n$ in g(co($R(A)$)). For every $n \in \mathbb{N}$ there exists $u_n \in$ co($R(A)$) such that

$$y_n(t) = \nu_0 + \int_a^t \phi^{-1}(\tau_{u_n} + u_n(r))dr, \quad \text{for all } t \in T,$$

where

$$u_n = \sum_{i=1}^{p(n)} \alpha_{n_i} v_{n_i} \text{ with } \alpha_{n_i} \in (0,1), v_{n_i} \in R(A), \quad \text{for all } i = 1, \ldots, p(n), \quad \text{and} \quad \sum_{i=1}^{p(n)} \alpha_{n_i} = 1.$$

Using (5),(9), hypothesis H_0 and the convexity of D, we obtain that the sequences $\{y_n\}_n$ and $\{y'_n\}_n$ are uniformly bounded on I and then, by continuity of ϕ^{-1} on $[-M, M]$ we deduce that $\{y_n\}_n$ is equicontinous on I.

Fixed $t_1, t_2 \in I$, by definition of u_n, we have

$$|u_n(t_1) - u_n(t_2)| \leq \sum_{i=1}^{p(n)} \alpha_{n_i} |v_{n_i}(t_1) - v_{n_i}(t_2)|$$

so, recalling that, for each $i = 1, \ldots, p(n)$ and every $n \in \mathbb{N}$, v_{n_i} is an element of $R(A)$, and using (5), we obtain

$$|u_n(t_1) - u_n(t_2)| \leq \sum_{i=1}^{p(n)} \alpha_{n_i} |\int_{t_1}^{t_2} \psi(s)ds|$$

from which, invoking the uniform continuity on I of the function $t \mapsto \int_a^t \psi(s)ds$ we conclude that the sequence $\{y_n'\}_n$ is equicontinuous on I. So by Arzelá-Ascoli theorem, the set $g(\text{co}(R(A))$ is relative compact in $C^1(I, \mathbb{R})$.

Now we consider the multivalued operator Γ: $C^1(I, \mathbb{R}) \to 2^{C^1(I, \mathbb{R})}$ defined as follows

$$\Gamma(x) = g(A(x)) \quad \text{for all } x \in C^1(I, \mathbb{R}).$$

Observe that, if $x \in C^1(I, \mathbb{R})$ is a fixed point of Γ, there exists $y \in A(x)$ and $f \in N(x)$ with the property that

$$x(t) = \nu_0 + \int_a^t \phi^{-1}\left(\tau_y + \int_a^r f(s)ds\right)dr \quad \text{for all } t \in I,$$

and so x is a solution of problem (3).

In order to prove that Γ has a fixed point, we denote by S the following convex and compact set

$$S = \text{co}\left(\overline{g(\text{co}(R(A)))}\right)$$

and we observe that the operator Γ: $S \to 2^S$ has closed graph, since it is $(\text{u.s.c.})_m$ (this is a simple consequence of the metric upper semicontinuity of A on S and the uniform continuity of g on $\text{co}(R(A))$ and it has closed values on S (cf. [8]).

Moreover since g is uniformly continuous on $\text{co}(R(A))$ and using Theorem 1 of [6] on the operator A, we deduce that for every $\epsilon > 0$ there exists a continuous single-valued function f: $S \to S$ with the property that

$$d^*(Grf, Gr\Gamma) \equiv \sup\{d(x, Gr\Gamma) : x \in Grf\} < \epsilon.$$

Now, taking into account Proposition 1 of [6], we obtain the existence of a fixed point for the multivalued operator Γ.

Finally note that if x is a solution of problem (3), from Claim 1 we have $\alpha(t) \leq x(t) \leq \beta(t)$ for all $t \in I$, so by Banach Lemma it follows

$$(\phi(x'(t)))' \in F(t, x(t), q_N(x'(t))) \quad \text{a.e. on } I.$$

Therefore, using Claim 2, we conclude that x is a solution of problem (I). Q.E.D.

We pass to consider now the more general problem with nonlinear boundary conditions:

$$\begin{aligned} &(\phi(x'(t)))' \in F(t, x(t), x'(t)), \quad \text{a.e. on } I \\ &g(x(a), x(b), x'(a), x'(b)) = 0 \\ &x(b) = h(x(a)) \end{aligned} \tag{II}$$

where g and h are functions on which we make the following assumptions:
B_1: g: $\mathbb{R} \times \mathbb{R} \times \mathbb{R} \times \mathbb{R} \to \mathbb{R}$ is such that

(i) g is continuous;
(ii) $\forall u, v, z \in \mathbb{R}$, the function $w \mapsto g(u, v, w, z)$ is not decreasing;
(iii) $\forall u, v, w \in \mathbb{R}$, the function $z \mapsto g(u, v, w, z)$ is not increasing.

B_2: h: $\mathbb{R} \to \mathbb{R}$ is continuous and not decreasing.

For the problem (II) a function $\alpha \in C^1(I, \mathbb{R})$ is said to be a *lower solution* if $\phi \circ \alpha' \in W^{1,1}(I, \mathbb{R})$ and

$$\begin{aligned} &F(t, \alpha(t), \alpha'(t)) \cap]-\infty, (\phi(\alpha'(t)))'] \neq \emptyset, \quad \text{a.e. on } I \\ &g(\alpha(a), \alpha(b), \alpha'(a), \alpha'(b)) \geq 0 \\ &\alpha(b) = h(\alpha(a)) \end{aligned}$$

In a similar way a function $\beta \in C^1(I, \mathbb{R})$ is said to be an *upper solution* for problem (II) if $\phi \circ \beta' \in W^{1,1}(I, \mathbb{R})$ and

$$\begin{aligned} &F(t, \beta(t), \beta'(t)) \cap [(\phi(\beta'(t)))', +\infty[\neq \emptyset, \quad \text{a.e. on } I \\ &g(\beta(a), \beta(b), \beta'(a), \beta'(b)) \leq 0 \\ &\beta(b) = h(\beta(a)) \end{aligned}$$

We prove the following existence result:

Theorem 2: *Let α and β be lower and upper solutions, respectively, for problem (II). If hypotheses H_0, $H(F)_1$ and B_1, B_2 hold and the hypotheses $H(F)_2$ is satisfied with respect to α and β, then problem (II) has at least one solution in the order interval $[\alpha, \beta]$.*

Proof: Let $c \in \mathbb{R}$ such that $\alpha(a) \leq c \leq \beta(a)$. For Theorem 1 there exists a solution x_c of the problem

$$\begin{aligned} &(\phi(x'(t)))' \in F(t, x(t), x'(t)), \quad \text{a.e. on } I \\ &x(a) = c, \quad x(b) = h(c) \end{aligned} \tag{10}$$

with the property that $\alpha(t) \leq x_c(t) \leq \beta(t), \forall t \in I$.

It is simple to see that if $c = \alpha(a)$ then $x_c'(a) \geq \alpha'(a)$ and $x_c'(b) \leq \alpha'(b)$ and so from B_1 we deduce that

$$g(x_c(a), x_c(b), x_c'(a), x_c'(b)) \geq 0. \tag{11}$$

In a similar way, if $c = \beta(a)$ we obtain that

$$g(x_c(a), x_c(b), x_c'(a), x_c'(b)) \leq 0. \tag{12}$$

In order to obtain the existence of a solution of the problem (II), it is sufficient to prove that there exists $c \in [\alpha(a), \beta(a)]$ such that $g(x_c(a), x_c(b), x_c'(a), x_c'(b)) = 0$. Indeed we assume that $g(x_c(a), x_c(b), x_c'(a), x_c'(b)) \neq 0, \forall c \in [\alpha(a), \beta(a)]$ and we consider the following subsets of $[\alpha(a), \beta(a)]$:

$$M_1 = \{c \in [\alpha(a), \beta(a)] : g(x_c(a), x_c(b), x_c'(a), x_c'(b)) < 0\}$$
$$M_2 = \{c \in [\alpha(a), \beta(a)] : g(x_c(a), x_c(b), x_c'(a), x_c'(b)) > 0\}.$$

It is evident that $[\alpha(a), \beta(a)] = M_1 \cup M_2$ and from (11) and (12) we have that $M_1 \neq \emptyset$ and $M_2 \neq \emptyset$. If we show that the sets M_1 and M_2 are closed then we obtain a contradiction with the fact that the interval $[\alpha(a), \beta(a)]$ is connected and so the proof is complete. So, in order to prove that M_1 is closed (the same is for M_2), let $(c_n)_n \subset M_1$ such that $c_n \to c_o$ and let x_n be a solution of the problem (10) when $c = c_n$. Consider now, the differential inclusion introduced in the proof of the previous theorem

$$(\phi(x'(t)))' \in \tilde{F}(t, x(t), q_N(\tau(x')(t))) + \arctan(u(t, x(t))), \quad \text{a.e. on } I \tag{13}$$

and let T: $C^1(I, \mathbb{R}) \to 2^{C^1(I, \mathbb{R})}$ the multivalued operator defined as

$$T(x) = \left\{ y \in C^1(I, \mathbb{R}) : y(t) = x(a) + \int_a^t [\phi^{-1}(\phi(x'(a)) + \int_a^r f(s)ds]dr : f \in N(x) \right\},$$

where N: $C^1(I, \mathbb{R}) \to 2^{L^1(I, \mathbb{R})}$ is, as in Theorem 1, the Nemytskii operator of the multifunction $(t, x, y) \to \tilde{F}(t, x, y) + \arctan(u(t, x))$. Observe that x is a solution of the equation (13) if and only if $x \in T(x)$ and so it is simple to see that $x_n \in T(x_n)$. Moreover from the monotonicity of ϕ^{-1} and from hypotheses $H(F)_2$, we obtain that the two sequences $(x_n)_n$ and $(x_n')_n$ are uniformly bounded on I and using the continuity of ϕ^{-1}, we deduce also the equicontinuity of the same two sequences. So, using Arzelá-Ascoli Theorem, we obtain that there exists a subsequence of $(x_n)_n$, denoted again with $(x_n)_n$, which converges in $C^1(I, \mathbb{R})$ to a function x_o.

Now observe that from the continuity of ϕ and from the fact that the Nemytskii operator N is weakly upper semicontinuous, the operator T has closed graph in $C^1(I, \mathbb{R}) \times C^1(I, \mathbb{R})$ and then $x_o \in T(x_o)$. Therefore x_o is a solution of the equation (13) and, since $x_n \in [\alpha, \beta]$ and $|x_n'(t)| \leq N, \forall t \in I$, then also $x_o \in [\alpha, \beta]$ and $|x_o'(t)| \leq N$, $\forall t \in I$. Taking into account the continuity of the function h we deduce that x_o is a solution of problem (10) when $c = c_o$.

Finally from the continuity of g we obtain that $g(x_o(a), x_o(b), x_o'(a), x_o'(b)) \leq 0$ but from our assumptions it must be $g(x_o(a), x_o(b), x_o'(a), x_o'(b)) < 0$ which implies that $c_o \in M_1$ and so M_1 is closed. Q.E.D.

Remark 1: We observe that the problem (II) contains as a particular case the periodic problem and that the previous theorem contains Theorem 2.1 of [5].

4. EXTREMAL SOLUTIONS

In this paragraph we show that the conclusion of Theorem 2 can be improved by proving that the problem (II) admits extremal solutions in the order interval $[\alpha, \beta]$ i.e., there exist a least solution $x_* \in [\alpha, \beta]$ and a greatest solution $x^* \in [\alpha, \beta]$ of problem (II), such that if $x \in [\alpha, \beta]$ is any other solution of problem (II), we have $x_*(t) \leq x(t) \leq x^*(t)$ for all $t \in I$.

Theorem 3: *Let α and β be lower and upper solutions, respectively, for problem (II) such that $\alpha \leq \beta$ in I. If hypotheses H_0, $H(F)_1$ and B_1, B_2 are satisfied and hypotheses $H(F)_2$ hold with respect to α and β, then problem (II) has extremal solutions in the order interval $[\alpha, \beta]$.*

Proof: Let S_1 be the set

$$S_1 = \{x \in [\alpha, \beta] : x \text{ is a solution of problem (II)}\}.$$

From theorem 2 we have that $S_1 \neq \emptyset$. First we will show that S_1 is an inductive set with respect to the order structure of $C^1(I, \mathbb{R})$ defined by

$$x \prec y \Leftrightarrow x(t) \leq y(t), \quad \forall t \in I.$$

To this end we denote by C a chian in S_1. Since $L^1(I, \mathbb{R})$ is a complete lattice and C is a bounded subset of $L^1(I, \mathbb{R})$, if $x = \sup C$, by Corollary IV.II.7 of [11], we can find a sequence $\{x_n\}_n \subset C$ such that $x = \sup_{n \in \mathbb{N}} \{x_n\}$ and $x \in L^1(I, \mathbb{R})$, then according to the monotone convergence theorem we obtain that $x_n \rightarrow x$ in $L^1(I, \mathbb{R})$.
For every $n \in \mathbb{N}$ the function x_n satisfies the differential inclusion

$$(\phi(x_n'(t)))' \in F(t, x_n(t), x_n'(t)), \quad \text{a.e. on } I,$$

so, arguing as in the previous proof, we establish the existence of a subsequence of $\{x_n\}_n$, denoted again by $\{x_n\}_n$, such that $x_n \rightarrow x$ in $C^1(I, \mathbb{R})$.
Therefore, from continuity of h and g, it follows that $x \in S_1$ and, invoking Zorn's Lemma, we infer the existence of a maximal element $x^* \in S_1$.

Now we will prove that S_1 is a directed set (i.e., if $x_1, x_2 \in S_1$ then there exists $x \in S_1$ such that $x_1 \prec x$ and $x_2 \prec x$). In this way we can conclude that x^* is unique and it is the greatest element of S_1 in the order interval $[\alpha, \beta]$.
To this end, let $x_1, x_2 \in S_1$ and put $x_3 = \max\{x_1, x_2\}$. Since $x_1, x_2 \in C^1(I, \mathbb{R})$ and $x_3 = (x_1 - x_2)_+ + x_2$, from Lemma 7.6 of [18], we have that $x_3 \in W^{1,1}(I, \mathbb{R})$ and

$$x_3'(t) = \begin{cases} x_1'(t) & \text{if } x_1(t) \geq x_2(t) \\ x_2'(t) & \text{if } x_1(t) \leq x_2(t), \end{cases}$$

a.e. on I.

First we suppose that $x_3 \in C^1(I, \mathbb{R})$; then, in each point in which x_1 and x_2 coincide, $x_1'(t)$ is equals to $x_2'(t)$ and so

$$x_3'(t) = \begin{cases} x_1'(t) & \text{if } x_1(t) > x_2(t) \\ x_2'(t) & \text{if } x_1(t) < x_2(t) \\ x_1'(t) & \text{if } x_1(t) = x_2(t) \end{cases}$$

for all $t \in I$. Let

$$I_0 = \{t \in I\colon \phi(x_1'(t)) = \phi(x_2'(t)) \quad \text{and} \quad (\phi(x_1'(t)))' \neq (\phi(x_2'(t)))'\}$$

and

$$I_1 = \{t \in I\colon \ \nexists(\phi(x_1'(t)))' \quad \text{or} \quad \nexists(\phi(x_2'(t)))'\};$$

from Banach Lemma we obtain that $m(I_0 \cup I_1) = 0$ and

$$(\phi(x_3'(t)))' = \begin{cases} (\phi(x_1'(t)))' & \text{if } x_1(t) \geq x_2(t) \\ (\phi(x_2'(t)))' & \text{if } x_1(t) \leq x_2(t) \end{cases}$$

for all $t \in I - (I_0 \cup I_1)$, from which we deduce that

$$(\phi(x_3'(t)))' \in F(t, x_3(t), x_3'(t)) \quad \text{a.e. on } I.$$

Now, using the boundary conditions for x_1 and x_2, we conclude that $x_3 \in S_1$ and x_3 is the element such that $x_1 \prec x_3$ and $x_2 \prec x_3$.

In the case in which $x_3 \notin C^1(I, \mathbb{R})$ we denote by $\tau_3\colon W^{1,1}(I, \mathbb{R}) \to W^{1,1}(I, \mathbb{R})$ the following truncation operator

$$\tau_3(x)(t) = \begin{cases} \beta(t) & \text{if } x(t) \geq \beta(t) \\ x(t) & \text{if } x_3(t) \leq x(t) \leq \beta(t) \\ x_3(t) & \text{if } x(t) \leq x_3(t), \end{cases}$$

for every $x \in W^{1,1}(I, \mathbb{R})$ and for all $t \in I$ and using again Lemma 7.6 of [18] we have that

$$\tau_3(x)'(t) = \begin{cases} \beta'(t) & \text{if } x(t) \geq \beta(t) \\ x'(t) & \text{if } x_3(t) \leq x(t) \leq \beta(t) \\ x_3'(t) & \text{if } x(t) \leq x_3(t), \end{cases}$$

for every $x \in W^{1,1}(I, \mathbb{R})$ and a.e. on I. We consider also the truncation function $q_N\colon \mathbb{R} \to \mathbb{R}$ introduced in the proof of Theorem 1 and the penalty function $u_3\colon I \times \mathbb{R} \to \mathbb{R}$ defined by

$$u_3(t, x) = \begin{cases} x - \beta(t) & \text{if } x \geq \beta(t) \\ 0 & \text{if } x_3(t) \leq x \leq \beta(t) \\ x - x_3(t) & \text{if } x \leq x_3(t), \end{cases}$$

for every $(t, x) \in I \times \mathbb{R}$.

Now let $F^*\colon I \times \mathbb{R} \times \mathbb{R} \to 2^{\mathbb{R}} - \{\emptyset\}$ the multifunction defined by

$$F^*(t, x, y) = \overline{\text{co}}\left(\bigcup_{i=1}^{3} F_i(t, x, y)\right), \quad \text{for all } (t, x, y) \in I \times \mathbb{R} \times \mathbb{R},$$

where $F_i: I \times \mathbb{R} \times \mathbb{R} \to 2^{\mathbb{R}}$, $i = 1, 2, 3$, denotes the multivalued applications so defined

$$F_1(t,x,y) = \begin{cases} F(t,x,y) & \text{if } x_3(t) \le x \le \beta(t) \\ \emptyset & \text{otherwise,} \end{cases}$$

$$F_2(t,x,y) = \begin{cases} F(t,x_3(t),y) \cap]-\infty, \tilde{z}(t)] & \text{if } x \le x_3(t) \text{ and } F(t,x_3(t),y) \cap]-\infty,\tilde{z}(t)] \neq \emptyset \\ \{\tilde{z}(t)\} & \text{if } x \le x_3(t) \text{ and } F(t,x_3(t),y) \cap]-\infty,\tilde{z}(t)] = \emptyset \\ \emptyset & \text{otherwise,} \end{cases}$$

$$F_3(t,x,y) = \begin{cases} F(t,\beta(t),y) \cap [(\phi(\beta'(t)))', +\infty[& \text{if } x \ge \beta(t) \text{ and } F(t,\beta(t),y) \cap [(\phi(\beta'(t)))', +\infty[\neq \emptyset \\ \{(\phi(\beta'(t)))'\} & \text{if } x \ge \beta(t) \text{ and } F(t,\beta(t),y) \cap [(\phi(\beta'(t)))', +\infty[= \emptyset \\ \emptyset & \text{otherwise,} \end{cases}$$

for all $(t,x,y) \in I \times \mathbb{R} \times \mathbb{R}$ and $\tilde{z} \in L^1(I,\mathbb{R})$ is the following function

$$\tilde{z}(t) = \min\{(\phi(x_1'(t)))', (\phi(x_2'(t)))'\}, \quad \text{a.e. on } I.$$

It is easy to check that F^* is a Caratheodory multifunction with nonempty, closed and convex values.
Consider now the following boundary value problem

$$\begin{aligned} &(\phi(x'(t)))' \in F^*(t,x(t),q_N(\tau_3(x)'(t))) + \arctan(u_3(t,x(t))), \quad \text{a.e. on } I \\ &g(x(a),x(b),x'(a),x'(b)) = 0, \\ &h(x(a)) = x(b). \end{aligned} \tag{14}$$

If we will prove the existence of a solution x of the problem (14) belonging to the interval $[x_3, \beta]$, since $x_3 \notin C^1(I,\mathbb{R})$, we can conclude that x is a really solution for problem (II), moreover $x_1 \prec x$ and $x_2 \prec x$ so S_1 is a directed set.
To see that, for every $c \in [x_3(a), \beta(a)]$, we introduce the Dirichlet problem

$$\begin{aligned} &(\phi(x'(t)))' \in F^*(t,x(t),q_N(\tau_3(x)'(t))) + \arctan(u_3(t,x(t)), \quad \text{a.e. on } I \\ &x(a) = c, \quad x(b) = h(c) \end{aligned} \tag{15}$$

and as in the proof of Claim 3 of Theorem 1, we establish that it has a solution x_c.
Now we will show that $x_c \in [x_3, \beta]$. From boundary conditions and by hypothesis B_2, it follows that

$$x_1(a) \le x_c(a) \le \beta(a) \quad \text{and} \quad x_1(b) \le x_c(b) \le \beta(b),$$

therefore, if there exists a point $t_0 \in (a,b)$ such that

$$(x_c - x_1)(t_0) = \min_{t \in I}(x_c - x_1)(t) < 0,$$

then it is possible to find an interval $[t_1, t_2] \subset [a, b]$ with $t_0 \in (t_1, t_2)$, in which $x_c(t) < x_1(t)$, $\forall t \in (t_1, t_2)$ and

$$(x_c - x_1)(t_1) = (x_c - x_1)(t_2) = 0.$$

Therefore

$$(\phi(x_c'(t)))' \in F_2(t, x_c(t), x_3'(t)) + \arctan((x_c - x_3)(t)), \quad \text{a.e. on } [t_1, t_2]$$

and so recalling the definition of F_2, we obtain

$$(\phi(x_c'(t)))' \le (\phi(x_1'(t)))' + \arctan((x_c - x_3)(t)) < (\phi(x_1'(t)))', \quad \text{a.e. on} [t_1, t_2]$$

which means that the function $\phi \circ x_c' - \phi \circ x_1'$ is decreasing on (t_1, t_2). Now, from this and by monotonicity and injectivity of ϕ, we have

$$(x_c - x_1)(t_2) < (x_c - x_1)(t_0) < 0$$

which is a contradiction with the existence of t_0, thus $x_c(t) \ge x_1(t)$ in I. In a similar way we show that $x_c(t) \ge x_2(t)$ and $x_c(t) \le \beta(t)$ on I. Hence we conclude that every solution x_c of (15) is located in the order interval $[x_3, \beta]$.

Finally proceeding as in the proof of Theorem 2, we establish the existence of a solution x in the interval $[x_3, \beta]$ for problem (14). So we have shown that S_1 is a directed set, which implies the existence of a maximal solution in $[\alpha, \beta]$.

In a similar way we can prove the existence of a minimal solution x_* for problem (II) in $[\alpha, \beta]$. Q.E.D.

5. PARTICULAR CASES

Now we focus our attention on the following problem

$$\begin{aligned} &(\phi(x'(t)))' \in F(t, x(t), x'(t)), \quad \text{a.e. on } I \\ &p(x(a), x'(a)) = 0 \\ &q(x(b), x'(b)) = 0 \end{aligned} \tag{III}$$

where the function p and q satisfy the following hypotheses:

B_3: p: $\mathbb{R} \times \mathbb{R} \to \mathbb{R}$ is such that

(i) p is continuous;
(ii) $\forall s \in \mathbb{R}$, the function $t \to p(s, t)$ is not increasing;

B_4: q: $\mathbb{R} \times \mathbb{R} \to \mathbb{R}$ is such that

(i) q is continuous;
(ii) $\forall s \in \mathbb{R}$, the function $t \to q(s, t)$ is not decreasing.

For the problem (III) a function $\alpha \in C^1(I, \mathbb{R})$ is said to be a *lower solution* if $\phi \circ \alpha' \in W^{1,1}(I, \mathbb{R})$ and

$$\begin{aligned} &F(t, \alpha(t), \alpha'(t)) \cap]-\infty, (\phi(\alpha'(t)))'] \ne \emptyset, \quad \text{a.e. on } I \\ &p(\alpha(a), \alpha'(a)) \le 0 \\ &q(\alpha(b), \alpha'(b)) \le 0. \end{aligned}$$

Anologously a function $\beta \in C^1(I, \mathbb{R})$ is said to be an *upper solution* for problem (III) if $\phi \circ \beta' \in W^{1,1}(I, \mathbb{R})$ and

$$F(t, \beta(t), \beta'(t)) \cap [(\phi(\beta'(t)))', +\infty[\neq \emptyset, \quad \text{a.e. on } I$$
$$p(\beta(a), \beta'(a)) \geq 0$$
$$q(\beta(b), \beta'(b)) \geq 0.$$

Note that problem (III) includes the Sturm-Liouville and Dirichlet problem. Indeed, since in these cases boundary conditions are respectively $c_0x(a) - d_0x'(a) = v_0$, $c_1x(b) + d_1x'(b) = v_1$ and $x(a) = v_0$, $x(b) = v_1$, it is sufficient to put $p(s,t) = c_0s - d_0t - v_0$, $q(s,t) = c_1s + d_1t - v_1$ in (III) to obtain the first and $p(t,s) = s - v_0$, $q(s,t) = s - v_1$ for the second.

It holds the following existence result:

Theorem 4: *Let α and β be lower and upper solutions, respectively, for problem (III) such that $\alpha \leq \beta$ in I. Assume that hypotheses H_0, $H(F)_1$, B_3, B_4 are satisfied and the hypotheses $H(F)_2$ hold with respect to α and β.*

Then problem (III) has at least one solution in the order interval $[\alpha, \beta]$.

Proof: Let $d \in \mathbb{R}$ such that $\alpha(b) \leq d \leq \beta(b)$. For every $c \in [\alpha(a), \beta(a)]$ we denote by $x(c, d; \cdot)$ a solution of the problem

$$\begin{aligned} &(\phi(x'(t)))' \in F(t, x(t), x'(t)), \quad \text{a.e. on } I \\ &x(a) = c, \quad x(b) = d \end{aligned} \tag{16}$$

with the property that $\alpha(t) \leq x(c, d; t) \leq \beta(t)$ for all $t \in I$. The existence of such solution follows immediately from Theorem 1.

If $c = \alpha(a)$, then $x'(c, d; a) \geq \alpha'(a)$ and using B_3 we obtain

$$p(x(c, d; a), x'(c, d; a)) \leq 0.$$

In a similar way, if $c = \beta(a)$ we have

$$p(x(c, d; a), x'(c, d; a)) \geq 0,$$

from which, proceeding as in the proof of Theorem 2, we deduce the existence of $c_d \in [\alpha(a), \beta(a)]$ with the property that

$$p(x(c_d, d; a), x'(c_d, d; a)) = 0.$$

In this way, to each $d \in [\alpha(b), \beta(b)]$ we associate a constant $\tilde{c}_d \in [\alpha(a), \beta(a)]$ defined as $\tilde{c}_d = \max\{c_d\}$.

If $d = \alpha(b)$, we have $x'(\tilde{c}_d, d; b) \leq \alpha'(b)$, hence, from B_4 it follows

$$q(x(\tilde{c}_d, d; b), x'(\tilde{c}_d, d; b)) \leq 0.$$

Analogously, if $d = \beta(b)$, we get

$$q(x(\tilde{c}_d, d; b), x'(\tilde{c}_d, d; b)) \geq 0.$$

Now, proceeding again as in the proof of Theorem 2, we prove the existence of $d_0 \in [\alpha(b), \beta(b)]$ such that the solution $x(c_0, d_0, \cdot)$ of problem (15) with $c_0 = \tilde{c_{d_0}}$, satisfies the condition:

$$q(x(\tilde{c}_0, d_0; b), x'(\tilde{c}_0, d_0; b)) = 0,$$

and this implies that this solution is really a solution of problem (III). Q.E.D.

Proceeding as in Theorem 3, it is possible to prove the existence of extremal solutions in the order interval $[\alpha, \beta]$ also for problem (III):

Theorem 5: *Let α and β be lower and upper solutions, respectively, for problem (III) such that $\alpha \leq \beta$ in I. If hypotheses H_0, $H(F)_1$ and B_3, B_4 are satisfied and hypotheses $H(F)_2$ hold with respect to α and β, then problem (III) has extremal solutions in the order interval* $[\alpha, \beta]$.

Finally we consider the following problem

$$\begin{aligned} &(\phi(x'(t)))' \in F(t, x(t), x'(t)), \quad \text{a.e. on } I \\ &q(x(a), x'(a)) = 0 \\ &p(x(b), x'(b) = 0 \end{aligned} \tag{IV}$$

where the function p and q satisfy the hypotheses B_3 and B_4.

We say that a function $\alpha \in C^1(I, \mathbb{R})$ is a *lower solution* for problem (IV) if $\phi \circ \alpha' \in W^{1,1}(I, \mathbb{R})$ and

$$\begin{aligned} &F(t, \alpha(t), \alpha'(t)) \cap]-\infty, (\phi(\alpha'(t)))'] \neq \emptyset, \quad \text{a.e. on } I \\ &q(\alpha(a), \alpha'(a)) \geq 0 \\ &p(\alpha(b), \alpha'(b)) \geq 0 \end{aligned}$$

and $\beta \in C^1(I, \mathbb{R})$ is an *upper solution* for problem (IV) if $\phi \circ \beta' \in W^{1,1}(I, \mathbb{R})$ and

$$\begin{aligned} &F(t, \beta(t), \beta'(t)) \cap [(\phi(\beta'(t)))', +\infty[\neq \emptyset, \quad \text{a.e. on } I \\ &q(\beta(a), \beta'(a)) \leq 0 \\ &p(\beta(b), \beta'(b)) \leq 0. \end{aligned}$$

Note that problem (IV) generalizes Neumann problem, indeed in this case the boundary conditions are $x'(a) = A$ and $x'(b) = B$, with $A, B \in \mathbb{R}$, so it is sufficient to put $q(t, s) = t - A$ and $p(t, s) = B - t$.

Also in this case we have the following:

Theorem 6: *Let α and β be lower and upper solutions, respectively, for problem (IV) such that $\alpha \leq \beta$ in I. If hypotheses H_0, $H(F)_1$ and B_3, B_4 are satisfied and hypotheses $H(F)_2$ holds with respect to α and β, then the problem (IV) has extremal solutions in the order interval* $[\alpha, \beta]$.

REFERENCES

[1] V. Barbu (1976). *Nonlinear semigroups and differential equations in Banach spaces.* Leyden, The Nertherlands: Noordhoff International Publishing.

[2] C. Berge (1959). *Espaces topologiques, Functions multivoques.* Paris: Dunod.

[3] L.E. Bobisud (1991). Steady-state turbolent flow with reaction, Rocky Mountain *J. Math.*, **21**, 993–1007.

[4] A. Cabada and R.L. Pouso (1997). Existence results for the problem $(\phi(u'))' = f(t,u,u')$ with periodic and Neumann boundary conditions, *Nonl. Anal. TMA*, **30**, 1733–1742.

[5] A. Cabada and R.L. Pouso (1999). Existence results for the problem $(\phi(u'))' = f(t,u,u')$ with nonlinear boundary conditions, *Nonl. Anal. TMA*, **35**, 221–231.

[6] A. Cellina (1969). Approximation of set-valued functions and fixed point theorems, *Annali di Mat.*, **82**, 3–24.

[7] C. De Coster (1994). On pairs of positive solutions for the one-dimensional p-Laplacian, *Nonl. Anal. TMA*, **23**, 669–681.

[8] K. Deimling (1992). *Multivalued differential equations*, Berlin: Walter de Gruyther.

[9] M. Del Pino, M. Elgueta and R. Manásevich (1989). A homotopic deformation along p of a Leray-Schauder degree result and existence for $(|u'|^{p-2}u')' + f(t,u) = 0$, $u(0) = u(T) = 0$, $p > 1$, *J. Differential Equations*, **80**, 1–13.

[10] M. Del Pino, R. Manaŝevich and A. Murua (1992). Existence and multiplicity of solutions with prescribed period for a second order quasilinear o.d.e, *Nonl. Anal. TMA*, **18**, 79–92.

[11] I.N. Dunford and J. Schwartz (1958). *Linear operators I*, New York: Wiley.

[12] J.R. Esteban and J.L. Vazquez (1986). On the equation of turbolent filtration in one-dimensional porus media, *Nonl. Anal. TMA*, **10**, 1303–1325.

[13] Ch. Fabry and D. Fayyad (1992). Periodic solutions of second order differential equations with a p-Laplacian and asymmetric nonlinearities, *Rend. Istit. Mat. Univ. Trieste*, **24**, 207–227.

[14] Ch. Fabry and P. Habets (1986). Upper and lower solutions for second order boundary value problems with nonlinear boundary conditions, *Nonl. Anal. TMA*, **10**, 985–1007.

[15] M. Frigon (1991). Problmes aux limites pour des inclusions différentialles sans condition de croissance, *Ann. Polon. Math.*, **LIV.1**, 69–83.

[16] M. Frigon (1995). Théorèmes d'existence de solutions d'inclusions différentielles, *Topological Methods in Differential Equations and Inclusions*, 51–87.

[17] M. Garciá-Huibodro, R. Manásevich and F. Zanolin (1995). Strongly nonlinear second-order ODE's with rapidly growing terms, *J. Math. Anal. Appl.*, **202**, 1–26.

[18] D. Gilbarg and N. Trudinger (1977). *Elliptic partial differential equations of second order*, New York: Springer-Verlag.

[19] M.A. Herrero and J.L. Vazquez (1982). On the propagation properties of a non linear degenerate parabolic equation, *Comm. Partial Differential Equation*, **7**, 1381–1402.

[20] C.J. Himmelberg (1975). Measurable relations, *Fund. Math.*, **87**, 53–72.

[21] R. Manásevich and J. Mawhin (1998). Periodic solutions for nonlinear system with p-Laplacian-like operators, *J. Differential Equations*, **145**, 367–393.

[22] R. Manásevich and F. Zanolin (1993). Time-mappings and multiplicity of solutions for the one-dimensional p-Laplacian, *Nonl. Anal. TMA*, **21**, 269–291.

[23] I.P. Natanson (1955). *The theory of functions of real variable*, New York: Ungar.

[24] D. O'Regan (1993). Some general principles and results for $(\phi(y'))' = qf(t,y,y')$, $0 < t < 1$, *SIAM J. Math. Anal.*, **24**, 648–668.

[25] D. O'Regan (1997). Existence theory for $(\phi(y'))' = qf(t,y,y')$, $0 < t < 1$, *Comm. Appl. Anal.*, **1**, 33–52.

[26] M. Palmucci and F. Papalini. Periodic and boundary value problems for second order differential inclusions, (to appear).

[27] T. Pruszko (1981). Topological degree methods in multi-valued boundary value problems, *Nonl. Anal. TMA*, **5**, 959–973.

[28] J. Wang and W. Gao (1997). Existence of solutions to boundary value problems for a nonlinear second order equation with weak Carathéodory functions, *Differential Equations and Dyn. Systems*, **5(2)**, 175–185.

[29] J. Wang, W. Gao and Z. Lin (1995). Boundary value problems for general second order equation and similarity solutions to the Rayleigh problem, *Tôhoku Math. J.*, **47**, 327–344.

[30] M. Zhang (1997). Nonuniform nonresonance at the first eigenvalue of the p-Laplacian, *Nonl. Anal. TMA*, **29**, 41–51.

SIMAA 4(2002) 403–409

23. Extensions of Monotone Sets*

Sehie Park

Department of Mathematics, Seoul National University, Seoul 151-742, Korea.
E-mail: shpark@math.snu.ac.kr

Abstract: Far-reaching generalizations of the extension theorem for monotone sets due to Debrunner and Flor are obtained. In fact, Browder's extension theorem involving Kakutani multimaps is extended to the one involving a large class of "better" admissible maps. Moreover, a sharpened version of the extension theorem for noncompact case due to Lassonde is also obtained. Finally, we mention another generalization of the Debrunner–Flor theorem due to the author.

1. INTRODUCTION

In 1964, Debrunner and Flor [4] proved an extension theorem for monotone sets. This generalized earlier works of Minty [9] and Grünbaum [6], which have interesting applications to nonlinear elliptic boundary value problems (see [1]) and monotone operator theory (for example, see [2,15]). Further generalized and sharpened forms of the extension theorem were obtained by Fan [5], Browder [3], Lassonde [8], and others.

Especially, Browder [3] obtained an extension theorem for monotone sets involving the Kakutani multimaps. In the present chapter, we extend this result to the one involving a large class of multimaps which were called "better" admissible by the present author [12]. Moreover, we sharpen the noncompact version of the extension theorem due to Lassonde [8]. Finally, we mention another generalization of the Debrunner–Flor theorem due to the author in a recent work [14].

2. PRELIMINARIES

In this chapter, t.v.s. means topological vector spaces.

Given two t.v.s. E and F, let $\langle , \rangle : F \times E \to \mathbf{R}$ be a bilinear pairing which is continuous on compact subsets of $F \times E$. This assumption is quite natural in most applications, since the natural pairing between a Hausdorff locally convex t.v.s. E and its dual space E^* equipped with the strong topology enjoys this property.

A subset $M \subset E \times F$ is said to be *monotone* if for any two points (u, w) and (u', w') in M, we have $\langle w - w', u - u' \rangle \geq 0$; see Debrunner and Flor [4].

* This research is partially supported by Research Institute of Mathematics, Seoul National University.

Browder [3] obtained the following extension theorem for monotone sets:

Lemma: (Theorem 8 [3]) *Let K be a compact convex subset of a t.v.s. E, and F a t.v.s. with a bilinear pairing $\langle , \rangle$: $F \times E \to \mathbf{R}$ which is continuous on compact subsets of $F \times E$. Let f: $K \to F$ be continuous and M a monotone subset of $K \times F$. Then there exists a $u_0 \in K$ such that*

$$\langle f(u_0) - w, u_0 - u \rangle \geq 0 \quad \text{for all } (u, w) \in M,$$

or equivalently, the set $M \cup \{(u_0, f(u_0))\}$ remains monotone.

This result sharpens corresponding result of Debrunner and Flor [4] for E locally convex and of Fan (Theorem 12 [5]) for F locally convex and quasi-complete.

3. BETTER ADMISSIBLE MULTIMAPS AND A COINCIDENCE THEOREM

A *multimap* (simply, a *map*) T: $X \multimap Y$ is a function from a set X into the power set 2^Y having nonempty values. Note that $y \in T(x)$ is equivalent to $x \in T^-(y)$ and, for $B \subset Y$, let $T^-(B) := \{x \in X : T(x) \cap B \neq \emptyset\}$ and $T^+(B) := \{x \in X : T(x) \subset B\}$. For topological spaces X and Y, a map T: $X \multimap Y$ is *upper semicontinuous* if $T^+(B)$ is open for each open set B in Y; *lower semicontinuous* if $T^+(B)$ is closed for each closed set B in Y; and *continuous* if it is upper and lower semicontinuous.

A *convex space* X is a convex set (in a vector space) equipped with a topology that induces the Euclidean topology on convex hulls of each finite subset of X. Such convex hulls are called *polytopes*; see Lassonde [8].

Let X and Y be topological spaces. An *admissible class* $\mathfrak{A}_c^\kappa(X,Y)$ of maps T: $X \multimap Y$ is one such that, for each compact subset K of X, there exists a map $\Gamma \in \mathfrak{A}_c(K,Y)$ satisfying $\Gamma x \subset Tx$ for all $x \in K$; where $\mathfrak{A}_c$ is consisting of finite composites of maps in $\mathfrak{A}$, and $\mathfrak{A}$ is a class of maps satisfying the following properties:

(i) $\mathfrak{A}$ contains the class $\mathbb{C}$ of (single-valued) continuous functions;
(ii) each $F \in \mathfrak{A}_c$ is upper semicontinuous and compact-valued; and
(iii) for each polytope P, each $F \in \mathfrak{A}_c(P,P)$ has a fixed point, where the intermediate spaces of composites are suitably chosen for each $\mathfrak{A}$.

Examples of $\mathfrak{A}$ are continuous functions $\mathbb{C}$, the Kakutani maps $\mathbb{K}$ (with convex values and codomains are convex spaces), the Aronszajn maps $\mathbb{M}$ (with R_δ values), the acyclic maps $\mathbb{V}$ (with acyclic values), the Powers map $\mathbb{V}_c$, the O'Neil maps $\mathbb{N}$ (continuous with values of one or m acyclic components, where m is fixed), the approachable maps $\mathbb{A}$ (in uniform spaces), admissible maps of Górniewicz, permissible maps of Dzedzej, and others. Further, $\mathbb{K}_c^\sigma$ due to Lassonde, $\mathbb{V}_c^\sigma$ due to Park *et al.*, and approximable maps $\mathbb{A}^\kappa$ due to Ben-El-Mechaiekh and Idzik are examples of $\mathfrak{A}_c^\kappa$. For the literature, see [11,12].

We now define a new "better" admissible class defined on a convex space X:

$$F \in \mathfrak{B}(X,Y) \Leftrightarrow \text{ for any polytope } P \text{ in } X \text{ and any } f \in \mathbb{C}(F(P),P),$$

$$f(F|_P): P \multimap P \text{ has a fixed point.}$$

Note that $\mathfrak{A}_c^\kappa \subset \mathfrak{B}$ and some examples of maps in $\mathfrak{B}$ not belonging to $\mathfrak{A}_c^\kappa$ were given in [13].

The following coincidence theorem is a particular form of (Theorem 1 [12]) and its proof is given here for completeness:

Theorem 1: *Let X be a convex space, Y a Hausdorff space, and T, S: $X \multimap Y$ maps satisfying*

(1) $T \in \mathfrak{B}(X,Y)$ *is compact;*
(2) *for each* $y \in T(X)$, $S^-(y)$ *is convex; and*
(3) $\{\text{Int}\, S(x) : x \in X\}$ *covers* $\overline{T(X)}$.

Then T and S have a coincidence point $x_0 \in X$; that is, $T(x_0) \cap S(x_0) \neq \emptyset$.

Proof: Since $\overline{T(X)}$ is compact and included in $\bigcup\{\text{Int}\, S(x) : x \in X\}$, there exists an $N = \{x_1, x_2, \ldots, x_n\} \subset X$ such that $\overline{T(X)} \subset \bigcup\{\text{Int}\, S(x) : x \in N\}$. Let $\{\lambda_i\}_{i=1}^n$ be the partition of unity subordinated to this cover of the Hausdorff compact space $\overline{T(X)}$, and $P = \text{co}\, N \subset X$. Define f: $\overline{T(X)} \to P$ by

$$f(y) = \sum_{i=1}^{n} \lambda_i(y)x_i = \sum_{i \in N_y} \lambda_i(y)x_i$$

for $y \in \overline{T(X)} \subset Y$, where

$$i \in N_y \;\Leftrightarrow\; \lambda_i(y) \neq 0 \;\Rightarrow\; y \in \text{int}\, S(x_i) \subset S(x_i).$$

Then $x_i \in S^-(y)$ for each $i \in N_y$. Clearly f is continuous and, by (2), we have $f(y) \in \text{co}\{x_i : i \in N_y\} \subset S^-(y)$ for each $y \in T(X)$. Since P is a polytope in X and $T \in \mathfrak{B}(X,Y)$, $(f|_{T(P)})(T|_P)$: $P \multimap P$ has a fixed point $x_0 \in P \subset X$. Since $x_0 \in (fT)(x_0)$ and $f^-(x_0) \subset S(x_0)$, we have $T(x_0) \cap S(x_0) \neq \emptyset$. This completes our proof.

Remark: For the subclass $\mathfrak{A}_c^\kappa$ of $\mathfrak{B}$, Theorem 1 is given earlier in [11].

4. EXTENSIONS OF MONOTONE SETS

We deduce the following equilibrium existence theorem from Theorem 1:

Theorem 2: *Let K be a compact convex subset of a t.v.s. E, K_1 a Hausdorff compact subset of a t.v.s. F, $T \in \mathfrak{B}(K, K_1)$ with closed graph, and $M \subset E \times F$. Let Φ: $E \times F \to \mathbf{R} \cup \{-\infty\}$ be a function such that*

(1) Φ *is u.s.c. on compact subsets of* $E \times F$;
(2) *for each* $x \in E$, $\Phi(x, \cdot)$ *is l.s.c. on compact subsets of* F;
(3) *for each* $w \in F$, $\Phi(\cdot, w)$ *is quasiconcave.*

Suppose that for each $y \in K_1$, there exists an $x \in K$ such that

$$\Phi(x - u, y - w) \geq 0 \quad \textit{for all } (u, w) \in M.$$

Then there exist a $u_0 \in K$ and a $w_0 \in T(u_0)$ such that

$$\Phi(u_0 - u, w_0 - w) \geq 0 \quad \textit{for all } (u, w) \in M.$$

Proof: For any $\varepsilon > 0$ and any nonempty finite subset N of M, we set

$$H_{(\varepsilon,N)} = \{(u_0, w_0) \in \mathrm{Gr}(T) : \Phi(u_0 - u, w_0 - w) \geq -\varepsilon \text{ for all } (u, w) \in N\}$$

and

$$\begin{aligned} H_0 &= \{(u_0, w_0) \in \mathrm{Gr}(T) : \Phi(u_0 - u, w_0 - w) \geq 0 \text{ for all } (u, w) \in M\} \\ &= \bigcap \big\{H_{(\varepsilon,N)} : \varepsilon > 0 \text{ and } N \text{ is a finite subset of } M\big\}. \end{aligned}$$

Then we have to show $H_0 \neq \emptyset$.

By (1), each $H_{(\varepsilon,N)}$ is a closed subset of Gr(T). The intersection of each finite family of such sets is also a set of the form $H_{(\varepsilon',N')}$ for some $\varepsilon' > 0$ and a finite subset N' of M. Therefore, in order to show $H_0 \neq \emptyset$, it suffices to show that each $H_{(\varepsilon,N)}$ is nonempty.

Choose a given $\varepsilon > 0$ and a nonempty finite subset N of M. Define a map S: $K \multimap K_1$ by

$$S(x) = \{y \in K_1 : \Phi(x - u, y - w) > -\varepsilon,\ (u, w) \in N\}$$

for $x \in X$. Then $S(x)$ is open in K_1 by (2). Moreover,

$$S^-(y) = \{x \in K : \Phi(x - u, y - w) > -\varepsilon,\ (u, w) \in N\}$$

is nonempty by hypothesis and convex by (3).

Now we apply Theorem 1. Then there exists a $(u_0, w_0) \in \mathrm{Gr}(T)$ such that $w_0 \in S(u_0)$; that is,

$$\Phi(u_0 - u, w_0 - w) > -\varepsilon \quad \text{for all } (u, w) \in N.$$

Therefore, $H_{(\varepsilon,N)}$ is nonempty. This completes our proof.

Remarks

1. In Theorem 2, instead of $T \in \mathfrak{B}(K, K_1)$ with closed graph, we can adopt $T \in \mathfrak{A}_c^\kappa(K, K_1)$ without affecting its conclusion.
2. In Theorem 2, since T has closed graph and K_1 is compact, T itself is actually u.s.c. with compact values.
3. For the subclass $\mathbb{C}$ of $\mathfrak{B}$, Theorem 2 reduces to Fan (Theorem 11 [5]), who assumed that F is locally convex and other restrictions.
4. For the subclass $\mathbb{K}$ of $\mathfrak{B}$, Theorem 2 reduces to Browder (Theorem 9 [3]), where F is locally convex.

The following is our theorem on extensions of monotone sets:

Theorem 3: *Let E be a t.v.s., F a Hausdorff t.v.s. with a bilinear pairing $\langle , \rangle$: $F \times E \to \mathbf{R}$ which is continuous on compact subsets of $F \times E$, K a compact convex subset of E, and K_1 a compact subset of F. Let $T \in \mathfrak{B}(K, K_1)$ have closed graph and M a monotone subset of $K \times F$. Then there exist a $u_0 \in K$ and a $w_0 \in T(u_0)$ such that*

$$\langle w_0 - w, u_0 - u\rangle \geq 0 \quad \textit{for all } (u, w) \in M.$$

Proof: We put $\Phi(x, w) = \langle w, x\rangle$ for $(x, w) \in E \times F$. Then Φ satisfies conditions (1)–(3) in Theorem 2. By Theorem 2, it suffices to show that for each $y \in K_1$, there exists an $x \in K$ such that

$$\langle y - w, x - u\rangle \geq 0 \quad \text{for all } (u, w) \in M.$$

Now, we define $f\colon K \to K_1$ by

$$f(v) = y \quad \text{for all } v \in K.$$

By applying Lemma to f, such an $x \in K$ exists. This completes our proof.

Remarks

1. In Theorem 3, we can replace $T \in \mathfrak{B}(K, K_1)$ with closed graph by $T \in \mathfrak{A}_c^\kappa(K, K_1)$.
2. For the subclass $\mathbb{C}$ of $\mathfrak{B}$, Theorem 2 reduces to Browder (Theorem 8 [3]) or Lemma.
3. Even for the subclass $\mathbb{K}$ of $\mathfrak{B}$, Theorem 2 improves Browder (Theorem 9 [3]), where F is assumed to be locally convex.

5. NONCOMPACT VERSIONS

The following is a noncompact version of Lemma:

Theorem 4: *Let X be a convex subset of a t.v.s. E, and F a t.v.s. with a bilinear pairing $\langle,\rangle\colon F \times E \to \mathbf{R}$ which is continuous on compact subsets of $F \times E$. Let $f\colon X \to F$ be a function which is continuous on compact subsets of X, and $M \subset X \times F$ a monotone subset satisfying the following compactness condition:*

(0) *There is a nonempty compact subset K of X such that for each nonempty finite subset N of X, there exists a compact convex subset L_N of X containing N such that for each $x \in L_N \backslash K$ there exists a $(u, w) \in M \cap (L_N \times F)$ with $\langle f(x) - w, x - u\rangle < 0$.*

Then there exists a $u_0 \in X$ such that

$$\langle f(u_0) - w, u_0 - u\rangle \geq 0 \quad \textit{for all } (u, w) \in M.$$

Proof: For each $(u, w) \in M$, consider the set

$$K_{(u,w)} = \{x \in K : \langle f(x) - w, x - u\rangle \geq 0\},$$

which is a closed subset of K. We show that $\{K_{(u,w)} : (u, w) \in M\}$ has the finite intersection property. Let $\{(u_1, w_1), \ldots, (u_n, w_n)\}$ be a finite subset of M. Then there exists a compact convex subset L_N of X containing $N = \{u_1, \ldots, u_n\}$ as in condition (0). Applying Lemma with L_N instead of K, we obtain a $u_0 \in L_N$ such that

$$\langle f(u_0) - w, u_0 - u\rangle \geq 0 \quad \text{for all } (u, w) \in M \cap (L_N \times F).$$

Now by condition (0), we should have $u_0 \in K$. Therefore,

$$\bigcap_{i=1}^{n} K_{(u_i,w_i)} \supset \bigcap\{K_{(u,w)} : (u, w) \in M \cap (L_N \times F)\} \supset \{u_0\} \neq \emptyset.$$

Since K is compact, we have

$$\bigcap\{K_{(u,w)} : (u,w) \in M\} \neq \emptyset.$$

This implies the conclusion.

Remarks

1. For $X = K$, Theorem 4 reduces to Lemma or Browder (Theorem 8 [3]).
2. Theorem 4 is a sharpened version of Lassonde (Theorem 2.7 [8]).
3. By adopting similar method, it is possible to establish noncompact versions of Theorems 2 and 3.

6. GENERALIZED KIRSZBRAUN THEOREMS

The well-known Kirszbraun theorem [7] asserts that a nonexpansive function from a finite domain in $\mathbf{R}^n$ to $\mathbf{R}^n$ can be extended to a larger domain including any arbitrarily chosen point so as to be nonexpansive. Extensions and variations of the theorem were obtained by Minty [9,10], Debrunner and Flor [4], and many others; for the literature, see [14].

Recently, the author [14] obtained a generalized Kirszbraun-Minty type inequality theorem which subsumes all of the above-mentioned results of the same sort. In this section, we introduce the main result of [14] for reader's convenience:

Theorem 5: *Let X be a vector space, Y a nonempty set, Φ: $X \times Y \times Y \to \mathbf{R}$ a function such that*

(0) *for each $y, y' \in Y$, $\Phi(\cdot, y, y')$ is finitely l.s.c. (that is, it is l.s.c. on any finite dimensional subspace of X).*

Let $(x_1, y_1), \ldots, (x_m, y_m) \in X \times Y$ and $y \in Y$ be given. Suppose that

(I) $\sum_{i=1}^{m} \lambda_i \Phi(x_i - x, y_i, y) \leq 0$ *for all* $x = \sum_{j=1}^{m} \lambda_j x_j$ *with* $\lambda_j \geq 0, \sum_{j=1}^{m} \lambda_j = 1$.

Then we have the following:

(II) *There exists an $x \in \mathrm{co}\{x_1, \ldots, x_m\}$ such that*

$$\Phi(x_i - x, y_i, y) \leq 0 \quad \textit{for each } 1 \leq i \leq m.$$

REFERENCES

[1] F.E. Browder (1963). The solvability of non-linear functional equations, *Duke Math. J.*, **30**, 557–566.

[2] F.E. Browder (1968). Nonlinear maximal monotone operators in Banach space, *Math. Ann.*, **175**, 89–113.

[3] F.E. Browder (1968). The fixed point theory of multi-valued mappings in topological vector spaces, *Math. Ann.*, **177**, 283–310.

[4] H. Debrunner and P. Flor (1964). Ein Erweiterungssatz für monotone Mengen, *Arch. Math.*, **15**, 445–447.

[5] Ky Fan (1966). Applications of a theorem concerning sets with convex sections, *Math. Ann.*, **163**, 189–203.

[6] B. Grünbaum (1962). A generalization of theorems of Kirszbraun and Minty, *Proc. Amer. Math. Soc.*, **13**, 812–814.

[7] M.D. Kirszbraun (1934). Über die zusammenziehende und Lipschitzsche Transformationen, *Fund. Math.*, **22**, 77–108.

[8] M. Lassonde (1983). On the use of KKM multifunctions in fixed point theory and related topics, *J. Math. Anal. Appl.*, **97**, 151–201.

[9] G.J. Minty (1962). On the simultaneous solution of a certain system of linear inequalities, *Proc. Amer. Math. Soc.*, **13**, 11–16.

[10] G.J. Minty (1967). On the generalization of a direct method of the calculus of variations, *Bull. Amer. Math. Soc.*, **73**, 315–321.

[11] Sehie Park (1994). Foundations of the KKM theory via coincidences of composites of upper semicontinuous maps, *J. Korean Math. Soc.*, **31**, 493–519.

[12] Sehie Park (1997). Coincidence theorems for the better admissible multimaps and their applications, World Congress of Nonlinear Analysts '96, Proceedings, *Nonlinear Anal.*, **30**, 4183–4191.

[13] Sehie Park (1997). Remarks on fixed point theorems of Ricceri. *Proc. '97 Workshop on Math. Anal. App.*, Pusan Nat. Univ., 1–9.

[14] Sehie Park, Generalized Kirszbraun-Minty type inequalities, (to appear).

[15] E. Zeidler (1986–1990). *Nonlinear Functional Analysis and Its Applications*, 5 volumes. New York: Springer-Verlag.

SIMAA 4(2002) 411–420

24. Convergence of Iterates of Nonexpansive Set-valued Mappings

Simeon Reich* and Alexander J. Zaslavski

Department of Mathematics, The Technion-Israel Institute of Technology, 32000 Haifa, Israel. E-mail: sreich@tx.technion.ac.il; ajzasl@tx.technion.ac.il

Abstract: We show that most nonexpansive set-valued mappings (in the sense of Baire's categories) are, in fact, contractive, and then we study the asymptotic behavior of their trajectories.

Key words and Phrases: Complete metric space, contractive mapping, fixed point, generic property, nonexpansive mapping, set-valued mapping, trajectory

1991 Mathematics Subject Classification: 47H04, 47H09, 47H10, 54C60

INTRODUCTION

Nonexpansive mappings have been studied extensively in recent years with many results and applications. See, for example, [3,9,10] and the references mentioned therein. Although the iterates of nonexpansive mappings do not converge in general, it is known that the iterates of contractive mappings converge in all complete metric spaces [14]. However, it is also known [6] that the iterates of most nonexpansive mappings (in the sense of Baire's categories) do converge to their unique fixed points. Recently, we have improved upon this result [19] by showing that most nonexpansive mappings are, in fact, contractive. Fixed point theory for set-valued contractive and nonexpansive mappings is more delicate and was studied by several authors. See, for example, [5,9,12,15,16] and the references mentioned there. Our purpose in the present chapter is to study set-valued contractive and nonexpansive mappings on bounded closed convex subsets of a Banach space. To this end, we first consider single-valued contractive sequences of mappings and prove a weak ergodic theorem for their infinite products (Section 1). In Section 2 we introduce star-shaped spaces and show that a generic nonexpansive sequence is, in fact, contractive. In Section 3 we breifly review an important instance of star-shaped spaces, namely, the class of hyperbolic spaces. Section 4 contains our main results. We show, in particular, that a generic nonexpansive set-valued mapping (acting on a bounded closed convex subset K of a Banach space) is contractive and that its iterates converge to a

* The work of Simeon Reich was partially supported by the Israel Science Foundation founded by the Israel Academy of Sciences and Humanities, by the Fund for the Promotion of Research at the Technion, and by the Technion VPR Fund – E. and M. Mendelson Research Fund.

unique closed invariant subset. This enables us to deduce some asymptotic properties of trajectories of such mappings. In the proofs we are able to use the framework of Section 2 because the space of closed subsets of K is, indeed, star-shaped. Such an approach, when a certain proprty is investigated for a whole space and not just for a single operator or sequence, turns out to be useful in many other areas of analysis [6,7,19–21].

1. CONTRACTIVE MAPPINGS

Let (X, ρ) be a complete metric space. An operator $A\colon X \to X$ is called nonexpansive if

$$\rho(Ax, Ay) \leq \rho(x, y) \quad \text{for all } x, y \in X.$$

We denote by $\mathfrak{A}$ the set of all nonexpansive operators $A\colon X \to X$. We assume that X is bounded and set

$$d(X) = \sup\{\rho(x, y) : \ x, y \in X\} < \infty.$$

We equip the set $\mathfrak{A}$ with the metric $\rho_{\mathfrak{A}}$ defined by

$$\rho_{\mathfrak{A}}(A, B) = \sup\{\rho(Ax, Bx) : \ x \in X\}, \ A, B \in \mathfrak{A}. \tag{1.1}$$

Clearly the metric space $(\mathfrak{A}, \rho_{\mathfrak{A}})$ is complete. Denote by $\mathcal{A}$ the set of all sequences $\{A_t\}_{t=1}^{\infty}$, where $A_t \in \mathfrak{A}$, $t = 1, 2, \ldots$. A member of $\mathcal{A}$ will occasionally be denoted by boldface $\mathbf{A}$. For the set $\mathcal{A}$ we define a metric $\rho_{\mathcal{A}}$ by

$$\rho_{\mathcal{A}}(\{A_t\}_{t=1}^{\infty}, \{B_t\}_{t=1}^{\infty}) = \sup\{\rho(A_t x, B_t x) : \ t = 1, 2, \ldots \text{ and } x \in X\}. \tag{1.2}$$

Clearly the metric space $(\mathcal{A}, \rho_{\mathcal{A}})$ is also complete.

A sequence $\{A_t\}_{t=1}^{\infty} \in \mathcal{A}$ is called contractive if there exists a decreasing function $\phi\colon [0, d(X)] \to [0, 1]$ such that

$$\phi(t) < 1 \quad \text{for all } t \in (0, d(X)] \tag{1.3}$$

and

$$\rho(A_t x, A_t y) \leq \phi(\rho(x, y))\rho(x, y) \quad \text{for all } x, y \in X \text{ and all integers } t \geq 1. \tag{1.4}$$

An operator $A \in \mathfrak{A}$ is called contractive if the sequence $\{A_t\}_{t=1}^{\infty}$ with $A_t = A$, $t = 1, 2, \ldots$, is contractive.

The notion of a contractive mapping as well as its modifications and applications were studied by many authors. See, for example, [1,2,11,15]. It is known [14] that the iterates of any contractive mapping converge to its unique fixed point. Our first theorem extends this result to infinite products.

Theorem 1.1: *Assume that the sequence $\{A_t\}_{t=1}^{\infty}$ is contractive and that $\epsilon > 0$. Then there exists a natural number N such that for each integer $T \geq N$, each mapping $h\colon \{1, \ldots, T\} \to \{1, 2, \ldots\}$ and each $x, y \in X$,*

$$\rho(A_{h(T)} \cdots A_{h(1)} x, A_{h(T)} \cdots A_{h(1)} y) \leq \epsilon. \tag{1.5}$$

Proof: There exists a decreasing function ϕ: $[0, d(X)] \to [0, 1]$ such that the inequalities (1.3) and (1.4) hold. Choose a natural number $N > 4$ such that

$$d(X)\phi(\epsilon)^N < \epsilon. \tag{1.6}$$

Assume that an integer $T \geq N$, a mapping h: $\{1, \ldots, T\} \to \{1, 2, \ldots\}$ and a pair of points $x, y \in X$ are given. We intend to show that (1.5) holds. Assume it does not. Then

$$\rho(x, y) > \epsilon \text{ and } \rho(A_{h(n)} \ldots A_{h(1)}x, A_{h(n)} \ldots A_{h(1)}y) > \epsilon, \; n = 1, \ldots, N. \tag{1.7}$$

It follows from (1.7) and (1.4) that

$$\rho(A_{h(1)}x, A_{h(1)}y) \leq \phi(\rho(x, y))\rho(x, y) \leq \phi(\epsilon)\rho(x, y)$$

and that for all integers $i = 1, \ldots, N - 1$,

$$\begin{aligned}&\rho(A_{h(i+1)}A_{h(i)} \ldots A_{h(1)}x, A_{h(i+1)}A_{h(i)} \ldots A_{h(1)}y)\\ &\quad\leq \phi(\epsilon)\rho(A_{h(i)} \ldots A_{h(1)}x, A_{h(i)} \ldots A_{h(1)}y).\end{aligned}$$

Together with (1.6) this implies that

$$\rho(A_{h(N)} \ldots A_{h(1)}x, A_{h(N)} \ldots A_{h(1)}y) \leq \phi(\epsilon)^N \rho(x, y) \leq d(X)\phi(\epsilon)^N < \epsilon,$$

a contradiction. This completes the proof of Theorem 1.1.

Corollary 1.1: *Assume that the sequence $\{A_t\}_{t=1}^{\infty}$ is contractive. Then*

$$\rho(A_{h(T)} \ldots A_{h(1)}x, A_{h(T)} \ldots A_{h(1)}y) \to 0 \text{ as } T \to \infty$$

uniformly in h: $\{1, 2, \ldots\} \to \{1, 2, \ldots\}$ and in $x, y \in X$.

We remark in passing that such results are called weak ergodic theorems in the population biology literature [4,13]. For more information on infinite products and their applications see [8].

We now show that Theorem 1.1 also implies the above-mentioned convergence result for a single contractive operator [14].

Theorem 1.2: *Let the operator $A \in \mathfrak{A}$ be contractive. Then there exists a unique point $x_A \in X$ such that $Ax_A = x_A$ and $A^n x \to x_A$ as $n \to \infty$, uniformly on X.*

Proof: By Theorem 1.1, for each $\epsilon > 0$ there exists a natural number N such that for each $x, y \in X$, $\rho(A^N x, A^N y) \leq \epsilon$. This implies that for each $x \in X$, the sequence $\{A_n x\}_{n=1}^{\infty}$ is a Cauchy sequence which converges to a point $x_A \in X$. This point is a fixed point of A and does not depend on x. It is not difficult to see that $A^n x \to x_A$ as $n \to \infty$ uniformly on X.

2. STAR-SHAPED SPACES

We will say that a complete metric space (X, ρ) is star-shaped if it contains a point $x_* \in X$ with the following property:

For each $x \in X$ there exists a mapping

$$t \to tx \oplus (1-t)x_*, \ t \in (0,1) \in X, \tag{2.1}$$

such that for each $t \in (0,1)$ and each $x, y \in X$,

$$\rho(tx \oplus (1-t)x_*, ty \oplus (1-t)x_*) \le t\rho(x,y) \tag{2.2}$$

and

$$\rho(tx \oplus (1-t)x_*, x) \le (1-t)\rho(x, x_*). \tag{2.3}$$

For each $A \in \mathfrak{A}$ and each $\gamma \in (0,1)$ define $A_\gamma \in \mathfrak{A}$ by

$$A_\gamma x = (1-\gamma)Ax \oplus \gamma x_*, \ x \in X. \tag{2.4}$$

For each $\mathbf{A} = \{A_t\}_{t=1}^\infty \in \mathcal{A}$ define $\mathbf{A}_\gamma = \{A_{\gamma t}\}_{t=1}^\infty$ by

$$A_{\gamma t}x = (1-\gamma)A_t x \oplus \gamma x_*, \ x \in X, \ t = 1, 2, \ldots \tag{2.5}$$

Theorem 2.1: *Assume that $\mathcal{B}$ is a closed subset of $\mathcal{A}$ such that for each $\mathbf{A} \in \mathcal{B}$ and each $\gamma \in (0,1)$ the sequence $\mathbf{A}_\gamma \in \mathcal{B}$. Then there exists a set $\mathcal{F}$ which is a countable intersection of open everywhere dense subsets of $\mathcal{B}$ (with the relative topology) such that each $\mathbf{A} \in \mathcal{F}$ is contractive.*

Proof: It follows from (2.3) that for each $\mathbf{A} = \{A_t\}_{t=1}^\infty \in \mathcal{B}$, each $\gamma \in (0,1)$ and each $x \in X$,

$$\rho(A_{\gamma t}x, A_t x) \le \gamma\rho(A_t x, x_*).$$

This implies that $\mathbf{A}_\gamma \to \mathbf{A}$ in $\mathcal{B}$ as $\gamma \to 0^+$ and that the set $\{\mathbf{A}_\gamma \colon \mathbf{A} \in \mathcal{B}, \ \gamma \in (0,1)\}$ is everywhere dense in $\mathcal{B}$.

Let $\mathbf{A} = \{A_t\}_{t=1}^\infty \in \mathcal{B}$ and $\gamma \in (0,1)$. The inequality (2.2) implies that

$$\rho(A_{\gamma t}x, A_{\gamma t}y) \le (1-\gamma)\rho(x,y) \tag{2.6}$$

for all $x, y \in X$ and all integers $t \ge 1$. For each integer $i \ge 1$, choose a positive number

$$\delta(\mathbf{A}, \gamma, i) < (4i)^{-1}d(X)\gamma \tag{2.7}$$

and define

$$U(\mathbf{A}, \gamma, i) = \{\mathbf{B} \in \mathcal{B} \colon \rho_{\mathcal{A}}(\mathbf{A}_\gamma, \mathbf{B}) < \delta(\mathbf{A}, \gamma, i)\}. \tag{2.8}$$

Let $i \ge 1$ be an integer. We will show that the following property holds:

P(1) For each $\mathbf{B} \in U(\mathbf{A}, \gamma, i)$, each $x, y \in X$ satisfying $\rho(x,y) \ge i^{-1}d(X)$ and each integer $t \ge 1$, the inequality $\rho(B_t x, B_t y) \le (1-\gamma/2)\rho(x,y)$ is valid.

Indeed, assume that $\mathbf{B} \in U(\mathbf{A}, \gamma, i)$, the points $x, y \in X$ satisfy

$$\rho(x,y) \ge i^{-1}d(X), \tag{2.9}$$

and that $t \geq 1$ is an integer. It follows from the definition of $U(\mathbf{A}, \gamma, i)$ (see (2.8), (2.7)), (2.6) and (2.9) that

$$\begin{aligned} \rho(B_t x, B_t y) &\leq \rho(A_{\gamma t} x, A_{\gamma t} y) + 2\delta(\mathbf{A}, \gamma, i) \\ &< 2\delta(\mathbf{A}, \gamma, i) + (1-\gamma)\rho(x, y) \leq (1-\gamma)\rho(x, y) + (2i)^{-1}\gamma d(X) \\ &\leq (1-\gamma)\rho(x, y) + 2^{-1}\gamma\rho(x, y) \leq (1-\gamma/2)\rho(x, y). \end{aligned}$$

Thus

$$\rho(B_t x, B_t y) \leq (1-\gamma/2)\rho(x, y). \tag{2.10}$$

Now define

$$\mathcal{F} = \cap_{i=1}^{\infty} \cup \{U(\mathbf{A}, \gamma, i) : \ \mathbf{A} \in \mathcal{B}, \ \gamma \in (0,1)\}. \tag{2.11}$$

Clearly $\mathcal{F}$ is a countable intersection of open everywhere dense subsets of $\mathcal{B}$ (with the relative topology). We claim that any $\mathbf{B} \in \mathcal{F}$ is contractive. To show this, assume that i is a natural number. There exist $\mathbf{A} \in \mathcal{B}$ and $\gamma \in (0,1)$ such that $\mathbf{B} \in U(\mathbf{A}, \gamma, i)$. By property P(1), for each $x, y \in X$ satisfying $\rho(x, y) \geq i^{-1}d(X)$ and each integer $t \geq 1$, the inequality (2.10) holds. Since i is an arbitrary natural number we conclude that $\mathbf{B}$ is contractive. Theorem 2.1 is proved.

Theorem 2.2: *Assume that $\mathfrak{B}$ is a closed subset of $\mathfrak{A}$ such that for each $A \in \mathfrak{B}$ and each $\gamma \in (0,1)$, the mapping $A_\gamma \in \mathfrak{B}$. Then there exists a set $\mathcal{F}$ which is a countable intersection of open everywhere dense subsets of $\mathfrak{B}$ (with the relative topology) such that each $A \subset \mathcal{F}$ is contractive.*

Proof: For each $A \in \mathfrak{B}$ denote by $Q(A)$ the sequence $\mathbf{A} = \{A_t\}_{t=1}^{\infty}$ with $A_t = A$, $t = 1, 2, \ldots$. Set

$$\mathcal{B} = \{Q(A) : \ A \in \mathfrak{B}\}.$$

It is easy to see that $\mathcal{B}$ is a closed subset of $\mathcal{A}$ and that for each $\mathbf{A} \in \mathcal{B}$ and each $\gamma \in (0,1)$, the sequence $\mathbf{A}_\gamma \in \mathcal{B}$. Now Theorem 2.2 follows from Theorem 2.1 and the equality

$$\rho_{\mathfrak{A}}(A, B) = \rho_{\mathcal{A}}(Q(A), Q(B)).$$

3. HYPERBOLIC SPACES

Let (E, ρ) be a metric space and let R^1 denote the real line. We say that a mapping c: $R^1 \to X$ is a metric embedding of R^1 into E if $\rho(c(s), c(t)) = |s - t|$ for all real s and t. The image of R^1 under a metric embedding will be called a metric line. The image of a real interval $[a, b] = \{t \in R^1 : a \leq t \leq b\}$ under such a mapping will be called a metric segment. Assume that (E, ρ) contains a family M of metric lines such that for each pair of distinct points x and y in E there is a unique metric line in M which passes through x and y. This metric line determines a unique metric segment joining x and y. We denote this segment by $[x, y]$. For each $0 \leq t \leq 1$ there is a unique point z in $[x, y]$ such that $\rho(x, z) = t\rho(x, y)$ and $\rho(z, y) = (1 - t)\rho(x, y)$. This point will be denoted by $(1 - t)x \oplus ty$. We will say that E, or more precisely (E, ρ, M), is a hyperbolic space if

$$\rho\left(\frac{1}{2}x \oplus \frac{1}{2}y, \frac{1}{2}x \oplus \frac{1}{2}z\right) \leq \frac{1}{2}\rho(y, z)$$

for all x, y and z in E. A set $K \subset E$ is called ρ-convex if $[x, y] \subset K$ for all x and y in K. It is clear that all normed linear spaces are hyperbolic. A discussion of more examples of hyperbolic spaces and in particular of the Hilbert ball can be found, for example, in [10,17,18]. We mention this class of spaces here because they provide us with examples of star-shaped spaces. Indeed, note that if (E, ρ, M) is a hyperbolic space, then

$$\rho((1 - t)x \oplus tz, (1 - t)y \oplus tw) \leq (1 - t)\rho(x, y) + t\rho(z, w)$$

for all x, y, z and w in E and $0 \leq t \leq 1$ (see [10:pp. 77 and 104] and [18]).

Therefore if X is a closed subset of E and there exists $x_* \in X$ such that for each $x \in X$,

$$tx \oplus (1 - t)x_* \in X, \; t \in (0, 1),$$

then it is not difficult to see that (X, ρ) is a star-shaped complete metric space.

4. SET-VALUED MAPPINGS

Assume that $(E, \|\cdot\|)$ is a Banach space, K is a nonempty bounded closed subset of E and there exists $\theta \in K$ such that for each $x \in K$,

$$tx + (1 - t)\theta \in K, \; t \in (0, 1).$$

We consider the star-shaped complete metric space K with the metric $\|x - y\|$, $x, y \in K$. Denote by $S(K)$ the set of all nonempty closed subsets of K. For $x \in K$ and $A \subset K$ set

$$\rho(x, A) = \inf\{\|x - y\| : \; y \in A\},$$

and for each $A, B \in S(K)$ let

$$H(A, B) = \max\{\sup_{x \in A} \rho(x, B), \; \sup_{y \in B} \rho(y, A)\}. \tag{4.1}$$

We equip the set $S(K)$ with the Hausdorff metric $H(\cdot, \cdot)$. It is well-known that the metric space $(S(K), H)$ is complete. Clearly $\{\theta\} \in S(K)$.

For each $A \in S(K)$ and each $t \in [0, 1]$ define

$$tA \oplus (1 - t)\theta = \{tx + (1 - t)\theta : \; x \in A\} \in S(K). \tag{4.2}$$

It is not difficult to see that the complete metric space $(S(K), H)$ is star-shaped.

Denote by $\mathfrak{A}$ the set of all nonexpansive operators T: $S(K) \to S(K)$. For the set $\mathfrak{A}$ we consider the metric $\rho_{\mathfrak{A}}$ defined by

$$\rho_{\mathfrak{A}}(T_1, T_2) = \sup\{H(T_1(A), T_2(A)) : A \in S(K)\},\ T_1, T_2 \in \mathfrak{A}. \tag{4.3}$$

Denote by $\mathcal{M}$ the set of all mappings T: $K \to S(K)$ such that

$$H(T(x), T(y)) \leq \|x - y\|,\ x, y \in K. \tag{4.4}$$

A mapping $T \in \mathcal{M}$ is called contractive if there exists a decreasing function ϕ: $[0, d(K)] \to [0, 1]$ such that

$$\phi(t) < 1 \quad \text{for all } t \in (0, d(K)] \tag{4.5}$$

and

$$H(T(x), T(y)) \leq \phi(\|x - y\|)\|x - y\| \quad \text{for all } x, y \in K. \tag{4.6}$$

Assume that $T \in \mathcal{M}$. For each $A \in S(K)$ denote by $\tilde{T}(A)$ the closure of the set $\cup\{T(x)$: $x \in A\}$ in the norm topology.

Proposition 4.1: *Assume that $T \in \mathcal{M}$. Then the mapping $\tilde{T}$ belongs to $\mathfrak{A}$.*

Proof: Let $A, B \in S(K)$. We will show that

$$H(\tilde{T}(A), \tilde{T}(B)) \leq H(A, B). \tag{4.7}$$

Given $\epsilon > 0$, there exist $x_1 \in \tilde{T}(A)$ and $x_2 \in \tilde{T}(B)$ such that

$$\max\{\rho(x_1, \tilde{T}(B)), \rho(x_2, \tilde{T}(A))\} + \epsilon/2 > H(\tilde{T}(A), \tilde{T}(B)). \tag{4.8}$$

We may assume that

$$\rho(x_1, \tilde{T}(B)) \geq \rho(x_2, \tilde{T}(A)).$$

Therefore

$$\rho(x_1, \tilde{T}(B)) + \epsilon/2 > H(\tilde{T}(A), \tilde{T}(B)). \tag{4.9}$$

We may also assume that $x_1 \in T(A)$. Then there exist $x_0 \in A$ such that $x_1 \in T(x_0)$ and $y_0 \in B$ such that

$$\|x_0 - y_0\| < \rho(x_0, B) + \epsilon/2 \leq H(A, B) + \epsilon/2.$$

Therefore the inequality (4.4) implies that

$$\rho(x_1, \tilde{T}(B)) \leq \rho(x_1, T(y_0)) \leq H(T(x_0), T(y_0)) \leq \|x_0 - y_0\| < H(A, B) + \epsilon/2.$$

Now (4.9) yields

$$H(\tilde{T}(A), \tilde{T}(B)) < H(A, B) + \epsilon.$$

Since ϵ is an arbitrary positive number we conclude that (4.7) holds. Proposition 4.1 is proved.

Proposition 4.2: *Assume that $T \in \mathcal{M}$. Then the mapping $\tilde{T}$ is contractive if and only if the mapping T is contractive.*

Proof: It is clear that T is contractive if $\tilde{T}$ is contractive. Assume now that the mapping T is contractive. Then there exists a decreasing function ϕ: $[0, d(K)] \to [0, 1]$ such that (4.5) and (4.6) hold.

Let $A, B \in S(K)$. We will show that

$$H(\tilde{T}(A), \tilde{T}(B)) \leq \max\{1/2, \phi(H(A,B)/4)\}H(A,B). \tag{4.10}$$

We may assume that $H(A, B) > 0$ and that

$$H(\tilde{T}(A), \tilde{T}(B)) > H(A,B)/2. \tag{4.11}$$

Let

$$\epsilon \in (0, H(A,B)/4). \tag{4.12}$$

By the definition of the Hausdorff metric, there exist $x_1 \in \tilde{T}(A)$ and $x_2 \in \tilde{T}(B)$ such that

$$\max\{\rho(x_1, \tilde{T}(B)), \rho(x_2, \tilde{T}(A))\} + \epsilon/2 > H(\tilde{T}(A), \tilde{T}(B)). \tag{4.13}$$

We may assume that

$$\rho(x_1, \tilde{T}(B)) \geq \rho(x_2, \tilde{T}(A)).$$

Therefore,

$$\rho(x_1, \tilde{T}(B)) + \epsilon/2 > H(\tilde{T}(A), \tilde{T}(B)). \tag{4.14}$$

We may also assume that $x_1 \in T(A)$. Then there exist $x_0 \in A$ such that $x_1 \in T(x_0)$ and $y_0 \in B$ such that

$$\|x_0 - y_0\| < \rho(x_0, B) + \epsilon/2 \leq H(A,B) + \epsilon/2. \tag{4.15}$$

Therefore (4.6) implies that

$$\begin{aligned} \rho(x_1, \tilde{T}(B)) \leq \rho(x_1, T(y_0)) \leq H(T(x_0), T(y_0)) \leq \phi(\|x_0 - y_0\|)\|x_0 - y_0\| \\ \leq \phi(\|x_0 - y_0\|)(H(A,B) + \epsilon/2). \end{aligned} \tag{4.16}$$

Combining this with (4.14) we see that

$$-\epsilon/2 + H(\tilde{T}(A), \tilde{T}(B)) < \phi(\|x_0 - y_0\|)(H(A,B) + \epsilon/2). \tag{4.17}$$

It follows from (4.4), (4.14), (4.11) and (4.12) that

$$\begin{aligned} \|x_0 - y_0\| \geq H(T(x_0), T(y_0)) \geq \rho(x_1, T(y_0)) \geq \rho(x_1, \tilde{T}(B)) \\ \geq -\epsilon/2 + H(\tilde{T}(A), \tilde{T}(B)) > -\epsilon/2 + H(A,B)/2 \geq H(A,B)/4. \end{aligned}$$

Thus

$$\|x_0 - y_0\| \geq H(A,B)/4.$$

Combining this last inequality with (4.17) we can deduce that

$$-\epsilon/2 + H(\tilde{T}(A), \tilde{T}(B)) < \phi(H(A,B)/4)(H(A,B) + \epsilon/2). \tag{4.17}$$

Since ϵ is an arbitrary positive number we conclude that

$$H(\tilde{T}(A), \tilde{T}(B)) \leq \phi(H(A,B)/4)(H(A,B)).$$

This completes the proof of Proposition 4.2.

We equip the set $\mathcal{M}$ with the metric $\rho_{\mathcal{M}}$ defined by

$$\rho_{\mathcal{M}}(T_1, T_2) = \sup\{H(T_1(x), T_2(x)) : x \in K\},\ T_1, T_2 \in \mathcal{M}. \tag{4.18}$$

It is not difficult to verify that the metric space $(\mathcal{M}, \rho_{\mathcal{M}})$ is complete.

For each $T \in \mathcal{M}$ set $P(T) = \tilde{T}$. It is easy to see that for each $T_1, T_2 \in \mathcal{M}$,

$$\rho_{\mathfrak{A}}(P(T_1), P(T_2)) = \rho_{\mathcal{M}}(T_1, T_2). \tag{4.19}$$

Denote

$$\mathfrak{B} = \{P(T) : T \in \mathcal{M}\}. \tag{4.20}$$

Clearly the metric spaces $(\mathfrak{B}, \rho_{\mathfrak{A}})$ and $(\mathcal{M}, \rho_{\mathcal{M}})$ are isometric.

For each $T \in \mathfrak{A}$ and each $\gamma > 0$ define

$$T_\gamma(A) = (1-\gamma)T(A) \oplus t\theta.$$

It is easy to see that $T_\gamma \in \mathfrak{A}$ for each $T \in \mathfrak{A}$ and each $\gamma > 0$, and moreover, $T_\gamma \in \mathfrak{B}$ if $T \in \mathfrak{B}$. Now we can apply Theorem 2.2 and obtain the following result.

Theorem 4.1: *There exists a set $\mathcal{F}$ which is a countable intersection of open everywhere dense subsets of $(\mathcal{M}, \rho_{\mathcal{M}})$ such that each $T \in \mathcal{F}$ is contractive.*

Theorem 1.2 and Proposition 4.2 imply the following result.

Theorem 4.2: *Assume that the operator $T \in \mathcal{M}$ is contractive. Then there exists a unique set $A_T \in S(K)$ such that $\tilde{T}(A_T) = A_T$ and $(\tilde{T})^n(B) \to A_T$ as $n \to \infty$, uniformly for all $B \in S(K)$.*

Let $T \in \mathcal{M}$. A sequence $\{x_n\}_{n=1}^N \subset K$ with $N \geq 1$ (respectively, $\{x_n\}_{n=1}^\infty \subset K$) is called a trajectory of T if $x_{i+1} \in T(x_i)$, $i = 1, \ldots, N-1$ (respectively, $i = 1, 2, \ldots$).

Theorem 4.2 leads to the following results.

Theorem 4.3: *Let the operator $T \in \mathcal{M}$ be contractive and let the set $A_T \in S(K)$ be as guaranteed by Theorem 4.2. Then for each $\epsilon > 0$ there exists a natural number n such that for each trajectory $\{x_i\}_{i=1}^n \subset K$ of T, $\rho(x_n, A_T) < \epsilon$.*

Theorem 4.4: *Let the operator $T \in \mathcal{M}$ be contractive and let the set $A_T \in S(K)$ be as guaranteed by Theorem 4.2. Then for each $\epsilon > 0$ there exists a natural number n such that for each $z \in K$ and each $x \in A_T$, there exists a trajectory $\{x_i\}_{i=1}^n \subset K$ of T such that $x_1 = z$ and $\rho(x_n, x) < \epsilon$.*

Corollary 4.1: *Let the operator $T \in \mathcal{M}$ be contractive and let the set $A_T \in S(K)$ be as guaranteed by Theorem 4.2. Then for each $x \in A_T$ there is a trajectory $\{x_i\}_{i=1}^{\infty} \subset A_T$ such that $x_1 = x$ and $\liminf_{i\to\infty} \|x_i - x\| = 0$.*

Corollary 4.2: *Let the operator $T \in \mathcal{M}$ be contractive and let the set $A_T \in S(K)$ be as guaranteed by Theorem 4.2. Assume that the set A_T is separable. Then for each $x \in A_T$ there is a trajectory $\{x_i\}_{i=1}^{\infty} \subset A_T$ such that $x_1 = x$ and for each $y \in A_T$, $\liminf_{i\to\infty} \|x_i - y\| = 0$.*

REFERENCES

[1] Ya. I. Alber and S. Guerre-Delabrière (1997). Principle of weakly contractive maps in Hilbert spaces, *New Results in Operator Theory and Its Applications, Operator Theory*, **98**, 7–22.
[2] Ya. I. Alber, S. Guerre-Delabrière and L. Zelenko (1998). The principle of weakly contractive mappings in metric spaces, *Comm. Appl. Nonlinear Analysis*, **5**(1), 45–68
[3] H.H. Bauschke, J.M. Borwein and A.S. Lewis (1997). The method of cyclic projections for closed convex sets in Hilbert space, *Recent Developments in Optimization Theory and Nonlinear Analysis, Contemporary Mathematics*, **204**, 1–38.
[4] J.E. Cohen (1979). Ergodic theorems in demography, *Bull. Amer. Math. Soc.*, **1**, 275–295.
[5] H. Covitz and S.B. Nadler Jr., (1970). Multi-valued contraction mappings in generalized metric spaces, *Israel J. Math.*, **8**, 5–11.
[6] F.S. De Blasi and J. Myjak (1976). Sur la convergence des approximations successives pour les contractions non linéaires dans un espace de Banach, *C. R. Acad. Sc. Paris*, **283**, 185–187.
[7] F.S. De Blasi and J. Myjak (1983). Generic flows generated by continuous vector fields in Banach spaces, *Adv. in Math.*, **50**, 266–280.
[8] J. Dye and S. Reich (1992). Random products of nonexpansive mappings, *Optimization and Nonlinear Analysis, Pitman Research Notes in Mathematics Series*, **244**, 106–118.
[9] K. Goebel and W.A. Kirk (1990). *Topics in Metric Fixed Point Theory*. Cambridge: Cambridge University Press.
[10] K. Goebel and S. Reich (1984). *Uniform Convexity, Hyperbolic Geometry, and Nonexpansive Mappings*. New York and Basel: Marcel Dekker.
[11] M.A. Krasnosel'skii and P.P. Zabreiko (1984). *Geometrical Methods of Nonlinear Analysis*. Berlin: Springer.
[12] T.-C. Lim (1974). A fixed point theorem for multivalued nonexpansive mappings in a uniformly convex space, *Bull. Amer. Math. Soc.*, **80**, 1123–1126.
[13] R.D. Nussbaum (1990). Some nonlinear weak ergodic theorems, *SIAM J. Math. Anal.*, **21**, 436–460.
[14] E. Rakotch (1962). A note on contractive mappings, *Proc. Amer. Math. Soc.*, **13**, 459–465.
[15] S. Reich (1972). Fixed points of contractive functions, *Boll. Un. Mat. Ital.*, **5**, 26–42.
[16] S. Reich (1978). Approximate selections, best approximations, fixed points, and invariant sets, *J. Math. Anal. Appl.*, **62**, 104–113.
[17] S. Reich (1993). The alternating algorithm of von Neumann in the Hilbert ball, *Dynamic Systems and Appl.*, **2**, 21–26.
[18] S. Reich and I. Shafrir (1990). Nonexpansive iterations in hyperbolic spaces, *Nonlinear Analysis: Theory, Methods and Applications*, **15**, 537–558.
[19] S. Reich and A.J. Zaslavski (2000). Almost all nonexpansive mappings are contractive, *C. R. Math. Rep. Acad. Sci. Canada*, **22**, 118–124.
[20] A.J. Zaslavski (1996). Dynamic properties of optimal solutions of variational problems, *Nonlinear Analysis: Theory, Methods and Applications*, **27**, 895–932.
[21] A.J. Zaslavski. Existence of solutions of optimal control problems without convexity assumptions, *Nonlinear Analysis: Theory, Methods and Applications*, accepted for publication.

SIMAA 4(2002) 421–423

25. A Remark on the Intersection of a Lower Semicontinuous Multifunction and a Fixed Set

Biagio Ricceri

Department of Mathematics, University of Catania, Viale A. Doria 6, 95125 Catania, Italy. E-mail: ricceri@dipmat.unict.it

Certainly, one of the most important notions in the theory of multifunctions is that of lower semicontinuity. Let us recall it. So, let X, Y be two topological spaces, $F\colon X \to 2^Y$ a multifunction and $x_0 \in X$. We say that F is lower semicontinuous at x_0 if, for every open set $\Omega \subseteq Y$, the condition $x_0 \in F^-(\Omega)$ implies that $x_0 \in \operatorname{int}(F^-(\Omega))$, where $F^-(\Omega) = \{x \in X : F(x) \cap \Omega \neq \emptyset\}$. We say that F is lower semicontinuous provided that it is so at each point of X.

In this very short chapter, we simply wish to present a contribution to the following question: if F is lower semicontinuous and C is a subset of Y, what additional assumptions ensure that the multifunction $x \to F(x) \cap C$ is lower semicontinuous too?

Of course, there are two answers which are immediate. Namely, if we further assume that either C is open or $F^-(y)$ is open for each $y \in C$, then the multifunction $x \to F(x) \cap C$ is lower semicontinuous.

Out of these two cases, the question assumes rather delicate aspects.

We point out that the lower semicontinuity of the metric projection, a classical issue in approximation theory, falls in the setting we are dealing with. Let us recall that, given a metric space (X, d) and a non-empty subset C of X, the metric projection on C is the multifunction $P_C\colon X \to 2^C$ defined by

$$P_C(x) = \{y \in C : d(x, y) = \operatorname{dist}(x, C)\}$$

for all $x \in X$.

We have:

Proposition 1: *Let X be a normed space ($X \neq 0$) and let C be a non-empty subset of X. For each $x \in X$, put*

$$Q_C(x) = \{y \in X : \| x - y \| = \operatorname{dist}(x, C)\}.$$

Then, the multifunction Q_C is lower semicontinuous.

Proof: For each $(x, y) \in X \times X$, put

$$f(x, y) = \| x - y \| - \operatorname{dist}(x, C).$$

Clearly, the function f is continuous. Moreover, for each $x \in X \backslash \bar{C}$, the function $f(x, \cdot)$ changes sign in X, has no local minima at which it vanishes, and is not constant on each non-empty open subset of X. Then, taking also into account that X is connected and locally connected, we can apply Theorem 1.1 of [4] to the restriction of f to $(X \backslash \bar{C}) \times X$. That result then ensures that the restriction of Q_C to $X \backslash \bar{C}$ is lower semicontinuous. Hence, since $X \backslash \bar{C}$ is open, Q_C is lower semicontinuous at each point of $X \backslash \bar{C}$. Now, let $x_0 \in \bar{C}$. Hence, $Q_C(x_0) = \{x_0\}$. Observe that, for each $\delta > 0$, each $x \in X$ satisfying $\| x - x_0 \| < \frac{\delta}{2}$ and each $y \in Q_C(x)$, one has

$$\| y - x_0 \| \leq \operatorname{dist}(x, C) + \| x - x_0 \| \leq 2 \| x - x_0 \| < \delta.$$

Clearly, this shows that Q_C is lower semicontinuous at x_0. △

Hence, on the basis of Proposition 1, the metric projection P_C is the intersection of a lower semicontinuous multifunction and the set C.

We now establish our result which has been inspired by [5] (see also [2], [6,7]). For a set A in a topological vector space, we denote by $\operatorname{ri}(A)$ the relative interior of A, that is the interior of A in $\operatorname{aff}(A)$, the affine hull of A.

Theorem 1: *Let X be a topological space; $x_0 \in X$; Y a topological vector space; $A \subseteq Y$, with $\operatorname{ri}(A) \neq \emptyset$; g: $X \to \operatorname{aff}(A)$ a function which is continuous at x_0; F: $X \to 2^Y$ a multifunction such that $F^-(y)$ is open for all $y \in g(x_0) - \operatorname{ri}(A)$.*

For each $x \in X$, put

$$G(x) = (g(x) - F(x)) \cap \operatorname{ri}(A).$$

Then, the multifunction G is lower semicontinuous at x_0.

Proof: Let $\Omega \subseteq Y$ be an open set such that $G(x_0) \cap \Omega \neq \emptyset$. Pick $y_0 \in F(x_0)$ so that $g(x_0) - y_0 \in \operatorname{ri}(A) \cap \Omega$. Since $y_0 \in g(x_0) - \operatorname{ri}(A)$, by assumption, the set $F^-(y_0)$ is a neighbourhood of x_0. Now, observe that, since $g(x_0) \in \operatorname{aff}(A)$, the set $y_0 + (\operatorname{ri}(A) \cap \Omega)$ is contained in $\operatorname{aff}(A)$ and, of course, is open there and contains $g(x_0)$. Consequently, since g is continuous at x_0 and $g(X) \subseteq \operatorname{aff}(A)$, the set

$$V = F^-(y_0) \cap g^{-1}(y_0 + (\operatorname{ri}(A) \cap \Omega))$$

turns out to be a neighbourhood of x_0. Finally, if $x \in V$, we have $y_0 \in F(x)$ as well as $g(x) - y_0 \in \operatorname{ri}(A) \cap \Omega$. That is, $g(x) - y_0 \in G(x) \cap \Omega$, and the proof is complete. △

Remark 1: We now comment the two leading assumptions of Theorem 1. First, if the range of g is not contained in $\operatorname{aff}(A)$, the result can fail. The simplest example is as follows. Take: $X = [0, 1]$, $x_0 = 0$, $Y = \mathbf{R}^2$, $A = [-1, 1] \times \{0\}$, $g(x) = (x, x^2)$, $F(x) = \{(0, 0)\}$ for all $x \in [0, 1]$. Clearly, in such a case, the multifunction $x \to (g(x) - F(x)) \cap \operatorname{ri}(A)$ is not lower semicontinuous at x_0. Likewise, the result is no longer true, in general, if we replace the assumption made on F by assuming merely that F is lower semicontinuous. In this connection, take: $Y = \mathbf{R}^2$, $X = A = [-1, 1] \times \{0\}$, $x_0 = (0, 0)$, $g(t, 0) = (0, 0)$ and

$$F(t, 0) = \begin{cases} \{(\frac{1}{2}, 0)\} & \text{if } t \in [-1, 0] \\ \operatorname{conv}(\{(0, 0), (1, t)\}) & \text{if } t \in]0, 1] \end{cases}$$

where "conv" stands for convex hull. The multifunction F (and so $-F$) is lower semicontinuous (see [1:pp. 408–409]). However, the multifunction $(t,0) \to -F(t,0) \cap \text{ri}(A)$ is not lower semicontinuous at x_0.

Here is an application of Theorem 1.

Theorem 2: *Let X be a non-empty, compact, convex, finite-dimensional set in a Hausdorff topological vector space Y, let g: $X \to \text{aff}(X)$ be a continuous function, and let Φ: $X \to 2^Y$ be a multifunction, with convex values, such that the set $\{x \in X: y \in g(x) - \Phi(x)\}$ is open in X for all $y \in g(X) - \text{ri}(X)$, and $\Phi(x) \cap \text{ri}(X) \neq \emptyset$ for all $x \in X$.*

Then, there exists $x^* \in \text{ri}(X)$ such that $x^* \in \Phi(x^*)$.

Proof: Apply Theorem 1, taking $A = X$ and $F(x) = g(x) - \Phi(x)$ for all $x \in X$. We then have that the multifunction $x \to \Phi(x) \cap \text{ri}(X)$ is lower semicontinuous. Moreover, by assumption, its values are non-empty finite-dimensional convex sets, and so, by Theorem 3.1‴ of [3], there exists a continuous function f: $X \to \text{ri}(X)$ such that $f(x) \in \Phi(x)$ for all $x \in X$. By the Brouwer theorem, there is $x^* \in X$ such that $x^* = f(x^*)$. Consequently, we have $x^* \in \Phi(x^*) \cap \text{ri}(X)$, and the proof is complete. △

REFERENCES

[1] P. Cubiotti (1993). Some remarks on fixed points of lower semicontinuous multifunctions, *J. Math. Anal. Appl.*, **174**, 407–412.

[2] P. Cubiotti and N. D. Yen (1997). A result related to Ricceri's conjecture on generalized quasi-variational inequalities, *Arch. Math. (Basel)*, **69**, 507–514.

[3] E. Michael (1956). Continuous selections. I, *Ann. of Math.*, **63**, 361–382.

[4] B. Ricceri (1982). Applications de théorèmes de semi-continuité inférieure, *C. R. Acad. Sci. Paris, Série I*, **294**, 75–78.

[5] B. Ricceri (1985). Un théorème d'existence pour les inéquations variationnelles, *C. R. Acad. Sci. Paris, Série I*, **301**, 885–888.

[6] B. Ricceri (1987). Existence theorems for nonlinear problems, *Rend. Accad. Naz. Sci. XL*, **11**, 77–99.

[7] B. Ricceri (1995). Basic existence theorems for generalized variational and quasi-variational inequalities, In F. Giannessi and A. Maugeri (Eds), *Variational inequalities and network equilibrium problems*, Plenum Press, 251–255.

SIMAA 4(2002) 425–429

26. Random Approximations and Random Fixed Point Theorems for Set-valued Random Maps

Naseer Shahzad

Department of Mathematics, King Abdulaziz University, P.O. Box 9028, Jeddah-21413, Saudi Arabia

1. INTRODUCTION

The theory of nonlinear analysis has come out as one of the momentous mathematical disciplines during the last 50 years. The study of random approximations and random fixed points of set-valued random operators is presently a very active field of research lying at the intersection of nonlinear analysis and probability theory. It has grown very rapidly and has many interesting applications in various fields.

The study of random fixed point theorems was initiated by the Prague school of probabilists in the 1950s. However, the interest in this subject was revived specially after the publication of the survey article by Bharucha-Reid [3] in 1976. On the other hand, the theory of random approximations has received further attention after the appearance of the chapters by Sehgal and Waters [13], Sehgal and Singh [12], and Papageorgiou [9]. Since then there has been remarkable progress in this area and many interesting results have appeared (see, e.g., [1,2,4,6,7,10,11,15–19]), etc. More recently, Shahzad [14] introduced condition (A) (see the definition below) and proved a very general random fixed point theorem for continuous set-valued random operators. In the present chapter, we consider a new condition which we will call condition (B) (see the definition below) and obtain a general random approximation theorem for d-continuous random operators, which gives us the stochastic analogues of several deterministic approximation theorems involving continuous mappings. Finally, we derive a random fixed point theorem for d-continuous set-valued random operators.

2. PRELIMINARIES

Let (Ω, Σ) be a measurable space and S a non-empty subset of a metric space $X = (X, d)$. We denote by 2^S the family of all subsets of S and by $C(S)$ the family of all non-empty closed subsets of S. A set-valued mapping $H\colon \Omega \to 2^S\backslash\{\emptyset\}$ is called measurable if, for each open subset U of $S, H^{-1}(U) = \{\omega \in \Omega : H(\omega) \cap U \neq \emptyset\} \in \Sigma$. A mapping $\xi\colon \Omega \to S$ is said to be a measurable selector of a measurable set-valued mapping $H\colon \Omega \to 2^S\backslash\{\emptyset\}$ if ξ is measurable and $\xi(\omega) \in H(\omega)$ for each $\omega \in \Omega$. A mapping $T\colon \Omega \times S \to 2^X\backslash\{\emptyset\}$ is called a set-valued random operator if, for any fixed $x \in S$, the map $T(\cdot, x) : \Omega \to 2^X\backslash\{\emptyset\}$ is measurable. A measurable mapping $\xi\colon \Omega \to S$ is said to be a random fixed point of a random operator $T: \Omega \times S \to 2^X \backslash \{\emptyset\}$ if $\xi(\omega) \in T(\omega, \xi(\omega))$ for each $\omega \in \Omega$. Let $T: S \to C(X)$ be a set-valued mapping. Then T is said to satisfy (1) condition (A) [14] if for any sequence $\{x_n\}$ in S and $D \in C(S)$ such that $d(x_n, D) \to 0$ and $d(x_n, T(x_n)) \to 0$ as $n \to \infty$, there exists an $x_0 \in D$ with $x_0 \in T(x_0)$; (2) condition (B) if for any sequence $\{x_n\}$ in S and $D \in C(S)$ such that $d(x_n, D) \to 0$ and $|d(x_n, T(x_n)) - \delta(S, T(x_n))| \to 0$, there exists an $x_0 \in D$ such that

$$d(x_0, T(x_0)) = \delta(S, T(x_0)),$$

where

$$d(x, D_1) = \inf\{d(x, y) : y \in D_1\}$$

and

$$\delta(D_1, D_2) = \inf\{d(x, y) : x \in D_1, y \in D_2\} \quad \text{for } D_1, D_2 \in C(S).$$

The map T is called (3) hemicompact [18] if each sequence $\{x_n\}$ in S has a convergent subsequence whenever $d(x_n, T(x_n)) \to 0$ as $n \to \infty$; (4) continuous if T is both lower semicontinuous and upper semicontinuous; (5) d-continuous if for all $y \in X$, $x \to d(y, T(x))$ is continuous; (6) h-continuous if it is continuous as a map from S into the metric space $(C(X), h)$, where h is the Hausdorff metric induced by the metric d. Note that an h-continuous set-valued mapping is d-continuous. The notions of continuity and h-continuity are equivalent when T is compact-valued. A random operator $T: \Omega \times S \to C(X)$ is said to be continuous (d-continuous, etc.) if the map $T(\omega, \cdot): S \to C(X)$ is so for each $\omega \in \Omega$.

3. MAIN RESULTS

In order to develop the proof of the main theorem, we need the following Lemmas.

Lemma 3.1: *Let S be a non-empty separable complete subset of a metric space $X = (X, d)$ and $T: S \to C(X)$ a d-continuous set-valued mapping. If $f: S \to \mathbb{R}$ and $g: S \to \mathbb{R}$ are defined by*

$$f(x) = d(x, T(x)) \quad \textit{and} \quad g(x) = \delta(S, T(x)),$$

then f is continuous and g is upper semicontinuous.

Proof: Let $\{x_n\}$ be a sequence in S such that $x_n \to x$. Then

$$\begin{aligned} |f(x_n) - f(x)| &\le |d(x_n, T(x_n)) - d(x, T(x_n))| \\ &\quad + |d(x, T(x_n)) - d(x, T(x))| \\ &\le d(x_n, x) + |d(x, T(x_n)) - d(x, T(x))| \\ &\longrightarrow 0. \end{aligned}$$

Thus f is continuous.

Now, take a countable dense subset $D = \{y_n\}$ of S and set $g_n(x) = d(y_n, T(x))$. Then, for each n, g_n is continuous. It further implies that $g(x) = \inf_n g_n(x)$ is upper semicontinuous (cf. [8:p. 200]).

The proof of the following lemma runs on the same line as that of Lemma 3 of [12]

Lemma 3.2: *Let S be a non-empty separable complete subset of a metric space $X = (X, d)$ and T: $\Omega \times S \to C(X)$ a set-valued random operator. Then, for each fixed $x \in S$, the mappings $f(\cdot, x)$ and $g(\cdot, x)$ defined by*

$$f(\omega, x) = d(x, T(\omega, x)) \quad \text{and} \quad g(\omega, x) = \delta(S, T(\omega, x))$$

are measurable.

Now, we are in a position to prove our main result.

Theorem 3.3: *Let S be a non-empty separable complete subset of a metric space $X = (X, d)$ and T: $\Omega \times S \to C(X)$ a d-continuous random operator satisfying condition (B). If the set*

$$H(\omega) = \{x \in S : d(x, T(\omega, x)) = \delta(S, T(\omega, x))\}$$

is non-empty for each $\omega \in \Omega$, then there exists a measurable mapping ξ: $\Omega \to S$ such that

$$d(\xi(\omega), T(\omega, \xi(\omega))) = \delta(S, T(\omega, \xi(\omega)))$$

for each $\omega \in \Omega$.

Proof: Let H: $\Omega \to 2^S$ be a mapping defined by

$$H(\omega) = \{x \in S : d(x, T(\omega, x)) = \delta(S, T(\omega, x))\}.$$

By hypothesis, $H(\omega) \neq \emptyset$ for each $\omega \in \Omega$.
Define a mapping h: $\Omega \times S \to \mathbb{R}$ by

$$h(\omega, x) = f(\omega, x) - g(\omega, x),$$

where

$$f(\omega, x) = d(x, T(\omega, x))$$

and

$$g(\omega, x) = \delta(S, T(\omega, x)).$$

From Lemmas 3.1 and 3.2, we conclude that $h(\cdot, x)$ is measurable for each $x \in S$ and $h(\omega, \cdot)$ is lower semicontinuous for each $\omega \in \Omega$. Clearly $h(\omega, x) \geq 0$ for each $(\omega, x) \in \Omega \times S$ and $H(\omega) = \{x \in S : h(\omega, x) \leq 0\}$. It further implies that $H(\omega)$ is complete for all $\omega \in \Omega$. For any non-empty closed subset D of S, let

$$L(D) = \bigcap_{n=1}^{\infty} \bigcup_{i=1}^{\infty} \left\{\omega \in \Omega : h(\omega, x_i) < \frac{2}{n}\right\},$$

where $\{x_i\}$ is a countable dense subset of D. We first show that $H^{-1}(D) = L(D)$. To see this, let $\omega \in L(D)$. Then, for each n, there exists $i(n)$ such that $h(\omega, x_{i(n)}) < \frac{2}{n}$, that is,

$$d(x_{i(n)}, T(\omega, x_{i(n)})) - \delta(S, T(\omega, x_{i(n)})) < \frac{2}{n}.$$

Since T satisfies condition (B), it follows that

$$d(x_0, T(\omega, x_0)) = \delta(S, T(\omega, x_0))$$

for some $x_0 \in D$. Thus $\omega \in H^{-1}(D)$. On the other hand, if $\omega \in H^{-1}(D)$, then there exists an $x_0 \in D$ such that $d(x_0, T(\omega, x_0)) = \delta(S, T(\omega, x_0))$. Since $\{x_i\}$ is dense in D, there is a subsequence $\{x_{i(n)}\}$ with $d(x_{i(n)}, x_0) < \frac{1}{n}$. From Lemma 3.1, we obtain

$$\begin{aligned} d(x_{i(n)}, T(\omega, x_{i(n)})) &< d(x_0, T(\omega, x_0)) + \frac{1}{n} \\ &= \delta(S, T(\omega, x_0)) + \frac{1}{n} \\ &< \delta(S, T(\omega, x_{i(n)})) + \frac{2}{n}. \end{aligned}$$

This means that $\omega \in L(D)$. Hence $H^{-1}(D) = L(D)$. Since $\{\omega \in \Omega : h(\omega, x) < \frac{2}{n}\} \in \Sigma$ for any $x \in S$, the mapping H is measurable. The result now follows from the Kuratowski and Ryll-Nardzewski selection theorem [5].

If $T(\omega, x) \cap S \neq \emptyset$ for all $\omega \in \Omega$ and all $x \in S$, then condition (B) coincides with condition (A) and we obtain at once the following random fixed point theorem.

Theorem 3.4: *Let S be a non-empty separable complete subset of a metric space $X = (X, d)$ and T: $\Omega \times S \to C(X)$ a d-continuous random operator satisfying condition* (A). *If the set*

$$H(\omega) = \{x \in S : x \in T(\omega, x)\}$$

is non-empty for each $\omega \in \Omega$, then T has a random fixed point.

Remark 3.5:

1. Let S be a non-empty compact subset of a normed space $X = (X, \|\cdot\|)$. If T: $\Omega \times S \to CK(X)$ is any continuous set-valued random operator, then T satisfies condition (B); here $CK(X)$ denotes the family of all non-empty compact convex subsets of X. To see this, fixed $\omega \in \Omega$. Let $\{x_n\} \subset S$ and $D \in C(X)$ be such that $d(x_n, D) \to 0$ and $|d(x_n, T(\omega, x_n)) - \delta(S, T(\omega, x_n))| \to 0$ as $n \to \infty$. Because S is compact, there exists a subsequence $\{x_m\}$ of $\{x_n\}$ such that $x_m \to x_0 \in S$. Lemma 2 of Sehgal and Singh [12] further implies that

$$d(x_0, T(\omega, x_0)) = \delta(T(\omega, x_0), S).$$

Since $x_0 \in D$, it follows that T satisfies condition (B). Hence Theorem 2 of Sehgal and Singh [12] follows immediately from Theorem 3.3.
2. Corollary 1 of Sehgal and Singh [12] and Theorem 2 – Corollary 6 of Beg and Shahzad [2] can also be obtained by applying Theorem 3.3.

3. Theorem 3.4 complements Theorem 2.1 of Shahzad [14]. We further remark that Theorem 2.1 of Shahzad [14] also holds for d-continuous random operators.
4. Every continuous hemicompact (in particular, condensing) random operator satisfies condition (A). It would be interesting to compare continuous hemicompact random operators to operators satisfying condition (B).

REFERENCES

[1] I. Beg and N. Shahzad (1999). Some random approximation theorems with applications, *Nonlinear Anal.*, **35**, 609–616.
[2] I. Beg and N. Shahzad (1995). Random fixed point theorems for multivalved inward random operators on Banach spaces, *Adv. Math. Sci. Appl.*, **5**, 31–37.
[3] A.T. Bharucha-Reid (1976). Fixed point theorems in probabilistic analysis, *Bull. Amer. Math. Soc.*, **82**, 641–657.
[4] S. Itoh (1979). Random fixed point theorems with an application to random differential equations in Banach spaces, *J. Math. Anal. Appl.*, **67**, 261–273.
[5] K. Kuratowski and C. Ryll-Nardzewski (1965). A general theorem on selectors, *Bull. Acad. Polon. Sci. Ser. Sci. Math. Astronom. Phys.*, **13**, 397–403.
[6] T.C. Lin (1995). Random approximations and random fixed point theorems for continuous 1-set-contractive random maps, *Proc. Amer. Math. Soc.*, **123**, 1167–1176.
[7] L.S. Liu (1997). Some random approximations and random fixed point theorems for 1-set-contractive random operators, *Proc. Amer. Math. Soc.*, **125**, 515–521.
[8] N.S. Papageorgiou (1992). Fixed points and best approximations for measurable multifunctions with stochastic domain, *Tamkang J. Math.*, **23**, 197–203.
[9] N.S. Papageorgiou (1986). Random fixed point theorems for measurable multifunctions in Banach spaces, *Proc. Amer. Math. Soc.*, **97**, 507–514.
[10] D. O'Regan (1998). A continuation type result for random operators, *Proc. Amer. Math. Soc.*, **126**, 1963–1971.
[11] D. O'Regan (1998). Fixed points and random fixed points for weakly inward approximable maps, *Proc. Amer. Math. Soc.*, **126**, 3045–3053.
[12] V.M. Sehgal and S.P. Singh (1985). On random approximations and a random fixed point theorem for set-valued mappings, *Proc. Amer. Math. Soc.*, **95**, 91–94.
[13] V.M. Sehgal and C. Waters (1984). Some random fixed point theorems for condensing operators, *Proc. Amer. Math. Soc.*, **90**, 425–429.
[14] N. Shahzad. Random fixed points of set valued maps, *Nonlinear Anal.*, (to appear).
[15] N. Shahzad (1996). Random fixed point theorems for various classes of 1-set-contractive maps in Banach spaces, *J. Math. Anal. Appl.*, **203**, 712–718.
[16] N. Shahzad and L.A. Khan (1999). Random fixed points of 1-set-contractive random maps in Frechet spaces, *J. Math. Anal. Appl.*, **231**, 68–75.
[17] N. Shahzad and S. Latif (1999). Random fixed points for several classes of 1-ball-contractive and 1-set-contractive random maps, *J. Math. Anal. Appl.*, **237**, 83–92.
[18] K.K. Tan and X.Z. Yuan (1994). Random fixed point theorems and approximation in cones, *J. Math. Anal. Appl.*, **185**, 378–390.
[19] H.K. Xu (1990). Some random fixed point theorems for condensing and nonexpansive operators, *Proc. Amer. Math. Soc.*, **110**, 395–400.

SIMAA 4(2002) 431–447

27. Existence Theorems for Two-variable Functions and Fixed Point Theorems for Set-valued Mappings

Wataru Takahashi

Department of Mathematical and Computing Sciences, Tokyo Institute of Technology, Ohokayama, Meguro-ku, Tokyo 152-8552, Japan. E-mail: wataru@is.titech.ac.jp

1. INTRODUCTION

Let X be a nonempty set and let F be a two-variable function of $X \times X$ into [0, 1]. Then an element $x_0 \in X$ is said to be a *fixed point* of F if $F(x_0, x_0) = 1$. Let T be a set-valued mapping of X into X and let F be the two-variable function of $X \times X$ into [0, 1] defined by

$$F(x, y) = 1_{Tx}(y) \quad \text{for every } x, y \in X,$$

where 1_A is the characteristic function for an arbitrary set A. Then we know that $x_0 \in X$ is a fixed point of T, i.e., $x_0 \in Tx_0$ if and only if $F(x_0, x_0) = 1$. There are well-known fixed point theorems for set-valued mappings. They are Nadler's fixed point theorem [13] for contractive mappings, Lim's fixed point theorem [11] for nonexpansive mappings, and Fan's fixed point theorem [7] for upper semicontinuous mappings. On the other hand, Takahashi [21] proved a nonconvex minimization theorem generalizing fixed point theorems for set-valued mappings. Recently, Kada *et al.* [9] introduced the concept of w-distances and improved Caristi's fixed point theorem [4], Ekeland's variational principle [6] and the nonconvex minimization theorem according to Takahashi [21] by using w-distances.

In this article, we deal with existence theorems for two-variable functions and fixed point theorems for set-valued mappings. In Section 2, we introduce the concept of w-distances in a metric space for proving existence theorems and fixed point theorems concerning set-valued mappings and give some useful examples. In Section 3, we prove an existence theorem for two-variable functions in a complete metric space which generalizes Nadler's fixed point theorem [13] and Takahashi's nonconvex minimization theorem [21], simultaneously. Further, we prove an existence theorem for two-variable functions in a locally convex topological vector space by applying Fan's fixed point theorem. In Section 4, we prove a fixed point theorem for set-valued mappings of contractive type by using w-distances. Then, using this, we prove a fixed point theorem which generalizes Nadler's fixed point theorem [13] and Edelstein's fixed point theorem [5] in an ε-chainable metric space. In Section 5, we give a fixed point theorem for nonexpansive set-valued mappings in a convex metric space which was proved by Shimizu and Takahashi [14]. Then, using this, we present another proof of Lim's fixed point theorem [11] in a uniformly convex Banach space by applying ultrafilters, without using the notion of regular sequences.

2. *W*-DISTANCES AND EXAMPLES

Throughout this article, we denote by $\mathcal{N}$ the set of positive integers and by $\mathcal{R}$ the set of real numbers. Let X be a metric space with metric d. Then a function $p\colon X \times X \to [0, \infty)$ is called a *w-distance* [9] on X if the following are satisfied:

(1) $p(x, z) \leq p(x, y) + p(y, z)$ for any $x, y, z \in X$;
(2) for any $x \in X$, $p(x, \cdot)\colon X \to [0, \infty)$ is lower semicontinuous;
(3) for any $\varepsilon > 0$, there exists $\delta > 0$ such that $p(z, x) \leq \delta$ and $p(z, y) \leq \delta$ imply $d(x, y) \leq \varepsilon$.

The metric d is a w-distance on X. We also know the following examples.

Example 2.1: Let X be a normed linear space with norm $\|\cdot\|$. Then a function $p\colon X \times X \to [0, \infty)$ *defined by*

$$p(x, y) = \|y\| \quad \text{for every } x, y \in X$$

is a w-distance on X.

Proof: Let $x, y, z \in X$. Then

$$p(x, z) = \|z\| \leq \|y\| + \|z\| = p(x, y) + p(y, z).$$

This implies (1). (2) is obvious. To show (3), let $\varepsilon > 0$ and put $\delta = \frac{\varepsilon}{2}$. Then if $p(z, x) \leq \delta$ and $p(z, y) \leq \delta$, we have

$$d(x, y) = \|x - y\| \leq \|x\| + \|y\| = p(z, x) + p(z, y) \leq \delta + \delta = \varepsilon.$$

This implies (3). □

Example 2.2: *Let X be a metric space and let T be a continuous mapping from X into itself. Then a function $p\colon X \times X \to [0, \infty)$ defined by*

$$p(x, y) = \max\{d(Tx, y), d(Tx, Ty)\} \quad \text{for every } x, y \in X$$

is a w-distance on X.

Proof: Let $x, y, z \in X$. Then if $d(Tx, z) \geq d(Tx, Tz)$, we have

$$\begin{aligned} p(x, z) &= d(Tx, z) \leq d(Tx, Ty) + d(Ty, z) \\ &\leq \max\{d(Tx, y), d(Tx, Ty)\} + \max\{d(Ty, z), d(Ty, Tz)\} \\ &= p(x, y) + p(y, z). \end{aligned}$$

In the other case, we have

$$\begin{aligned} p(x, z) &= d(Tx, Tz) \leq d(Tx, Ty) + d(Ty, Tz) \\ &\leq \max\{d(Tx, y), d(Tx, Ty)\} + \max\{d(Ty, z), d(Ty, Tz)\} \\ &= p(x, y) + p(y, z). \end{aligned}$$

This implies (1). Since T is continuous, it is clear that for any $x \in X$, $p(x, \cdot)$: $X \to [0, \infty)$ is lower semicontinuous. Let $\varepsilon > 0$ and put $\delta = \frac{\varepsilon}{2}$. Then if $p(z, x) \leq \delta$ and $p(z, y) \leq \delta$, we have $d(Tz, x) \leq \delta$ and $d(Tz, y) \leq \delta$. Therefore

$$d(x, y) \leq d(Tz, x) + d(Tz, y) \leq 2\delta = \varepsilon.$$

This implies (3). □

Example 2.3: Let F be a bounded and closed subset of a metric space X. Assume that F contains at least two points and c is a constant with $c \geq \delta(F)$, where $\delta(F)$ is the diameter of F. Then a function p: $X \times X \to [0, \infty)$ defined by

$$p(x, y) = \begin{cases} d(x, y) & \text{if } x, y \in F, \\ c & \text{if } x \notin F \text{ or } \quad y \notin F \end{cases}$$

is a w-distance on X.

Proof: If $x, y, z \in F$, we have

$$p(x, z) \;=\; d(x, z) \;\leq\; d(x, y) + d(y, z) \;=\; p(x, y) + p(y, z).$$

In the other case, we have

$$p(x, z) \;\leq\; c \;\leq\; p(x, y) + p(y, z).$$

Let $x \in X$. If $\alpha \geq c$, we have $\{y \in X : p(x, y) \leq \alpha\} = X$. Let $\alpha < c$. If $x \in F$, then $p(x, y) \leq \alpha$ implies $y \in F$. So, we have

$$\{y \in X : p(x, y) \leq \alpha\} \;=\; \{y \in X : d(x, y) \leq \alpha\} \cap F.$$

If $x \notin F$, we have $\{y \in X : p(x, y) \leq \alpha\} = \emptyset$. In each case, the set $\{y \in X : p(x, y) \leq \alpha\}$ is closed. Therefore $p(x, \cdot) : X \to [0, \infty)$ is lower semicontinuous. Let $\varepsilon > 0$. Then there exists $n_0 \in \mathcal{N}$ such that $0 < \frac{\varepsilon}{n_0} < c$. Let $\delta = \frac{\varepsilon}{2n_0}$. Then $p(z, x) \leq \delta$ and $p(z, y) \leq \delta$ imply $x, y, z \in F$. So, we have

$$d(x, y) \;\leq\; d(x, z) + d(y, z) \;=\; p(z, x) + p(z, y) \;\leq\; \frac{\varepsilon}{2n_0} + \frac{\varepsilon}{2n_0} \;=\; \frac{\varepsilon}{n_0} \;\leq\; \varepsilon. \qquad \square$$

Let $\varepsilon \in (0, \infty]$. A metric space X with metric d is called *ε-chainable* [5] if for every $x, y \in X$ there exists a finite sequence $\{u_0, u_1, \ldots, u_k\}$ in X such that $u_0 = x$, $u_k = y$ and $d(u_i, u_{i+1}) < \varepsilon$ for $i = 0, 1, \ldots, k-1$. Such a sequence is called an *ε-chain* in X linking x and y.

Example 2.4: Let $\varepsilon \in (0, \infty]$ and let X be an ε-chainable metric space with metric d. Then a function p: $X \times X \to [0, \infty)$ defined by

$$p(x, y) = \inf\left\{ \sum_{i=0}^{k-1} d(u_i, u_{i+1}) : \{u_0, u_1, \ldots, u_k\} \text{is an } \varepsilon\text{-chain linking } x \text{ and } y \right\}$$

is a w-distance on X.

Proof: Note that p is well-defined because X is ε-chainable. Let $x, y, z \in X$ and let $\eta > 0$ be arbitrary. Then there exist ε-chains $\{u_0, u_1, \ldots, u_k\}$ linking x and y and $\{v_0, v_1, \ldots, v_l\}$ linking y and z such that

$$\sum_{i=0}^{k-1} d(u_i, u_{i+1}) \leq p(x, y) + \eta \quad \text{and} \quad \sum_{i=0}^{l-1} d(v_i, v_{i+1}) \leq p(y, z) + \eta.$$

Since $\{u_0, u_1, \ldots, u_k, v_1, v_2, \ldots, v_l\}$ is an ε-chain linking x and z, we have

$$\begin{aligned} p(x, z) &\leq \sum_{i=0}^{k-1} d(u_i, u_{i+1}) + \sum_{i=0}^{l-1} d(v_i, v_{i+1}) \\ &\leq p(x, y) + p(y, z) + 2\eta. \end{aligned}$$

Since $\eta > 0$ is arbitrary, we have $p(x, z) \leq p(x, y) + p(y, z)$.

Let us prove (2). Let $x, y \in X$ and let $\{y_n\}$ be a sequence in X with $y_n \to y$. Choose $n_0 \in \mathcal{N}$ such that $d(y, y_n) < \varepsilon$ for every $n \geq n_0$. Let $\eta > 0$ be arbitrary and let $n \geq n_0$. Then there exists an ε-chain $\{u_0, u_1, \ldots, u_k\}$ linking x and y_n such that

$$\sum_{i=0}^{k-1} d(u_i, u_{i+1}) \leq p(x, y_n) + \eta.$$

Since $d(y, y_n) < \varepsilon$, $\{u_0, u_1, \ldots, u_k, y\}$ is an ε-chain linking x and y. So, we have

$$\begin{aligned} p(x, y) &\leq \sum_{i=0}^{k-1} d(u_i, u_{i+1}) + d(y_n, y) \\ &\leq p(x, y_n) + \eta + d(y_n, y) \end{aligned}$$

and hence

$$p(x, y) \leq \liminf_{n \to \infty} p(x, y_n) + \eta.$$

Since $\eta > 0$ is arbitrary, we have

$$p(x, y) \leq \liminf_{n \to \infty} p(x, y_n).$$

This implies that $p(x, \cdot)$ is lower semicontinuous. We show (3). Let $\varepsilon > 0$ and $\delta = \frac{\varepsilon}{2}$. Then, if $p(z, x) \leq \delta$ and $p(z, y) \leq \delta$, we have

$$\begin{aligned} d(x, y) &\leq d(x, z) + d(z, y) \\ &= d(z, x) + d(z, y) \\ &\leq p(z, x) + p(z, y) \leq \varepsilon. \end{aligned}$$

So, p is a w-distance on X. □

The following lemma was proved in [9].

Lemma 2.5: *Let X be a metric space with metric d and let p be a w-distance on X. Let $\{x_n\}$ and $\{y_n\}$ be sequences in X, let $\{\alpha_n\}$ and $\{\beta_n\}$ be sequences in $[0, \infty)$ converging to 0, and let $x, y, z \in X$. Then the following hold:*

(1) *If $p(x_n, y) \leq \alpha_n$ and $p(x_n, z) \leq \beta_n$ for any $n \in \mathcal{N}$, then $y = z$. In particular, if $p(x, y) = 0$ and $p(x, z) = 0$, then $y = z$;*
(2) *if $p(x_n, y_n) \leq \alpha_n$ and $p(x_n, z) \leq \beta_n$ for any $n \in \mathcal{N}$, then $\{y_n\}$ converges to z;*
(3) *if $p(x_n, x_m) \leq \alpha_n$ for any $n, m \in \mathcal{N}$ with $m > n$, then $\{x_n\}$ is a Cauchy sequence;*
(4) *if $p(y, x_n) \leq \alpha_n$ for any $n \in \mathcal{N}$, then $\{x_n\}$ is a Cauchy sequence.*

Proof: We first prove (2). Let $\varepsilon > 0$ be given. From the definition of w-distance, there exists $\delta > 0$ such that $p(u, v) \leq \delta$ and $p(u, z) \leq \delta$ imply $d(v, z) \leq \varepsilon$. Choose $n_0 \in \mathcal{N}$ such that $\alpha_n \leq \delta$ and $\beta_n \leq \delta$ for every $n \geq n_0$. Then we have, for any $n \geq n_0$, $p(x_n, y_n) \leq \alpha_n \leq \delta$ and $p(x_n, z) \leq \beta_n \leq \delta$ and hence $d(y_n, z) \leq \varepsilon$. This implies that $\{y_n\}$ converges to z. It follows from (2) that (1) holds. Let us prove (3). Let $\varepsilon > 0$ be given. As in the proof of (2), choose $\delta > 0$ and then $n_0 \in \mathcal{N}$. Then for any $n, m \geq n_0 + 1$

$$p(x_{n_0}, x_n) \leq \alpha_{n_0} \leq \delta \quad \text{and} \quad p(x_{n_0}, x_m) \leq \alpha_{n_0} \leq \delta$$

and hence $d(x_n, x_m) \leq \varepsilon$. This implies that $\{x_n\}$ is a Cauchy sequence. As in the proof of (3), we can prove (4). □

3. EXISTENCE THEOREMS FOR TWO-VARIABLE FUNCTIONS

Let X be a metric space with metric d. Then, for any $x \in X$ and any nonempty subset A of X, we define

$$d(x, A) = \inf\{d(x, y) : y \in A\}.$$

It is easy to see that for each $x, y \in X$ and each nonempty subset A of X, $d(x, A) \leq d(x, y) + d(y, A)$. Moreover, we observe that for every $x \in X$ and every sequence $\{K_n\}$ of nonempty subsets of X, $\lim_{n \to \infty} d(x, K_n) = 0$ if and only if there exists a sequence $\{x_n\}$ of elements of X such that $x_n \in K_n$ and $\lim_{n \to \infty} d(x_n, x) = 0$. We denote by $CB(X)$ the class of all nonempty bounded closed subsets of X. For any $A, B \in CB(X)$, H denotes the Hausdorff metric with respect to d, i.e.,

$$H(A, B) = \max\{\sup_{u \in A} d(u, B), \sup_{v \in B} d(v, A)\}.$$

We see at once that for each $x \in X$ and each $A, B \in CB(X)$, $d(x, B) \leq H(A, B)$ if x belongs to A; see, for instance [20]. A set-valued mapping T of X into $CB(X)$ is said to be *k-contractive* if there exists a real number k with $0 \leq k < 1$ such that

$$H(Tx, Ty) \leq kd(x, y) \quad \text{for every } x, y \in X.$$

Lemma 3.1: *Let (X, d) be a complete metric space and let $\{K_n\}$ be a sequence of nonempty closed subsets of X such that $K_n \supset K_{n+1}$ for all $n \in \mathcal{N}$. Assume that there exist a sequence $\{u_n\}$ of elements of X and a w-distance p on X such that*

$$\sup_{x,y\in K_n} \max\{p(u_n,x),p(u_n,y)\} \to 0$$

as $n \to \infty$. Then $\bigcap_{n=1}^{\infty} K_n$ consists of one point.

Proof: Put $\delta(K_n) = \sup\{d(x,y) : x, y \in K_n\}$ for every $n \in \mathcal{N}$. Then it is sufficient to show that $\delta(K_n) \to 0$ as $n \to \infty$. Let $\varepsilon > 0$ be given. The definition of a w-distance guarantees the existence of $\delta > 0$ such that $p(u,w) \leq \delta$ and $p(u,z) \leq \delta$ imply $d(w,z) \leq \varepsilon$. Therefore, there exists $n_0 \in \mathcal{N}$ such that for all $n \geq n_0$, $\sup_{x,y\in K_n} \max\{p(u_n,x),\ p(u_n,y)\} \leq \delta$. Then, it follows that for any $n \geq n_0$ and any $x, y \in K_n$, $p(u_n,x) \leq \delta$ and $p(u_n,y) \leq \delta$. This implies that for any $n \geq n_0$ and any $x, y \in K_n$, $d(x,y) \leq \varepsilon$. Hence, we have proved that for any $\varepsilon > 0$, there exists $n_0 \in \mathcal{N}$ such that for all $n \geq n_0$, $\delta(K_n) \leq \varepsilon$. □

Using Lemma 3.1, Amemiya and Takahashi [2] proved the following theorem for two-variable functions.

Theorem 3.2: *Let X be a complete metric space, let f: $X \to (-\infty, \infty]$ be a proper, lower semicontinuous function bounded from below and let F be a two-variable function on X. Assume that there exists a w-distance p on X such that for each $x \in X$, there is $y \in X$ with $F(x,y) = 1$ and $f(y) + p(x,y) \leq f(x)$. Then there exists $x_0 \in X$ such that $F(x_0,x_0) = 1$ and $p(x_0,x_0) = 0$.*

Proof: Let u be an element of X such that $f(u) < \infty$ and put $u_0 = u$. Then we set

$$S_1 = \{w \in X : f(w) + p(u_0,w) \leq f(u_0)\}.$$

We shall construct a sequence $\{S_n\}$ of subsets of X by induction, starting with S_1. Suppose that $S_n \subset X$ ($n \in \mathcal{N}$) is known. Then take $u_n \in S_n$ such that

$$f(u_n) \leq \inf_{z\in S_n} f(z) + \frac{1}{n}$$

and we set

$$S_{n+1} = \{w \in X : f(w) + p(u_n,w) \leq f(u_n)\}.$$

Then, since u_n belongs to S_n for any $n \in \mathcal{N}$, it is clear that for each $n \in \mathcal{N}$,

$$f(u_n) \leq f(u_n) + p(u_{n-1},u_n) \leq f(u_{n-1}),$$

so that, the sequence $\{f(u_n)\}$ of elements of $\mathcal{R}$ is convergent. We claim that $\bigcap_{n=1}^{\infty} S_n$ consists of one point. Indeed, S_n are obviously nonempty closed subsets of X for all $n \in \mathcal{N}$, and it follows that for any $n \in \mathcal{N}$,

$$\begin{aligned} w \in S_{n+1} \Rightarrow\ & f(w) + p(u_n,w) \leq f(u_n) \text{ and} \\ & f(w) + p(u_{n-1},w) \leq f(w) + p(u_{n-1},u_n) + p(u_n,w) \\ \Rightarrow\ & f(w) + p(u_{n-1},w) \leq f(u_n) + p(u_{n-1},u_n) \leq f(u_{n-1}) \\ \Rightarrow\ & w \in S_n, \end{aligned}$$

which implies that $S_n \supset S_{n+1}$. Moreover, putting $\alpha_n = f(u_{n-1}) - \{f(u_n) - \frac{1}{n}\}$, we infer that for any $n \in \mathcal{N}$ and any $w \in S_n$,

$$p(u_{n-1},w) \leq f(u_{n-1}) - f(w) \leq f(u_{n-1}) - \inf_{z\in S_n} f(z) \leq \alpha_n$$

and thus, that $\sup_{w,z\in S_n}\max\{p(u_{n-1},w),p(u_{n-1},z)\}\le\alpha_n\to 0$ as $n\to\infty$. Therefore, we deduce from Lemma 3.1 that $\bigcap_{n=1}^{\infty}S_n$ consists of one point. Let $\bigcap_{n=1}^{\infty}S_n=\{x_0\}$ and we set

$$S=\{w\in X: f(w)+p(x_0,w)\le f(x_0)\}.$$

Then there exists $w\in S$ such that $F(x_0,w)=1$. Further, we observe that for each $n\in\mathcal{N}$, $S\subset S_n$. In fact, we have that for any $n\in\mathcal{N}$,

$$\begin{aligned} w\in S &\Rightarrow f(w)+p(x_0,w)\le f(x_0) \text{ and}\\ &\quad f(w)+p(u_{n-1},w)\le f(w)+p(u_{n-1},x_0)+p(x_0,w)\\ &\Rightarrow f(w)+p(u_{n-1},w)\le f(x_0)+p(u_{n-1},x_0)\le f(u_{n-1})\\ &\Rightarrow w\in S_n. \end{aligned}$$

This implies that $S=\{x_0\}$ and hence, that $F(x_0,x_0)=1$ and $p(x_0,x_0)=0$. This completes the proof. □

We also have the following corollary proved in [9], which generalizes the result of Takahashi [21].

Corollary 3.3: *Let X be a complete metric space and let $f\colon X\to(-\infty,\infty]$ be a proper, lower semicontinuous function bounded from below. Assume that there exists a w-distance p on X such that for each $x\in X$ with $\inf_{u\in X}f(u)<f(x)$, there is $y\in X$ with $y\ne x$ and $f(y)+p(x,y)\le f(x)$. Then there exists $x_0\in X$ such that $f(x_0)=\inf_{u\in X}f(u)$.*

Proof: Suppose that $\inf_{u\in X}f(u)<f(x)$ for all $x\in X$. Let

$$Sx=\{w\in X: w\ne x\}$$

for every $x\in X$ and define a two-variable function F on X by

$$F(x,y)=1_{Sx}(y)$$

for every $x,y\in X$. Then we deduce from assumption that for each $x\in X$, there is $y\in X$ with $F(x,y)=1$ and $f(y)+p(x,y)\le f(x)$. Therefore, by Theorem 3.2, we have $x_0\in X$ such that $F(x_0,x_0)=1$, that is, $x_0\in Sx_0$. This is a contradiction. Hence, there exists $x_0\in X$ such that $f(x_0)=\inf_{u\in X}f(u)$. □

The following theorem is due to Nadler [13]. We shall give another proof by using Theorem 3.2.

Theorem 3.4: *Let (X,d) be a complete metric space and let T be a k-contractive set-valued mapping of X into $CB(X)$. Then, T has a fixed point in X, i.e., there exists $x_0\in X$ such that $x_0\in Tx_0$.*

Proof: Let F be a two-variable function on X defined by

$$F(x,y)=1_{Tx}(y)\quad\text{for all } x,y\in X.$$

Suppose that $d(x,Tx)>0$ for all $x\in X$. Choose $\varepsilon>0$ with $\varepsilon<\frac{1}{k}-1$. Then for each $x\in X$, there exists $y\in Tx$ such that

$$d(x,y)\le(1+\varepsilon)d(x,Tx).$$

Then it follows that

$$\begin{aligned} d(y,Ty) &\le H(Tx,Ty) \\ &\le kd(x,y) \\ &\le k(1+\varepsilon)d(x,Tx) \end{aligned}$$

and therefore we have that

$$\begin{aligned} d(x,Tx) - d(y,Ty) &\ge \frac{1}{1+\varepsilon}d(x,y) - kd(x,y) \\ &= \left(\frac{1}{1+\varepsilon} - k\right)d(x,y). \end{aligned}$$

Hence it follows that

$$F(x,y) = 1,$$

$$f(y) + d(x,y) \le f(x),$$

where f is a continuous real valued function on X defined by

$$f(x) = \left(\frac{1}{1+\varepsilon} - k\right)^{-1} d(x,Tx)$$

for all $x \in X$. Then by Theorem 3.2, there exists $x_0 \in X$ such that $F(x_0,x_0) = 1$, that is, $d(x_0,Tx_0) = 0$. This is a contradiction. □

Next, we obtain a fixed point theorem for two-variable functions on a locally convex topological vector space by using Fan's fixed point theorem [7].

Theorem 3.5: *Let X be a compact convex subset of a locally convex topological vector space and let F be an upper semicontinuous two-variable function of $X \times X$ into* $[0,1]$. *Suppose that for each $x \in X$, there exists $y \in X$ such that $F(x,y) = 1$ and $F(x,\cdot)$ is convex. Then there exists $x_0 \in X$ such that $F(x_0,x_0) = 1$.*

Proof: Define a set-valued mapping T on X by

$$Tx = \{y \in X : F(x,y) = 1\}$$

for all $x \in X$. Then for each $x \in X$, Tx is a nonempty, closed and convex subset of X. Let

$$G(T) = \{(x,y) \in X \times X : F(x,y) = 1\}.$$

Then $G(T)$ is a closed subset of $X \times X$. In fact, let $(x_\alpha, y_\alpha) \to (x,y)$ and $(x_\alpha, y_\alpha) \in G(T)$. Then we have that

$$1 \ge F(x,y) \ge \inf_\alpha \sup_{\beta \ge \alpha} F(x_\beta, y_\beta) = 1$$

and therefore $(x, y) \in G(T)$. This implies that $G(T)$ is closed in $X \times X$. Hence, T is upper semicontinuous on X. Then, by Fan's fixed point theorem, we obtain $x_0 \in X$ such that $x_0 \in Tx_0$, that is, $F(x_0, x_0) = 1$. This completes the proof. □

We can also show that Theorem 3.5 is a generalization of Fan's fixed point theorem. Let X be a compact convex subset of a locally convex topological vector space and let T be an upper semicontinuous set-valued mapping from X into X such that for each $x \in X$, Tx is a nonempty, closed, convex subset of X. Define a two-variable function F on X by

$$F(x, y) = 1_{Tx}(y)$$

for all $x, y \in X$. Then F is upper semicontinuous. Indeed, if $\alpha \leq 0$, then the set

$$\{(x, y) \in X \times X : F(x, y) \geq \alpha\}$$

is obviously $X \times X$; if $\alpha > 1$, then the set is empty. Moreover, if $0 < \alpha \leq 1$, we have that $\{(x, y) \in X \times X : F(x, y) \geq \alpha\} = G(T)$ is closed. Therefore F is upper semicontinuous. Then by Theorem 3.5, there exists $x_0 \in X$ such that $F(x_0, x_0) = 1$, that is, $x_0 \in Tx_0$.

4. FIXED POINT THEOREMS FOR CONTRACTIVE MAPPINGS

In this section, we first prove a fixed point theorem for set-valued mappings of contractive type by using w-distances. Then, using this, we prove a fixed point theorem which generalizes Nadler's fixed point theorem [13] for set-valued mappings and Edelstein's fixed point theorem [5] in an ε-chainable metric space. Let X be a metric space with metric d. A set-valued mapping T from X into itself is called *weakly contractive* or *p-contractive* if there exist a w-distance p on X and $r \in [0, 1)$ such that for any $x_1, x_2 \in X$ and $y_1 \in Tx_1$ there is $y_2 \in Tx_2$ with $p(y_1, y_2) \leq rp(x_1, x_2)$.

Theorem 4.1: *Let X be a complete metric space and let T be a set-valued p-contractive mapping from X into itself such that for any $x \in X$, Tx is a nonempty closed subset of X. Then there exists $x_0 \in X$ such that $x_0 \in Tx_0$ and $p(x_0, x_0) = 0$.*

Proof: Let p be a w-distance on X and let $r \in [0, 1)$ such that for any $x_1, x_2 \in X$ and $y_1 \in Tx_1$, there exists $y_2 \in Tx_2$ with $p(y_1, y_2) \leq rp(x_1, x_2)$. Fix $u_0 \in X$ and $u_1 \in Tu_0$. Then there exists $u_2 \in Tu_1$ such that $p(u_1, u_2) \leq rp(u_0, u_1)$. Thus, we have a sequence $\{u_n\}$ in X such that $u_{n+1} \in Tu_n$ and $p(u_n, u_{n+1}) \leq rp(u_{n-1}, u_n)$ for every $n \in \mathcal{N}$. For any $n \in \mathcal{N}$, we have

$$p(u_n, u_{n+1}) \leq rp(u_{n-1}, u_n) \leq r^2 p(u_{n-2}, u_{n-1}) \leq \cdots \leq r^n p(u_0, u_1)$$

and hence, for any $n, m \in \mathcal{N}$ with $m > n$,

$$\begin{aligned} p(u_n, u_m) &\leq p(u_n, u_{n+1}) + p(u_{n+1}, u_{n+2}) + \cdots + p(u_{m-1}, u_m) \\ &\leq r^n p(u_0, u_1) + r^{n+1} p(u_0, u_1) + \cdots + r^{m-1} p(u_0, u_1) \\ &\leq \frac{r^n}{1-r} p(u_0, u_1). \end{aligned}$$

By Lemma 2.5, $\{u_n\}$ is a Cauchy sequence. Hence $\{u_n\}$ converges to a point $v_0 \in X$. Fix $n \in \mathcal{N}$. Since $\{u_m\}$ converges to v_0 and $p(u_n, \cdot)$ is lower semicontinuous, we have

$$p(u_n, v_0) \le \liminf_{m\to\infty} p(u_n, u_m) \le \frac{r^n}{1-r} p(u_0, u_1). \qquad (*)$$

By hypothesis, we also have $w_n \in Tv_0$ such that $p(u_n, w_n) \le rp(u_{n-1}, v_0)$. So, we have, for any $n \in \mathcal{N}$,

$$p(u_n, w_n) \le rp(u_{n-1}, v_0) \le \frac{r^n}{1-r} p(u_0, u_1).$$

By Lemma 2.5, $\{w_n\}$ converges to v_0. Since Tv_0 is closed, we have $v_0 \in Tv_0$. For such v_0, there exists $v_1 \in Tv_0$ such that $p(v_0, v_1) \le rp(v_0, v_0)$. Thus, we also have a sequence $\{v_n\}$ in X such that $v_{n+1} \in Tv_n$ and $p(v_0, v_{n+1}) \le rp(v_0, v_n)$ for every $n \in \mathcal{N}$. So, we have

$$p(v_0, v_n) \le rp(v_0, v_{n-1}) \le \ldots \le r^n p(v_0, v_0).$$

By Lemma 2.5, $\{v_n\}$ is a Cauchy sequence. Hence $\{v_n\}$ converges to a point $x_0 \in X$. Since $p(v_0, \cdot)$ is lower semicontinuous, $p(v_0, x_0) \le \liminf_{n\to\infty} p(v_0, v_n) \le 0$ and hence $p(v_0, x_0) = 0$. Then, for any $n \in \mathcal{N}$,

$$p(u_n, x_0) \le p(u_n, v_0) + p(v_0, x_0) \le \frac{r^n}{1-r} p(u_0, u_1).$$

So, using ($*$) and Lemma 2.5, we obtain $v_0 = x_0$ and hence $p(v_0, v_0) = 0$. □

Let X be a metric space with metric d and let T be a single-valued mapping from X into itself. Then T is called *weakly contractive* or *p-contractive* if there exist a w-distance p on X and $r \in [0, 1)$ such that $p(Tx, Ty) \le rp(x, y)$ for every $x, y \in X$. In the case of $p = d$, T is called *contractive*. The following theorem is a direct result of Theorem 4.1.

Theorem 4.2: *Let X be a complete metric space. If a mapping T from X into itself is weakly contractive, then T has a unique fixed point $x_0 \in X$. Further the x_0 satisfies $p(x_0, x_0) = 0$.*

Proof: Let p be a w-distance and let $r \in [0, 1)$ be such that $p(Tx, Ty) \le rp(x, y)$ for every $x, y \in X$. Then from Theorem 4.1, there exists $x_0 \in X$ with $Tx_0 = x_0$ and $p(x_0, x_0) = 0$. If $y_0 = Ty_0$, then we have

$$p(x_0, y_0) = p(Tx_0, Ty_0) \le rp(x_0, y_0)$$

and hence $p(x_0, y_0) = 0$. So, by $p(x_0, x_0) = 0$ and Lemma 2.5, we have $x_0 = y_0$. □

Remark: From [16], we also have that if X is a metric space and every weakly contractive mapping of X into itself has a fixed point, then X is complete. We know a metric space X which X is not complete and every contractive mapping of X into itself has a fixed point.

Let X be a metric space with metric d. A mapping T from X into $CB(X)$ is said to be (ε, σ)*-uniformly locally contractive* [5] if there exists $\sigma \in [0, 1)$ such that

$H(Tx, Ty) \le \sigma d(x, y)$ for every $x, y \in X$ with $d(x, y) < \varepsilon$. In particular, T is said to be *contractive* when $\varepsilon = \infty$. The following theorem was proved by Suzuki and Takahashi [16].

Theorem 4.3: *Let $\varepsilon \in (0, \infty]$ and let X be a complete and ε-chainable metric space with metric d. Suppose that a mapping T from X into $CB(X)$ is (ε, σ)-uniformly locally contractive. Then there exists $x_0 \in X$ with $x_0 \in Tx_0$.*

Proof: Define a function p from $X \times X$ into $[0, \infty)$ as follows:

$$p(x, y) = \inf\left\{\sum_{i=0}^{k-1} d(u_i, u_{i+1}) : \{u_0, u_1, \ldots, u_k\} \text{ is an } \varepsilon\text{-chain linking } x \text{ and } y\right\}.$$

From Example 2.4, p is a w-distance on X. We prove that T is p-contractive. Choose a real number r such that $\sigma < r < 1$. Let $x_1, x_2 \in X, y_1 \in Tx_1$ and $\eta > 0$. Then there exists an ε-chain $\{u_0, u_1, \ldots, u_k\}$ linking x_1 and x_2 such that

$$\sum_{i=0}^{k-1} d(u_i, u_{i+1}) \le p(x_1, x_2) + \eta.$$

Put $v_0 = y_1$. Since T is (ε, σ)-uniformly locally contractive, there exists $v_1 \in Tu_1$ such that

$$d(v_0, v_1) \le rd(u_0, u_1) < r\varepsilon \le \varepsilon.$$

In a similar way, we define an ε-chain $\{v_0, v_1, \ldots, v_k\}$ linking y_1 and v_k such that $v_i \in Tu_i$ for every $i = 0, 1, \ldots, k$ and

$$d(v_i, v_{i+1}) \le rd(u_i, u_{i+1}) < \varepsilon$$

for every $i = 0, 1, \ldots, k-1$. Putting $y_2 = v_k$, since $y_2 \in Tx_2$ and $\{v_0, v_1, \ldots, v_k\}$ is an ε-chain linking y_1 and y_2, we have

$$\begin{aligned} p(y_1, y_2) &\le \sum_{i=0}^{k-1} d(v_i, v_{i+1}) \\ &\le \sum_{i=0}^{k-1} rd(u_i, u_{i+1}) \\ &\le rp(x_1, x_2) + r\eta \\ &< rp(x_1, x_2) + \eta. \end{aligned}$$

Since $\eta > 0$ is arbitrary, we have $p(y_1, y_2) \le rp(x_1, x_2)$. So, T is a *p-contractive* set-valued mapping from X into itself. Theorem 4.1 now gives the desired result. □

Theorem 3.4 (Nadler's fixed point theorem) is a direct consequence of Theorem 4.3. In fact, we may assume that there exists $\sigma \in [0, 1)$ such that $H(Tx, Ty) \le \sigma d(x, y)$ for every $x, y \in X$. Since T is (∞, σ)-uniformly locally contractive and X is ∞-chainable, using Theorem 4.3, we obtain the desired result. Edelstein's theorem [5] is also a direct result of Theorem 4.3.

Corollary 4.4: *Let $\varepsilon \in (0, \infty]$ and let X be a complete and ε-chainable metric space with metric d. Suppose that a mapping T from X into itself is (ε, σ)-uniformly locally contractive. Then T has a unique fixed point.*

5. FIXED POINT THEOREMS FOR NONEXPANSIVE MAPPINGS

In this section, we prove Lim's fixed point theorem [11] for nonexpansive set-valued mappings in a uniformly convex Banach space by applying ultrafilters, without using the notion of regular sequences. Before proving it, we give a theorem in a convex metric space which was proved by Shimizu and Takahashi [14]. Let X be a metric space with metric d and let $CB(X)$ be the family of all nonempty bounded closed subsets of X. We also denote by $K(X)$ the family of all nonempty compact subsets of X. Then, a set-valued mapping T of X into $CB(X)$ is said to be *nonexpansive* if for every $x, y \in X$,

$$H(Tx, Ty) \leq d(x, y),$$

where H is the Hausdorff metric on $CB(X)$. A set-valued mapping T of X into $CB(X)$ is said to have the *almost fixed point property* in X if

$$\inf_{x \in X} d(x, Tx) = 0,$$

where $d(x, A) = \inf\{d(x, a) : a \in A\}$. The following theorem obtained by Shimizu and Takahashi [14] generalizes Kijima's result [10] for single-valued mappings.

Theorem 5.1: *Let X be a bounded metric space with metric d, and suppose that for each pair $x, y \in X$, there exists $z \in X$ such that*

$$d(z, u) \leq \frac{1}{2} d(x, u) + \frac{1}{2} d(y, u) \quad \textit{for all } u \in X. \tag{**}$$

Let T be a nonexpansive mapping of X into $K(X)$. Then, T has the almost fixed point property in X.

Proof: Suppose $\inf_{x \in X} d(x, Tx) = 2\delta > 0$. Let ε be an arbitrary positive number. Then, there exists $x_0 \in X$ such that

$$d(x_0, Tx_0) \leq 2\delta + \varepsilon.$$

Since Tx_0 is nonempty, there exists $y_0 \in Tx_0$ such that

$$d(x_0, y_0) \leq 2(\delta + \varepsilon).$$

Define inductively sequences $\{x_n\}$ and $\{y_n\}$ in X with $y_n \in Tx_n$ as follows. Suppose x_k and y_k with $y_k \in Tx_k$ are known. Then, by (**), we choose $x_{k+1} \in X$ such that

$$d(x_{k+1}, u) \leq \frac{1}{2} d(x_k, u) + \frac{1}{2} d(y_k, u) \quad \text{for all } u \in X.$$

Since Tx_{k+1} is nonempty and compact, we choose $y_{k+1} \in X$ such that

$$y_{k+1} \in Tx_{k+1} \quad \text{and} \quad d(y_k, y_{k+1}) = d(y_k, Tx_{k+1}).$$

Then we have

$$d(y_k, y_{k+1}) \leq d(x_k, x_{k+1}).$$

In fact, we know

$$\begin{aligned} d(y_k, y_{k+1}) = d(y_k, Tx_{k+1}) &\leq \sup_{y \in Tx_k} d(y, Tx_{k+1}) \\ &\leq H(Tx_k, Tx_{k+1}) \\ &\leq d(x_k, x_{k+1}). \end{aligned}$$

We first show that for each nonnegative integer k,

$$d(x_k, x_{k+1}) \leq \delta + \varepsilon \tag{5.1}$$

and

$$d(x_k, y_k) \leq 2(\delta + \varepsilon).$$

If $k = 0$, we have

$$\begin{aligned} d(x_0, x_1) &\leq \frac{1}{2} d(x_0, x_0) + \frac{1}{2} d(x_0, y_0) \\ &= \frac{1}{2} d(x_0, y_0) \leq \delta + \varepsilon \end{aligned}$$

and

$$d(x_0, y_0) \leq 2(\delta + \varepsilon).$$

Suppose $d(x_k, x_{k+1}) \leq \delta + \varepsilon$ and $d(x_k, y_k) \leq 2(\delta + \varepsilon)$. Then we have

$$\begin{aligned} d(x_{k+1}, y_{k+1}) &\leq d(x_{k+1}, y_k) + d(y_k, y_{k+1}) \\ &\leq \frac{1}{2} d(x_k, y_k) + d(x_k, x_{k+1}) \\ &\leq (\delta + \varepsilon) + (\delta + \varepsilon) \\ &\leq 2(\delta + \varepsilon) \end{aligned}$$

and

$$d(x_{k+1}, x_{k+2}) \leq \frac{1}{2} d(x_{k+1}, y_{k+1}) \leq \delta + \varepsilon.$$

Therefore, (5.1) holds for every nonnegative integer k.

Next we show that for all nonnegative integers k and n,

$$d(x_k, y_{k+n}) \geq (n + 2)(\delta + \varepsilon) - 2^{n+1} \varepsilon \tag{5.2}$$

Let $n = 0$. Then we have

$$d(x_k, y_k) \geq d(x_k, Tx_k) \geq 2\delta.$$

So, suppose (5.2) holds for a nonnegative integer n and every nonnegative integer k. Then we have

$$\begin{aligned} d(x_k, y_{k+n+1}) &\geq 2d(x_{k+1}, y_{k+n+1}) - d(y_k, y_{k+n+1}) \\ &\geq 2d(x_{k+1}, y_{k+n+1}) - \sum_{i=0}^{n} d(y_{k+i}, y_{k+i+1}) \\ &\geq 2d(x_{k+1}, y_{k+n+1}) - \sum_{i=0}^{n} d(x_{k+i}, x_{k+i+1}) \\ &\geq 2\{(n+2)(\delta+\varepsilon) - 2^{n+1}\varepsilon\} - (n+1)(\delta+\varepsilon) \\ &= (2n+4)(\delta+\varepsilon) - 2^{n+2}\varepsilon - (n+1)(\delta+\varepsilon) \\ &= (n+3)(\delta+\varepsilon) - 2^{n+2}\varepsilon. \end{aligned}$$

Therefore, (5.2) holds for all nonnegative numbers k and n. Let m be a nonnegative integer and set $\varepsilon = \frac{\delta}{2^m}$. Then, as in the previous proof, we choose sequences $\{x_n^m\}$ and $\{y_n^m\}$ in X such that

$$d(x_0^m, Tx_0^m) \leq 2\delta + \frac{\delta}{2^m}$$

and

$$d(x_k^m, y_{k+n}^m) \geq (n+2)\left(\delta + \frac{\delta}{2^m}\right) - 2^{n+1}\frac{\delta}{2^m}$$

for nonnegative numbers k and n. Particularly, we have

$$\begin{aligned} d(x_0^m, y_m^m) &\geq (m+2)\left(\delta + \frac{\delta}{2^m}\right) - 2^{m+1}\frac{\delta}{2^m} \\ &= m\delta + \frac{m}{2^m}\delta + \frac{\delta}{2^{m-1}} \\ &> m\delta. \end{aligned}$$

This implies

$$\lim_{m\to\infty} d(x_0^m, y_m^m) = +\infty,$$

which contradicts boundedness of X. □

The following is a fixed point theorem for nonexpansive set-valued mappings in a compact convex metric space.

Theorem 5.2: *Let X be a bounded metric space as in Theorem 5.1. Let T be a nonexpansive mapping of X into $K(X)$ such that the closure $\overline{T(X)}$ of $T(X) = \cup_{x\in X} Tx$ is compact. Then, there exists $x_0 \in X$ such that $x_0 \in Tx_0$.*

Proof: By Theorem 5.1, there exists a sequence $\{x_n\}$ in X such that $d(x_n, Tx_n) \to 0$ as $n \to \infty$. Since for each positive integer n, Tx_n is nonempty and compact, there exists $y_n \in X$ such that $y_n \in Tx_n$ and

$$d(x_n, y_n) = d(x_n, Tx_n).$$

Since $\{y_n\} \subseteq \overline{T(X)}$ and $\overline{T(X)}$ is compact, there exists a subsequence $\{y_{n_i}\}$ of $\{y_n\}$ converging to an element $x_0 \in X$. From $\lim_i d(x_{n_i}, y_{n_i}) = 0$, we also have $\lim_i x_{n_i} = x_0$. On the other hand, since

$$|d(x_{n_i}, Tx_0) - d(x_{n_i}, Tx_{n_i})| \leq H(Tx_{n_i}, Tx_0),$$

we have

$$\begin{aligned} d(x_0, Tx_0) &\leq d(x_0, x_{n_i}) + d(x_{n_i}, Tx_0) \\ &\leq d(x_0, x_{n_i}) + d(x_{n_i}, Tx_{n_i}) + H(Tx_n, Tx_0) \\ &\leq 2d(x_0, x_{n_i}) + d(x_{n_i}, Tx_{n_i}) \end{aligned}$$

and hence $d(x_0, Tx_0) = 0$. This implies that the set of fixed points of T is nonempty. □

Let X be a nonempty set. A nonempty family $\mathcal{F}$ of subsets of X is called a *filter* on X if it has the following properties: (1) $\phi \notin \mathcal{F}$; (2) if $A \subset B$ and $A \in \mathcal{F}$, then $B \in \mathcal{F}$; (3) if $A, B \in \mathcal{F}$, then $A \cap B \in \mathcal{F}$. If $\mathcal{F}_1$ and $\mathcal{F}_2$ are filters on X with $\mathcal{F}_1 \subset \mathcal{F}_2$, then we say that $\mathcal{F}_2$ is *finer* than $\mathcal{F}_1$. A filter $\mathcal{U}$ on X is called an *ultrafilter* if there is no filter on X which is strictly finer than $\mathcal{U}$. A nonempty class $\mathcal{B}$ of subsets of X is called a *filterbase* on X if it has the following properties: (1) $\phi \notin \mathcal{B}$; (2) for any A_1 and A_2 in $\mathcal{B}$, there exists A_3 in $\mathcal{B}$ such that $A_3 \subset A_1 \cap A_2$. If $\mathcal{B}$ is a filterbase on X, then

$$\mathcal{F} = \{A \subset X : B \subset A,\ B \in \mathcal{B}\}$$

is a filter on X. In this case, $\mathcal{B}$ is said to be a *base* of $\mathcal{F}$ or to *generate* $\mathcal{F}$. Let X be a topological space and let $\mathcal{B}$ be a filterbase on X. Then $\mathcal{B}$ is said to *converge* to a point x in X or to have a *limit* x in X if for any neighbourhood V of x, there is a set A in $\mathcal{B}$ such that $A \subset V$. If $\mathcal{U}$ is an ultrafilter on a compact set X, then $\mathcal{U}$ has a limit in X. Let $\mathcal{U}$ be an ultrafilter on a set X and P be a mapping of X into a set D. Then $P(\mathcal{U})$ is a filterbase on D and it generates an ultrafilter on D. In fact, it is obvious that since $\mathcal{U}$ is an ultrafilter on X, then $P(\mathcal{U})$ is a filterbase on D. Let

$$\mathcal{B} = \{B \subset D : P(A) \subset B \text{ for some } A \in \mathcal{U}\}$$

and let $\mathcal{K}$ be a filter on D with $\mathcal{K} \supset \mathcal{B}$. If $K \in \mathcal{K}$, then $P^{-1}K \in \mathcal{U}$ or $P^{-1}K^c \in \mathcal{U}$, where K^c is the complement of K. Suppose $A = P^{-1}K^c \in \mathcal{U}$. Then $P(A) = P(P^{-1}K^c) \subset K^c$, and hence $K^c \in \mathcal{B}$. This is a contradictions. So, $P^{-1}K \in \mathcal{U}$. Since $P(P^{-1}K) \subset K$, we have $K \in \mathcal{B}$ and hence $\mathcal{K} = \mathcal{B}$. This implies that $\mathcal{B}$ is an ultrafilter on D.

Let E be a uniformly convex Banach space and let $\mathcal{B}$ be a filterbase on E which contains at least one bounded subset of E. Then we define

$$r(x, \mathcal{B}) = \inf_{A \in \mathcal{B}} \sup_{y \in A} \|x - y\| = \lim_{A \in \mathcal{B}} \sup_{y \in A} \|x - y\|$$

for every $x \in E$. Since for every $x, y \in E$, $|r(x,\mathcal{B}) - r(y,\mathcal{B})| \le \|x-y\|$, the real-valued function $r(\cdot,\mathcal{B})$ on E is continuous. Further, $r(\cdot,\mathcal{B})$ is convex. We also have that if $\|x_n\| \to \infty$, then $r(x_n,\mathcal{B}) \to \infty$. So, if X is a closed convex subset of E, there exists a unique $u_0 \in X$ such that

$$r(u_0,\mathcal{B}) = \min_{x\in X} r(x,\mathcal{B}).$$

The following theorem is due to Lim [11].

Theorem 5.3: *Let X be a bounded closed and convex subset of a uniformly convex Banach space. If T is a nonexpansive set-valued mapping which assigns to each point of X a nonempty compact subset of X, then T has a fixed point in X.*

Proof: By Theorem 5.1, there exists a sequence $\{x_n\}$ in X such that $d(x_n, Tx_n) \to 0$ as $n \to \infty$. For every positive integer n, define

$$A_n = \{x_n, x_{n+1}, \ldots\}.$$

Then $\{A_n\}$ is a filterbase on X and generates a filter $\mathcal{F}$ on X. So, we know that there is an ultrafilter $\mathcal{U}$ finer than $\mathcal{F}$. Clearly we have

$$\inf_{A\in\mathcal{U}} \sup_{x\in A} d(x,Tx) = 0.$$

So, there exists a unique element $u_0 \in X$ such that

$$r(u_0,\mathcal{U}) = \inf_{x\in X} r(x,\mathcal{U}).$$

Since for each $x \in X$, Tx is nonempty and compact, we obtain elements $Sx \in Tx$ and $Px \in Tu_0$ such that

$$\|x - Sx\| = d(x,Tx) \quad \text{and} \quad \|Sx - Px\| = d(Sx, Tu_0).$$

Thus, we have got a mapping $P\colon X \to Tu_0$. We know that $P(\mathcal{U})$ is a filterbase on Tu_0 and the filter generated by $P(\mathcal{U})$ is an ultrafilter on Tu_0. Since Tu_0 is compact, $P(\mathcal{U})$ has a limit p_0 in Tu_0. So, we have

$$\begin{aligned}
r(p_0,\mathcal{U}) &= \inf_{A\in\mathcal{U}} \sup_{x\in A} \|p_0 - x\| \le \inf_{A\in\mathcal{U}} \sup_{x\in A}\{\|p_0 - Px\| + \|Px - Sx\| + \|Sx - x\|\} \\
&= \inf_{A\in\mathcal{U}} \sup_{x\in A}\{\|p_0 - Px\| + d(Sx, Tu_0) + d(x,Tx)\} \\
&\le \inf_{A\in\mathcal{U}} \sup_{x\in A}\{\|p_0 - Px\| + H(Tx, Tu_0) + d(x,Tx)\} \\
&\le \inf_{A\in\mathcal{U}} \sup_{x\in A}\{\|p_0 - Px\| + \|x - u_0\| + d(x,Tx)\} \\
&= \inf_{A\in\mathcal{U}} \sup_{x\in A} \|x - u_0\| = r(u_0,\mathcal{U}).
\end{aligned}$$

So, we have $u_0 = p_0 \in Tu_0$. This completes the proof. □

REFERENCES

[1] S. Amemiya and W. Takahashi (2000). Generalization of shadows and fixed point theorems for fuzzy sets, *Fuzzy Sets and Systems*, **114**, 469–476.

[2] S. Amemiya and W. Takahashi. Fixed point theorems for fuzzy mappings in complete metric spaces, *Fuzzy Sets and Systems*, (to appear).

[3] F.E. Browder (1969). The fixed point theory of multi-valued mappings in topological vector spaces, *Math. Ann.*, **177**, 283–301.

[4] J. Caristi (1976). Fixed point theorems for mappings satisfying inwardness conditions, *Trans. Amer. Math. Soc.*, **215**, 241–251.

[5] M. Edelstein (1961). An extension of Banach's contraction principle, *Proc. Amer. Math. Soc.*, **12**, 7–10.

[6] I. Ekeland (1979). Nonconvex minimization problems, *Bull. Amer. Math. Soc.*, **1**, 443–474.

[7] K. Fan (1952). Fixed point and minimax theorems in locally convex topological linear spaces, *Proc. Nat. Acad. Sci. USA*, **38**, 121–126.

[8] K. Fan (1969). Extensions of the two fixed point theorems of F.E. Browder, *Math. Z.*, **112**, 234–240.

[9] O. Kada, T. Suzuki and W. Takahashi (1996). Nonconvex minimization theorems and fixed point theorems in complete metric spaces, *Math. Japonica*, **44**, 381–391.

[10] Y. Kijima (1992). A fixed point theorem for nonexpansive self-maps of a metric space with some convexity, *Math. Japonica*, **37**, 707–709.

[11] T.C. Lim (1974). A fixed point theorem for multivalued nonexpansive mappings in a uniformly convex Banach space, *Bull. Amer. Math. Soc.*, **80**, 1123–1126.

[12] N. Mizoguchi and W. Takahashi (1989). Fixed point theorems for multivalued mappings on complete metric spaces, *J. Math. Anal. Appl.*, **141**, 177–188.

[13] S.B. Nadler, Jr. (1969). Multi-valued contraction mappings, *Pacific J. Math.*, **30**, 475–488.

[14] T. Shimizu and W. Takahashi (1992). Fixed point theorems in certain convex metric spaces, *Math. Japonica*, **37**, 855–859.

[15] T. Shimizu and W. Takahashi (1996). Fixed points of multivalued mappings in certain convex metric spaces, *Topol. Methods Nonlinear Anal.*, **8**, 197–203.

[16] T. Suzuki and W. Takahashi (1996). Fixed point theorems and characterizations of metric completeness, *Topol. Methods Nonlinear Anal.*, **8**, 371–382.

[17] W. Takahashi (1970). A convexity in metric space and nonexpansive mappings I, *Kōdai Math. Sem. Rep.*, **22**, 142–149.

[18] W. Takahashi (1976). Nonlinear variational inequalities and fixed point theorems, *J. Math. Soc. Japan*, **28**, 168–181.

[19] W. Takahashi (1984). Fixed point theorems for families of nonexpansive mappings on unbounded sets, *J. Math. Soc. Japan*, **36**, 543–553.

[20] W. Takahashi (1988). *Nonlinear Functional Analysis (Japanese)*, Kindaikagakusha, Tokyo.

[21] W. Takahashi (1991). Existence theorems generalizing fixed point theorems for multivalued mappings. In J.B. Baillon and M. Théra (Eds), *Fixed Point Theory and Applications*, Pitman Research Notes in Mathematical Series, **252**, 397–406.

SIMAA 4(2002) 449–460

28. An Extension Theorem and Duals of Gale-Mas-Colell's and Shafer-Sonnenschein's Theorems

Kok-Keong Tan* and Zhou Wu

Department of Mathematics and Statistics, Dalhousie University, Halifax, Nova Scotia, Canada B3H 3J5

Abstract: In this chapter, we shall first prove a Tietze-Dugundji extension theorem for upper semicontinuous correspondences with non-empty compact star-shaped values. By applying this extension theorem, Eilenberg and Montgomery (1946) fixed point theorem, Ky Fan's (1972) minimax inequality, Fan-Glicksberg (1952) fixed point theorem, or an improvement of an extension theorem of Pruszko (1997), we shall study the duals of the Gale-Mas-Colell's (1975, 1979) and Shafer-Sonnenschein's (1975) Theorems where the correspondences are upper semicontinuous instead of being lower semicontinuous. Some applications to equilibrium existence theorems for qualitative games are also given.

Keywords: Extension theorem, star-shaped, abstract economy, qualitative game, equilibrium

1. INTRODUCTION

If X is a set, we shall denote by 2^X the family of all subsets of X. We begin with the following fixed point theorem which was first formulated by Browder [4] as Theorem 1 but which is equivalent to an earlier result of Fan [10] stated as Lemma 4. It is now often called Fan-Browder fixed point theorem.

Theorem 1.1: *Let X be a non-empty compact convex subset of a Hausdorff topological vector space E, and $F: X \to 2^X$ be a multifunction with non-empty convex values such that for each $x \in X, F^{-1}(x)$ is open in X. Then F has a fixed point.*

We remark here that in Theorem 1.1, the requirement for the space E to be Hausdorff is not necessary as can be seen in Theorem 3′ of Ding and Tan [6].

We also remark that a multifunction $F: X \to 2^X$ with the property that $F^{-1}(x)$ is open in X for each $x \in X$ is necessarily lower semicontinuous.

In studying game theory and mathematical economics, we often need to study a family of correspondences from the product space into each of its factor spaces. In 1975 and 1979, Gale and Mas-Colell [12,13] proved the following result which improves Theorem 1.1 when the space E is finite-dimensional:

* The author was partially supported by NSERC of Canada under grant A-8096.

Theorem 1.2: *Let I be a finite index set. For each $i \in I$, let X_i be a non-empty compact convex subset of $\mathbb{R}^{n_i}$ and $F_i: X := \Pi_{j\in I} X_j \to 2^{X_i}$ be a lower semicontinuous correspondence with convex values. Then there exists $x \in X$ such that for each $i \in I$ either $x_i \in F_i(x)$ or $F_i(x) = \emptyset$.*

Originally, in [12], the correspondences in the above theorem were assumed to have open graphs instead of lower semicontinuous. In [13], Gale and Mas-Colell gave the above form while commenting that the proof of the above theorem was the same as the original one but applied Michael's [19] selection theorem instead.

In 1995, Deguire and Lassonde [5] proved the following theorem which partially generalizes Theorem 1.2:

Theorem 1.3: *Let I be any index set. For each $i \in I$, let X_i be a non-empty compact convex subset of Hausdorff locally convex space E_i and $F_i: X := \Pi_{j\in I} X_j \to 2^{X_i}$ be a correspondence with convex values such that for each $y_i \in X_i, F_i^{-1}(y_i)$ is open in X. Suppose for each $x \in X$, there exists $i \in I$ such that $F_i(x) \neq \emptyset$. Then there exist $x \in X$ and $i \in I$ such that $x_i \in F_i(x)$.*

In 1975, Shafer and Sonnenschein [21] proved the following equilibrium existence theorem for abstract economies:

Theorem 1.4: *Let I be finite and let $(X_i, F_i, P_i)_{i\in I}$ be an abstract economy such that for each $i \in I$,*

(1) *$X_i \subset \mathbb{R}^{n_i}$ is non-empty, compact and convex;*
(2) *F_i is a continuous correspondence with non-empty compact convex values;*
(3) *GrP_i is open in $X \times X_i$;*
(4) *$x_i \notin \mathrm{co}(P_i(x))$ for all $x \in X$.*

Then $(X_i, F_i, P_i)_{i\in I}$ has an equilibrium $x \in X$; i.e., for each $i \in I, x \in F_i(x)$ and $F_i(x) \cap P_i(x) = \emptyset$.

The following well-known fixed point theorem was proved independently by Fan [9] and Glicksberg [14] (where in finite-dimensional case, it is known as Kakutani [16] fixed point theorem):

Theorem 1.5: *Let X be a non-empty compact convex subset of a Hausdorff locally convex space $E, F: X \to 2^X$ be an upper semicontinuous multifunction with non-empty closed convex values. Then F has a fixed point.*

Theorem 1.5 can be thought as a dual of Theorem 1.1 in the sense that Theorem 1.1 is a fixed point theorem for lower semicontinuous multifunctions while Theorem 1.5 is a fixed point theorem for upper semicontinuous multifunctions.

In this chapter, we shall first improve Ma's generalization [18] of Dugundji's Tietze extension theorem [7] for upper semicontinuous correspondences with non-empty compact star-shaped values. By applying this extension theorem, Eilenberg-Montgomery fixed point theorem [8], Ky Fan's minimax inequality [11], Fan-Glicksberg fixed point theorem [9,14], or an improvement of an extension theorem of Pruszko [20], we shall study the duals of the above Theorems 1.2, 1.3 and 1.4 where the correspondences are upper semicontinuous instead of being lower semicontinuous. Some applications to equilibrium existence theorems for qualitative games are also given.

2. AN EXTENSION THEOREM

A set X in a linear space is said to be *star-shaped* if there exists $x_0 \in X$ such that for any $x \in X, tx_0 + (1-t)x \in X$ for all $t \in [0,1]$. Such an x_0 is called a *center* of the star-shaped set X. If (X, d) is a metric space, $x \in X, A, C$ are non-empty subsets of X and $r > 0, B_d(x;r) := \{y \in X : d(x,y) < r\}, B_d(A;r) := \{y \in X : d(y,a) < r \text{ for some } a \in A\}$, $d(A) =$ the diameter of $A, d(x,A) := \inf\{d(x,a) : a \in A\}$ and $d(A,C) := \inf\{d(a,c) : a \in A, c \in C\}$. If X is a topological space and $A \subset X$, then $\text{int}_X(A)$ denotes the interior of A in $X, \text{cl}_X(A)$ denotes the closure of A in X and $\partial_X(A)$ denotes the boundary of A in X.

Proposition 2.1: *Let X, Y be two star-shaped sets in a linear space, then $X+Y$ is also a star-shaped set.*

Proof: Let x_0 be a center of X and y_0 be a center of Y. Let $z_0 = x_0 + y_0 \in X + Y$. For any $z = x + y \in X + Y$ and $t \in [0,1], tz_0 + (1-t)z = [tx_0 + (1-t)x] + [ty_0 + (1-t)y] \in X + Y$. So $X + Y$ is also a star-shaped set. □

The proof of the following result (e.g., see [1:p. 108]) is easy and is thus omitted:

Lemma 2.2: *If (X, d) and (Y, ρ) are metric spaces and $F: X \to 2^Y$ is a non-empty and compact-valued correspondence, then F is upper semicontinuous at $x_0 \in X$ if and only if for each $\epsilon > 0$, there exists $\delta > 0$ such that $F(x) \subset B_\rho(F(x_0); \epsilon)$ for all $x \in B_d(x_0; \delta)$.*

We frequently refer to the following result, which is Proposition 11 on page II.34 in Bourbaki [3]:

Lemma 2.3: *Let E be a metrizable locally convex space. The topology of E can be defined by a metric that is invariant under translations and for which the open balls are convex.*

We now improve Ma's generalization [18] of the Dugundji's Tietze extension theorem [7] for compact-convex valued upper semicontinuous correspondences to compact star-shaped valued upper semicontinuous correspondences.

Theorem 2.4: *Let (X, d) be a metric space, M be a non-empty closed subset of X, E be a metrizable locally convex space and $F: M \to 2^E$ be an upper semicontinuous correspondence with non-empty compact star-shaped values. Then there exists an upper semicontinuous correspondence $\tilde{F}: X \to 2^E$ with non-empty compact star-shaped values such that $\tilde{F}|_M = F$ and $\tilde{F}(x) \subset \text{co}(F(M))$ for each $x \in X$. Moreover, if F is single-valued (respectively, convex-valued), the extension $\tilde{F}$ is also single-valued (respectively, convex-valued).*

Proof: (1) By Lemma 2.3, the topology on E can be induced by a metric ρ on E which is invariant under translations and for which the open balls are convex.

For each $y \in X\backslash M$, let U_y be an open ball in $X\backslash M$ with center at y and $d(U_y) < d(U_y, M)$. Thus $\{U_y : y \in X\backslash M\}$ is an open cover of $X\backslash M$. Since X is a metric space, there is a partition of unity $\{f_y : y \in X\backslash M\}$ on $X\backslash M$ subordinated to the covering $\{U_y : y \in X\backslash M\}$; that is, $f_y: X\backslash M \to [0,1]$ is continuous for each $y \in X\backslash M$ and is zero outside U_y while each $x \in X\backslash M$ has an open neighborhood $V(x)$ in $X\backslash M$ such that all but a finite number of f_y are identically zero on $V(x)$ and

$$\sum_{y \in X\backslash M} f_y(x) = 1, \text{ for all } x \in X\backslash M.$$

For each $y \in X \backslash M$, choose any $m_y \in M$ such that $d(m_y, U_y) < 2d(M, U_y)$. Now define $\tilde{F}: X \to 2^E$ by

$$\tilde{F}(x) = \begin{cases} F(x), & \text{if } x \in M; \\ \sum_{y \in X \backslash M} f_y(x) F(m_y), & \text{if } x \in X \backslash M. \end{cases}$$

By Proposition 2.1, $\tilde{F}(x)$ is a non-empty compact star-shaped subset of $\text{co}(F(M))$ for each $x \in X$.

(2) Clearly $\tilde{F}$ is upper semicontinuous at each point in $\text{int}_X(M)$. We shall first show that $\tilde{F}$ is upper semicontinuous at each point in $X \backslash M$.

Indeed, let $x_0 \in X \backslash M$ be given. Then we can find an open neighborhood $V(x_0)$ of x_0 in $X \backslash M$ (since M is closed in X) such that all but only a finite number of $f_y, y \in X \backslash M$, are identically zero. We denote the latter as $f_{y_1}, \ldots, f_{y_n}$. For any $\epsilon > 0$, since each $F(m_{y_i})$ is compact (and hence bounded), we can find $\delta > 0$ such that $ry \in B_\rho(0, \epsilon/n)$ for all $y \in \cup_{i=1}^n F(m_{y_i})$ and for all $|r| < \delta$.

Since each $f_{y_i}(x)$ is continuous, there exists an open neighborhood $V'(x_0)$ of x_0 in $X \backslash M$ such that for each $i \in \{1, \ldots, n\}, |f_{y_i}(x) - f_{y_i}(x_0)| < \delta$ for all $x \in V'(x_0)$. Let $V''(x_0) = V'(x_0) \cap V(x_0)$, then $V''(x_0)$ is an open neighborhood of x_0 in $X \backslash M$. Let $i \in \{1, \ldots, n\}$ be arbitrarily fixed. For each $x \in V''(x_0)$ and any $y \in F(m_{y_i})$, we have

$$\rho(f_{y_i}(x)y, f_{y_i}(x_0)y) = \rho((f_{y_i}(x) - f_{y_i}(x_0))y, 0) < \epsilon/n.$$

It follows that

$$f_{y_i}(x)y \in f_{y_i}(x_0)F(m_{y_i}) + O(0, \varepsilon/n),$$

so that

$$f_{y_i}(x)F(m_{y_i}) \subset f_{y_i}(x_0)F(m_{y_i}) + O(0, \varepsilon/n).$$

Therefore,

$$\sum_{i=1}^n f_{y_i}(x)F(m_{y_i}) \subset \sum_{i=1}^n f_{y_i}(x_0)F(m_{y_i}) + O(0, \varepsilon);$$

i.e., $\tilde{F}(x) \subset \tilde{F}(x_0) + O(0, \varepsilon)$ for all $x \in V''(x_0)$. This implies that $\tilde{F}$ is upper semicontinuous at x_0.

(3) To complete the proof that $\tilde{F}$ is upper semicontinuous on X, it remains to show that $\tilde{F}$ is also upper semicontinuous at each point in $\partial_X(M)$.

Let $x_0 \in \partial_X(M)$ be given, then $\tilde{F}(x_0) = F(x_0)$. By Lemma 2.2, since F is upper semicontinuous, for each $\epsilon > 0$, there exists $\delta_1 > 0$ such that $F(x) \subset F(x_0) + B_\rho(0, \epsilon) = \tilde{F}(x_0) + B_\rho(0, \epsilon)$ for all $x \in B_d(x_0, \delta_1) \cap M$.

If $x \in X \backslash M$ and $f_y(x) \neq 0$ for some $y \in X \backslash M$, then $x \in U_y$. Applying the triangle inequality yields

$$d(m_y, x) \leq d(m_y, U_y) + d(U_y) \leq 3d(M, U_y) \leq 3d(x_0, x);$$

it follows that

$$d(m_y, x_0) \leq d(m_y, x) + d(x, x_0) \leq 4d(x_0, x).$$

Take $\delta_2 = \delta_1/4$. For any $x \in B_d(x_0, \delta_2) \cap (X \backslash M)$, if $f_y(x) \neq 0$ for some $y \in X \backslash M$, we have $d(m_y, x_0) \leq 4d(x_0, x) < 4\delta_2 = \delta_1$. Hence $F(m_y) \subset \tilde{F}(x_0) + B_\rho(0, \epsilon)$ as $m_y \in M$.

(3.1) For any $x \in B_d(x_0, \delta_2) \cap (X \backslash M)$, we have $\sum_{y \in X \backslash M} f_y(x) = 1$ and only a finite number of $f_y(x)$ are not zero.

Note that $B_\rho(0, \epsilon)$ is convex, we have

$$\sum_{y \in X \backslash M} f_y(x) F(m_y) \subset \sum_{y \in X \backslash M} f_y(x) F(x_0) + B_\rho(0, \epsilon),$$

or

$$\tilde{F}(x) \subset \tilde{F}(x_0) + B_\rho(0, \epsilon)$$

for each $x \in B_d(x_0, \delta_2) \cap (X \backslash M)$.

(3.2) For each $x \in B_d(x_0, \delta_2) \cap M \subset B_d(x_0, \delta_1) \cap M$, we have

$$\tilde{F}(x) = F(x) \subset F(x_0) + B_\rho(0, \epsilon) = \tilde{F}(x_0) + B_\rho(0, \epsilon).$$

By (3.1) and (3.2), for each $x \in B_d(x_0, \delta_2)$, we have $\tilde{F}(x) \subset \tilde{F}(x_0) + B_\rho(0, \epsilon)$. Thus F is upper semicontinuous at each $x_0 \in \partial_X(M)$.

Finally, if F is single-valued (respectively, convex-valued), from the construction of its extension $\tilde{F}, \tilde{F}$ is also single-valued (respectively, convex-valued). □

We remark that Theorem 2.4 is a partial generalization of Theorem 2.1 of Ma [18]: our correspondence F may have star-shaped values (instead of convex values) but our space E is required to be metrizable.

3. EQUILIBRIA OF QUALITATIVE GAMES AND ABSTRACT ECONOMIES

Let us recall the definition of an acyclic space. A compact metrizable space X is said to be *acyclic* if

(1) X is non-empty;
(2) the homology groups $H_q(X)$ vanish for $q > 0$;
(3) the reduced 0-th homology group $\tilde{H}_0(X)$ vanishes.

Obviously non-empty compact convex or star-shaped sets in a metrizable locally convex space are acyclic.

The following is the Eilenberg–Montgomery fixed point theorem [8]:

Lemma 3.1: *Let X be an acyclic absolute neighborhood retract and $F: X \to 2^X$ be an upper semicontinuous correspondence such that for every $x \in X$, the set $F(x)$ is acyclic. Then F has a fixed point.*

We shall need the following simple fact whose proof is omitted:

Lemma 3.2: *Let X and Y be topological spaces and $F: X \to 2^Y$ be upper semi-continuous. Then the set $\{x \in X : F(x) \neq \emptyset\}$ is a closed subset of X.*

The following result is a dual of Theorem 1.2:

Theorem 3.3: *Let I be countable. For each $i \in I$, let X_i be a non-empty compact convex subset of the metrizable locally convex space E_i and $F_i: X := \Pi_{j\in I} X_j \to 2^{X_i}$ be an upper semicontinuous correspondence with closed star-shaped values. Then there exists $\hat{x} \in X$ such that for each $i \in I$, either $\hat{x}_i \in F_i(\hat{x})$ or $F_i(\hat{x}) = \emptyset$.*

Proof: Note that $E := \Pi_{i\in I} E_i$ is also a locally convex space when equipped with the product topology. Since I is countable and each E_i is metrizable, E is also metrizable. By the Tychonoff theorem, X is a compact subset of E. Obviously, X is convex. By Dugundji's Tietze extension theorem [7] (i.e., the single-valued case of our Theorem 2.4) and (2.18) of Borsuk [2:p. 103], X is an absolute retract and hence an absolute neighborhood retract. Moreover, X is also acyclic. For each $i \in I$, let $C_i = \{x \in X: F_i(x) \neq \emptyset\}$. By Lemma 3.2, C_i is a closed subset of X. Define $\tilde{F}_i: X \to 2^{X_i}$ as follows:

(1) If $C_i = \emptyset$, let $\tilde{F}_i(x) = X_i$ for all $x \in X$;
(2) If $C_i = X$, let $\tilde{F}_i(x) = F_i(x)$ for all $x \in X$;
(3) If C_i is a proper non-empty subset of X, by Theorem 2.4, there exists an upper semicontinuous correspondence $\tilde{F}_i: X \to 2^{X_i}$ with non-empty closed star-shaped values such that $\tilde{F}_i(x) = F_i(x)$ for all $x \in C_i$.

Define $F: X \to 2^X$ by $F = \Pi_{i\in I}\tilde{F}_i$. Then by Lemma 3 of Fan [9], F is an upper semicontinuous correspondence with non-empty closed star-shaped values. By Lemma 3.1, there exists $\hat{x} \in X$ such that $\hat{x} \in F(\hat{x})$. Now for each $i \in I$, if $F_i(\hat{x}) \neq \emptyset$, then $\tilde{F}_i(\hat{x}) = F_i(\hat{x})$ which in turn implies $\hat{x}_i \in F_i(\hat{x})$. □

As an immediate consequence of Theorem 3.3, we have the following equilibrium existence theorem of a qualitative game:

Theorem 3.4: *Let $(X_i, P_i)_{i\in I}$ be a qualitative game, where the set I of players is countable. For each $i \in I$, let X_i be a non-empty compact convex subset of a metrizable locally convex space E_i. Suppose that for each $i \in I, P_i : X := \Pi_{j\in I} X_j \to 2^{X_i}$ is an upper semicontinuous correspondence with closed star-shaped values and $x_i \notin P_i(x)$ for all $x \in X$. Then $(X_i, P_i)_{i\in I}$ has an equilibrium $\hat{x} \in X$; i.e., $P_i(\hat{x}) = \emptyset$ for all $i \in I$.*

Remark 3.5: Note that in the Eilenberg–Montgomery fixed point theorem [8], the values of the correspondence are assumed to be acyclic only. We wonder whether the sum of two acyclic sets is acyclic. If it were true, we could replace the star-shaped values with acyclic ones in Theorem 2.4 and hence Theorem 3.3 and Theorem 3.4 could be improved in this way. However, the following simple example given by K. Johnson shows that even the sum of two contractible sets is not necessarily an acyclic set:

Example 3.6: Let two figures in the plane have the shapes of C and I, where the height of I is equal to that of the gap of C. Obviously they are contractible and hence they are acyclic. But the sum of them is a set with its homology group $H_1 \neq 0$. □

Nevertheless, we have the following question:

Question 1: Let $I = \{1, \ldots, n\}$. For each $i \in I$, let X_i be an acyclic absolute neighborhood retract and $F_i: X := \Pi_{j=1}^n X_j \to 2^{X_i}$ be an upper semicontinuous correspondence with acyclic or empty-set values. Does there exist $x \in X$ such that for each $i \in I$, either $x_i \in F_i(x)$ or $F_i(x) = \emptyset$?

We shall need the following result which is Proposition 1 of Tarafdar, Watson and Yuan [26]:

Lemma 3.7: *Let X be a topological space and Y be a normal topological space. If $F: X \to 2^Y$ is upper semicontinous, then the correspondence $\text{cl}\, F : X \to 2^Y$, defined by $\text{cl}\, F(x) = \text{cl}_Y(F(x))$ for each $x \in X$, is also upper semicontinuous.*

Lemma 3.8: *Let (X, d) be a metric space and Y be a non-empty compact convex subset of a metrizable locally convex space E. Suppose that $G: X \to 2^Y$ is an upper semicontinuous correspondence, then $T: X \to 2^Y$, defined by $T(x) := \text{cl}_Y(\text{co}(G(x)))$ for each $x \in X$, is also upper semicontinuous.*

Proof: Since G is upper semicontinuous, by Lemma 3.7, the correspondence $F := \text{cl}\, G$ is also upper semicontinuous. Note that for each $x \in X, T(x) = \text{cl}_Y(\text{co}(G(x))) = \text{cl}_Y(\text{co}(\text{cl}_Y(G(x)))) = \text{cl}_Y(\text{co}(F(x)))$.

Since E is a metrizable locally convex space, by Lemma 2.3, the topology of E can be defined by a metric ρ that is invariant under translations and for which the open balls are convex.

Let $\epsilon > 0$ and $x_0 \in X$. Since F is upper semicontinuous, by Lemma 2.2, there exists $\delta > 0$ such that for each $x \in B_d(x_0, \delta), F(x) \subset F(x_0) + B_\rho(0, \epsilon/2)$. Now take any $z \in \text{co}F(x)$. Then there exist $z_1, \ldots, z_n$ in $F(x)$ and non-negative numbers $\lambda_1, \ldots, \lambda_n$ with $\sum_{i=1}^n \lambda_i = 1$ such that $z = \sum_{i=1}^n \lambda_i z_i$. Choose $y_1, \ldots, y_n$ in $F(x_0)$ with $z_i \in y_i + B_\rho(0, \epsilon/2)$ for all $i = 1, \ldots, n$. Then we have

$$\sum_{i=1}^n \lambda_i z_i \in \sum_{i=1}^n \lambda_i y_i + \sum_{i=1}^n \lambda_i B_\rho(0, \epsilon/2).$$

Since $B_\rho(0, \epsilon)$ is convex, we have

$$\sum_{i=1}^n \lambda_i z_i \in \sum_{i=1}^n \lambda_i y_i + B_\rho(0, \epsilon/2)$$

or $\text{co}F(x) \subset \text{co}F(x_0) + B_\rho(0, \epsilon/2)$. Thus $\text{cl}_Y(\text{co}F(x)) \subset \text{cl}_Y(\text{co}F(x_0)) + B_\rho(0, \epsilon)$, i.e., $T(x) \subset T(x_0) + B_\rho(0, \epsilon)$. By Lemma 2.2 again, T is upper semicontinuous. □

Theorem 3.9: *Let $(X_i, P_i)_{i \in I}$ be a qualitative game, where the set I of players is countable. For each $i \in I$, let X_i be a non-empty compact convex subset of a metrizable locally convex space E_i. Suppose that for each $i \in I, P_i : X := \Pi_{j \in I} X_j \to 2^{X_i}$ is an upper semicontinuous correspondence such that $x_i \notin \text{cl}_{X_i}(\text{co}P_i(x))$ for each $x \in X$. Then $(X_i, P_i)_{i \in I}$ has an equilibrium.*

Proof: By Lemma 3.8, the correspondence $T_i: X \to 2^{X_i}$, defined by $T_i(x) = \text{cl}_{X_i}(\text{co}(P_i(x)))$ for each $x \in X$, is also upper semicontinuous. By hypothesis, for each $x \in X, x_i \notin T_i(x)$. Therefore by Theorem 3.4, the qualitative game $(X_i, T_i)_{i \in I}$ has an equilibrium $\hat{x} \in X$; i.e., $T_i(\hat{x}) = \emptyset$ for all $i \in I$. It follows that $P_i(\hat{x}) = \emptyset$ for all $i \in I$. Thus $(X_i, P_i)_{i \in I}$ has an equilibrium. □

Let E be a topological vector space. We shall denote by E' the continuous dual of E, by $\langle w, x \rangle$ the pairing between E' and E for $w \in E'$ and $x \in E$ and by $Re\langle w, x \rangle$ the real part of $\langle w, x \rangle$.

First we note that the same proof of Lemma 1 of Shih and Tan [22] gives the following slightly more general formulation:

Lemma 3.10: *Let X be a topological space, E be a topological vector space and $S: X \to 2^E$ be upper semicontinuous such that for each $x \in X, S(x)$ is non-empty and bounded.* Then for each $p \in E'$, *the map $f_p: X \to \mathbb{R}$ defined by $f_p(x) = \sup_{y \in S(x)} \mathrm{Re}\langle p, y\rangle$ for all $x \in X$ is upper semicontinuous.*

The following minimax inequality is equivalent to the celebrated Ky Fan minimax inequality [11]:

Lemma 3.11: *Let E be a topological vector space, X be a non-empty compact convex subset of E and $f: X \times X \to \mathbb{R}$ be such that*

(1) *for each $x \in X$, $f(x, x) \leq 0$;*
(2) *for each fixed $x \in X, y \longrightarrow f(x, y)$ is lower semicontinuous on X;*
(3) *for each $y \in X$, $x \longrightarrow f(x, y)$ is quasi-concave on X.*

Then there exists $\hat{y} \in X$ such that $f(x, \hat{y}) \leq 0$ for all $x \in X$.

The following result is a dual of Theorem 1.3:

Theorem 3.12: *Let I be any (countable or uncountable) index set. For each $i \in I$, let X_i be a non-empty compact convex subset of the Hausdorff locally convex space E_i, $F_i: X := \Pi_{j \in I} X_j \to 2^{X_i}$ be an upper semicontinuous correspondence with closed convex values. Then there exists $\hat{x} \in X$ such that for each $i \in I$, either $\hat{x}_i \in F_i(\hat{x})$ or $F_i(\hat{x}) = \emptyset$.*

Proof: For each $i \in I$, let $\pi_i: X \to X_i$ be the projection map. Suppose that the conclusion is false. Let $x \in X$ be arbitrarily given. Then there exists $i \in I$ such that $F_i(x) \neq \emptyset$ and $x_i \notin F_i(x)$. By the Hahn-Banach separation theorem, there exists $p_i \in E_i'$ such that

$$\mathrm{Re}\langle p_i, x_i\rangle > \sup_{y_i \in F_i(x)} \mathrm{Re}\langle p_i, y_i\rangle.$$

By Lemma 3.10, the set

$$V_{p_i} = \{x \in X: \mathrm{Re}\langle p_i, \pi_i(x)\rangle > \sup_{y_i \in F_i(x)} \mathrm{Re}\langle p_i, y_i\rangle\}$$

is open in X.

Thus $\{V_{p_i}: p_i \in E_i', i \in I\}$ is an open covering of X.

Since X is compact, we may assume without loss of generality that there exist $p_i^1, \ldots, p_i^{j_i} \in E_i'$, for $i = 1, \ldots, n \in I$ such that $X = \bigcup_{i=1}^n \bigcup_{k=1}^{j_i} V_{p_i^k}$.

Let $\{f_i^k: k = 1, \ldots, j_i, i = 1, \ldots, n\}$ be a partition of unity subordinated to the open covering $\{V_i^k: k = 1, \ldots, j_i, i = 1, \ldots, n\}$. Define $\phi: X \times X \to \mathbb{R}$ by

$$\phi(x, y) = \sum_{i=1}^{n} \sum_{k=1}^{j_i} f_i^k(y) \mathrm{Re}\langle p_i^k, \pi_i(y) - x_i\rangle$$

for each $x, y \in X$. Since for each fixed $x \in X$, $y \mapsto \phi(x, y)$ is continuous and for each fixed $y \in X$, $x \mapsto \phi(x, y)$ is concave, by Lemma 3.11, there exists $\bar{y} \in X$ such that $\phi(x, \bar{y}) \leq 0$ for all $x \in X$.

Now since there is at least one i such that $F_i(\bar{y}) \neq \emptyset$ (see the beginning of this proof), we can take $\bar{x}$ as follows: let $\bar{x}_i$ be any point in $F_i(\bar{y})$ if it is non-empty and be any point in

X_i if $F_i(\bar{y})$ is empty. We shall show that $\phi(\bar{x}, \bar{y}) > 0$. Indeed, if $f_i^k(\bar{y}) > 0$, then $\bar{y} \in V_{p_i^k}$ which in turn implies that $\text{Re}\langle p_i, \pi_i(\bar{y})\rangle - \text{Re}\langle p_i, \bar{x}_i\rangle > 0$ since $\bar{x}_i \in F_i(\bar{y})$. Since $\sum_{i=1}^n \sum_{k=1}^{j_i} f_i^k(\bar{y}) = 1$, there must be some $f_i^k(\bar{y}) > 0$. Hence $\phi(\bar{x}, \bar{y}) > 0$ which is a contradiction. □

As an immediate consequence of Theorem 3.12, we have the following equilibrium existence theorem of a qualitative game:

Theorem 3.13: *Let $(X_i, P_i)_{i\in I}$ be a qualitative game. For each $i \in I$,* let *X_i be a non-empty compact convex subset of a locally convex space E_i. Suppose that for each $i \in I$, $P_i : X := \Pi_{j\in I} X_j \to 2^{X_i}$ is an upper semicontinuous correspondence with closed convex values and $x_i \notin P_i(x)$ for all $x \in X$. Then $(X_i, P_i)_{i\in I}$ has an equilibrium $\hat{x} \in X$; i.e., $P_i(\hat{x}) = \emptyset$ for all $i \in I$.*

We next give an existence theorem of equilbria for a qualitative game in which the preferences are majorized by upper semicontinuous correspondences. Let us recall some concepts and notations which were introduced by Tan and Yuan [24]:

Let X be a topological space, Y a non-empty subset of a vector space E. Let $\theta : X \to E$ be a single-valued mapping and $\phi : X \to 2^Y$ be a correspondence. Then $\phi : X \to 2^Y$ is said to be of class $\mathcal{U}_\theta$ if (a) for each $x \in X, \theta(x) \notin \phi(x)$ and (b) ϕ is upper semicontinuous with closed and convex values in Y; (2) ϕ_x is a *$\mathcal{U}_\theta$-majorant* of ϕ at x if there is an open neighborhood $N(x)$ of x in X and $\phi_x : N(x) \to 2^Y$ such that (a) for each $z \in N(x)$, $\phi(z) \subset \phi_x(z)$ and $\theta(z) \notin \phi_x(z)$ and (b) ϕ_x is upper semicontinuous with closed and convex values; (3) ϕ is said to be *$\mathcal{U}_\theta$-majorized* if for each $x \in X$ with $\phi(x) \neq \emptyset$, there exists a *$\mathcal{U}_\theta$-majorant* ϕ_x of ϕ at x.

The following result is Theorem 2.1 of Tan and Yuan [24]:

Lemma 3.14: *Let X be a paracompact space and Y be a non-empty normal subset of a topological vector space E. Let $\theta : X \to E$ and $P : X \to 2^Y\backslash\{\emptyset\}$ be $\mathcal{U}_\theta$-majorized. Then there exists a correspondence $\phi : X \to 2^Y\backslash\{\emptyset\}$* of class $\mathcal{U}_\theta$ *such that* $P(x) \subset \phi(x)$ for each $x \in X$.

Here we shall only deal with the case $X = \Pi_{i\in I} X_i$ and $\theta = \pi_i : X \to X_i$, the projection from X onto X_i. In the following, we shall write $\mathcal{U}$ instead of $\mathcal{U}_\theta$.

Theorem 3.15: *Let $(X_i, P_i)_{i\in I}$ be a qualitative game, where the set I of players is countable. For each $i \in I$, let X_i be a non-empty compact convex subset of a metrizable locally convex space E_i and $P_i : X := \Pi_{j\in I} X_j \to 2^{X_i}$ be $\mathcal{U}$-majorized such that the set $C_i := \{x \in X : P_i(x) \neq \emptyset\}$ is closed in X. Then $(X_i, P_i)_{i\in I}$ has an equlibrium.*

Proof: Since I is countable, the space $E := \Pi_{j\in I} E_j$ is a metrizable locally convex space when it is equipped with the product topology.

Let $i \in I$ be arbitrarily fixed. Note that C_i is paracompact. Since $P_i : C_i \to 2^{X_i}\backslash\{\emptyset\}$ is *$\mathcal{U}$-majorized*, by Lemma 3.14, there exists a correspondence $\phi_i : C_i \to 2^{X_i}\backslash\{\emptyset\}$ of class $\mathcal{U}$such that $P_i(x) \subset \Phi_i(x)$ for all $x \in X$. By Theorem 2.4, there exists an upper semicontinuous correspondence $\Phi_i : X \to 2^{X_i}\backslash\{\emptyset\}$ with compact star-shaped values such that $\Phi_i|_{C_i} = \phi_i$.

Now define $\Phi : X \to X$ by

$$\Phi(x) = \Pi_{i\in I} \Phi_i(x) \quad \text{for all } x \in X.$$

Then Φ is upper semicontinuous (by Lemma 3 of Fan [9]) with non-empty star-shaped values. By Lemma 3.1, Φ has a fixed point $\hat{x} \in X$. For each $i \in I$, we must have $\hat{x}_i \notin C_i$ for otherwise $\hat{x}_i \in \Phi_i(\hat{x}) = \phi_i(x)$ which would contradict the assumption that ϕ_i is of class $\mathcal{U}$. Thus for each $i \in I, \phi_i(\hat{x}) = \emptyset$ so that $P_i(\hat{x}) = \emptyset$. Therefore $\hat{x}$ is an equilibrium of $(X_i, P_i)_{i\in I}$. □

Remark 3.16: Theorem 3.12 generalizes the well-known Fan-Glicksberg fixed point theorem [9,14]. However, we know that the Himmelberg fixed point theorem [15] is a non-compact generalization of the Fan-Glicksberg fixed point theorem. We post the following:

Question 2: Let I be any index set. For each $i \in I$, let X_i be a non-empty convex subset of Hausdorff locally convex space E_i, D_i be a non-empty compact subset of X_i and $F_i: X := \Pi_{j\in I}X_j \to 2^{D_i}$ be an upper semicontinuous correspondence with closed convex values. Does there exist $x \in D := \Pi_{j\in I}D_j$ such that for each $i \in I$, either $x_i \in F_i(x)$ or $F_i(x) = \emptyset$?

Let (X, d) be a metric space, E be a topological vector space and $F: X \to 2^E$ be a correspondence with non-empty closed convex values. Then F is *completely continuous* if for each bounded subset B of X, the set $\text{cl}_E(\cup_{x\in B} F(x))$ is compact.

Pruszko [20] proved an extension theorem for an upper semicontinuous correspondence which is dominated by a completely continuous correspondence with non-empty closed convex values in a normed space. By applying Lemma 3.7, Lemma 3.8 and by modifying his proof (especially we have to show that $\tilde{F}(x_0)$ is the Hausdorff limit of the sequence $(A_n)_{n=1}^{\infty}$ while each set A_n need not be compact), we find that Pruszko's result holds in metrizable locally convex spaces as well. We shall state it below without proof.

Theorem 3.17: *Let M be a non-empty closed subset of a metric space X, E be a metrizable locally convex space, $F: M \to 2^E$ be upper semicontinuous with non-empty closed convex values and $\phi: X \to 2^E$ be continuous with non-empty closed convex values such that $F(y) \subset \phi(y)$ for each $y \in M$. Further suppose that the closure of the set $\phi(N) := \cup_{x\in N}\phi(x)$ in E is compact for any bounded subset N of X. Then there exists an upper semicontinuous correspondence $\tilde{F}: X \to 2^E$ with non-empty closed convex values such that $\tilde{F}|_M = F$ and $\tilde{F}(x) \subset \phi(x)$ for each $x \in X$.*

The following equilibrium existence theorem of an abstract economy is a dual of Theorem 1.4:

Theorem 3.18: *Suppose that I is countable. Let $(X_i, F_i, P_i)_{i\in I}$ be an abstract economy such that for each i*

(i) X_i is a non-empty compact convex subset of a metrizable locally convex space E_i;
(ii) $F_i: X := \Pi_{j\in I}X_j \to 2^{X_i}$ is a continuous correspondence with non-empty compact convex values;
(iii) $P_i: X \to 2^{X_i}$ is upper semicontinuous.
(iv) $x_i \notin \text{cl}_{X_i}(\text{co}(P_i(x)))$ for all $x \in X$.

Then $(X_i, F_i, P_i)_{i\in I}$ has an equilibrium.

Proof: Note that since I is countable, the space $E := \Pi_{j\in I}E_j$ is a metrizable locally convex space when it is equipped with the product topology. Also note that X is a compact and convex subset of E.

Fix an arbitrary $i \in I$. Define $G_i : X \to 2^{X_i}$ by

$$G_i(x) = F_i(x) \cap \mathrm{cl}_{X_i}(\mathrm{co}(P_i(x))) \quad \text{for each } x \in X.$$

Since P_i is upper semicontinuous, by Lemma 3.8, the correspondence $T_i : X \to 2^{X_i}$ defined by $T_i(x) = \mathrm{cl}_{X_i}(\mathrm{co}(P_i(x)))$ for each $x \in X$ is also upper semicontinuous. Furthermore, since F_i is upper semicontinuous, G_i is also upper semicontinuous by Lemma 2.2 of Tan and Yuan [25]. Now let $M_i := \{x \in X : G_i(x) \neq \emptyset\}$, then M_i is closed in X by Lemma 3.2. Note that for each $x \in M_i$, we have $G_i(x) \subset F_i(x)$. By Theorem 3.17, there exists an upper semicontinuous correspondence $\tilde{G}_i(x)$ with non-empty closed convex values such that $\tilde{G}_i(x)|_{M_i} = G_i(x)$ and $\tilde{G}_i(x) \subseteq F_i(x)$ for each $x \in X$.

Now define $G : X \to 2^X$ by $G(x) = \Pi_{i\in I}\tilde{G}_i(x)$ for each $x \in X$, then G is upper semicontinuous by Lemma 3 of Fan [9] with non-empty closed convex values. By the Fan-Glicksberg fixed point theorem [9,14], G has a fixed point $x \in X$. Obviously, x is an equilibrium of $(X_i, F_i, P_i)_{i\in I}$. □

Remark 3.19: As pointed out in Tan and Wu [23] that the statement and proof of Theorem 1 of Kim and Lee [17] were not correct. Thus our Theorem 3.18 is not a special case of Kim and Lee's result. For other equilibrium existence theorem of abstract economies with upper semicontinuous constraint and preference correspondences, we refer to Tan and Wu [23].

Finally we shall post the following:

Question 3: In Theorem 3.17, can the "metrizable" assumption on the spaces E be dropped?

Question 4: In Theorem 3.18, can the "metrizable" assumption on the spaces E_i be dropped?

REFERENCES

[1] J.P. Aubin and I. Ekeland (1984). *Applied Nonlinear Analysis*. New York: John Wiley & Sons.

[2] K. Borsuk (1967). *Theory of Retracts*. Warszawa: PWN – Polish Scientific Publisher.

[3] N. Bourbaki (1987). *Topological Vector Spaces*. New York: Springer-Verlag.

[4] F.E. Browder (1968). The fixed-point theory of multi-valued mappings in topological vector spaces, *Mathematische Annalen*, **177**, 283–301.

[5] P. Deguire and M. Lassonde (1995). Familles sélectantes, *Topological Methods in Nonlinear Analysis*, **5**, 261–269.

[6] X.P. Ding and K.K. Tan (1992). A minimax inequality with applications to existence of equilibrium point and fixed point theorems, *Colloquium Mathematicum*, **63**, 233–247.

[7] J. Dugundji (1951). An extension of Tietze's Theorem, *Pacific Journal of Mathematics*, **1**, 353–367.

[8] S. Eilenberg and D. Montgomery (1946). Fixed point theorems for multi-valued transformations, *American Journal of Mathematics*, **68**, 214–222.

[9] K. Fan (1952). Fixed-point and minimax theorems in locally convex topological linear spaces, *Proceedings of the National Academy of Sciences of the USA*, **38**, 121–126.

[10] K. Fan (1961). A generalization of Tychonoff's fixed point theorem, *Mathematische Annalen*, **142**, 305–310.

[11] K. Fan (1972). A minimax inequality and applications, Inequalities III, O. Shisha (Ed.). New York: Academic Press.

[12] D. Gale and A. Mas-Colell (1975). An equilibrium existence theorem for a general model without ordered preferences, *Journal of Mathematical Economics*, **2**, 9–15.

[13] D. Gale and A. Mas-Colell (1979). Corrections to an equilibrium existence theorem for a general model without ordered preferences, *Journal of Mathematical Economics*, **6**, 297–298.

[14] I. Glicksberg (1952). A further generalization of the Kakutani fixed-point theorem with applications to Nash equlibrium points, *Proceedings of the American Mathematical Society*, **3**, 170–174.

[15] C.J. Himmelberg (1972). Fixed points of compact multifunctions, *Journal of Mathematical Analysis and Applications*, **38**, 205–207.

[16] S. Kakutani (1941). A generalization of Brouwer's fixed point theorem, *Duke Mathematical Journal*, **8**, 457–459.

[17] W.K. Kim and K.H. Lee (1997). Existence of equilibrium and separation in generalized games, *Journal of Mathematical Analysis and Applications*, **207**, 316–325.

[18] T.-M. Ma (1972). Topological degree of set-valued compact fields in locally convex spaces, *Dissnertationes Mathematica*, **92**, 1–47.

[19] E. Michael (1956). Continuous selections I, *Annals of Mathematics*, **63**, 361–382.

[20] T. Pruszko (1997). Completely continuous extensions of convex-valued selectors, Nonlinear Analysis Theory, *Methods and Applications*, **27**, 781–784.

[21] W. Shafer and H. Sonnenschein (1975). Equilibrium in abstract economies without ordered preferences, *Journal of Mathematical Economics*, **2**, 345–348.

[22] M.H. Shih and K.K. Tan (1985). Generalized quasi-variational inequalities in locally convex topological spaces, *Journal of Mathematical Analysis and Applications*, **108**, 333–343.

[23] K.K. Tan and Z. Wu (1997). A note on abstract economies with upper semicontinuous correspondences, *Applied Math. Letters*, **11**, 21–22.

[24] K.K. Tan and X.Z. Yuan (1993). Equilibria of generalized games with U-majorized preference correspondences, *Economics Letters*, **41**, 379–383.

[25] K.K. Tan and X.Z. Yuan (1994). Maximal elements and equilibria for U-majorized preferences, *Bulletin of Australian Mathematical Society*, **49**, 47–54.

[26] E. Tarafdar, P. Watson and X.Z. Yuan (1997). Jointly measurable selections of condensing Carathéodory set-valued mappings and its applications to random fixed points, *Nonlinear Analysis Theory, Methods and Applications*, **28**, 39–48.

SIMAA 4(2002) 461–467

29. Iterative Algorithms for Nonlinear Variational Inequalities Involving Set-valued *H*-Cocoercive Mappings

Ram U. Verma

Mathematical Sciences Division, International Publications, USA, 12046 Coed Drive, Suite A-29 Orlando, Florida 32826, USA

Abstract: Consider the following class of nonlinear variational inequality (NVI) problems, whose solvability is based on an iterative procedure characterized by a variational inequality: Determine an element $x^* \in K$ and $u^* \in T(x^*)$ such that

$$\langle u^*, x - x^* \rangle \geq 0 \quad \text{for all } x \in K,$$

where T: $K \to P(H)$ is a multivalued mapping from a real Hilbert space H into $P(H)$, the power set of H, and K is a nonempty closed convex subset of H. The iterative procedure is characterized as a nonlinear variational inequality, that is, for any arbitrarily chosen initial point $x^0 \in K$ and $u^0 \in T(x^0)$, we have

$$\langle u^k + x^{k+1} - x^k, x - x^{k+1} \rangle \geq 0 \quad \textit{for all } x \in K \textit{ and, for } u^k \in T(x^k) \textit{ and for } k \geq 0,$$

which is equivalent to a projection equation

$$x^{k+1} = P_K[x^k - u^k],$$

where P_K denotes the projection of H onto K. This extends the existing results to the case of a class of nonlinear variational inequalities involving multivalued α-H-cocoercive mappings in a Hilbert space setting.

Keywords: Nonlinear variational inequalities, α-H-cocoercive mappings, projection equations, projection method, extragradient method

AMS Subject Classification: 49J40, 90C20

1. INTRODUCTION

Recently He [4–7] applied a projection-contraction method to the solvability of a class of linear complementarity problems leading to convex quadratic programming problems and pointed out that the preliminary numerical experiments showed that the method could be very efficient for large sparse problems, while the applications of projection-contraction, extragradient, and projection-type methods to the solvability of nonlinear monotone variational inequalities are given in the context of mathematical programming by He [4], Korpelevich [9], Pang and Chan [14], Marcotte and Wu [10], Solodov and Tseng [15], and others. We note that the convergence for most of these methods is based on the estimates computed using iterative schemes, which are characterized, as usual, by the corresponding projection equations, while Marcotte and Wu [10] constructed the iterative procedure – a sort of new – characterized by a variational inequality leading to the convergence of the projection method, though the estimate does not coincide with the estimate found applying the iterative scheme represented by a projection equation. It is worth noting that an iterative scheme characterized by a variational inequality seems to be applicable only to the solvability of a variational inequality problems involving cocoercive mappings.

We plan to consider the approximation-solvability of a class of nonlinear variational inequalities involving multivalued α-H-cocoercive mappings in a real Hilbert space setting, which extends the variational inequality problems studied by Marcotte and Wu [10] and others. However, we establish another estimate – simple yet possibly an alternative to that of Marcotte and Wu [10] – based on an iterative scheme generated by a corresponding projection equation.

Let H be a real Hilbert space with inner product $<\cdot,\cdot>$ and norm $\|\cdot\|$. Let $C(H)$ denote the family of all nonempty compact subsets of H. Let T: $K \to C(H)$ be a multivalued mapping and K a closed convex subset of H. We consider a class of nonlinear variational inequality (NVI) problems: find an element $x^* \in K$ and $u^* \in T(x^*)$ such that

$$\langle u^*, x - x^* \rangle \geq 0 \quad \text{for all } x \in K. \tag{1.1}$$

Next, we consider the construction of the iterative algorithm upon which the convergence estimate depends.

Iterative Algorithm 1.1: For an arbitrary element $x^0 \in K$, we choose $u^0 \in T(x^0)$, and set

$$\langle \rho u^0 + x^1 - x^0, x - x^1 \rangle \geq 0 \text{ for all } x \in K, \tag{1.2}$$

where $\rho > 0$ is a constant.

Then (1.2) is equivalent to a projection equation

$$x^1 = P_K[x^0 - \rho u^0], \tag{1.3}$$

where P_K is the projection of H onto K

Since $u^0 \in T(x^0) \in C(H)$, there exists, by Nadler [11], an element $u^1 \in T(x^1)$ such that

$$\| u^0 - u^1 \| \leq H(T(x^0), T(x^1)).$$

By induction we can generate iterative sequences $\{x^k\}$ and $\{u^k\}$ as follows:

$$\begin{aligned} & u^k \in T(x^k), \\ & \| u^{k+1} - u^k \| \leq H(T(x^{k+1}), T(x^k)), \text{ and} \\ & \langle \rho u^k + x^{k+1} - x^k, x - x^{k+1} \rangle \geq 0 \quad \text{for all } x \in K \quad \text{and} \quad k \geq 0, \end{aligned} \tag{1.4}$$

where $H(\cdot,\cdot)$ denotes the *Hausdorff metric* defined by

$$H(A,B) = \max\left\{\sup_{a\in A} d(a,B), \sup_{b\in B} d(b,A)\right\}.$$

Here $d(a,B) = \inf_{b\in B} d(a,b)$ is the distance of the point a from the set B for $A, B \subset H$.

The iterative procedure (1.4) can be characterized as a projection equation

$$x^{k+1} = P_K[x^k - \rho u^k] \quad \text{for } k \geq 0. \tag{1.5}$$

A mapping T: $H \to C(H)$ is called r-strongly monotone if for all $x, y \in H$, we have

$$\langle u - v, x - y\rangle \geq r \,\|x - y\|^2 \quad \text{for } u \in T(x) \quad \text{and} \quad v \in T(y),$$

where $r > 0$ is a constant.

The mapping T is said to r-H-expanding if there exists a constant $r > 0$ such that

$$H(T(x), T(y)) \geq r\,\|x - y\|, \quad \text{for all } x, y \in H.$$

T is called an H-expanding mapping if $r = 1$.

A mapping T: $H \to C(H)$ is said to be β-H-Lipschitz continuous if

$$H(T(x), T(y)) \leq \beta\,\|x - y\| \quad \text{for all } x, y \in H,$$

where $\beta \geq 0$ is a constant. When $\beta = 1$, T is called an H-nonexpansive mapping.

A mapping T: $H \to C(H)$ is said to be α-H-cocoercive if for all $x, y \in H$, we have

$$\langle u - v, x - y\rangle \geq \alpha[H(T(x), T(y))]^2 \quad \text{for all } x, y \in H,$$

where $\alpha > 0$ is a constant.

Every α-H-cocoercive and H-expanding mapping is α-strongly monotone, and every α-H-cocoercive mapping T implies that

$$\langle u - v, x - y\rangle \geq \alpha\,\|u - v\|^2.$$

Clearly, every α-H-cocoercive mapping is $(1/\alpha)$-H-Lipschitz continuous for $\alpha > 0$.

Lemma 1.1: *For all $v, w \in H$, we have*

$$\|v\|^2 + \langle v, w\rangle \geq -(1/4)\,\|w\|^2.$$

Lemma 1.2: *Let $v, w \in H$. Then we have*

$$\langle v, w\rangle = (1/2)[\|v + w\|^2 - \|v\|^2 - \|w\|^2].$$

Lemma 1.3: ([3]) *Let K be a nonempty subset of a real Hilbert space H, and T: $K \to P(H)$ a multivalued mapping. Then the NVI problem (1.1) has a solution (x^*, u^*) if and only if x^* is a fixed point of the mapping F: $K \to P(K)$ defined by*

$$F(x) = \bigcup_{u\in T(x)} \{P_K[x - \rho u]\} \quad \text{for all } x \in K,$$

where $\rho > 0$ is a constant.

2. NONLINEAR VARIATIONAL INEQUALITIES

In this section we intend to present the main results on the solvability of the NVI problem (1.1) involving multivalued α-H-cocoercive mappings in a Hilbert space setting.

Theorem 2.1: *Let H be a real Hilbert space and T: $K \to C(H)$ an α-H-cocoercive mapping from a nonempty closed convex subset K of H into $C(H)$. Suppose that x^* and u^* form a solution of the NVI problem (1.1). Then the following conclusions hold:*

(i) *The sequences $\{x^k\}$ and $\{u^k\}$ generated by the iterative algorithm (1.4) satisfy the estimate*

$$\| x^{k+1} - x^* \|^2 \leq \| x^k - x^* \|^2 - [1 - (\rho/2\alpha)] \| x^k - x^{k+1} \|^2 .$$

(ii) *The sequences $\{x^k\}$ and $\{u^k\}$ converge to x^* and u^*, respectively, for $0 < \rho < 2\alpha$, a solution of the NVI problem (1.1).*

Proof: Before we show that the sequences $\{x^k\}$ and $\{u^k\}$ generated by the iterative algorithm (1.4) converge, respectively, to x^* and u^*, we need to establish the estimate in (i). Since x^{k+1} satisfies the iterative algorithm (1.4), we have for a constant $\rho > 0$ that

$$\langle \rho u^k + x^{k+1} - x^k, x - x^{k+1} \rangle \geq 0 \quad \text{for all } x \in K \quad \text{and for } u^k \in T(x^k). \tag{2.1}$$

Since x^* and u^* form a solution of the NVI problem (1.1), we have for constant $\rho > 0$ that

$$\langle \rho u^*, \ x - x^* \rangle \geq 0. \tag{2.2}$$

Replacing x by x^* in (2.1) and x by x^{k+1} in (2.2), and adding, we obtain

$$\begin{aligned} 0 &\leq \langle \rho(u^k - u^*), x^* - x^{k+1} \rangle + \langle x^{k+1} - x^k, x^* - x^{k+1} \rangle \\ &= -\rho\langle [u^k - u^*, x^k - x^* \rangle - \rho\langle u^k - u^*, x^{k+1} - x^k \rangle + \langle x^{k+1} - x^k, x^* - x^{k+1} \rangle \\ &\leq -\rho\alpha H(T(x^k), T(x^*))^2 - \rho\langle u^k - u^*, x^{k+1} - x^k \rangle) + \langle x^{k+1} - x^k, x^* - x^{k+1} \rangle \\ &\leq -\rho\alpha\{\| u^k - u^* \|^2 + (1/\alpha)\langle u^k - u^*, x^{k+1} - x^k \rangle\} + \langle x^{k+1} - x^k), x^* - x^{k+1} \rangle \end{aligned} \tag{2.3}$$

Setting $v = u^k - u^*$ and $w = (1/\alpha)[x^{k+1} - x^k]$ in Lemma 1.1, we obtain

$$-\{\| u^k - u^* \|^2 + (1/\alpha)\langle u^k - u^*, x^{k+1} - x^k \rangle\} \leq (1/4\alpha^2) \| x^{k+1} - x^k \|^2 . \tag{2.4}$$

Applying (2.4) to (2.3), we get

$$0 \leq (\rho/4\alpha) \|x^{k+1} - x^k \|^2 + \langle x^{k+1} - x^k, x^* - x^{k+1} \rangle \tag{2.5}$$

Taking $v = x^{k+1} - x^k$ and $w = x^* - x^{k+1}$ in Lemma 1.2, and applying to (2.5), we have

$$0 \leq (\rho/4\alpha) \| x^{k+1} - x^k \|^2 + (1/2)[\| x^* - x^k \|^2 - \| x^{k+1} - x^k \|^2 - \| x^* - x^{k+1} \|^2].$$

It follows that

$$\| x^{k+1} - x^* \|^2 \leq \| x^k - x^* \|^2 - [1 - (\rho/2\alpha)] \| x^{k+1} - x^k \|^2 . \tag{2.6}$$

It follows from (2.6) for $0 < \rho < 2\alpha$ that

$$\text{either } \lim_{k\to\infty} \| x^k - x^* \| = 0, \text{ or } \lim_{k\to\infty} \| x^k - x^{k+1} \| = 0.$$

Assume that the first alternative holds. Then the sequence $\{x^k\}$ converges to x^* and

$$\lim_{k\to\infty} \| x^k - x^{k+1} \| = 0 \text{ as well.}$$

Since T is $(1/\alpha)$-H-Lipschitz continuous, it implies by Algorithm 1.1 that

$$\| u^k - u^* \| \leq H(T(x^k), T(x^*)) \leq (1/\alpha) \| x^k - x^* \| \to 0,$$

that means, $u^k \to u^*$.

On the other hand, assume that the second alternative holds, that is,

$$\lim_{k\to\infty} \| x^k - x^{k+1} \| = 0.$$

Let x' be a cluster point of the sequence $\{x^k\}$. Then there exists a subsequence $\{x^{ki}\}$ such that $\{x^{ki}\}$ converges to x' since the left hand term of (2.6) is bounded. Finally, the mapping F in Lemma 1.3 in light of (1.5) is a contraction for $0 < \rho < 2\alpha$, and hence x' is a fixed point of F. As a result, the entire sequence $\{x^k\}$ converges to x' and $\{u^k\}$ converges to u' by the $(1/\alpha)$-H-Lipschitz continuity of T and by the construction of Algorithm 1.1.

Next, in the following theorem, we obtain an alternative to the estimate of Theorem 2.1, which is simple, yet easy for the convergence of the approximate solutions.

Theorem 2.2: *Let T: $K \to C(H)$ be an α-H-cocoercive mapping. Suppose that the sequences $\{x^k\}$ and $\{u^k\}$ are generated by the projection equation (1.5) and, $x^* \in K$ and $u^* \in T(x^*)$ form a solution of the NVI problem (1.1). Then we have:*

(*i*) *The estimate*

$$\| x^{k+1} - x^* \|^2 \leq [1 - (2\rho/\alpha)(1 - (\rho/2\alpha))] \| x^k - x^* \|^2 .$$

(*ii*) *The sequences $\{x^k\}$ and $\{u^k\}$ converge, respectively, to x^* and u^*, which form a solution of the NVI problem (1.1) for $\rho < 2\alpha$.*

Proof: Since P_K is nonexpansive and T is α-H-cocoercive and hence $(1/\alpha)$-H-Lipschitz continuous, we have

$$\begin{aligned}
\| x^{k+1} - x^* \|^2 &= \| P_K[x^k - \rho u^k] - P_K[x^* - \rho u^*] \|^2 \\
&\leq \| x^k - x^* \|^2 - 2\rho\langle u^k - u^*, x^k - x^* \rangle + \rho^2 \| u^k - u^* \|^2 \\
&\leq \| x^k - x^* \|^2 - 2\rho\alpha H(T(x^k, T(x^*))^2 + \rho^2 H(T(x^k), T(x^*))^2
\end{aligned}$$

$$
\begin{aligned}
&= \| x^k - x^* \|^2 + (\rho^2 - 2\rho\alpha) H(T(x^k), T(x^*))^2 \\
&\le \| x^k - x^* \|^2 + (\rho^2 - 2\rho\alpha)/(\alpha^2) \| x^k - x^* \|^2 \\
&= [1 - (2\rho/\alpha)(1 - \rho/2\alpha)] \| x^k - x^* \|^2 .
\end{aligned}
$$

In light of this estimate, $\{x^k\}$ converges to x^* for $\rho < 2\alpha$. And since $u^* \in T(x^*)$, it implies by Iterative Algorithm 1.1 that there exists $u^k \in T(x^k)$ such that

$$\| u^k - u^* \| \le H(T(x^k), T(x^*)) \le (1/\alpha) \| x^k - x^* \| \to 0.$$

This completes the proof.

REFERENCES

[1] C. Baiocchi and A. Capelo (1984). Variational and Quasivariational Inequalities. New York: Wiley & Sons.

[2] X.P. Ding (1994). A new class of generalized strongly nonlinear quasivariational inequalities and quasicomplementarity problems, *Indian J. Pure Appl. Math.*, **25**(11), 1115–1128.

[3] J.S. Guo and J.C. Yao (1992). Extension of strongly nonlinear quasivariational inequalities, *Appl. Math. Letters*, **5**(3), 35–38.

[4] B.S. He (1992). A projection and contraction method for a class of linear complementarity problems and its applications, *Applied Math. Optim.*, **25**, 247–262.

[5] B.S. He (1994). A new method for a class of linear variational inequalities, *Math. Programming*, **66**, 137–144.

[6] B.S. He (1994). Solving a class of linear projection equations, *Numer. Math.*, **68**, 71–80.

[7] B.S. He (1997). A class of projection and contraction methods for monotone variational inequalities, *Applied Math. Optim.*, **35**, 69–76.

[8] D. Kinderlehrer and G. Stampacchia (1980). *An Introduction to Variational Inequalities*, New York: Academic Press.

[9] G.M. Korpelevich (1976). The extragradient method for finding saddle points and other problems, *Matecon*, **12**, 747–756.

[10] P. Marcotte and J.H. Wu (1995). On the convergence of projection methods, *J. Optim. The. Appl.*, **85**, 347–362.

[11] S.B. Nadler (1969). Multi-valued contraction mappings, *Pacific J. Math.*, **30**(2), 475–488.

[12] M.A. Noor (1998). An implicit method for mixed variational inequalities, *Appl. Math. Lett.*, **11**(4), 109–113.

[13] M.A. Noor (199). An extragradient method for general monotone variational inequalities, *Adv. Nonlinear Var. Inequal.*, **2**(1), 25–31.

[14] J.S. Pang and D. Chan (1982). Iterative methods for variational and complementarity problems, *Math. Programming*, **24**, 284–313.

[15] M.V. Solodov and P. Tseng (1996). Modified projection-type methods for monotone variational inequalities, *SIAM J. Control Optimiz.*, **34**(5), 1814–1830.

[16] R.U. Verma (1997). A fixed point theorem involving Lipschitzian generalized pseudocontractions, *Proc. Royal Irish Acad.*, **97A**, 83–86.

[17] R.U. Verma (1997). Nonlinear variational and constrained hemivariational inequalities involving relaxed operators, *ZAMM*, **77**(5), 387–391.

[18] R.U. Verma (1998). Generalized pseudocontractions and nonlinear variational inequalities, *Publicationes Math. Debrecen*, **33**(1–2), 23–28.

[19] R.U. Verma (1998). On generalized KKM type selections in generalized H-spaces, *Math. Sci. Res. Hot-Line*, **2**(12), 1–11.

[20] R.U. Verma (1998). R-S-G-H-KKM theorems and their applications, *Math. Sci. Res. Hot-Line*, **2**(12), 19–27.

[21] R.U. Verma (1998). Some minimax inequalities on generalized H-convex sections, *Math. Sci. Res. Hot-Line*, **2**(1), 27–32.

[22] R.U. Verma (1998). An iterative algorithm for a class of nonlinear variational inequalities involving generalized pseudocontractions, *Math. Sci. Res. Hot-Line*, **2**(5), 17–21.

[23] R.U. Verma (1999). Strongly nonlinear quasivariational inequalities, *Math. Sci. Res. Hot-Line*, **3**(2), 11–18.

[24] R.U. Verma (1999). R-KKM selections applied to minimax inequalities, *Adv. Nonlinear Var. Inequal.*, **2**(1), 33–42.

[25] R.U. Verma (1999). Intersection theorems on generalized H-convex sections, *Adv. Nonlinear Var. Inequal.*, **2**(1), 11–24.

[26] R.U. Verma (1999). R-KKM theorems and their applications to minimax inequalities, *Adv. Nonlinear Var. Inequal.*, **2**(1), 55–63.

[27] R.U. Verma (1999). Minimax inequalities based on generalized KKM type selections, *Adv. Nonlinear Var. Inequl.* **2**(1), 65–71.

[28] R.U. Verma (1999). On generalized nonlinear variational inequalities on G-H-spaces, *Math. Sci. Res. Hot-Line*, **3**(1), 1–5.

[29] R.U. Verma. RKKM mapping theorems and variational inequalities, *Proc. Royal Irish Acad.*, (to appear).

[30] R.U. Verma. A class of nonlinear variational inequalities involving pseudomonotone operators, *J. Appl. Math. Stochastic Anal.*, (to appear).

[31] R.U. Verma (1999). A class of quasivariational inequalities involving cocoercive mappings, *Adv. Nonlinear Var. Inequal.*, **2**(2), (to appear).

[32] E. Zeidler (1986). Nonlinear Functional Analysis and its Applications I, New York: Springer-Verlag.